Periodic Table of the Elements with Atomic Weights*

Transition elements

1 H 1.0079																		2 He 4.00260
3 Li 6.941	4 Be 9.01218											5 B 10.81	6 C 12.011	7 N 14.0067	8 O 15.9994	9 F 18.99840	10 Ne 20.179
11 Na 22.98977	12 Mg 24.305											13 Al 26.98154	14 Si 28.086	15 P 30.97376	16 S 32.06	17 Cl 35.453	18 Ar 39.948
19 K 39.098	20 Ca 40.08	21 Sc 44.9559	22 Ti 47.90	23 V 50.9414	24 Cr 51.996	25 Mn 54.9380	26 Fe 55.847	27 Co 58.9332	28 Ni 58.70	29 Cu 63.546	30 Zn 65.38	31 Ga 69.72	32 Ge 72.59	33 As 74.9216	34 Se 78.96	35 Br 79.904	36 Kr 83.80
37 Rb 85.4678	38 Sr 87.62	39 Y 88.9059	40 Zr 91.22	41 Nb 92.9064	42 Mo 95.94	43 Tc (97)	44 Ru 101.07	45 Rh 102.9055	46 Pd 106.4	47 Ag 107.868	48 Cd 112.40	49 In 114.82	50 Sn 118.69	51 Sb 121.75	52 Te 127.60	53 I 126.9045	54 Xe 131.30
55 Cs 132.9054	56 Ba 137.34	57 La 138.9055	72 Hf 178.49	73 Ta 180.9479	74 W 183.85	75 Re 186.207	76 Os 190.2	77 Ir 192.22	78 Pt 195.09	79 Au 196.9665	80 Hg 200.59	81 Tl 204.37	82 Pb 207.2	83 Bi 208.9804	84 Po (209)	85 At (210)	86 Rn (222)
87 Fr (223)	88 Ra 226.0254	89 Ac (227)	104 (Rf)	105 (Ha)	106												

Lanthanides

58 Ce 140.12	59 Pr 140.9077	60 Nd 144.24	61 Pm (145)	62 Sm 150.4	63 Eu 151.96	64 Gd 157.25	65 Tb 158.9254	66 Dy 162.50	67 Ho 164.9304	68 Er 167.26	69 Tm 168.9342	70 Yb 173.04	71 Lu 174.97
90 Th 232.0381	91 Pa 231.0359	92 U 238.029	93 Np 237.0482	94 Pu (244)	95 Am (243)	96 Cm (247)	97 Bk (247)	98 Cf (251)	99 Es (254)	100 Fm (257)	101 Md (258)	102 No (255)	103 Lw (260)

Actinides

*1973 values based on $^{12}C = 12$.

Chem One

Chem One

Second Edition

JÜRG WASER
Formerly, Department of Chemistry, California Institute of Technology

KENNETH N. TRUEBLOOD
Department of Chemistry, University of California, Los Angeles

CHARLES M. KNOBLER
Department of Chemistry, University of California, Los Angeles

McGraw-Hill Book Company

New York   St. Louis   San Francisco   Auckland   Bogotá   Hamburg
Johannesburg   London   Madrid   Mexico   Montreal   New Delhi   Panama
Paris   São Paulo   Singapore   Sydney   Tokyo   Toronto

Chem One

34567890 RMRM   8321

This book was set in Century Schoolbook by York Graphic Services, Inc. The editors were Jay Ricci, Donald C. Jackson, and Sibyl Golden; the designer was Janet Durey Bollow; the production supervisor was Charles Hess. New drawings were done by Judith L. McCarty. Rand McNally & Company was printer and binder.

Library of Congress Cataloging in Publication Data

Waser, Jürg.
    Chem one.

    Bibliography: p.
    Includes index.
    1. Chemistry.   I. Trueblood, Kenneth N., joint
author. II. Knobler, Charles M.,   date   joint author.
III. Title.
QD31.2.W36   1980   540   79-15670
ISBN 0-07-068432-4

Contents in Brief

Contents

29 Organic Chemistry 643

30 Biochemistry 678

Appendixes

Preface

"Chem One" is a text for a first-year college or university course for students planning careers in the physical and life sciences and in engineering. This second edition is a considerable revision of the first edition that was published four years ago.

We have attempted to avoid the pitfalls of half-truth and misleading oversimplification. A book with such a goal need not, however, *begin* each topic on an advanced level. We have tried, through careful definitions, rather full explanations, and many worked examples with parallel exercises, to provide the means by which a student who is willing to work can master the fundamental concepts and be able to apply them. Those whose preparation has been inadequate should find that the "Study Guide" prepared for this edition by Emily Maverick will make the task easier.

Certain concepts are encountered a number of times, first in an introductory fashion in the early chapters, then again in greater depth and sophistication in the middle sections of the book, and finally still again as they are used in the discussions of descriptive chemistry that comprise the final eight chapters. In addition to the pedagogical advantages that accompany repeated exposure at increasing depth, including a better overall perspective on the unifying themes of chemistry, this scheme makes it possible for students to do significant laboratory experiments during the early stages of the course. Thus errors, stoichiometry, and gases are treated in detail in Chapters 3 and 4, and discussions of the periodic table, bonding, three-dimensional structure, and reactions in aqueous solution are introduced in an elementary fashion in the first third of the book. Chapters 11 through 13 contain a detailed treatment of chemical equilibrium to facilitate the early introduction of quantitative laboratory work (such as titrations) that requires an understanding of equilibria in solution. The detailed discussion of quantum theory, atomic and molecular structure, and the periodic table is taken up in the middle of the book (Chapters 14 through 18), but these chapters may readily be assigned as a unit either earlier (e.g., after Chapter 7 or Chapter 10) or later (after Chapter 22). The structure of the book allows several other variations in order of presentation as well. For courses in which laboratory experiments dealing with solutions are begun as early as the fourth or fifth week, Chapters 4 (gases) and 7 may be deferred, allowing Chapters 8 and 9 to be covered earlier.

We have again used the SI system primarily, because those being trained now will doubtless use SI units throughout most of their working lives. However, we have retained the atmosphere, the torr, and the angstrom, and we have not completely given up the milliliter in favor of the identical cubic centimeter because students will use glassware calibrated in milliliters. Table headings, labels on the axes of graphs, and some equations are written in "slash" notation. In this convention, dimensioned quantities are divided by their units to give dimensionless numbers. Thus, we write $\log (P/\text{atm})$, which represents the logarithm of a number, the pressure in atmospheres divided by the unit atmosphere. Each of the SI conventions is explained when it is introduced and the SI system is discussed in Chapter 1 and, in detail, in Appendix A.

In any book as comprehensive as this there are topics or sections that can be omitted or abridged to suit the tastes of the instructor and the students as well as the limitations of time. We have attempted to relegate the more advanced and least essential material to the latter parts of many chapters so that they may most readily be skipped. Some sections of the book can be left for the student to read and need not be discussed extensively in class. Chapter 7, for example, can stand on its own as an introduction to structure. With selective omissions, a lecturer could have time to devote at least three weeks to organic chemistry and biochemistry, which comprise the final two chapters, if this is felt desirable. Although the latter are often found in second-year courses, there are many students who do not continue beyond the first year. It seems to us essential that any general chemistry book provide such students with some exposure to these important areas of modern chemistry, even if they are not covered in formal course work.

The following are among the more significant changes from the first edition:

1. The number of worked examples has been increased to 165 and each is now accompanied by a parallel exercise with which students can test their understanding.

2. The number of end-of-chapter problems and questions has been increased by about one-third, with particular emphasis on straightforward, relatively simple, problems.

3. Each chapter concludes with a summary and a list of new terms and concepts, with a page reference to the introduction of each term or concept so that it may be reviewed in context rather than as a disconnected entity.

4. Many sections of the book have been completely rewritten, and much of the material has been reorganized, with the number of chapters increased by three although the text as a whole is somewhat shorter. There are new chapters on oxidation and reduction (Chapter 10) and nuclear chemistry (Chapter 28), and an early overview of the periodic table (in Chapter 2).

5. Derivations using calculus notation explicitly have now been put into Appendix C, with only the results in the text itself, in the chapters dealing with thermodynamics and kinetics.

6. The term *formality* and the related symbol F have been abandoned because so few instructors distinguish between formality and molarity. The terms

equivalent weight and *normality* have also been removed because they are useful chiefly in a laboratory context and can readily be learned there, as needed. (The principles behind them remain, however; for example, in Problems 9-14 through 9-17, 10-9, and 10-10.)

In this revision we have benefited greatly from the advice, suggestions, and criticism of many students and colleagues at UCLA and elsewhere, and especially from those of David Adams, Robert Allendoerfer, Daniel Atkinson, Kyle Bayes, James Espensen, Jerry Kasper, Ed Lingafelter, Sam Markowitz, Emily Maverick, George Miller, Verner Schomaker, Bernice Segal, Arden Slotter, and Charles West. We owe a great deal to them, as we do to those others who contributed to the first edition in various ways, including Jay M. Anderson, Bill Benjamin, John P. Chesick, Deirdre Devereux, Ed Friedrich, Jenny Glusker, James B. Ifft, Daniel Kivelson, Caroline Lanford, Richard Marsh, James D. McCullough, Kathy North, Julian L. Roberts, Jr., Raymond J. Suplinskas, Judy Swain, Julie Swain, Robert Weiss, and Alan Wingrove.

We have appreciated the professionalism and talent of Janet Bollow, the designer; of Judith McCarty, who prepared the new figures; of John Hannon, who copyedited the manuscript; and of Donald Jackson, Sibyl Golden, and Charles Hess, who have supervised the various stages of editing and production with patience and encouragement. Finally, we owe a very special debt to Delna Jacobs, whose skill and unfailing good spirits in typing the revisions of revisions of revisions has made our task much easier and more pleasant.

Jürg Waser
Kenneth N. Trueblood
Charles M. Knobler

To the Student

The material in this book varies widely in difficulty. Some topics may seem essentially a review of high school chemistry, whereas others will be new to all students and sometimes rather abstract. Many topics are treated a number of times so that you can become familiar with them while working with them. They are first encountered at an elementary and qualitative level, are later developed in more detail, often quantitatively, and finally turn up again when they are utilized in systematizing and explaining the great body of chemical facts called "descriptive chemistry".

Not all of the material in the book is easy to grasp. We and most of our colleagues in chemistry had to struggle with many parts of it when we were learning it and have to think about some aspects of it carefully even now. We have, however, tried to smooth the way for you. New words are italicized when they are first defined, and the key terms and concepts new to each chapter are listed at the end of the chapter, with a specific page reference that will help you find the place where each is introduced and discussed, in context, in the chapter. The extensive index should also be helpful in this connection. Each chapter has a summary at the end; make it a practice to read this after you have finished studying the chapter to be sure that you have grasped all of the essential points. You may also find it helpful to get an overview by scanning the summary before studying the chapter.

Many of the figures have extensive legends which are intended to clarify both the figures and the accompanying text. Make it a habit to read them carefully. You will find many worked examples in most chapters, designed to show you how to apply principles and methods to specific situations. Each example is accompanied by a parallel exercise, which you should be able to do without difficulty if you have followed through and understood the example. Each chapter concludes with many problems and questions that will help you develop your ability to work with chemical concepts.

Frequent cross-references tie together related concepts and facts found in different sections of the book. Anyone learning a new subject finds it hard at times to see interrelationships and to appreciate general principles; the cross-references are intended to help provide a broader perspective. Appendixes A and B are self-contained essays on special topics (Units and Chemical Nomenclature) that you may want to consult a number of times. Appendix C presents derivations by the methods of calculus of some of the equations used in chemical thermodynamics (Chapters 19 and 20) and in the discussion of the rates of chemical reactions (Chapter 22). Appendix D contains tables of data that you will need to refer to often. Appendix E gives answers to all exercises and to many odd-numbered end-of-chapter problems and questions. Frequently used tables are to be found inside the front and back covers.

The "Study Guide" that accompanies the book should prove helpful, especially for those students whose background in chemistry is weak and in those places in the text where some background in physics or mathematics is essential for a thorough understanding. The "Solutions and Supplementary Material" manual that also accompanies the text provides detailed solutions and answers for every end-of-chapter problem and question. In addition, it contains supplementary material on some topics that may be of interest to those students who find their curiosity piqued but their questions not entirely answered by the material in the text.

We suggest that when the going gets tough, as it will at times, you give the more difficult material a rest after a first cursory reading. Follow up later with a second, more careful, study, jotting down key words and concepts and frequently closing the book for quick mental reviews. Retrace the steps of derivations and check the details of the worked examples. Work or rework the exercises. Don't worry if at first you understand some new topic only partially and even have some wrong ideas about it. Often an initial false start that is later corrected helps to clarify something, because it gives a perspective not available to someone who has not thought about the topic at all.

Try to retain a critical attitude at all times. Don't accept anything stated here, or elsewhere, simply on the basis of the apparent authority of the source. Apply your powers of reasoning as much as you can; search for internal consistency. We have tried to avoid errors but it is unlikely that we have caught them all.

Learning is a lonely pursuit and takes a good deal of discipline. Yet all these sober words of caution and advice should not obscure the fact that chemists really *enjoy* chemistry. We know that you can too—and we hope you will.

Jürg Waser
Kenneth N. Trueblood
Charles M. Knobler

Chem One

Introduction

"Those sciences are vain and full of errors that are not born from experiment, the mother of all certainty, and that do not end with one clear experiment."
LEONARDO da VINCI

"According to convention there is a sweet and a bitter, a hot and a cold, and according to convention there is a color. In truth there are atoms and a void."
DEMOKRITOS, GREEK PHILOSOPHER, FIFTH CENTURY B.C.

1-1 Chemistry as One of the Natural Sciences

The Natural Sciences The development of modern science during the last three centuries has been at once so broad and so deep that a great deal of learning and thought is required for even the best of human minds to encompass a portion of it. This is in large part because progress in science is cumulative—science builds on and extends what has been observed and understood earlier. It grows by the interplay of experimental observations, imaginative reasoning, predictions based on this reasoning, and experiments to test the predictions. As the body of systematized observations and generalizations about the natural world has increased, it has been artificially subdivided into "different" scientific fields, and these have in turn been partitioned, all because human life is too short and the mind too limited to be able to learn all that has been observed and postulated.

The different natural sciences were once considered to be physics, chemistry, biology, geology, and astronomy, but now the subdivisions and extensions of these fields have become far more numerous. Not only do they include areas in which several of these disciplines overlap, such as geophysics (the physics of the earth) and biochemistry (the chemistry of living things) but, increasingly now, they encompass areas in which the methods of many of the classical natural sciences are integrated and brought to bear on a particular portion of our world. For example, oceanography includes the related studies of the physical, chemical, biological, and geological aspects of the ocean, and planetary science involves a correlated study in all the foregoing disciplines of the members of our solar system.

Chemistry Chemistry is concerned with the composition, structure, and properties of substances, the transformations of these substances into others by reactions, and the different kinds of energy changes that accompany these reactions. In the main, the chemist is interested in the properties and reactions of matter as it commonly exists on the earth, and most of this book deals with this aspect of chemistry. However, astronomical observations and space exploration give us every reason to believe that the general principles discussed are more widely applicable.

Since the field of chemistry covers an enormous range of activities, it is, in turn, subdivided loosely into many branches. This division is done in several distinct ways, depending on the focus of interest. It may be based on the kinds of substances that are involved; for example, organic chemistry deals with compounds containing the element carbon (of which there are literally millions) and inorganic chemistry deals with substances that do not contain this element. However, the distinctions are not clear cut. For example, the carbonate minerals are regarded as inorganic even though they contain carbon.

Another classification scheme focuses on the types of operations and reactions that one may perform. Two of the earliest major areas of chemistry were analytical chemistry, the determination of the identity and the proportions of the components of a compound or of a mixture, and synthetic chemistry, the creation of one substance from others. Still another way of subdividing the field of chemistry is on the basis of its overlap with other fields—thus one finds references to physical chemistry, biochemistry, geochemistry, cosmochemistry, and so on. Finally, the contrasting phrases "theoretical chemistry" and "descriptive chemistry" are often used to refer to the concepts, principles, and theories on the one hand and to the experimentally observed facts on the other.

Many of the experimental methods of modern chemistry were developed originally by men and women regarded as physicists—for example, the methods of spectroscopy and those of structure determination by the diffraction of X rays. Similarly, modern theoretical chemistry is based chiefly on thermodynamics, statistical mechanics, and quantum mechanics, all originally regarded as fields of theoretical physics. The interaction has not all been one-sided, however; many of the ideas important in modern physics came originally from chemistry. The atomic nature of matter was first recognized quantitatively by John Dalton (1766–1844) and was part of chemists' thinking for about a century before most physicists accepted it without reservation. The interactions of chemistry with the other nominally distinct branches of science, most notably biology and geology, have also been numerous and important.

A word of caution is needed about the vocabulary of chemistry. You will encounter many new terms, most of which have quite precisely defined meanings. Not only must you learn these meanings carefully so that you understand how to use the terms and how they differ from related ones, but sometimes you must avoid confusion with popular usage of the same words that is usually (though not always) broader and vaguer. Thus, the terms *energy, work, heat,* and *force* have more restricted meanings in physics and chemistry than in everyday speech; on the other hand, the words *salt* and *alcohol* have more general meanings, referring to classes of substances rather than to specific ones. Sometimes a definition is mathematical and is used in a mathematical way. A few scientific terms have multiple meanings, and while each meaning is quite precise, only the context makes it possible to decide which of them applies in any particular case. For example, the term *neutral* may refer to the absence of an excess of positive or negative charge, or to the absence of an excess of acid or base. Such possible ambiguities will often be pointed out, but you should be alert for them.

1-2 Units and Dimensions of Physical Quantities

Handling Units in Calculations Units as well as numbers are needed to specify many physical quantities, such as length, time, or mass. If a distance is said to be 12.5, that is not enough—the unit of distance or length must also be specified, for example, feet or meters or miles. Physical quantities with which units are associated are said to have *dimensions* that can be established from the equations that relate the quantities to others considered to be "fundamental", such as length, time, and mass. For example, the dimensions of an area are length squared (length^2), those of a volume are length cubed (length^3), and those of a speed are length time^{-1}. Dimensions are characteristic attributes of physical quantities, whereas the units with which the quantities are measured depend on the particular scale or standard relative to which the measurement is made. Thus a length might be measured in centimeters, meters, kilometers, or miles, and a speed in meters per second (m s^{-1}) or feet per hour (foot hour^{-1}). Pure numbers, such as the ratio of two quantities with the same dimensions, are termed *dimensionless*.

When giving the value of some quantity that has dimensions, one must give both the numerical value and the unit in which that value is expressed. If we represent the volume of some solid by V and the solid occupies a volume of 5.02 cm^3, we write

$$V = 5.02 \text{ cm}^3$$

In any calculations involving any quantity, both the numerical value *and the units* are handled by the ordinary rules of arithmetic and algebra. Thus if the solid just referred to has a mass m of 12.72 g, its *density* ρ (Greek rho), defined as the ratio of mass to volume, is given by

$$\rho = \frac{m}{V} = \frac{12.72 \text{ g}}{5.02 \text{ cm}^3} = 2.53 \text{ g cm}^{-3} \qquad (1\text{-}1)$$

In working problems, always label every quantity with its units and carry the units along in each step of every calculation. Following this practice minimizes errors and makes it easier to remember some definitions correctly.

☐ **Volume from Mass and Density** Find the volume V of a block of wood that has a mass m of 10.3 g and a density ρ of 0.97 g cm^{-3}.

Example 1-1

Solution The volume must have dimensions (length^3). Thus, even if you do not recall the definition of density, you know that the relation to be used between the quantities given (mass and density) must be such that the answer (volume) comes out with the dimensions length^3. The correct result is obtained by dividing the mass by the density, as is apparent from rearrangement of (1-1):

$$V = \frac{m}{\rho} = \frac{10.3 \text{ g}}{0.97 \text{ g cm}^{-3}} = 10.6 \text{ cm}^3$$

This answer does have the dimensions of volume, the units being cubic centimeters.

Suppose that you had forgotten the definition of density, and had multiplied m by ρ

$$m\rho = 10.3 \text{ g} \times 0.97 \text{ g cm}^{-3} = 10.0 \text{ g}^2 \text{ cm}^{-3}$$

This result cannot be correct. It does not have the proper units and cannot be a volume. ∎

Exercise 1-1

☐ Suppose that the price of gasoline is 70 cent gal^{-1}. What is the price in dollar ton^{-1}? The (metric) ton is defined as 1 ton = 10^3 kg, the density of gasoline is 0.70 g cm^{-3}, and 1 gal = 3.79 liter. ∎

SI Units   Most units used in this text are the internationally adopted SI units (Système Internationale d'Unités), adopted in 1960 by the General Conference on Weights and Measures, the international organization that defines metric units. These units have been in use at the U.S. National Bureau of Standards since 1964. Seven of the units of the system, such as the meter (m), the kilogram (kg), the second (s), and the mole (mol), are termed *basic*. The reason is that all other units are derived from the basic units, either by multiplying or dividing the basic units by factors of 10, 100, or (preferably) 1000, or by combining the basic units. Examples of multiples or fractions are the kilometer (1 km = 1000 m), the centimeter (1 cm = 10^{-2} m), the gram (1 g = 10^{-3} kg), and the milligram (1 mg = 10^{-6} kg). Examples of combinations are the volume units m^3 and cm^3, and the units of speed km s^{-1} and m s^{-1}. Not all units used in this text are SI units, for example, the pressure units atmosphere (atm) and torr (1 atm = 760 torr) (see Sec. 4-1). We also often use the volume units liter and milliliter (1 ml = 10^{-3} liter = 1 cm^3).

More details on the SI system are given in the first section of Appendix A, which you should study.

Dimensionless Numerical Values of Physical Quantities   It is customary to regard the symbol for a physical quantity, such as the volume V or mass m, as representing the quantity including its dimensions. The symbol represents the product

Physical quantity = numerical value × unit

The symbol used for the physical quantity does not and should not imply any *particular* choice of units. On the other hand, if the symbol for the physical quantity is divided by an appropriate unit, the result is a pure number that represents the value of the quantity in the units considered. For example, $m/\text{g} = 50$ implies a mass of 50 g and $V/\text{cm}^3 = 40$ implies a volume of 40 cm^3. The coordinate axes of graphs are often labeled by such pure numbers, as are the columns in tables of numerical values. A column in a table might be headed d/mm; numbers in that column would then represent values of d in millimeters. An equivalent heading would be 1000 d/m, obtained from d/mm by multiplying numerator and denominator by 1000; numbers given would still represent distances in millimeters, an entry of 5 implying

$$1000 \, d/\text{m} = 5 \qquad \text{or} \qquad d = 5 \times 10^{-3} \text{ m} = 5 \text{ mm}$$

The notation used in the preceding paragraph for expressing ratios $\left(\text{written, for}\right.$ example, either as V/cm^3 or $\left.\dfrac{V}{\text{cm}^3}\right)$ provides a convenient way to express conversion factors between different units. From the equality 1 inch = 25.4 mm, for example, we can extract the relation

$$\frac{25.4 \text{ mm}}{1 \text{ inch}} = 1$$

a relation that permits conversion from inches to millimeters or vice versa.

☐ **Unit Conversion** The thickness of a plastic swimming-pool cover is $d = 6 \times 10^{-3}$ inch ("6 mil"). What is d in millimeters?

Example 1-2

Solution We use the relationship between inch and millimeter in the form 25.4 mm/inch = 1. Since multiplication by unity causes no change, it follows that

$$6 \times 10^{-3} \text{ inch} = 6 \times 10^{-3} \text{ inch} \times \frac{25.4 \text{ mm}}{\text{inch}} = \underline{0.15 \text{ mm}} \quad \blacksquare$$

☐ From the relations $3600 \text{ s}/1 \text{ hour} = 1$, $1.609 \text{ km}/1 \text{ mile} = 1$, and $1000 \text{ m}/1 \text{ km} = 1$, calculate the speed of an automobile traveling at 55 mile hour^{-1} in the units km hour^{-1} and m s^{-1}. ■

Exercise 1-2

1-3 Substances, Mixtures, and States of Aggregation

Physical and Chemical Properties The simplest way to describe a substance is in terms of its properties. Its *physical properties,* such as color, density, melting point, and solubility, are those that do not depend on its reaction with other substances (or with itself). Physical properties usually describe the response of a substance to external influences, such as temperature changes or pressure changes. Examples are the change in volume of a substance as the result of temperature or pressure changes and the reflection of specific colors when a substance is illuminated with white light. The *chemical properties* of a substance, on the other hand, are those that relate to its behavior in the presence of other substances, in particular to its reactions with other substances (or with itself) to form new substances with different properties, or its failure to react under particular conditions. Examples are the reaction of sulfur, a yellow solid, with the gas oxygen to form the irritating gas sulfur dioxide and the reaction of the green gas chlorine with the silvery metal sodium to form the colorless solid sodium chloride (table salt).

There is no sharp line between chemical and physical properties. The interaction of a substance with light may not only reveal its color but also cause chemical reaction, as it does in a photographic emulsion; the flow of electric current may cause chemical changes, and a sharp blow may cause explosives to detonate (a chemical change). Similarly, the dissolving of a substance in water is often accompanied by chemical interaction between the dissolving particles and the water molecules.

5

Pure Substances and Mixtures The term *substance* (or *pure substance*) is used to refer to a sample of matter that has distinct physical and chemical properties and a definite composition. A few of the materials that we deal with in everyday life are pure substances—for example, sugar, salt, copper, and silver.

Most familiar materials—milk, vinegar, brass, air, dirt, concrete—are mixtures of various substances.[1] Mixtures do not have a definite composition. As a result they do not have one or more of the well-defined properties of pure substances. For example, they may not have sharp melting or boiling temperatures, or they may appear to the naked eye to have several distinct ingredients. Even if they *appear* to be composed of a single substance, it may be possible to separate them into substances with different properties. This might be accomplished, for instance, by heating or by treatment with a suitable solvent.

Mixtures can be categorized as either homogeneous—uniform in properties throughout the sample—or heterogeneous. Homogeneous mixtures are called *solutions* (Chaps. 8 and 9). The implications of the terms *homogeneous* and *heterogeneous* are discussed further in the next section.

Macroscopic and Microscopic Viewpoints Explanations and interpretations of chemical phenomena involve two distinct points of view, the macroscopic and the microscopic. The macroscopic world is that part of the physical world that is directly apparent to our senses. In common usage it includes all phenomena that can be observed by the naked eye. In chemistry and related fields, however, all those phenomena that are not on an atomic or molecular scale are considered *macroscopic,* while *microscopic* (Greek: *makros,* large; *mikros,* small) phenomena are those that occur at the atomic and molecular level.

One of the chief objectives of chemistry is to find explanations for macroscopic phenomena on the microscopic level, that is, in terms of the properties and interactions of atoms and molecules. Thus the interplay between macroscopic and microscopic viewpoints is considered throughout this text.

Phases A piece of matter is called *homogeneous* if its macroscopic properties (e.g., density, or speeds of light and sound in it) are the same throughout. If it consists of macroscopic regions that have different properties and are separated by macroscopically sharp boundaries, it is said to be *heterogeneous.* The homogeneous parts of such a system are called *phases* (Greek: *phainein,* to appear). For example, a piece of granite consists of three principal solid phases—quartz, feldspar, and mica. A pitcher of water with ice cubes contains the liquid and solid phases of water, while the air above the water, including some water vapor, is a third, gaseous phase.

In some respects a one-phase region may not be strictly homogeneous. The density of a gas or a liquid is different at points that differ in height, and other properties may also be different. These differences are often small. It is important to note that the properties in a one-phase region change continuously and slowly with position, whereas they change abruptly when a boundary to a new phase is crossed.

Granite

---

[1] Even "pure" substances—distilled water, refined sugar, electrolytically purified copper—inevitably have traces of contaminants, although perhaps only a few parts of impurity per million parts of the substance itself. In special cases even higher purities are obtainable. Germanium used in transistors is routinely produced with only 1 atom of impurity in 10 million atoms of germanium.

□ **Classification of Common Materials** (*a*) Is seawater a pure substance or a mixture?

Example 1-3

Solution Seawater is easily separated, by evaporation of the water in it, into pure water and a salty residue. Hence it is a mixture.

(*b*) Are the following homogeneous or heterogeneous: sand from a beach or from the desert; soil from a field; a snowball; tap water; a crystal of sugar or salt?

Solution The only sure way to answer a question of this kind is by experiment (that is, by careful observation). You may be familiar with all these things already and be able to answer with assurance; if not, we suggest that you examine some of them carefully. When you do you will probably observe the following things: Most <u>sands</u> are very obviously <u>heterogeneous</u>, being composed of distinct particles of quite different materials, with varying color, hardness, and other properties; <u>soil</u> from almost any field is <u>heterogeneous</u>, composed of some decaying vegetable matter, sand, small stones of various kinds, small live organisms, and all sorts of other things; a <u>snowball</u> is composed primarily of ice crystals (snow) but usually contains some liquid water and thus is <u>heterogeneous</u>; tap water, although almost always a mixture, is usually <u>homogeneous</u>, unless the water contains sediment of some kind; a <u>crystal of sugar or salt</u> is <u>homogeneous</u>, although the crystals available to you may be so small and your methods so limited that you cannot readily be sure of that. ■

□ Classify as a pure substance or a mixture: (*a*) distilled water, (b) a piece of wood. Classify as homogeneous or heterogeneous: (*c*) copper wire, (*d*) maple syrup, and (*e*) chocolate chip ice cream. ■

States of Aggregation The three common states of aggregation of matter are the solid, the liquid, and the gaseous states. Liquid and solid states are also called condensed states or condensed phases. Their densities are usually much greater than those of gases, and the density of a solid is usually somewhat greater than that of the corresponding liquid phase.

□ **Density of Solid and Gas** A 5.00-cm³ piece of solid carbon dioxide (dry ice) weighs 7.80 g. (*a*) What is its density?

Example 1-4

Solution Using the definition of density (1-1) we write

$$\text{Density solid carbon dioxide} = \rho = \frac{m}{V} = \frac{7.80\text{ g}}{5.00\text{ cm}^3} = \underline{1.56\text{ g cm}^{-3}}$$

(*b*) At 0°C and a pressure of 1 atm gaseous carbon dioxide has a density of $1.98 \times 10^{-3}\text{ g cm}^{-3}$. Under these conditions what is the volume in liters occupied by 7.80 g gaseous carbon dioxide?

Solution

$$V = \frac{m}{\rho} = \frac{7.80\text{ g}}{1.98 \times 10^{-3}\text{ g cm}^{-3}}$$

$$= 3.94 \times 10^3\text{ cm}^3 \times \frac{1\text{ liter}}{1000\text{ cm}^3} = \underline{3.94\text{ liter}}$$

Note that the definition 1 liter = 1000 cm³ has been used to convert the answer from cubic centimeters to liters. ■

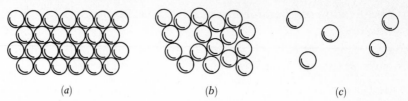

(a) (b) (c)

Figure 1-1 Atomic Aspects of (a) a Crystalline Solid, (b) a Liquid, and (c) a Gaseous Phase
The spheres represent molecules or atoms. In (a) the arrangement is regular, and adjacent molecules are in contact with each other; in (b) it is irregular but there is still contact between adjacent molecules; in (c) the arrangement is irregular; most molecules are distant from each other.

Exercise 1-4

☐ (a) What is the density of a 15.20-g piece of brass, the volume of which is 1.810 cm^3? (b) What is the volume of 21.50 g of mercury, the density of which is 13.55 g cm^{-3}? ∎

It is instructive to compare the macroscopic and microscopic descriptions of the three states (Fig. 1-1). A solid has the macroscopic property of resisting attempts to change its original shape and volume. A liquid is also difficult to compress but it assumes the shape of as much of its container as it fills.[2] A gas fills all the space available to it uniformly, as long as effects of gravitation are negligible.

On the microscopic level we start with a description of gases. The molecules of a gas move about freely, with distances between them that are large compared to their size—for example, under ordinary conditions of temperature and pressure, about 10 molecular diameters. The forces between the molecules are small, and, except for many mutual collisions, the molecules move about almost independently at high speeds, preferring no part of the container over another and filling it uniformly. The pressure exerted by the gas is the result of the numerous collisions of its molecules with the container walls.

On the other hand, the molecules of a solid are packed closely and oscillate about fixed positions, moving back and forth over increasingly larger distances as the temperature rises. Only occasionally do molecules in a solid break loose and interchange places. Solids are not easily compressed because of the strong repulsion at small distances between molecules, and the resistance to other changes in shape is due to the forces that hold the molecules close together.

The packing of molecules in liquids is somewhat looser, and the molecules slip by each other continually.[3] There is still sufficient attraction for the molecules to cling together, and strong repulsive forces arise when they are pushed even closer, as when the liquid is compressed. Thus it is understandable that a liquid resists attempts to change its volume, but not its shape.

In crystalline solids the molecules are arranged with great regularity. Except for minor flaws, this regularity extends throughout the crystal (Figs. 1-2a and 1-3) and thus affects its macroscopic properties. There are also noncrystalline solids, called *amorphous solids.* Their molecular arrangement is more or less random (Fig. 1-2b).

---

[2] Drops of liquids, and larger amounts in the absence of gravitation, tend to assume a spherical shape because of *surface tension,* that is, because of forces that strive to make the surface area of the liquid a minimum.

[3] The unceasing motion of the molecules in a liquid is indicated by the random motion of tiny particles suspended in a liquid, first observed with a microscope by the English botanist Robert Brown in the early nineteenth century and therefore called brownian motion.

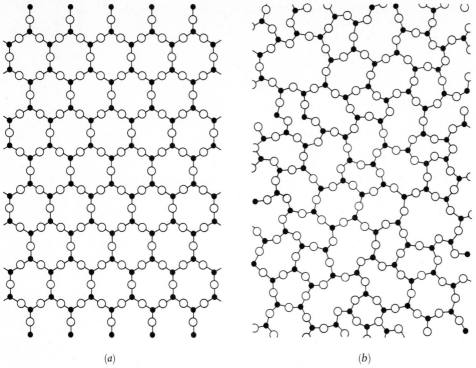

(a)   (b)

Figure 1-2　Two-Dimensional Models for (a) a Crystalline Substance and (b) an Amorphous Substance

In the crystalline substance the arrangement is regular and persists over long distances. The amorphous substance has an irregular arrangement. Less schematic representations of a crystalline solid are given in Fig. 1-3.

Figure 1-3　The Structure of a Molecular Crystal

The molecule is tetracyanoethylene, depicted in terms of a ball-and-stick model in (a). All atoms of the molecule lie in one plane. The arrangement of the molecules in the crystal is shown in (b). In (c) the molecules are represented with their appropriate relative sizes ("packing" radii). The molecule in the lower right of the layer closest to the viewer has been omitted in (c) to give a better view of the packing.

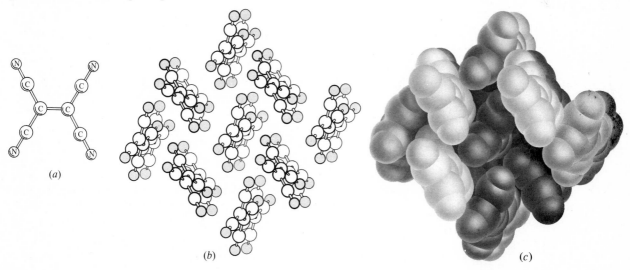

(a)   (b)   (c)

9

The idea that there are elementary components, or elements, common to all substances is at least as old as the fifth century B.C. Leukippos and Demokritos developed at that time an *atomistic* (Greek: *a*, not; *tomos*, divisible) theory in which all substances were regarded as composed of indivisible and indestructible particles (atoms) which differed from each other in shape, weight, relative position, and orientation but did not differ internally. Differences in the properties of substances were considered to arise from the differing configurations and combinations of atoms. Thus, a liquid such as water would be composed of smooth and round atoms that rolled over one another readily, while a solid such as iron would be composed of atoms that differed from those of water in having a jagged, rough shape, which would permit them to cling together to form a solid body.

These early atomic ideas survive today, in greatly modified form, but they were partly in the shadow of an alternative theory for many centuries. This view of the composition of matter retained the concept of a few elements but regarded them as qualities rather than as indivisible material particles. It was initiated by Anaxagoras and Empedokles, also in the fifth century B.C., and was developed and supported by the immense authority of Aristotle. According to this notion, all matter was continuous—that is, indefinitely divisible—and composed of the same fundamental material. Differences in properties resulted from the presence of differing amounts of two sorts of antagonistic principles: dry-wet and hot-cold. It is not clear why just these principles were regarded as fundamental; the fact that they relate only to the sense of touch and not to the other senses suggests that only the sense of touch was deemed trustworthy. The four classical elements were composed of combinations of pairs of these principles: fire (hot and dry), air (hot and wet), water (cold and wet), and earth (cold and dry).

The alchemists of the Middle Ages and later centuries adopted a similar set of elements, although their "principles", such as metallic behavior, solubility, and combustibility, were different. Their continuing efforts to transmute common metals such as lead into rare ones such as gold are often regarded now with moderate contempt or amusement, but from the standpoint of a theory that views distinct substances as differing only in the proportion of different qualities, such conversions might be accomplished merely by appropriate changes in this proportion. There is no intrinsic reason why this sort of change should be regarded as less likely than some of the remarkable syntheses done routinely in the laboratory nowadays—until one understands more about the true nature of the chemical elements and the principles governing their interconversion.

The atomic theory became established unambiguously only after the brilliant insight of John Dalton in the early nineteenth century had been supported and developed by subsequent convincing experimental evidence. Dalton's essential contribution was to emphasize the possibility of deducing by experiment the relative weights of different atoms and the combining proportions of these atoms in different compounds. Many of his conjectures about chemical formulas were incorrect—chiefly because he did not acknowledge the possibility that the fundamental "particle" of an element, which we now call a molecule, might contain more than one atom (e.g., in modern terminology, his formulas for gaseous oxygen and hydrogen were O and H rather than O_2 and H_2). However, this shortcoming was soon corrected by Avogadro and Cannizzaro, and with improved analytical techniques and data the atomic theory was eventually put on a firm quantitative basis, although not without some controversy.

We shall consider in Chap. 3 the quantitative relations among the components of substances and among the reactants and products involved in chemical changes. The remainder of this chapter is devoted to a discussion of some current ideas about atoms and molecules, together with an account of a few of the

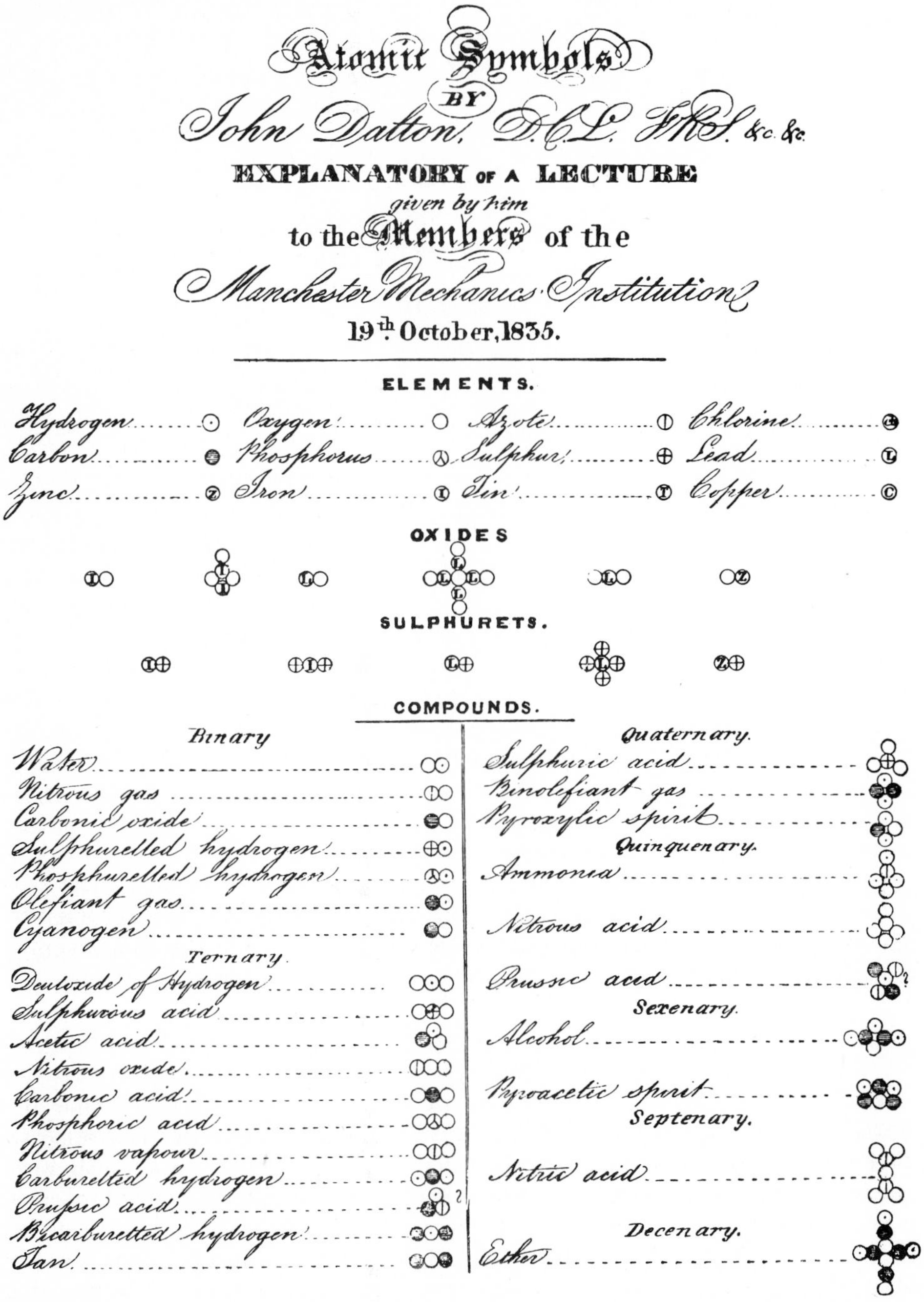

Atomic Symbols

BY

John Dalton, D.C.L. F.R.S. &c. &c.

EXPLANATORY OF A LECTURE

given by him

to the Members of the

Manchester Mechanics Institution

19th October, 1835.

ELEMENTS.

Hydrogen	Oxygen	Azote	Chlorine
Carbon	Phosphorus	Sulphur	Lead
Zinc	Iron	Tin	Copper

OXIDES

SULPHURETS.

COMPOUNDS.

Binary		Quaternary.	
Water		Sulphuric acid	
Nitrous gas		Binoliefiant gas	
Carbonic oxide		Pyroxylic spirit	
Sulphuretted hydrogen		Quinquenary.	
Phosphuretted hydrogen		Ammonia	
Olefiant gas			
Cyanogen		Nitrous acid	
Ternary.			
Deutoxide of Hydrogen		Prussic acid	
Sulphurous acid		Sexenary.	
Acetic acid		Alcohol	
Nitrous oxide			
Carbonic acid		Pyroacetic spirit	
Phosphoric acid		Septenary.	
Nitrous vapour			
Carburetted hydrogen		Nitric acid	
Prussic acid			
Bicarburetted hydrogen		Decenary.	
Tan		Ether	

Fascimile of Some of Dalton's Symbols

11

critical experiments that helped to establish them. The details of atomic and molecular structure are considered in later chapters.

1-5 The Emergence of Modern Concepts about the Atom

In the present-day view, atoms have a structure and are made up of constituents. While "constituents of atoms"—parts of the indivisible—is literally a contradiction in terms, this merely illustrates again the way in which the implications of scientific concepts change. Dalton's idea of atoms as the fundamental building blocks of the chemist has survived, but these atoms are now recognized to have components that can be separated from one another under appropriate conditions.

Rudimentary ideas concerning the structure of atoms developed during the nineteenth century, but it was not until the end of that century and the start of the present one that these ideas took definite form and the modern picture began to emerge. Important contributions to the development of atomic theory included the discovery of the electron by J. J. Thomson in 1897, the development of the nuclear model of the atom by Ernest Rutherford in 1911, and the application of quantum ideas to atomic structure by Niels Bohr in 1913. Our current views on atomic structure emerged only after the elaboration of modern quantum theory in the mid-1920s and the discovery of the neutron in 1932. The first of these developments is discussed in this chapter; the later ones are considered in Chaps. 14 and 15.

The Electron It has been known since ancient times that rubbing certain substances, such as amber and glass, with fur or wool makes them attract light objects such as bits of paper. Objects touched by rubbed amber or glass repel each other. These attractions and repulsions, termed electrical phenomena (Greek: *elektron,* amber), were explained in the early part of the eighteenth century by attributing them to the presence in matter of two kinds of fluid: positive and negative electricity. It was postulated that there is an attraction between portions of fluid of opposite sign and a repulsion between fluids of the same sign. A substance would normally contain the fluids in equal amounts, so that their properties canceled. The fluids could be separated by friction, however, and all the observed electrical phenomena could be explained as arising from an excess of either positive or negative electricity.

In 1827, Georg Ohm likened the flow of electrical fluid to the flow of water and used this analogy to give precise meaning to the concepts quantity of electricity, electric current, and electromotive force. By this time, a number of experimenters had shown that the passage of electric current through solutions caused chemical changes. For example, water is decomposed when electricity is passed through it (Fig. 1-4), hydrogen being produced at one electrode, i.e., at one of the wires dipping into the liquid, and oxygen at the other.

Michael Faraday coined the name *electrolysis* for the process of decomposing substances by electricity, and his quantitative investigations of the phenomenon gave the first experimental indication that electricity might exist as discrete particles. Faraday showed that the volumes of hydrogen and oxygen gas liberated when a current passes through water are directly proportional to the quantity of electricity. Thus, if matter was atomic in nature, a given quantity of electricity liberated a specific number of atoms. This suggested that there must be some fundamental unit of electricity associated with every atom.

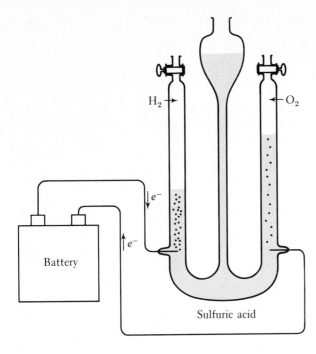

Figure 1-4   An Electrolysis Experiment
When an electric current passes through water that has been
made conducting by the addition of a small amount of acid, gaseous hydrogen is produced at the electrode at which the electrons
enter, and gaseous oxygen is formed at the other electrode. The
volume of hydrogen produced is always double the volume of oxygen, and the volumes liberated are directly proportional to the
quantity of electricity that passes through the solution. Electrolysis is discussed in Chap. 21.

Convincing experimental evidence for the particulate nature of electricity was
obtained only in the last decade of the nineteenth century as the outgrowth of
four decades of study of the electrical conductivity of gases. These studies were
made chiefly in glass tubes of various designs, usually evacuated to low pressures.
Rays were found to emanate from the cathode (the negative electrode) of a tube
containing gas at a low pressure to which a sufficiently high voltage (about 10^4 V)
had been applied to cause conductivity (Fig. 1-5). It was shown that these
cathode rays carry negative electricity, but there were conflicting views on
whether the rays consisted of particles or waves. J. J. Thomson in 1897 hypothe-
sized that they were particles, for which he adopted the name proposed a few
years earlier by Stoney, *electrons*. Thomson designed experiments for measuring
the ratio of their charge (the quantity of electricity they possess) to their mass,
e/m, by two different methods, the simpler of which involved balancing opposing
deflections of the rays by electrostatic and magnetic fields (Fig. 1-6). He found the
same value of e/m *whatever the gas present in the tube;* this strongly suggested
the presence of just one kind of negative particle.

The value of e/m found by Thomson was higher by about 10^3 than that for the
simplest charged particles previously known—atoms of hydrogen with a single
positive charge, now called *protons*. This implied either that the charge of the
electron is much higher than that of the proton, or that its mass is much smaller,
or some combination of these possibilities. R. A. Millikan's determination of the
electronic charge in 1911 permitted separate evaluation of e and m for the
electron; it was established that the mass of the electron is about $\frac{1}{1800}$ that of the
proton and its charge the same in magnitude as that of the proton. Millikan's
experiment also established unambiguously that electric charge occurs in discrete
(that is, indivisible) units and that an electric current is thus a flow of charged
particles, for example, electrons in a wire or ions (charged atoms or molecules) in
solution.

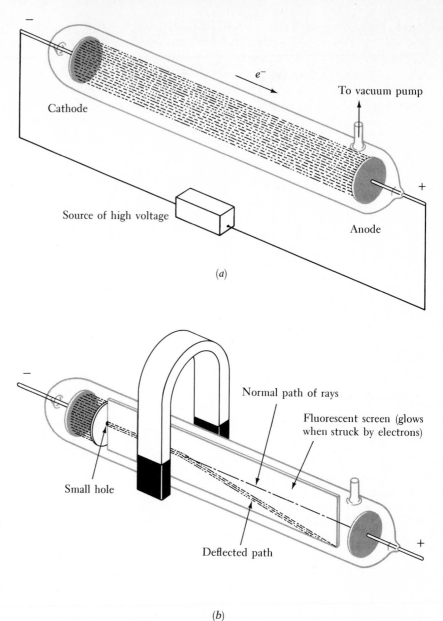

(a)

Figure 1-5 Simple Cathode-Ray Tubes

(a) When a high voltage is applied to the electrodes of an evacuated tube, a *cathode ray* (actually a stream of electrons) issues from the negative electrode (the cathode). (b) The cathode ray can be deflected by magnetic fields. (It can also be deflected by electric fields.) (*Adapted from Konrad Krauskopf and Arthur Beiser, "The Physical Universe", McGraw-Hill Book Co., Inc. Copyright 1960 by McGraw-Hill Book Co., Inc.*)

(b)

X Rays and Radioactivity The next steps in the development of our modern picture of the atom were a direct outgrowth of three remarkable discoveries that came during the last few years of the last century. The first two were experimental—the discoveries of X rays and of radioactivity. The third was Planck's quantum hypothesis and its later development, the theoretical implications of which marked a major turning point in the history of physics and of our world. These implications were so revolutionary that they were unacceptable to many who had been trained in nineteenth-century physics.

The discovery of X rays in 1895 by W. K. Röntgen was purely accidental. Röntgen was investigating the discharge of electricity through gases in an evacuated tube, as so many had before him. In the room with the equipment he also had a paper screen that glowed when exposed to ultraviolet light. Röntgen noticed that this screen emitted light even when various objects were between it and the cathode-ray tube and sought to explain this chance observation. He found that some extremely penetrating radiation was coming from the point on the cathode-ray tube where the cathode rays struck the glass wall and caused a weak glow (fluorescence). In a very short time he established many of the important properties of this radiation, which he termed X rays, including its diagnostic value in medicine. However, it was nearly two decades before the exact nature of X rays as very short wavelength electromagnetic radiation, similar in many ways to visible light, was clearly understood.

Röntgen's discovery caused great excitement, and considerable skepticism as well, and many physicists immediately turned to a study of these mysterious rays. Antoine Henri Becquerel, a professor in Paris, supposing that there might be some connection between the fluorescence of the glass wall of the cathode-ray tube and the production of X rays, undertook a systematic study of all minerals that fluoresced when exposed to sunlight.

Among the substances Becquerel studied was uranium potassium sulfate. His technique was to try to detect the possible X rays by placing the salt on a photographic plate that was carefully wrapped in thick black paper and exposing the combination to bright sunlight for several hours. He found that some very penetrating radiation was indeed present and would even pass through thin sheets of aluminum or copper. However, during a period of bad weather, while storing some photographic plate and salt combinations in the dark, *before* he had an opportunity to expose them to sunlight, he found that the salt emitted the

Figure 1-6 A Cathode-Ray Tube

In an evacuated glass tube similar to a television picture tube, electrons are emitted from a hot wire, accelerated, and shaped into a beam by a set of positively charged, perforated disks (*A*). The beam may be deflected toward the reader to a larger or smaller degree (*D*) by increasing or decreasing the charges on the vertical plates (*B*) or by moving the magnet (*C*) in or out. By quantitative experiments J. J. Thomson determined an approximate value of the ratio e/m for electrons. Similar measurements can be made for other charged particles.

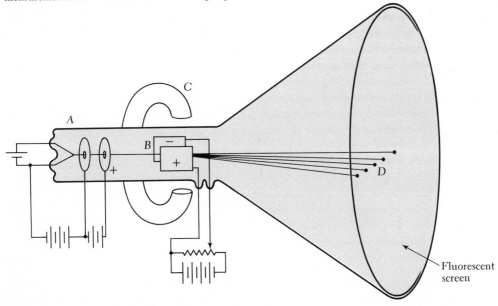

Fluorescent screen

penetrating radiation even when it had not been exposed to sunlight. Further experiments showed that the radiation was characteristic of uranium in any form or in any of its compounds. It persisted undiminished for months. The name *radioactivity* was applied to this phenomenon by one of Becquerel's students, Marie Curie, who (with her husband Pierre) discovered several radioactive elements, including radium and polonium.

Becquerel's discovery, like Röntgen's, was essentially an accidental one, resulting in this instance from the pursuit of a false hypothesis. However, the way in which each of these men carefully noted and recognized the implications of his apparently anomalous initial observations illustrates well the truth of Louis Pasteur's dictum that "chance favors the prepared mind only".

Ernest Rutherford, a young New Zealand physicist trained under Thomson at Cambridge, was among many who turned their attention to radioactivity. He and others soon found three distinct types of radiation from radioactive substances, termed initially alpha, beta, and gamma rays. Alpha rays were shown by Rutherford to consist of streams of doubly charged helium atoms, He^{2+}. Beta rays were identified as streams of electrons, like cathode rays. Gamma rays were found to be similar to X rays in their properties. Rutherford and F. Soddy succeeded in establishing the fact that radioactivity was accompanied by the spontaneous change of the radioactive atoms into different atoms (see Chap. 28), an idea that was hard for many to accept because of the supposed immutability of atoms.

The Nuclear Atom The model of the atom considered most plausible around 1910 was one proposed by Thomson in which the positive electricity was spread uniformly throughout a sphere inside which the electrons moved in circular orbits. From considerations of various properties of gases and the densities of solids, the effective diameters of atoms were known to be a few angstroms (Å) ($1Å \equiv 10^{-10}$ m); the number of electrons in any atom was approximately known from experiments with X rays.

Rutherford did not trust this model. He and his collaborators had been probing atomic structure by bombarding atoms of metals, such as silver, gold, and platinum in thin foils, with alpha particles (Fig. 1-7). The fraction of particles scattered was measured for each angle, and it was found that while most of the alpha particles passed through the foil undeflected a few of them were deflected through large angles, some even in the backward direction (that is, through angles greater than 90°). The number of these strongly scattered alpha particles increased in proportion to the thickness of the gold foil.

Rutherford's results could not be explained on the basis of Thomson's model, which pictured the positive charge of the gold atoms as spread quite *uniformly* through the foil. The large scattering angles for a few alpha particles could be understood only if the positive charge and most of the mass of each atom were concentrated in a very small region, which Rutherford termed the *nucleus* (Latin: *nucula,* small nut). Occasional backward scattering of alphas could then be explained as a result of a more or less direct collision of each of these alphas with a nucleus. Scattering of most of the alphas through small or zero angles could be explained as a result of electrostatic repulsion between alphas and nuclei at greater distances. Any electrons present in the atoms would not scatter alpha particles noticeably, because the mass of an electron is only about $\frac{1}{7500}$ that of an alpha particle. Expressed in terms of the scattering from a (hypothetical) foil consisting of a single layer of gold atoms, about 1 alpha particle in 10^8 was scattered back near the direction from which it had come. From this result,

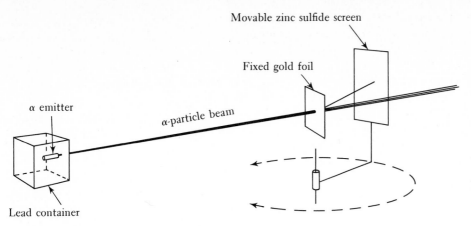

Movable zinc sulfide screen

Fixed gold foil

α emitter

α-particle beam

Lead container

Figure 1-7 Rutherford's Scattering Experiment
A beam of α particles is made to pass through a gold foil. The particles can be observed by the light flashes they produce when striking a fluorescent screen. The screen can be rotated about the center of the gold foil so that the angular dependence of the scattering of the α particles can be determined.

Rutherford was able to estimate that the area of the nucleus was about 10^{-8} times that of the entire atom, and thus the diameter of the nucleus was about 10^{-4} times that of the entire atom, or about 10^{-4} Å. Furthermore, he was able to deduce that the positive charge on the nucleus was about half of what is called the atomic weight. Since the atom as a whole was neutral, this was also the number of electrons in the atom.

These results formed the basis of Rutherford's atom model, in which a massive, central, positively charged nucleus is surrounded by Z orbiting electrons, similar to a sun surrounded by planets. The nucleus contains almost the entire mass of the atom and has a positive charge of Z electronic units, exactly neutralizing the negative charge of the Z electrons. This model, however, presented formidable difficulties. It can be proved that no static arrangement of positive and negative charges is stable, and so the electrons had to be in motion around the nucleus. It was well known from the laws of classical physics, however, that a charged particle moving in a circular or similar path should continuously radiate energy. Thus atoms should constantly lose energy and eventually the electrons should fall into the nucleus and the atom should collapse—yet it was known that atoms did not continuously radiate and that almost all atoms were stable indefinitely. The resolution of this conflict between experimental results and theoretical predictions came with Bohr's quantum theory of atomic structure in 1913 and its modification and extension in the quantum mechanics of the mid-1920s (Chap. 15).

Isotopes and the Neutron After Thomson determined the charge-to-mass ratio for the electron, he applied a similar technique to the determination of this ratio for many positively charged atoms and molecules (positive ions). His method was extended by F. W. Aston and others, and several high-precision instruments called *mass spectrometers* have been developed for measuring the charge-to-mass ratio for almost any charged particle (ion) that can be formed in a gas. One mass spectrometer is described in Fig. 1-8. Since the charge on the particle, expressed in units of the electronic charge, is always a small integer, usually 1 or 2, the mass of the ion can thus be determined with high precision.

When the first mass spectrometer was used with a pure sample of the gaseous element neon, it was discovered that there were two distinct atoms present, with masses approximately 20 and 22 times that of the hydrogen atom. It was soon found that many other chemical elements consisted of atoms with different masses but essentially identical chemical properties. The term *isotope* (Greek:

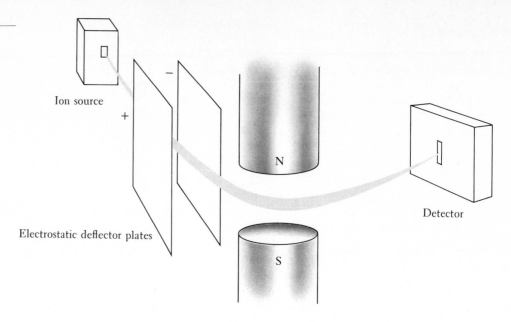

Ion source

Electrostatic deflector plates

Detector

Magnetic deflector

Figure 1-8   A Mass Spectrometer
The atoms or molecules of interest are ionized and accelerated by electric fields in the ion source. They are deflected from a straight-line path by an electric field and then by a magnetic field, both arranged in such a way that the total deflection does not depend on the velocity of these ions but just on the ratio ne/m_{ion} of their charge to their mass where n is a small integer and e is the magnitude of the charge on the electron. The ions are finally collected and counted with a detector. Photographic plates were first used for detection, but electronic means are now much more common. A variable magnetic field is used so that ions of different ne/m_{ion} successively pass through the slit of the detector as the magnetic field is changed.

isos, equal; *topos,* place) was coined to apply to these different atoms that occupy the same place in the classification of the elements according to their chemical properties, the periodic table (Sec. 2-4).

In 1932, James Chadwick, a British physicist working in Rutherford's laboratory in Cambridge, demonstrated the existence of the neutron, a neutral particle with mass approximately equal to that of the hydrogen atom. The neutron had actually been suggested by Rutherford in 1920, but many experiments to detect it during the 1920s had been unsuccessful. Chadwick's discovery made possible the present view that a nucleus "contains" protons and neutrons and that isotopes of a given element differ only in the number of neutrons they contain, as described in more detail in the next chapter. Some properties of the "elementary" particles electron, proton, and neutron are summarized in Table 1-1.

**Table 1-1
Some Properties of
Elementary Particles**

	MASS		CHARGE	
	kg	Atomic mass* units, u	Coulombs, C	Electronic charge units
Electron	9.1095×10^{-31}	5.4858×10^{-4}	-1.602×10^{-19}	-1
Proton	1.67265×10^{-27}	1.00728	$+1.602 \times 10^{-19}$	$+1$
Neutron	1.67495×10^{-27}	1.00867	0	0

*See Sec. 2-1.

Summary

The primary concerns of chemistry are the composition, structure, and properties of substances, the transformations of these substances into others by reactions, and the different kinds of energy changes that accompany these reactions. Units as well as numbers are needed to specify many physical quantities. Those physical quantities with which units are associated are said to have dimensions. Pure numbers, such as the ratio of two quantities with the same dimensions, are termed dimensionless.

Most of the units in this text are SI units, which are based on the meter, the kilogram, the second, and four other fundamental units (Appendix A). In calculations involving any quantity, both the numerical value and the units are handled by the rules of arithmetic and algebra. The cancellation of units often provides a useful check in the solution of problems and in the conversion of numerical values from one system of units to another.

The physical properties of a substance are those such as color and density that do not depend on its reaction with other substances. Chemical properties relate to the reactions of a substance with itself or other substances to form new substances with different properties. Pure substances have distinct physical and chemical properties and definite composition. Mixtures lack one or more of the well-defined properties of pure substances. For example, mixtures are usually of variable composition. The properties of homogeneous mixtures (solutions) are uniform throughout. A heterogeneous sample contains regions that have different properties and are separated by sharp boundaries.

Phenomena that occur on the atomic or molecular scale are termed microscopic; those that are directly apparent to our senses are called macroscopic. The macroscopic differences in properties that characterize the three common states of aggregation of matter—solid, liquid, and gaseous—can be explained on the microscopic level in terms of the closeness and regularity of packing of the constituent atoms and molecules. Molecules in crystalline solids are arranged with great regularity; in amorphous solids the molecular arrangement is more or less random.

Atoms consist of a positively charged nucleus surrounded by electrons, particles of negative charge. The constituents of the nucleus are protons (hydrogen atoms with a single positive charge) and neutrons (uncharged particles comparable in mass to the proton). The mass of the electron is about $\frac{1}{1800}$ that of the proton. Atoms are electrically neutral because the number of electrons is equal to the nuclear charge.

Three distinct types of emission from radioactive nuclei are known: α rays, doubly charged helium nuclei, He^{2+}; β rays, electrons; and γ rays, similar to X rays.

Terms and Concepts

Problems and Questions

1-1 Definitions of Terms (*a*) What is the distinction between a physical property and a chemical property? Give examples to illustrate your answer. (*b*) Can a sample of a pure substance be heterogeneous? Can a sample of an impure substance be homogeneous? Explain, citing examples.

1-2 Definitions and Distinctions Distinguish carefully between the following terms, giving examples of each to illustrate the distinction:
(*a*) Quantities with dimensions and dimensionless quantities.
(*b*) Units and dimensions.
(*c*) Macroscopic and microscopic points of view.

1-3 Phases and Substances Define the terms phase and substance. How many phases and how many substances are present in each of the following?
(*a*) A sample of water in a sealed container at 0°C, with ice, liquid water, and water vapor all present.
(*b*) A solution of sugar in water at 25°C, with no undissolved sugar present. Disregard any air space that may be present above the solution.
(*c*) A sample of air, containing nitrogen, oxygen, argon, carbon dioxide, and gaseous water.

1-4 Density and Volume The density of carbon tetrachloride, CCl_4, at 20°C is 1.595 g cm^{-3}. What is the volume of 375.0 g of CCl_4 at 20°C?

1-5 Density of a Solid The volume of a solid may be measured indirectly in terms of the mass of a liquid of known density that it displaces from a previously full container. An irregular solid of mass 5.72 g was found to displace 2.50 g of a liquid whose density was 1.04 g cm^{-3}. What is the density of the solid?

1-6 Volume of Flask To determine the volume of a flask, the flask was first weighed dry and was then filled to a calibration mark with water of known temperature (and therefore known density) and weighed again. The empty flask weighed 38.365 g, the filled flask 63.411 g. The density of the water used was 0.99823 g ml^{-1}. What was the volume of the flask?

1-7 Bus Routes The distance covered by a bus on a certain route is 29 mile and the bus covers this route 25 times in a 20-hour day. What is the average speed of the bus during this period, in km $hour^{-1}$? Set the problem up so that appropriate units cancel (1 mile = 1.609 km).

1-8 Speed of Mariner IV En route to Mars, Mariner IV covered 1.34×10^8 mile in the 228 days required for the journey. What was its average speed in mile $hour^{-1}$ and in km s^{-1}? Set the problem up so that appropriate units cancel.

1-9 Fuel Economy In some countries the fuel economy of an auto is measured in terms of the liters of gasoline needed to cover 100 km. Suppose your car gets 25 mile to the gallon. What is its fuel economy in liter $(100 \text{ km})^{-1}$? Set the problem up so that appropriate units cancel (1 gal = 3.79 liter; 1 mile = 1.609 km).

1-10 Physical and Chemical Properties Classify each of the following observations or phenomena as reflecting chiefly physical properties or chemical properties of the materials concerned:

Knives made of stainless steel (*a*) are resistant to corrosion but (*b*) tend to lose their cutting edges through frequent use.
Common household illumination (*c*) in the eighteenth century was by use of candles but (*d*) now involves chiefly electric light bulbs.
The care of swimming pools involves (*e*) pumping the pool water through a filter and (*f*) the action of chlorine and other disinfectants on bacteria.

1-11 Historically Important Experiments Describe briefly the importance of each of the following experiments or experimental observations in the development of atomic theory:
(*a*) Scattering of alpha particles by metal foil.
(*b*) Measurement of charge-to-mass ratio of different positive ions in a mass spectrometer.
(*c*) The fact that the charge-to-mass ratio for cathode rays is independent of the nature of the gas in the cathode-ray tube.

1-12 Important Concepts (*a*) What is an electric current? (*b*) How do a crystalline solid and an amorphous solid differ? (*c*) In what respects are a proton and a neutron similar and in what respects do they differ? (*d*) How are the charges on a proton and an electron related and what are the approximate relative masses of these particles?

Atoms, Molecules, and the Periodic Table

2

"The majority of English chemists represent the atomic weight of carbon by 6, that of oxygen by 8, and that of sulphur by 16. Dr. Frankland would double the atomic weight of carbon, but would retain the old atomic weights of oxygen and sulphur. Mr. Griffin, who lays claim to priority in doubling the atomic weights of carbon and oxygen, ridicules the notion of doubling that of sulphur. Dr. Williamson, Mr. Brodie, and myself have for a long time advocated the doubling of all three.

For silicon Thomson took 7.12, assuming the chloride to be SiCl; Gmelin prefers $SiCl_2$ and 14.25, and Berzelius 21.37 and $SiCl_3$; whereas in my opinion the balance of argument is in favour of $SiCl_4$ and 28.50."

W. ODLING, 1859

". . . I have tried to base a system on the magnitudes of the atomic weights of the elements. My first attempt in this respect was the following: I chose the substances with the smallest atomic weights and arranged them according to the sizes of their atomic weights. This showed that there existed a periodicity in the properties of these simple substances and that even according to their atomicity[1] the elements followed one another in the arithmetical sequence of their atomic weights."

D. I. MENDELEEV, 1869

In this chapter we continue the discussion of atoms and elements begun in Chap. 1, progressing then to a discussion of the atomic weight scale, molecules and compounds, and the important concept of the mole. The final sections of the chapter provide an introduction to the periodic table, which is an arrangement of the elements in a systematic way that reflects many similarities among and trends in their chemical properties. We return to a more detailed consideration of the periodic table in Chap. 16.

2-1 Atoms and Atomic Weights

Atoms and Chemical Elements The charge Z on the nucleus of an atom is termed the *atomic number*. The modern definition of a chemical element is that it is a substance that contains only atoms of the same atomic number. There are now 106 elements known, each represented by a one-letter or two-letter symbol (H, O, Cl) representing important initial letters in the name that was in common use for the element when the symbol was adopted. Most elements have unique names and symbols. An exception is the simplest element, hydrogen, for which the three isotopes have different symbols and names, as discussed later. The names and symbols are listed inside the front cover.

---

[1] The word *atomicity,* a literal translation from the Russian, was used by many chemists at that time for the concept later termed *valence.*

The nucleus of each atom is composed of protons and neutrons (except for the isotope of hydrogen of atomic weight 1, whose nucleus consists of a single proton). The sum of the numbers of neutrons and protons in a nucleus is called the *mass number A*. Since the proton charge is not lost when neutrons and protons are together in a nucleus, the nucleus of an atom with atomic number Z contains Z protons and $(A-Z)$ neutrons. Neutrons and protons are called *nucleons*.

The nucleus of an atom is surrounded by Z electrons, so that the entire atom is electrically neutral. The chemical behavior of an atom is determined almost exclusively by these electrons. It is, in fact, mainly the outermost of these electrons, also known as *valence electrons,* that are responsible for the chemical properties, as discussed in later chapters. To a very small extent, the chemical properties depend on the mass number A, but this influence of A on the chemistry of a particular element can usually be neglected. The chemical identity of an atom is thus determined almost entirely by its atomic number Z.

Isotopes   An atomic species of given atomic number Z and specified mass number A is called a *nuclide*. It is often represented by adding a preceding superscript A and subscript Z to the chemical symbol "Ch" of the atom:

$$^A_Z\text{Ch}$$

Thus $^{12}_6\text{C}$ represents a carbon atom whose nucleus contains 6 protons and 6 neutrons. It is often called carbon 12. The subscript 6 is redundant, since the same information is contained in the symbol C. It is not easy to remember the atomic numbers of all elements, however, so explicit listing of Z as a subscript is often useful. For the lighter elements ($Z \leq 20$), A is about twice Z; for the heavier elements, A is somewhat greater than twice Z. *Isotopes* have the same atomic number (Z) but different mass numbers (A).

The simplest element, hydrogen, has three isotopes, ^1_1H, ^2_1H, and ^3_1H, the last of which is radioactive. The relative difference in mass among these isotopes is the largest known; that is, for no other element are the ratios of isotopic masses as large as $3:1$ and $2:1$ or even $3:2$. The corresponding chemical effects of these mass differences are thus the largest known. For this reason, separate names and symbols are sometimes used for the isotopes: hydrogen 2 is called deuterium, D; hydrogen 3, tritium, T; and hydrogen 1, occasionally protium. The nucleus of deuterium is called the deuteron, and the tritium nucleus is called the triton; in addition to one proton they contain one and two neutrons, respectively.

There are two stable isotopes of carbon, ^{12}C and ^{13}C. One of its radioactive isotopes, ^{14}C, is important for radiocarbon dating (Chap. 28). The natural abundances of the isotopes of hydrogen and carbon are listed in Table 2-1. These are number percents, so that, for example, of 100,000 representative hydrogen nuclei 99,984 are ^1_1H and 16 are ^2_1H.

Atomic Weight and Isotopes   Dalton emphasized the possibility of determining the relative weights of different atoms from precise data on the composition and reacting proportions of different substances, and a great deal of effort was devoted to this task during the nineteenth century and much of the present one. The term *atomic weight* (AW) is customarily used to designate the relative weight of an atom, referred to a common standard that is defined by setting the atomic weight of the most abundant form of carbon atom at exactly 12.

Table 2-1
Natural Abundances of
Isotopes of Hydrogen and
Carbon

	Percent abundance	Atomic weight		Percent abundance	Atomic weight
$^1_1$H	99.984	1.0078	$^{12}_6$C	98.89	12.0000*
$^2_1$H (=D)	0.016	2.0141	$^{13}_6$C	1.11	13.0034
$^3_1$H (=T)	10^{-16}†	3.0160	$^{14}_6$C	10^{-10}†	14.0032

*The atomic weight of ^{12}C is by definition exactly 12.

†Tritium and carbon 14 occur naturally, despite their continuous radioactive decay at a rate that is rapid on a geologic time scale, because they are continually formed in the upper atmosphere by the reaction of neutrons with ^{14}N and ^{16}O. The neutrons are formed by the interaction of cosmic-ray protons with these same atmospheric constituents. Tritium is also formed during thermonuclear explosions, and it may be detected in atmospheric and surface waters in much larger proportions for many years after an explosion in the atmosphere.

Thus, the atomic weight of the ^{16}O isotope is 15.995, which implies the relation

$$\frac{\text{Weight of one } ^{16}\text{O atom}}{\text{Weight of one } ^{12}\text{C atom}} = \frac{15.995}{12.000}$$

On the same scale, the simplest element, hydrogen, has an atomic weight of very nearly 1. Since weight is proportional to mass if all measurements are made under the same conditions of gravity, there is usually no distinction made between the terms *atomic mass* and *atomic weight* referred to this dimensionless scale. We shall normally use the more common term, atomic weight. The unit of mass normally used for individual atoms is the atomic mass unit, u, exactly one-twelfth the mass of a carbon 12 atom. Since the mass in grams of a ^{12}C atom can be determined only by experiment, its value is subject to refinement. A change in this value will not alter the definition of the atomic mass unit, but it will change the number of atomic mass units per gram.

If a naturally occurring element consists of a mixture of isotopes, as most do, the atomic weight of this element as determined by chemical methods represents the average weight of the different isotopes, weighted in proportion to their abundance. The average weight is observed because typical chemical experiments involve very large numbers of atoms and molecules. On the other hand, atomic weights determined with the mass spectrometer represent the weights of the individual isotopes of which the different elements are composed. Although many chemical atomic weights (listed inside the front cover) are nonintegral, the weights of individual isotopes are very nearly integral, i.e., whole numbers.

□ **Atomic Weight from Natural Abundances** Silicon consists of three stable isotopes, whose atomic weights and relative abundances are as follows:

Example 2-1

	Atomic weight	Percent abundance
^{28}Si	27.98	92.2
^{29}Si	28.98	4.7
^{30}Si	29.97	3.1

What is the (chemical) atomic weight of silicon?

Solution The atomic weight is the weighted average of the individual values:

$$\text{Atomic weight Si} = \frac{27.98(92.2) + 28.98(4.7) + 29.97(3.1)}{92.2 + 4.7 + 3.1}$$

$$= \underline{28.09} \ \blacksquare$$

Exercise 2-1

☐ Calculate the (chemical) atomic weight of boron from the following natural abundances and atomic weights for its two stable isotopes, given in parentheses: ^{10}B (19.6 percent, 10.013), ^{11}B (80.4 percent, 11.009). ■

We speak of *chemical* atomic weights when this situation requires emphasis. The accuracy with which the atomic weight of a given element is known reflects not only the quality of the experimental work on which the value is based but also the fact that the natural abundances of the different isotopes vary slightly, depending on the source of the element or compound considered. It is fortunate that for most elements natural abundances do not vary enough to affect usual weight relations in chemical reactions.

2-2 Molecules and the Mole

Molecules and Compounds Discrete uncharged groupings of atoms, tied together by specific forces in a definite arrangement, are called *molecules*. The arrangement need not be rigid, but there must be a clear distinction between atoms that belong to the molecule and those that do not. The atoms in a particular molecule may be either the same (H_2, O_2, P_4, S_8) or different.

A chemical compound contains at least two elements. The defining characteristic is that a compound contains different elements in a *definite* ratio, independent of the way in which the compound was prepared. The chemical composition of a compound is expressed by formulas (H_2O, Na_2SO_4, $C_{12}H_{22}O_{11}$) that indicate the relative numbers of atoms of different elements present in the substance. The definite composition may be a consequence of the compound's existing as molecules, for example, H_2O or CO_2. However, other compounds are composed of *ions*, that is, charged atoms or molecules, and the definite composition of such a compound results from its overall neutrality, which implies that it contains appropriate numbers of ions of opposite charge to make the charges balance. Thus sodium nitrate, $NaNO_3$, contains equal numbers of Na^+ and NO_3^- ions (the names and formulas of common ions are listed in Appendix B), and ammonium sulfate, $(NH_4)_2SO_4$, contains two ammonium ions (NH_4^+) for every sulfate ion (SO_4^{2-}). These compounds, which are crystalline solids at ordinary temperatures, have properties that depend on those of their component ions; no molecules are present, that is, it is not possible to associate a given sodium ion with a given nitrate ion, or some pair of ammonium ions with a given sulfate ion.

Molecular Weight and Formula Weight The concept of relative weights of atoms can be extended to molecules and compounds.

The molecular weight of a molecule (abbreviated MW) is the sum of the atomic weights of the atoms it contains. We normally use the term *molecular weight* only with reference to compounds that consist of molecules. More generally, the term *formula weight* (FW) will be used, defined as the sum of the atomic weights of all atoms shown in the formula of the compound. Some examples follow.

Example 2-2

☐ **Molecular Weight** Carbon dioxide exists as discrete CO_2 molecules in the solid, liquid, and gaseous states. The molecular weight and the formula weight are thus identical:

$$AW\ C + 2\ AW\ O = 12.01 + 2(16.00) = \underline{44.01}\ \blacksquare$$

☐ What are the molecular and formula weights of ammonia, a gas that consists of molecules of formula NH_3? ■

Example 2-3

☐ **Formula Weight** Sodium sulfate, Na_2SO_4, in the crystalline form or when melted, consists only of Na^+ and SO_4^{2-} ions, in the ratio 2:1. Similarly, in aqueous solutions (solutions in water) there are no Na_2SO_4 molecules, but only Na^+ and SO_4^{2-} ions, associated fleetingly with various numbers of water molecules, as discussed in Chap. 9. Under ordinary circumstances, no Na_2SO_4 molecules exist, and the term *molecular weight* is inappropriate. The formula weight of Na_2SO_4 is

$$2\ AW\ Na + AW\ S + 4\ AW\ O = 2(22.99) + 32.06 + 4(16.00) = \underline{142.04}\ \blacksquare$$

☐ What is the formula weight of $Ca_3(PO_4)_2$, which consists of Ca^{2+} and PO_4^{3-} ions? ■

Example 2-4

☐ **Water of Crystallization**[2] When an aqueous solution of sodium sulfate is evaporated, crystals of $Na_2SO_4 \cdot 10H_2O$ appear below 32.4°C, and crystals of anhydrous (water-free) Na_2SO_4 appear above this temperature. The formula weight of $Na_2SO_4 \cdot 10H_2O$ is

$$FW\ Na_2SO_4 + 10\ FW\ H_2O = 142.04 + 10(18.02) = \underline{322.2}$$

The water molecules in these crystals are typical of what is called water of crystallization, and the compound exists only in the crystalline state. ■

☐ What is the formula weight of $CaSO_4 \cdot \frac{1}{2}H_2O$ (plaster of paris)? ■

Avogadro's Number and the Mole Most chemical experiments involve macroscopic amounts of substances, and it is therefore extremely useful to define a macroscopic quantity that always contains the same number of microscopic particles such as atoms, molecules, or ions. This unit is called the *mole* (SI symbol, mol),[3] and the number of particles it contains is called *Avogadro's number,* for which the symbol N_A will be used. Avogadro's number N_A is *defined* as the number of atoms contained in exactly 12 g of ^{12}C. The *value* (as opposed to the definition) of N_A is 6.022×10^{23}. Further, 1 mol of a chemical species (atoms, molecules, or ions) is defined as the quantity that contains N_A particles, as indicated by the formula given (such as H, H_2, NO_2, SO_4^{2-}). Moreover, if *one formula unit* of a compound is defined as that quantity that contains exactly the

---

[2]The dot (·) is used in formulas to indicate that the components shown are associated in a definite proportion, without any implication about the nature of the exact molecular or ionic species present or the way in which they are combined.

[3]The unusually lengthy symbol *mol* was presumably selected because further contraction of *mole* might lead to confusion. Its usage is parallel to that of other symbols such as m or kg. No plural exists and the symbol is usually employed with numerals, as in "3 mol" or "5 kg", in contrast to "three moles" or "five kilograms". *Mol* must not be taken as an abbreviation for molecule(s).

number of different species indicated by the formula of the compound, then *one mole* of the compound contains N_A formula units. Thus 1 mol $Al_2(SO_4)_3$ contains $2N_A$ Al^{3+} ions and $3N_A$ $SO_4{}^{2-}$ ions. Sometimes the term *mole* is used to refer to N_A particles that are not all of the same kind. For example, 1 mol of air contains Avogadro's number of "air molecules", with appropriate contributions from the component gases, chiefly N_2 and O_2.

This *number aspect of the mole* should be carefully noted and remembered: a mole is a specific number of things, just as a dozen is, although the number in a mole is almost incomprehensibly large.[4] When we speak of a dozen objects, their shape, weight, or color is unimportant. There are 12 objects, and in 10 dozen there are 120. Similarly, 1 mol refers to 6.022×10^{23} of the objects considered, and 10 mol to 6.022×10^{24}. Although our main concern will be with moles of atoms, molecules, ions, or formula units, we shall have occasion to refer to a mole of electrons (which is called the *faraday*, a fundamental unit in electrochemistry).

A second aspect of the mole is of considerable practical value in chemical calculations. *The weight in grams of a mole of any atomic, molecular, or ionic species, or of a mole of formula units, is numerically the same as the weight of that species or formula unit on the atomic weight scale.* This is a necessary consequence of the definition of the atomic weight scale and of Avogadro's number, N_A. Since the mass of one atom of ^{12}C is defined to be exactly 12 u, where u is 1 atomic mass unit,[5] and since, by definition, N_A atoms of ^{12}C weigh exactly 12 g, it follows that

$$12N_A \, \text{u} = 12 \text{ g} \qquad \text{and} \qquad N_A \, \text{u} = 1 \text{ g}$$

or
$$1 \, \text{u} = \frac{1}{N_A} \text{ g} = \frac{1}{6.022 \times 10^{23}} \text{ g} = 1.6606 \times 10^{-24} \text{ g}$$
$$= 1.6606 \times 10^{-27} \text{ kg}$$

This relation and Table 2-2 show that Avogadro's number is a *scale factor* for conversion from atomic mass units to grams. Table 2-2 also illustrates the advantage of using atomic mass units when individual atoms or molecules are considered, because the masses in macroscopic units, such as grams, are inconveniently small. For the large numbers of atoms and molecules that are normally used in chemical experiments, the mole is a convenient unit because it represents a macroscopic quantity, and more specifically because for any molecular compound the weight of *one mole of molecules,* expressed in grams, is numerically equal to the molecular weight of the molecule. Moreover, recalling that *one*

---

[4] The sizes of very large and very small numbers are hard to comprehend, especially when they are expressed as powers of 10. For example, 1 Å is 10^{-10} m; this implies that it takes 250 million Å, a number well beyond everyday comprehension, to make up 1 inch. It is particularly difficult to appreciate the enormous size of Avogadro's number and the infinitesimal dimensions of atoms and atomic nuclei. The following examples are illustrative:

The volume of N_A dust grains, represented by tightly packed cubes each 0.010 mm on a side, is that of a cube more than 840 m (about a half mile) on a side.

Were we to mark the molecules in 100 ml (about $\frac{1}{3}$ cup) of water and mix them with the entire volume of the oceans (1.37×10^9 km^3), each 100 ml of the mixture would still contain over 200 of the original molecules.

Imagine trying to see with your unaided eye a large beachball and a sand grain resting on the surface of the moon. Their images at that distance are, respectively, about the same size as those of an iodine atom and its nucleus viewed from a distance of 20 cm.

[5] The atomic mass unit, sometimes abbreviated amu, is also called the dalton, abbreviated d.

Table 2-2
Molecular and Molar
Quantities

	WEIGHT OF ONE PARTICLE		WEIGHT OF A MOLE OF PARTICLES	
Particle*	Atomic mass units	Grams	Atomic mass units	Grams
^{12}C atom	12†	1.9927×10^{-23}	7.2264×10^{24}	12†
C atom	12.01115	1.9945×10^{-23}	7.2331×10^{24}	12.01115
O atom	15.9994	2.6568×10^{-23}	9.6348×10^{24}	15.9994
O_2 molecule	31.9988	5.3132×10^{-23}	19.2697×10^{24}	31.9988
CO_2 molecule	44.0100	7.3082×10^{-23}	26.5028×10^{24}	44.0100

* An average over the natural isotopic abundances is implied unless specific isotopes are indicated by superscripts.

† By definition.

formula unit of a compound is defined as that quantity that contains exactly the numbers of different species indicated by the formula of the compound, then the weight of *one mole of formula units,* expressed in grams, is numerically equal to the formula weight of the compound.

The value of Avogadro's number necessarily depends both on the choice of the standard for the atomic weight scale and on the choice of the gram as the unit of mass for comparison. For example, the basic SI unit of mass is the kilogram (kg), and the scale factor from atomic mass units to kilograms would be 1000 times Avogadro's number. Hence, a factor 10^{-3} is required when the mass of 1 mol is to be expressed in SI mass units. For example, the masses of 1 mol each of ^{12}C and ^{1}H atoms are, respectively, 12×10^{-3} kg and 1.008×10^{-3} kg.

Note that 1 mol H *atoms* weighs 1.008 g, and 1 mol H_2 *molecules* weighs 2.016 g. If no particular molecular species is indicated, as in "a mole of hydrogen", the inference is that the stable species at room temperature is meant. This still requires you to know the chemical formula of the element; under normal conditions, hydrogen is a diatomic molecule, H_2, and so "a mole of hydrogen" normally means 6.022×10^{23} hydrogen molecules, or 2.016 g of the element. It is important to learn the formulas of simple substances as quickly as possible so as to know what is implied by a mole of such substances. Also note that 1 mol Na_2SO_4 (142.04 g) contains 1 mol SO_4^{2-} and 2 mol Na^+, as does 1 mol (or 322.19 g) $Na_2SO_4 \cdot 10H_2O$ (which contains, in addition, 10 mol water).

You should memorize the names and symbols of the elements with atomic numbers 1 through 38 at once, and also should learn soon those for elements 47 through 56 and 78 through 83.

□ **Formula Weights and Moles** The following questions should be answered with respect to *each* of these three species: O (an oxygen atom), O_2 (an oxygen molecule), and O_3 (an ozone molecule).

Example 2-5

(*a*) What is the formula weight?

Solution The atomic weight of oxygen is 16.0 and the formula weight is defined as the sum of the atomic weights of all the atoms in the formula. Thus the formula weights for O, O_2, and O_3 are 16.0, 32.0, and 48.0, respectively.

(*b*) What is the weight in grams of 1.00 mol?

Solution The weight in grams of 1 mol is numerically the same as the formula weight. Thus 1.00 mol of O weighs 16.0 g, 1.00 mol of O_2 weighs 32.0 g, and 1.00 mol of O_3 weighs 48.0 g.

(c) How many oxygen atoms are there in one dozen of each species?

Solution Since one dozen is just 12 of the species represented by the formula, one dozen O contains just <u>12</u> atoms of oxygen, while one dozen O_2 contains $2 \times 12 = \underline{24}$ atoms of oxygen, and one dozen O_3 contains $3 \times 12 = \underline{36}$ oxygen atoms.

(d) How many oxygen atoms are there in 1.00 mol?

Solution Since 1 mol contains 6.0×10^{23} of the species represented by the formula, 1.00 mol O contains $\underline{6.0 \times 10^{23}}$ oxygen atoms, while 1.00 mol O_2 contains $2 \times 6.0 \times 10^{23} = 12.\overline{0 \times 10^{23}} = \underline{1.20 \times 10^{24}}$ oxygen atoms, and 1.00 mol O_3 contains $3 \times 6.0 \times 10^{23} = \underline{1.80 \times 10^{24}}$ oxygen atoms.

(e) Give the number of moles that contain 36×10^{23} oxygen atoms.

Solution We can use the results from (d), writing each of them in terms of moles of species/number of oxygen atoms:

$$\frac{1.00 \text{ mol O}}{6.0 \times 10^{23} \text{ O atoms}} \times 36 \times 10^{23} \text{ O atoms} = \underline{6.0} \text{ mol O}$$

$$\frac{1.00 \text{ mol } O_2}{12.0 \times 10^{23} \text{ O atoms}} \times 36 \times 10^{23} \text{ O atoms} = \underline{3.0} \text{ mol } O_2$$

$$\frac{1.00 \text{ mol } O_3}{18.0 \times 10^{23} \text{ O atoms}} \times 36 \times 10^{23} \text{ O atoms} = \underline{2.0} \text{ mol } O_3$$

Note that we have labeled each factor with its units, and canceled units when they occur to the same power in the numerator *and* the denominator. ∎

Exercise 2-5

☐ Calculate the following for the three species S, S_2, S_6: (a) formula weight, (b) weight in grams of 0.50 mol, (c) the number of S atoms in one gross (144) of each species, (d) the number of S atoms in 2.00 mol of each species, (e) the number of moles that contain 6.0×10^{24} S atoms. ∎

Example 2-6

☐ **Calculations Using Molecular Weight and Density** Calculate the following quantities for the compound difluorohexane, $C_6H_{12}F_2$, a liquid with density 0.90 g ml^{-1}:

(a) *Grams in 2.00 mol:* The molecular weight is $6(12.0) + 12(1.0) + 2(19.0) = 122.0$. Hence 2.00 mol corresponds to

$$2.00 \text{ mol} \times \frac{122.0 \text{ g}}{\text{mol}} = \underline{244 \text{ g}}$$

(b) *Volume occupied by 1.00 mol:* Density information must be used here. We know from (a) that 1 mol weighs 122.0 g. Hence the volume occupied by 1 mol is[6]

$$\frac{122.0 \text{ g}}{0.90 \text{ g ml}^{-1}} = \underline{136 \text{ ml}}$$

---

[6] Note that the answer is given to three significant figures even though one factor in the preceding step is given only to two significant figures. The reason is that if the last digit in this factor, 0.90, were different by 1 this would be a change of 1 part in 90, or about 1 percent. The answer should reflect a comparable percentage uncertainty, which is the case with 136 ml, whereas retention of only two significant figures as with 1.4×10^2 ml would imply an uncertainty of 1 part in 14, or 7 percent. Significant figures are discussed in detail at the start of Chap. 3.

(c) *Weight in grams that contains 3.0×10^{23} hydrogen atoms:* One method
of solving problems of this kind is to multiply the original quantity by
factors that connect related quantities, in units chosen so that in the end
there is cancellation of all units except those desired for the answer. The
factors are derived from information given in the problem, such as (1 mol
compound)/(12 mol hydrogen atoms), which follows from the formula
$C_6H_{12}F_2$ or from definitions, such as (1 mol atoms)/(6.022×10^{23} atoms).
Thus

$$3.0 \times 10^{23} \text{ hydrogen atoms} \times \frac{1 \text{ mol hydrogen atoms}}{6.0 \times 10^{23} \text{ hydrogen atoms}}$$

$$\times \frac{1 \text{ mol compound}}{12 \text{ mol hydrogen atoms}} \times \frac{122.0 \text{ g compound}}{1 \text{ mol compound}} = \underline{5.1 \text{ g}} \text{ compound}$$

In other words, the different factors convert the number of hydrogen atoms
given initially, in turn, into moles of hydrogen atoms, moles of compound,
and grams of compound.

   Another method of setting up problems of this kind is by use of proportions. The number of hydrogen atoms, 3.0×10^{23}, represents $\frac{1}{2}$ mol. The
formula of the compound shows that one molecule of it contains 12 hydrogen atoms, so 1 mol contains 12 mol hydrogen atoms. Thus, if x represents
the moles of compound desired,

$$\frac{x}{0.50 \text{ mol H atoms}} = \frac{1 \text{ mol compound}}{12 \text{ mol H atoms}}$$

$x = \frac{1}{24}$ mol compound, which weighs

$$\frac{122 \text{ g}}{\text{mol}} \times \frac{1}{24} \text{ mol} = \underline{5.1 \text{ g}}$$

(d) *Weight in grams of 10 molecules:* Since 1 mol weighs 122.0 g, 1 molecule
weighs $(122.0/N_A)$g, and 10 molecules weigh $(1220/N_A)$g $= (1220/6.022$
$\times 10^{23})$g $= \underline{2.026 \times 10^{-21} \text{ g}}$.

(e) *Weight that contains 6.0 g carbon:* Using the first method of part (c), we
find

$$6.0 \text{ g carbon} \times \frac{1 \text{ mol C atoms}}{12.0 \text{ g C}} \times \frac{1 \text{ mol compound}}{6 \text{ mol C atoms}} \times \frac{122.0 \text{ g compound}}{1 \text{ mol compound}}$$
$$= \underline{10.2 \text{ g}} \text{ compound}$$

To utilize the proportion method, we first note that since the atomic weight
of C is 12.0, 6.0 g represents 0.50 mol C atoms. One mole of the compound
contains 6 mol C atoms; thus, if x is moles compound,

$$\frac{x}{0.50 \text{ mol C atoms}} = \frac{1 \text{ mol compound}}{6 \text{ mol C atoms}}$$

and $x = 0.083$ mol compound, which amounts to

$$0.083 \text{ mol} \times \frac{122 \text{ g}}{\text{mol}} = \underline{10.2 \text{ g}} \quad \blacksquare$$

☐ Calculate the following for hexachlorodisilane, Si_2Cl_6, a colorless liquid
of density 1.58 g ml^{-1} at 0°C: (a) grams in 5.00 mol, (b) volume occupied by

Exercise 2-6

2.00 mol at $0°C$, (*c*) weight in grams that contains 1.80×10^{24} Cl atoms, (*d*) weight in grams of 15 molecules, (*e*) weight that contains 20.0 g Si. ∎

2-3 The Periodicity of Chemical Properties

Throughout the first half of the nineteenth century, and even somewhat earlier, many chemists were searching for patterns among the numerical values of atomic weights and of the combining weights of substances—that is, the relative weights of two substances that combine with each other in a chemical reaction. Many of the early attempts at finding a quantitative systematization of the chemical properties of the elements were discarded, and even ridiculed, because they seemed too limited and often led to values of atomic and molecular weights contradictory to the sets of apparently accurate atomic weights then being determined. Only a few of the chemists of the time recognized that it was not atomic weights themselves, but rather relative combining weights, that were being established quite accurately; the factors (small whole numbers or simple fractions, such as $\frac{2}{3}$) needed to convert these into a scale of relative atomic weights were usually quite uncertain.[7]

By the 1830s most chemists regarded the atoms of different elements, to the extent that they accepted the atomic hypothesis at all, as totally distinct, unrelated to one another. Nonetheless some of their contemporaries pursued the search for order, and gradually the existence of different groups of elements with similar properties was recognized. By the mid-1850s, these included the *alkaline-earth metals* (magnesium, calcium, strontium, and barium); the *halogens* (fluorine, chlorine, bromine, and iodine); the *alkali metals* (lithium, sodium, and potassium); the group oxygen, sulfur, selenium, and tellurium; and the group nitrogen, phosphorus, arsenic, antimony, and bismuth.

After Cannizzaro had, in 1858, clarified once and for all the distinction between atoms and molecules and had, with the help of Avogadro's law (Sec. 4-3), established approximately correct atomic weights for the majority of the elements then known, renewed attempts were made to find some pattern when the elements were arranged in order of increasing atomic weight. One of the most significant of these was the proposal in 1863 by the 25-year-old English chemist, John Newlands, of the *law of octaves*. Newlands noticed that when the elements were arranged in order of ascending atomic weight, there was a regular recurrence of properties among the elements of lower weight, every element resembling the eighth element following it, like the eighth note in an octave of music. The law broke down completely when applied to the heavier elements and was derided and ridiculed as too fantastic to be worthy of serious consideration. However, only 6 years later, Dmitri Mendeleev in Russia and Lothar Meyer in Germany independently and conclusively showed that there was indeed a periodicity in chemical and physical properties, although a more elaborate one than that suggested by Newlands.

Mendeleev's papers were especially convincing; he boldly disregarded some of the atomic weights then accepted, because they led to positions for the elements concerned that were not in accord with his periodic system, and he also left

---

[7] The state of confusion in the late 1850s with regard to the atomic weights of even the most common elements is indicated clearly in the quotation from Odling that opens this chapter.

certain gaps in his table for elements still to be discovered. By considering carefully the chemical properties of related elements, he was able to predict with uncanny accuracy the chemical and physical properties of several of these "missing" elements and their compounds. Within two decades, three of these elements (scandium, gallium, and germanium) had been discovered and found to behave just as he had predicted (Table 2-3). Later, additional elements that he had predicted were discovered or were prepared artificially, including technetium, rhenium, and polonium.

Mendeleev's periodic law was accepted almost at once because he presented the evidence in favor of it so convincingly, but he and others regarded it merely as an empirical generalization based on observation rather than arising from a knowledge of some underlying principles. Since the existence of atomic nuclei, and thus of nuclear charge and atomic number, was unsuspected, the initial basis for the ordering of the elements was their atomic weights rather than their atomic numbers. While atomic weights are indeed fundamental quantities that could even then be measured with precision, they are valuable in relation to the periodic system only because elements with larger atomic numbers *usually* also have larger atomic weights. However, for a few pairs of elements (Te and I, Co and Ni, and, after the discovery of argon in the 1890s, Ar and K), the order of atomic weights was (and still is) such that the element of higher atomic weight belongs first in the periodic table. The wisdom of considering the chemical properties of the elements as more fundamental than their atomic weights when these inconsistencies occurred should be particularly noted, for this bold step was at first regarded as unsound by some persons with less imagination and insight than Mendeleev, Meyer, and their followers. It is not uncommon in the development of science that the recognition of new connections and relationships requires at first disregarding some of what at the time seems relevant evidence. It was only after Moseley in 1913 and 1914 found a way to measure nuclear charge that this quantity, the atomic number, was recognized and accepted as the true basis for the ordering of the elements in the periodic table. For each of the pairs Te–I, Co–Ni, and Ar–K, the element listed first and placed first in the periodic table has indeed the lower atomic number despite the fact that it has the higher

Ekasilicon (Es) (predicted in 1871)	Germanium (Ge) (discovered in 1886)
Atomic weight 72 (average of the atomic weights of the neighbors, Ga and As)	Atomic weight 72.6
Dark gray metal, mp higher than that of Sn, perhaps 800°C; density 5.5 g cm^{-3}	Gray metal, mp 958°C; density 5.36 g cm^{-3}
Heating in air will yield a white powder, EsO_2, of density 4.7 g cm^{-3}; lower oxides may also exist	Heating in air yields GeO_2, white; density 4.70 g cm^{-3}; GeO is also known
Action of acids such as HCl will be slight; Es will resist attack by alkalies such as NaOH	Neither HCl nor NaOH dissolves Ge, but concentrated HNO_3 does
Hydrated EsO_2 will be soluble in acids and easily reprecipitated	$Ge(OH)_4$ dissolves in dilute acid and reprecipitates upon dilution or addition of base
Like Sn, Es will form a volatile, liquid chloride, $EsCl_4$; bp somewhat below 100°C and density 1.9 g cm^{-3}	$GeCl_4$ is a volatile liquid; bp 83°C and density 1.88 g cm^{-3}
Like Sn, Es will form a yellow sulfide, EsS_2, insoluble in water but soluble in ammonium sulfide solution	GeS_2 is white, insoluble in water and dilute acids but readily soluble in ammonium sulfide solution

Table 2-3
Comparison of Some of Mendeleev's Predictions for Ekasilicon with Observed Properties of Germanium

atomic weight. Shortly thereafter, the discovery of isotopes made it clear that atoms of a given element might vary in mass and thus that the atomic weight could not be of primary significance in determining chemical properties, which are nearly identical for all the different isotopes of a given element.

2-4   The Modern Periodic Table

Descriptive Comments   Many schemes have been proposed for effectively displaying the periodicity in chemical and physical properties. One modern form of the periodic table is shown inside the front cover and in Fig. 2-1. A slightly altered version, with a few features added to emphasize the interrelationships of the elements that compose the different groups or families, is presented in Fig. 16-1.

Whatever the form of the periodic table, it is designed to group together elements with similar properties. No arrangement is entirely satisfactory, since the lengths of the periods (that is, the number of elements included before similar properties recur) are not constant, becoming greater as atomic number increases. However, the arrangements shown in Fig. 2-1 and in Fig. 16-1 reveal many of the important relationships.

The horizontal rows in these charts represent the *periods*. Each period ends with one of the noble gases (inert gases), the atoms of which are so stable that the gases are monatomic and form compounds only with difficulty if at all. Their atomic numbers are 2, 10, 18, 36, 54, and 86; their names are helium (He), neon (Ne), argon (Ar), krypton (Kr), xenon (Xe), and radon (Rn). The lengths of the periods beyond He thus represent the differences in the atomic numbers of these unusually stable elements and are, successively, 8, 8, 18, 18, and 32. The last

Figure 2-1   One Form of the Periodic Table of the Elements

period, starting after Rn, would presumably have 32 elements also if it were not incomplete; at present, the element with highest known atomic number is that with $Z = 106$, whereas the last element in that period would have $Z = 86 + 32 = 118$. Except for the first very short period (hydrogen and helium), every period begins with an alkali metal.

The columns in the table contain elements with related properties, and several columns in the longer periods may be related to a given column in the shorter ones. The columns containing the elements of the two short eight-element periods (Li to Ne, $Z = 3$ to 10; Na to Ar, $Z = 11$ to 18) are numbered IA to VIIA, with 0 for the noble-gas column. The elements in column 0 and in the A columns are often called the *representative elements*. In addition, there are ten more columns, seven of which are headed IB to VIIB. The similarity in properties between elements in A and B columns with the same Roman numeral (IA and IB, IIA and IIB, and so on) is at best only very weak. The three remaining columns in the 18-element periods are grouped as one triple column, designated VIII.

Following Pauling, we use the term *congeners* for those elements that form a closely related group, such as those in column 0 or in any column A in Fig. 2-1—as, for example, the alkali metals, the noble gases, or the halogens. In addition, the elements in each column B (such as Cu, Ag, and Au) or those in any one of the three columns of the triple column VIII (such as Ni, Pd, and Pt) are also termed congeners. The marked similarity in properties of the elements in any A column is exemplified by the fact that for those elements that readily form monatomic ions, the charges on those ions are the same within any column and are easily correlated with the column. For example, the alkali metals (column IA) all form ions of charge $+1$, the elements in column IIA form ions of charge $+2$, those in column VIA ions of charge -2, and those in column VIIA ions of charge -1. These facts, and the other correlations in properties reflected in the periodic table, can be explained in terms of the electronic structure of atoms (Chaps. 15 and 16).

Inspection of the 18-element periods shows that after columns IA and IIA come columns IIIB to VIIB, the three columns VIII, and then IB and IIB. The elements in the nine columns IIIB through IB are often referred to as the *transition elements;* the elements in the last column, IIB, have been called the *posttransition elements.* The last six columns of the long periods contain congeners of the elements in the short periods.

The two series of 14 elements grouped separately at the bottom of the chart comprise portions of the two 32-element periods, the second of which is still incomplete. Neither of these groups has congeners among the lighter elements. The *lanthanides* (elements 58 to 71) follow the element lanthanum (La) and resemble it in many ways, as their name implies; these elements are also sometimes called, especially in the older literature, the *rare-earth* elements. The *actinides* (elements 90 to 103) occupy a similar position in the next period, following actinium (Ac). Somewhat surprisingly, the actinides do not closely resemble the lanthanides. While most of the lanthanides are very similar to one another in chemical properties, the actinides differ from each other in significant ways, just as the transition elements do. Most of the actinides—all those beyond uranium (U)—occur on earth only as man-made elements, having been synthesized only during the last four decades. The nuclei of these elements decompose too rapidly for them to have survived since the creation of the earth, even if they were present then.

Summary

The atomic number Z is the charge on the nucleus of an atom. The mass number A is equal to the number of protons and neutrons in the nucleus; there are $A - Z$ neutrons. Isotopes are atoms with the same Z but different A. The atomic weight (AW) of an atom is its weight relative to the arbitrary choice of the weight 12 for the most abundant isotope of carbon, ^{12}C.

Molecules are discrete, uncharged groupings of atoms tied together by specific forces in a definite arrangement. Ions are charged particles resulting from the loss or gain of electrons by atoms or molecules. Chemical compounds may consist of neutral molecules, such as H_2O or CO_2, or of ions in such numbers that their charges balance, as in K^+Br^-, $Ca^{2+}(SO_4)^{2-}$, or $Mg^{2+}(Cl^-)_2$.

The molecular weight (MW) of a molecule is the sum of the atomic weights of the atoms it contains. The formula weight (FW) is similarly defined for any compound, whether it consists of molecules or ions.

Avogadro's number (N_A) is, by definition, the number of atoms in exactly 12 g of ^{12}C; its value is approximately 6×10^{23}. The mole is a quantity of anything that contains N_A units—atoms, molecules, formula units, eggs. The weight in grams of a mole of atoms, molecules, ions, or formula units is numerically equal to the weight of that species or formula unit on the atomic weight scale.

The periodic table is an arrangement of the elements in order of increasing atomic number, with rows (or periods) of 2, 8, 8, 18, 18, 32, . . . elements, starting with H. Elements in the same column are called congeners, exhibit characteristic similarities in chemical properties, and may have special group names: noble gases, alkali metals, alkaline-earth metals, halogens. Other characteristic groups are the transition elements, located in the middle portions of the periods of 18 or 32 elements, and the lanthanides and actinides, found in the periods of 32 elements.

Terms and Concepts

Problems and Questions

2-1 Composition of Atoms and Ions How many protons, how many neutrons, and how many electrons are there in each of the following species: ^2H, $^3H^+$, ^{40}K, $^{37}Cl^-$, ^{17}O?

2-2 Definitions Define the following terms: nucleon, nuclide, atomic number, mass number, isotope, atomic weight.

2-3 Definitions (a) Define each of the following terms carefully: atom, molecule, element, compound, ion. (b) Give examples of molecules that are compounds and of molecules that are not compounds.

2-4 Old Atomic Weight Scale The atomic weight scale in a chemistry text published in 1823 was based on assigning oxygen the value 100. What is the formula weight of nitric acid, HNO_3, on this scale?

2-5 Atomic Weight from Mass-spectrometer Data Nitrogen consists of two stable isotopes, whose atomic weights and relative abundances were determined with a mass spectrometer to be: ^{14}N, 14.0031, 99.63 percent; ^{15}N, 15.0001, 0.37 percent. What is the (chemical) atomic weight of nitrogen?

2-6 Natural Abundance Sulfur consists primarily of three isotopes ^{32}S, ^{33}S, and ^{34}S, with atomic weights 31.97, 32.97, and 33.97, respectively. The percent abundance of ^{32}S is 95.00, and that of ^{33}S is 0.76. Show that these data are consistent with the fact that the atomic weight of sulfur is 32.06.

2-7 Abundances from Isotopic Weights and Atomic Weight Chlorine consists of two isotopes, ^{35}Cl with atomic weight 34.97 and ^{37}Cl with atomic weight 36.97. The atomic weight of chlorine is 35.45. What are the relative abundances of the isotopes?

2-8 Isotopic Abundance Element Y has three naturally occurring isotopes of atomic weight $(A - 4)$, A, and $(A + 1)$. If the atomic weight of natural Y is just A, what are the maximum and minimum percent abundances of the heaviest isotope consistent with this information?

2-9 Elements (a) Give the chemical symbols for the following elements: sodium, magnesium, potassium, cobalt, argon, tin, bromine, nickel, silver, iron, calcium, sulfur. (b) Give the names of the elements corresponding to the following symbols: Ne, Cl, P, Cr, Cu, Na, C, Mg, Sr, Sn, Mn, Zn, Si.

2-10 Determination of Avogadro's Number Rutherford and Boltwood found in 1911 that the steady alpha radiation that issues from a standard preparation of radium corresponds to 27.8×10^{-3} mg He for 1 g of the preparation in 1 year. (Each alpha particle picks up two electrons as it travels through the air and becomes a helium atom.) It had also been found, by counting the light flashes produced by the impact on a zinc sulfide screen of the individual alpha particles that issue from such a preparation, that 1 g of the standard preparation emits 13.8×10^{10} alpha particles per second. Assume that Avogadro's number is not known. Calculate (a) the number of alpha particles emitted by the sample in one year, (b) the mass of one He atom (in grams), and (c) the number of He atoms in 1 mol (4.00 g) of He, i.e., Avogadro's number.

2-11 The Size of Avogadro's Number

The Walrus and the Carpenter
  Were walking close at hand:
They wept like anything to see
  Such quantities of sand:
"If this were only cleared away,"
  They said, "it would be grand!"

"If seven maids with seven mops
  Swept it for half a year,
Do you suppose," the Walrus said,
  "That they could get it clear?"
"I doubt it," said the Carpenter,
  And shed a bitter tear.

     "Through the Looking Glass,"
             Lewis Carroll

The sands of the beach are usually regarded as countless, but in fact the number of grains of sand on an average beach is considerably fewer than the number of gas molecules in 1 cm^3 of ordinary air. Suppose that there are, on the average, 60 grains of sand per mm^3 on a typical beach. If the beach is 100 m wide and the sand is 10 m deep, how long a strip of beach is needed to contain N_A grains of sand? The circumference of the earth at the equator is 4×10^4 km. How many times would this strip of beach stretch around the earth at the equator?

2-12 The Ark-Mole (a) Suppose that there is another inhabited planet, Arko, in a distant galaxy, with intelligent beings for whom a convenient unit of mass is the ark. This unit is so defined that 1.00 ark equals 70.0 g. The life of these beings is, like ours, tied to the chemistry of carbon. Since they worship the number 5, they have chosen to assign the most abundant isotope of carbon (which we call carbon 12) an atomic weight of exactly 5.00. . . . What is the analog of Avogadro's number on Arko, that is, the number of molecules in an ark-mole? (b) Assume the information given in (a), and calculate the molecular weight of CO_2 on the scale used by the inhabitants of Arko.

2-13 Number of Molecules in a New Mole Suppose that the atomic weight of ^{12}C were taken to be 5.000 and that a mole were defined as the number of atoms in 5.000 kg carbon 12, or more generally as the number of atoms in 1 "kilogram atomic weight" of any element. Calculate the number of molecules in a mole under these conditions.

2-14 Formulas for Compounds (*a*) Give the formulas of the following compounds: sulfuric acid, ammonia, calcium chloride, nitric acid, lead sulfide, silver oxide, sodium carbonate, aluminum bromide. Consult Appendices B-1 and B-2 if necessary. (*b*) Give names of compounds corresponding to the following formulas: $(NH_4)_2SO_4$, KI, $Sr(NO_3)_2$, HCl, $ZnCO_3$, H_2O, $NaClO_3$, $LiHCO_3$, BaO_2, HNO_2.

2-15 Quantitative Implications of Chemical Formulas *para*-Dichlorobenzene, $C_6H_4Cl_2$, consists of molecules that form a solid with density 1.53 g ml^{-1}. Calculate for this substance: (*a*) the molecular weight; (*b*) weight of compound that contains 200 molecules; (*c*) weight of compound that contains 15.0 mol hydrogen atoms; (*d*) volume of compound that contains 3.0×10^{22} chlorine atoms.

2-16 Quantitative Implications of Chemical Formulas Sodium carbonate decahydrate, $Na_2CO_3 \cdot 10H_2O$, is an ionic solid with density 1.44 g ml^{-1}. When heated to $50°C$, it is converted to the monohydrate, $Na_2CO_3 \cdot H_2O$. Calculate the following quantities with respect to sodium carbonate decahydrate: (*a*) the formula weight (FW); (*b*) moles of sodium ion in 46 g of the compound; (*c*) weight of oxygen in 100.0 g compound; (*d*) number of water molecules given off when 5.0 ml of compound is heated to $50°C$.

2-17 Quantitative Implications of Chemical Formulas 4-Nitrobenzoic acid, $C_7H_5NO_4$, is a colorless solid with density 1.55 g ml^{-1}. Calculate for this substance: (*a*) the molecular weight; (*b*) weight of hydrogen in 200 g compound; (*c*) weight of compound containing 200 carbon atoms; (*d*) moles of compound in 100 ml compound.

2-18 Quantitative Implications of Chemical Formulas (*a*) The compounds phosphorus trioxide and phosphorus pentoxide are often given the formulas P_2O_3 and P_2O_5. What are the formula weights corresponding to these two formulas? (*b*) These compounds exist as molecules in the vapor state and in some crystals, with the formulas P_4O_6 and P_4O_{10}, respectively. Give the following with respect to these two molecules: (i) molecular weight; (ii) moles of oxygen atoms in 100 g; (iii) weight of compound containing 62 g phosphorus; (iv) weight of compound containing 1.0×10^{23} oxygen atoms.

2-19 Quantitative Implications of Chemical Formulas Chlorobenzaldehyde is a solid that melts at $47°C$. Its molecular formula is C_7H_5OCl, $MW = 140$, and density $= 1.25 \text{ g ml}^{-1}$. Calculate the following quantities with respect to this compound: (*a*) moles of compound in 80 g compound; (*b*) volume of 1.50 mol compound; (*c*) total number of atoms in 28 g compound; (*d*) weight of compound containing as many hydrogen atoms as 90 g water.

2-20 Quantitative Implications of Chemical Formulas Calculate the following quantities for cryolite, Na_3AlF_6: (*a*) grams in 2.10 mol; (*b*) moles of fluorine atoms in 7.0 g compound; (*c*) weight of sodium in 100 g compound; (*d*) number of atoms in 4.0 g compound; (*e*) number of moles that contain 2.00 g Al.

2-21 Quantitative Implications of Chemical Formulas Give the following quantities for the compound $C_9H_{12}N_2$, a solid with density 1.25 g ml^{-1}: (*a*) grams in 3.00 mol; (*b*) volume occupied by 2.50 mol; (*c*) weight that contains 7.0 g nitrogen; (*d*) weight in grams that contains 1.0×10^{23} atoms of hydrogen; (*e*) weight in grams of five molecules.

2-22 Periodic Table (*a*) What is implied by the term *periodic* in the phrase "periodic classification of the elements"? What do the different periods have in common? How do they differ? (*b*) What is meant by the term *congener*? Give (in order of increasing atomic number) three cogeners of each of the following elements: He, Li, C, N, O, F, Mg.

Stoichiometry: The Quantitative Relationships Implied by Chemical Formulas and Equations

3

"It often seems that the practical strength of the scientific method is not so much that science provides the right answer as that the scientist has learnt how to calculate the margins of error that surround his tentative answer. . . . The scientist is not so much a man who likes to get all the answers right as a man with a very clear view of error, and what errors mean, and when they are important."
DAVID WILSON IN *THE LISTENER*, 1967

"We must recognize an invisible hand that holds the balance in the formation of compounds. A compound is a substance to which Nature assigns fixed ratios; it is, in short, a being which Nature never creates other than balance in hand, *pondere et mensura*."
JOSEPH LOUIS PROUST, 1799

When the basic assumptions of the atomic theory are used to interpret the changes that occur in a chemical reaction, quantitative[1] relationships among the constituents of a compound or the substances participating in the reaction can be deduced. The term *stoichiometry* (Greek: *stoicheion,* element; *metron,* measure) is applied to these relationships. This chapter is devoted to stoichiometry; it begins with a discussion of errors in measurement and ways of carrying out calculations so as to avoid giving a false impression of the uncertainty inherent in experimental results.

3-1 Errors and Significant Figures

All quantitative science is based on measurements, but measurements are usually afflicted by errors, which must be taken into account. Several strategies may be used to serve this purpose. One of them is to repeat the measurements many times. The results are seldom identical, and an analysis of the way they are

---

[1]The terms *quantitative* and *qualitative* are used in a contrasting sense in many discussions in the sciences. The use of the adjective *quantitative* implies a concern with measurement of some property numerically and usually rather precisely. *Qualitative* suggests instead a concern with distinctive characteristics or qualities, but is sometimes used, as in "a qualitative estimate", to suggest a very approximate (rather than a precise) estimate of amount, perhaps good within only a factor of 2 or 3, as in a forecast of the size of a crop.

scattered yields information about their reliability. This analysis is called the statistical treatment of data. Another approach is to measure the same quantity by several widely different methods. The closeness of the results is an indication of their trustworthiness. For most routine problems, such as the determination of the chlorine content of a sample, the first procedure is almost always used. One method executed in triplicate is often satisfactory, provided the method is well established and its limitations are known and adequate. However, for quantities of central importance, such as Avogadro's number, many elaborate and differing types of experiments are used.

An important aspect of the handling of data is that the quantity of primary interest is usually not what is measured directly, so that intermediate computations are required. This raises two questions. One concerns the relation of the uncertainty in the quantity of interest to the errors in the original measurements. In other words, how do the errors propagate through the equations that connect the quantity of interest with the data? The other question concerns the accuracy with which intermediate calculations should be carried out. To do justice to the data, the accuracy of the calculations should be sufficiently high that no new errors are introduced, but excessive accuracy is pointless and inefficient.

A related issue concerns ways of expressing the uncertainty in an experimentally determined quantity. In a rough way this is done by the conventions of significant figures, to be discussed shortly. We now consider some of these questions, beginning with definitions of the concepts of error, accuracy, and precision.

Error, Accuracy, and Precision   The *error* of an observation is the difference between the observation and the actual, or true, value of the quantity observed. The *accuracy* of a set of observations is the difference between the average of the values observed and the true value of the observed quantity. The *precision* of a set of measurements is a measure of the range of values found, that is, of the reproducibility of the measurements. The meanings of accuracy and precision are thus quite different. A set of observations may, for example, have high precision and low accuracy at the same time (Fig. 3-1).

Errors are often classified as either systematic or random. *Systematic errors* may be caused by fundamental flaws in the experimental equipment (e.g., the experimenter is unaware that his stopwatch runs fast) or by inadequate under-

Figure 3-1   Different Possible Combinations of High and Low Precision and Accuracy

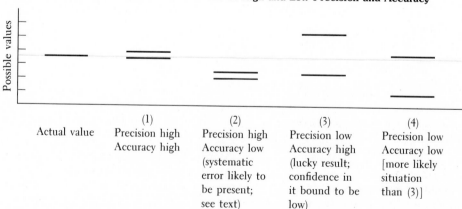

standing of the theory underlying the measurement (e.g., unwarranted neglect of the buoyant effect of air on weights and on an object being weighed). Sometimes systematic errors have their source in the observer. He may, for example, be prejudiced by prior information, as when he is repeating thermometer readings and finds it impossible to forget estimates of fractional degrees made just a moment earlier. Similarly, his estimate of a tenth of a division may be different when he is reading close to one of the etched marks and when he reads halfway between marks. Systematic errors usually do not average out, even if the observations are repeated many times.

Random errors are accidental errors that vary in a completely unreproducible way from measurement to measurement. The concept of randomness can best be explained by the use of examples: the sequence of heads and tails in tossing a perfect coin is said to be random. Random errors find their origin, at least in part, in the limited precision of instrument readings (e.g., the reading of a ruler with engraved scale marks of finite widths) and in the degree to which external conditions (e.g., the temperature of the ruler) are controlled. (Both effects may cause systematic errors as well.)

Random errors can be treated statistically, which makes it possible to relate the precision of an experimental result to the precision with which each of the experimental variables (e.g., pressure, temperature) is known. An error analysis also shows the weakest link in the chain of observations that leads to the determination of a certain quantity, the step that needs to be improved first if the precision of the result is to be improved.

Systematic errors are frequently difficult to discover. They may usually be detected by going outside the framework of the original measurements, for example, by measuring the quantity of interest in several fundamentally different ways. Close agreement of the results gives confidence in the unlikelihood of systematic errors, but the detection of such errors is never certain.

An illustration of the effects of unsuspected systematic errors is provided by the determination of the electronic charge by Millikan. His value for e was dependent on the viscosity of air in a way that he understood, and he considered carefully all the different precise measurements of the viscosity that had been made. Millikan then estimated that the value he used could "scarcely be in error by more than one part in two thousand". Taking into account this limit of error and those in his own work, he estimated that his value of e, 1.5924×10^{-19} C (coulomb), was in error by no more than 0.1 percent. Yet the currently accepted value is about 0.6 percent higher, 1.6022×10^{-19} C, with an estimated uncertainty of less than 0.01 percent. Almost all the discrepancy between Millikan's value and that currently accepted is attributable to an error in the viscosity of air, which was about 0.5 percent low, although it was believed to be about 10 times more precise than that. Millikan estimated his precision quite correctly; it was a systematic error that made his value far less accurate than he thought.

Precision and Significant Figures Precision is discussed in two distinct ways. *Absolute* precision refers to the actual uncertainty in a quantity, in whatever decimal place the least significant digit occurs. For example, if we assume an uncertainty of 1 in the least significant digit, the absolute precision of 2.47×10^4 is 1×10^2 and that of 0.0943 is 1×10^{-4}. *Relative* precision implies expressing the uncertainty as a fraction of the quantity of interest (or as a percentage of it). If the length of a rod is given as 5.1 m with an uncertainty of 0.1 m, the absolute precision is 0.1 m and the relative precision is 0.1 m/5.1 m, or 0.02, or 2 percent.

39

Relative precision is always a dimensionless quantity, the units cancelling out.

In general, results of observations should be reported in such a way that the last digit given is the only one whose value may be in doubt. The digits that constitute the result, excluding *leading* zeros, are then termed *significant figures*. The number of significant figures is the same in 23.5 mg and 0.0235 g; this is reasonable, for the uncertainty in a result should not depend on the units in which the answer is expressed. Note that there is some uncertainty about the number of significant figures in a weight reported to be 2350 kg, because it is unclear whether the final zero is significant or not. The matter is clarified by reporting the weight either as 2.350×10^3 kg or 2.35×10^3 kg, whichever applies.

The concept of significant figures does not apply to quantities known to be integers; for example, in a molecular formula, such as C_6H_{14}, the number of figures in each subscript does not imply that the composition is uncertain. We can assume that there are exactly 6 mol carbon atoms per mole of compound. Similar considerations apply to integral numbers of objects; the integers are regarded as perfectly precise.

There is no general agreement about the degree of uncertainty associated with the least significant digit given, that is, with the last digit on the right. It is sometimes assumed that the implied uncertainty is one unit of the decimal position of the least significant digit. According to this convention,[2] which we shall use, a weight given only as 2.350 kg is uncertain by 0.001 kg.

The number of significant figures with which a quantity is given expresses roughly its *relative precision*. Thus, if we assume an uncertainty of one or two units in the least significant digit, the quantities 201 and 0.0198 are each precise to 1 or 2 parts in about 200, or about 1 percent. However, it must be noted that when a value such as 98.3 mm is given with three significant figures, its relative precision is comparable to that of a value such as 103.7 mm, given with four significant figures. If we associate an uncertainty of 1 in the last place with each quantity, each is given to a precision of about 1 part in 1000, although the number of significant figures differs. Despite minor shortcomings of this kind, the system of significant figures is often adequate and useful.

In statements of experimental procedures or of numerical problems, more significant figures are sometimes implied than are given. For example, the meanings of "1 liter" and "10 g" *may* well be those of "1.000 liter" and "10.000 g". Such implications can usually be established from the context, and confusion should seldom arise once it is understood that under such circumstances the "1" in "1 liter" does not necessarily imply a one-digit significance.

"Mount Whitney Elevation 14,496.811 ft." Is the precision meaningful?

3-2 Precision in Calculations

There are several useful rules for carrying out calculations so that the precision of the answer properly reflects the precision of the data used.

Rounding Off This operation consists of discarding digits that are not significant and adjusting the residue appropriately, as deduced from the following simple rules:

---

[2] This practice is rather restrictive; more generally, in the absence of any specific statement, the uncertainty in the last digit might be regarded as $\sqrt{10}$, that is, ± 3.

1. Discard the unwanted digits.

2. (a) Increase the last retained digit by one unit (*round up*) if the most significant discarded digit is larger than 5 or if it is equal to 5 but followed by other digits some of which are different from zero (for example, 94.3507 rounds to 94.4 if only three significant digits are warranted).

   (b) Leave the last retained digit unchanged (*round down*) if the most significant digit discarded is less than 5 (for example, 94.348 rounds to 94.3). Note that rounding should be done in one step; rounding one digit at a time would give the wrong answer, 94.348 to 94.35 to 94.4 [see (c)].

   (c) If the digits discarded are a 5 followed only by zeros (or by nothing), round to the nearest even digit. This last convention provides that, in the long run, the total number of rounding-up operations is the same as the number of rounding-down operations so that averages are unaffected. Thus, 4.55 and 4.65 would both be rounded to 4.6 if the final 5 were not a significant digit.

Addition and Subtraction First look for the term of lowest absolute precision, that is, the term for which the absolute uncertainty is largest. This determines the precision warranted for the result, and it is meaningless and misleading to give the result with significant figures implying an absolute precision greater than that of this term. An illustration is given in Table 3-1. Note that rounding all terms to the lowest absolute precision before performing the calculation is undesirable because rounding errors may accumulate in the subsequent computations and affect the figures finally retained. Instead, before adding or subtracting, round in a preliminary fashion the terms given with higher precision, retaining one or two places beyond the precision warranted for the result. Then perform the additions or subtractions, and finally round the result.

Multiplication and Division The situation here is like that discussed for addition and subtraction except that *relative* rather than absolute precisions are involved. The result of any multiplication or division must not be expressed with a relative precision much higher than that of the term involved that has the lowest relative precision. However, to prevent the accumulation of rounding errors, terms given with higher relative precision should be rounded in a preliminary fashion only. During the intermediate calculations, retain one or two figures

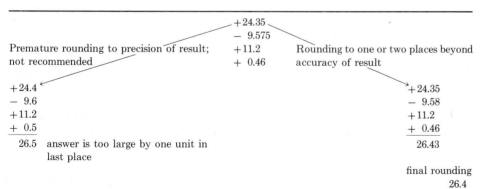

**Table 3-1
Combination of Terms
with Different Absolute
Precision in Addition
and Subtraction**

41

beyond the relative precision warranted by the least precise quantity and perform the rounding as the last step.

□ **Multiplication and Division**   Find $(438.807 \times 5.2)/4.03$.

Solution   Multiplying first and retaining three significant figures, we obtain $2.28 \times 10^3/4.03 = 5.66 \times 10^2 \approx 5.7 \times 10^2$. The end result is not given as 6×10^2 because a relative precision of 1 part in 57 is closer to 1 part in 52, the uncertainty of the term with least relative precision, than is 1 part in 6. Note that retaining only two significant figures in the intermediate calculations yields $2.3 \times 10^3/4.0 = 5.75 \times 10^2 \approx 5.8 \times 10^2$, provided the rounding-off procedure described earlier is followed. ■

Exercise 3-1

□ Find $1.55/(45.64 \times 46.7)$. ■

General Intermediate Computations   In computations that are more involved than those just illustrated, calculate intermediate results to one or two places beyond the precision warranted by the data and round off only the final results. There are some circumstances in which one must be extremely careful about premature rounding off. The following discussion illustrates the danger of incorrectly neglecting small terms in intermediate steps because these terms seem insignificant. Whenever you are tempted to neglect a "small" quantity, remember always to ask: "Small compared with what? What effect does neglecting this quantity have on the final answer?"

Example 3-2

□ **Square Root of a Quantity Close to Unity**   A certain problem calls for the evaluation of the expression

$$x = \sqrt{1 + 4.0 \times 10^{-14}} - 1 \tag{3-1}$$

Find x.

Solution   It may be argued that the significant figures of 1 under the square-root sign suggest that the term 4.0×10^{-14} is of no concern and that to keep it would imply too high precision. Deletion of this term leads, however, to $x = 0$, which is wrong. The term must therefore be retained. One method of evaluating the square root is by use of the approximation

$$\sqrt{1 + y} \approx 1 + \frac{y}{2}$$

which is valid[3] for very small values of y. This leads to

$$\sqrt{1 + 4.0 \times 10^{-14}} = 1 + 2.0 \times 10^{-14} \tag{3-2}$$

(Note that a typical small electronic calculator does not retain enough decimal places to allow the square root in (3-2) to be evaluated directly and that to find this square root in a table would have required a table of much

---

[3]To prove this relation, square both sides: $1 + y \approx 1 + 2(y/2) + y^2/4$. Note that y^2 is 16×10^{-28} when y is 4.0×10^{-14}.

higher accuracy than is usually available.) Inserting the value of the square root into (3-1) gives

$$x = 1 + 2.0 \times 10^{-14} - 1 = \underline{2.0 \times 10^{-14}} \quad \blacksquare$$

☐ Find the value of $(16 + 3.7 \times 10^{-9})^{1/2}$. ■

Exercise 3-2

If extensive algebraic developments are required, they should be carried through to the final result symbolically and computations should be performed only at the end. Computation of intermediate results may cause errors, because terms that should be identical may become slightly different through rounding off and may no longer cancel in the end as they should.

3-3 Chemical Formulas

Chemical formulas have already been used in Chap. 2. Several types of formulas serve to designate chemical compounds, each type conveying different kinds of information. We consider here some of the most important types of formulas, and in the following section we represent chemical reactions by equations between such formulas. The quantitative implications of formulas and equations are examined further in Sec. 3-6 and those of equations are considered in Sec. 3-7.

Empirical formulas (also called *simplest formulas*) express the results of elemental analysis in the simplest way possible. For stoichiometric compounds the atoms involved carry the smallest integral subscripts possible, so that the different subscripts have no common factor. If a substance consists of molecules, there is no implication that the empirical formula corresponds to the correct number of atoms contained in one molecule. The molecule may correspond to the empirical formula or any integral multiple of it. Examples are given in the second column of Table 3-2.

Molecular formulas give the correct atomic composition of molecules. No information about the structure of the molecule is implied, except in very simple cases such as He, H_2, or NO (nitric oxide). To be consistent, ionic compounds should be given *ionic formulas,* for example, Na^+Cl^- and $Ca^{2+}SO_4^{2-}$, but the plus and minus signs are awkward and are usually left out. Just as with molecules, the total number of atoms in an ion need not be that implied by the empirical formula. For example, sodium peroxydisulfate has the empirical formula $NaSO_4$, but the ionic formula is $(Na^+)_2S_2O_8^{2-}$ or $Na_2S_2O_8$ (see also Fig. 3-2).

The empirical formula can be calculated from a knowledge of the chemical composition alone. The molecular formula cannot, however, be deduced from the

Name	Empirical formula	Molecular or ionic formula	Structural formula (topological)
Phosphorus	P	P_4 (white P) P_∞ (red, black P)	See Fig. 3-2
Sodium chloride	NaCl	Na^+Cl^- or NaCl	Ionic; see Fig. 6-1
Phosphorus pentoxide	P_2O_5	P_4O_{10}	See Fig. 3-2
Ammonium peroxydisulfate	H_4NO_4S	$(NH_4^+)_2S_2O_8^{2-}$ or $(NH_4)_2S_2O_8$	See Fig. 3-2
Acetylene	CH	C_2H_2	$H-C\equiv C-H$

Table 3-2
Different Types of Chemical Formulas

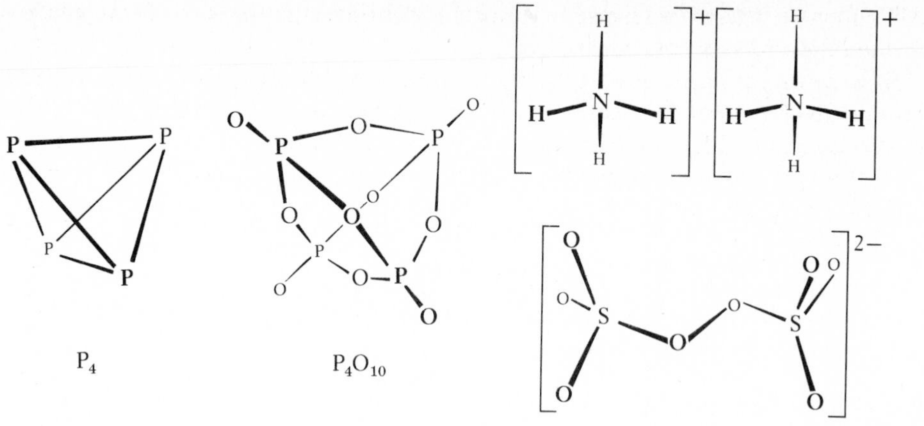

P_4

P_4O_{10}

$(NH_4)_2S_2O_8$

Figure 3-2   Examples of Three-Dimensional Structural Formulas
The four P atoms in P_4 are at the corners of a tetrahedron. This is also true for P_4O_{10}, but here the P atoms are linked by O atoms that are positioned just outside the centers of the tetrahedral edges. An additional O atom is linked to each P atom. In red phosphorus and black phosphorus the P atoms are bonded together in three-dimensional networks. $(NH_4)_2S_2O_8$ consists of tetrahedral NH_4^+ ions and an $S_2O_8^{2-}$ ion in which two tetrahedral SO_4 groups are joined by an O—O bond. The lines between atoms represent covalent bonds (Sec. 6-4).

empirical formula. More information is needed, such as the molecular weight, which then can be compared with the formula weight corresponding to the empirical formula. Direct structure determinations, for example by X-ray diffraction, also provide molecular and ionic formulas. Examples of molecular and ionic formulas are given in the third column of Table 3-2.

Example 3-3

□ **Empirical and Molecular Formulas**   The compounds benzene and acetylene both have the empirical formula CH. The molecular weight of benzene is 78, and that of acetylene is 26. Give the molecular formulas of the compounds.

Solution   The formula weight corresponding to the empirical formula CH is $12 + 1 = 13$. To determine the molecular formulas we divide the molecular weights by this formula weight:

$$\text{Number of formula units in benzene} = \frac{78}{13} = 6$$

$$\text{Number of formula units in acetylene} = \frac{26}{13} = 2$$

Thus the molecular formula of benzene is $(CH)_6$ or $\underline{C_6H_6}$, and that of acetylene is $(CH)_2$ or $\underline{C_2H_2}$. ■

Exercise 3-3

□ Glucose has the empirical formula CH_2O and the molecular weight 180. What is its molecular formula? ■

When desired, *states of aggregation* may be indicated by letters in parentheses, as in $Br_2(g)$, $Br_2(l)$, and $Br_2(s)$, referring to gaseous bromine, liquid bromine, and solid bromine, respectively. Similarly, the symbol (*aq*) may be used, as in

$Al^{3+}(aq)$ or $Cl_2(aq)$, to indicate a species dissolved in water to form what is termed an aqueous solution (Chap. 9).

Structural formulas give partial or complete information on the detailed atomic arrangement in a compound. There are several types of structural formulas, some of which are discussed here. All of them reveal, at least by implication, the way in which the atoms are linked. Some show no more than that, while others reveal the detailed three-dimensional spatial arrangement of the atoms.

The formula

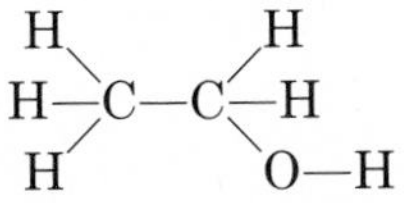

$$\begin{array}{c} H \qquad\quad H \\ H-C-C-H \\ H \qquad\quad O-H \end{array}$$

shows how the atoms in the ethanol (ethyl alcohol) molecule are linked. It also implies some features of the three-dimensional arrangement to one who knows that four single bonds to a carbon atom are directed approximately toward the corners of a regular tetrahedron and two single bonds to an oxygen atom make an angle of about 109° with one another. This structural formula is often given in the abbreviated form CH_3CH_2OH (chiefly for typographical convenience), which indicates the linkage of the atoms but says little about the spatial arrangement. Other examples of the same types of formulas are

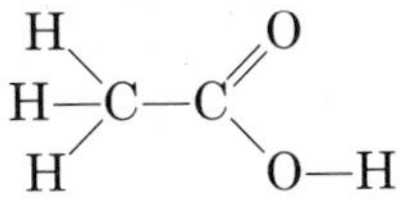

$$\begin{array}{c} H \qquad\quad O \\ H-C-C \\ H \qquad\quad O-H \end{array} \qquad \text{or} \qquad CH_3COOH$$

for acetic acid and

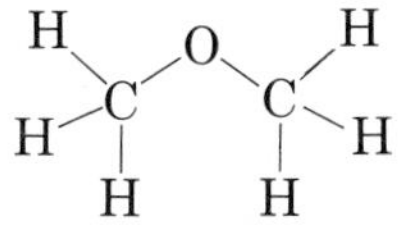

$$\begin{array}{c} H \qquad O \qquad H \\ C \qquad\quad C \\ H \qquad H \qquad H \\ H \qquad\quad H \end{array} \qquad \text{or} \qquad CH_3OCH_3$$

for dimethyl ether.

Ethyl alcohol and dimethyl ether have different molecular structures and, as a result, they have very different properties. Note, however, that these compounds have the same molecular formula, C_2H_6O. This phenomenon is called *isomerism* (Greek: *iso,* same; *meros,* part), and compounds with the same elemental composition and molecular weight but different structures are called *isomers* of one another. There are many forms of isomerism, some of which will be encountered in later chapters.

☐ **Structural and Molecular Formulas** Lactic acid has the structural formula

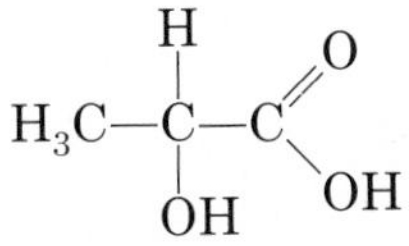

Example 3-4

$$\begin{array}{c} H \qquad\quad O \\ H_3C-C-C \\ OH \qquad OH \end{array}$$

What is the molecular formula?

Solution The only information the molecular formula supplies is how many atoms of each kind the molecule contains. By counting atoms we find 3 C atoms, 6 H atoms, and 3 O atoms. The formula sought is thus $C_3H_6O_3$. ∎

☐ The vitamin niacin has the structural formula

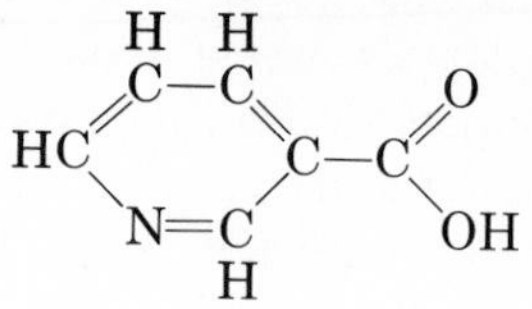

What is the molecular formula? ■

3-4   Chemical Equations

The reactions of chemical substances may be represented by chemical equations that show the formulas and relative numbers of the reactants and products involved. Equations should, by definition, be balanced: they should show the same number of atoms of a given kind on each side, and the net charge should be the same on each side. The condition, sometimes slavishly insisted upon, that all coefficients in an equation must be whole numbers is not essential; coefficients may be fractional or may have a common factor.

Example 3-5

☐ **Balancing Equations by Inspection**   Balance the following equations:

(a) $Al + H_2SO_4 \longrightarrow Al_2(SO_4)_3 + H_2$
(b) $C_3H_8O + O_2 \longrightarrow CO_2 + H_2O$

Solution   (a) On the right-hand side of Equation (a) the formula units Al and SO_4 appear in the ratio $2:3$. Thus there must be 2 Al for every 3 SO_4 on the left-hand side of the equation:

$$2Al + 3H_2SO_4 \longrightarrow Al_2(SO_4)_3 + H_2$$

To complete the balancing we must have 3 H_2 on the right:

$$\underline{2Al + 3H_2SO_4 \longrightarrow Al_2(SO_4)_3 + 3H_2}$$

(b) The simplest procedure is to balance first an atomic species that appears in only one compound on each side of the equation. We balance first with respect to C:

$$C_3H_8O + O_2 \longrightarrow 3CO_2 + H_2O$$

Hydrogen also appears in only one compound on each side. There are 8 H in C_3H_8O so that hydrogen can be balanced by putting 4 H_2O on the right:

$$C_3H_8O + O_2 \longrightarrow 3CO_2 + 4H_2O$$

With 3 CO_2 and 4 H_2O on the right, there are 10 oxygen atoms on the right, and thus there must be 10 on the left. There is already one atom of O present in each molecule of C_3H_8O.

$$\underline{C_3H_8O + \tfrac{9}{2}O_2 \longrightarrow 3CO_2 + 4H_2O} \quad ■$$

Exercise 3-5

☐ Balance the equations

(a) $Ca(OH)_2 + H_3PO_4 \longrightarrow Ca_3(PO_4)_2 + H_2O$
(b) $C_6H_{12}O_6 + O_2 \longrightarrow CO_2 + H_2O$ ■

Many equations are easy to balance, once the important reactants and products are known. For example, when dissolved in water, silver nitrate, $AgNO_3$, and sodium sulfate, Na_2SO_4, react to form solid silver sulfate, Ag_2SO_4, which precipitates from the solution. An unbalanced equation for the reaction is

$$AgNO_3 + Na_2SO_4 \longrightarrow Ag_2SO_4 + \cdots \qquad (3\text{-}3)$$

which may be balanced to

$$2AgNO_3 + Na_2SO_4 \longrightarrow Ag_2SO_4 + 2NaNO_3 \qquad (3\text{-}4)$$

The arrow is often replaced by an equal sign for typographical reasons, but this should be done only when the equation is balanced. Throughout this book, we use arrows. Methods of balancing equations in some less obvious situations are discussed in Chap. 10.

The reaction considered in (3-3) and (3-4) is actually one between the ions Ag^+ and SO_4^{2-} rather than between $AgNO_3$ and Na_2SO_4. It is therefore usually written as an *ionic equation:*

$$2Ag^+ + 2NO_3^- + 2Na^+ + SO_4^{2-} \longrightarrow Ag_2SO_4(s) + 2Na^+ + 2NO_3^- \quad (3\text{-}5)$$

or
$$2Ag^+ + SO_4^{2-} \longrightarrow Ag_2SO_4(s) \qquad (3\text{-}6)$$

The ions NO_3^- and Na^+ remain unchanged and may thus be omitted, in the same way that the H_2O molecules present in the solution are also not shown in the equation. The same basic reaction would occur between any two water-soluble substances that contain, respectively, Ag^+ and SO_4^{2-}; for example, Na_2SO_4 could be replaced by K_2SO_4. Note that (3-6) implies the relation

$$\frac{\text{Mol } Ag^+ \text{ reacted}}{\text{Mol } SO_4^{2-} \text{ reacted}} = \frac{2}{1}$$

In (3-5) and (3-6) the total of the charges on each side is zero, but it need not be, as long as it is the *same* on each. For example, in the balanced equation

$$8H^+ + MnO_4^- + 5Fe^{2+} \longrightarrow Mn^{2+} + 5Fe^{3+} + 4H_2O$$

the total of the charges on either side is $+17 [= (8 \times 1) - 1 + (5 \times 2) = +2 + (5 \times 3)]$.

While certain ions may not be shown in a reaction equation, it must be remembered that their presence is nevertheless required to keep the solution electrically neutral. Thus, SO_4^{2-} ions may not be introduced into a solution without some positive ions such as Na^+, K^+, or Mg^{2+}. Such extraneous ions must, of course, not participate in the reaction considered (otherwise they would have to be shown in the reaction equation).

The balanced chemical equation shows the overall results of a chemical reaction and provides quantitative information about the relative amounts of reactants and products concerned. It is often referred to as the *stoichiometric equation* for the reaction. Such an equation, however, provides no information about the detailed sequence of ionic and molecular processes that actually occur as the reactants combine to form the products.

3-5   General Principles of Stoichiometry

The term *stoichiometry* denotes the quantitative relationships among the constituents of a chemical formula or a balanced equation for a chemical reaction. The principles of stoichiometry are no more complicated than those of everyday calculations based on proportions in shopping. The following facts are of principal importance:

1. All stoichiometric calculations involve *whole units* on the atomic level: atoms, molecules, ions, or atomic groups.
2. Mass[4] is conserved in chemical reactions. Thus, for each kind of atom, the total weight of the products must be the same as the total weight of the reactants.
3. As discussed in Chap. 2, for practical reasons these units are scaled up to macroscopic dimensions by multiplication by Avogadro's number. One therefore deals with moles instead of with individual atoms, molecules, and ions.
4. The proportions by weight of the different constituents in a compound or of the different reactants and products in a particular reaction are definite and characteristic of the compound or the reaction.

Our concern here is to illustrate and clarify the principles involved in stoichiometric calculations rather than to develop a set of procedures for solving different types of problems. Consequently, no formal rules are given; instead, the working of stoichiometric problems is demonstrated by examples. It is sometimes possible, and tempting, to use "plug-in" methods, but such rote methods are dangerous when new situations are encountered. A grasp of the overall principles—which involve no more than is stated in (1) to (4) above—will prove far more useful. One way to develop this mastery is to consider alternative methods of solution; a variety of approaches is always possible.

The solution of any problem can be developed in a series of steps. For a beginner, this sequence of steps is often somewhat lengthy, but as insight, experience, and confidence develop, some steps can usually be combined and the whole solution considerably shortened. However, this process should not be rushed. Difficulties can be avoided by:

1. Remembering that scientific terms have precise definitions. Learn and use these definitions properly, including such common terms as fraction and percent.
2. Appreciating the implications of both the number aspect and the weight aspect of the mole (Sec. 2-2).
3. Being careful in the use of significant figures to express the precision of an answer. This is especially necessary when an electronic calculator is used, the apparent precision of the answer usually being higher than warranted by the data input. For example, although your calculator may give 51.05148600 when 5.24 is multiplied by 9.74265, the proper scientific answer is 51.1. An imprecise or inaccurate answer can also arise because of premature rounding. In most of the examples and problems presented here, an answer good to 0.2 to 0.5 percent is adequate for the precision of the data given and is all that is

---

[4]We are not differentiating between the terms *mass* and *weight;* see "Study Guide".

needed for an understanding of the principles involved. However, the accuracy of the customary methods of chemical analysis is usually such that higher precision is often justified and indeed is necessary if the full implications of the analysis are to be appreciated.

Extensive practice with problems of various kinds is essential for the development of an understanding of chemical principles. We will discuss some typical problems, which may be classified into two general categories: those dealing with the interconversion of chemical formulas and percentage compositions, and those dealing with weight relations in chemical reactions. A number of problems fall in both categories.

3-6   Chemical Formulas and Percentage Composition

One of the simplest applications of stoichiometry is the calculation of the composition of a compound in weight percent from its formula and atomic weight information.

☐ **Composition in Weight Percentage**   Find the composition of nitric acid, HNO_3, on a weight basis. The weight percentage of a component A in a sample of matter is defined as

$$\text{Weight percentage } A = \frac{\text{weight of } A}{\text{total weight}} \times 100$$

Example 3-6

Solution   The method used to compute the answer is indicated by the steps shown in Table 3-3. The atomic weights of H, N, and O are, respectively, 1.01, 14.01, and 16.00. Therefore, the weights of the components in 1 mol HNO_3 are as indicated in the second column of the table. The total of these entries is equal to the molecular weight of HNO_3. The percentage composition is given in the fourth column. ■

☐ Find the composition of sulfuric acid, H_2SO_4, on a weight basis. ■

Exercise 3-6

The deduction of an empirical (or simplest) formula from percentage composition information is the reverse of the calculation illustrated in Example 3-6. However, it usually presents appreciably greater difficulties to the beginner and so, instead of starting with a chemical example, we present first an illustration involving something more familiar than atoms.

Table 3-3
Weight-Percentage Composition of HNO₃

Number of moles of atoms	Weight/g	Weight fraction	Weight percentage
1 H	1.01	$\frac{1.01}{63.02} = 0.016$	1.6
1 N	14.01	$\frac{14.01}{63.02} = 0.222$	22.2
3 O	$3 \times 16.00 = 48.00$	$\frac{48.00}{63.02} = 0.762$	76.2
SUMS:	63.02 (molecular weight)	1.000	100.0

Example 3-7

□ **Empirical Formula of Postage** Twenty-four percent of the value of the postage on a package is represented by 15¢ stamps (designated Fn), 16 percent by 20¢ stamps (Ty), and 60 percent by 50¢ stamps (Fy). What are the smallest possible numbers of stamps of each kind present?

Solution The solution is found in two steps, with results collected in Table 3-4. The information provided is shown in the first two columns. The first step is to find the numbers of stamps that would correspond to a convenient assumed total postage, such as 100¢. These relative numbers are in the third column; for example, the number of 15¢ stamps that corresponds to 24¢ is $\frac{24}{15}$. The final step is to scale the relative numbers until they are all integers, since no fractional numbers of stamps can be used. This is done by trial. For example, as a first step we may divide all the relative numbers by the smallest: $\frac{1.6}{0.8} = 2.0$, $\frac{0.8}{0.8} = 1.0$, $\frac{1.2}{0.8} = 1.5$. Multiplication by 2 then yields the smallest integral numbers of stamps, given in the last column.

The "empirical formula" that corresponds to the specified postage is $Fn_4Ty_2Fy_3$. It represents a monetary value of $(4 \times 15) + (2 \times 20) + (3 \times 50)$¢ $= 250$¢ that corresponds to the formula weight of a compound. Note that any integral multiple of the formula ($Fn_8Ty_4Fy_6$, $Fn_{24}Ty_{12}Fy_{18}$, and so on) and of the total value (500¢, 1500¢, and so on) would lead to the same percentage distribution of the stamps. On the basis of the information given it is impossible to determine the actual postage used. ∎

Exercise 3-7

□ Forty percent of the value of a small sum of money is represented by quarters (Q), 48 percent by dimes (D), and 12 percent by nickels (N). What are the smallest possible numbers of coins of each kind? ∎

The example just worked is parallel to the problem faced by a chemist who wishes to find the chemical formula from the analysis of a compound:

Example 3-8

□ **Chemical Formula from Elemental Composition** The elemental composition of a substance is as indicated by the first and third columns of Table 3-5. What is the empirical formula of the compound?

Solution If a quantity of 100 u of substance is considered, the numbers in the third column of the table represent the weights of the different kinds of atoms in atomic mass units. The *relative* numbers of atoms are found by dividing these weights by the atomic weights, as shown in the fourth column. The smallest *integral* numbers of atoms, given in the fifth column, are found by appropriate scaling—that is, dividing by the largest common factor among the relative numbers of atoms, here 1.32, which gives the smallest integers for the numbers of the different kinds of atoms. Note that

Table 3-4
Postage on Parcel

Kind of stamp	Percent of value	Relative numbers of stamps	Smallest integral numbers of stamps
15¢ (Fn)	24	$\frac{24}{15} = 1.6$	4
20¢ (Ty)	16	$\frac{16}{20} = 0.8$	2
50¢ (Fy)	60	$\frac{60}{50} = 1.2$	3

Table 3-5
Empirical Formula of a
Chemical Compound

Kind of atom	AW	Weight percentage	Relative numbers of atoms	Smallest integers
N	14.0	36.8	$\frac{36.8}{14.0} = 2.63$	2
H	1.0	5.3	$\frac{5.3}{1.0} = 5.3$	4
C	12.0	15.8	$\frac{15.8}{12.0} = 1.32$	1
S	32.1	42.1	$\frac{42.1}{32.1} = 1.31$	1

because of experimental uncertainties in the percentage composition data, some numbers may vary by as much as about 1 percent from integral values.

The *empirical* formula of the compound is thus N_2H_4CS. The actual formula cannot be established from the data given. It could be any integral multiple of the empirical formula. A compound of composition that fits the data given is ammonium thiocyanate, NH_4SCN, containing the ions NH_4^+ and SCN^-.

The same problem can also be worked in macroscopic units. If the scaling is by Avogadro's number, then 100 g rather than 100 u of substance is assumed, and each number in the fourth column of Table 3-5 gives the number of *moles* of atoms rather than the number of atoms in the assumed sample. However, the sample might be 100 ounces, 100 tons, or any arbitrary number of any size unit; the procedure of finding the simplest integers relating the numbers of different kinds of atoms present remains the same. ■

☐ What is the empirical formula of a compound with the following composition in weight percentage: Co 13.0, K 25.9, N 18.6, O 42.5? ■

In the example just given, the starting point is the elemental composition, which is information that generally does not follow from chemical analysis without computations. However, to find the formula of a compound it is not necessary to work out the elemental composition; the analytical data may be used directly as the starting point. Before considering an example to illustrate this point (Example 3-15 below), we first discuss some simple examples of weight relations in chemical reactions.

3-7 Quantities of Substances Participating in Reactions

One of the simplest applications of stoichiometry to chemical reactions is to find out the quantities of substances that react and that are produced. These quantities may be expressed in moles, weights, volumes, or in still other ways. It is always assumed, unless stated otherwise, that the reactions go to completion, that is, that they proceed until one or more of the starting materials is used up. Even though this may not always be true, the assumption is a reasonable and simplifying one, and corrections for incomplete reaction can be made.

☐ **Quantities of Reactants and Products** When carbon monoxide is heated with water to a high temperature, hydrogen and carbon dioxide are produced:

Example 3-9

$$CO + H_2O \longrightarrow H_2 + CO_2 \qquad (3\text{-}7)$$

How many grams of H_2 and how many moles of CO_2 would be formed if 56 g CO reacted with an excess of water? How many grams of water would be consumed?

Solution Remember that any chemical equation, such as (3-7), applies to moles of reactants as well as to individual atoms, molecules, ions, or other species represented by the formulas. The coefficients do *not* represent relative numbers of grams, or ounces, or tons—because the weight of a mole of one substance usually differs from the weight of a mole of another, just as the weight of a dozen eggs usually differs from the weight of a dozen oranges. The balanced equation (3-7) indicates that 1 mol CO reacts with 1 mol H_2O to form 1 mol H_2 and 1 mol CO_2. Perhaps the easiest way to solve this problem is to calculate the number of moles of CO at the start and then use the simple mole relationships implied by the equation. The molecular weight of CO is $12.0 + 16.0 = 28.0$, and thus 56 g CO is

$$56 \text{ g CO} \times \frac{1 \text{ mol CO}}{28.0 \text{ g CO}} = 2.0 \text{ mol CO} \qquad (3\text{-}8)$$

Hence if all the CO reacts, it will use up 2.0 mol H_2O and form 2.0 mol H_2 and 2.0 mol CO_2. To calculate the weights of hydrogen and water involved we need their molecular weights, 2.0 for H_2 and 18.0 for H_2O. Hence the weight of H_2 formed is

$$2.0 \text{ mol } H_2 \times \frac{2.0 \text{ g } H_2}{\text{mol } H_2} = 4.0 \text{ g } H_2$$

and the weight of water consumed is

$$2.0 \text{ mol } H_2O \times \frac{18.0 \text{ g } H_2O}{\text{mol } H_2O} = 36 \text{ g } H_2O \quad \blacksquare$$

Exercise 3-9

□ Ammonia, NH_3, and oxygen, O_2, react when heated in the presence of platinum to form gaseous water and nitric oxide, NO:

$$4NH_3 + 5O_2 \longrightarrow 6H_2O + 4NO$$

How many grams of H_2O and how many moles of NO would be formed if 51 g NH_3 reacted with an excess of O_2? How many grams of O_2 would be consumed? ∎

Example 3-10

□ **Quantities of Substances Produced** When crystals of potassium permanganate ($KMnO_4$) are heated to several hundred degrees, oxygen, solid potassium manganate (K_2MnO_4), and solid manganese dioxide (MnO_2) are formed by the reaction

$$2KMnO_4(s) \longrightarrow K_2MnO_4(s) + O_2(g) + MnO_2(s) \qquad (3\text{-}9)$$

How many grams of MnO_2 result when 10.0 g $KMnO_4$ is heated?

Solution The balanced equation (3-9) shows that in this reaction 2 mol $KMnO_4$ produces 1 mol MnO_2. This problem may be solved in a manner parallel to that used for Example 3-9, but we shall use an alternative, though equivalent, approach. Let x represent the grams of MnO_2 that are formed. Then, by (3-9), the following proportion exists:

$$\frac{\text{FW MnO}_2}{\text{2FW KMnO}_4} = \frac{x}{10.0 \text{ g}}$$

so that

$$x = 10.0 \text{ g} \times \frac{\text{FW MnO}_2}{\text{2FW KMnO}_4}$$

The formula weights are 86.9 for MnO_2 and 158.0 for $KMnO_4$. Inserted,

$$x = 10.0 \text{ g} \times \frac{86.9}{2(158.0)} = \underline{2.75 \text{ g}}$$

Note that the units work out properly, to grams, as they should. ■

□ How many grams of Fe_3O_4 are produced when 50.0 g Fe_2O_3 reacts with an excess of the gas carbon monoxide according to the equation

Exercise 3-10

$$3Fe_2O_3 + CO \longrightarrow 2Fe_3O_4 + CO_2 \; ■$$

□ **Analysis by Means of Weight Loss in Known Reaction** When solid sodium bicarbonate, $NaHCO_3$, is heated to about $300°C$ it decomposes completely to form solid sodium carbonate, Na_2CO_3, and gaseous carbon dioxide and water. A mixture of sodium bicarbonate and sodium carbonate weighing 200 g was heated until all the $NaHCO_3$ present had been converted to sodium carbonate. The final weight of the mixture was 180 g. What was the percentage of sodium carbonate in the original mixture, before it was heated?

Example 3-11

Solution An essential first step is to write a balanced equation for the reaction taking place:

$$2NaHCO_3(s) \longrightarrow Na_2CO_3(s) + CO_2(g) + H_2O(g)$$

We use the symbols (s) and (g) to emphasize which substances are solid and which are gaseous. For every 2 mol $NaHCO_3$ that react, 1 mol CO_2 and 1 mol H_2O are lost. Thus for every 2 mol $NaHCO_3$ that react there is a weight loss of 62 g (44 g CO_2 + 18 g H_2O). We can therefore write

$$\frac{\text{Wt lost}}{\text{mol NaHCO}_3 \text{ reacting}} = \frac{62 \text{ g}}{2 \text{ mol NaHCO}_3}$$

or

$$\text{mol NaHCO}_3 \text{ reacting} = (\text{wt lost}) \times \frac{2 \text{ mol NaHCO}_3}{62 \text{ g}}$$

In this example the weight loss is $(200 - 180)\text{g} = 20$ g. Hence

$$\text{Wt NaHCO}_3 \text{ reacting} = 20 \text{ g} \times \frac{2 \text{ mol NaHCO}_3}{62 \text{ g}} \times \frac{84 \text{ g NaHCO}_3}{\text{mol NaHCO}_3}$$

$$= \frac{20 \times 2 \times 84}{62} \text{ g NaHCO}_3 = 54 \text{ g NaHCO}_3$$

Since all the $NaHCO_3$ in the original 200-g sample reacted, the percentage of $NaHCO_3$ in this sample is

$$100 \times \frac{\text{wt NaHCO}_3}{\text{wt of sample}} = 100 \times \frac{54}{200} = \underline{27 \text{ percent}} \; ■$$

Exercise 3-11

☐ Lime, CaO, is produced by heating limestone, $CaCO_3$, the reaction being

$$CaCO_3 \longrightarrow CaO + CO_2$$

When 200 g of a mixture of $CaCO_3$ and CaO was heated until all the $CaCO_3$ was converted to CaO, the loss in weight was 25 g. What was the percentage of CaO in the original mixture? ■

More involved considerations are needed when one or several of the reactants are not completely used up at the end of the reaction. The reactant that runs out first and thereby limits the extent of the reaction must first be identified.

Example 3-12

☐ **Extent of a Reaction Limited by a Reactant** The metal zinc, Zn, reacts with a solution of sulfuric acid, H_2SO_4, in water to form gaseous hydrogen, H_2, and a solution of zinc sulfate, $ZnSO_4$.

(a) Suppose that 0.50 mol Zn reacts with a solution containing 0.70 mol H_2SO_4 until one of the reactants is used up. How much of which reactant will remain and how much of each product will be formed?

Solution We first write a balanced equation for the reaction

$$Zn + H_2SO_4 \longrightarrow H_2 + ZnSO_4 \tag{3-10}$$

The molar proportions for reaction of zinc and sulfuric acid are seen to be 1:1. Because we have 0.50 mol Zn, 0.50 mol H_2SO_4 is needed to react with it. We have, however, 0.70 mol H_2SO_4 present in the solution originally, more than is needed. Hence $0.70 - 0.50 = \underline{0.20 \text{ mol sulfuric acid left unreacted}}$, and the reaction of 0.50 mol Zn with 0.50 mol H_2SO_4 produces $\underline{0.50 \text{ mol } H_2}$ and $\underline{0.50 \text{ mol } ZnSO_4}$.

(b) Suppose that 50 g Zn is allowed to react with a solution containing 50 g H_2SO_4. What weight of hydrogen will be produced, and what weight of which reactant will be left unreacted?

Solution One way to solve this problem is to convert the weights of each reactant into moles and follow the method used in (a). The atomic weight of Zn is 65.4, so 50 g Zn is 50 g Zn $\times$ (1 mol Zn/65.4 g Zn) = 0.76 mol Zn. The formula weight of H_2SO_4 is 98.1, so 50 g H_2SO_4 is 50 g H_2SO_4 $\times$ (1 mol H_2SO_4/98.1 g H_2SO_4) = 0.51 mol H_2SO_4. Thus there is an excess of zinc; after 0.51 mol Zn has reacted with the 0.51 mol H_2SO_4, all the acid is used up, and no further reaction can occur. The molecular weight of hydrogen is 2.02; hence the weight of hydrogen formed is

$$0.51 \text{ mol } H_2SO_4 \times \frac{1 \text{ mol } H_2}{\text{mol } H_2SO_4} \times \frac{2.02 \text{ g } H_2}{\text{mol } H_2} = \underline{1.03 \text{ g } H_2}$$

The weight of *unreacted* zinc will be

$$(0.76 - 0.51) \text{ mol Zn} \times \frac{65.4 \text{ g Zn}}{\text{mol Zn}} = \underline{16 \text{ g Zn}} \quad ■$$

Exercise 3-12

☐ Iron sulfide, FeS, reacts with a solution of hydrogen chloride, HCl, in water to form gaseous hydrogen sulfide, H_2S, and a solution of ferrous chloride, $FeCl_2$. (a) Write a balanced equation for the reaction. (b) How much of which reactant will remain and how much of each product will be formed when 0.50 mol FeS and 0.70 mol HCl react until one of the reactants is used up? (*c*) When 50 g FeS reacts with a solution containing 50 g HCl until one of the reactants is used up, what weight of H_2S will be produced and what weight of which reactant will be left unreacted? ■

□ **Extent of a Reaction Limited by a Reactant** In the thermite process, used to produce iron for certain welding applications, powdered aluminum is allowed to react with ferric oxide, Fe_2O_3, to produce metallic iron, Fe, and aluminum oxide, Al_2O_3. The reaction gives off so much heat that the iron produced is, under appropriate conditions, melted and flows readily to the joint to be welded.

Example 3-13

(*a*) Suppose that 0.100 mol Al is allowed to react with 0.100 mol Fe_2O_3 until one or the other is used up. How many moles of iron will be produced, what weight of Al_2O_3 will be produced, and how many moles of which reactant will remain unreacted?

Solution The balanced equation is

$$2Al + Fe_2O_3 \longrightarrow 2Fe + Al_2O_3 \tag{3-11}$$

Initially equal numbers of moles of Fe_2O_3 and Al are present, but the number of moles of Fe_2O_3 reacting is only half as great as that of Al. Thus the number of moles of Fe_2O_3 left unreacted is

$$0.100 \text{ mol } Fe_2O_3 \text{ initially} - \frac{1 \text{ mol } Fe_2O_3}{2 \text{ mol Al}} \times 0.100 \text{ mol Al reacting}$$

$$= 0.100 - 0.050 = \underline{0.050 \text{ mol } Fe_2O_3 \text{ unreacted}}$$

Because 0.100 mol Al reacts, $\underline{0.100 \text{ mol Fe is formed.}}$
    Finally, for every 2 mol Al reacting, 1 mol Al_2O_3 is formed. Thus

$$\text{Wt } Al_2O_3 = 0.100 \text{ mol Al} \times \frac{1 \text{ mol } Al_2O_3}{2 \text{ mol Al}} \times \frac{102 \text{ g } Al_2O_3}{\text{mol } Al_2O_3} = \underline{5.1 \text{ g } Al_2O_3}$$

(*b*) Suppose that 1.00 ton of Al is allowed to react with 1.00 ton of Fe_2O_3 until one of them is used up. How many tons of iron will be produced?

Solution First find out which is the limiting reagent, that is, whether the aluminum or the iron oxide is used up. Then find the weight of iron produced when 1.00 ton of that reagent reacts in this particular reaction.
    It is not necessary to convert tons to grams and then to moles. This is a possible approach, but there is a simpler way. The important point is that any chemical equation implies not only the relative numbers of moles of reactants and products involved, but also, *indirectly,* the *relative proportions by weight.* It does not matter whether these weights are expressed in atomic weight units, in grams, in tons, or in any other units, as long as the units are the same for each reactant and product.
    Specifically, (3-11) says that two Al atoms react with each formula unit of Fe_2O_3. The relative weights that react are then

$$\frac{\text{Wt Al}}{\text{Wt } Fe_2O_3} = \frac{2 \text{ AW Al}}{\text{FW } Fe_2O_3} = \frac{54}{160}$$

This ratio is independent of the units of weight; 54 g of Al will react with

160 g Fe_2O_3, and 54 ton of Al will react with 160 ton of Fe_2O_3. We may rewrite this expression as

$$\text{Wt Al reacting} = \tfrac{54}{160} \text{ wt } Fe_2O_3 \text{ reacting} \qquad (3\text{-}12)$$

Hence 1.00 ton of Fe_2O_3 will react with just $\tfrac{54}{160}$ (1.00) ton of Al = 0.34 ton Al. When we start with 1.00 ton of each, all the ferric oxide will react and (1.00 − 0.34) = 0.66 ton Al will be left over unreacted. How many tons of iron will be produced? For every formula unit of Fe_2O_3 we get two Fe atoms; thus

$$\frac{\text{Wt } Fe_2O_3 \text{ reacting}}{\text{Wt Fe produced}} = \frac{\text{FW } Fe_2O_3}{2 \text{ AW Fe}} = \frac{160}{2 \times 56} = 1.43$$

$$\text{or} \quad \text{Wt Fe produced} = \frac{\text{Wt } Fe_2O_3 \text{ reacting}}{1.43} = \frac{1.00 \text{ ton}}{1.43} = \underline{0.70 \text{ ton}} \quad \blacksquare$$

Exercise 3-13

☐ In the production of the metal chromium, the ore chromite, $FeCr_2O_4$, is allowed to react with carbon, the products being Fe, Cr, and the gas carbon monoxide, CO. (*a*) Suppose that 0.100 mol $FeCr_2O_4$ and 0.100 mol C react until one or the other is used up. How many moles of iron will be produced, what weight of chromium will be produced, and how many moles of which reactant will remain unreacted? (*b*) How many tons of chromium are produced when 1.00 ton $FeCr_2O_4$ reacts with 1.00 ton C until one of them is used up? ■

Example 3-14

☐ **Partially Completed Reaction** When solid silver oxide, Ag_2O, is heated for a long enough time at about 200°C it decomposes to silver, Ag, and gaseous O_2, which escapes into the atmosphere.

(*a*) A sample of Ag_2O is heated until it is entirely decomposed to Ag. What is the fractional weight loss?

Solution We start by writing a balanced equation relating reactants and products

$$Ag_2O(s) \longrightarrow 2Ag(s) + \tfrac{1}{2} O_2(g) \qquad (3\text{-}13)$$

The formula weight of Ag_2O is 2(107.87) + 16.00 = 231.74. Thus the weight of O_2 escaping for every 231.74 g Ag_2O that decomposes is

$$\text{Wt } O_2 = 1 \text{ mol } Ag_2O \times \frac{0.5 \text{ mol } O_2}{\text{mol } Ag_2O} \times \frac{32.00 \text{ g } O_2}{\text{mol } O_2} = 16.00 \text{ g } O_2$$

The fractional weight loss is then

$$\tfrac{16.00}{231.74} = \underline{0.0690}$$

or 6.9 percent.

(*b*) Suppose that 100.0 lb of Ag_2O is heated for a rather short time and, after cooling, the residue is found to weigh 97.1 lb. What fraction of the Ag_2O decomposed? How much Ag is in the residue?

Solution Here we are dealing with an incomplete reaction, for if all the Ag_2O had decomposed, the weight would have decreased by 6.9 percent, which is 6.9 lb. Instead it decreased by only 2.9 lb.

We might say that the fraction of the Ag_2O that decomposed must be

2.9/6.9 = 0.42, or 42 percent—and indeed this is correct. But how do we *know* that that is right? Use of this relationship is equivalent to saying that the loss in weight is directly proportional to the weight of Ag_2O decomposed, and indeed this is a necessary consequence of the weight relationships implied by Equation (3-13):

$$\frac{\text{Wt } Ag_2O \text{ decomposed}}{\text{Wt loss}} = \frac{231.74}{16.00} = 14.48$$

or

$$\text{Wt } Ag_2O \text{ decomposed} = 14.48 \times \text{wt loss} \qquad (3\text{-}14)$$

Thus the weight of Ag_2O decomposed is 14.48×2.9 lb = 42 lb, which, as a fraction of 100 lb, is 0.42.

Equation (3-13) also permits us to calculate readily the weight of Ag formed:

$$\text{Wt Ag formed} = \frac{2 \text{ FW Ag}}{\frac{1}{2} \text{ FW } O_2} \times \text{wt loss as } O_2$$

$$= 2 \times \frac{107.87}{16.00} \times 2.9 \text{ lb}$$

$$= \underline{39 \text{ lb}}$$

As a check, let us list the final amounts:

Reacted Ag_2O	42 lb
Unreacted Ag_2O	58 lb
Resulting Ag	39 lb
O_2 produced	3 lb

The last three items add to 100 lb, as they should. ∎

Exercise 3-14

☐ At $1300°C$ Fe_3O_4 slowly reacts with O_2 to form Fe_2O_3. (*a*) A sample of Fe_3O_4 is heated until it has been changed completely into Fe_2O_3. What is the fractional weight gain? (*b*) When a 100.0-lb sample of Fe_3O_4 reacts with O_2 the weight increase is 2.1 lb. What fraction of the sample has reacted? How many pounds of Fe_2O_3 have been formed? ∎

Example 3-15

☐ **Chemical Formula from Analytical Results** A compound containing only P and S was analyzed by converting the sulfur to $BaSO_4$ and the phosphorus to $Mg_2P_2O_7$ (magnesium pyrophosphate). The amounts of these solids were determined by weighing, and in a typical analysis the results were, per gram of compound, 3.1254 g $BaSO_4$ and 2.0443 g $Mg_2P_2O_7$. What is the empirical formula of the compound?

Solution Instead of computing the percentages of S and P in the compound and proceeding as in Example 3-8, we calculate the moles of S atoms and of P atoms in 1.000 g compound directly from the data given. Thus, because 1 mol $BaSO_4$ (233.40 g) contains 1 mol S, 3.1254 g $BaSO_4$ contains the following number of moles of S:

$$\text{mol S} = 3.1254 \text{ g } BaSO_4 \times \frac{1 \text{ mol } BaSO_4}{233.40 \text{ g } BaSO_4} \times \frac{1 \text{ mol S}}{\text{mol } BaSO_4}$$

$$= 1.339 \times 10^{-2} \text{ mol S}$$

Similarly, 1 mol $Mg_2P_2O_7$ (222.54 g) contains 2 mol P, and thus the number of moles of P atoms contained in $Mg_2P_2O_7$ is

$$\text{mol P} = 2.0443 \text{ g } Mg_2P_2O_7 \times \frac{1 \text{ mol } Mg_2P_2O_7}{222.54 \text{ g } Mg_2P_2O_7} \times \frac{2 \text{ mol P}}{\text{mol } Mg_2P_2O_7}$$
$$= 1.837 \times 10^{-2} \text{ mol P}$$

The ratio of moles of P to S is thus $1837:1339 = 1.372$. The smallest integers with a ratio near this number are 4 and 3 $(4:3 = 1.333)$. The integers 11 and 8 are in a ratio that is even closer to 1.372 $(11:8 = 1.375)$, but these integers are suspiciously large, and it is likely that the difference between 1.333 and 1.372 is caused by experimental inaccuracy. The empirical formula would thus be $\underline{P_4S_3}$.

To check the accuracy of the analysis, it is useful to find the actual weights of P and S and to see whether they add up to the sample weight:

$$\text{Wt S} = 3.1254 \text{ g } BaSO_4 \times \frac{1 \text{ mol } BaSO_4}{233.40 \text{ g } BaSO_4} \times \frac{1 \text{ mol S}}{\text{mol } BaSO_4} \times \frac{32.06 \text{ g S}}{\text{mol S}}$$
$$= \underline{0.4293 \text{ g S}}$$

$$\text{Wt P} = 2.0443 \text{ g } Mg_2P_2O_7 \times \frac{1 \text{ mol } Mg_2P_2O_7}{222.54 \text{ g } Mg_2P_2O_7} \times \frac{2 \text{ mol P}}{\text{mol } Mg_2P_2O_7}$$
$$\times \frac{30.97 \text{ g P}}{\text{mol P}} = \underline{0.5690 \text{ g P}}$$

SUM OF WEIGHTS   0.9983 g

The weights of sulfur and phosphorus found by analysis do not quite add to the 1.000-g sample weight. Thus, the analysis is demonstrably not very accurate, and the simpler interpretation of the analytical results, leading to the formula P_4S_3 (rather than $P_{11}S_8$), is the more reasonable. A clearer distinction between these two possibilities could be provided by more accurate analyses. ■

Exercise 3-15

□ A compound containing only P and S was analyzed in the way just described, yielding 9.40 g $BaSO_4$ and 2.56 g $Mg_2P_2O_7$ per gram of compound. What is the empirical formula of the compound? ■

Example 3-16

□ **Chemical Formula by Determination of Quantities of Products Formed in Combustion** An unknown compound contains only carbon, hydrogen, and oxygen. When 100.0 g of the compound is burned with an excess of oxygen, 39.2 g water and 95.6 g carbon dioxide are formed. What is the simplest formula of the unknown substance?

Solution When the compound reacts with oxygen to form CO_2 and H_2O, all the carbon and hydrogen must come from the compound itself. We can use the data given to calculate the numbers of moles of carbon atoms and hydrogen atoms in the original sample, in a manner analogous to that used to find the moles of phosphorus and sulfur atoms in the preceding example. However, to determine the oxygen content of the original sample we must also know the *weight* of the carbon and hydrogen in it; the weight of original compound not accounted for as C or H must be oxygen. The weight of oxygen is then converted to moles of oxygen atoms, and finally the empirical formula of the compound is found from the relative numbers of moles of hydrogen, carbon, and oxygen atoms that the sample contains.

The calculation of moles of hydrogen atoms and carbon atoms is summarized in the following table. The number of moles of hydrogen atoms is equal to twice the number of moles of water formed, while the number of moles of carbon atoms is equal to that of carbon dioxide. An excess of oxygen is assumed to be present during combustion, since if there were insufficient oxygen, some of the carbon would have been converted to CO rather than CO_2.

Element	Combustion product	Weight of combustion product	Moles of combustion product	Moles of atoms in 100.0-g sample	Weight of element
H	H_2O	39.2 g	$\frac{39.2}{18.02} = 2.18$	4.36 mol H	4.4 g
C	CO_2	95.6 g	$\frac{95.6}{44.01} = 2.17$	2.17 mol C	26.1 g

The 100.0-g sample contains a total of 30.5 g carbon and hydrogen, and thus must contain 69.5 g oxygen, the only other constituent. This corresponds to $\frac{69.5}{16.00} = 4.34$ mol oxygen atoms. Finally, then, we know that the sample contains C, H, and O atoms in the ratio $2.17:4.36:4.34$; thus, the empirical formula of the compound is $\underline{CH_2O_2}$. The molecular formula may be the same, or any integral multiple of it. (The molecule might be, for example, formic acid, HCOOH.) ■

□ Burning 100.0 g of a compound containing only C, H, and O with an excess of oxygen yielded 220 g CO_2 and 120 g H_2O. What is the simplest formula of the compound? ■

Exercise 3-16

Summary

The magnitudes of random errors may be assessed from the results of multiple measurements, while the existence of systematic errors is usually hard to establish and may remain hidden. The number of significant figures with which a result is reported should be an indication of the estimated error of the result. The precision of intermediate calculations and the rounding of intermediate results should take this estimated error into account. In addition and subtraction, the precision of an answer is governed by the lowest absolute precision of the numbers treated; in multiplication and division, the lowest relative precision determines the relative precision of the answer.

The simplest chemical formulas (also called empirical formulas) represent usually just the composition of a compound, while more elaborate molecular, ionic, and structural formulas (which may be three-dimensional) yield information about structure. Compounds with the same elemental composition and molecular weight may have different structures, a phenomenon called isomerism.

A chemical reaction may be represented by a combination of chemical formulas in a chemical equation. In a balanced equation, the number of atoms of a given kind is the same on each side, as is the overall charge. Balanced equations permit the calculation of the amounts of compounds that may be used up or formed in chemical reactions. The simplest formula of a compound can be determined experimentally from weight relationships.

Terms and Concepts

Problems and Questions

3-1 Precision of Calculations Express the results of the following arithmetic operations with the appropriate precision:

(a) 429×0.0020
(b) $3.7 \times 0.025 - 0.2370$
(c) $(4.30 + 0.1031)(101 - 88.1)$
(d) $(1037 - 952)/(0.2790 - 0.2687)$
(e) $(5300)(0.014000)$

3-2 Appropriate Choice of Numbers (a) A certain measurement given as 14.3487 g was later found to be 0.12 percent too large. What is the correct value? (b) The population of a town was about 151,000 at a certain date. In the following months 1036 babies were born, 739 persons died, 310 people moved into the town, and 479 moved away. What number would you choose to describe the population after these changes? What is the percentage change of the population?

3-3 Percentages (a) What is 2.3 percent of 7452? (b) What is the result of increasing 1372 and 13,728 by 4.1 percent?

3-4 Conversion of Units Convert the values of the quantities in the following statements into SI units. The answers should reflect the precision implied by such statements, some of which are the kind of common statement that is not very precise. (Conversion factors: 1.609 km = 1 mile; 1 inch = 2.54 cm; 1 gal = 3.785 liter; and 1 Å = 10^{-10} m.) (a) The mountains are about 14,000 feet high. (b) The altitude of the top of Mt. Whitney is 14,497 feet. (c) The swimming pool contains about 130,000 gal of water. (d) The Moffat tunnel in Colorado is 6.21 mile long. (e) The circumference of the earth is 25,000 mile. (f) The distance between the H atom and the O atom in a water molecule is 0.958 Å. (g) The speed limit on certain city streets is 25 mile hour^{-1}.

3-5 Balancing Equations Balance the following chemical equations in which all the reactants and products involved are given:

(a) $H_3BO_3 \longrightarrow H_4B_6O_{11} + H_2O$
(b) $CaCN_2 + H_2O \longrightarrow CaCO_3 + NH_3$
(c) $C_7H_{16} + O_2 \longrightarrow CO_2 + H_2O$
(d) $CO + Fe_3O_4 \longrightarrow FeO + CO_2$
(e) $Fe_3O_4 + H_2 \longrightarrow Fe + H_2O$
(f) $Ca + H_2O \longrightarrow Ca(OH)_2 + H_2$
(g) $SiO_2 + HF \longrightarrow SiF_4 + H_2O$

3-6 Formula of an Oxide A certain oxide of iodine contains 24.0 percent oxygen. What is the empirical formula of the oxide?

3-7 Formulas of Compounds (a) A 0.849-g sample of calcium reacted with hydrogen to yield 0.892 g of a compound. (b) A 0.467-g sample of lead reacted with bromine to yield 0.827 g of a compound. What are the simplest formulas of the compounds?

3-8 Simplest Formula A certain oxide of phosphorus contains 56.3 percent oxygen and an even number of oxygen atoms. What is its simplest formula?

3-9 Formula of a Compound A compound contains 66.9 percent Ag, 15.8 percent V, and the remainder O. What is its simplest formula?

3-10 Simplest Formula A certain compound contains only carbon, hydrogen, and oxygen. If it contains 47.4 percent carbon and if there is one oxygen atom present for every four hydrogen atoms, what is its simplest formula?

3-11 Simplest Formula A 10.0-g sample of a compound of C, H, and N contains 17.7 percent N and 3.8×10^{23} atoms of H. What is its simplest formula?

3-12 Simplest Formula and Molecular Formula A certain oxide of carbon is analyzed and found to contain 50.0 percent carbon. A very rough determination of the molecular weight shows it to be 300, within a possible error of about 10 percent. (*a*) What is the simplest formula of the oxide consistent with the analysis (without considering the molecular weight)? (*b*) What is the molecular formula of the oxide?

3-13 Simplest Formula and Molecular Formula A certain solid compound contains 31.8 percent nitrogen and 13.6 percent hydrogen; the rest is carbon. (*a*) What is the simplest formula of this compound? (*b*) Suppose that you know that 2 mol of the compound contain 56 g nitrogen. What is the molecular formula of the compound?

3-14 Simplest Formula Water may be shown by analysis to consist of 11.2 percent hydrogen and 88.8 percent oxygen. Suppose you believed that an atom of hydrogen weighed one-twelfth as much as an atom of oxygen. What would be the simplest formula for water consistent with these supposed relative weights and with the analytical data?

3-15 Implications of Incorrect Assumed Formula It is known by analysis that hydrogen peroxide, a compound containing only hydrogen and oxygen, contains 5.9 percent hydrogen. Suppose you believed its formula to be H_4O_3. What ratio of the atomic weight of oxygen to that of hydrogen is consistent with the analysis and with this assumption? (Do *not* use known atomic weights in solving this problem.)

3-16 Atomic Weights from Analytical Data Let X and Y represent two unknown elements that form a compound with oxygen of formula $X_8Y_4O_6$. If 53.6 g of this compound contains 19.2 g X and 7.2×10^{23} atoms of oxygen, what are the atomic weights of X and Y?

3-17 Formula Weight from Analytical Data A certain stable, nonvolatile chemical substance, X, may readily be crystallized from solution as a dihydrate $X \cdot 2H_2O$. When 5.00 g of the dihydrate is heated gently, it loses all its water and leaves a residue of 4.00 g pure X. What is the formula weight of X?

3-18 Atomic Weight from Chemical Composition A compound of the formula K_nXCl_m contains 16.1 percent K, 40.2 percent X, and 43.7 percent Cl. Compute possible atomic weights of X.

3-19 Simplest Formula from Analytical Data When 24.0 g of an unknown gaseous compound containing only carbon and hydrogen was burned in excess oxygen, 36.0 g water and 73.4 g CO_2 were formed. What is the simplest formula of the compound? Show how the answer could have been obtained from the weight of only one product, water or CO_2.

3-20 Formula of a Compound Compound X is known to contain only the elements C, H, and S. When it is burned in excess O_2, the products consist exclusively of CO_2, H_2O, and SO_2. A 2.35-g sample of X was burned in excess O_2; 2.20 g of CO_2 and 1.35 g of H_2O were formed. What is the empirical formula of X?

3-21 Simplest Formula and Molecular Formula A certain compound, X, contains only C, H, and N. Analysis shows that it contains 60.0 percent C. When 2.00 g of the compound is burned, all the hydrogen is converted to water weighing 0.90 g. (*a*) What is the weight of each element in a 2.00-g sample of X? (*b*) What is the simplest formula of X? (*c*) Suppose an experiment shows that, within a precision of 5 percent, the molecular weight of X is 155. What is the molecular formula of X?

3-22 Atomic Weight from Chemical Composition A certain metal forms a compound with bromine that contains 71.4 percent Br and 28.6 percent metal. What conclusions can be drawn about the atomic weight of the metal if it is assumed that one atom of the metal combines with n atoms of bromine, where n is some positive integer?

3-23 Atomic Weight from Chemical Composition An oxide of element X contains 24.2 percent oxygen. (*a*) Find at least one possible value of the atomic weight of X. (It would be preferable to find the general set of values.) (b) The atomic number of X is 33. What is a likely value for its atomic weight?

3-24 Rule of Dulong and Petit The *rule of Dulong and Petit*, which holds approximately for solid elements with atomic weights above about 30, states that the heat capacity of a mole of atoms of a solid element is about 26 J mol^{-1} K^{-1}. (The heat capacity of a substance is defined as the energy that must be absorbed by the substance to raise its temperature by one kelvin[5].) This rule proved of great value during the first half of the nineteenth century, since it permitted choosing correct atomic weights among the several that were possible on chemical grounds. For example, the chemical composition of a metallic oxide can be used to calculate the atomic weight of the metal only if the empirical formula of the oxide is known; different possible formulas yield different possible atomic weights for the metal that are related by the ratios of simple integers, such as 2:1 or 3:2. Thus a rule for choosing an approximate value of the atomic weight, such as that of Dulong and Petit, was of great help; it permitted choosing among different possible formulas and thus finding the precise atomic weight.

A pure red oxide of iron contains 69.9 percent iron and 30.1 percent oxygen. The *specific heat* of iron (i.e., the heat capacity per gram of iron) is 0.48 J g^{-1} K^{-1}. Assume that the atomic weight of oxygen is known to be 16.00 and find the simplest formula of the oxide and the precise atomic weight of iron.

3-25 Rule of Dulong and Petit An element A has a heat capacity per gram of 1.02 J g^{-1} K^{-1}. The chloride of A contains 25.53 percent A. With the help of the rule of Dulong and Petit (Prob. 3-24), calculate the precise atomic weight of A.

3-26 Atomic Weight from Analytical Data When 5.50 g of a certain metal M reacts with 1.50 g fluorine, a compound MF_3 is formed. (a) What is the atomic weight of M? (b) What weight of M is needed to prepare 2 mol of the fluoride MF_3?

3-27 Atomic Weight and Simplest Formula A certain metal M forms a bromide of formula MBr_4. When 1.15 g MBr_4 is treated with an excess of

[5] See p. 68.

silver nitrate, all the bromide is converted to AgBr weighing 1.88 g. (a) What is the atomic weight of M? (b) If M forms an oxide containing 85.4 percent M, what is the empirical formula of the oxide?

3-28 Formula of an Organic Compound A 0.464-g sample of a compound containing the elements C, H, O, and S was burned completely yielding 0.969 g CO_2 and 0.198 g H_2O. All the S in a separate 0.586-g sample was converted to $BaSO_4$ (by fusion with Na_2O_2 and reaction with $BaCl_2$ solution) producing 1.082 g $BaSO_4$. What is the empirical formula of the compound?

3-29 Formula Weight and Percentage of One Component A compound of bromine contains 83.6 percent Br. Give possible values of the formula weight.

3-30 Composition of Fertilizer A fertilizer contains 73 percent of its total nitrogen in the form of KNO_3 and 27 percent as $(NH_4)_2SO_4$. What is the weight ratio of these two salts in this fertilizer?

3-31 Production of Chlorine Oxygen can react with hydrochloric acid, HCl, to produce Cl_2 and water. How many moles of HCl are needed to produce (a) 0.75 mol of Cl_2; (b) 100.0 g of Cl_2?

3-32 Quantity of Product Formed Sulfur dioxide, SO_2, reacts with NaOH to form Na_2SO_3. (a) What is the name of Na_2SO_3? (b) How many grams of Na_2SO_3 could be formed from the SO_2 produced by burning a 6.10-g sample of $C_4H_{10}S_2$ if SO_2 were the only sulfur-containing product formed when the compound was burned?

3-33 Quantities of Reactants and Products How many moles of H_3PO_4 are needed to convert 112 g CaO into $Ca_3(PO_4)_2$ if no other product containing calcium is formed?

3-34 Quantities of Reactants and Products How many moles of H_2SO_4 are needed to convert 51 g Al_2O_3 into $Al_2(SO_4)_3$ if no other product containing aluminum is formed?

3-35 Quantities of Reactants and Products Cyclopentane, C_5H_{10}, is a liquid with density 0.75 g ml^{-1}. How many moles of CO_2 can be formed by burning 200 ml cyclopentane with excess oxygen?

3-36 Quantities of Reactants and Products
Ethylene bromide, $C_2H_4Br_2$, is a liquid with density 2.18 g ml^{-1}. At high temperatures all the bromine in the molecule will react with lead (Pb) to form $PbBr_4$. What is the maximum weight of $PbBr_4$ that can be formed from 20.0 ml $C_2H_4Br_2$?

3-37 Quantities of Reactants and Products
Some fuel oils contain significant quantities of sulfur. When the oil is burned, the sulfur is oxidized to SO_2:

$$S + O_2 \longrightarrow SO_2$$

In a major city, 465 ton SO_2 is emitted by power plants each day. If the SO_2 comes from the combustion of fuel oil that contains 3.0 percent sulfur by weight, how many tons of fuel oil are burned per day?

3-38 Maximum Yield of Product Calculate the maximum amount of silver that could be obtained from 10.0 g of a mixture that contains 60 percent $AgCl$, 20 percent AgI, and 20 percent Na_2SO_4.

3-39 Quantities of Reactants and Products: Limiting Reagent Magnesium reacts with oxygen to form MgO but does not react with neon. Suppose 0.180 g magnesium is allowed to react completely with 0.250 g of a mixture of neon and oxygen that contains 60.0 percent oxygen by weight. Calculate the weight of each substance that will be present after reaction has occurred.

3-40 Proportions of Reactants and Products: Limiting Reagent Zinc and iodine form only one compound, ZnI_2. If 13 g zinc and 60 g iodine are sealed in a tube and heated until reaction is complete, what weights of what substances will be in the tube?

3-41 Proportions of Reactants and Products
Sodium nitrite, $NaNO_2$, can be prepared by the reduction of sodium nitrate with lead, the reaction being

$$NaNO_3 + Pb \longrightarrow NaNO_2 + PbO$$

Suppose that you wish to use a 10 percent excess of lead over the theoretical quantity needed. How much lead and how much sodium nitrate would you take to make 10.0 lb $NaNO_2$?

3-42 Composition of a Mixture Potassium hydrogen sulfate, $KHSO_4$, when heated strongly, forms potassium pyrosulfate, $K_2S_2O_7$, and water. The water is volatile and escapes. (a) Write a balanced equation for this reaction. (b) Calculate the percentage of $KHSO_4$ in a mixture if a 50-g sample of the mixture lost 1.8 g when heated strongly. Assume that no other reaction takes place and that no other components of the mixture are volatile.

3-43 Composition of a Mixture Ethanol, C_2H_5OH, reacts with oxygen by burning to form carbon dioxide and water according to the following equation:

$$C_2H_5OH + 3O_2 \longrightarrow 2CO_2 + 3H_2O$$

Sodium hydroxide, $NaOH$, reacts with CO_2 as follows:

$$2NaOH + CO_2 \longrightarrow H_2O + Na_2CO_3$$

When a 200-g sample of a mixture containing ethanol was burned with an excess of air and the resulting CO_2 was allowed to react with $NaOH$, 212 g Na_2CO_3 was formed. The mixture contained no other substance that could form Na_2CO_3. What was the percentage of ethanol in the mixture?

3-44 Compensation of Weight Loss To counterbalance the weight loss of fuel used in the operation of a dirigible, the water contained in the exhaust gases is partly condensed whereas the carbon dioxide is allowed to escape. Exactly what percentage of the water produced need be condensed to achieve weight balance if the chemical composition of the fuel is represented by the formula C_nH_{2n}, with n any (positive) integer?

3-45 Mixture of Silver Halides A mixture of pure $AgCl$ and pure $AgBr$ contains 60.4 percent silver. What is the percentage of bromine? What is the error estimate of the answer if the error estimate of the percentage of silver is 0.1 (absolute) percent?

3-46 Composition of a Mixture When calcium carbonate, $CaCO_3$, is heated to about $900°C$, it decomposes to CaO and CO_2. The gaseous CO_2 escapes; the CaO is not further changed. Calcium sulfate, $CaSO_4$, is stable and unchanged under these conditions.

A mixture of $CaCO_3$ and $CaSO_4$ that initially weighed 300 ton was heated at $900°C$ until all the CO_2 had been driven off and the weight had become constant. The loss in weight was 99 ton. (*a*) What was the percentage of $CaCO_3$ in the original mixture? (*b*) What was the percentage of Ca in the final product?

3-47 Composition of a Mixture A mixture of calcium carbonate and carbon is heated strongly in air. The calcium carbonate decomposes completely to calcium oxide and carbon dioxide, and the carbon is oxidized by the air to carbon dioxide. If the total weight of the carbon dioxide formed is equal to the weight of the original mixture, what is the percentage of carbon in the original mixture?

3-48 Erroneous Analysis A certain mineral contains just 20.0 percent calcium. A student was given this mineral as an unknown and was asked to determine the content of calcium. The procedure used was to dissolve the sample, precipitate the calcium as calcium oxalate monohydrate, $CaC_2O_4 \cdot H_2O$, and heat this until it decomposed entirely to CaO, with attendant formation of CO, CO_2, and H_2O. The CaO was then to be weighed. However, the student did not heat the sample sufficiently and succeeded only in converting the oxalate to calcium carbonate, but he did not know this and assumed he was weighing the oxide. If he made no other errors, what was the percentage of Ca he reported?

The Behavior
of Gases

"The particles of air are in contact with each other; yet they do not fit closely in every part, but void spaces are left between them, as between the grains of sand on a seashore. (These grains must be imagined to correspond to the particles of air, and the air between the grains of sand to the void spaces between the particles of air.) Hence, when any force is applied to it, the air is compressed and—contrary to its nature—is forced into the vacant spaces by the pressure exerted on its particles. When the force is withdrawn, however, the air returns again to its former position on account of the elasticity of its particles; in this, it resembles horn shavings and sponge which, if compressed and then released, return to their original position and volume."

HERO OF ALEXANDRIA, "PNEUMATIKA", AROUND A.D. 60[1]

In this chapter we shall be concerned with the nature and properties of gases. To a very considerable degree one can discuss the gaseous state both qualitatively and quantitatively without considering the specific nature of the molecules of which the gas is composed. Our concern is primarily with an "ideal" or "perfect" gas instead of with any particular real gas. However, the properties of some actual gases are considered near the end of Sec. 4-3.

After a discussion of pressure units and of energy and temperature, we consider ideal gases and apply the ideal equation of state to numerous situations. The kinetic-molecular theory, which explains the properties of gases, is considered in Secs. 4-4 and 4-5.

4-1   Introduction

Speculation about the nature of the gaseous state occupied philosophers of science for many centuries. Some of the ancient speculations had considerable merit, but it was only after the quantitative experiments of Robert Boyle in the seventeenth century, and those of his successors during the next two centuries, that the modern kinetic-molecular picture of a gas evolved. As the name implies (Greek: *kinesis*, motion), this picture is a microscopic-level view of the consequences of the independent motion of molecules in gases. According to this model, the molecules of a gas "fill" the space allotted to them by their ever-present, extremely rapid motion, much as space is "occupied" by a moving swarm of bees. The fraction of space occupied by the molecules themselves is small, often negligibly small. The pressure exerted on the walls of a vessel containing the gas is caused by the many collisions of the molecules with the wall. The temperature

---

[1]Quoted by Toulmin and Goodfield, "The Architecture of Matter", Harper & Row, Publishers, Inc., New York, 1962.

of a gas (and more generally, of any phase) is a measure of the average kinetic energy associated with the continuous random motion of its molecules and atoms. During the nineteenth century a remarkably successful theory based on this simple model was developed that permits quite precise quantitative predictions of many different properties of gases (Sec. 4-4). Furthermore, to the extent that some properties of real gases deviate from the predictions of the theory, the deviations are often understandable in terms of the assumptions used in formulating the model. Thus, appropriate modifications of the model can be made and the theory can be improved in a systematic way. This kind of interplay of model, theory, and experimental observation is the essence of modern science.

The interrelations of the properties of any phase, be it solid, liquid, or gas, may be expressed quantitatively in its *equation of state*. This is often written as a mathematical relation between the pressure, the volume, the temperature, and the mass of material involved, although other variables may be used. The ideal gas has a particularly simple equation of state, a fact that is of considerable practical importance since the behavior of most real gases under ordinary conditions approximates the ideal within a few percent. Better equations for real gases exist but are more complicated. There is no generally applicable ideal equation of state for liquids or for solids.

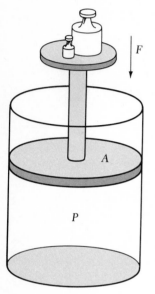

Pressure and Pressure Units   Pressure is defined as force per unit area. For example, the pressure P on a gas or liquid trapped by a piston in a cylinder is given by the force F exerted by the piston divided by its cross-sectional area A: $P = F/A$. A device to measure the pressure of a gas is the manometer, shown in Fig. 4-1. The difference between the levels of the liquid in the two branches of the U-tube can be related to the pressure exerted by the gas. A liquid often used in the manometer is mercury.

The pressure unit in the SI system is the *pascal* (Pa), which is defined as a

Figure 4-1   Two Manometers

In either of the two versions shown, the U-shaped part of the glass tube contains a liquid, usually mercury, and the level difference h is an indication of the pressure of the gas in a vessel to which the left side of the manometer is connected. The *open* manometer (a) is open to the atmosphere on the right and measures the pressure difference between vessel and outside atmosphere. This difference is also called the *gauge pressure*. In the example shown, P_1 is greater than atmospheric pressure, the difference being measured by h_1.

In the *closed* manometer (b) the space above the liquid at the closed end is evacuated and the instrument shows the actual pressure; P_2 is thus measured directly by h_2. It is assumed that the pressure exerted by the vapor of the manometer fluid in the evacuated space is negligible. This is a reasonable assumption for mercury at room temperature.

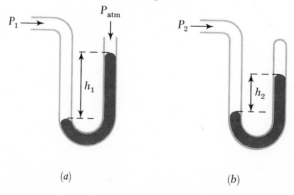

(a)   (b)

force of one newton per square meter ($N\,m^{-2}$ or $kg\,m^{-1}s^{-2}$) (Appendix A).
Other pressure units in chemistry include the *atmosphere* (atm) and the *torr*,
by definition related to the pascal by the equations

$$1\ atm = 1.01325 \times 10^5\ Pa$$

and

$$1\ atm = 760\ torr$$

The average pressure exerted by the earth's atmosphere at sea level is about
1 atm. The torr, named after Torricelli, the inventor of the barometer, differs by
less than 2 parts in 10^7 from the older pressure unit millimeters of mercury (mm
Hg), defined as one millimeter of level difference in a mercury manometer at $0°C$
and standard gravity.

A pressure unit that is approximately equal to the atmosphere, but is
more simply related to the newton per square meter, is the bar: 1 bar =
$10^5\ N\,m^{-2} = 0.987$ atm. In engineering applications the unit pound per square
inch (lb $inch^{-2}$), is often used. It is the pressure exerted by one pound on an
area of one square inch at the earth's surface; 1 atm corresponds to

$$14.696\ lb\ inch^{-2} \approx 14.7\ lb\ inch^{-2}$$

4-2 Energy and Temperature

Energy and Energy Units   Energy is an important attribute of physical systems.
There are different kinds of energy. For example, a moving object of mass m and
speed v has an energy of motion, called *kinetic energy*, ε_k, given by

$$\varepsilon_k = \frac{mv^2}{2}$$

Another kind of energy is *potential energy*. As the word *potential* implies, this
energy is stored energy, for example, the energy in a coiled spring or the gravitational energy associated with the height of an object.

In a system isolated from its environment, the sum of the kinetic and potential
energies remains constant. Kinetic energy can be converted into potential energy,
for example, by using it to coil a spring or lift an object. Conversely, potential
energy can be transformed into kinetic energy, for example, by allowing a spring
to uncoil or an object to fall, as well as in numerous other ways.

Work and *heat* are used to describe quantities of energy transferred. For
example, the energy gained by a weight when it is lifted is the work performed in
lifting the weight. When two objects at different temperatures are in contact,
energy is transferred from the hot to the cold object. Energy transferred in this
way is called heat. A detailed discussion of work and heat is given in Chap. 19.
The SI units of energy are the joule (J), which is the basic unit of energy, and
the kilojoule (kJ), equal to 1000 J. Other units still widely used by scientists are
the calorie (cal) and the kilocalorie (kcal), 1000 cal. By international agreement
the calorie (or more precisely the thermochemical calorie) is now by definition
4.1840 joules:

$$1\ cal = 4.1840\ J$$

Confusion may arise because the "calorie" of nutritionists is the kilocalorie of
chemists. Sometimes the abbreviation Cal with a capital C is used to designate
kilocalories.

Temperature and Temperature Scales Although our initial impression of temperature is merely the sensing of hot and cold, we can measure temperatures precisely with thermometers. A thermometer makes use of some standard substance with properties that change in a known and reproducible way when temperature is changed. For example, the expansion of mercury or alcohol with increasing temperature is the basis of temperature measurement with common thermometers suitable for use at ordinary temperatures.

The use of thermometers is based on the important experimental fact that two objects in contact with each other will eventually come to the same temperature. In measuring the temperature of any object A with a thermometer B, we leave B in contact with A for a sufficient time and assume that B is so small relative to A that the temperature of A is not significantly affected by contact with the thermometer B.

A number of temperature scales are used in different applications. The two most commonly used in chemistry are the *Kelvin scale* and the *Celsius (centigrade) scale*. To define a temperature scale, two fixed points of reference are needed; these fix both the size of the temperature unit and the zero of the scale. The SI temperature scale is the Kelvin scale, the unit of which is the *kelvin*, designated by K. This temperature unit was formerly called the degree Kelvin, abbreviated °K, a usage that is still often found. The Kelvin scale is an absolute scale, because its zero point is the lowest temperature that can be reached theoretically ("the absolute zero of temperature"). The second fixed point on this scale is the temperature at which pure water freezes.

The Celsius scale is tied by definition to the Kelvin scale. Temperatures on it are designated by "degree Celsius" (°C). By definition,

$$\text{Temperature in K} = \text{temperature in °C} + 273.15 \text{ K}$$

With this assignment the ice point, at which water freezes when saturated with air at 1 atm, is 273.15 K or 0.00°C, and the boiling point of water at 1 atm pressure is 373.15 K or 100.00°C. We designate temperature on the Celsius scale by t and on the Kelvin or absolute scale by T, as is widely customary, and usually use the approximation

$$T \approx t + 273 \text{ K}$$

Another temperature scale in wide current use is the Fahrenheit scale. Its relation to the Celsius scale is

$$\text{Temperature in °F} = \tfrac{9}{5}(\text{temperature in °C}) + 32\text{°F}$$

The three temperature scales are shown in Fig. 4-2.

4-3 Ideal Gases

A very useful concept for the discussion of the properties of gases is that of the ideal gas. As discussed in Sec. 4-4, on the molecular level an ideal gas corresponds to a situation in which there are no forces between the molecules and in which the molecules themselves take up no space. The equation of state for such a gas, which turns out to be mathematically very simple, is often a good approximation to the properties of real gases at low pressures. *All* real gases become more nearly ideal as the pressure is lowered at a given temperature; ideality is enhanced also by any other change in conditions for which the volume occupied by a given

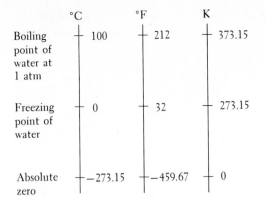

Figure 4-2 **Different Temperature Scales**

number of molecules grows larger. The effects of intermolecular forces become smaller as the distances between the molecules become larger, and at the same time the volume occupied by the gas molecules themselves becomes a smaller fraction of the total volume occupied by the gas.

The Ideal-Gas Law The equation of state for an ideal gas is

$$PV = nRT \tag{4-1}$$

where P = pressure of the gas
   V = volume
   T = absolute temperature
   n = number of moles of gas
   R = universal constant called the *gas constant*

The numerical value of the gas constant depends on the choice of units for P, V, and T.

   Unless stated otherwise, gas pressures will be given in atmospheres. Volumes will usually be expressed in liters, and absolute temperatures in kelvins. The units of n are moles, but this does not imply that the gas considered must be pure; as discussed in Sec. 2-2, a mole of air contains a total of N_A molecules, where N_A is Avogadro's number, with the different components present in proportion to their relative abundances. Many gas mixtures, including air, behave like nearly ideal gases under some conditions. For a gas mixture, n is simply the total number of gas molecules divided by N_A, or, what amounts to the same thing, the sum of the numbers of moles of the component gases.

   The value of the gas constant appropriate to the most common units is

$$R = 0.082057 \text{ liter atm mol}^{-1}\,\text{K}^{-1}$$

The units are necessarily the same as those of PV/nT. If some of the variables in the ideal-gas equation are in other units, the numerical value of R must be adjusted appropriately.

□ **Change of Units of the Gas Constant** R What is the value of R when the volume is expressed in cubic centimeters and the pressure is in torr?

Example 4-1

Solution Multiply or divide by conversion factors in such a way that cancellation yields the units sought:

$$R = 0.082057 \text{ liter atm mol}^{-1}\,\text{K}^{-1} \times \frac{1000 \text{ cm}^3}{\text{liter}} \times \frac{760 \text{ torr}}{\text{atm}}$$

$$= \underline{6.236 \times 10^4 \text{ cm}^3 \text{ torr mol}^{-1}\,\text{K}^{-1}} \quad ■$$

Exercise 4-1

☐ What is the value of R when the volume is expressed in cubic feet (foot^3) and the pressure is in lb inch^{-2}? (1 foot^3 = 28.3 liter; 14.7 lb inch^{-2} = 1 atm) ■

We now turn to a consideration of some applications of (4-1) under special conditions, with different variables held constant. Historically, these special cases were discovered from experiments and are associated with the names of their discoverers. In the mid-nineteenth century the information contained in these separate laws was embodied into the generalized law given in (4-1).

Boyle's Law In the seventeenth century, Robert Boyle, one of the first men to appreciate the importance of quantitative experimentation, established the fact that the product of the pressure and the volume of a fixed quantity of gas is constant at a given temperature; that is,

$$PV = \text{const} \qquad (\text{if } T = \text{const}, n = \text{const}) \qquad (4\text{-}2)$$

This relationship, Boyle's law, is seen to be merely a special case of the ideal gas law. Boyle's law is illustrated in the following example and in Fig. 4-3.

Example 4-2

☐ **Boyle's Law** A sample of gas occupies 100 liter at 3.0 atm pressure. What pressure is needed to bring it to a volume of 20 liter at the same temperature?

Solution Using the subscripts 1 and 2 for the initial and final states, respectively, we have $P_1 V_1 = P_2 V_2$. Thus

$$P_2 = \frac{P_1 V_1}{V_2} = \frac{100 \text{ liter} \times 3.0 \text{ atm}}{20 \text{ liter}} = \underline{15 \text{ atm}} \quad ■$$

Exercise 4-2

☐ A sample of gas occupies 500 foot^3 at 200 torr. What is the volume when the pressure is changed to 800 torr at the same temperature? ■

The Law of Gay-Lussac and Charles Another special case results when the quantity of gas and the pressure are kept constant, so that, in terms of (4-1),

$$\frac{V}{T} = \text{const} \qquad (\text{for } P = \text{const}, n = \text{const}) \qquad (4\text{-}3)$$

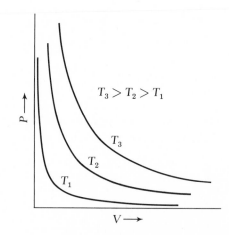

Figure 4-3 Volume and Pressure of an Ideal Gas at Constant Temperature
For a fixed amount of an ideal gas, the relationship between volume and pressure at constant temperature is represented by a hyperbola. Each of the curves shown represents Boyle's law at a given temperature.

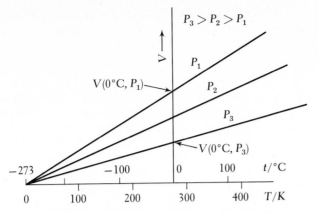

Figure 4-4 Volume of a Fixed Amount of an Ideal Gas at Constant Pressure
Each of these lines represents the law of Gay-Lussac and Charles at a particular pressure. At constant pressure, the volume of a fixed amount of an ideal gas is linearly related to its temperature and is proportional to its Kelvin temperature. Thus the volume of an ideal gas can be used as a "thermometer" to indicate absolute temperature. As T goes toward zero, so does the volume V of an ideal gas. There is no known substance that behaves like an ideal gas near $0\,K$, because all real gases eventually condense. Nevertheless, most gases exhibit the linear relation of volume and temperature shown here with reasonably high precision when the temperature is not too low and the pressure not too high. Temperature measurements near $0\,K$ would be difficult or impossible if ideal-gas behavior were the only basis of the absolute scale.

This law was discovered in an alternative form by Gay-Lussac and Charles in the late eighteenth century, and it played a key role in the later establishment of an absolute temperature scale. They found that the volume of a fixed quantity of a gas at constant pressure was a linear function of its temperature (Fig. 4-4). The relation between volume and temperature found by Gay-Lussac and Charles can be given the form

$$V = a + bt = a\left(1 + \frac{b}{a}t\right) \tag{4-4}$$

where t denotes the temperature in °C. The constant a is merely the volume of the gas when $t = 0°C$, which we shall call V_0, and the constant b/a turns out to be approximately the same for all gases, about $1/273$ per °C or $(b/a) = 1/(273°C)$. Thus (4-4) may be rewritten as

$$V = V_0\left(1 + \frac{t}{273°C}\right) = V_0\left(\frac{273 + t/°C}{273}\right) \tag{4-5}$$

It is evident from Fig. 4-4 that the volume of a gas (provided it did not condense to a liquid or solid as it was cooled) would become zero at about $-273°C$. This suggested to Lord Kelvin the introduction of a new temperature scale with degrees the same size as those of the Celsius scale but with a different zero point. By the substitution $T/K = t/°C + 273$ into (4-5), the law of Gay-Lussac and Charles can be expressed in the form

$$V = V_0\frac{T}{273\ K}$$

or

$$\frac{V}{T} = \frac{V_0}{273\ K} = \text{const}$$

which is just (4-3). In this context, the "absolute" aspect of T lies in the fact that its introduction permits an (idealized) description of the thermal expansion of any gas by the same simple proportionality, expressed in (4-3). The relationship between (4-3) and (4-4) is illustrated in Fig. 4-4, where the scale shows kelvins as well as degrees Celsius.

The law of Gay-Lussac and Charles shows that the volume and absolute temperature of an ideal gas are directly proportional to each other when P and n are kept constant. Indeed, this proportionality may be used as the basis for measuring absolute temperature. The one aspect of the Kelvin scale that is not absolute is the size of its degree, but this has recently been defined by international agreement (as discussed in Sec. 4-2) in conjunction with the Celsius scale.

Constancy of PV/T for a Fixed Number of Moles of Gas   If only the number of moles of gas remains constant in (4-1), then we have

$$\frac{PV}{T} = \text{const} \qquad (\text{when } n = \text{const}) \tag{4-6}$$

or
$$\frac{P_1 V_1}{T_1} = \frac{P_2 V_2}{T_2} \qquad (\text{when } n = \text{const}) \tag{4-7}$$

This is a very useful general relationship when one has a fixed number of moles of gas; its utility is illustrated in the following example.

Example 4-3

□ **Effect of Simultaneous Change of Volume and Temperature**   A sample of air at 20°C is obtained by filling a 100-ml tube to a pressure of 1.00 atm. A second sample of air is taken later, at 5°C in a 150-ml tube. To what pressure should the second sample tube be filled to provide the same quantity of air as taken in the first sampling?

Solution   The problem is equivalent to the calculation of the effect of changes in volume and temperature on the pressure of a fixed quantity of gas; that is, $n = P_1 V_1/RT_1 = P_2 V_2/RT_2$, or

$$P_2 = P_1 \frac{V_1}{V_2} \frac{T_2}{T_1}$$

The required pressure is given by

$$P_2 = 1.00 \text{ atm} \times \frac{278 \text{ K}}{293 \text{ K}} \times \frac{100 \text{ ml}}{150 \text{ ml}} = \underline{0.63 \text{ atm}}$$

Note that this expression may be considered as involving the multiplication of the initial pressure by two factors: (1) the temperature ratio, applied in such a way that a reduced temperature corresponds to a decreased pressure (in other words, this factor must be less than unity here); (2) the ratio of volumes, applied in such a way that a larger volume corresponds to a lower pressure, so that this factor must also be less than unity here. The units in the ratios cancel; hence the answer is dimensionally correct. This is not a complete check of the correctness of the answer, because the dimensions would still be correct if one or both ratios had inadvertently been turned upside down. ■

Exercise 4-3

□ A sample of a gas at 50°C and 700 torr has a volume of 250 ml. What is the volume of the gas if temperature and pressure are changed to 25°C and 800 torr? ■

It must be kept in mind that each of the special laws so far discussed is *valid only if n is constant*. There are two reasons why n might vary when the conditions of a system are changed: (1) gas may be added or removed without chemical reaction, as by vaporization or condensation; (2) gas molecules may be added or removed by chemical reaction. For example, nitrogen tetroxide, N_2O_4, dissociates partially into NO_2 when the pressure on it is decreased:

$$N_2O_4(g) \longrightarrow 2NO_2(g)$$

Since each molecule of N_2O_4 is replaced by two molecules of NO_2 upon dissociation, there is a change in n. This second possibility is discussed in Chap. 11 in Example 11-3 and Sec. 11-4.

Avogadro's Law   When P and T are constant, V is proportional to n,

$$V = \frac{nRT}{P} = n \times \text{const} \tag{4-8}$$

with the constant the same for all ideal gases. This means that *at the same* T *and* P *equal volumes of two ideal gases contain the same number* n *of moles* and therefore also the *same number* nN$_A$ *of molecules*. This is called *Avogadro's law*.

In discussions of gas volumes, the standard temperature 0°C and pressure 1 atm, abbreviated STP, are often used. The *standard molar volume*,[2] i.e., the volume of 1 mol of ideal gas at STP, is

$$\widetilde{V}_{STP} = \frac{V}{n} = \frac{RT}{P}$$

$$= 0.082057 \text{ liter atm mol}^{-1} \text{ K}^{-1} \times \frac{273.15 \text{ K}}{1.00000 \text{ atm}} \tag{4-9}$$

$$= 22.414 \text{ liter mol}^{-1}$$

For real gases, the molar volume at STP may deviate from this value. The following values are illustrative. The molar volumes given apply at standard conditions; the percentage deviations are from the ideal value:

Gas	He	H$_2$	N$_2$	O$_2$	CH$_4$	CO$_2$	NH$_3$
$\widetilde{V}$/liter mol^{-1}	22.425	22.431	22.402	22.392	22.366	22.258	22.076
Deviation/%	+0.05	+0.08	−0.05	−0.09	−0.21	−0.70	−1.5

As is evident from the table, deviations in either direction may occur. They reflect both molecular size and intermolecular forces. Gases consisting of small molecules containing only one kind of atom of low atomic number, such as He, H$_2$, and Ne, deviate at most a few tenths of a percent over a wide range of pressures at ordinary temperatures. As the atomic number and the molecular size and complexity increase, the deviations may be larger. Nonetheless, the ideal-gas law is a remarkably good approximation for real gases in many situations.

An important consequence of Avogadro's law is that volume composition and mole composition of an ideal gas are identical.

☐ **Volume and Mole Composition of a Gas**   A 22.4-liter sample of a gas at STP contains by volume 20 percent ethylene (C_2H_4), 50 percent methane   **Example 4-4**

---

[2] A tilde ($\sim$) is used to indicate values that pertain to 1 mol; thus $\widetilde{V}$ is the volume of 1 mol of substance.

(CH_4), and 30 percent nitrogen (N_2). How many moles of carbon atoms are contained in this gas sample?

Solution   Since 22.4 liter at STP corresponds to 1.00 mol gas, the quantities of component gases are 0.20 mol C_2H_4, 0.50 mol CH_4, and 0.30 mol N_2. Therefore, the number of moles of carbon atoms is equal to

$$2 \times 0.20 + 0.50 = \underline{0.90 \text{ mol}} \quad \blacksquare$$

Exercise 4-4

☐ An 11.2-liter sample of a gas at STP contains by volume 30 percent ammonia (NH_3), 35 percent nitrogen (N_2), and 35 percent carbon dioxide (CO_2). How many moles of nitrogen atoms are contained in this gas sample? ∎

Historical Importance of Avogadro's Law   John Dalton postulated that atoms were the smallest particles of elements and wrote formulas for water and ammonia that corresponded to HO and NH. He considered the hypothesis that equal volumes of gases contain equal numbers of atoms but rejected it because it did not appear to fit certain observed relationships between volumes of reacting gases. For example, two volumes of hydrogen and one of oxygen combine (above 100°C) to give two volumes of water vapor whereas one volume of oxygen and one of nitrogen form two volumes of nitric oxide (NO). Dalton did not consider that in elemental substances several atoms might be combined into one molecule, as in H_2 and O_2. Amedeo Avogadro recognized this possibility and postulated in 1811 that equal volumes of gases contained equal numbers of *molecules,* not atoms. He saw, for example, that $2x$ hydrogen atoms (with $x =$ any integer) might constitute a hydrogen molecule H_{2x} and similarly $2x$ oxygen atoms might constitute a molecule O_{2x}. If water had the formula $H_{2x}O_x$, reaction between hydrogen and oxygen would then be described by

$$2H_{2x} + O_{2x} \longrightarrow 2H_{2x}O_x$$

which would, by his law of volumes, explain the experimental observation that two volumes of hydrogen and one of oxygen result in two volumes of water vapor. It would also give oxygen the atomic weight 16 relative to an atomic weight of 1 for hydrogen, rather than 8 as implied by the formula HO for water. There was no way at that time to determine the value of x, which was taken to be 1 for simplicity's sake. In a similar way the equation

$$N_2 + O_2 \longrightarrow 2NO$$

would reproduce correctly the volumes of nitrogen and oxygen that combine to form NO, and again a common factor, here set equal to 1, remained undetermined.

The idea of molecules composed of atoms of the *same kind* seemed unreasonable to Avogadro's contemporaries, and many decades passed before it was accepted. In the meantime many new elements were discovered and their compounds analyzed, particularly by Jöns Jakob Berzelius, who also introduced our system of chemical symbols. Avogadro's principle was applied to the new elements and their compounds by Stanislao Cannizzaro, who successfully obtained in this way a consistent set of atomic weights and set the stage for the discovery of the periodic system in 1869 by Dmitri Mendeleev and, independently, by Lothar Meyer (Sec. 2-3).

Molecular Weight of Gas Molecules The ideal-gas law can be given a form that shows an important connection between the density ρ of a gas and the molecular weight of its molecules. Let m be the weight of a sample of ideal gas and let $\tilde{M}$ be the weight per mole of gas, so that the molecular weight is $MW = \tilde{M}/(g\ mol^{-1})$. The number of moles of the gas considered is

$$n = \frac{m}{\tilde{M}} \qquad (4\text{-}10)$$

so that

$$PV = \frac{mRT}{\tilde{M}} \qquad (4\text{-}11)$$

We note that it is possible to isolate on one side the combination m/V, which is just the gas density ρ:

$$\rho = \frac{m}{V} = \tilde{M}\frac{P}{RT} \qquad (4\text{-}12)$$

The density of a gas at a given T and P is thus proportional to the molecular weight.

□ **Density of a Gas** What is the approximate density of CO at 700 torr and 30°C?

Example 4-5

Solution The molecular weight of CO is $12.0 + 16.0 = 28.0$. Using the approximation that CO behaves like an ideal gas, we have, by (4-12),

$$\rho = \frac{P\tilde{M}}{RT} = \frac{700\ torr \times (1\ atm/760\ torr) \times 28.0\ g\ mol^{-1}}{0.0821\ liter\ atm\ mol^{-1}\ K^{-1} \times 303\ K}$$
$$= \underline{1.037\ g\ liter^{-1}}$$

with the least significant digit very uncertain. ∎

□ What is the approximate density of NH_3 at 1.03 atm and 20°C? ∎

Exercise 4-5

We may for convenience rewrite (4-12) by solving for $\tilde{M}$,

$$\tilde{M} = \rho\frac{RT}{P} \qquad (4\text{-}13)$$

and calculate the molecular weight from the density of a gas at a given temperature and pressure.

Neither (4-12) nor (4-13) need be memorized. They can always be derived from the ideal-gas equation, $PV = nRT$. The following two examples are applications of (4-13).

□ **Molecular Weight from Vapor Density** The density of ethyl alcohol vapor, C_2H_5OH, at 76°C and 625 torr is found to be 1.373 g liter^{-1}. What molecular weight corresponds to this vapor density?

Example 4-6

Solution By (4-13)

$$\tilde{M} = \frac{\rho RT}{P} = \frac{1.373\ g\ liter^{-1} \times 0.0821\ liter\ atm\ mol^{-1}\ K^{-1} \times 349\ K}{625\ torr \times (1\ atm/760\ torr)}$$
$$= \underline{47.8\ g\ mol^{-1}}$$

The sum of the atomic weights is 46.1, smaller than the value just found by about 3.5 percent. The difference is due to deviations from ideality, which are to be expected for a vapor close to the pressure at which it condenses. The vapor pressure of ethanol at 76°C is 698 torr and the normal boiling point is 78°C (see pp. 100–101). ∎

Exercise 4-6

☐ The density of methyl chloride gas, CH_3Cl, is reported to be 2.308 g $liter^{-1}$ at 0°C and 1.000 atm. What molecular weight corresponds to this value? ∎

Example 4-7

☐ **Molecular Aggregation from Vapor Density** Arsenious trioxide, of composition As_2O_3, forms a vapor of density 5.66 g $liter^{-1}$ at 743 torr and 571°C. What is the approximate molecular weight, and what is the molecular formula of this compound?

Solution We proceed as in the previous example and find

$$\widetilde{M} = \frac{5.66 \text{ g liter}^{-1} \times 0.0821 \text{ liter atm mol}^{-1} \text{ K}^{-1} \times 844 \text{ K}}{743 \text{ torr} \times (1 \text{ atm}/760 \text{ torr})}$$

$$= \underline{401 \text{ g mol}^{-1}}$$

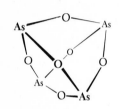

The sum of the atomic weights for As_2O_3 is 197.8, which is about half the value found. Arsenious oxide thus exists as As_4O_6, with molecular weight 395.6, at least under these conditions in the vapor phase. This has been confirmed by electron diffraction studies of the vapor, which also yield information about the structure. ∎

Exercise 4-7

☐ Phosphorus pentoxide, of composition P_2O_5, forms a vapor of density 1.94 g $liter^{-1}$ at 200 torr and 200°C. What is the approximate molecular weight, and what is the molecular formula of this compound? ∎

Stoichiometry Involving Gases When gases are involved in stoichiometric considerations, the best procedure is to find the number of moles of gas directly rather than to calculate weights.

Example 4-8

☐ **Volume of Gas by Stoichiometry** Iron reacts with oxygen to form Fe_2O_3. What volume of oxygen at 800 torr and 25°C reacts quantitatively with exactly 1 g iron?

Solution The reaction equation is

$$4Fe + 3O_2 \longrightarrow 2Fe_2O_3$$

Therefore, 4 mol iron reacts with 3 mol oxygen; the number of moles of oxygen required is

$$\frac{1 \text{ g Fe}}{55.85 \text{ g mol}^{-1}} \times \frac{3 \text{ mol O}_2}{4 \text{ mol Fe}} = 0.01343 \text{ mol O}_2$$

The volume of this number of moles of oxygen at 25°C and 800 torr is

$$V = \frac{nRT}{P} = \frac{0.01343 \text{ mol} \times 0.0821 \text{ liter atm mol}^{-1} \text{ K}^{-1} \times 298 \text{ K}}{800 \text{ torr} \times (1 \text{ atm}/760 \text{ torr})}$$

$$= \underline{0.312 \text{ liter oxygen}} \quad ∎$$

□ Silver reacts with chlorine to form AgCl. What volume of chlorine at 700 torr and 20°C reacts quantitatively with 1.00 g silver? ∎

In the next example the volume of a gas produced in a reaction is given, and the problem posed is to interpret this volume stoichiometrically.

Example 4-9

□ **Analysis of Impure Ore** A 5.00-g sample of crude ZnS ore was heated in air, a process called roasting. At the completion of the reaction 1.560 liter SO_2 gas at 200°C and 750 torr pressure had been produced and all ZnS converted to ZnO. What is the minimum volume of air (containing 20 mol percent O_2) at STP needed for the reaction? If the ore contains no S except in the form of ZnS, what is the percentage of ZnS in the ore?

Solution First, we find the number of moles of SO_2 formed:

$$n = \frac{PV}{RT} = \frac{750 \text{ torr} \times (1 \text{ atm}/760 \text{ torr}) \times 1.560 \text{ liter}}{0.0821 \text{ liter atm mol}^{-1}\,\text{K}^{-1} \times 473 \text{ K}}$$

$$= 39.6 \times 10^{-3} \text{ mol}$$

To answer the first part of the question we consider the reaction equation

$$\text{ZnS} + \tfrac{3}{2}O_2 \longrightarrow \text{ZnO} + SO_2$$

and note that $\tfrac{3}{2}$ mol O_2 is used up for each mol SO_2 formed. Thus, the number of moles of air required is

$$39.6 \times 10^{-3} \text{ mol } SO_2 \times \frac{\tfrac{3}{2}\text{ mol } O_2}{\text{mol } SO_2} \times \frac{1 \text{ mol air}}{0.20 \text{ mol } O_2} = 0.30 \text{ mol air}$$

where the third factor is equivalent to the statement that air contains 20 mol percent oxygen. At STP the air takes up a volume of

$$0.30 \text{ mol} \times \frac{22.4 \text{ liter}}{\text{mol}} = \underline{6.7 \text{ liter}}$$

For the second part of the question we note that, according to the reaction equation, 1 mol SO_2 corresponds to 1 mol ZnS, or $(65.4 + 32.1) \text{ g} = 97.5 \text{ g ZnS}$. Thus the quantity of gas calculated earlier corresponds to

$$(39.6 \times 10^{-3} \text{ mol } SO_2)\frac{1 \text{ mol ZnS}}{1 \text{ mol } SO_2} \times \frac{97.5 \text{ g ZnS}}{1 \text{ mol ZnS}} = 3.86 \text{ g ZnS}$$

In a 5.00-g sample this is

$$\frac{3.86}{5.00} \times 100 = \underline{77.2 \text{ percent}} \;\; \blacksquare$$

□ Sulfur dioxide, SO_2, can be obtained by roasting the iron ore pyrite, FeS_2, the iron being transformed to Fe_2O_3. A 10.00-g sample of impure pyrite yielded 6.18 liter of sulfur dioxide gas at 250°C and 700 torr pressure. After writing the equation for the reaction of oxygen with FeS_2, calculate the minimum volume of air at STP needed for the reaction. What is the percentage of FeS_2 in the ore if there is no other source of SO_2 present? ∎

In problems that are concerned with the volumes of reacting gases, it is simplest to remember Avogadro's law: total numbers of moles and gas volumes are proportional to each other at constant T and P.

Example 4-10

□ **Reaction of Gases** To 50 ml of a mixture of CO and CO_2 is added 40 ml O_2, all at 20°C and 740 torr. The mixture is heated until reaction has converted the carbon monoxide completely to carbon dioxide. When the original temperature and pressure are reestablished, the gas mixture is found to occupy 75 ml. What was the volume composition of the original gas?

Solution Note that, although the temperature was not constant, the original temperature was restored at the end. The final pressure was also the same as that originally. Hence we can say that the numbers of moles at the start and the end are proportional to the volumes at the start and the end.

Let there be x ml CO and $(50 - x)$ ml CO_2 in the original mixture. The reaction equation is

$$CO + \tfrac{1}{2}O_2 \longrightarrow CO_2$$

and thus x ml CO uses up $(x/2)$ ml O_2 and produces x ml CO_2. At the end there is 50 ml CO_2 [that is, $(50 - x)$ ml unchanged plus x ml produced by oxidation of CO] and $(40 - x/2)$ ml O_2, which combined must total 75 ml: $50 + 40 - (x/2) = 75$. Thus, $x = 30$. The original gas mixture therefore consists of <u>30 ml CO and 20 ml CO_2</u>. ■

Exercise 4-10

□ To 50 ml of a mixture of CO and CH_4 is added 100 ml O_2, all at 20°C and 700 torr. The mixture is heated until reaction has converted the CO and the CH_4 completely to CO_2 and H_2O. When the original temperature and pressure are reestablished and the water is removed, the resulting gas mixture is found to occupy 75 ml. What was the volume composition of the original gas? Be sure to write balanced equations for the reactions. ■

The Behavior of Real Gases Even though real gases behave very nearly like ideal gases under many conditions, they may show large deviations from ideal

Figure 4-5 Behavior of Real Gases
The ratio $P\widetilde{V}/RT$ plotted against P approaches 1 as P goes to zero because at very low pressure all gases behave ideally, so $P\widetilde{V} = RT$. The horizontal dashed line corresponds to an ideal gas. The deviations of the curves from this horizontal line are a measure of the deviations from ideal-gas behavior. Graph (a) extends from 0 to 1000 atm and exhibits large deviations. For pressures that do not exceed a few atmospheres the deviations are small, as shown in graph (b), which shows pressures up to 10 atm. Graph (b) is a greatly expanded view of the low-pressure portion of graph (a).

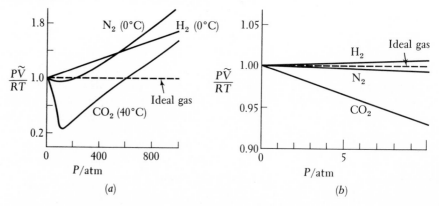

behavior at low temperature, particularly when a gas is close to condensation. The product $P\widetilde{V}$ is constant for an ideal gas at constant temperature and has the value RT. Thus the ratio $P\widetilde{V}/RT$ has the value 1 for an ideal gas at any temperature. Figure 4-5 shows a graph of $P\widetilde{V}/RT$ versus P for some *real* gases, and it is apparent that there are sizable deviations from ideal-gas behavior when the pressure exceeds a few atmospheres.

Dalton's Law of Partial Pressures   As indicated at the start of the chapter, the ideal-gas law applies as well to mixtures as it does to pure gases. Indeed, it was with a mixture, ordinary air, that Boyle, Gay-Lussac, and Charles did their early experiments. Suppose we have a mixture of ideal gases that is itself ideal, occupying a volume V at a temperature T. If we denote the different components of the mixture by subscripts $i = 1, 2, 3, \ldots$, we may write the ideal-gas law, (4-1), for each:

$$P_1 = n_1 \frac{RT}{V}$$

$$P_2 = n_2 \frac{RT}{V} \tag{4-14}$$

and generally
$$P_i = n_i \frac{RT}{V}$$

with the pressure P_i being the pressure that each component would exert if it occupied the total volume V individually. Suppose that we now add these pressures, symbolizing this addition by the symbol ΣP_i, with Σ, the Greek letter sigma, implying summation:

$$\Sigma P_i = \Sigma n_i \frac{RT}{V} = \frac{RT}{V} \Sigma n_i = \frac{nRT}{V} \tag{4-15}$$

where we have taken RT/V as a common factor in front and have used the fact that the total moles, n, in the gas mixture is the sum of the moles of component gases, $n = \Sigma n_i$. Since, by (4-1), the total pressure $P = nRT/V$, we have

$$\Sigma P_i = P \tag{4-16}$$

For an ideal-gas mixture, *the total pressure is thus equal to the sum of the pressures that each component gas would exert if it were present alone.* The pressures P_i are called the *partial pressures* of the different component gases of the mixture, and the law just expressed is called *Dalton's law.* An alternative way of stating this law, and one that is of considerable utility, can be deduced by combining (4-14), (4-15), and (4-16) to give

$$\frac{P_i}{P} = \frac{n_i}{n} \tag{4-17}$$

The partial pressure of component i is thus related to the total pressure in the following simple way:

$$P_i = \frac{n_i}{n} P = x_i P \tag{4-18}$$

where
$$x_i = \frac{n_i}{n} \tag{4-19}$$

is called the *mole fraction* of component i. It represents the fraction of the total moles (or molecules) present that consists of component i.

While we have just seen that in an ideal-gas mixture each component gas behaves as if it were present by itself, this is only approximately true for a mixture of *real* gases. Here the partial pressure is again defined as the pressure each component would exert if it were present by itself, but the total pressure may no longer be equal to the sum of the partial pressures. In other words, while the total pressure may be considered to be the sum of contributions by the component gases, these contributions may be different from the pressures the components would exert when alone. The reason is, of course, that there are interactions between the components of a mixture. The components no longer behave as if they were present alone. However, such deviations from Dalton's law are usually small and we shall neglect them.

An example of the application of Dalton's law concerns the density of a gas mixture.

Example 4-11

□ **Density of a Gas Mixture** What is the density at 700 torr and 27°C of a gas containing 30.0 mol percent ethane (C_2H_6) and 70.0 mol percent carbon dioxide?

Solution Since we are free to consider whatever quantity of gas is most convenient, we choose 1.000 mol of gas, consisting of 0.300 mol C_2H_6 and 0.700 mol CO_2. The molecular weight of C_2H_6 is 30.0 and that of CO_2 is 44.0. The total weight is therefore $m = (0.300 \times 30.0 + 0.700 \times 44.0)$ g $= 39.8$ g. To find the volume we assume that the mixture behaves like an ideal gas, so that

$$V = \frac{nRT}{P} = \frac{1.000 \text{ mol} \times 0.0821 \text{ liter atm mol}^{-1} \text{ K}^{-1} \times 300 \text{ K}}{700 \text{ torr} \times (1 \text{ atm}/760 \text{ torr})} = 26.7 \text{ liter}$$

The density is

$$\rho = \frac{m}{V} = \frac{39.8 \text{ g}}{26.7 \text{ liter}} = \underline{1.49 \text{ g liter}^{-1}}$$

Note that the mass of 1 mol of the gas mixture is 39.8 g, so that the gas behaves as if its molecular weight were 39.8. This value is called the *apparent molecular weight* of the gas mixture. ■

Exercise 4-11

□ What is the density at 750 torr and 20°C of a gas mixture containing 20.0 mol percent H_2 and 80.0 mol percent CH_4? ■

Another application shows how volume percentage in a gas mixture may be converted into weight percentage.

Example 4-12

□ **The Composition of Dry Air** The average composition by volume of dry air is, restricting ourselves to the three major constituents, N_2, 78.0 percent; O_2, 21.0 percent; and Ar, 1.0 percent. What is the composition by weight?

Solution Again we consider 1 mol of gas. The first row of Table 4-1 shows the moles of components in 1 mol of air. Their sum is 1.000 as shown in the last column. On the second line are the weights of the components, obtained by multiplication by the respective molecular weights, 28.0, 32.0, and 40.0.

Table 4-1
Composition of 1 mol Dry Air

	N_2	O_2	Ar	Σ
n_i/mol	0.780	0.210	0.010	1.000
m_i/g	$0.780 \times 28.01 = 21.85$	$0.210 \times 32.00 = 6.72$	$0.010 \times 39.95 = 0.40$	28.97
Weight fraction	$\frac{21.85}{28.97} = 0.754$	$\frac{6.72}{28.97} = 0.232$	$\frac{0.40}{28.97} = 0.014$	1.000
Weight percentage	75.4	23.2	1.4	100.0

The sum of these weights, 28.97 g (rounded to 29.0 g) is the weight of 1 mol of dry air. The apparent molecular weight of dry air is thus 29.0. On the third line are the weight fractions, obtained by division by 28.97. Multiplication by 100.0 converts to the percentages shown on the last line. ■

□ The gas emitted by a volcano has the following typical composition, in volume percentage: H_2O, 55; SO_2, 25; CO_2, 20. What is the composition by weight of the gas, and what is the average molecular weight? ■

Exercise 4-12

Despite popular belief that on a humid day the air is "heavier" because it seems more oppressive, air containing water vapor has a lower density and apparent molecular weight than dry air at the same temperature, if pressure is kept constant, because water has molecular weight 18, considerably less than that of the other constituents. On the other hand, the presence of appreciable carbon dioxide, with molecular weight 44, will raise the density of air.

4-4   Kinetic Theory and the Ideal-Gas Law

Introductory Remarks   We turn now from the macroscopic properties of gases to the microscopic picture of a gas—the kinetic theory of gases, developed during the nineteenth century chiefly by Rudolf Clausius, James Clerk Maxwell, and Ludwig Boltzmann. The aim of kinetic theory is to explain the detailed behavior of gases on the basis of elementary principles of physics. The theory is convincingly successful; we consider here a highly simplified version that still contains the important ideas. The theory leads to the ideal-gas law and gives information about molecular velocities.

The Model   Recall that the molecules of a gas are in rapid and continuous motion and that they are, on the average, widely separated. Despite the fact that a gas is "mostly empty space", there are about 3×10^{19} molecules per cubic centimeter at ordinary temperature and pressure, and any given molecule collides about 10^{10} times per second with other molecules in the gas. Molecules also collide with the walls of the container and these collisions are responsible for the pressure exerted by the gas.

As a result of collisions with other molecules, the motion of any given molecule consists of a zigzag sequence of many short straight-line segments. It might appear that because of this complicated motion any microscopic theory of gases must be hopelessly complex. This is not so. We can to a very good approximation neglect the intermolecular collisions and depict the molecules as moving in straight paths interrupted only by collisions with the *walls*. We are able to neglect collisions between molecules because the role of a molecule whose

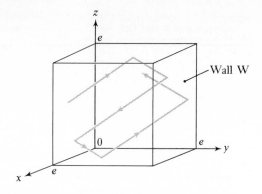

Figure 4-6   Molecule Moving in a Cubic Container
The molecule shown collides eventually with all six walls of the cube. In the text we consider first a molecule that moves in the y direction and thus collides just with the wall W and that opposite it. We then simulate the effect of the random motions of all the molecules by letting one-third of them move in the x direction, one-third in the y direction, and one-third in the z direction.

velocity has been changed by such a collision is in essence assumed by another molecule whose velocity has, in turn, just been suitably changed. This results from the fact that the number of molecules in even a small gas sample is enormous, and the number of intermolecular collisions in even a short time interval is very large.

Molecular Collisions with the Wall   The collisions between the gas molecules and the container walls, which are responsible for the pressure exerted by the gas, are assumed to be elastic. This means that the molecules rebound from the wall without any loss in speed. We calculate now the number of these collisions per second and the force they impart to the wall.

For simplicity, assume 1 mol of gas to be enclosed in a cube of edge e. Figure 4-6 depicts the path of one molecule as it moves through the cubic container. First, consider the collisions of a molecule traveling in a direction perpendicular to the wall at the right of the cube, which we shall call W. Between any two such collisions the molecule travels all the way to the opposite wall and back to W. How long does it take the molecule to travel that distance? The distance traveled is $2e$, twice the cube edge. Thus the time between collisions with W is the distance traveled divided by the speed v in a direction perpendicular to the wall, $t = 2e/v$, and the number of collisions that the molecule makes with W per second is $1/t = v/2e$.

Next we calculate the force imparted to the wall by these collisions. In each collision the velocity of the molecule in a direction perpendicular to the wall is changed from $+v$ to $-v$, a total change of $2v$. The rate of change of speed is called acceleration, and by Newton's second law (Study Guide) force and acceleration are related by

$$\text{Force} = \text{mass} \times \text{acceleration}$$
$$= \text{mass} \times \text{rate of change of velocity}$$

The force exerted on the wall by each molecule is therefore equal to its mass times its velocity change per collision, $m(2v)$, times the number of collisions per second, $v/2e$.

$$\text{Force per molecule} = 2mv\frac{v}{2e} = \frac{mv^2}{e} \qquad (4\text{-}20)$$

To evaluate the total number of collisions with the wall, we note that if a macroscopic sample of a gas is to remain stationary and uniform, there can be no preferred direction for the motion of molecules. We can, for simplicity, assume that one-third of the molecules move back and forth in the direction perpendicular to W, and one-third in each of the directions perpendicular to the other two pairs of walls. (This simplification leads to results in agreement with much more realistic models.) Because molecules have different speeds, it is necessary to replace v^2 in (4-20) by $\langle v^2 \rangle$, the average of the squares of all molecular speeds.

The total force on W is given then by

$$\text{Total force on W} = \text{force per molecule} \times \text{number of molecules}$$

$$= \frac{m\langle v^2 \rangle}{3e} \times N_A = F \tag{4-21}$$

The Pressure on the Wall The collisions of the molecules with the wall are seen macroscopically as a pressure exerted by the gas on the wall. This pressure P is a force per area and, since the area A of the wall W is e^2, we have

$$P = \frac{F}{A} = \frac{N_A m\langle v^2 \rangle/3e}{e^2} = \frac{N_A m\langle v^2 \rangle}{3e^3} \tag{4-22}$$

Introducing the cube volume, $V = e^3$, and rearranging, we get

$$PV = \frac{N_A m\langle v^2 \rangle}{3} \tag{4-23}$$

The average translational kinetic energy[3] of a molecule is given by

$$\langle \epsilon_k \rangle = \frac{m\langle v^2 \rangle}{2} \tag{4-24}$$

Using this relation in (4-23), we get

$$PV = \frac{N_A m\langle v^2 \rangle}{3} = \frac{(2/3)N_A m\langle v^2 \rangle}{2} = (2/3)N_A\langle \epsilon_k \rangle \tag{4-25}$$

A rigorous derivation leads to the same result. The equation does not depend on the size or shape of the container or on the assumption that the collisions between molecules and wall are elastic. It does, however, depend on the following two assumptions:

1. The volume of the molecules themselves is zero. The space accessible to each molecule is the entire volume V of the container and is thus not decreased by the presence of the other molecules.
2. There are no forces between the molecules.

These two assumptions are interrelated, because if the volume of the molecules themselves is not zero, this implies that a repulsive force arises when two molecules attempt to occupy the same space. (If molecules are represented by "hard" spheres, i.e., spheres with no "give" whatever, the force would arise on contact only but would then be infinite.)

We have assumed in deriving (4-25) that the molecular motion is completely random, and we pointed out that the quantity $\langle \epsilon_k \rangle$ in (4-24) represents the *average* of the molecular translational kinetic energy. The pressure exerted by a macroscopic sample of a gas represents the average effect of the innumerable nearly simultaneous collisions of gas molecules with the walls. In air at ordinary

---

[3]That is, the kinetic energy associated with the motion of the molecule as a whole but not including energy due to rotation or that associated with vibration of the atoms.

temperature and pressure there are about 10^{24} collisions with each square centimeter of the walls during each second, but there are fluctuations as a result of the random motions of the molecules. The reason that these fluctuations are not noticeable at ordinary pressures is that random fluctuations in a large number N are on the average equal to $N^{1/2}$, so the average *relative* fluctuation is $N^{1/2}/N = 1/N^{1/2}$. Thus if the pressure-measuring device has an area of 1 cm², the expected fluctuation in pressure from one second to the next is of the order of 1 part in 10^{12}, far too little to detect. If on the other hand the area is greatly reduced, the fluctuations are noticeable. *Brownian motion* of fine dust particles in air results from this kind of effect—the collisions of gas molecules on opposite sides of the dust particle do not balance out, and hence the particle is buffeted about and moves in a jerky, random fashion. Relative fluctuations can also be magnified by reducing the pressure or, with delicate instrumentation, by reducing the time required to measure it.

The Ideal-Gas Equation   Let us compare Equations (4-23) and (4-25) with the ideal-gas equation for 1 mol of gas, $P\widetilde{V} = RT$. It is known experimentally that the ideal-gas equation describes approximately the behavior of a real gas (the lower the density of the gas, the better the approximation). This is precisely the condition under which the assumptions leading to (4-23) and (4-25) are expected to be correct, because at low densities the molecules are far apart so that both the forces between molecules and the molecular volumes are of little account.

If (4-25) is to be identified with the ideal-gas equation, its right side must be equal to RT, since we have chosen our cube to contain just 1 mol of gas. Thus

$$\tfrac{2}{3}N_A\langle\epsilon_k\rangle = RT \tag{4-26}$$

where we should remember that $\langle\epsilon_k\rangle$ is the *average* translational kinetic energy of one gas molecule. This energy must therefore equal

$$\langle\epsilon_k\rangle = \frac{m\langle v^2\rangle}{2} = \frac{(3/2)RT}{N_A} = \frac{3}{2}kT \tag{4-27}$$

where

$$k = \frac{R}{N_A} \tag{4-28}$$

is the gas constant per molecule, also called the *Boltzmann constant*. The average translational kinetic energy per gas molecule is thus proportional to the temperature, and the proportionality factor is *independent of the kinds of molecules* that are being considered. Thus, at a given temperature the average translational kinetic energy of the molecules H_2, O_2, Hg, and C_2H_6 in the gaseous state is the same in spite of differences in their masses and sizes. While this conclusion has been reached here by a combination of theoretical and experimental results, the more refined kinetic theory of gases furnishes this result by theoretical arguments alone.

The kinetic theory of gases also provides a basis for Avogadro's law. Molecules of an ideal gas fill the space available to them by virtue of their rapid motion and not by taking up the space themselves. The volume filled by the gas at a given temperature and pressure is thus not affected by the nature of the molecules but only by their number.

Equation (4-26) makes it apparent that RT is an energy per mole, a fact that is perhaps surprising, considering that the units we have so far favored for the gas constant R have been liter atmospheres per mole per kelvin (liter atm mol⁻¹ K⁻¹).

In other words, liter atmospheres are energy units—the product of volume and pressure represents an energy. Consider the relation

$$\text{Volume} \times \text{pressure} = \text{volume} \times \text{force/area}$$

Since volume/area has the dimensions of a length or a distance, volume × pressure has the same dimensions as distance × force. These are the dimensions appropriate to mechanical work (Study Guide), so that liter atmospheres are indeed energy units.

To convert liter atmospheres to joules, we note that 1 liter is by definition exactly 1000 cm³ while 1 atm is 1.01325×10^5 N m⁻². Thus

$$1 \text{ liter atm} = 1000.000 \text{ cm}^3 \times \frac{1 \text{ m}^3}{10^6 \text{ cm}^3} \times 1.01325 \times 10^5 \text{ N m}^{-2}$$

$$= 101.325 \text{ N m} = 101.325 \text{ J} \tag{4-29}$$

Division by 4.1840 J cal⁻¹ converts this into calories:

$$1 \text{ liter atm} = 24.217 \text{ cal} \tag{4-30}$$

The most useful units for the gas constant R depend on the circumstances. They may be liter atmospheres, ergs, joules, or calories—all multiplied by mol⁻¹ K⁻¹. In SI units, J mol⁻¹ K⁻¹,

$$R = 0.082057 \text{ liter atm mol}^{-1} \text{ K}^{-1} \times 101.325 \text{ J liter}^{-1} \text{ atm}^{-1}$$
$$= 8.314 \text{ J mol}^{-1} \text{ K}^{-1}$$

Table 4-2 gives R in four different units. Additional energy-conversion factors are given in Appendix A, Tables A-4 and A-5.

The Boltzmann constant k may be obtained in different units by dividing any of these values by Avogadro's number. Thus,

$$k = \frac{R}{N_A} = \frac{8.314 \text{ J mol}^{-1} \text{ K}^{-1}}{0.6022 \times 10^{24} \text{ mol}^{-1}}$$
$$= 1.381 \times 10^{-23} \text{ J K}^{-1} = 3.300 \times 10^{-24} \text{ cal K}^{-1} \tag{4-31}$$

where the units mol⁻¹ have been associated with N_A.

In the foregoing all molecules were considered to be of the same kind, but generalization to gas mixtures is straightforward and leads to the result that in a gas mixture the *average translational kinetic energy of all molecules is the same*, as given by (4-27), irrespective of their masses.

The Root-Mean-Square Speed Equation (4-27) contains information about the average (or mean) speed of gas molecules. The quantity directly involved is $\langle v^2 \rangle$, the average of the square of the speed, for which

$$\langle v^2 \rangle = \frac{3kT}{m} \tag{4-32}$$

Multiplying numerator and denominator on the right by Avogadro's number and

**Table 4-2
The Gas Constant R in
Several Units**

0.08206	liter atm mol⁻¹ K⁻¹
8.314	J mol⁻¹ K⁻¹
8.314×10^7	erg mol⁻¹ K⁻¹
1.987	cal mol⁻¹ K⁻¹

replacing $N_A k$ by the gas constant R and $N_A m$ by $\widetilde{M}$ we find

$$\langle v^2 \rangle = \frac{3RT}{\widetilde{M}} \tag{4-33}$$

The square root of the average (or mean) of the square of any quantity is called the *root-mean-square* value, abbreviated rms. The rms value is always greater than the average value unless all the numbers being averaged happen to be the same. For gas molecules, $\langle v^2 \rangle^{1/2}$, the rms speed, exceeds the average speed only by about 8 percent so that the value of the root-mean-square speed,

$$v_{\mathrm{rms}} = \sqrt{\frac{3RT}{\widetilde{M}}} \tag{4-34}$$

gives a good indication of the average speed of gas molecules.

Example 4-13

□ **The Root-Mean-Square Speed of H_2 at 1000 K** What is v_{rms} of H_2 at 1000 K?

Solution To apply (4-34) we use R in SI units, $R = 8.31$ J mol^{-1} K^{-1}, and note that the mass per mole for H_2 is 2.016×10^{-3} kg mol^{-1}. Thus

$$\langle v^2 \rangle = \frac{3RT}{\widetilde{M}} = \frac{3 \times 8.31 \text{ J mol}^{-1} \text{ K}^{-1} \times 1000 \text{ K}}{2.02 \times 10^{-3} \text{ kg mol}^{-1}}$$

$$= 1.23 \times 10^7 \text{ J kg}^{-1}$$

$$= 1.23 \times 10^7 \text{ J kg}^{-1} \times \frac{1 \text{ kg m}^2 \text{ s}^{-2}}{1 \text{ J}}$$

$$= 1.23 \times 10^7 \text{ m}^2 \text{ s}^{-2}$$

Here we have used the definition of the joule, $1 \text{ J} = 1 \text{ kg m}^2 \text{ s}^{-2}$, to convert the answer to convenient units. Taking the square root yields

$$v_{\mathrm{rms}} = \underline{3.5 \times 10^3 \text{ ms}^{-1}}$$

$$= 3.5 \times 10^3 \text{ ms}^{-1} \times \frac{1 \text{ km}}{10^3 \text{ m}} \times \frac{3600 \text{ s}}{\text{hour}}$$

$$= \underline{12.6 \times 10^3 \text{ km hour}^{-1}}$$

or in more familiar units

$$v_{\mathrm{rms}} = 12.6 \times 10^3 \text{ km hour}^{-1} \times \frac{1 \text{ mile}}{1.609 \text{ km}}$$

$$= \underline{7.8 \times 10^3 \text{ mile hour}^{-1}} \quad \blacksquare$$

Exercise 4-13

□ (*a*) What is v_{rms} of CO_2 at 800 K? (*b*) At what temperature is v_{rms} of He just 1000 mile hour^{-1}? ■

Table 4-3 contains additional values of v_{rms}. These rms molecular speeds may seem surprisingly large, but it must be remembered that under ordinary conditions a gas molecule does not move in a straight line for more than a very short time because it suffers innumerable collisions with other gas molecules. Thus the mixing of two gases at ordinary pressures solely as a result of their intrinsic molecular motion (rather than as a consequence of bulk stirring by convection or by mechanical means) is a comparatively slow process despite the high speeds of individual molecules.

Table 4-3
Root-Mean-Square Speeds
for Some Gases

Gas	Temperature/K	Molecular weight	v_{rms}/km s^{-1}
H$_2$	273 1000	2.02	1.84 3.51
O$_2$	273 1000	32.00	0.461 0.883
H$^+$	10^7 (in stars)	1.01	497
Hg	273 1000	200.6	0.184 0.353

The average of the distances through which a gas molecule moves between collisions is called the *mean free path*. It is of the order of 10^{-7} m for oxygen at 1 atm. At a given temperature the mean free path is inversely proportional to the pressure, so that for O$_2$ at 10^{-7} atm, a moderately good vacuum easily obtainable in the laboratory, it has increased to 1 m.

Because such a large fraction of the space occupied by a gas is empty, the chances of collision are relatively small as the molecules move about. Consequently, the mean free path for gas molecules is much greater than the average distance between molecules. To find the average distance between molecules at STP, consider that 1 mol of gas contains 6.02×10^{23} molecules and occupies 22.4 liter $= 22.4 \times 10^{-3}$ m^3. The average volume per molecule is then

$$\frac{22.4 \times 10^{-3} \text{ m}^3 \text{ mol}^{-1}}{6.02 \times 10^{23} \text{ mol}^{-1}} = 3.72 \times 10^{-26} \text{ m}^3$$

If we imagine each molecule to be in a small cube, the distance between the centers of these cubes, which is then the average distance between molecules, is the cube root of the average volume per molecule, or 3.3×10^{-9} m. Since the mean free path of a gas such as O$_2$ is of the order of 10^{-7} m at STP, the distance traveled by a molecule between collisions under these conditions is about 30 times the average distance between molecules.

□ **Mean Free Path** The pressure in the tail of a comet is about 10^{-12} torr. What would be the approximate mean free path of an oxygen molecule in the tail of such a comet? If this value is taken as representative of other molecules present, and if a typical molecular velocity is taken as 1 km s^{-1}, estimate the time between collisions for an average molecule. What bearing has this rough calculation on the fact that comets are often observed to contain molecules that are much too reactive to persist for an appreciable time under ordinary conditions on earth?

Example 4-14

Solution A pressure of 10^{-12} torr is equivalent to about 10^{-15} atm: 10^{-12} torr (1 atm/760 torr) $\approx 10^{-15}$ atm. Since the mean free path of O$_2$ is about 10^{-7} m at 1 atm, and since the mean free path is inversely proportional to pressure, the mean free path of O$_2$ in a comet's tail must be about $(10^{-7}$ m$)(1$ atm$/10^{-15}$ atm$) = 10^8$ m or 10^5 km. If a typical velocity is 1 km s^{-1}, then the time it takes a molecule to travel 10^5 km must be 10^5 s, or about 30 hour. Hence molecules that are created in the tails of comets (for example, by the action of sunlight on other substances present) may persist for many days just because they meet nothing else with which they can react—quite unlike what would happen to them at much higher pressures. ■

Exercise 4-14

☐ What is the average time between collisions for O_2 molecules at STP? What is the average number of collisions per second? ■

Inspection of (4-34) shows that v_{rms} is directly proportional to $\sqrt{T}$ and inversely proportional to $\sqrt{\widetilde{M}}$. The same dependence on T and $\widetilde{M}$ is shown by phenomena that are related to the motion of gas molecules, such as *diffusion,* the motion of one kind of molecule through an assemblage of others as a result of random motion, and *effusion,* the streaming of a gas into a vacuum through a fine hole in a thin wall. These two phenomena obey laws of the same form, even though the basic molecular mechanisms are quite different. Both laws were discovered by Thomas Graham, his law of diffusion in 1831 and his law of effusion in 1846. These laws state that for two gases, 1 and 2, the *times* required for passage of equal numbers of molecules past some reference point under the same conditions are in the ratio

$$\frac{t_1}{t_2} = \sqrt{\frac{\widetilde{M}_1}{\widetilde{M}_2}} \tag{4-35}$$

It is thus possible to compare the molecular weights of two gases by comparing their rates of effusion.

Example 4-15

☐ **Molecular Weight from Effusion Time** A gas is confined in a chamber with a tiny hole in one wall. The number of moles of gas passing through the hole in a given time—i.e., the molar rate of effusion of the gas—at STP is 1.41 that for oxygen under the same conditions. (*a*) What is the molecular weight of the gas? (*b*) What weight of gas will escape in the time that 32 mg of oxygen escapes?

Solution (*a*) Since the *rate* of effusion is 1.41 times as great, the *time* for effusion of equal numbers of moles must be 1.41 times smaller. Thus, with $\widetilde{M}_1 = 32.0 \text{ g mol}^{-1}$ (for O_2) and $t_2/t_1 = 1/1.41$ in (4-35), we have

$$\widetilde{M}_2 = \widetilde{M}_1 \times \frac{t_2{}^2}{t_1{}^2} = \frac{32.0 \text{ g mol}^{-1}}{1.41^2} = \underline{16 \text{ g mol}^{-1}}$$

(*b*) Let n_1 and n_2 denote the numbers of moles of the gases that escape and w_1 and w_2 the weights. Then

$$n_2 = 1.41\, n_1$$

Recall that $w = n\widetilde{M}$, hence

$$\frac{w_2}{\widetilde{M}_2} = 1.41 \times \frac{w_1}{\widetilde{M}_1}$$

Solving for w_2 gives

$$w_2 = 1.41\, w_1 \times \frac{\widetilde{M}_2}{\widetilde{M}_1}$$

From part (a) we have $\widetilde{M}_2/\widetilde{M}_1 = 1/2$, so

$$w_2 = 1.41 \times 32 \text{ mg} \times \tfrac{1}{2} = \underline{23 \text{ mg}}$$

(The gas in question was methane, CH_4. Note that the first step in almost any problem is to see what one can infer qualitatively so that one can check the reasonableness of the answer. The fact that the gas effuses more rapidly than oxygen implies at once that it must have a lower molecular weight, which is consistent with the answer. It is all too easy to get ratios inverted when doing problems like this one.) ■

□ What is the number of moles of CO_2 that escapes through a tiny hole during the time that 1.00 mol N_2 escapes under identical conditions? What weight of CO_2 will escape in the time that 28 mg of N_2 escapes? ■

Exercise 4-15

Diffusion can, under appropriate conditions, be used to separate gases of different molecular weight. If a mixture of oxygen and methane is allowed to diffuse through a porous wall, the methane passes through more rapidly because of its smaller molecular weight. Thus the mixture that has penetrated the wall is enriched in methane and there is a depletion of methane in the original mixture. There are actually better methods for the separation of these two gases, based on their chemical properties or their different solubilities in liquids, but the present method is useful for one class of mixtures difficult to separate in other ways: mixtures of isotopic molecules. The classical case is the separation of ^{235}U and ^{238}U in the form of their fluorides, UF_6, which are gaseous at ordinary conditions. Since the masses of $^{235}UF_6$ and $^{238}UF_6$ are almost identical, the effect is very small. In fact the ratio of the two diffusion times is

$$\frac{t_1}{t_2} = \left(\frac{235 + 6 \times 19}{238 + 6 \times 19}\right)^{1/2} = \left(\frac{349}{352}\right)^{1/2} = 0.9957$$

To be effective, the separation procedure has to be repeated many times, which is achieved by diffusion through a sequence of thousands of porous barriers. To a first approximation, the effect of s barriers is to change the ratio 0.9957 to $(0.9957)^s$.

The effect of 1000 porous barriers on the separation of $^{235}UF_6$ and $^{238}UF_6$ by diffusion is to change the ratio to $(0.9957)^{1000}$. To find the value of this number (call it f) we use logarithms: $\log f = \log (0.9957)^{1000} = 1000 \log 0.9957 = -1.87$. Thus $f = 1.34 \times 10^{-2} \approx 1/74$. Having 1000 barriers is thus seen to be a great improvement over having only one barrier since the ratio of diffusion times is changed by a *factor of 74* (as though the molecular weights differed by a factor of 74^2 or almost 5500!). An elaborate experimental design, first used at Oak Ridge, Tennessee, during World War II, is needed to put this method in effect.

4-5 The Maxwell-Boltzmann Distribution

So far in the discussion of the kinetic gas theory we have made the simplifying assumptions that all molecules have the same speed and that they collide only with the walls. As we have already pointed out, in reality there are very many collisions between molecules. In most of these one molecule gains energy at the expense of another. It is therefore possible that after several collisions some molecules may have acquired energies much higher than the average, while others will be left with much lower energies. In any sample of a gas, then, there is a distribution of molecular energies and speeds.

If the gas is kept at constant temperature, the distribution does not change with time: the fraction of molecules with energies in a given range remains constant. This situation is completely analogous to the time-independence of the density of a gas. If we were to focus on a small volume of a gas we would observe molecules entering and leaving. The number of molecules within the volume remains constant, however, because molecules that leave are replaced by others that enter. Similarly, any molecule that lies within a given narrow energy range is likely to be knocked out of this range by a collision. It is, however, replaced by another molecule that has just received or lost enough energy in a collision to bring it into the range considered.

What fraction of the molecules are in a given energy range for a gas at a given temperature? This question was answered toward the end of the nineteenth century by theoretical investigations of Maxwell and Boltzmann. Closely related to the question just asked is what fraction of the molecules have a given range of speeds. The equation that gives the fraction of molecules that is associated with any given range of speeds is termed the *Maxwell-Boltzmann speed distribution*.

The Maxwell-Boltzmann Distribution of Speeds of Gas Molecules Although the distribution of molecular speeds was first derived from the kinetic theory of gases, it has become possible with modern techniques to determine the distribution experimentally. The results of some measurements on thallium vapor are shown in Fig. 4-7. Molecular speed is plotted along the horizontal axis; the quantity plotted in the vertical direction is proportional to the fraction of molecules with a given speed. The general shape of this curve is typical of the speed distributions of all gases. Relatively small fractions of the molecules have very low or very high speeds, although such distributions are unsymmetric and

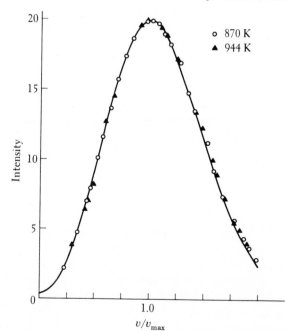

Figure 4-7 Speed Distribution of Thallium Atoms Obtained by Molecular Beam Measurements

The data represented by circles and triangles were measured at different temperatures. However, when v/v_{max} is used as horizontal variable rather than v (where v_{max} is the speed observed most often), the same curve, calculated from the kinetic theory of gases, fits all data points.

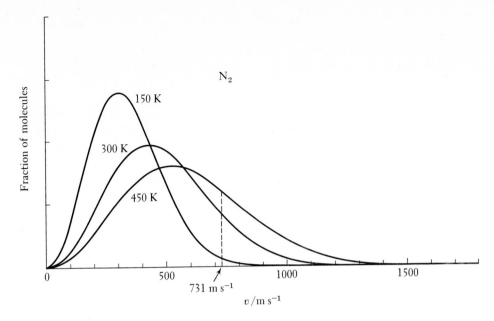

Figure 4-8 Maxwell-Boltz-mann Distribution for N$_2$ at Three Temperatures
The areas underneath the curves to the right of the vertical dashed line at 731 m s^{-1} represent the fractions 0.007, 0.112, and 0.261 of the entire areas, in the order of increasing temperature. These fractions represent the fractions of molecules with speeds equal to or above 731 m s^{-1}.

have tails that extend to high speeds. The line drawn through the experimental points has been calculated from kinetic theory. The agreement is well within experimental error, and many similar experiments have confirmed the accuracy of the theory.

The calculated curves for N$_2$ shown in Fig. 4-8 illustrate the striking effect of temperature on the speed distribution. Most important is the increase of the fraction of fast molecules as the temperature is raised. Since the total area underneath each curve must be unity (i.e., the sum over all speeds of the fractions with given speeds must total 1), a flattening of the curves must accompany the rise in the high-temperature tail with an increase in temperature. The maximum in the curve (corresponding to the most probable speed) therefore becomes lower. For chemical purposes, the molecules that have a speed greater than a certain minimum are of special interest because usually only molecules with an excess of energy may react upon collision with other molecules. The differences in the areas underneath the curves to the right of the vertical line at 731 m s^{-1} in Fig. 4-8 illustrate the increase with temperature of the fraction of molecules with speeds equal to or in excess of 731 m s^{-1}, which is twice the rms speed of N$_2$ at 150 K. The implications of this increase are discussed further in Sec. 22-4.

Summary

Pressure is defined as force per unit area. Common units of pressure are the atmosphere, the torr, and the pascal, the SI unit. Kinetic energy is energy associated with motion; stored energy is called potential energy. Kinetic energy can be converted into potential energy and potential energy into kinetic energy. The SI unit of energy is the joule (J). Work and heat are forms of energy transfer.

Temperature, which we sense as the quality of hotness and coldness, can be measured quantitatively with thermometers. It is proportional to the average kinetic energy associated with random motion of molecules and atoms. The two temperature scales most commonly used in chemistry are the Kelvin and Celsius scales.

Under many conditions gases conform to the

ideal-gas equation, $PV = nRT$, which reduces to simpler forms in special cases, such as Boyle's law: $P_1V_1 = P_2V_2$ (constant n, T); the law of Charles and Gay-Lussac: $V_1/T_1 = V_2/T_2$ (constant P, n); and $P_1V_1/T_1 = P_2V_2/T_2$ (constant n). At constant P and T, equal volumes of two ideal gases contain the same number of molecules (Avogadro's law). Consequently, under these conditions there is a proportionality between the density of a gas and its molecular weight. The ideal-gas law applies as well to mixtures as it does to pure gases. A consequence is Dalton's law: the total pressure of a mixture is equal to the sum of the partial pressures. The partial pressure of each component is the pressure it would exert if it were present alone. Even though real gases behave like ideal gases under many conditions, they may deviate significantly from ideal behavior, particularly when they are close to condensation.

The simple kinetic gas theory, in which it is assumed that molecules are in constant random motion, occupy negligible volume, and do not interact except by collision, provides a basis for the ideal-gas equation and leads to the following conclusions: (1) the average translational kinetic energy of a gas molecule is $\frac{3}{2}kT$, where k is the gas constant per molecule, $k = R/N_A$; (2) the root-mean-square speed v_{rms} of a molecule is proportional to $(T/\tilde{M})^{1/2}$, where $\tilde{M}$ is the mass per mole. The dependence of v_{rms} on T and $\tilde{M}$ is shown also by phenomena such as diffusion and effusion that are related to the motion of gas molecules.

In any sample of gas there is a distribution of molecular energies and speeds. The fraction of molecules with speeds in a given range is given by the Maxwell-Boltzmann speed distribution. Relatively small fractions of the molecules have very low or very high speeds. The distribution is asymmetric, however, and has a tail that extends to very high speeds. A striking effect of an increase in the temperature of a gas is the marked growth in the fraction of molecules with speeds very much higher than the most probable speed.

Terms and Concepts

Problems and Questions

4-1 Conversion of Temperature and Pressure Units Convert the temperatures and pressures in the following statements into degrees Celsius and atmospheres, respectively. (*a*) Normal body temperature is 98.6°F. (*b*) The surface temperature of the sun is about 11,000°F. (*c*) The barometric pressure was 738 torr. (*d*) A recommended tire pressure is 24 lb inch^{-2}.

4-2 Temperature Scales The inhabitants of the very cold planet Cryos base their temperature scale on the properties of grain alcohol, ethanol.

Their temperature unit is termed the degree frigid (°f) and the zero of their scale is the melting point of ethanol, −117°C. The boiling point of ethanol, 78°C, is assigned the value 100°f. What are the values of the freezing and boiling points of water in degrees frigid?

4-3 Boyle's Law In this question assume the temperature and the number of moles of gas to remain constant:
(a) If a certain quantity of helium occupies 4.00 liter at a pressure of 1.00 atm, what volume will it occupy at a pressure of 50 atm?
(b) A certain quantity of hydrogen is expanded from a volume of 2.00 to 150 foot³. If the pressure was originally exactly 1 atm, what is the final pressure, in atmospheres and in torr?

4-4 Charles' Law If a fixed weight of gas is heated at constant pressure from 20 to 100°C, by what factor will the volume increase?

4-5 Gas Law A certain reaction was allowed to happen four different times, and the volume of one particular gaseous product was measured each time. The pressure of the gas was always the same, but the weight of this gas may have varied.
(a) The first time the volume was 140 ml and the temperature was 25°C. The weight of the gas was found to be 0.350 g.
(b) The second time the volume was 160 ml and the temperature was still 25°C. What was the weight of the gas?
(c) The third time the volume was 140 ml and the temperature was 15°C. What was the weight of gas?
(d) The fourth time the volume was 140 ml and the weight of gas was 0.250 g. What was the temperature?

4-6 Gas Law How many moles of gas are contained in 3.00 liter at 25°C and 1.00 atm? If the gas is SO_2, what is its weight? What is its density?

4-7 Gas Law How many moles of water vapor are present in a room with a volume of 50 m³ (= 50 × 10³ liter) if the temperature is 20°C and the pressure of the water vapor in the room is 15 torr? How many grams of water is this?

4-8 Gas Law How many moles of "air" are present in a large lecture room with a volume of 3.0 × 10³ m³ at a temperature of 25°C if the atmospheric pressure is 755 torr?

4-9 Gas Law What weight of neon will occupy, at 20°C and 740 torr, the same volume that 2.00 mol of oxygen occupies at 0°C and 760 torr?

4-10 Gas Law What is the pressure in a sealed 4.00-liter flask containing 4.80 g of the gas CH_4 (methane) at 27°C?

4-11 Gas Law The volume of a certain amount of a gas is doubled at 20°C. To what temperature would the gas have to be heated in order to restore the original pressure? Give the answer in kelvins and in degrees Celsius.

4-12 Gas Law If 20.0 g of a gas occupies 5.00 liter at a pressure of 1.50 atm and 27°C, what volume will 35.0 g of this gas occupy at 77°C and a pressure of 720 torr?

4-13 Compression of Gas A gas occupying 5.000 liter at 100°C and 1.000 atm is compressed, which raises the temperature to 150°C, the final pressure being 9.500 atm. What is the volume of the gas?

4-14 Pressure Units In some experiments at high temperature, the metal gallium has been used as a manometer fluid because it has a lower vapor pressure than mercury. The density of $Ga(l)$ is 6.1 g cm⁻³; that of $Hg(l)$, 13.6 g cm⁻³. How high is a column of gallium equivalent to 1 atm?

4-15 Application of Gas Laws A 224-liter steel reaction vessel is filled with hydrogen at 100°C and 1.00 atm. Calculate (a) the density of the gas; (b) the volume of the same quantity of hydrogen at 0°C and 1.50 atm; (c) the pressure of an equal weight of Ne in the same vessel at 100°C; (d) the moles of H_2 that must be removed if the pressure is to remain 1.00 atm when the vessel is heated to 200°C; (e) the volume of oxygen, measured at 20°C and 740 torr, needed to react with all the hydrogen to form water.

4-16 Heating Air Suppose that air were not soluble in body fluids and that you inhaled 2.00 liter of air at 1.00 atm and −20°C and then held your breath so that the air was warmed to body temperature (37°C). If there were no change in the volume available to the gas, what would be the change in pressure? Conversely, if the pressure of the gas were held constant, what would be the increase in volume?

4-17 Application of Gas Laws A tank is filled with 1000 g N_2 at 0°C and 16.0 atm. The tank is then heated to 50°C and the valve is opened. What is the total weight of the nitrogen that escapes if the external pressure is 1.0 atm and the temperature is kept at 50°C?

4-18 Molar Volume One of the commonest errors of beginning students of chemistry is to assert that the volume occupied by a mole of a substance is 22.4 liter, with no further qualification. Under what conditions is this true? (Be neither too general nor too specific in your answer.)

4-19 Molecular Weight The weight of 950.0 ml of a gas at 40°C and 765 torr is found to be 2.83 g. What is the molecular weight of the gas?

4-20 Density of Gas The density of a gas is 3.74 g liter^{-1} at 740 torr and 20°C. What is the density at 35°C and 763 torr?

4-21 Stratosphere The portion of the atmosphere at altitudes greater than 10 km, where the temperature ceases to fall with increasing altitude, is called the stratosphere. At an altitude of 40 km the temperature is about −55°C and the pressure about 2 torr. How many molecules are in 1 ml of air at these conditions?

4-22 Pressure Exerted by Weight A cubic block of iron 5.0 cm on a side exerts a pressure P on the smooth horizontal area on which it is resting. The density of iron is 7.8 g cm^{-3} and the acceleration due to gravity g is 9.8 m s^{-2}. What is the force, $F = mg$, exerted on the surface, and what is the pressure P in pascals and in atmospheres?

4-23 Volume and Cost of Natural Gas Natural gas is a mixture of low-molecular-weight hydrocarbons such as methane (CH_4) and propane (C_3H_8). This mixture of gases is an excellent fuel and is stored prior to use in large natural underground reservoirs of fixed volume and relatively unchanging temperature or in large tanks on the surface of the ground. (a) Suppose that a supply of natural gas is stored in an underground reservoir of volume 6.0×10^5 m^3 at a pressure of 3.6 atm and a temperature of 17°C. How many storage tanks of volume 2.0×10^4 m^3 could be filled with this gas at 7°C and 1.20 atm? (b) Natural gas is normally sold to consumers by volume. Suppose a consumer obtains some gas in the winter, at an average temperature

of 3°C and a pressure of 1.00 atm, and additional gas in the spring, at an average temperature of 17°C and 1.00 atm. If the price per unit volume is the same in winter and spring, will a given amount of money buy more, fewer, or the same number of moles of gas in winter as in spring? Explain your reasoning clearly, and indicate the percentage difference, if any.

4-24 Formula of an Oxide (a) A certain oxide of phosphorus contains 56.3 percent oxygen. What is its simplest empirical formula? (b) If the density of the vapor of this oxide is 1.95 g liter^{-1} when measured at 540°C and 350 torr, what is the molecular formula of the oxide?

4-25 Molecular Weight A certain gas, X, has a density of about 3.6 g liter^{-1} at 10°C and a pressure of 745 torr. (a) What is the approximate molecular weight of X? (b) If X is known to contain just 44.0 percent fluorine, how many fluorine atoms are there in a molecule of X, and what is a more precise value of the molecular weight of X than that obtained in (a)?

4-26 Molecular Formula A certain compound contains 60.0 percent carbon, 5.0 percent hydrogen, and 35.0 percent nitrogen. At 149 torr and 21°C its vapor has a density of 1.30 g liter^{-1}. What is the molecular formula of the compound?

4-27 Molecular Formula and Atomic Weight A 9.8-g sample of the gaseous oxide of a (hypothetical) element X occupies a volume of 2.00 liter at 27°C and a pressure of 1.00 atm. (a) What is the approximate molecular weight of the gaseous oxide? (b) If the oxide contains 40.0 percent oxygen, how many atoms of oxygen are there in each molecule of the oxide? (c) Give two values of the atomic weight of X consistent with the information given.

4-28 Molecular Formula (a) An unknown gaseous hydrocarbon (that is, a compound containing only C and H) has a density of about 3.5 g liter^{-1} at 746 torr and 100°C. What is its approximate molecular weight? (b) When this gas is burned with an excess of oxygen, the products being carbon dioxide and water, the weight of water formed is just the same as the weight of hydrocarbon at the start. What is the empirical formula of the unknown hydrocarbon? (c) What is the molecular formula of the compound and what is its exact molecular weight?

4-29 Octane and Air Mixture Octane (C_8H_{18}) is vaporized at 25°C and 1.00 atm into 1.00 liter air (containing 21 vol percent oxygen). What is the maximum weight of octane that can be vaporized if the vapor is to be burned completely to H_2O and CO_2? What is the volume of the mixture of air and octane vapor at 25°C and 1.00 atm?

4-30 Volume Composition of a Gas Given a gas mixture of the following volume composition: 20 parts He, 20 parts N_2, 50 parts NO, 50 parts N_2O. How many grams of nitrogen are there in 200 liter of the mixture at 0°C and 760 torr?

4-31 Number of Atoms in Molecules Two elements A and B in the gaseous state may react in the following proportions. Three volumes of A react completely with two volumes of B to produce six volumes of a new gas that is definitely one compound, not a mixture. What can you say from these data about the number of atoms contained (a) in the molecules of element A, (b) in the molecules of element B, (c) in the molecules of the new compound?

4-32 Limestone Analysis A 1.00-g sample of a limestone produces 219 ml CO_2 at 30°C and 750 torr pressure when treated with HCl. What percentage of the mineral is $CaCO_3$ if this is the only carbonate in the mineral?

4-33 Stoichiometry of a Gas Reaction A 100-ml sample of O_2 is added to 50 ml of a mixture of CO and C_2H_6 (at the same temperature and pressure) and the entire mixture is heated. The CO and C_2H_6 burn completely to CO_2 and H_2O; some of the O_2 may be left unreacted. After the initial temperature and pressure are restored and the H_2O has been removed entirely, the gas volume is 85 ml. Find the volume percentages of CO and of C_2H_6 in the original mixture (before the O_2 was added).

4-34 Gas Composition from Its Density A mixture of CO and CO_2 has a density of 1.82 g liter^{-1} at STP. What weight fraction of the gas is CO?

4-35 Deviation from Ideal-Gas Law At 50 atm and 50°C, 10.0 g CO_2 occupies 85 ml. What is the percentage deviation from the ideal-gas law?

4-36 Mixture of NO_2 and N_2O_4 Under ordinary conditions the chemical substance with simplest formula NO_2 is actually a mixture of two gaseous species, NO_2 and N_2O_4. At 45°C and 1.00 atm total pressure, the density of a sample containing only these substances is 2.56 g liter^{-1}. (a) What is the partial pressure of each component? (b) What is the percentage by weight of each component?

4-37 Mixture of Gases A 150-liter steel reaction vessel is filled with a mixture containing equal parts by weight of fluorine and argon at 100°C and 2.00 atm total pressure. (a) If the temperature is changed to 50°C, what will be the resulting total pressure? (b) What is the partial pressure of each component at 50°C? (c) What is the density of the mixture under the original conditions? (d) What is the ratio of the rms speed of an argon atom to that of a fluorine molecule under the conditions described in (a)?

4-38 Analysis of a Mixture of Gases When acetylene, C_2H_2, reacts with hydrogen, it reacts completely to form ethane, C_2H_6, if there is sufficient hydrogen present. Suppose that at a given temperature and pressure, 10.0 volumes of a mixture of hydrogen and acetylene react to form 6.0 volumes of a gas that contains ethane and may also contain some unreacted hydrogen, but no acetylene. (a) Write a balanced equation for the reaction of hydrogen and acetylene. (b) State Avogadro's law and indicate its relevance to this question. (c) What is the mole fraction of acetylene in the original mixture? (d) What is the mole fraction of ethane in the gas present at the end?

4-39 Van der Waals Equation A relatively simple equation that approximates the behavior of real gases when they show moderate deviations from ideal-gas behavior was put forward by J. D. van der Waals in 1873. This equation has the form

$$\left(P + \frac{a}{\widetilde{V}^2}\right)(\widetilde{V} - b) = RT$$

where a and b are constants that depend on the gas being described. The term $a/\widetilde{V}^2$ is added to P to account for the attraction between molecules, and the molar volume is reduced by b to correct for the volume occupied by the molecules. For SO_2, a and b are 6.7 atm liter^2 mol^{-2} and 0.056 liter mol^{-1}, respectively. Use the van der Waals equation to calculate the pressure at which a mole of SO_2

occupies a volume of 22,414 cm^3 at 0°C. The experimental value of this pressure is 0.976 atm.

4-40 Helium-Nitrogen Mixture A mixture of helium and nitrogen occupies a certain volume at a temperature of 27°C and a pressure of 4.00 atm. The same sample is allowed to expand to a volume exactly five times the original volume, and the temperature is then changed until the pressure is just 2.00 atm. (*a*) What is the final temperature (°C)? (*b*) If the gas mixture contains just 33 percent helium by weight, what is the final partial pressure of nitrogen? (*c*) What is the ratio of the average molecular velocity of helium to that of nitrogen in the mixture?

4-41 Mixture of Gases A flask of volume 8.2 liter contains 4.0 g hydrogen, 0.50 mol oxygen, and sufficient argon so that the partial pressure of argon is 2.00 atm. The temperature is 127°C. (*a*) What is the density of the mixture in the flask? (*b*) What is the total pressure in the flask? (*c*) What is the mole fraction of hydrogen? (*d*) Suppose that a spark is passed through the flask so that the following reaction occurs until one reactant is entirely used up:

$$2H_2(g) + O_2(g) \longrightarrow 2H_2O(g)$$

What will the pressure be when the temperature returns to 127°C?

4-42 Comparison of Two Gases Suppose that you have two flasks of the same volume and at the same temperature, one containing gas A and the other gas B. The density of A is just four times that of B in these flasks. Assume that the gases are ideal. What can you conclude about the ratio of the pressures of A and B in the flasks? (If additional information is needed, state precisely what it is and show just how you would use it.)

4-43 Gaseous Diffusion and Molecular Velocities (*a*) A certain molecule X is observed to diffuse 0.50 times as rapidly at 300 K as does oxygen at 400 K. What is the molecular weight of X? (*b*) At what temperature will the rms speed of a hydrogen molecule be the same as that of a molecule of Cl_2 at 100°C?

4-44 Comparison of Oxygen and Helium Imagine two boxes of equal size at the same temperature and pressure, one containing oxygen and the other helium. Compare the contents of the boxes as quantitatively as possible in the following respects: (*a*) number of molecules, (*b*) total mass, (*c*) average kinetic energy of molecules, (*d*) average number of impacts per second on walls of box, (*e*) average force exerted on wall by each collision.

Liquids and Phase Changes

"While the behavior of molecules in a gas can be described mathematically with very great precision, leading to elegant and simple laws, the picture of molecular motion in liquids is unpleasant and untidy. The molecules forming a liquid can be compared to Japanese beetles caught in abundance in a patented trap, or to the worms in a fisherman's can. They crawl and wiggle around each other, they constantly come to clinches like two inexperienced boxers, and they do not permit a physicist to say anything simple or reasonable about them. We can build simple theories and formulate simple mathematical laws only when things themselves are simple, and the motion of molecules in liquids is certainly not that."

GEORGE GAMOW[1]

In the previous chapter, we considered the gaseous state and the kinetic-molecular theory. This chapter, which continues and extends that discussion, is concerned with the nature of the liquid state and of the transitions between the gaseous, liquid, and solid forms of a substance (phase changes). The emphasis is on pure substances; solutions are dealt with in Chaps. 8 and 9.

Liquids play a central role in our lives and in the world about us. The most familiar liquid, water, is essential to all forms of life that we know of, and we use numerous other liquids for a variety of purposes—gasoline, alcohol, mercury, different oils, and many more. Melted solids are involved in volcanic eruptions and other mountain-building processes that shape the surface of the earth. Transformations between solids, liquids, and gases are part of our daily experience too—for example, the melting of ice, evaporation of water from oceans, lakes, and rivers, condensation of water vapor in the atmosphere to form clouds and rain and snow, vaporization of gasoline in the engines of automobiles. The aim of this chapter is to illuminate some of the common features of these different phase changes.

5-1 The Liquid State

Although the nature and structure of liquids are rather well understood qualitatively, no good quantitative theories of the liquid state have been developed, despite intensive efforts over many years by able theoreticians. The reasons are that (1) the molecules of a liquid are close together, so that the essential simplifying assumptions of the kinetic theory of gases—negligible intermolecular forces and negligible molecular volumes—are totally inapplicable, and (2) the molecules are not packed in an ordered array, so that the symmetry that characterizes the crystalline state and makes it relatively easy to treat theoretically is also lacking.

---

[1]G. Gamow, "Matter, Earth, and Sky", p. 204, Prentice-Hall, Inc., Englewood Cliffs, N.J., 1958.

Structure of Liquids It is extremely difficult to design experiments that will give direct information about the microscopic structure of liquids, and many of our ideas about this subject come from calculations made with computers. In essence the computer is used to imitate or simulate nature in order to permit a detailed microscopic picture of a model of a liquid to be obtained at any instant. In a typical computer simulation experiment, the paths of about a thousand molecules in a cubic volume about 10 molecular diameters on a side are calculated. At the start of the calculation the molecules are given random kinetic energies whose average is consistent with the desired temperature. The positions and velocities of all the molecules are then calculated as they change with time because of collisions and less direct interactions. Throughout the study, the interactions of each molecule with all other molecules in the volume are computed according to some specified intermolecular potential-energy function. When a molecule leaves the volume, the computer program generates an identical molecule that enters from the opposite side, so as to avoid spurious effects that would result from collisions with the walls.

The basic reliability of the simulation technique can be checked in a number of ways. For example, in a properly designed simulation the system reaches an equilibrium condition in which the molecular speeds follow the Maxwell-Boltzmann distribution, which holds for liquids as well as gases. There is little doubt that the results obtained from computer studies exhibit the essential characteristics of real liquids and can therefore be used to gain insight into the microscopic properties of the liquid state. The basic ideas that have emerged are these: to a good approximation, the structure of a simple liquid, such as argon or methane, is determined primarily by the repulsive forces between molecules; the attractive forces act only as a "glue" that holds the fluid together. Thus, when structure is considered, one might picture liquid argon as a collection of equal-size marbles. At high densities (i.e., when the marbles are confined to a small volume) the marbles have little mobility. They pack into an ordered solid structure that exhibits a *long-range order* typical of crystals, extending over a great many marbles in all directions (Fig. 5-1a). However, if the density is decreased slightly, by about 10 percent, the highly ordered structure breaks down, and the marbles can move past each other (Fig. 5-1b). For almost all real substances, melting is accompanied by a similar decrease in density and increase in mobility, in agreement with this "hard-sphere" picture.

There are some exceptions to this generalization, however, the most notable being water, which at the melting temperature is about 8 percent *more* dense than ice. This is a consequence of the fact that at ordinary pressures the structure of ice is quite open (Fig. 23-4), with only 4 nearest-neighbor water molecules about any given molecule instead of the maximum of 12 that is possible for an ordered arrangement of spheres.

Figure 5-1 Effect of Density Change on Regularity of Packing

(*a*) Ordered structure at high density; (*b*) disordered structure at slightly lower density.

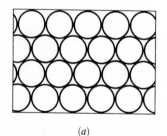
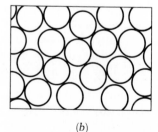

(*a*)   (*b*)

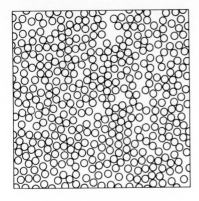

Figure 5-2   Computer Simulation of a Liquid
The figure can be thought of as a "snapshot" that records the positions of the rapidly moving molecules at one instant in time.

Any structural view of a liquid must take into account that liquids are dynamic systems. The relative positions of the molecules in a liquid are constantly changing. If we could film the molecular motion (with an ultramicroscopic and ultra-high-speed camera), we could examine each frame of the film to determine at a given instant the geometric arrangement of the molecules—the liquid structure. This information is available from computer simulations.

A typical "frame" for a monatomic liquid is shown in Fig. 5-2. Note that there are regions in which the ordered structure typical of a crystal seems to be approached, but in many places the order is destroyed because one or more atoms are missing. The order in the liquid is of *short range*, that is, as seen from any given atom, the order extends only over a few neighboring atoms. Many atoms are surrounded by a first shell of neighbors in a nearly regular arrangement, but the regularity of packing with respect to a given "central" atom rapidly decreases as the distance from that atom is increased. Remember, also, that the frame shown in the figure is an instantaneous view. In another instant the positions of the atoms will have altered and the arrangement around any given atom will have changed.

5-2   Phase Transitions and Vapor Pressure

Vapors and Gases   All liquids and solids are to some extent volatile; that is, they tend to evaporate to form gases. For most solids, the pressure of the vapor so formed is extremely small at ordinary temperatures, but it is significant for some solids as it is for most liquids. The term *vapor* is frequently used to refer to a gas that can readily be condensed to a liquid or solid; thus, one usually speaks of water vapor, sodium vapor, naphthalene vapor, and so on. However, there is *no difference between a vapor and a gas;* every vapor is a gas and every gas can be condensed under appropriate conditions. It is merely an accident of our terrestrial environment that we happen to refer to gaseous water, sodium, and naphthalene as *vapors* and to gaseous oxygen and argon as *gases*.

Melting and Freezing   At temperatures very close to absolute zero, all substances are solid, with the exception of helium, which remains liquid at the lowest temperatures unless sufficient pressure is applied (about 26 times normal atmospheric pressure). When the temperature of a solid is raised, the average energy of molecular motion increases. Eventually the motion becomes sufficiently extensive to destroy the close association that keeps the molecules near fixed positions, and the solid *melts*. Crystalline and amorphous solids (Sec. 1-3, Fig. 1-2) differ in their melting behavior. For crystalline solids melting occurs at a very definite temperature, the *melting point* (mp), while for amorphous solids the transition to the liquid state is gradual. It begins with a softening of the solid at a temperature that cannot be measured or specified precisely. As the temperature is raised, an amorphous solid gradually begins to flow as a thick liquid. With further increase in temperature the *viscosity* of the liquid, that is, its resistance to flow, usually decreases because the increasing agitation of the molecules more readily overcomes their mutual attractions.

Conversely, when the temperature of a liquid is lowered sufficiently it freezes. If the resulting solid is crystalline, the transition occurs at a well-defined temperature, the *freezing point*, identical with the melting point. The liquid may also turn into an amorphous solid, the transition taking place over a temperature interval. When this happens the resulting solids are often called *glasses*, because ordinary glass behaves in this way. Other examples of glasses are lava, pitch, resin, and many common plastics.

Some liquids that normally freeze at definite temperatures may nevertheless be cooled below their freezing points by cooling them in smooth and absolutely clean vessels protected against vibrations. Water may be *supercooled* in this way to about $-20°C$. Jarring the vessel or introducing a speck of impurity causes sudden freezing. Crystallization is also aided greatly by wall irregularities on which the phase change may begin to take place. Best of all is the presence of some crystals of the actual substance to be crystallized; these are called *seed crystals*, and only a very few small ones are needed. The properties of a supercooled liquid are a smooth continuation of the properties the liquid possesses at higher temperatures. If freezing does not occur, there are no discontinuities when the temperature passes through the normal freezing point (see Figs. 5-7 and 5-8).

Many solid compounds can occur in several distinct crystalline forms, and transitions between these different solid phases can be observed under appropriate conditions. Transitions between solid phases are usually very slow.

The Vapor Pressure of Liquids A key to the understanding of the relation between a liquid and its vapor is the pressure exerted by the vapor. Consider a closed vessel containing only a liquid and its vapor (Fig. 5-3). The vessel is surrounded by a constant-temperature bath. As already noted, temperature is a measure of the average energy of the random motion of the molecules. Some molecules have just this average energy, but molecules with both greater and less

Figure 5-3 Liquid and Vapor in Closed Container
A closed vessel is partly filled by a liquid, and the temperature is maintained at a value T by a surrounding bath of constant temperature. Given enough liquid and sufficient time, a vapor phase establishes itself above the liquid, with a pressure that is a property of the liquid and depends upon the temperature. Molecules in the liquid and near the surface may break loose and escape into the vapor phase if they have sufficient energy. Molecules in the vapor phase that strike the liquid surface may be captured and stay with the liquid. Another gas may be present in the vapor space without affecting the liquid-vapor equilibrium.

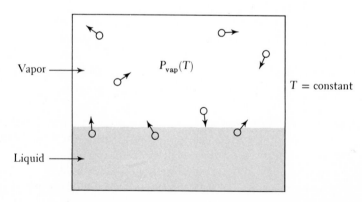

than average energy are also present. Some molecules near the surface of the liquid always have sufficient energy to escape the attraction of the other molecules in the liquid—they *evaporate*. Conversely, vapor molecules that strike the liquid surface are usually retained; that is, they *condense*. When equal numbers of molecules evaporate and condense per unit time, we say that there is equilibrium.

At each temperature this equilibrium corresponds to a definite pressure $P_{vap}(T)$ of the vapor.[2] It is called the *vapor pressure* of the liquid, and it increases with increasing temperature. When the pressure of the vapor phase is below this equilibrium value, more molecules evaporate than condense, continually increasing the pressure in the gas phase. When the gas pressure is above $P_{vap}(T)$, more molecules condense than evaporate, thus lowering the gas pressure. Any energy liberated or absorbed during such condensation or evaporation is taken up or furnished by the constant-temperature bath. At equilibrium, the rates of evaporation and condensation are in balance and no further net change occurs; all visible aspects of condensation and evaporation cease. This balance of opposing rates is the essential feature of all *dynamic* equilibria, of which this is a typical example.

---

[2]We use the symbol T in parentheses here to stress the fact that the vapor pressure is strongly dependent on the temperature.

Figure 5-4 Liquid and Vapor at Constant Pressure
A cylinder is partly filled by liquid, and the remaining volume is taken up by the vapor of the same substance. The temperature is maintained by an external bath at a constant value T, and a piston that forms a tight seal exerts a constant pressure P. If P is larger than the vapor pressure $P_{vap}(T)$, vapor keeps condensing, and the piston moves down until it is in direct contact with the liquid. If P is less than $P_{vap}(T)$, liquid keeps evaporating, and the piston moves up until all liquid is gone (provided the cylinder is large enough). If P is equal to $P_{vap}(T)$, a state of dynamic equilibrium exists no matter what the position of the piston is, and the piston does not move in or out. If the force on the piston is increased or decreased very slightly, the piston can be slowly moved in or out, and the system adjusts by evaporation of liquid or condensation of vapor, the pressure remaining constant. These considerations apply as well to an equilibrium between a solid and its vapor.

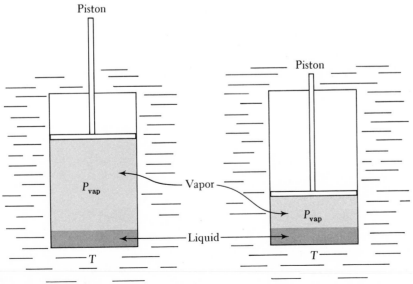

The system illustrated in Fig. 5-3 is at constant volume and temperature. What happens if the volume of the vapor space is changed slowly while the pressure is kept substantially constant? This situation is discussed in Fig. 5-4. Because of the dynamic nature of the equilibrium between the liquid and its vapor, the system adjusts to the volume change (as long as both phases are present) while maintaining the pressure of the vapor at the equilibrium value $P_{vap}(T)$ that corresponds to the temperature of the bath.

The vapor pressure of a pure liquid does *not* depend on the amount of liquid or vapor present, nor on the area of the interface between the two phases. It is necessary, however, that some of each phase be present to ensure that there is an equilibrium between them. The vapor pressure is a characteristic property of the liquid.

Figure 5-5 shows some typical curves representing the vapor pressure as a function of temperature. The concept of vapor pressure is not tied to the presence of a closed vessel. In a vessel open to the outside, some vapor molecules escape continually and an equilibrium can no longer be established—the liquid keeps evaporating.

When the temperature of a liquid is raised to the point where the vapor pressure equals the outside pressure, vapor bubbles begin to form and the liquid boils. The vapor now has sufficient pressure to push the air away. The *normal boiling point* (bp) of a liquid is defined as the temperature at which its vapor pressure equals one atmosphere (760 torr; Fig. 5-5).

If air or other gases are also present in the vapor phase, the total pressure P_{tot} in the gas phase may be considered to be the sum of two contributions: the pressure caused by the vapor molecules and that due to the other gas molecules. The equilibrium between a liquid and its vapor leads to the same vapor pressure as in the absence of such gases, but $P_{vap}(T)$ is now a partial pressure rather than the total pressure. When the partial pressure of the vapor is less than the vapor pressure P_{vap} of the liquid, evaporation exceeds condensation; if it is larger than P_{vap} (e.g., because the temperature has just been lowered), condensation exceeds evaporation. Equilibrium exists when the partial pressure of the vapor in the gas phase equals P_{vap}.

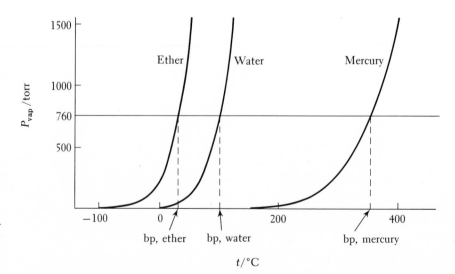

Figure 5-5 Vapor-Pressure Curves for Diethyl Ether, Water, and Mercury

Note the increase of the vapor pressure with temperature. The abbreviation bp stands for the normal boiling point, i.e., the boiling temperature at 1 atm pressure.

The Vapor Pressure of Solids A solid also has a vapor pressure. Usually there are molecules near the surface of a solid that have sufficient energy to escape and form a vapor phase, and in a closed container a dynamic equilibrium is achieved. The vapor pressure of a solid is an inherent property, depending only on the temperature and the kind of substance. It increases with the temperature, like the vapor pressure of a liquid. A solid may thus evaporate directly, without melting, a process called *sublimation*. For many solids the vapor pressure is so small at normal temperatures that effectively no evaporation takes place. However, some solids have measurable vapor pressures, that of ice at 0°C being 4.6 torr or about 0.006 atm. Although this might at first seem a very low pressure, those who are familiar with cold climates know that snow often disappears from the ground and frozen wash on the clothesline dries by sublimation even when the temperature never rises above the melting point of ice.

A few solids have vapor pressures that reach 1 atm at temperatures below their melting points. The temperature for which the vapor pressure of a solid is 1 atm is called the *normal sublimation point*. At this temperature, the vapor has sufficient pressure to push away the surrounding atmosphere, but of course evaporation of such a solid also occurs below this temperature, just as it does for a liquid below its normal boiling point. Solid carbon dioxide (dry ice) is an example of a substance that normally sublimes rather than melts. Its sublimation point is $-78°C$.

An Important Distinction It is essential to recognize that, in the absence of the liquid or solid, the pressure exerted by the molecules of a vapor at a given temperature T may have any value from zero up to the vapor pressure at that temperature, $P_{vap}(T)$. As has been stressed, the vapor pressure is a property characteristic of an equilibrium between a vapor and some other phase (or phases) at a particular temperature. In the presence of vapor alone or under nonequilibrium conditions, the *pressure of the vapor* of any substance is not necessarily the same as, and is usually less than, the *vapor pressure*. For example, on a relatively dry day at (say) 25°C (77°F), the pressure of water vapor in the atmosphere may be only 10 torr while the vapor pressure of water at that temperature is 24 torr, so that the *relative humidity* is $\frac{10}{24}(100) = 42$ percent. When the relative humidity is very high, the pressure of water vapor in the atmosphere approaches the vapor pressure. The pressure of the vapor may also be above the vapor pressure; such vapor is called supercooled or supersaturated and eventually condenses, as we discuss later.

Vapor Pressure and Partial Pressure When a gas is collected over water (Fig. 5-6), the total pressure includes a contribution from the water vapor. The pressure that would be exerted by the gas alone can be calculated by the use of Dalton's law (Equation 4-16).

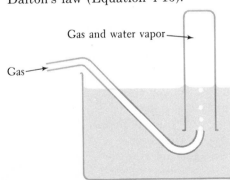
Gas and water vapor
Gas

Figure 5-6 Collecting a Gas over Water
The inverted cylinder is completely filled with water before any gas is collected. At the end, to find the pressure due to the gas, a correction must be applied to allow for the partial pressure of the water vapor in the gas collected. The gas is usually saturated with water vapor, and the total pressure of the gas collected (P_{tot}) is equal to the sum of the vapor pressure of water (P_{vap}) at the temperature considered and the pressure (P_{gas}) the gas would exert at the same volume and temperature if the water vapor were removed. The cylinder may need to be lowered or raised to ensure that the levels of water inside and outside it are equal. Otherwise a correction is needed to allow for the pressure difference caused by the difference in levels.

103

Example 5-1

☐ **Collecting a Gas over Water**   A gas Y was collected above water (Fig. 5-6) at 18°C and an outside pressure of 745 torr. Its volume was 234.0 ml. How many moles of Y were collected?

Solution   According to Dalton's law, the total pressure is given by

$$P_{\text{tot}} = P_Y + P_{H_2O}$$

and here P_{H_2O} is the vapor pressure of water at 18°C because the space above the water is saturated with water vapor, i.e., contains water vapor at its vapor pressure. The vapor pressure of water at 18°C is 15.5 torr (Table D-1, Appendix D), so that the pressure of Y is $745 - 15.5 = 729.5$ torr. (Even though it is not significant, the final digit is retained during the calculation, as discussed in Sec. 3-1.) The number of moles of Y is then given by

$$n_Y = \frac{P_Y V}{RT}$$

$$= \frac{729.5 \text{ torr} \times 234.0 \text{ ml} \times (1 \text{ liter}/1000 \text{ ml})}{62.4 \text{ liter torr mol}^{-1} \text{ K}^{-1} \times 291 \text{ K}}$$

$$= \underline{9.40 \times 10^{-3} \text{ mol}} \qquad ■$$

Exercise 5-1

☐ A gas was collected above water at 20°C ($P_{\text{vap}} = 17.5$ torr) and at total pressure 750 torr. Its volume was 455.2 ml. After the gas was dried, it was stored at 25°C and 720 torr. What was its volume then?   ■

Stability of Phases   When equilibrium exists between any two condensed phases, for example, between a liquid and a solid phase, the vapor pressure of the two phases must be identical. Otherwise the phase with the higher vapor pressure would evaporate continually and condense into the other phase and no equilibrium would be reached. In the end only the phase with the lower vapor pressure would exist. These considerations apply equally well when there are two different solid phases in equilibrium with each other, as happens under appropriate conditions for a number of substances (for example, different forms of ice under high pressure, or sulfur at about 96°C and 1 atm).

When more than one condensed phase of a substance can exist at a given temperature and the phases are not in equilibrium with each other, that is, they do not have the same vapor pressure, then the phase with the lower or lowest vapor pressure is called the *stable* phase, all others being termed *metastable* phases. For example, supercooled water (that is, liquid water below 0°C) has a higher vapor pressure than ice at the same temperature (see Fig. 5-7). The term "metastable" implies that the unstable phase exists long enough to be observed, because the rate at which it changes into the stable phase is very low.

The boiling of liquids is often associated with a delay effect, *superheating,* similar to the supercooling of liquids. The smooth boiling of liquids is enhanced by specks of impurities or rough spots on the container walls, both of which further the formation of vapor bubbles. In the absence of either, vapor bubbles may not form at once, although the boiling point has been reached and even passed. Conversely, vapor may be supercooled in the absence of suspended solid or liquid particles, the pressure of the vapor then exceeding the vapor pressure.

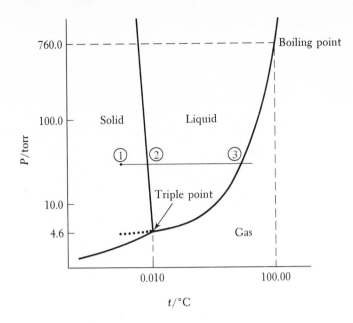

Figure 5-7 Schematic Phase Diagram for Water
This diagram describes the stable states of water at various combinations of temperature (T) and pressure (P) and makes it possible to predict what changes will occur as T and P are changed from any initial conditions. Consider, for example, point 1, corresponding to conditions ($-25°C$, 30 torr) under which water is solid (ice). Suppose that T is increased at constant P; this change in conditions corresponds to moving horizontally toward the right in the diagram. When the boundary between the solid and liquid phases is crossed, at point 2, the ice is transformed into liquid water. Similarly, at point 3 the boundary between the liquid and gaseous phases is crossed and the water is vaporized. The dots represent the vapor pressure of the supercooled liquid; see text.

Phase Diagrams for Pure Substances The ranges of temperature and pressure in which the different phases of a pure substance are stable may be represented graphically by a *phase diagram*. Each substance has its own characteristic diagram; those for water and carbon dioxide are shown in Figs. 5-7 and 5-8. These diagrams show whether the most stable state at any specified temperature and pressure is a gas, a liquid, or a solid, or whether several phases may be equally stable and thus in equilibrium with each other. The area in which the gaseous state is most stable extends to low pressures and high temperatures, whereas that representing the solid state extends to high pressures and low temperatures. The stability range for the liquid state forms a wedge between the other two regions.

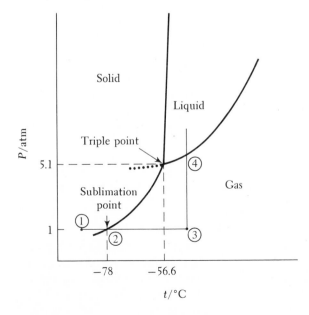

Figure 5-8 Schematic Phase Diagram for Carbon Dioxide
A noteworthy aspect of this diagram is that the triple point lies above atmospheric pressure. It takes at least 5.1 atm to liquefy carbon dioxide, and at atmospheric pressure only solid and gaseous carbon dioxide can exist. For example, when carbon dioxide at $-90°C$ and 1 atm (point 1) is heated at constant pressure, it sublimes at $-78°C$ (point 2). Gaseous carbon dioxide at $-50°C$ and 1 atm (point 3) may be liquefied at constant temperature by an increase in pressure (point 4). Solid carbon dioxide is often called *dry ice.*

The boundary line between any two areas gives the temperatures and pressures at which there is equilibrium between the two phases corresponding to these areas. Thus, the boundary line between the liquid and gaseous regions is the vapor-pressure curve for the liquid, like those shown in Fig 5-5, and the boundary curve between the solid and gaseous regions is the vapor-pressure curve for the solid. The equilibrium curve between liquid and solid is almost vertical; the deviations from the vertical have been exaggerated in Figs. 5-7 and 5-8 to make them visible.

The boundaries of the three regions meet at a point; this *triple point* indicates the conditions under which all three phases may exist in equilibrium with each other. The dotted continuation of the liquid-vapor curve into the area of the solid (in each diagram) shows the vapor pressure of the supercooled liquid. The vapor pressure of a supercooled liquid is greater than that of the corresponding crystalline solid because the solid is the more stable phase below the triple point. There is no corresponding extension of the vapor-pressure curve for the solid above the triple point; pure crystals always melt sharply at the melting point and cannot be superheated.

Example 5-2

□ **Interpretation of a Phase Diagram** A portion of the phase diagram of a hypothetical substance A is shown in Fig. 5-9. (*a*) What are the stable phases of A under the following conditions: 20°C, 0.1 atm; 60°C, 0.7 atm; 30°C, 0.6 atm; 25°C, 0.3 atm? (*b*) Describe any phase changes that occur when 1 mol of A, initially at 60°C and 0.4 atm, is cooled to 10°C at constant pressure.

Solution (*a*) Each point on the phase diagram can be identified by its temperature and pressure coordinates. The point 20°C, 0.1 atm lies within the gaseous region; A is therefore a gas under these conditions. The point 60°C, 0.7 atm lies within the liquid portion of the diagram; hence A is entirely liquid. The point 30°C, 0.6 atm lies on the line separating the liquid and solid regions of the diagram. Liquid and solid A are in equilibrium under these conditions. The three phase boundaries intersect at 25°C, 0.3 atm. This is a triple point: solid, liquid, and gas are in equilibrium. (*b*) The path followed can be indicated by a horizontal line at 0.4 atm extending from 60 to 10°C. Initially, A is gaseous. The boundary between the gaseous and liquid regions is reached at about 50°C; here liquid and gas are in equilibrium. At a lower temperature A is entirely liquid. At about 27°C, the liquid-solid boundary is reached, and solid A appears, initially in equilibrium with the liquid. At lower temperatures the sample is entirely solid. ■

Exercise 5-2

□ Use the phase diagram of substance A, Fig. 5-9, to answer the following: (*a*) What are the stable phases of A under the following conditions: 10°C, 0.2 atm; 40°C, 0.2 atm; 10°C, 0.1 atm; 70°C, 0.7 atm? (*b*) Describe any phase changes that occur when 1 mol of A initially at 30°C, 0.1 atm is brought to 30°C, 0.7 atm at constant temperature. ■

The Critical Point At temperatures near the triple point, liquid water and the vapor with which it is in equilibrium are easily distinguishable. For example, at 10°C the density of the liquid is 10^5 times that of the vapor. Under these conditions liquid water is much more viscous than water vapor: the ratio of the viscosities is 100.

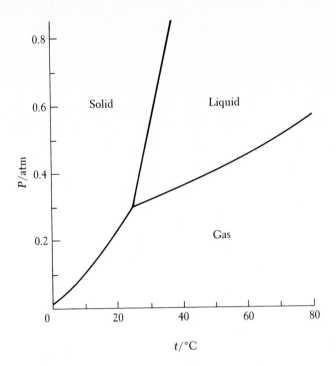

Figure 5-9

As the temperature is increased, the liquid and vapor in equilibrium become more nearly alike. At the normal boiling point, 100°C, the liquid is 1600 times denser than the vapor, and the ratio of viscosities has fallen to 20. Other properties show a similar trend with temperature. The two phases in equilibrium become identical in their properties at 374°C. The vapor pressure at this temperature is 218.3 atm and the density (of both liquid and vapor) is 0.325 g cm^{-3}. This point, which is called the *critical point* of water, is the termination of the vapor-pressure curve (Fig. 5-10). Above the critical temperature, 374°C, it is not possible to speak of distinct liquid and vapor phases. When water vapor at a temperature greater than 374°C is compressed, there is no discontinuous change in properties until solid begins to form,[3] that is, until the melting curve is reached. Sometimes, rather than referring to liquid or vapor phases above the critical point, the term *fluid* phase is used.

All liquids have critical points. In other words all vapor-pressure curves, that is, all phase boundaries between liquid and vapor, end at a finite temperature and pressure. The critical temperature, pressure, and density are characteristic properties of each substance. Critical parameters for several common substances are given in Table 5-1. With pure substances, critical points are found only for vapor-liquid equilibria. There do not appear to be comparable terminations of solid-liquid or solid-solid phase boundaries.

Energy Requirements in Phase Transitions   There is molecular[4] motion in all solids, which consists of vibrations of the molecules about fixed positions and is

---

[3] At 375°C this occurs when the pressure reaches about 10^5 atm.

[4] In this discussion the terms *molecule* and *molecular* refer to any of the basic units that comprise a solid—atoms, ions, or molecules.

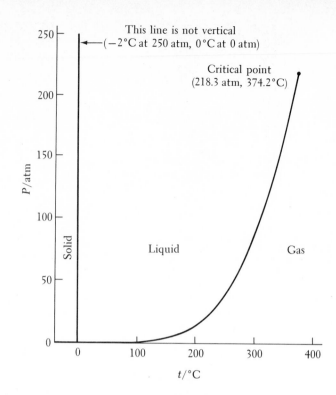

Figure 5-10 **Phase Diagram of H₂O Showing Critical Point**

present even at very low temperatures. This vibrational motion in solids has been simulated in computer studies (Fig. 5-11a). As the temperature of the solid is increased, the amplitude of the motion increases. When the temperature of a solid reaches a certain value, the motion becomes sufficiently great that the solid melts to a liquid (Fig. 5-11b) and an equilibrium can be established between solid and liquid. Like all equilibria, this equilibrium represents a state of balance. Here the balance is between a tendency for the energy to decrease (corresponding to the stronger molecular interactions in the solid) and a tendency toward disorder (corresponding to the more random arrangement in the liquid). The disordering tendency becomes more important with increasing temperature, as we shall see when we consider the thermodynamics of equilibria (Chap. 20). The melting of a solid is virtually always accompanied by absorption of energy.

Since the distribution of kinetic energy among the molecules of a liquid follows the Maxwell-Boltzmann law, the average kinetic energy is $\frac{3}{2}kT$ per molecule or $\frac{3}{2}RT$ per mole, just as in the vapor. The process of vaporization corresponds, at ordinary temperature and pressure, to an increase in the average intermolecular distance by a factor of about 10, since the volume increases by about 10^3. Because

Table 5-1
Critical Temperature, Pressure, and Density for Some Common Substances

Substance	Critical temperature/K	Critical pressure/atm	Critical density/g cm⁻³
He	5.3	2.26	0.0693
H₂	33.3	12.8	0.0310
N₂	126.1	33.5	0.3110
O₂	153.4	49.7	0.430
CO₂	304.2	73.0	0.460
NH₃	405.6	111.5	0.235
H₂O	647.4	218.3	0.325

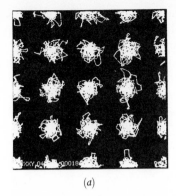 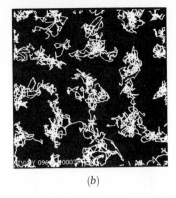

(a)   (b)

Figure 5-11 Atomic Motion in Solids and Liquids

The two figures show computer-simulated projections of the trajectories of 32 rigid spheres during the time it takes them to make 3000 collisions inside a cubic box. In (a) a solid is simulated;
in (b), a liquid. In the solid each sphere moves in a well-defined region; in the liquid the regions
of individual motion are much less well-defined, and a sphere may move from one region to another. The reason that in the liquid there are regions of individual motion at all is that the
space taken up by one sphere cannot also be occupied by another; the spheres are impenetrable,
as is essentially true also of atoms and molecules. (*B. J. Alder and T. E. Wainwright, Scientific
American, p. 113, October 1959.*)

this increased separation is associated with an increase in the potential energy,
energy must be supplied to vaporize a liquid. The amount of energy required
varies widely for different liquids, depending on the strength of the interactions
among the particles of which the liquid is composed.

The fact that energy is absorbed during evaporation and released during
condensation is of great practical importance. Evaporative cooling helps maintain the body temperature of many animals, including man, and considerable
atmospheric heating and turbulence accompany the condensation of water vapor
during heavy rain storms.

Summary

The microscopic structure of liquids is characterized by short-range order. Each molecule is surrounded by a shell of nearest neighbors in a somewhat regular arrangement, but the order around a
molecule decreases rapidly beyond the nearest-
neighbor distance. Liquids are dynamic systems;
the arrangement of molecules does not remain
fixed. There is long-range order in crystalline solids. When the temperature of a crystalline solid is
raised to a certain point, the long-range order is
destroyed by the vigor of molecular motion, and
the solid melts. If the temperature of an amorphous solid is raised there is a gradual transition to
the liquid state.

When a liquid partially fills a closed vessel, a
dynamic equilibrium is set up in which the number
of molecules evaporating from the surface is equal
to the number of vapor molecules that condense.
At each temperature this equilibrium corresponds
to a definite pressure of the vapor called the vapor
pressure of the liquid. The vapor pressure is a
characteristic property of a liquid. Solids also have
vapor pressures; the evaporation of a solid is called
sublimation.

When equilibrium exists between any two condensed phases, the vapor pressure of the two
phases must be identical. If two (or more) condensed phases of a substance exist at a given temperature and the phases are not in equilibrium
with each other, the phase with the lower vapor
pressure is the stable phase, and the other is called
a metastable phase. At equilibrium the substance
will exist solely in the stable phase.

The phase diagram of a pure substance indicates
the conditions of temperature and pressure under
which its solid, liquid, and gas phases are stable.

The boundaries between the different regions of stability represent transitions between states of aggregation, transitions such as melting (freezing), vaporization (condensation), and sublimation (condensation). A triple point is a point at which the boundaries of three regions meet.

At the melting point of a substance, the solid and liquid phases are in equilibrium with each other. The boiling point of a liquid, defined as the temperature at which the vapor pressure of a liquid equals the applied pressure, similarly corresponds to equilibrium between liquid and vapor. At the sublimation point of a solid there is equilibrium between solid and vapor. The distinction between the gas and liquid forms of a substance is lost at the critical point, which is the termination of the vapor pressure curve.

Energy must be supplied to vaporize a liquid or to melt a solid. Conversely, energy is liberated when a vapor condenses or when a liquid freezes.

Terms and Concepts

Problems and Questions

5-1 Differences between Solids and Liquids (a) Describe in detail the differences between the microscopic structure of a typical substance just below and just above the melting point. Include in your description the average distance between molecules, the long- and short-range order, the degree of molecular motion, and the average potential energy per molecule. (b) Explain how the differences in microscopic properties of solids and liquids are reflected in differences in the macroscopic properties of solids and liquids.

5-2 Differences between Liquids and Gases Compare and contrast the microscopic properties of a liquid and its vapor. Include in your discussion those properties specified in Prob. 5-1. Explain how these differences are reflected in the macroscopic properties.

5-3 Internal Pressure in Liquids The van der Waals equation (Prob. 4-39) describes the behavior of a substance in the liquid as well as in the gaseous state. In this equation the pressure P is increased by $a/\widetilde{V}^2$, a term that represents an internal pressure that would prevail in the fluid even in the absence of an external pressure. This internal pressure is quite large for liquids. Estimate it for N_2 and SO_2 by evaluating $a/\widetilde{V}^2$ for these two substances at their normal boiling points. The respective densities and temperatures are 0.81 g cm^{-3} ($-196°C$) for N_2 and 1.43 g cm^{-3} ($-73°C$) for SO_2. Values of a are 1.38 and 6.7 atm liter^2 mol^{-2}, respectively.

5-4 Definitions (a) Give concise definitions of the following terms: melting point, sublimation, superheating, boiling point, vapor pressure. (b) Distinguish clearly between *vapor pressure* and *pressure of a vapor*.

5-5 Vapor Pressure and Stability The vapor pressure of liquid benzene at a given temperature can be computed from the empirical formula

$$\log \frac{P}{\text{torr}} = \frac{-1784.8}{T/\text{K}} + 7.962$$

and that of solid benzene from

$$\log \frac{P}{\text{torr}} = \frac{-2309.7}{T/\text{K}} + 9.846$$

(a) Which of the phases of benzene, solid or liquid, is the more stable at 10°C? (b) At what temperature are solid, liquid, and vapor in equilibrium?

5-6 Temperature Variation of Vapor Pressure With reference to the vapor pressure curves shown in Fig. 5-5, speculate on how the vapor pressure of a pure liquid could be used as a thermometer. Which of the three liquids described in the figure would make the most sensitive vapor-pressure thermometer at 50°C?

5-7 Relative Humidity Under which conditions is the pressure of water vapor in the atmosphere greater, a day when the temperature is 18°C and the relative humidity is 100 percent, or one when the temperature is 35°C and the relative humidity is 40 percent? (The vapor pressure of water at various temperatures is given in Table D-1.)

5-8 Gas Collected over Water A sample of nitrogen is collected over water at 30°C at which temperature the vapor pressure of water is 32 torr. The total pressure of the gas is 656 torr, and the volume is 606 ml. How many moles of nitrogen does the sample contain?

5-9 Vapor Pressure of Unknown Liquid A volume of just 4.00 liter oxygen was collected over a certain liquid at 25°C and a pressure of 750 torr. When the vapor of the liquid was removed from the oxygen by appropriate treatment and the oxygen volume was again measured, this time at STP, it was found to be 3.00 liter. Calculate the vapor pressure of the liquid; assume that the oxygen was initially saturated with vapor.

5-10 Mixture of Oxygen and Water Vapor A mixture of oxygen and water vapor at 90°C and 738 torr was cooled to 20°C. Some liquid water condensed from the sample during the cooling; at 20°C, the vapor pressure of water is 18 torr. The pressure of the gas after cooling was the same as that at the start; the volume of the gas had decreased to 0.500 of what it had been originally. What was the final pressure of the oxygen, and what was it initially (at 90°C)? (If you need this information, the vapor pressure of water at 90°C is 526 torr.)

5-11 Humidity of Air On a humid day in the tropics when the air was at 35°C, a 20.0-liter air sample was collected at a pressure of 1.00 atm. The air sample was dried by passing it over a salt that would remove all the moisture; the weight of the water removed was 0.72 g. (a) How many moles of water were removed from the air? (b) What was the pressure of water in the sample of air before it had been dried? (c) What relative humidity does this correspond to?

5-12 Density of Wet Gas What is the density of a gas that consists of CO_2 saturated with H_2O at 40°C and a total pressure of 700 torr? Vapor pressures for water may be found in Appendix D, Table D-1.

5-13 Relative Humidity (a) How many moles of water, and what weight of water, are present in 1.00 m^3 of air if the relative humidity is 80 percent at 26°C and 760 torr? (b) Air of 80 percent relative humidity at 10°C is heated at constant pressure to 30°C. What is the relative humidity at 30°C? (c) If the air in (b) had been heated to 30°C at constant volume, what would the final relative humidity have been?

5-14 Water Vapor The air in a large room, which has a volume of $8.0 \times 10^2 \text{ m}^3$, is at 20°C and a pressure of 735 torr. Use data from Table D-1 as needed. (a) If 10.2 kg of $H_2O(g)$ is present in the room, what is the partial pressure of water vapor in the room? (b) Suppose that an object at a temperature of 5°C were placed in the room under the conditions described in (a). Would water vapor condense as liquid on the surface of that object? Explain your answer.

5-15 Phase Diagram of CO_2 Refer to Fig. 5-8. Suppose that a sample of carbon dioxide at −80°C and 0.1 atm is (1) compressed at constant temperature to a pressure of 7 atm, then (2) is heated at constant pressure to a temperature of −50°C, and then (3) is held at constant temperature while the pressure is lowered to 3 atm. Describe any phase changes that will occur during each of the stages 1, 2, and 3. Indicate whether they come about at some particular temperature and pressure (if so, what?) or over a range.

5-16 Phase Diagram Figure 5-12 is a phase diagram for a hypothetical substance X that has two solid forms, 1 and 2; solid 2 is stable at higher

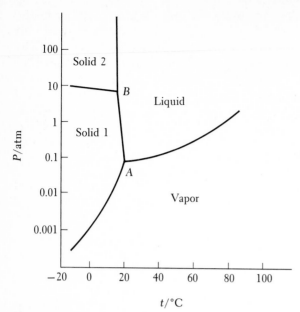

Figure 5-12

pressures, as indicated. (*a*) Explain the significance of point *A* (each letter refers to the point of intersection of the lines adjacent to it) and point *B*. (*b*) Describe any phase changes that would occur, giving the approximate temperature and pressure of each, if the following sequence of steps were followed: (i) Some vapor at 60°C and 0.001 atm is compressed at constant temperature to 100 atm and cooled to −10°C; the pressure is then gradually reduced until it is 10^{-4} atm. (ii) Some solid 1 is gradually heated at a constant pressure of 0.01 atm from −20°C until the temperature is 60°C.

5-17 Interpretation of Phase Diagram of Water Imagine that a beaker of water supercooled to −5°C is placed next to a beaker of ice, also at −5°C, and the two are isolated in a closed box. If the temperature of the box is kept at −5°C and the water does not freeze, what change *will* take place eventually? (Under proper conditions water can be kept supercooled for a long time.) Explain, with the help of the phase diagram for water, Fig. 5-7.

5-18 Interpretation of Phase Diagram The schematic phase diagram for water, Fig. 5-7, shows

that at room temperature and a pressure of 1 atm water is in its liquid form. We know from experience, however, that there is water vapor above the surface of water in a glass at these conditions of temperature and pressure. Explain this apparent contradiction.

5-19 Cooling by Evaporation Explain in terms of heat effects during phase changes why a hot humid day is usually far more uncomfortable than a dry day at the same temperature.

5-20 Phase Diagram A certain substance A has the following properties:

(i)   normal boiling point is 250°C;
(ii) melting point is 100°C, independent of pressure;
(iii) vapor pressure at the triple point is 10 torr;
(iv) vapor pressure at 150°C is 20 torr; and
(v)   vapor pressure at 50°C is 0.5 torr.

(*a*) Sketch an approximate phase diagram for A, using for convenience a pressure scale similar to that in Fig. 5-7 and a temperature scale ranging from 0 to 300°C. Indicate the experimental points by an x in a circle ($\widehat{x}$) and label the regions of stability of the solid, the liquid, and the gaseous forms of A.

(*b*) Suppose that a sample of A, originally at 50°C and a pressure of 1 torr, is warmed gradually at constant pressure until its temperature is 250°C, then the pressure is raised at a temperature of 250°C until the pressure is 100 torr, and then the sample is cooled at a constant pressure of 100 torr until its temperature is 25°C. Indicate the path of these changes by a dashed line (————) on your diagram and indicate by numbers (1, 2, 3, . . .) on the diagram the places on this path at which phase changes would be expected to occur. Indicate whether each of these phase changes (refer to it by the number you have used) would occur at a specific temperature and pressure or over a range, and approximately what the corresponding pressure(s) and temperature(s) would be.

(*c*) Under what conditions, if any, may the pressure exerted by the vapor of any substance be substantially less than the vapor pressure of that substance at the same temperature?

Interatomic and Intermolecular Forces

6

"There are therefore Agents in Nature able to make the Particles of Bodies stick together by very strong Attractions. And it is the Business of Experimental Philosophy to find them out."
ISAAC NEWTON, 1717

"We shall say that there is a chemical bond between two atoms or groups of atoms in case that the forces acting between them are such as to lead to the formation of an aggregate with sufficient stability to make it convenient for the chemist to consider it as an independent molecular species."
LINUS PAULING,[1] 1939

This chapter deals in an introductory way with the forces that bind atoms together in molecules and compounds—chemical bonds—and the weaker forces that are responsible, at low enough temperatures, for the condensation of all gases into liquids and solids. Chemical bonds are considered in greater depth in Chaps. 17 and 18.

6-1 Background

The existence of solids and liquids suggests that there are attractive forces between molecules. Similarly, it follows from the existence of distinct chemical compounds containing definite proportions of different kinds of atoms that there are forces holding these atoms together in stable aggregates. This led nineteenth-century chemists to the idea of valency, an expression of the number of other atoms with which a given atom can combine in a given compound. Much of the chemical research during that century was devoted to establishing the formulas of compounds in an effort to understand the tendency of atoms to combine with one another, and many theories were evolved about the nature of the bonds—i.e., the forces—between atoms and thus the nature of combining power or valency. By the end of the century it was recognized that some substances (for example, sodium chloride) were composed of oppositely charged ions with mutual electrostatic attraction. It was only after the development of the Rutherford-Bohr model of the atom, however, that a clear idea of the nature of the forces in nonionic compounds began to develop, initially through the work of G. N. Lewis, an American chemist.

The forces that hold atoms together in molecules, and molecules together in

[1] From Linus Pauling, "The Nature of the Chemical Bond", 3d ed., Cornell University Press, Ithaca, N.Y., 1960. Copyright 1939 and 1940, third edition © 1960, by Cornell University.

larger aggregates, are all electrical in origin. These forces vary widely in strength; some are referred to as chemical bonds or chemical interactions, others as physical interactions. The distinction is, however, somewhat artificial, and, because it is not sharply defined there are some differences in usage. We use the term *chemical interaction,* or *chemical bond,* to refer to an interaction leading to a molecular or ionic species that persists sufficiently long to be detected, at least by instrumental methods. These interactions discussed at length below and in later chapters, include ionic and covalent bonds, hydrogen bonds, and some dipole interactions. Attractive interactions that do not lead to species that persist long enough for detection are termed *physical.*

There are two types of strong chemical bonds, the ionic bond and the covalent bond. Typical energies of interaction—energies required to separate atoms held together by such bonds—are in the range 150 to 400 or more kJ mol^{-1}. A much weaker but very important chemical bond is the hydrogen bond, with an interaction energy of some 5 to 40 kJ mol^{-1}. For comparison, the average kinetic energy of a gas molecule, $\frac{3}{2}RT$, is of the order of 4 kJ mol^{-1} at room temperature.

6-2 Electrons in Atoms

During the last half of the nineteenth century and the early years of this one, extensive studies of the light absorbed and emitted by atoms (Chap. 14) gave detailed information about the energies associated with the electrons in an atom. It was found that atomic electrons can be grouped broadly into two classes—*inner* and *outer* electrons—depending on the energy needed to remove them from an atom. This energy is called the *ionization energy* because removal of an electron from a neutral atom results in the formation of a positively charged ion. Typical ionization energies of outer electrons are many hundreds of kJ mol^{-1} (Table 16-4). Inner electrons have higher ionization energies—they are relatively more difficult to remove.

Within any period of the periodic table (Fig. 2-1), the ionization energy is least for the element in column IA (hydrogen and the alkali metals—lithium, sodium, potassium, and the others) and greatest for the element in column 0 (the noble, or inert, gases—helium, neon, argon, and the others). Elements within a given column of the table have similar chemical reactivities; within any period the noble gas is invariably the least reactive. These observations, and many others, gradually led chemists to a realization that *chemical interaction* involves the *outer electrons* of atoms and to a cardinal generalization that there are certain particularly stable arrangements of these electrons, those of the noble gases, that are found again and again. These arrangements are called *completed shells* or *closed shells.* The interpretation of these closed shells in terms of the electronic arrangement in atoms is discussed in Chap. 16. For our present purposes only the following points are important:

1. The outermost closed shell for each noble gas except He contains eight electrons; that for the He atom has only two electrons because this atom has only two electrons.
2. For atoms of elements in the A columns, the number of electrons in the outermost incomplete shell, beyond the shell of the next lower noble gas, is the same as the number of the column. Thus, the alkali metals (column IA)

have one electron in the outer shell, the elements in column IIA have two electrons in the outer shell, those in column IIIA have three outer electrons, and so on.

3. When atoms combine to form molecules and compounds, each atom tends to surround itself with a completed outer shell of electrons, since this corresponds to an especially stable arrangement. For example, an alkali metal atom loses its single outer electron, forming an ion M^+, and an atom in column IIA loses its two outer electrons, forming M^{2+}. Since atoms of the halogens (column VIIA) have seven outer electrons, they need just one more to form a completed shell of eight and thus can form stable ions of charge -1 by gaining one electron. Atoms can also get completed shells by *sharing* electrons, as well as by gaining or losing them, as we shall illustrate shortly.

6-3 Ionic Bonds

The force of attraction between oppositely charged ions, that is, anions and cations,[2] is commonly referred to as the *ionic bond*. This force is due to the mutual electrostatic (coulomb) attraction of the opposite charges (Appendix A). Consequently the energy ϵ of such a bond varies directly as the product of the charges $\mathbf{Q}_1$ and $\mathbf{Q}_2$ of the ions concerned, and inversely as the distance r between their centers:[3]

$$\epsilon = \frac{\mathbf{Q}_1 \mathbf{Q}_2}{\alpha r} \tag{6-1}$$

Since the distance between an anion and a cation when they are in contact is equal to the sum of their radii, the energy required to break a bond between two small ions is considerably greater than that required to break a bond between two larger ions with charges of the same magnitude. For example, the energy of the bond between Li^+ and F^-, two relatively small ions, is about 570 kJ mol^{-1}, which is about 70 percent larger than the energy of the bond between the significantly larger Cs^+ and I^- ions, about 340 kJ mol^{-1}.

Oppositely charged ions may form ionic crystals, in which the ions are packed together in highly regular three-dimensional arrays that are held together chiefly by ionic bonds. For example, crystals of sodium chloride contain equal numbers of Na^+ and Cl^- ions in a regular pattern (Fig. 6-1) in which each Na^+ has six Cl^- neighbors and each Cl^- has six Na^+ neighbors. While electrostatic forces are attractive between ions of opposite charge, they are repulsive between ions of the same charge. Thus, in the sodium chloride crystal the Na^+ ions repel each other and the Cl^- also repel each other. In the interlaced arrangement of Na^+ ions and Cl^- ions, however, the sum of all attractive forces is larger than that of the repulsive forces, so that the solid is stable.

---

[2] The names *anion* for a *negatively* charged ion and *cation* for a *positively* charged ion are derived from the names of the electrodes to which the respective ions move in an electrolysis cell (Fig. 1-4). Anions are attracted to the positive electrode, the anode, and cations to the negative electrode, the cathode.

[3] The α has to do with units. In the SI system the units for r are meters (m) and those for $\mathbf{Q}_1$ and $\mathbf{Q}_2$ are coulombs (C), which with $\alpha = 1.113 \times 10^{-10}$ C^2 J^{-1} m^{-1} yields joules (J) as energy units (Appendix A).

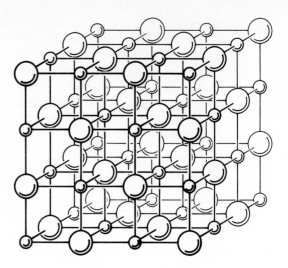

Figure 6-1 The Arrangement of Sodium and Chloride Ions in a Crystal of Sodium Chloride
The larger spheres represent Cl^-, the smaller ones Na^+. This is a skeletal rather than a packing model of the structure (the contrast between these two ways of representing structures is shown in Fig. 1-3). The only purpose of the lines connecting the spheres is to aid in the visualization of geometric relationship, and they must not be mistaken for covalent bonds. The radii of the spheres representing the ions have been made relatively small so that the arrangement of the ions in the structure can be visualized. This arrangement is harder to see in a packing model, such as that in Table 7-1.

In addition to these electrostatic forces that cause long-range attractions and repulsions, there are also forces of repulsion that become noticeable only when the distance between the two ions (or indeed two neutral atoms) becomes less than a critical value (Fig. 6-2). These forces increase rapidly the closer the ions get pushed together. Without this repulsion ionic crystals would collapse.

It is important to note that sodium chloride crystals do not contain individual NaCl molecules, but only Na^+ and Cl^- ions. Similarly, molten sodium chloride or an aqueous solution of sodium chloride contains only Na^+ and Cl^- ions, and no NaCl molecules. Only NaCl vapor, existing at high temperatures, contains NaCl "molecules", a "molecule" actually being a tightly associated combination of an Na^+ and a Cl^- ion, more appropriately called an *ion pair*. It might thus be better to write Na^+Cl^- rather than NaCl, but the second of these forms is preferred for simplicity.

The names and formulas of many common ions are given in Appendix B.

6-4 Covalent Bonds

Uncharged atoms can also attract one another strongly. The chemical bond formed under these circumstances results from the *sharing* of one or more pairs of electrons by the two atoms. The forces involved are electrostatic in origin, but the principle of sharing electrons can be fully understood only with the help of quantum mechanics, and further discussion is deferred until Chap. 18.

A shared pair of electrons constitutes a *covalent bond*. In diagrams depicting the structure of molecules, each shared pair of electrons in a covalent bond is conventionally represented by a line between the two bonded atoms, for example, H—H or H—Cl. There may be two or three such pairs shared between two atoms, giving rise to what is called a *double* or *triple* bond. The *unshared* or *lone* electron pairs on each atom are often of chemical importance as well, because they can partake in chemical reactions to form new covalent bonds with approaching atoms and because they have a strong influence on the geometric arrangement of

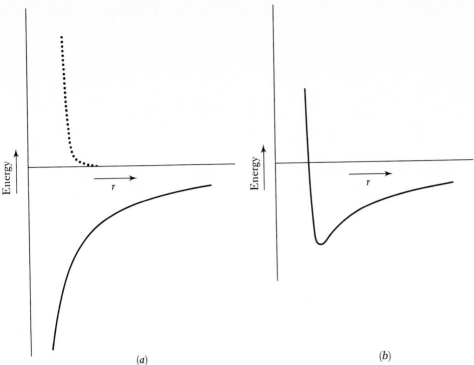

Figure 6-2 Interaction Energies of Oppositely Charged Ions as a Function of the Distance between Their Centers

(a) Electrostatic energy of attraction (solid) and energy of repulsion (dotted); (b) net interaction energy, the sum of the two curves in (a).

The lower curve in (a) is a hyperbola, represented analytically by Equation 6-1. The energy varies as $1/r$ and becomes increasingly negative with decreasing r because the ions have charges of opposite sign. The curve falls as r decreases. Zero energy is associated with an infinite distance between charges. The upper curve in (a) rises as r decreases and becomes extremely steep at short distances.

Curve (b) represents the actual interaction energy of a pair of ions such as Na^+ and Cl^-. The force acting on an ion is in the direction of downward slope, i.e., in the direction of lower potential energy. In the absence of kinetic energy, the minimum of the potential energy curve represents a stable state. At this distance, attraction and repulsion just balance—the force acting on the ion is zero. This separation corresponds to the average distance (about which the ions would oscillate) in an ion pair, Na^+Cl^-. This is about 10 percent less than the average distance in an NaCl crystal because of the modifying effects of the other surrounding ions. The usual estimates of ionic radii are derived from measurements on ionic crystals, because these are the stable forms of ionic substances at ordinary temperatures.

molecules. It is customary to represent unshared outer electrons by dots,[4] as in the formulas

$$H-\ddot{\underset{\displaystyle H}{O}}: \qquad H-\ddot{\ddot{F}}: \qquad H-\ddot{\underset{\displaystyle H}{N}}-H \qquad :\ddot{O}=C=\ddot{O}: \qquad H-C\equiv N: \qquad (6\text{-}2)$$

The bonds that hold together the atoms in polyatomic ions are predominantly

---

[4] Some chemists also use pairs of dots to represent covalent bonds, as in $H:\ddot{\ddot{F}}:$ or $H:H$.

covalent also—for example, the bonds between the nitrogen and hydrogen atoms in the ammonium ion, NH_4^+, and those beween the sulfur and oxygen atoms in the sulfate ion, SO_4^{2-}:

$$
\begin{bmatrix} & H & \\ & | & \\ H & -N- & H \\ & | & \\ & H & \end{bmatrix}^+ \quad
\begin{bmatrix} & :\overset{..}{\underset{}{O}}: & \\ :\overset{..}{\underset{..}{O}} & -S- & \overset{..}{\underset{..}{O}}: \\ & :\overset{..}{\underset{..}{O}}: & \end{bmatrix}^{2-}
\qquad (6\text{-}3)
$$

Formulas of the kind illustrated in (6-2) and (6-3) are called *electron-dot* or *Lewis* formulas and are explained more fully in Chap. 17.

Ionic and covalent bonds represent the two simplest and most important kinds of chemical forces. A number of contrasts between them are worthy of emphasis. Because the ionic bond is the result of simple electrostatic attraction between charged particles, it is nonspecific; that is, it does not depend on the chemical nature of the ions involved but only on their charges and the distance between them. There is no limit to the number of such bonds a given ion can form, provided only that there is room around it for all the ions of opposite charge.

On the other hand, covalent bonds are highly specific. They vary considerably in strength and in other respects with the detailed nature of the atoms involved and even to some extent with the nature of the other bonds formed by these atoms. Furthermore, covalent bonds are highly directional; if two atoms are bonded covalently to a third atom, the angle between them depends strongly, in a quite predictable way, on the electronic structure of the atom to which they are bonded. In contrast, since the electric field around a spherical ion is spherically symmetrical, ionic bonds are not directional. Finally, the number of covalent bonds that can be formed by a given atom is limited. For example, hydrogen normally forms only one covalent bond and carbon normally forms four.

Pure ionic and pure covalent bonds represent ideal extremes. Most ionic bonds (e.g., the Na^+Cl^- bonds in sodium chloride crystals) have some covalent character and most covalent bonds (e.g., the H-O bonds in H_2O) have some ionic character.

6-5   Van der Waals Forces

Weak attractive forces exist between uncharged atoms or molecules even in the absence of any tendency for formation of covalent bonds. Perhaps the most obvious evidence is the fact that all gases condense to liquids and eventually solidify at sufficiently low temperature and sufficiently high pressure. Accurate measurements show that even in the gaseous state all substances behave in a way that indicates the presence of weak intermolecular forces. These weak forces have long been known collectively as *van der Waals forces* because the Dutch physicist J. D. van der Waals was the first to obtain an estimate of their magnitude in gases by modification of the ideal gas law.

One type of van der Waals force arises when the center of the positive charges of a molecule does not coincide with the center of the negative charges. Such molecules are said to have electric dipole moments (Fig. 6-3). The forces between molecules with permanent dipoles (*polar* molecules) depend strongly on the relative orientations and positions of these molecules (Fig. 6-4) and are of very short range compared to the range of electrostatic forces between ions.

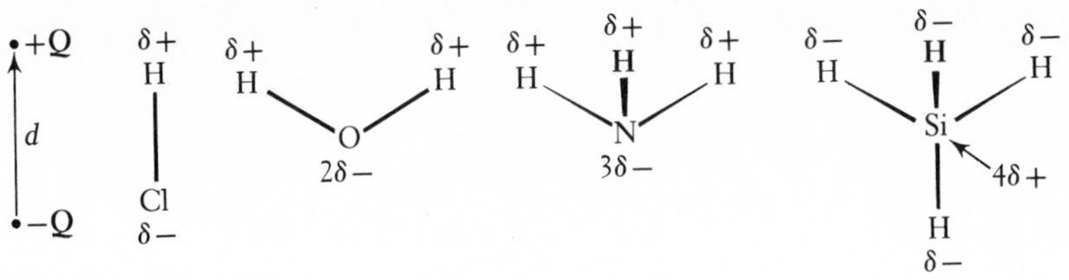

$\delta+$ $+Q$ $\delta+$ $\delta+$ $\delta+$ $\delta+$ $\delta+$ $\delta+$ $\delta-$ $\delta-$ $\delta-$

Figure 6-3 Dipoles

A simple dipole (left) and several molecular examples are shown. The symbols $\delta+$ and $\delta-$ indicate positive and negative charges associated with the atoms in question. They are equal to each other in magnitude for any given molecule here, but they may differ from molecule to molecule. In HCl the separated positive and negative charges are associated with the two atoms. In H_2O the center of the positive charges is halfway between the two H atoms. In NH_3 the H atoms form the base of a relatively flat pyramid with N at the top (shown here inverted); the center of the positive charges is at the center of the base. In SiH_4 (silane) the H atoms surround the Si tetrahedrally and the center of the positive charges coincides with that of the negative charges. The molecule of silane is therefore nonpolar, even though each bond in the molecule is a dipole. This situation is not uncommon. In all the other molecules shown, the directions of the dipoles are the same as in the dipole on the far left consisting of the two point charges $+Q$ and $-Q$ separated by the distance d. The dipole moment μ is the product of Q and d.

Even individual atoms and molecules with no permanent dipoles or other permanent charge distributions interact with other atoms and molecules, quite apart from covalent bond formation. These interactions occur chiefly because other atoms and molecules, when close enough, can deform the electronic charge distribution of the given atom or molecule, *inducing* dipole moments in them (Fig. 6-5). The attractive force between a pair of such induced dipoles is sometimes called the *dispersion force* or the *London force,* after F. London, the man who first explained it quantitatively. London forces increase with the number of electrons in the atoms or molecules concerned and with the ease with which an electric field can deform the electronic charge distribution. The composite of these two qualities is termed *polarizability*.[5]

---

[5]In the simplest case the polarizability of an atom or molecule is a proportionality constant that relates the dipole moment induced in an atom or molecule by an electric field to the electric field strength (see Study Guide). The name *dispersion force* has its origin in the circumstance that polarizability is the basis not only of dispersion forces but also of the refractive index of transparent substances, which in turn governs the separation of white light into its colored spectral components by prisms. The last phenomenon was called *dispersion* of light by Isaac Newton.

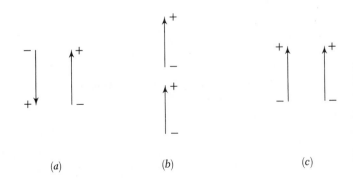

(a)   (b)   (c)

Figure 6-4 Dipoles with Different Relative Orientations
In (a) the positive end of one dipole is close to the negative end of the other while like-charged ends are further removed from each other. The overall result is attraction between the dipoles. A similar situation holds for (b). In (c) charges of equal sign are close to each other and charges of opposite sign are farther apart. The two dipoles repel each other.

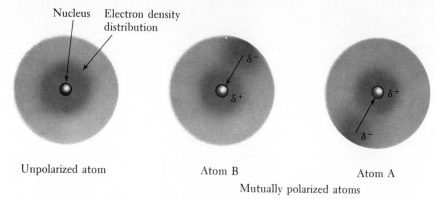

Nucleus   Electron density
           distribution

Unpolarized atom   Atom B   Atom A

Mutually polarized atoms

Figure 6-5   London Dispersion Forces (Highly Schematic Drawing)
On the left is an unpolarized atom with a spherically symmetric electronic charge distribution. On the right, one of the two atoms, such as atom A, has become polarized through a momentary fluctuation that causes the electronic-charge distribution to deviate from spherical symmetry, creating a momentary dipole, symbolized by the arrow. This dipole affects the charge distribution of atom B and induces a dipole in atom B in such a direction that the two dipoles attract each other. This charge distribution is rapidly followed by a sequence of similar distributions with dipoles that are always oriented so as to attract each other. The net result is weak attraction, the London dispersion force. (In these schematic drawings, the relative size of the nucleus has been greatly exaggerated.)

Since the boiling point of a liquid is a rough measure of the attractive forces between the molecules it contains, the magnitudes of the London forces between molecules without permanent dipole moments can be assessed roughly by comparison of their normal boiling points (Fig. 6-6). London forces are a general property of matter because of its electronic makeup. The force of attraction is effective only for very short distances and is large between atoms containing many electrons. The mutual attraction of molecules or parts of molecules is

Figure 6-6   Boiling Points and Molecular Weights of Some Nonpolar Molecules
Only an approximate correlation of boiling points and molecular weights is to be expected, since the intermolecular forces depend strongly on the distances between pairs of atoms in different molecules and on the ease with which the electronic structure of these atoms can be deformed (the polarizability of the atoms). The polarizability of fluorine and its compounds is relatively low because fluorine atoms bind their electrons particularly tightly, and thus nonpolar fluorine compounds have lower boiling points and higher volatility than other substances of similar molecular weight. A plot of boiling point against *numbers of electrons* is very similar to the curve shown here, since the number of electrons present increases regularly with increasing molecular weight.

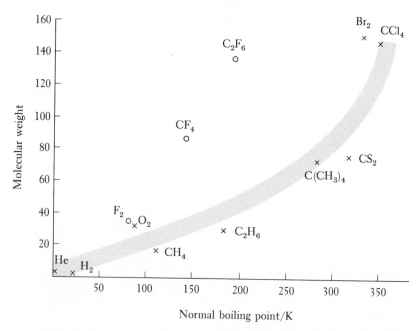

strongly shape-dependent because the number of close atomic approaches between the molecules considered depends on their shapes.

At sufficiently short distances all atoms repel one another. Without this
repulsion, solid and liquid matter as we know it would collapse to a state of much
higher (or infinite) density. The balance between attractive and repulsive forces
implies that a graph of the energy of interaction of a pair of atoms as a function of
their distance apart shows a minimum. Figure 6-7 shows a comparison among the
energies characteristic of typical covalent, ionic, and London interactions. One
important point is that the repulsive portions of these curves are extremely steep.
This means that when atoms or ions approach one another more closely than a
minimum distance, they behave as though they were almost incompressible, since
a small decrease in distance leads to a great increase in energy. The position of

Figure 6-7 Interaction Energies between Atoms

Typical energy-versus-distance curves are shown for three situations, a covalent bond (H—H), an
ionic bond (Na⁺Cl⁻) (compare Fig. 6-2), and a van der Waals interaction (indicated by Ar---Ar).
At large separations all curves rise slowly with *increasing* distance, signifying attraction. At short
distances they rise sharply with *decreasing* distance, corresponding to strong repulsion. All curves
exhibit a minimum, representing the position of balance between attraction and repulsion. The
curves for covalent and ionic interactions are similar to each other except that the ionic interaction extends appreciably further than the covalent interaction. The minimum of the van der Waals
curve is very shallow and can be clearly seen only in the inset on the upper right, in which
the ordinate has been magnified by 100. The minimum of the van der Waals curve is usually
too shallow to permit the formation of molecules. Furthermore, it occurs at a larger distance than
do the minima for covalent or ionic interactions for atoms of similar atomic number. The strongly
repulsive portions of all three curves are similar because basically they have the same origin, the
mutual interpenetration of the electron clouds of the two atoms concerned.

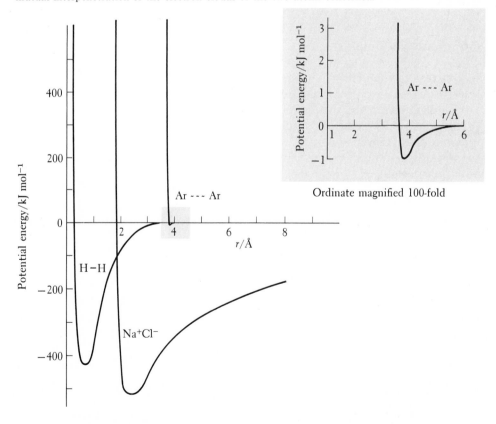

Ordinate magnified 100-fold

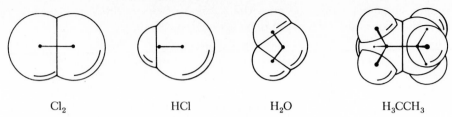

| Cl_2 | HCl | H_2O | H_3CCH_3 |

Figure 6-8 Van der Waals Models of Simple Molecules
Centers of atoms are indicated by dots. The spheres approximately represent the space taken up by the atoms, in the sense that there is strong repulsion when nonbonded atoms approach each other so closely that the spheres begin to interpenetrate. A van der Waals representation of molecules is also shown in Fig. 1-3c.

the minimum in the energy depends on the size of the atom and on whether the interaction is of the covalent, ionic, or van der Waals type.

In considering the van der Waals interaction of two atoms that are not joined by a chemical bond, it is a good approximation to regard the atoms as nearly hard and slightly sticky spheres, with characteristic radii called the van der Waals radii (Fig. 6-8). For example, molecules in crystals pack in such a way that atoms of different molecules rarely are closer to each other than the sum of their van der Waals radii. One speaks of a *van der Waals contact* when the distance between two nonbonded atoms is approximately equal to the sum of their van der Waals radii. These radii are in the range 1 to 2 Å. Typical values are H, 1.1 Å; O, 1.4 Å; N, 1.5 Å; C, 1.6 Å; Cl, 1.8 Å; Br, 2.0 Å; I, 2.2 Å. Characteristic radii for covalent bonding and ionic bonding can also be assigned to different atoms (Chap. 17).

6-6 Hydrogen Bonds

In contrast to the nonspecific van der Waals forces, which depend only on the relative positions and polarizabilities of the atoms involved but not otherwise on their individual nature, the *hydrogen bond* is a specific kind of attractive chemical interaction. Its existence was first postulated in 1920 by M. L. Huggins (in an undergraduate thesis) and by W. M. Latimer and W. H. Rodebush to explain the surprisingly high boiling point of water, a molecule with only 10 electrons and a molecular weight of 18 (Fig. 6-9). Evidently there are strong attractive forces between water molecules. Many other properties of water are also unusual and can be explained by the existence of such forces, as can similar unusual properties of some other substances, all of which contain hydrogen atoms. Studies of these substances, by a variety of methods, in the solid, liquid, and gaseous states and in solution have provided detailed information about the unique interaction called the hydrogen bond.

A hydrogen bond involves an arrangement of three atoms X—H---Y (Fig. 6-10), with the symbol --- representing this bond. Other atoms are usually bonded to X and Y, which may be identical and must exhibit a strong tendency to attract electrons.[6] Atom Y must have an unshared pair of electrons. The hydrogen atom is covalently bonded to atom X and points in the general direction of atom Y. As a result of its attraction for electrons, atom X pulls the shared pair in

---

[6] Such atoms are said to be strongly electronegative. See Chap. 16.

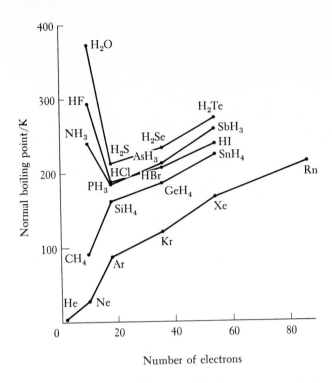

Figure 6-9 Boiling Points of the Noble Gases and of Some Compounds of Hydrogen

The curves for the noble gases and the compounds of the formula XH_4 are similar, the compounds of formula XH_4 boiling some 80 to 100 K higher than the noble gases with the same total number of electrons. The situation for the compounds of formula XH_3, H_2X, and HX is similar, with the exception of NH_3, H_2O, and HF, each of which has an abnormally high boiling point (particularly H_2O), signifying strong intermolecular forces.

the X-H bond toward itself, leaving the hydrogen atom with a residual positive charge and relatively unshielded by electrons. Atom Y is endowed with a residual negative charge because of its strong tendency to attract electrons. The electrostatic attraction between the positive hydrogen atom and the negative atom Y, particularly its unshared electron pair, is what constitutes the hydrogen bond. There may also be a *small* contribution of covalent bonding between atoms H and Y. Atom X is referred to as the donor atom since it donates the hydrogen atom for the hydrogen bond; atom Y is called the acceptor atom.

The most effective donors and acceptors for hydrogen bonds are the atoms F, O, and N, which may be covalently bonded to still other atoms to form molecules or ions (for example, CH_3OH, OH^-, OH_3^+, NH_4^+, NH_2^-, SO_4^{2-}). An outright negative charge enhances the power of an atom to act as an acceptor, which not only makes F^- an excellent acceptor but also permits Cl^- and Br^- to act in the same capacity. A few other atoms, for example, S, may also be involved in weak hydrogen bonds.

The energy of hydrogen bonds is between 5 and 40 kJ mol^{-1}, or about 5 to 10

Figure 6-10 The Hydrogen Bond

The hydrogen atom attached to atom X in (*a*) points to and is closer to atom Y than would correspond to a normal van der Waals contact. Atoms X and Y may be identical but need not be. A specific example of a hydrogen bond is shown in (*b*), where the oxygen atom of the water molecule is hydrogen bonded to the nitrogen atom of the ammonia molecule, which has an unshared electron pair (see also Fig. 6-11).

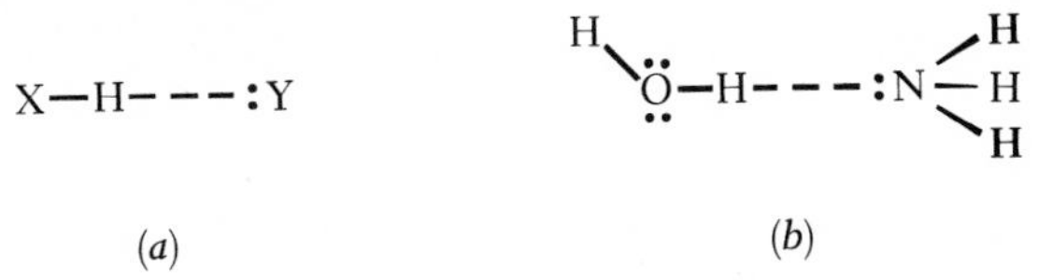

$$X—H---:Y$$

(*a*)

(*b*)

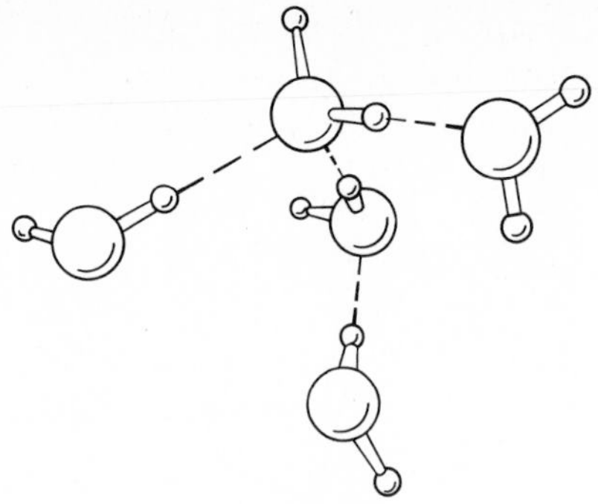

Figure 6-11 Hydrogen-Bonded Water Molecules in Liquid Water or Ice
Each oxygen atom is covalently bonded to two hydrogen atoms at about 1 Å to form a water molecule. The two hydrogen atoms are in general directed toward and are close to the oxygen atoms of two neighboring molecules (at about 1.8 Å), forming hydrogen bonds. In ice, two additional hydrogen bonds are formed to a given oxygen atom by hydrogen atoms of neighboring water molecules, so that each water molecule is surrounded tetrahedrally by four others, to each of which it is tied by a hydrogen bond. When ice melts, some of the hydrogen bonds are broken, more or less randomly, and the average number of hydrogen bonds formed by each water molecule decreases somewhat.

percent of that of a typical covalent bond. The H- - -Y distance is usually about 0.5 to 1 Å smaller than the sum of the van der Waals radii of H and Y, attesting to the strength of the interaction.

Figure 6-11 is a schematic representation of hydrogen bonding in water and ice. As shown, the hydrogen atoms lie closer to the oxygen atoms to which they are covalently bonded. This is the usual situation, but there are a few examples of very strong hydrogen bonds between fluorine atoms, and between oxygen atoms, in which the hydrogen atom is halfway between the other two atoms. The roles of donor and acceptor atoms then become indistinguishable.

Hydrogen bonds are of great importance in every area of chemistry. For example, they play a dominant role in determining the physical behavior of ice and liquid water, and they exert a major influence in the detailed three-dimensional architecture of protein molecules and of the strands of DNA, deoxyribonucleic acid. Hence the details of many of the most essential processes occurring in organisms are controlled by hydrogen bonds.

The preceding discussion of van der Waals forces and hydrogen bonds has been presented in terms of interactions between different molecules (*inter*molecular interactions). It applies equally well to interactions between different parts of the same molecule when these parts are in appropriate relative positions. *Intra*molecular hydrogen bonds and van der Waals forces are particularly significant in the chemistry of large molecules, and the important interactions in proteins and DNA just mentioned are almost entirely intramolecular.

Summary

The forces that hold together atoms or groups of atoms in molecules and compounds are called chemical bonds. These bonds involve the outer electrons of atoms, those electrons for which the ionization energy is relatively small. The outer-electron arrangements of the noble gases are called completed, or closed, shells. Each closed shell consists of eight electrons (with the exception of that for helium, which has only two). When atoms combine to form molecules and compounds, each atom tends to surround itself with a completed outer shell.

The force of attraction between oppositely charged ions is commonly referred to as the ionic bond. Oppositely charged ions may form ionic crystals in which the ions are packed together in highly regular three-dimensional arrays held together chiefly by ionic bonds. Such crystals do not contain individual molecules.

A shared pair of electrons constitutes a covalent bond. Two bonded atoms may share two or three electron pairs giving rise to a double or triple bond. Unshared, or lone, pairs are of chemical importance as well because they can partake in chemical reactions to form new bonds. It is customary in structural formulas to represent covalent bonds by lines and unshared outer electrons by dots.

Weak attractive forces exist between uncharged atoms or molecules even in the absence of covalent bonds. The boiling point of a liquid is a rough measure of the attractive forces between the molecules it contains. Van der Waals forces arise from the interactions between molecular dipoles. In polar molecules, the dipoles arise from a permanent separation of charge; dipoles in nonpolar molecules arise from fluctuations in the distribution of charge. The hydrogen bond is a specific interaction between certain bonded hydrogen atoms and atoms such as F, O, and N.

Terms and Concepts

Problems and Questions

6-1 Types of Bonding Distinguish between ionic and covalent bonds, and give examples of compounds held together by each type of bond.

6-2 Polar Molecules What is meant by the term *polar molecule?* Give an example and show how it fits your definition.

6-3 Intermolecular Forces Coulomb interactions are often referred to as *long-range* forces, and dispersion interactions are termed *short-range* forces. The energies of the former interactions vary as $1/r$, those of the latter as $1/r^6$, with r the distance between the centers of the atoms. Compare the falloff with distance of these two types of interactions by computing for each the ratio of the energy of attraction at 2, 3, 4, and 5 molecular diameters to that at 1 molecular diameter.

6-4 Forces between Nonpolar Molecules What evidence is there that forces exist between *non-polar* molecules? On what do these forces depend? Why are they so sensitive to molecular shape?

6-5 Coulomb Interactions Assume that the ions Li^+ and F^- behave like point charges. Calculate (a) the coulomb energy between them at a distance of 2.0 Å and (b) the energy for 1 mol of Li^+, F^- pairs. (Charge on the electron $= -1.6 \times 10^{-19}$ C.)

6-6 Van der Waals Radius At 20°C and a pressure of 1 atm, 1 mol of argon gas occupies a volume of 24 liter. Estimate the van der Waals radius for an argon atom from the onset of the repulsive part of the potential energy curve in Fig. 6-7 and calculate the fraction of the gas volume that consists of argon atoms.

6-7 Comparison of Physical Properties In each of the following pairs of substances, choose the one that you would expect to have the *higher* value of the property mentioned. Explain briefly why this should be so.

(a) Solubility in water: $C_2H_4(OH)_2$; C_4H_{10}
(b) Melting point: SiH_4; LiF
(c) Boiling point: CCl_4; CBr_4

6-8 Boiling Points of Noble Gases Give an explanation for the observed order of increase of the boiling points of the noble gases. Does this interpretation explain also the variation of boiling points of the hydrogen halides (HF, HCl, HBr, and HI)? Explain why or why not.

6-9 Properties of Substances In each of the following pairs of substances, indicate which should have the *lower* value of the property mentioned and explain in a few words the reasons for your choice.

(*a*) Vapor pressure at $-100°C$: F_2, Cl_2; Ar, Xe; CCl_4, $GeCl_4$

(*b*) Normal boiling point: CH_3CH_2OH, CH_3Cl; NH_3, PH_3

6-10 Hydrogen Bonding Would you expect the hydrogen-bonding tendencies of deuterium fluoride to be appreciably greater than, about the same as, or appreciably less than those of hydrogen fluoride? Explain.

6-11 Hydrogen Bonds Ammonia is both a donor and an acceptor of hydrogen bonds. Indicate by a sketch an interaction of two ammonia molecules that illustrates this fact. Make it clear what is implied by the term *hydrogen bond*.

A Classification of Chemical Substances

"A remarkable feature of the material world is the variety of substances with which we are familiar in everyday life. . . . The reason for this extraordinary variety of materials lies, then, not so much in the number of different elements available but in the fact that atoms can link up to form arrangements ranging in complexity from simple molecules consisting of only two atoms to groupings so complicated that even today we have only the most rudimentary ideas of their structures."
A. F. WELLS, 1956[1]

This chapter describes a classification of chemical substances that is based on the nature of the interactions between the atomic or molecular units of which the substance is composed. Brief descriptions of nonstoichiometric compounds, clathrates, and liquid crystals are also presented.

7-1 Elements and Compounds Composed of Molecules

Many elements are composed of isolated molecules that contain one or more atoms. Examples include the monatomic noble gases (He, Ne, Ar, Kr, Xe, and Rn); various diatomic species such as H_2, O_2, and the halogens (F_2, Cl_2, Br_2, I_2, and At_2); and polyatomic molecules such as P_4 and S_8. Similarly, many compounds consist of molecules that are all the same kind, the overall composition of the compound then being the same as that of one of its molecules.[2]

We can, somewhat arbitrarily, divide this category into *volatile* and *nonvolatile* substances according to the magnitudes of their vapor pressures under ordinary conditions. There is no sharp boundary between these groups; a normal vapor pressure of 0.1 torr ($\sim 10^{-4}$ atm) or greater might be taken as a criterion of volatility, but the exact value chosen is unimportant. Typical volatile substances include sulfur dioxide (SO_2), ethyl alcohol (C_2H_5OH), and iodine, which are respectively gaseous, liquid, and solid under normal conditions. Figure 1-3 shows the crystal structure of another such compound (tetracyanoethylene). Typical nonvolatile molecular substances are sucrose (ordinary sugar, $C_{12}H_{22}O_{11}$) and substances with high molecular weight (e.g., polyethylene, C_nH_{2n} with n in the thousands), as well as more complicated molecules such as proteins.

---

[1] Quoted from A. F. Wells, "The Third Dimension in Chemistry", Oxford University Press, New York, 1956.

[2] Or better, of one of its *average* molecules, taking the existence of isotopes of the elements into account. On a weight basis the elemental composition of such a compound therefore varies slightly if the isotopic composition of the elements is changed. The composition is exactly constant when considered on the basis of *numbers* of atoms of different elements involved.

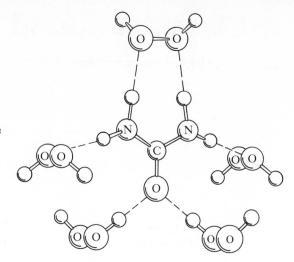

Figure 7-1 A Portion of the Structure of a Crystal of $(NH_2)_2CO \cdot H_2O_2$
The urea [$(NH_2)_2CO$] molecule in the center is involved in six hydrogen bonds to five hydrogen peroxide molecules, four of which are shown almost end-on (the small unlabeled spheres represent hydrogen atoms). Four of the six hydrogen bonds are of the type N—H---O and two are of the type O—H---O. All other urea molecules in the crystal have the same environment. Each hydrogen peroxide molecule is in turn involved in six hydrogen bonds of the same types to five different urea molecules (only one of which is shown here for each H_2O_2 molecule).

In the solid state, substances composed of molecules are often soft; in the liquid or solid form they are usually transparent or translucent and often colorless. They show no metallic luster and do not conduct electricity. In both the solid and the liquid states, the molecules are held together by van der Waals forces and hydrogen bonds. The degree of volatility reflects the energy characteristic of the interactions among neighboring molecules. If there are many weak van der Waals attractions per molecule (as in polyethylene) or several strong hydrogen bonds (as in sugar), the energy required to remove a molecule from the solid or liquid is substantially greater than the thermal energies available under ordinary conditions and the substance is nonvolatile. On the other hand, if there are comparatively few and weak intermolecular forces, the molecules can easily break loose and the substance is volatile. Since there are no ions present and the molecules do not transfer electrons to one another readily, no charge carriers are present; consequently these substances do not conduct electricity.

Crystals that contain only discrete molecules are called *molecular crystals*. The molecules in the crystal need not all be identical; sometimes two or more kinds of molecules may pack together in an ordered and energetically favorable array. For example, urea, $(NH_2)_2CO$, and hydrogen peroxide, H_2O_2, crystallize together to form a stable crystal containing a 1:1 ratio of the components, $(NH_2)_2CO \cdot H_2O_2$ (Fig. 7-1). Frequently these *molecular compounds* exist only in the crystalline state. This may be a consequence of some accidentally favorable mode of packing—the shapes may be complementary so that mutually attractive portions of different molecules fit together readily. Hydrogen bonds are frequently involved. If the mutual attraction of the different components is strong enough, the molecular compound may persist in the liquid state, in solution, and even, rarely, in the gaseous state.

7-2 Covalent Solids

In many crystalline substances, the atoms are linked by covalent bonds into one-dimensional, two-dimensional, and three-dimensional arrays, which are often referred to as chains, networks, and frameworks, respectively. Each of these extends, at least ideally, throughout the structure in one, two, or three dimensions. Because these arrays do extend throughout the crystal, and thus normally involve millions of atoms, they are often said to be infinite in extent, since the only limitation on their size is that crystals are finite in size.

What is the minimum number of neighboring atoms to which each atom must be bonded in the different kinds of arrays? If atoms of a particular kind form no bonds, the resulting crystals necessarily contain single atoms and thus are simple molecular crystals. If an atom can have only one bonded neighbor, it necessarily forms diatomic molecules, which also yield molecular crystals (Fig. 7-2a). Hydrogen and the halogens fall in this class. With two bonded neighbors per atom, there are two possible alternatives: an infinite chain or a ring (Fig. 7-2b). Sulfur and the elements below it in the periodic table belong, in their elemental state, in this category and exhibit both possibilities. Networks or layer structures become possible when each atom is bonded to three neighbors, but isolated molecules can also exist (e.g., the tetrahedron of Fig. 7-2c). Both possibilities illustrated in Fig.

Figure 7-2 Some Different Kinds of Zero-, One-, Two-, and Three-Dimensional Arrays

In (a), each atom forms bonds to only one neighbor; a molecular crystal containing diatomic molecules results (for example, I_2). In (b), each atom forms bonds to two neighbors; infinite chains or rings are produced (for example, S_n or Se_n, S_8 or Se_8, S_6). In (c), each atom is bonded to three neighbors: a two-dimensional network or layer structure results (for example, As_n) or a tetrahedron (P_4 or As_4). In (d), each atom is bonded to four neighbors, and a three-dimensional framework structure is formed [for example, C(diamond), ZnS, SiC, or BN]. In (e), another type of chain structure is shown, with some of the atoms in the chain forming more than two bonds but without covalent cross-links between different chains [for example, $(SiO_3)_n^{2n-}$, $(SO_3)_n$, $(CH_2O)_n$].

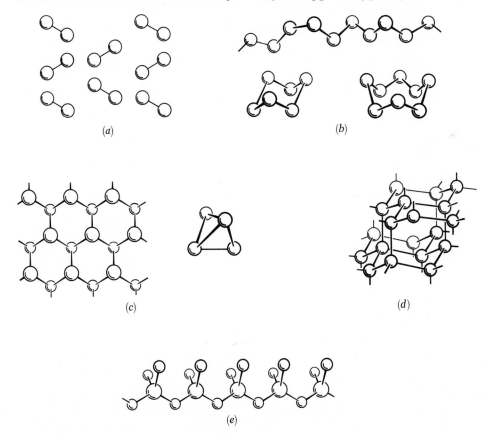

7-2c occur for arsenic and phosphorus. Their vapors and crystalline white phosphorus contain tetrahedral molecules, while crystals of arsenic, black phosphorus, and the elements below them in the periodic table consist of puckered layers. The layer is puckered rather than planar because there is an unshared electron pair on each atom; this effect is discussed further in Chap. 18.

When all atoms can form bonds to four neighbors, a three-dimensional (framework) structure is the usual result (Fig. 7-2d). If all atoms are assumed to be identical, Fig. 7-2d depicts the structure of diamond, a form of carbon; if it is assumed that there are equal numbers of two different kinds of atoms (Fig. 7-6), Fig. 7-2d represents the structure of a great variety of substances, including the abrasive carborundum (SiC) and the common mineral sphalerite (a form of ZnS).

The examples given in Fig. 7-2b to d involve no bonds other than those required to link the one-, two-, and three-dimensional arrays together. Figure 7-2e illustrates a chain with some additional atoms joined to the atoms composing the chain, but these additional atoms form no further covalent bonds, so that the chains are not linked together covalently to form a layer or a three-dimensional structure. Silicate chains of this type exist with composition $(SiO_3)_n^{2n-}$, with n a large integer. They occur, for example, in the mineral diopside, $CaMg(SiO_3)_2$, in which the chains are held together in the crystal by electrostatic attraction to the Ca^{2+} and Mg^{2+} ions interspersed in the structure. One form of sulfur trioxide, $(SO_3)_n$, also exists as chains like those depicted in Fig. 7-2e; these chains are held together in crystals by van der Waals forces.

An example of a crystal structure built up from separate chains weakly linked to one another ($CuCl_2$) is shown in Fig. 7-3. Typical layer structures are depicted in Figs. 7-4 and 7-5. Figure 7-4 illustrates individual layers and their mode of stacking in the structures of one form of boron nitride, BN, and of graphite, a form of carbon; Fig. 7-5 shows a single layer of the $CdCl_2$ structure. It is instructive to compare the arrangement of the atoms in the $CdCl_2$ layer with that in Fig. 7-2c. The only difference between them is that in the $CdCl_2$ structure, which is common to many salts with the general formula MX_2, an additional atom has been added below the center of each of the six-membered rings in the layers of Fig. 7-2c. This kind of relationship between two structures—interstices in one being filled systematically to form the other—is a common one, especially in ionic crystals. Figure 7-6 shows the three-dimensional framework of the sphalerite structure.

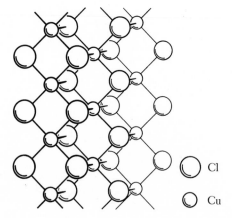

Figure 7-3 The Structure of Crystals of Copper Chloride, CuCl$_2$

Chains containing copper atoms surrounded by four chlorine atoms at the corners of an approximate square constitute the primary structural feature. Each chlorine atom is joined to two copper atoms at an angle of nearly 90°. The chains run vertically in the diagram. They are tied together by weaker bonds between copper and chlorine atoms in neighboring chains, in a direction nearly perpendicular to the page. Each copper atom is thus surrounded by an octahedron (Fig. 7-7) of chlorine atoms, two of which, on opposite sides of the copper atom, are about 25 percent more distant and much more weakly bonded than the others. The illustration shows this bonding for the chain in the middle only. The structure of CuCl$_2$ illustrates the difficulties of rigid classification; while it is primarily a chain structure, it definitely shows aspects of a three-dimensional framework as well.

Cl

Cu

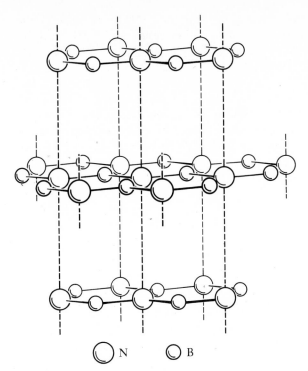

N ⚪   B ⚪

Figure 7-4   The Crystal Structures of Boron Nitride (BN) and Graphite

Crystals of the common form of BN consist of layers in which boron and nitrogen atoms are covalently linked into a planar arrangement of regular hexagons, the corners of which are alternately occupied by boron and nitrogen atoms. The layers are held together in a parallel array by van der Waals forces. A common type of graphite, a form of carbon, has this same structure, the corners of the hexagons being occupied by covalently linked carbon atoms. In other specimens of graphite the stacking of the layers is less regular than that shown here; the structures of these graphites are thus termed partially disordered. The structures of other forms of BN and carbon are illustrated in Figs. 7-2d and 7-6.

In all these crystals the chains or the layers may be considered to be molecules, and the entire crystal may be so considered in the case of a three-dimensional network. However, these molecules are infinitely large in the sense that they are of limited size only because of limitations in material available. It is better to restrict the term *molecule* to particles of a *finite size,* a size *that is always the same* no matter how much material is available, as is the case for the molecules H_2O and $C_{12}H_{22}O_{11}$ and even for giant protein molecules.

Crystals containing the kinds of arrangements of atoms discussed in this section are referred to as *covalent crystals* because predominantly covalent forces are operative throughout the crystal, in one dimension for chains, in two for layers, and in three for frameworks. Such compounds are characterized by high

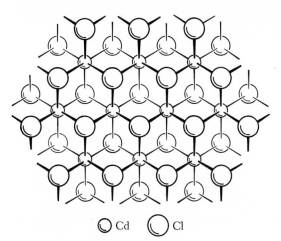

⚪ Cd   ⚪ Cl

Figure 7-5   The Structure of a Layer in a Crystal of Cadmium Chloride, $CdCl_2$

Cadmium chloride crystals are composed of layers held to each other by van der Waals forces. One of these layers is represented here. Each cadmium atom in the central plane of the layer is bonded to six chlorine atoms, three above the central plane and three below it. The chlorine atoms are at the six corners of an octahedron (Fig. 7-7) with the cadmium atom at the center. Each chlorine atom is bonded to three cadmium atoms. The layers are stacked parallel to one another to form the crystal and are held together by van der Waals forces. Other substances exhibiting the same crystal structure include NiI_2, $FeCl_2$, $CoCl_2$, $NiCl_2$, $ZnBr_2$, $MnCl_2$, and $MgCl_2$.

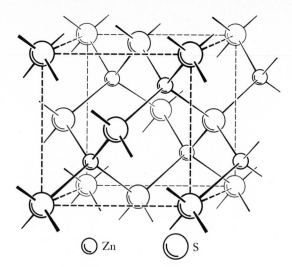

Figure 7-6   The Crystal Structure of Sphalerite, ZnS

Sphalerite is a mineral of composition ZnS. Each zinc atom is surrounded tetrahedrally by four sulfur atoms, and conversely, so that a
three-dimensional framework is created. A number of other compounds, among them BeS, BeSe, BeTe, ZnSe, CdTe, HgS, HgSe, HgTe,
CuCl, CuBr, CuI, and SiC, have the same structure. By application of
very high pressure and temperature, boron nitride can be transformed
from the structure shown in Fig. 7-4 to the present structure. Diamond, a form of carbon, has a similar structure except that all atoms
are carbon atoms. Graphite (Fig. 7-4) can be transformed into diamond
by applying very high pressure and temperature.

melting and boiling points, and they frequently decompose before they melt or
sublime. Electrical conductivity is normally extremely low in the solid state, and
solubility in water is also very low unless there is chemical reaction with water.
For example, the infinite chains in $(SO_3)_n$ react with water, the chains breaking
apart to form individual molecules of sulfuric acid, H_2SO_4.

Covalent crystals are usually translucent and show no metallic luster. Those
based on a three-dimensional framework are often very hard and brittle; diamond
and a second form of boron nitride (borazon), with the structure shown in Fig.
7-6, are the hardest substances known. Examination of the structure suggests the
reason: to fracture the crystal, it is necessary to break many strong covalent
bonds no matter in what direction one applies a force. Almost as hard is silicon
carbide (SiC, carborundum), which exists in many forms with closely related
structures, one of which is that of Fig. 7-6.

Crystals based on two-dimensional networks tend to cleave in layers (e.g.,
graphite, $CdCl_2$, mica). Crystals composed of chainlike structures cleave in strings
(e.g., asbestos, a silicate that contains chains like those of Fig. 7-2e, joined
together in pairs).

Covalent solids need not be crystalline; often they are amorphous (Sec. 1-3).
Most natural and synthetic fibers contain atoms covalently linked in chains with
only weak forces between the chains (nylon, cellulose, DNA). In amorphous
silica, SiO_2, each Si atom is tetrahedrally surrounded by four O atoms and each O
links two Si together, which accounts for the composition. The substance is not
crystalline, however, because there is no long-range order. It is, in fact, a typical
glass.

7-3   High Polymers

An important class of covalent compounds is that of high polymers. These consist
of extremely large molecules, with molecular weights of 10^4 to 10^5 or even higher.
They are made by combination of many small molecules, called *monomers*. The
monomer molecules used to make a given polymer may all be identical or may be

of two kinds that will suitably react with one another. Most polymers are mixtures of molecules that vary chiefly in size. Consequently in many ways polymers behave like pure compounds, but they seldom have sharp melting points. The common name *plastics* is often applied to many of these substances. Typical synthetic polymers include synthetic rubber, nylon, polystyrene, polyethylene (polythene), polytetrafluoroethylene (Teflon), and polymethylmethacrylate (Lucite or Plexiglas) (Chap. 29).

Even elements may exist in polymeric form. An example is lambda sulfur (plastic sulfur), which consists of long chains of sulfur atoms and has rubberlike properties. Many natural substances are polymeric, including cellulose, starch, and rubber. Furthermore, the myriad of different kinds of protein molecules and nucleic acid molecules that exist in nature can be regarded in many ways as polymers, although there are more different kinds of monomers from which any individual large molecule is constructed than for most other polymers, and these individual units are arranged in highly definite sequences in each protein or nucleic acid.

7-4 Metals

About 70 percent of all elements are metallic, and many mixtures and compounds also show certain *metallic properties*. They have a characteristic luster and are opaque in all but the thinnest layers. They are good conductors of electricity and heat. There are no chemical changes when metals conduct electricity, in contrast to the behavior of ionic compounds (see below). These properties are associated with the presence of very mobile electrons, which not only move through the metal easily but can also be made to pass into and out of it readily. Many metallic substances, especially when pure, are soft and can be worked easily—rolled or hammered into sheets (malleable) and drawn into wires (ductile). Mercury is the only metallic element that is liquid at ordinary temperature; it freezes at $-39°C$.

There are several distinct and useful points of view concerning the nature of the bonding in metals, considered in some detail in Chap. 25. In one view, the atoms are linked into three-dimensional arrays by bonds that are similar to covalent bonds but also permit the shifting of electrons throughout the crystal. In this sense metallic solids are related to the three-dimensional covalent solids just discussed. In the other predominant picture, the metal is regarded as consisting of a framework of positive ions surrounded by a sea of very mobile electrons.

An *alloy* is a metallic substance containing at least two elements.[3] It may be liquid or solid, homogeneous or heterogeneous. If it is homogeneous, it may be an intermetallic compound such as $NaZn_{13}$, Al_9Co_2, or Fe_5Zn_{21}. However, some homogeneous crystalline alloys are of variable composition because some of the atoms can easily be replaced by others without separation of a second phase. Other solid alloys are mixtures containing small crystals of several different metallic substances in various proportions and for this reason have no fixed composition or fixed properties.

---

[3] See Sec. 25-2.

Many crystalline substances are composed of ions. The chemical composition of such crystals is usually definite and constant. There are two reasons for this: electrical neutrality requires a fixed ratio between the numbers of ions of different charge, and the packing requirements in the crystal usually do not permit substitution of one ion by another of the same charge. This is not an invariable rule (Sec. 7-7), but it is a very good one for most substances.

Ionic compounds lose their identity when dissolved in water. For example, a solution of potassium nitrate, KNO_3, contains no KNO_3 molecules but rather only K^+ and NO_3^- ions; its characteristics are due to the ions K^+ and NO_3^-, not to molecules of KNO_3. The KNO_3 is said to be completely *dissociated* in aqueous solution (even though the solid also contains no KNO_3 molecules, only ions). A solution made by dissolving amounts of NaCl and KNO_3 chosen to make the numbers of K^+, Na^+, Cl^-, and NO_3^- ions equal cannot be distinguished from a solution in which appropriate amounts of $NaNO_3$ and KCl are dissolved. These solutions contain only Na^+, K^+, Cl^-, and NO_3^- ions. Which of the four possible compounds NaCl, KCl, $NaNO_3$, or KNO_3 will crystallize first when a solution containing these ions is evaporated depends on the relative numbers of ions present and on the temperature. For example, with equal numbers of ions present, KNO_3 crystallizes first at temperatures near $0°C$.

The physical characteristics of ionic crystals are high melting points and very high boiling points. Ionic substances show no metallic luster, and they are brittle and relatively hard. They conduct electricity when molten or dissolved, because

Figure 7-7 Examples of Complex Ions: $PtCl_6^{2-}$, $Al(H_2O)_6^{3+}$, and $Zn(OH)_4^{2-}$

In $PtCl_6^{2-}$, the chlorine atoms surround the platinum atom octahedrally, and in a similar way the aluminum atom in $Al(H_2O)_6^{3+}$ is surrounded octahedrally by the oxygen atoms of the six water molecules. In other complexes a central atom may be surrounded by different numbers of atoms in different arrangements. Thus in $Zn(OH)_4^{2-}$ the zinc atom is surrounded tetrahedrally by four oxygen atoms of OH groups; in most Ni^{2+} complexes in which the nickel atom has four near neighbors, they occupy the corners of a square about the nickel atom. Many other complexes, with a variety of structural arrangements, are known.

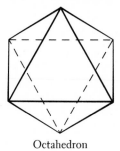

Octahedron

Tetrahedron

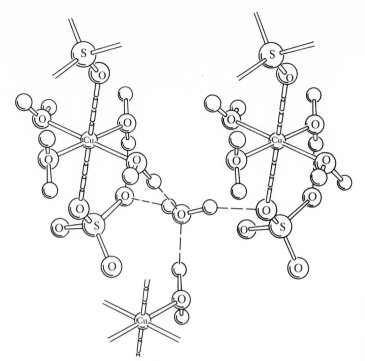

Figure 7-8 The Structural Arrangement of Ions and Molecules in a Crystal of CuSO₄•5H₂O

The crystal structure is built up by a periodic repetition of the components indicated in the formula; a portion of it is depicted here.

of the movement of ions, and chemical reaction takes place at the electrodes where the current enters and leaves. Ionic substances are at least slightly, and often appreciably, soluble in water. The solubility may be very small but it is almost always measurable, while the solubility in water of metals and covalent solids is with few exceptions immeasurably small. (The solubilities of substances that form molecular crystals are often large.)

Crystalline ionic solids may contain molecules in addition to ions. A case in point is the frequent occurrence of water of crystallization (Example 2-4, Sec. 2-2). To illustrate, evaporation of a solution of sodium carbonate below $32°C$ leads to crystals of the formula $Na_2CO_3 \cdot 10H_2O$. This compound exists because it is possible to pack Na^+, CO_3^{2-}, and H_2O in the correct proportions in a favorable way. Another such solid is the mineral carnallite, which is sometimes given the formula $MgCl_2 \cdot KCl \cdot 6H_2O$. It is called a double salt, but this name and the foregoing formula are misleading because the compound does not contain $MgCl_2$ and KCl but rather Mg^{2+}, K^+, Cl^-, and H_2O in the ratio $1:1:3:6$. Carnallite owes its existence solely to the accident that a favorable mode of packing exists for these particles in these proportions. The formula $MgKCl_3 \cdot 6H_2O$ is therefore preferable.

A number of compounds contain ions that are made up of several atoms linked by covalent bonds, such as NO_3^-, PO_4^{3-}, and NH_4^+. Another important polyatomic ion of this sort is present in the substance K_2PtCl_6, which contains K^+ but not Pt^{4+} and Cl^- as might be expected. Instead, Pt and Cl are linked by covalent bonds to form ions, $PtCl_6^{2-}$ (Fig. 7-7). This hexachloroplatinate ion is an example of a *complex ion,* of which a great many exist (Chaps. 9 and 27).

Frequently part or all of the water of crystallization is covalently bound to a central metal atom with the formation of a complex ion; such a hydrated ion is $[Al(H_2O)_6]^{3+}$ (Fig. 7-7). In crystals of $CuSO_4 \cdot 5H_2O$, each copper atom (Fig. 7-8) is at the center of a square formed by the oxygen atoms of four water molecules and is tied to these oxygen atoms by covalent bonds. (In addition, there is an oxygen atom of a sulfate group directly above and another sulfate oxygen directly below each copper atom.) The fifth water molecule in the formula lies between two oxygen atoms of sulfate groups and two oxygen atoms of water molecules, tying them together with hydrogen bonds. The structures of many other salt hydrates show similar features. The combination $Cu(H_2O)_4^{2+}$ is known to persist even in aqueous solution, as do other hydrated ions.

7-6 Summary of the Different Types of Substances

Table 7-1 is a compilation of the different types of substances that have been discussed. It lists characteristic properties and gives examples. While this classification is valuable, it must be realized that there are no sharp boundaries between the categories, and many substances do not fit into the scheme.

For example, crystals of pyrite (FeS_2) and of galena (PbS) are exceptions. They show metallic luster but are poor conductors of electricity and are hard; they are based on three-dimensional, largely covalent arrays of atoms. Many compounds formed by certain metals (such as titanium, vanadium, and tungsten) with boron, carbon, or nitrogen (compounds called borides, carbides, or nitrides) are metallic in character, showing luster and high electrical conductivity; yet they are extremely hard and brittle and have very high melting points and low chemical reactivity. Oxide crystals like quartz (a form of SiO_2) often show ionic as well as covalent character.

Other substances may belong to several categories, depending on the temperature and other circumstances. Aluminum chloride as a solid has a covalent, layered structure. It exists as $Al(H_2O)_6^{3+}$ and Cl^- ions in solution but is molecular in the melt (Al_2Cl_6), in the vapor (Al_2Cl_6 and $AlCl_3$), and in organic solvents.

7-7 Some Other Classes of Substances

Nonstoichiometric Compounds Compounds of fixed composition, such as those consisting of identical molecules, are called *stoichiometric* compounds. In solid compounds not made up of identical molecules it sometimes happens that enough of the constituent ions or molecules are of the wrong kind or are missing, that is, there are enough vacancies or other irregularities in the crystal lattice (lattice defects), to affect the macroscopic properties. If there is an excess of ions of one kind, the crystal gains or loses as many electrons as are necessary to maintain electrical neutrality. The composition of such crystals depends on the details of the irregularities and may lie within a certain range rather than remain constant. It has proved useful to employ the term *compound* for these substances as long as they retain properties characteristic of the stoichiometric substance. Such compounds of variable composition[4] are called *nonstoichiometric* compounds.

Stoichiometric compounds are sometimes called *daltonides* and nonstoichiometric compounds *berthollides,* after the English chemist Dalton and the French

---

[4]We are not concerned here with slight variations in weight composition caused by different isotopic composition. What matters is the composition by numbers of atoms of different elements.

Table 7-1
Types of Chemical Substances

Types	Macroscopic properties	Microscopic properties	
Molecules	Gases or liquids and solids of mp and bp below ~400°C; electrical insulators if solid; often soft, no luster, translucent (at least in thin layers), often soluble	Finite molecules of small or intermediate size	H H *Com...* CO_2, ..., NH_3, HCl, CH_4 (methane), C_6H_6 (benzene), P_4O_{10} (phosphorus pentoxide)
Covalent solids	Crystalline or amorphous; high mp and bp; insulators; brittle and hard, translucent, no luster, insoluble	Atoms held in one-, two-, or three-dimensional arrays by covalent forces Diamond	*Elemental:* C (diamond), Si, P (black), Se (chains) *Compounds:* SiC, BN (borazon), AlB, ZnS, $CdCl_2$, $CuCl_2$
High polymers	Solids or viscous liquids; often soft and flexible; no luster; when soluble, solution viscous; plastics	Long chains or networks of covalently bonded atoms; usually mixtures of molecules of same type but different size	*Elemental:* S (chains, amorphous), Se *Compounds (often mixtures):* starch, cellulose, rubber, nylon, Teflon, polystyrene, polyethylene
Metals	Highly conducting in solid or liquid state; metallic luster, opaque even in thin layers, insoluble	Highly mobile electrons	*Elemental:* Na, Cu, Fe, Cr *Compounds:* $NaZn_{13}$, Al_9Co_2, Fe_5Zn_{21}
Ionic compounds	Hard and brittle crystals; high mp and bp; water-soluble although at times in trace amounts; electrically conducting in melts and solutions; translucent, no luster	Crystals consisting of ions—for example, NaCl: Na^+, smaller spheres; Cl^-, larger spheres*	*Elemental:* none *Compounds:* NaCl, CaF_2, KNO_3, $(NH_4)_2SO_4$, K_2PtCl_6

*The illustration is taken from a paper published by William Barlow in 1898. Barlow proposed the sodium chloride structure and four others from considering the highly symmetric packing of spheres of equal and unequal sizes. His proposal came before the discovery of X rays and before it could therefore be proved that the structures he proposed actually occurred in nature.

chemist Berthollet. Dalton was a strong proponent of the idea that compounds always have a constant composition; Berthollet opposed this idea equally strongly, in part by arguments later shown to be in error by the French chemist Proust. Berthollet believed that the composition of compounds depended on their history, that is, on their mode of preparation.

Many metal oxides and sulfides are berthollides, including the oxides of iron. Stoichiometric compounds exist with composition Fe_2O_3 and Fe_3O_4, which can be written $Fe_{0.67}O$ and $Fe_{0.75}O$. However, FeO is not stable at ordinary temperatures. The material called "ferrous oxide" is not FeO but rather exists with a range of compositions between about $Fe_{0.84}O$ and $Fe_{0.94}O$. The fact that there is a slightly smaller number of iron ions than of oxide ions means that some of the iron ions must be Fe^{3+} instead of Fe^{2+} in order to maintain charge balance.

Intermetallic compounds are also usually berthollides. Their nonstoichiometric composition is due mainly to replacement in the crystal lattice of atoms of the appropriate kind by wrong atoms. For example, the ideal composition of gamma brass is Cu_5Zn_8 or $CuZn_{1.60}$, but the gamma-brass phase, as it is called, extends over the range $CuZn_{1.58}$ to $CuZn_{1.65}$. The boundaries of such ranges depend on the temperature, and the numbers are quoted just to give an idea of magnitudes.

Clathrate Compounds   In recent years many examples of a novel type of solid compound have been discovered. In these compounds, called *clathrates* (Latin: *clathratus,* encaged, enclosed by crossbars), ions or molecules form a three-dimensional array with regularly spaced cavities in which still other molecules are

Figure 7-9   The Structure of a Clathrate Crystal

The framework of hydrogen-bonded water molecules consists of nearly regular pentagonal dodecahedra linked together by additional water molecules, some oxygen atoms of which are shown. The polygons that border the spaces between linked dodecahedra have two hexagonal and six pentagonal faces; they are not shown explicitly. All hydrogen bonds, indicated either explicitly or by lines in the figure, are about the same length as in ordinary ice. The two kinds of polyhedra act as chambers, entrapping atoms (such as Ar, Kr, and Xe) or small molecules (such as CH_4 or H_2S). The figure shows H_2S molecules. Clathrates involving many different frameworks of water molecules are known.

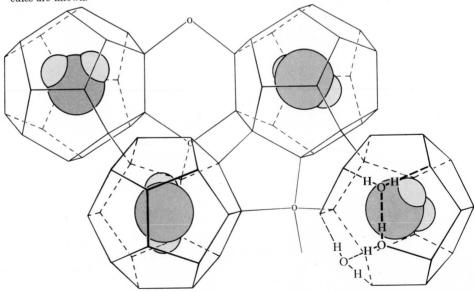

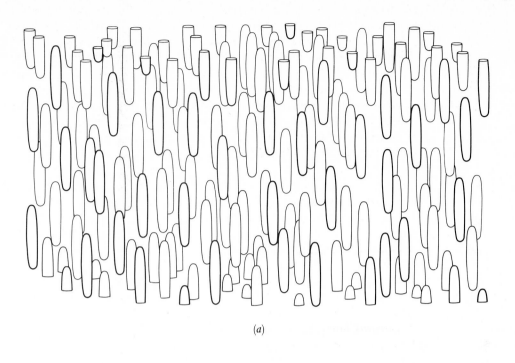

(a)

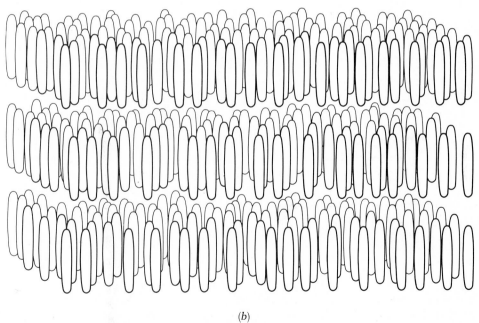

(b)

Figure 7-10 Schematic Representations of Some Liquid Crystals
In (a) the molecules are aligned but can rotate and slip past each other parallel to their long axes.
In (b) freely rotating molecules are arranged in layers, which can move freely past each other.
More complex arrangements are also known.

trapped. In most of the clathrates so far studied, the framework is held together by hydrogen bonds, but in a few it is covalently bonded at least in two dimensions. Among the most common clathrates are those in which hydrogen-bonded water molecules make up the cages, enclosing molecules like Cl_2, CH_4, and even inert gas atoms such as Xe. Gas hydrates of definite composition are formed, for example, $Cl_2 \cdot 8H_2O$, $Xe_4 \cdot 23H_2O$, and $(CH_4)_4 \cdot 23H_2O$ (Fig. 7-9). Clathrates are stabilized by the mutual van der Waals attraction of the molecules enclosed in the cage and the molecules forming the cage. If many of the cavities are vacant, the structure is usually unstable, and most gas hydrates are at best rather unstable because of the ease with which the hydrogen bonds linking the framework together can be broken.

Liquid Crystals When some solids composed of long rodlike or platelike molecules melt, the liquid formed exhibits partial long-range order. For example, above the melting point certain rodlike molecules remain aligned parallel to one another like matches in a long box (Fig. 7-10*a*). Each of the molecules can rotate around its axis, and the molecules can slip past each other so the molecular arrangement in the liquid is essentially random in two directions. The one-dimensional order, which can be detected by optical measurements, is destroyed when the liquid undergoes a sharp "melting" transition at a higher temperature. The order is restored on cooling.

Phases that have properties intermediate between those of liquids and crystals are called *liquid crystals*. In some liquid crystalline phases the molecules are arranged in planes (Fig. 7-10*b*), while in others even more ordered arrangements are found. However, all liquid crystal phases flow readily. Transitions from one liquid crystal phase to another are also observed.

Once considered merely a scientific curiosity, liquid crystals are now used extensively in electronic displays. In one such device a small quantity of a clear liquid crystal is sandwiched between two electrodes, one of which is transparent. When a small voltage is applied to the electrodes, small impurity ions move through the liquid crystal, collide with the large molecules, and break up the ordered structure, leaving only small ordered regions. The turbulence in these small regions makes the fluid appear more opaque. Numbers in a display can be produced when different combinations of liquid crystal devices are switched from the clear to the opaque state.

Summary

It is possible to classify chemical substances by the nature of the interactions between the atomic and molecular units of which the substances are composed. The macroscopic properties reflect the strength and nature of the forces between the microscopic units. Elements and compounds composed of covalently bonded molecules constitute one class. In both the solid and liquid states these substances are held together by weak interactions—van der Waals forces and sometimes hydrogen bonds. Covalent solids are linked by covalent bonds into one-, two-, or three-dimensional arrays. High polymers, an important class of covalent compounds, consist of extremely large molecules made by combination of many small molecules called monomers. Metallic properties—characteristic luster, good electrical and heat conduction—are found in many elements, mixtures, and compounds. These properties are associated with the presence of very mobile electrons, which move through the metal easily and can be made to pass into or out of it readily. Alloys are metallic substances that contain at least two elements.

Crystalline substances made up of ions are called ionic compounds. Such compounds lose their identity when dissolved in water because only their constituent ions are present in the solution. Some crystalline ionic solids contain molecules such as water in addition to ions.

Other classes of substances include nonstoichiometric compounds, clathrate compounds, and liquid crystals. Nonstoichiometric compounds have variable composition, a result of vacancies or other irregularities in the crystal lattice. In clathrates, ions or molecules form three-dimensional arrays with regularly spaced cavities in which other molecules are trapped. Liquid crystals are fluid phases that exhibit partial long-range order and therefore have properties intermediate between those of liquids and crystals.

Terms and Concepts

Problems and Questions

7-1 Volatility What is a *volatile* substance? What is the relation between volatility and the strength of intermolecular forces?

7-2 Molecular Crystals Define the term *molecular crystal* and give two examples. What are some characteristic physical properties of molecular crystals?

7-3 Molecular Crystals Benzene molecules consist of covalently bonded carbon and hydrogen atoms, yet benzene forms molecular crystals. Explain.

7-4 Covalent Solids Explain the relation between the number of bonds that can be formed by a typical atom in a crystal and the possibility of forming linear, two-dimensional, and three-dimensional network structures.

7-5 Covalent Crystals Define the term *covalent crystal*. What are the characteristic physical properties of covalent crystals?

7-6 Metallic Properties Many familiar household items are made of metal. Which characteristic properties make metals the materials of choice for the fabrication of the following items: pots; foil wrapper; jewelry; electrical wiring; light-bulb filaments?

7-7 Distinguishing Ionic and Molecular Crystals Although large crystals of sugar (rock candy) and large crystals of salt (rock salt) have different geometric shapes, they look much the same to the untrained observer. What physical tests other than taste might be performed to distinguish between these two crystalline substances?

7-8 Distinguishing Types of Solids You are presented with samples of ionic, covalent, molecular, and metallic solids. The samples are unlabeled and you must distinguish them. Describe a series of physical tests you could perform to categorize the samples.

7-9 Distinguishing Types of Solids Describe the chief differences in structure and properties between molecular crystals and covalent network solids. Indicate the reasons for these differences.

7-10 Reasons for Volatility Give a structural interpretation of the fact that all the halogens are relatively volatile—that is, have an appreciable vapor pressure at ordinary temperatures—while carbon and silicon are not.

7-11 Arrays of Atoms What is the smallest number of neighbors to which a given atom is bonded in arrays of dimension 0, 1, 2, and 3?

7-12 Classification of Structures Classify the structures illustrated in Figs. 7-3, 7-4, 7-5, and 7-6 as exhibiting one-, two-, or three-dimensional arrays, explaining the reasons for your choices.

7-13 Properties of Substances For each of the following pairs of substances, indicate which substance should have the *higher* value of the property mentioned, explaining briefly the reasons for your choice: (*a*) melting point: KBr, CBr_4; (*b*) solubility in nonpolar solvents: I_2, graphite; (*c*) hardness: S_8, ZnS.

7-14 Nonstoichiometric Oxide of Iron The substance referred to as FeO, which has the sodium chloride structure (Fig. 6-1), typically contains 76.8 percent Fe. (*a*) If its formula is written as Fe_xO, what is the value of x? (*b*) If it is written as FeO_y, what is the value of y? (*c*) If iron oxide contains only Fe^{2+} ions and oxide ions, its composition is FeO; if it contains only Fe^{3+} ions and oxide ions, its composition is Fe_2O_3. If the oxide considered here contains Fe^{2+}, Fe^{3+}, and O^{2-}, what is the ratio of Fe^{2+} to Fe^{3+}?

Solutions

"One molecule of a fixed, non-saline substance, in dissolving in 100 molecules of any volatile liquid, diminishes the vapor pressure of the liquid by a nearly constant fraction of its value."
F. M. RAOULT, 1886–1887

8

Solutions in liquid solvents are the concerns of this chapter and the next one. The solvent properties of many liquids, and especially the remarkable properties of water, make possible homogeneous conditions for carrying out chemical reactions. Such conditions are enormously advantageous because they ensure uniform and intimate contact of the reactants. Most biochemical reactions in our bodies and in other organisms occur in aqueous solution. The great majority of the important industrial chemical processes used to prepare synthetic plastics and fibers, medicinal chemicals, photographic supplies, petroleum products, ultrapure materials for transistors, and numerous other products of modern civilization involve liquid solutions at various stages.

This chapter is concerned primarily with the general properties of liquid solutions, the emphasis being on nonionic solutions. Chapter 9 deals with common and important kinds of solutions in water, including solutions of acids, bases, and salts.

8-1 Solutions

Solutions are homogeneous mixtures. They are intimate mixtures on the molecular level and may exist in all three states of aggregation. Liquid solutions are the most familiar and are our primary concern here. However, since all gases mix homogeneously with one another in any proportions, all mixtures of gases can be classified as solutions. Solid solutions also exist, not only for amorphous solids but for crystals as well, because in many crystals some of the original atoms, ions, or molecules may be replaced randomly by foreign atoms, ions, or molecules without destroying the regular structure characteristic of the original substance.

Composition of Solutions The predominant component of a solution is referred to as the *solvent,* the minor components as the *solutes.* These terms are somewhat arbitrary when the components are present in equal or nearly equal amounts, but this is not a common situation. The proportions of the components may always be varied over at least a narrow range, with an accompanying continuous variation in the properties of the solution. The *concentration* of a solution is a quantitative measure of the relative amounts of solute and solvent present—for example moles of solute per liter of solution. Any particular solution has a definite composition, which may be specified in many different ways. For example, dry air contains 21.0 percent oxygen by volume, which means that 100.0 liter of air can be reproduced by combining 21.0 liter of oxygen with 79.0 liter of the remaining components of air (chiefly nitrogen), all at the same temperature and pressure. Note the difference between weight and volume percentages. There is 23.2 wt percent oxygen in air, so that 100.0 g air contains 23.2 g oxygen. A percentage composition without further specification usually implies weight percentage, because the weights of the components of a solution are additive whereas the volumes often are not. Composition by weight is therefore easier to establish.

Example 8-1

□ **Volume Percentage and Weight Percentage** A solution of alcohol (ethanol) at 4°C is prepared by adding enough water to 31.73 g alcohol to make a total of 100.00 ml. The density of pure alcohol at 4°C is 0.7932 g ml^{-1} and that of the solution is 0.9510 g ml^{-1}.

(a) *What is the volume percentage of alcohol in the solution?*

Solution The volume of alcohol can be calculated from the weight and density:

$$\text{Volume alcohol} = \frac{31.73 \text{ g}}{0.7932 \text{ g ml}^{-1}} = 40.00 \text{ ml}$$

The total volume of solution is 100.00 ml; hence

$$\text{Volume percentage alcohol} = \frac{40.00 \text{ ml}}{100.00 \text{ ml}} \times 100 = \underline{40.00}$$

(b) *What is the weight percentage of alcohol?*

Solution The total weight of the solution can be calculated from the volume and density:

$$\text{Weight solution} = 100.00 \text{ ml} \times 0.9510 \text{ g ml}^{-1} = 95.10 \text{ g}$$

The weight percentage of alcohol is then

$$\text{Weight percentage alcohol} = \frac{31.73 \text{ g}}{95.10 \text{ g}} \times 100 = \underline{33.36} \blacksquare$$

Exercise 8-1

□ The density of pure acetic acid is 1.049 g ml^{-1}, and the density of a solution prepared by diluting 333 g of acetic acid to 1.00 liter is 1.026 g ml^{-1}. (a) What is the volume percentage of acetic acid in the solution? (b) Calculate the weight percentage of acetic acid. ■

Saturated Solutions When a solution of a substance is in contact with additional pure solute, the dissolving of more solute depends upon the concentration of the solute already in solution. A solution is said to be *saturated* when there is equilibrium between dissolved solute and any excess solute. This equilibrium is a dynamic one, like those involving vapor pressure discussed in Sec. 5-2. At equilibrium the rates at which the solute goes into the solution and comes out of it are equal, so that there is no net change in the amount of dissolved material. The concentration of solute in a saturated solution is called the *solubility* of the substance; the amount of excess undissolved solute present does not affect the solubility. Two liquids may be mutually soluble (*miscible*) in all proportions, as, for example, are ethanol and water. For such solutions, the term *solubility* has no meaning since the solution cannot be saturated.

Supersaturation   It is sometimes possible to prepare solutions that contain more solute than they would under equilibrium conditions. For example, the temperature of a saturated solution of a solid may be changed carefully to a temperature at which the solubility is smaller. When this is done without agitation in a vessel with smooth walls and in the absence of specks of solid impurities, the excess of the solute often fails to crystallize. Such solutions are called *supersaturated*. The phenomenon is in many respects similar to those of superheating and supercooling. Scratching the container walls or introducing a tiny crystal of the solute (or occasionally even a speck of *any* solid) usually causes the excess solute to crystallize. To give an example, a saturated solution of $Na_2SO_4 \cdot 10H_2O$ contains 92.8 g of this substance per 100 g water at 30°C and 13.8 g at 5°C, but when a filtered solution saturated at 30°C is carefully cooled, no solid precipitates and the solution is supersaturated. If a tiny seed crystal of the solute is dropped into the supersaturated solution, however, crystallization begins immediately.

Solutions of gases in liquids can also be supersaturated. The solubility of all gases increases with increasing pressure and decreases with increasing temperature. Consequently when the temperature of a saturated solution is raised, or the pressure of gas above the solution is decreased, the amount of dissolved solute exceeds the solubility under the new conditions. If the solution is not disturbed, it may remain supersaturated for some time—but if it is shaken, it usually releases the excess gas almost explosively, as anyone knows who has ever shaken a warm, freshly opened bottle of a carbonated drink.

Volatile Solutes   When the solute is a volatile solid or liquid, the partial pressure of the solute must be the same for all phases in equilibrium with each other. For example, its partial pressure above a saturated solution in equilibrium with pure solute must be the same as that of the pure solute (Fig. 8-1). The reason is that solute and saturated solution may interact through the vapor phase as well as by direct contact. Equilibrium can exist only if it covers all possible paths of

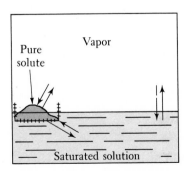

Figure 8-1   A Saturated Solution of a Volatile Substance
The wire basket contains excess solid that is partly immersed in the solution. At equilibrium the solution is saturated and no further solute dissolves. The existence of equilibrium implies that the vapor pressure of the solute above the *pure* solute and above the *saturated solution* must be the same.

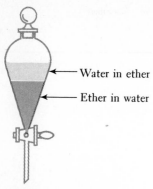

← Water in ether

← Ether in water

Figure 8-2 Saturated Solutions of Water in Ether (above) and Ether in Water (below) in a Separatory Funnel

When equal volumes of water and ether [more properly diethyl ether, $(C_2H_5)_2O$] are shaken together until each of these liquids is saturated with the other, two layers separate. The upper layer is a solution of water in ether containing, at 20°C, about 1.2 wt percent water. The lower layer is a solution of ether in water containing about 6.5 wt percent ether at 20°C. The separatory funnel permits convenient separation of two immiscible liquids.

interaction. If the vapor pressure of the pure solute (Sec. 5-2) were higher than its partial pressure above the saturated solution, the solute would evaporate and condense in the solution. For example, the vapor pressure of solid iodine at 25°C is 0.31 torr, and this is also the partial pressure of iodine above a saturated solution of iodine in water, or in any other solvent, at 25°C.

It is even more instructive to consider a saturated solution of ether in water, which may be obtained by shaking the two liquids together. When equilibrium has been reached and the shaking is stopped, two layers separate (Fig. 8-2). The aqueous layer at the bottom is saturated with ether, and the ether layer on top is saturated with water. The compositions of these two liquid phases are very different, although each is in equilibrium with the other and with the vapor phase, which is a mixture of gaseous water and gaseous ether.

8-2   Concentration Units

One significant advantage of working with solutions in carrying out chemical reactions is that precise and accurate measurements of liquid volumes can readily be made. Thus it is often easy, with concentration units expressed in terms of the amount of solute per unit *volume* of solution, to study the stoichiometry of chemical reactions or to make measurements of reaction rates or positions of equilibrium. Concentration units based on the weight of the solution or its components are also sometimes used, particularly in the study of what are called the "colligative" properties of solutions (Sec. 8-3).

Molarity   The most common unit of concentration, expressed as amount of solute per unit *volume* of *solution,* is molarity, which is symbolized by square brackets around the chemical formula of the solute and is defined as follows:

$$[A] = \frac{\text{number of moles of solute A}}{\text{volume of } solution \text{ in liters}} \tag{8-1}$$

The units of molarity are thus moles of *solute* per liter of *solution,* abbreviated mol liter^{-1} or just M. For example, a solution that contains 0.1000 mol A liter^{-1} is said to be 0.1000 molar in A, or 0.1000 M in A.

Example 8-2

☐ **Molarity of Sucrose Solution**   Calculate the molarity of an aqueous solution that contains 158.9 g sucrose ($C_{12}H_{22}O_{11}$) per liter of solution.

Solution   The molecular weight of sucrose is 342.3 so that 158.9 g sucrose represents 158.9 g/342.3 g mol^{-1} = 0.464 mol. The solution is therefore 0.464 M . ∎

Exercise 8-2

☐ Calculate the molarity of a solution that contains 29.8 g methanol, CH_3OH, per liter of solution. ∎

Example 8-3

☐ **Interconversion of Concentration Units**   The solubility of sodium chloride is listed in a handbook as 36 g per 100 ml water at 0°C. The density of the resulting solution is 1.21 g ml^{-1}, and the density of water at this temperature is 1.00 g ml^{-1}. What is the molarity of the solution?

Solution To calculate the molarity one must know the number of moles of solute per liter of *solution*. The weight of 1 liter of solution can be calculated from the density:

$$\text{Weight of 1 liter solution} = 1000 \text{ ml} \times 1.21 \text{ g ml}^{-1}$$
$$= 1.21 \times 10^3 \text{ g}$$

This weight includes both solute and solvent. To calculate the fraction of the weight attributable to NaCl, we make use of the solubility. There are 36 g NaCl in every 100 ml water, or 36 g NaCl in every 100 g water because the density of water is 1.00 g ml^{-1} at this temperature.

$$\text{Weight fraction NaCl} = \frac{36 \text{ g}}{(36 + 100) \text{ g}} = 0.26$$

For 1 liter of solution,

$$\text{Weight NaCl in 1 liter} = 1.21 \times 10^3 \text{ g} \times 0.26 = 3.2 \times 10^2 \text{ g}$$

Finally, the formula weight of NaCl is 35.5 + 23.0 = 58.5 so that

$$\text{Moles NaCl in 1 liter} = \frac{3.2 \times 10^2 \text{ g}}{58.5 \text{ g mol}^{-1}} = 5.5 \text{ mol}$$

This is, by definition, the molarity, and the solution is 5.5 M . ■

□ A solution prepared by dissolving 251 g of sodium nitrate, NaNO$_3$, in 1 liter of water has a density of 1.14 g ml^{-1} at 20°C. What is the molarity of the solution? ■

Exercise 8-3

Note that since NaCl is essentially completely dissociated into Na$^+$ and Cl$^-$ in aqueous solution, the concentrations of Na$^+$ and Cl$^-$ in Example 8-3 are *each* 5.5 M. There are no NaCl molecules in the solution. Similarly, in the NaNO$_3$ solution of Exercise 8-3, the concentrations of Na$^+$ and NO$_3^-$ are each 2.69 M.

□ **Ion Concentrations** What are the concentrations of Ca^{2+}, H$^+$, and Cl$^-$ in a solution if 500 ml of the solution is prepared from 0.200 mol CaCl$_2$·2H$_2$O and 0.150 mol HCl? Both CaCl$_2$ and HCl are fully dissociated into ions in aqueous solution.

Example 8-4

Solution The molar concentration of the solution, expressed in terms of CaCl$_2$·2H$_2$O and HCl, is

$$\frac{0.200 \text{ mol CaCl}_2 \cdot 2\text{H}_2\text{O}}{0.500 \text{ liter}} = 0.400 \text{ } M$$

$$\frac{0.150 \text{ mol HCl}}{0.500 \text{ liter}} = 0.300 \text{ } M$$

An identical solution could be prepared by starting with 0.200 mol of the anhydrous salt, CaCl$_2$, rather than the dihydrate, adding 0.150 mol HCl and enough water to make the volume 0.500 liter. The solution can thus also be described as 0.400 M in $CaCl_2$. Both $CaCl_2$ and HCl are fully ionized, so that $[Ca^{2+}] = \underline{0.400\ M}$, $[H^+] = \underline{0.300\ M}$, and $[Cl^-] = 2 \times 0.400 + 0.300 = \underline{1.100\ M}$. ■

Exercise 8-4

□ What are the concentrations of Ca^{2+}, Cl^-, and NO_3^- in a solution prepared by dissolving 0.100 mol $CaCl_2 \cdot 6H_2O$ and 0.120 mol $Ca(NO_3)_2$ in water and diluting to 10.0 liter? Both compounds are completely dissociated into ions in aqueous solution. ■

Volume concentration units have many practical advantages, but one significant disadvantage is that they are temperature-dependent. As the temperature changes, the volume of solution changes while the quantity of solute remains constant, so that their ratio, the concentration, also changes. Consequently, careful temperature control is needed in precise quantitative volumetric work.

Weight Concentration Units Concentrations that are expressed as a ratio of weights (or as numbers proportional to weights, such as numbers of moles) do not vary with temperature, because weights are essentially temperature-independent. Among useful weight concentration units are:

1. $\quad$ Molality $= \dfrac{\text{number of moles of solute}}{\text{weight of } \textit{solvent} \text{ in kilograms}} = m \qquad (8\text{-}2)$

Concentrations expressed in this way will be denoted by m. They are sometimes referred to as mola*l* concentrations, the final letter distinguishing them from mola*r* concentrations, defined in (8-1) above. It is important to note that the denominator is the kilograms of solvent, not the kilograms of the entire solution.

2. $\quad$ Weight percentage $= \dfrac{\text{weight of solute}}{\text{weight of } \textit{solution}} \times 100 \qquad (8\text{-}3)$

A useful feature of weight concentrations is that the weights of solute and solvent are additive, their sum being the weight of the solution. It is only rarely and accidentally true that the volume of a solution is the sum of the volumes of solute and solvent. No knowledge of densities is required to find molality from the weight percentage of a solution, but to transform a weight concentration to a volume concentration does require such knowledge.

Example 8-5

□ **Molal and Molar Concentrations** The density of the 0.464 M sucrose solution described in Example 8-2 is 1.059 g cm^{-3}. What is its molality?

Solution $\quad$ One liter of the solution weighs

$$1000\ \text{cm}^3 \times 1.059\ \text{g cm}^{-3} = 1.059 \times 10^3\ \text{g} = 1.059\ \text{kg}$$

and contains 158.9 g sucrose and therefore

$$(1.059 - 0.159)\ \text{kg} = 0.900\ \text{kg H}_2\text{O}$$

The molality, that is, the number of moles of sucrose per kilogram of water, is therefore

$$\text{Molality sucrose} = 158.9\ \text{g} \times \frac{1\ \text{mol}}{342.3\ \text{g}} \times \frac{1}{0.900\ \text{kg}}$$

$$= \underline{0.516\ \text{mol kg}^{-1}}\ \blacksquare$$

□ The density of a 0.930 M solution of methanol in water is 0.9930 g ml^{-1}. **Exercise 8-5**
What is the molal concentration of methanol? ■

Mole Fraction The mole fraction, defined in Equation (4-19) for gas mixtures, is also a useful concentration unit for solutions. Let the molecular or ionic species in a solution be designated by the index i, this index running over all species present, including the solvent. The solution then consists of n_1 moles of species 1, n_2 moles of species 2, and so on if there is more than one solute. Then

Total number of moles present: $n = \Sigma n_i$ (8-4)

Mole fraction of component i: $x_i = \dfrac{n_i}{n}$

The sum of all mole fractions is necessarily unity:

$$\Sigma x_i = 1 \qquad (8\text{-}5)$$

□ **Interconversion of Concentration Units** A certain solution of HNO_3 in **Example 8-6**
water contains 30.0 percent by weight of HNO_3 and has a density of 1.180 g ml^{-1}. Calculate its composition in terms of molarities, molalities, and mole fractions, assuming that in this solution the acid is 84 percent dissociated,[1] i.e., that 16 percent exists as HNO_3 and 84 percent as H^+ and NO_3^-.

Solution (*a*) To calculate the molarity, we proceed much as in Example 8-3. To find the number of moles HNO_3(FW = 63.0) in 1000 ml solution, we first use the density of the solution to calculate its weight,

Weight of 1000 ml solution = 1000 ml $\times$ 1.180 g ml^{-1} = 1180 g (8-6)

and then use the percentage composition to calculate the weight of HNO_3 in this weight of solution,

$$\text{Weight of } HNO_3 \text{ in 1000 ml} = \frac{\text{wt percent } HNO_3}{100} \times \text{wt of solution}$$
$$= 0.300 \times 1180 \text{ g} = 354 \text{ g} \qquad (8\text{-}7)$$

Finally,

$$\text{mol } HNO_3 \text{ per liter} = \frac{354 \text{ g}}{63.0 \text{ g mol}^{-1}} = \underline{5.62 \text{ mol}}$$

which would be numerically equal to the molarity if HNO_3 were undissociated. The actual molar concentrations are $[H^+] = [NO_3^-] = 0.84 \times 5.62 = \underline{4.7\,M}$ and $[HNO_3] = 0.16 \times 5.62 = \underline{0.9\,M}$.
(*b*) To calculate the molality, we must know the number of moles of each component per kg H_2O (the solvent in this solution). From part (*a*) we know that for each 1180 g solution there are 354 g HNO_3, and thus, by difference, 826 g H_2O. Thus there are

$$\frac{354 \text{ g } HNO_3}{63.0 \text{ g mol}^{-1}} \times \frac{1}{0.826 \text{ kg } H_2O} = 6.80 \text{ mol } HNO_3 \text{ per kg } H_2O$$

---

[1] In less concentrated solutions, for example, 1 M or less, HNO_3 is almost completely dissociated.

If the HNO_3 did not dissociate the molality would therefore be 6.80. The actual molalities of both H^+ and NO_3^- are $0.84 \times 6.80 = \underline{5.7 \text{ molal}}$, while HNO_3 is $0.16 \times 6.80 = \underline{1.1 \text{ molal}}$.

(c) Because the number of moles of H^+, NO_3^-, and HNO_3 per kg H_2O have already been calculated in (b), we need find only the number of moles of H_2O per kilogram to be able to compute the mole fractions:

$$\frac{1000 \text{ g}}{18.02 \text{ g mol}^{-1}} = 55.5 \text{ mol } H_2O$$

The total number of moles in the solution is then $5.7 + 5.7 + 1.1 + 55.5 = 68.0$, and the mole fractions are

$$x_{H^+} = x_{NO_3^-} = \frac{5.7}{68.0} = \underline{0.084}$$

$$x_{HNO_3} = \frac{1.1}{68.0} = \underline{0.016}$$

$$x_{H_2O} = \frac{55.5}{68.0} = \underline{0.816}$$

The sum of the mole fractions is 1.000, as it must be. ■

Exercise 8-6

☐ A solution of H_2SO_4 in water contains 5.00 percent by weight of H_2SO_4 and has a density of 1.032 g ml^{-1}. The H_2SO_4 is completely dissociated into HSO_4^- and H^+,

$$H_2SO_4 \longrightarrow H^+ + HSO_4^-$$

and HSO_4^- is in turn 2.1 percent dissociated by the reaction

$$HSO_4^- \longrightarrow H^+ + SO_4^{2-}$$

In other words, 2.1 percent of the acid exists as SO_4^{2-} and 97.9 percent as HSO_4^-. Two H^+ are produced for each SO_4^{2-} in solution and one for each HSO_4^-. Calculate the molarities, the molalities, and the mole fractions for these three species. ■

8-3 Colligative Properties of Solutions

There are certain properties of any given solution that are quantitatively interrelated and are therefore known as *colligative properties* (Latin: *colligare,* to tie together). They include the lowering of the solvent vapor pressure, the elevation of the boiling point, the lowering of the freezing point, and the osmotic pressure. A closely related property is the pressure dependence of gas solubilities. The magnitude of each of these effects depends on the concentration and temperature of the solution, the pressure, and other variables.

Some of the colligative properties of solutions are described and illustrated in the following pages. In general, the quantitative laws stated below are only approximately valid for most solutions, but they become increasingly applicable as solute concentrations decrease, that is, for dilute solutions. The terms *dilute* and *concentrated* are often used in referring qualitatively to the concentration of solute in a solution; like solute and solvent, they are only relative terms, with no

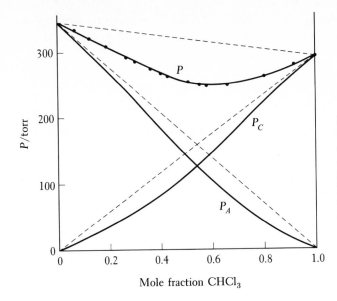

Figure 8-3 Vapor Pressure of a Solution
The graph shows the partial and total pressures for solutions
of chloroform and acetone as a function of mole fraction at
35°C. Full lines signify measured pressures: P_C = partial pressure of chloroform, P_A = partial pressure of acetone, P = total
pressure. Dashed lines represent the partial and total pressures
calculated from Raoult's law [Equation (8-8)]. Note that for
high concentrations of chloroform P_C approaches the Raoult's
law value. Similarly, Raoult's law is a good approximation for
P_A at high concentrations of acetone. When, as in this case,
the total pressure is less than the ideal-solution value, we
speak of negative deviations from Raoult's law. Positive deviations from Raoult's law, that is, $P > P_{ideal}$, are also found.

absolute meaning that holds in all circumstances. A solution containing a few
percent of a solute is considered dilute for many purposes.

Raoult's Law The vapor pressure of a solution usually lies between the vapor
pressures of its pure components. An approximate relation describing the partial
pressure of the solvent P_1 is Raoult's Law

$$P_1 = P_1^* x_1 \qquad (8\text{-}8)$$

where P_1^* is the vapor pressure of the pure solvent at the temperature in question
and x_1 is the mole fraction of the solvent. A solution for which Raoult's law holds
exactly, over the entire concentration range from solute mole fraction zero to
unity, is said to be *ideal*. No change in volume occurs upon dissolving the solute
in an ideal solution, nor is there any release or absorption of energy.

If the solute in an ideal solution is also volatile, Raoult's law applies to it as
well, and indeed if there are several volatile solutes, it applies to all of them. Even
for nonideal solutions (8-8) applies to the solvent reasonably well, at least when
the solution is dilute, but there are often marked deviations in more concentrated
solutions (Fig. 8-3).

The vapor pressure of a solution is the sum of the partial pressures of the
components. This relationship is important when one is considering the boiling
point of a solution. A solution will boil when the *total* pressure of its vapor equals
the applied pressure, usually atmospheric pressure. If a dissolved solute is *nonvolatile,* its partial pressure is zero and the vapor pressure of the solution is
simply the partial pressure of the solvent. The vapor pressure of the solvent has
effectively been lowered by the addition of a nonvolatile solute. Sometimes this
lowering of the vapor pressure, that is, $P_1^* - P_1$, is of interest. In a binary ideal solution, that is, an ideal solution with two components,

$$P_1^* - P_1 = P_1^* - x_1 P_1^* = P_1^*(1 - x_1) = P_1^* x_2 \qquad (8\text{-}9)$$

since $x_2 = 1 - x_1$. Thus the lowering of the vapor pressure of the solvent by a
dissolved solute is proportional to the mole fraction of the solute in an ideal

151

solution, and approximately so in a real dilute solution. In the presence of several solute species, $1 - x_1$ is the *sum* of the mole fractions of all solute species and x_2 in the last expression in (8-9) must therefore be replaced by this sum.

The following example describes a situation in which both components of a solution are volatile and there is complete miscibility for all proportions of the components.

Example 8-7

□ **Vapor of a Mixture** At 80°C the vapor pressure of benzene (C_6H_6) is 753 torr and that of toluene (C_7H_8) is 290 torr. What is the total vapor pressure and what is the composition of the vapor over a solution of $\frac{1}{3}$ mol of benzene in $\frac{2}{3}$ mol of toluene? Assume that Raoult's law applies to both components of the mixture.

Solution The mole fraction of benzene is $\frac{1}{3}$ so that by Raoult's law

$$P_{\text{benzene}} = \tfrac{1}{3}(753) = 251 \text{ torr}$$

Similarly, $x_{\text{toluene}} = \frac{2}{3}$ and therefore

$$P_{\text{toluene}} = \tfrac{2}{3}(290) = 193 \text{ torr}$$

The total vapor pressure is the sum of the two partial pressures, $P = 251 + 193 = \underline{444 \text{ torr}}$. The composition of the vapor is, in mole fractions,

$$x_{\text{benzene}} = \frac{P_{\text{benzene}}}{P} = \frac{251}{444} = \underline{0.565}$$

$$x_{\text{toluene}} = \frac{P_{\text{toluene}}}{P} = \frac{193}{444} = \underline{0.435} \blacksquare$$

Exercise 8-7

□ Consider a mixture of the same components at the same temperature, such that the *vapor* composition is given by $x_{\text{benzene}} = x_{\text{toluene}} = 0.500$. What is the mole fraction of benzene in the liquid? ∎

Boiling-Point Elevation The boiling point of a solution of a nonvolatile solute lies above that of the pure solvent. The reason is that, by Raoult's law, at any given temperature the partial pressure of the solvent in a solution is lower than that of the pure solvent. Consequently it is necessary to heat the solution to a somewhat higher temperature than the normal boiling point of the solvent, T_b^*, to raise its vapor pressure to 1 atm (Fig. 8-4). For a dilute solution, the increase in the boiling point, ΔT_b, is proportional to the molality of the solute, m:

$$\Delta T_b = T_b - T_b^* = k_b m \qquad (8\text{-}10)$$

The constant k_b is called the *molal boiling-point elevation;* typical values are given in Table 8-1. We shall treat (8-10) as an empirical law, but k_b is related to the boiling temperature of the solvent, its heat of vaporization, and its molecular weight. When several nonvolatile solute species are present, their effect on the boiling point is additive.

Freezing-Point Depression The freezing point of a solution, T_f, is below that of the pure solvent, T_f^*, provided that, as is usually true, the crystals formed as the solution freezes are pure solvent and not a solid solution containing solute. The

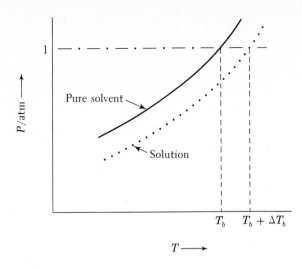

Figure 8-4   Boiling-Point Elevation of a Solution of a Nonvolatile Solute

The solid curve represents the vapor pressure of the pure solvent as a function of temperature; the solvent boils at T_b. The dotted curve, somewhat lower at any given temperature, represents the vapor pressure of the solution. If the solute is nonvolatile, the vapor pressure above the solution necessarily reaches 1 atm (and thus the solution boils) only at a temperature above T_b. The new boiling point is $T_b + \Delta T_b$, with ΔT_b given by (8-10).

lowering of the freezing point is proportional to the molality of the solute, m:

$$\Delta T_f = T_f - T_f^* = -k_f m \qquad (8\text{-}11)$$

The constant k_f is called the *molal freezing-point lowering;* typical values are given in Table 8-1. As with boiling-point elevation, the effect is additive when several solute species are present.

Figure 8-5 illustrates schematically the reason for the freezing-point depression of a solution when the solid solvent that separates is pure. If the solute and the solvent form a solid solution (also called mixed-crystal formation), the situation is more complex.

□ **Molecular Weight from Boiling-Point Elevation**   When 5.0 g of an unknown organic compound is dissolved in 100 g benzene, the boiling point of the benzene rises 0.65 K above its normal value. What is the approximate molecular weight of the unknown substance?

Example 8-8

Solution   This solution is sufficiently dilute that (8-10) should be a reasonable approximation. Since k_b for benzene is 2.64 K kg mol⁻¹, the molal concentration of the unknown solution is

$$m = \frac{\Delta T_b}{k_b} = \frac{0.65\ \text{K}}{2.64\ \text{K kg mol}^{-1}}$$
$$= 0.25\ \text{mol kg}^{-1}\ \text{or}\ 0.25\ \text{molal}$$

Since there is 0.25 mol per kilogram of solvent, or 0.025 mol per 100 g

BOILING-POINT ELEVATION [EQUATION (8-10)]		FREEZING-POINT LOWERING [EQUATION (8-11)]		Table 8-1
Substance	k_b/(K kg mol⁻¹)	Substance	k_f/(K kg mol⁻¹)	**Some Boiling-Point-Elevation and Freezing-Point-Lowering Constants**
H_2O	0.52	H_2O	1.86	
Ethanol (C_2H_5OH)	1.23	Naphthalene ($C_{10}H_8$)	6.8	
Carbon disulfide (CS_2)	2.34	Benzene (C_6H_6)	5.1	
Benzene (C_6H_6)	2.64	Phenanthrene ($C_{14}H_{10}$)	12.0	
Toluene ($C_6H_5CH_3$)	3.37	Camphor ($C_{10}H_{16}O$)	40	

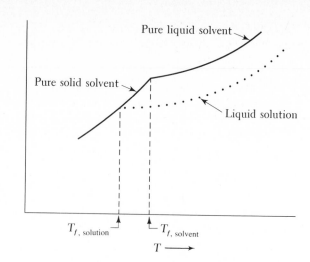

Figure 8-5 Freezing-Point Lowering of a Solution from Which Pure Solid Solvent Separates

The solid curve represents the vapor pressure of the pure solvent as a function of temperature. The portions of the curve representing the vapor pressure of the liquid and that of the solid meet at the normal freezing point, since at that temperature these two phases are in equilibrium and their vapor pressures must thus be the same (see Fig. 5-7 or Fig. 5-8). The freezing point of the solution is also necessarily the temperature at which the vapor pressure of solvent above the solution is the same as that of the solvent above the crystallized solid phase. Since the vapor pressure of the solution (represented by the dotted curve) is lower than that of the pure solvent, it is necessary to go to a lower temperature to make it equal to that of the solid solvent.

solvent, we know that the 5.0-g sample represents 0.025 mol. Hence the weight per mole is

$$\widetilde{M} = \frac{5.0 \text{ g}}{0.025 \text{ mol}} = \underline{2.0 \times 10^2 \text{ g mol}^{-1}}$$

no higher precision being warranted. ∎

Exercise 8-8

□ An unknown weight of a compound with MW 111 was dissolved in 90.0 g carbon disulfide. The solution boiled 0.89 K higher than pure carbon disulfide. What was the weight of the compound? ∎

Example 8-9

□ **Molecular Weight from Freezing-Point Depression** When 25.0 mg of a naturally occurring organic substance of unknown structure, just isolated from the bark of a West African tree known to possess unusual medicinal properties, was dissolved in 1.00 g camphor, the melting point of the camphor fell by 2.0 K. What is the approximate molecular weight of the unknown substance?

Solution Applying (8-11) with the value of k_f for camphor equal to 40, we have

$$m = -\frac{\Delta T_f}{k_f} = \frac{2.0 \text{ K}}{40 \text{ K kg mol}^{-1}} = 0.050 \text{ mol kg}^{-1}$$

Since the solution contains 25 mg of unknown per gram of camphor, it contains 25 g per kilogram of camphor. Thus 25 g of the unknown is 0.050 mol, so that the weight per mole is

$$\widetilde{M} = \frac{25 \text{ g}}{0.050 \text{ mol}} = \underline{5.0 \times 10^2 \text{ g mol}^{-1}}$$

no higher precision being warranted. ∎

Exercise 8-9

□ When 20.0 mg of an unknown substance was dissolved in 20.0 g benzene the melting point of the benzene was lowered by 0.055 K. What is the approximate molecular weight of the unknown compound? ∎

Henry's Law In a dilute solution, the equilibrium concentration (that is, the solubility) of a dissolved gas A is proportional to its partial pressure:

$$[A] = k_H P_A \qquad (8\text{-}12)$$

The Henry's law proportionality constant, k_H, depends on the nature of the gas and the solvent. The increase of gas solubility with increasing pressure is responsible for the difficulties that deep-sea divers experience with "the bends" if they rise too rapidly to the surface. The atmosphere they breathe at appreciable depths is at relatively high pressure, and if they rise too rapidly the excess dissolved gas does not equilibrate fast enough with the gradually lowered pressure in their lungs. Consequently the fluids in their tissues are supersaturated with gas, which separates as fine (and usually very painful and dangerous) bubbles throughout their body tissues. One way to minimize this effect is to dilute the essential oxygen in a diver's atmosphere with a gas that has very low solubility even at high pressure, such as helium (rather than with nitrogen as in air).

Henry's law applies, at least approximately, to dilute solutions not only of gases but also of most volatile solutes. For the latter it is more natural to say that the partial vapor pressure of the solute A is proportional to its concentration, rather than the other way around as in (8-12). Furthermore it is often more satisfactory to use the mole fraction of A rather than its molarity, thus leading to

$$P_A = k'_H x_A \qquad (8\text{-}13)$$

Equation (8-13) is approximately equivalent to (8-12) for dilute solutions, for which the mole fraction and the molarity are approximately proportional to one another.

Since Henry's law may be stated in several ways, using different concentration units, it is especially important to note the units of the Henry's law constants in any tabulation before using such constants.

Osmotic Pressure Suppose a solution is separated from the pure solvent by a membrane that is permeable to solvent molecules but not to solute species (Fig. 8-6); such a membrane is called *semipermeable*. If both solution and pure solvent are under the same (usually atmospheric) pressure, it is observed that more solvent molecules pass through the semipermeable membrane in the direction of the solution than in the reverse direction, so that the solution becomes continuously more dilute. This may be interpreted as resulting from the fact that the vapor pressure of the pure solvent is greater than its partial pressure above the solution. For the two phases to be in equilibrium, the solvent vapor pressure must be the same in each. The vapor pressure of the solvent in the solution can be raised without changing the temperature by applying a higher external pressure to the solution, because it happens that the vapor pressure of a liquid is raised by applied pressure. The excess pressure on the solution, at which there is equilibrium so that there is no net passage of solvent through the diaphragm in either direction, is called the *osmotic pressure,* here designated Π (Greek capital pi). For an ideal solution, or any sufficiently dilute solution,

$$\Pi V = n_2 RT \qquad (8\text{-}14)$$

where V = volume of solution
n_2 = total number of moles of solute present
R = gas constant
T = absolute temperature

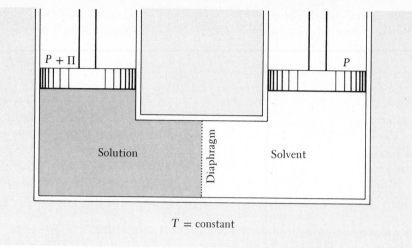

$P + \Pi$ P

Solution   Diaphragm   Solvent

$T = \text{constant}$

Figure 8-6 Osmotic Pressure
The solvent and the solution are separated by a diaphragm that is permeable to the solvent and impermeable to the solute. Solvent molecules constantly pass through the diaphragm in both directions. If the same pressure were applied to both pistons, the system would not be at equilibrium because the partial pressure of the solvent on the solution side would be lower than the vapor pressure of the pure solvent. As a result, solvent would flow through the diaphragm from right to left. The partial pressure of the solvent can be increased by applied pressure. Thus, if sufficient additional pressure is applied to the left-hand piston to increase the partial pressure of the solvent in the solution until it equals the vapor pressure of the pure solvent, net movement of solvent through the diaphragm can be prevented. The additional pressure required to achieve this condition is called the osmotic pressure, Π.

If the osmotic pressure Π is expressed in atmospheres, V in liters, and T in kelvins, then the appropriate value of R is 0.08206 liter atm mol^{-1} K^{-1}. Because (8-14) is identical in form with the ideal-gas law, a great deal of effort was expended during the last century in trying to explain the osmotic pressure law in terms of some sort of ideal-gas behavior of the solute molecules, but such interpretations have little significance. Osmotic pressure is not exerted by solute molecules; it is a pressure that must be applied to the solution to achieve equilibrium with pure solvent.

Measurements of osmotic pressure are of considerable value in providing estimates of the molecular weights of very large molecules, including both natural and synthetic polymers.

Example 8-10

□ **Molecular Weight from Osmotic Pressure** The osmotic pressure of a benzene solution containing 5.0 g polystyrene per liter was found to be 7.6 torr at 25°C. What is the approximate molecular weight of this sample of polystyrene?

Solution After substituting $\Pi = 7.6 \text{ torr}/(760 \text{ torr atm}^{-1}) = 0.0100$ atm, $V = 1.00$ liter, $n_2 = 5.0 \text{ g}/\widetilde{M}$ (where $\widetilde{M}$ is the weight per mole of the sample), and $T = 298$ K,

$$\widetilde{M} = \frac{5.0 \text{ g}(0.0821 \text{ liter atm mol}^{-1} \text{ K}^{-1})(298 \text{ K})}{(0.0100 \text{ atm})(1.00 \text{ liter})}$$
$$= \tfrac{122}{0.0100} \text{ g mol}^{-1} = \underline{12.2 \times 10^3 \text{ g mol}^{-1}} \quad \blacksquare$$

□ A solution of a polymer of MW 49×10^3 has an osmotic pressure of 6.9×10^{-3} atm at 25°C. How many grams of the polymer are dissolved in 100 ml solution? ■

Osmotic phenomena are responsible for maintaining a balance of fluids in the body, particularly in the functioning of the kidney. The contents of body cells, such as red blood corpuscles, are normally at osmotic equilibrium with the surrounding blood plasma. If red blood cells are put in distilled water, they swell and eventually burst. This does not happen in a salt solution that has the same osmotic pressure as blood plasma. Such a solution is known as *isotonic* (Greek: *isos*, equal; *tonos*, tension) saline solution.

Osmosis also plays a role in one method of desalting seawater and other brines. When salt water is separated from fresh water by a membrane that permits only the passage of water, an excess pressure on the salt water equal to its osmotic pressure II is required to prevent its dilution by the fresh water. If a pressure larger than II is applied to the solution (e.g., by a piston), pure water is squeezed out. The semipermeable membrane acts in effect as a molecular filter. This procedure is called reverse osmosis. The average salt content of ocean water corresponds to an osmotic pressure of about 22 atm, and reverse osmosis of ocean water therefore requires pressures in excess of this value. Lower pressures suffice for less salty water.

Comparison of Magnitudes of Colligative Properties The relative magnitudes of the effects manifested in the different colligative properties are instructive. The relative vapor-pressure lowering is usually very small. For example, in a 1.00 wt percent aqueous solution of sucrose ($C_{12}H_{22}O_{11}$, MW = 342) the mole fraction of solute is 5.3×10^{-4}, so that the vapor pressure of water is lowered by 5.3×10^{-2} percent. At 25°C this means a pressure drop of 0.013 torr out of a total pressure of 23.8 torr. The calculated boiling-point increase is 0.015 K and the calculated freezing-point lowering, 0.054 K. Both temperature changes can easily be measured, although special thermometers are needed. The osmotic pressure is calculated to be 0.71 atm at 25°C, a pressure that can easily be measured without special equipment.

8-4 Colloidal Solutions

Very small particles suspended in a liquid or in a gas may settle to the bottom only very slowly or may not even settle at all over long periods of time. Suspensions of fine particles that do not settle are called *colloidal suspensions* or *colloids;*[2] if the medium in which they are suspended is a liquid, they are often referred to as colloidal solutions. At ordinary temperatures the suspended particles execute brownian motion, a random, rapid, vibratory motion that never ceases and prevents the particles from settling.

The particles in many colloidal solutions are so small that they are not even visible with a high-powered microscope, and special means must be used to

---

[2]The term *colloid* comes from the Greek: *kolla*, glue. Many glues are made from the gelatinlike material collagen, which is the main constituent of fibrous connective tissue in animals. Gelatin is a typical colloid.

observe them. One such method is to use the Tyndall effect. A beam of light passing through a colloidal solution forms a trace that is visible from the side. This effect is observed when a beam of strong light is visible in hazy or smoke-filled air. The beam can be seen because the suspended particles, even though they are too small to be observed directly as individual particles, scatter some of the light sideways. Information about the size and shape of colloidal particles can be obtained by careful studies of the pattern of light scattered by a colloidal solution.

Because of their size, colloidal particles cannot pass through the pores of some membranes, such as cellophane, collodion, or parchment paper. This fact is often used to free a colloidal suspension of small molecules and ions that may happen to be dissolved in the suspending liquid. The suspension is placed in a cellophane or other suitable bag immersed in a bath of solvent that is frequently renewed. The dissolved small ions and molecules pass through the pores into the solvent and the colloidal solution remains in the bag. This process is called *dialysis* (Greek: *dia,* through; *lyein,* to loosen). Both natural and artificial kidneys function essentially by dialysis, small molecules and some salts being eliminated from the body in the urine and large molecules being retained.

The behavior of colloidal solutions is in many respects similar to that of ordinary solutions, and there is no sharp boundary between the two. The suspended particles may be solid or liquid aggregates of many molecules that are themselves insoluble or are soluble to a very small degree in the suspending liquid. They may also be single large molecules. In this case the suspension is in effect an intimate mixture of solute and solvent particles on the molecular level, just like any other solution. However, the fact that the solute particles are so much larger than the solvent molecules makes it sometimes desirable to classify the solution as colloidal. In other words, the distinction between ordinary and colloidal solutions is best based on the criterion of *size* of solute particles rather than on whether the solute particles are molecules or aggregates of molecules, even though the size distinction is not sharp and clear-cut. Particles with dimensions in the range 10^2 to 10^5 Å (10^{-5} to 10^{-2} mm) are often considered colloidal.

Many proteins and other large molecules form colloidal solutions, as do some simpler compounds and even some elements. For example, a colloidal suspension of gold in water can be prepared. At a particle size that has been found to yield long-lasting colloidal suspensions, this solution has a beautiful red color. One such suspension prepared by Michael Faraday has been kept apparently unchanged for more than 100 years. At other particle sizes, the suspension may be yellow, blue, or violet. The color is caused by interference (Sec. 14-1) between the light waves scattered from different parts of the particles and therefore depends on the particle size. Colloidal suspensions of gold in glass cause a deep ruby color also, at a dilution of about 1 part by weight of gold to 50,000 parts of glass. Colloidal metal sulfides and other precipitates are often obtained inadvertently (and to the despair of the beginning analyst) when the sulfides or other compounds of metal ions are precipitated.

The term *colloid* is used to refer to more than colloidal solutions. The characteristic that all colloidal systems have in common is that some substance, finely divided in at least one dimension, is dispersed through a second one. Water is a common medium for dispersion, but some colloids are solid, such as the ruby glass mentioned above. The dispersion medium may also be a gas; fogs and smokes are often colloidal suspensions of fine particles of liquids and solids, respectively, in air. Foams and other thin films are properly regarded as colloidal too, when one of their dimensions is in the colloidal range. Surface effects play a major role in colloidal phenomena because any finely divided substance has a surface area greatly increased over that which it would have if not so divided. Molecules and ions are attracted to any surface by van der Waals and coulomb forces, and if the surface is very large, the numbers of particles so attracted may also be large. Thus, this phenomenon of fixation of particles on surfaces (*adsorption*) plays an important part in many colloidal systems.

8-5   Separation of Substances

Highly purified substances are essential in many different fields of research and in industry and commerce as well, since traces of impurities sometimes significantly alter physical and chemical properties. The essence of purification is the separation of a substance from others with similar properties. The most important methods for separation include filtration, fractional crystallization, distillation, and chromatography.

Filtration   A filter permits the separation of material suspended in a fluid by virtue of differences in particle size. Macroscopic particles larger than a certain minimum size do not pass through the filter. The role of the filter may also be played by a semipermeable membrane, considered in the discussions of osmosis and colloids.

Fractional Crystallization   This term implies a separation of one of several solutes from a solution by crystallization in relatively pure form. Separation by fractional crystallization may be accomplished by slowly lowering the temperature of a solution or by evaporating the solvent until crystals form. Which of the different possible crystalline substances will form first depends upon the temperature of the solution, upon the concentrations of the various solutes present, and upon their solubilities at different temperatures. The temperature variation of the solubility often differs markedly from one compound to another, even for substances that are very similar chemically. For example, the solubility of KCl is more than twice as great at $100°C$ as at $0°C$ while that of NaCl increases by only about 11 percent over this same temperature interval.

Crystallization is a highly selective process, so that even in the presence of many impurity molecules or ions only a particular species usually enters a particular growing crystal and the impurities remain in solution. The reason for this is structural: each molecule or ion has a particular size and shape, and sometimes also particular ways by which it may interact strongly or specifically with its environment, for example, by hydrogen bonding or by ionic forces. Other molecules or ions cannot normally fit into the spaces in the crystal that are tailored for this particular species, nor can they interact in the specific way it does. On the other hand, if another molecule or ion is very similar in size and shape to one of the structural units of the crystal, a solid solution or *mixed crystal* may form and purification by fractional crystallization is ineffective.

Distillation and Related Methods   Among the most effective and widely used separation methods are those based upon differences in vapor pressure. The term *distillation* implies converting a liquid or solid phase into a gas and then condensing it again by cooling. With a mixture of nonvolatile and volatile substances in which only the nonvolatile materials are of interest, a separation may be achieved by evaporation alone (without recondensation). For example, salt is obtained by evaporation of seawater and other brines, and in the technique of freeze-drying, a solvent (usually water) is removed by sublimation from a solution frozen in a thin layer.

It often happens that two or more desired components of a mixture to be separated are volatile and may even have similar vapor pressures. Separation by distillation can still be achieved as long as the vapor is richer in one of the components than is the liquid. Consider, for example, a liquid mixture of N_2 and O_2. The vapor in equilibrium with a boiling sample of the liquid is always richer than the liquid in N_2, which has a higher vapor pressure than O_2 at any temperature. If the vapor is condensed it therefore yields a liquid richer in N_2 than the original liquid. Distillation of the condensed vapor will lead to still further enrichment in N_2. Such a procedure of successive redistillations is called *fractional distillation*. There are ways of doing it in one continuous step using an appropriately designed *distillation column,* rather than in many distinct steps.

Since the vapor is richer in N_2 than is the boiling liquid from which it is removed, the liquid left behind must necessarily be richer in O_2 than it was initially. As distillation continues, the remaining liquid becomes progressively richer in O_2.

Distillation is an extremely important industrial and laboratory method with many highly sophisticated variations. One of its most important applications is as a stage in the refining of petroleum. Each year, more than 700 billion liters of crude oil are fractionated into various grades of gasoline, kerosene, oils, and other products in the United States alone.

Chromatography While the operations of filtration, crystallization, and distillation have been known and used for millennia, many other important separation methods are relatively modern. Perhaps the most versatile of these is chromatography. The method was originated in 1906 by the Russian botanist Tswett to separate and isolate plant pigments; his term *chromatography* (Greek: *chroma,* color) is somewhat misleading because the method is not limited to colored substances. The great potential of chromatographic methods first began to be realized in the 1930s and 1940s.

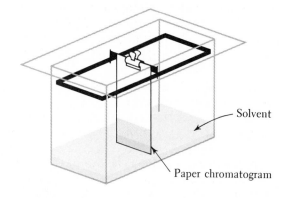

Figure 8-7 Paper Chromatography

A drop of the mixture to be separated is placed near one end of a strip of filter paper, which is then suspended in a jar in such a way that the mixture to be separated is at the lower end of the strip. Enough liquid solvent of an appropriate composition is placed in the bottom of the jar so that it just touches the lower edge of the paper strip, and the jar is then closed. The solvent rises in the paper by capillary action, and the components of the mixture, migrating at different rates up the paper strip, gradually separate. If they are colored, their positions are visible. If not, they can be detected in various ways; for example, they might be visible when viewed in ultraviolet light or they might react with an appropriate reagent to give a colored product.

Solvent

Paper chromatogram

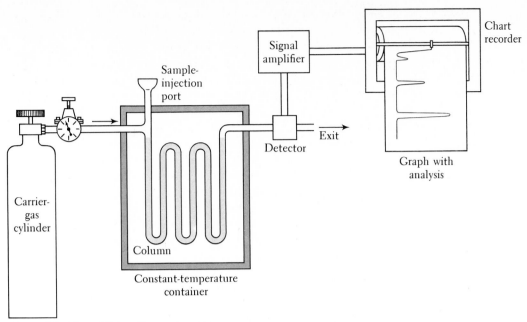

Figure 8-8 Vapor-Phase Chromatography
The mixture to be separated flows as a vapor in an inert *carrier gas,* e.g., helium, through a "column", a tube filled with finely divided solid that has been coated with a film of a liquid such as a high-boiling, essentially nonvolatile, hydrocarbon or silicone (Sec. 24-3) oil. As the mixture of vapor and carrier gas passes slowly through the column, a gradual separation of the components of the mixture occurs. The separated fractions are detected and may be quantitatively estimated as they emerge from the column. One common detector is a device that ionizes each component and measures the resulting current; it is capable of detecting as little as 10^{-10} g s^{-1}.

In one form of chromatography, essentially that originated by Tswett, a solution of the mixture to be separated is poured into a vertical glass tube packed with a powdered solid, such as aluminum oxide (Al_2O_3) or silica gel (especially prepared SiO_2), capable of adsorbing the substances to be separated. Initially the components of the mixture are adsorbed in the top portion of the packed column. Then a suitable solvent is allowed to flow slowly through the column. The substances to be separated move gradually down the column because they are at times dissolved in the flowing solvent and at other times adsorbed on the column. The rate at which a given solute moves depends on a delicate balance between its interactions with the solvent and with the column. With appropriately chosen solvent and adsorbent, the components of the mixture emerge in well-separated portions at the bottom of the column, and their solutions can be collected in different containers.

An important variant of the general method is *paper chromatography,* in which the role of the solid adsorbent is played by a strip of filter paper (Fig. 8-7). Some disadvantages of using paper, such as the fact that it is hard to be sure that different batches have the same properties, can be overcome by using *thin-layer chromatography* (TLC), in which a glass plate coated with adsorbent is used instead of paper. Both paper chromatography and TLC are of great value in the analysis of a great variety of complex mixtures; they are rapid methods that require only very modest equipment and are applicable to microgram quantities. One of the most sensitive current methods of separation and detection is *vapor-phase chromatography* (VPC) (Fig. 8-8). Vapor-phase chromatography can be used in the quantitative analysis of a sample of a mixture weighing as little as 10^{-6} or 10^{-7} g. The volatility of the sample to be analyzed need not be high, and it can be increased by carrying out the experiment at an elevated temperature.

Chromatography of liquid solutions under high pressure (HPLC) is now used extensively for separating complex mixtures. It has the advantage that volatilization is not required, and it can be done on a large scale; it may, however, consume large quantities of expensive solvents.

The common feature of all chromatographic methods is that a fluid mobile phase containing the sample to be analyzed passes over a stationary phase with which the components of the mixture can interact reversibly. As long as different components are distributed differently between the mobile and immobile phases at any moment, they will migrate with different speeds and hence can be separated. Chromatography has played a major role in many of the spectacular advances in chemistry and biochemistry in recent decades just because it provides convenient, versatile, sensitive, and highly specific methods of analysis.

Summary

A solution is a homogeneous mixture. The concentration of a solution is a quantitative measure of the relative amounts of solute and solvent present. Concentration units include volume percentage and weight percentage, molarity, molality, and mole fraction. A solution is said to be saturated when there is equilibrium between dissolved solute and any excess solute. Supersaturated solutions contain more solute than they would under equilibrium conditions.

A number of properties of dilute solutions can be related to the concentration by simple quantitative laws. The partial pressure of the solvent is given by Raoult's law: $P_1 = P_1^* x_1$ and the solubility of gases by Henry's law: $P_A = k_H' x_A$. The boiling point of a dilute solution of a nonvolatile solute is higher than that of the pure solvent, and the freezing point, if pure solvent crystallizes, is lower than that of the pure solvent. These changes in temperature can be expressed in terms of the molality of the solute and constants characteristic of the solvent: $\Delta T_b = k_b m$ and $\Delta T_f = -k_f m$. The osmotic pressure (Π), the pressure that must be exerted on a solution to maintain equilibrium when the solution is separated from pure solvent by a membrane through which only the solvent can pass, is given by $\Pi V = n_2 RT$.

Colloidal solutions are suspensions of fine particles that do not settle. Particles with dimensions in the range 10^2 to 10^5 Å are often considered colloidal. A beam of light passing through such a solution may be visible from the side (Tyndall effect). A colloidal solution may be freed of small molecules and ions by allowing them to pass through the fine pores of a membrane (dialysis). Other types of colloids may also be prepared. Surface effects, e.g., adsorption, are important in all colloidal phenomena because of the large surface-to-volume ratio of finely divided matter.

Among the methods of separation and purification are filtration and fractional crystallization; the effectiveness of the latter depends on differences in solubilities and on the temperature variation of solubilities. Separation by distillation requires a difference in composition between the mixture to be distilled and the vapor in equilibrium with it. In chromatography, substances to be separated are distributed between a mobile phase and a stationary phase over which the mobile phase flows at a slow but steady rate.

Terms and Concepts

Problems and Questions

8-1 Gaseous Solution The average composition by volume of a mixture of gases is Ar, 25.0 percent; H_2, 49.1 percent; and N_2, 25.9 percent. What is the mole fraction of N_2 in the mixture? Why?

8-2 Concentration Units A solution prepared by dissolving 66.0 g urea, $(NH_2)_2CO$, in 950 g water has a density of 1.018 g ml^{-1}. Urea does not dissociate in water. Calculate the concentration of urea in (*a*) weight percentage; (*b*) mole fraction; (*c*) molarity; (*d*) molality.

8-3 Hydrobromic Acid Solution (*a*) Gaseous hydrogen bromide is very soluble in water. What volume of the gas, measured at 20°C and 750 torr, is needed to prepare 300 ml of 4.0 *M* hydrobromic acid? (*b*) If the density of the resulting solution is 1.10 g ml^{-1}, what is the mole fraction of water in the solution? Note that HBr is completely dissociated into H^+ and Br^-.

8-4 Solution of Sodium Carbonate A 14.0 percent solution of Na_2CO_3 in water has a density of 1.146 g ml^{-1}. Eighty-five milliliters of this solution was evaporated to dryness. (*a*) What weight of dry Na_2CO_3 was obtained? Suppose that the residue had the composition $Na_2CO_3 \cdot 10H_2O$; what weight of it would be present? (*b*) What is the molarity of the solution? (*c*) What is the molality?

8-5 Density of Ethanol Solution A 9.9 *M* solution of ethanol (C_2H_5OH) in water contains just 50.0 percent ethanol by weight. What is the density of the solution?

8-6 Ethanol Solution A certain aqueous solution contains 13.0 wt percent ethanol, C_2H_5OH, and has a density of 0.96 g ml^{-1}. What is the concentration of ethanol in grams per liter and what is its molarity?

8-7 Solubility of Different Forms of the Same Substance Some substances can exist in two modifications, one more stable than the other. Which modification must have the lower vapor pressure? Which must have the lower solubility in any solvent?

8-8 Fructose Solution A certain aqueous solution of fructose ($C_6H_{12}O_6$) contains 40.0 percent by weight of fructose and has density 1.180 g ml^{-1} at 20°C. (*a*) What is the concentration of fructose in grams per liter of solution? (*b*) What is the molarity of fructose in this solution? (*c*) What is the molality of fructose? (*d*) What is the mole fraction of water?

8-9 Perchloric Acid Solution Suppose that you need some pure perchloric acid and wish to prepare it by mixing ordinary commercial concentrated aqueous $HClO_4$ (density 1.67 g ml^{-1}, 70.6 percent perchloric acid by weight) with 300 g of 6.1 percent (by weight) solution of Cl_2O_7 in $HClO_4$ that contains no water. Cl_2O_7 reacts with water to form $HClO_4$. What volume of the concentrated aqueous $HClO_4$ should you add, and how many grams of pure perchloric acid would you obtain?

8-10 Glycerol Solution The mole fraction of water is 0.884 in a 4.8 *M* aqueous solution of glycerol, $C_3H_5(OH)_3$. (*a*) What is the percentage by weight of glycerol in the solution? (*b*) What is the approximate freezing point of the solution?

8-11 Molecular Weight by Freezing-Point Lowering The molal freezing-point-lowering constant for benzene is 5.1 K kg mol^{-1}. When 0.196 g compound A is dissolved in 20.0 g benzene, the freezing point decreases by 0.51 K. (*a*) What is the approximate molecular weight of A? (*b*) If compound A contains just 25.0 percent carbon (by weight), what is the exact molecular weight of A? Explain any discrepancy from the value obtained in (a).

8-12 Automobile Radiator Ethylene glycol, CH_2OHCH_2OH, is a common antifreeze liquid for use, mixed with water, in automobile radiators. (a) What should be the approximate molality of solute in an aqueous solution if the solution is to freeze at a temperature no higher than $-20°C$ ($-4°F$)? (b) What volume of ethylene glycol (density $1.11\ g\ cm^{-3}$) should be added to 30 liter water (density $1.00\ g\ cm^{-3}$) to give the molality calculated in (a)? (c) What would be the approximate boiling point of this solution at 1 atm?

8-13 Solution of Mercuric Chloride When a 47.5-g sample of mercuric chloride, $HgCl_2$, is dissolved in 1000 g H_2O, the boiling point of the solution is raised by 0.085 K. What conclusions can be drawn about the dissociation of $HgCl_2$ in solution?

8-14 Boiling-Point Elevation A solution of 0.318 g of a compound in 35.4 g benzene has a boiling point 0.026 K higher than that of pure benzene. The compound contains 49.3 wt percent arsenic. (a) What is the molecular weight of the compound? (b) How many As atoms does one molecule of the compound contain?

8-15 Concentration Units A certain aqueous solution contains 9.0 wt percent ethanol (MW = 46) and has a density of 0.95 g ml^{-1}. (a) What is the concentration of ethanol in grams per liter and what is its molarity? (b) If this same solution contains one (and only one) other solute, Y, and freezes at $-10.2°C$, what is the sum of the molalities of ethanol and Y? (c) If the solution contains 10.0 wt percent of Y, what is the molality of ethanol, what is the molality of Y, and what is the molecular weight of Y?

8-16 Molal Boiling-Point Elevation When 5.12 g naphthalene ($C_{10}H_8$) is dissolved in 100 g CCl_4, the boiling point of the CCl_4 is raised 2.00 K. What is the molal boiling-point-elevation constant for CCl_4?

8-17 Temperature Variation of Solubility The solubility of $(NH_4)_2SO_4$ in water is 706 g liter^{-1} of water at 0°C and 1033 g liter^{-1} at 100°C. Suppose that 500 g of the salt is added to 200 ml water at 100°C and that the mixture is stirred until no more

of the salt dissolves and is then filtered at 100°C. Assume that no water evaporates and that equilibrium conditions are attained. (a) How much $(NH_4)_2SO_4$ is retained by the filter? (b) If the solution passing through the filter is chilled to 0°C, how much of the dissolved salt crystallizes out? (c) If you did not take account of the difference in density of water at 0°C and 100°C (1.000 and 0.958 g cm^{-3}, respectively), by how much is your answer to (b) in error?

8-18 Solubility of a Gas The solubility of carbon dioxide in water is 335 mg per 100 g water at 0°C when the partial pressure of carbon dioxide is 760 torr. How many milligrams of carbon dioxide will dissolve in 1.00 kg water at 0°C when the partial pressure of carbon dioxide is 5.0 atm? What is the freezing point of this solution?

8-19 Solubility of Oxygen At 20°C the Henry's law constant for the solubility of oxygen in water is $k'_H = 3.88 \times 10^4$ atm. How many grams of oxygen will dissolve in 1 liter of water that is in equilibrium with air at 1 atm (21 percent by volume oxygen)?

8-20 Freezing Point of Water At 0°C, 23.54 cm^3 nitrogen at a partial pressure of 1 atm will dissolve per liter of water. Under the same conditions, 48.89 cm^3 of oxygen will dissolve. Calculate the amount by which the freezing point of air-free water is lowered by being saturated with air. (Composition of air: 79 vol percent N_2, 21 vol percent O_2.)

8-21 Molecular Weight from Freezing-Point Lowering A 6.0-g sample of a solute dissolved in 100 g H_2O lowers the freezing point of water by 1.02 K. Calculate the apparent molecular weight of the solute.

8-22 Molecular Weight of a Protein The osmotic pressure at 25°C of a solution containing 1.35 g protein in 100 ml water was found to be 9.9×10^{-3} atm. What molecular weight would account for this datum?

8-23 Relationship between Osmotic Pressure and Freezing-Point Lowering Calculate the approximate osmotic pressure, at 25°C, of an aqueous solution that freezes at $-0.035°C$.

8-24 Vapor-Pressure Lowering A 4.0-g sample

of a substance X is dissolved in 156 g benzene (MW = 78) and found to lower the vapor pressure of benzene from 200.0 to 196.4 torr. (*a*) What are the mole fractions of benzene and of X? (*b*) What is the molecular weight of X?

8-25 Vapor Pressure of the Solution of a Volatile Solute The following data apply at 25°C. Solid iodine has a vapor pressure of 0.31 torr. Chloroform (liquid), $CHCl_3$, has a vapor pressure of 199.1 torr. In a saturated solution of iodine in chloroform, the mole fraction of iodine is 0.0147. (*a*) What is the partial pressure of iodine vapor at equilibrium with such a saturated solution? (*b*) Assuming Raoult's law, what is the total vapor pressure of this solution?

8-26 Raoult's Law for a Mixture of Volatile Liquids At 80°C the vapor pressure of benzene (C_6H_6) is 753 torr and that of toluene (C_7H_8) is 290 torr. What is the total pressure and what is the composition of the vapor over a solution of exactly $\frac{2}{3}$ mol benzene and exactly $\frac{1}{3}$ mol toluene? Assume Raoult's law for both components of the solution.

8-27 Heptane-Octane Mixture The liquids heptane (MW = 100) and octane (MW = 114) form a nearly ideal solution. At 40°C, heptane has a vapor pressure of 100 torr and octane a vapor pressure of 40 torr. Suppose that a solution is prepared by mixing equal weights of these two liquids. (*a*) What is the mole fraction of each component in this solution? (*b*) What is the partial pressure of each component over this solution at 40°C? (*c*) Suppose that some of the vapor described in (*b*) is condensed to a liquid. What would be the mole fraction of each component in this liquid, and what would be the vapor pressure of each component above this liquid at 40°C?

8-28 Distillation of an Ideal Solution Liquids A and B form an ideal solution. The vapor pressures of pure A and pure B at 100°C are 300 and 100 torr, respectively. Suppose that the vapor above a solution composed of 1.00 mol A and 1.00 mol B at 100°C is collected and condensed. This condensate is then heated to 100°C and the vapor is again condensed to form liquid X. What is the mole fraction of A in X?

8-29 Distillation of an Ideal Solution Liquids A and B form an ideal solution. At 75°C the vapor pressures of pure A and pure B are 100 torr and 200 torr, respectively. (*a*) What is the pressure of the vapor in equilibrium with a solution prepared from 3.00 mol A and 1.00 mol B? What is the mole fraction of A in this vapor? (*b*) Suppose that the vapor described in (*a*) is condensed to a liquid and the resulting liquid is heated again to 75°C. What will be the mole fraction of A in the vapor in equilibrium with this liquid at 75°C?

8-30 Equilibration of Two Solutions through the Vapor Two open 1-liter beakers, A and B, containing aqueous solutions of NaCl are placed together in a box, which is then sealed tightly. Initially there are 600 ml solution in A and 300 ml in B. Consider two situations: (*a*) the concentration of the solution in A is initially twice that in B; (*b*) the concentration in A is initially half that in B. Assume that equilibrium is established by waiting sufficiently long (perhaps several days). What will be the relation between the volumes of the solutions in A and B for case (*a*) and for case (*b*)? Explain clearly.

8-31 Equilibration of Solutions Imagine that two 1-liter beakers, A and B, each containing an aqueous solution of fructose (a nonvolatile sugar with MW = 180) are placed together in a box, which is then sealed. (The concentrations of the solutions are not necessarily the same.) The temperature remains constant at 26°C. Initially there is 600 ml of solution in A and 100 ml of solution in B.

As the solutions stand in the sealed box, their volumes change slowly for a while. When they stop changing, beaker A contains 400 ml and beaker B contains 300 ml. It is then determined that the solution in A is 1.5 *M* in fructose and has a density of 1.10 g ml^{-1}.

(*a*) What is the molar concentration of fructose in the solution in beaker B at the end? Explain.
(*b*) Calculate the concentration of fructose in the solution in A at the start.
(*c*) Calculate the concentration of fructose in the solution in B at the start.
(*d*) The vapor pressure of pure water at 26°C is 25.2 torr. What is the pressure of water vapor in the box at the end, after the volumes have stopped changing?

Acids, Bases, and Ionic Solutions

9

Aqueous solutions merit special attention. Water is a good solvent for a great variety of compounds, ranging from ionic substances like common salt to covalent compounds such as sugar, ammonia, and hydrogen chloride, and even to giant protein molecules. An understanding of the chemical properties of aqueous solutions is essential to an understanding of many different fields of chemistry. Among the most common reactions in solution are those in which protons are exchanged between the reactants—acid-base reactions. Another fundamental class of reactions is that in which electrons are exchanged—oxidation-reduction reactions, which are considered in Chap. 10. The chemical equilibria involving acids, bases, and salts are taken up quantitatively in Chaps. 12, 13, and 27.

9-1 Acids, Bases, and Salts

Early Views The alchemists and earliest chemists used the term *acid* (Latin: *acidus,* pungent, sharp) to refer to any of a variety of substances that had certain qualities and chemical properties in common, such as a characteristic sour, slightly sharp (acidic) taste, reaction with some metals to produce hydrogen, and the ability to cause certain color changes in various vegetable dyes. The first modern chemist, Lavoisier, noted two centuries ago that the oxides of sulfur, phosphorus, nitrogen, and carbon all produced acidic solutions. After his discovery of the element common to all these compounds, he chose the name *oxygen* for this element on the basis of the misconception that all acids contained oxygen (Greek: *oxys,* sharp, pungent, acid; *gennan,* to form). However, it was soon recognized that acids containing no oxygen existed and that the acid property was related to hydrogen rather than oxygen. In 1815, Davy and Dulong defined an acid as a substance containing hydrogen that could be replaced by a metal, a point of view later emphasized by Liebig as well.

As the concept of acids was developing, several other important classes of compounds were also recognized, including bases and salts. Solutions of typical bases (or alkalies) could be obtained by soaking the ashes left after burning various plants and filtering out the insoluble material. "Potash" crystallized from such a solution was soon recognized as potassium carbonate, K_2CO_3, and the element potassium owes its name to this process. Bases have characteristic properties in common, just as acids do; they have a somewhat bitter taste, and produce particular colors with various vegetable dyes (different from the colors produced by acids).

The modern view of the nature of solutions of acids, bases, and salts in water was first proposed in 1887 by Svante Arrhenius (and was greeted with much skepticism before the evidence in favor of it was recognized as overwhelming). This theory of *ionic dissociation* asserts that acids, bases, and salts are partially or completely dissociated into ions in aqueous solution. When dissolved in water, an acid HX yields H^+ and a charged ion, X^-. Thus, for example, a solution of hydrochloric acid contains H^+ and Cl^- and a solution of nitric acid contains H^+ and NO_3^-. The bases NaOH, KOH, and $Ca(OH)_2$ yield OH^- ions and the positive ions Na^+, K^+, and Ca^{2+}, respectively. Arrhenius therefore defined an acid as a substance that yields H^+ ions when it is dissolved in water and defined a base as a subtance that yields OH^- ions when it is dissolved in water. Acids and bases react to form *salts*[1] and water, the distinctive properties of both the acid and the base thereby disappearing, as for example

$$2HNO_3 + Ca(OH)_2 \longrightarrow Ca(NO_3)_2 + 2H_2O \qquad (9\text{-}1)$$

or, writing this as an ionic equation balanced with respect to charge as well as species,

$$2H^+ + 2NO_3^- + Ca^{2+} + 2OH^- \longrightarrow Ca^{2+} + 2NO_3^- + 2H_2O \qquad (9\text{-}2)$$

This characteristic reaction between an acid and a base is called *neutralization*. Since the Ca^{2+} and NO_3^- ions remain unchanged, they may be omitted and the equation for the neutralization reaction can be written simply as

$$H^+ + OH^- \longrightarrow H_2O \qquad (9\text{-}3)$$

Strong and Weak Electrolytes Substances whose solutions conduct electric current by migration of ions are called *electrolytes* (Fig. 9-1). Arrhenius recognized that the reason some acids and bases are much poorer electrolytes than others is that they are only slightly dissociated into ions whereas the others are more completely dissociated. The terms *strong* and *weak* are used to distinguish qualitatively between electrolytes that are almost completely dissociated and those that are dissociated only to a small extent. As with other qualitative terms, there is no sharp boundary between strong and weak electrolytes, but in practice this causes little difficulty. More quantitative ways of describing the strength of acids and bases are discussed in Sec. 12-2.

It is important not to confuse the relative solubility of an electrolyte with its strength. For example, acetic acid is extremely soluble in water but is a weak electrolyte. On the other hand, barium hydroxide is not very soluble but what

---

[1] The term *salt* originates from common table salt (sodium chloride, NaCl), which can be produced by the reaction of hydrochloric acid and sodium hydroxide

$$NaOH + HCl \longrightarrow NaCl + H_2O$$

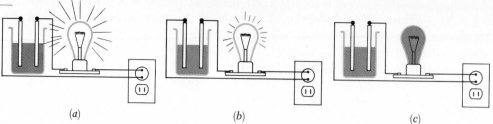

Figure 9-1 Equipment for Demonstrating the Conductivity of Solutions
When the electrodes are immersed in the liquid, the brightness of the bulb is an indication of the number of ions present in solution. (*a*) Strong electrolyte, (*b*) weak electrolyte, (*c*) nonelectrolyte.

does dissolve dissociates almost completely so that barium hydroxide is a strong electrolyte in water. Similarly, the present use of the terms strong and weak should not be confused with popular usage; hydrofluoric acid, HF, is a weak acid but it is extremely corrosive and will even dissolve glass.

Pure water is a very weak electrolyte, as it contains relatively few ions and is a poor conductor of electricity. Very sensitive instruments are needed to detect its conductivity. If an acid, a base, or a salt is dissolved in water, the conductivity becomes greater, and it increases with increasing concentration and with the extent of dissociation of the electrolyte involved.

Strong and Weak Acids In aqueous solutions, HCl, HBr, HI, HNO_3, and $HClO_4$ (perchloric acid) are essentially completely dissociated, according to the reaction

$$HA \longrightarrow H^+ + A^- \tag{9-4}$$

and are thus strong acids. Acids such as HF (hydrofluoric acid), HNO_2 (nitrous acid), and HCN (hydrocyanic acid) are known as weak acids because they do not dissociate completely except in solutions of very low concentration. Thus an aqueous solution of HF contains both undissociated HF molecules and the ions H^+ and F^-.

Acidic Hydrogen Atoms Many important acids containing carbon are classified as organic acids. An example is acetic acid, which is somewhat weaker than nitrous acid and has the structural formula

$$\begin{array}{c} H \\ | \\ H-C-C \\ | \\ H \end{array} \begin{array}{c} O \\ \diagup \\ \diagdown \\ O-H \end{array}$$

Organic acids

As mentioned in Sec. 3-3, this formula is frequently written CH_3COOH for typographical convenience; note, however, that there is no O-O bond in the molecule. The hydrogen atom that ionizes is the one joined to the oxygen atom, as is true for almost all oxygen-containing acids. The group of atoms —COOH is present in many organic acids and is called the *carboxyl group*. We abbreviate acetic acid by the symbol HOAc, Ac thus indicating the atomic grouping CH_3CO. In abbreviated form, the ionic dissociation of acetic acid is indicated by the equation HOAc → H^+ + OAc$^-$, which in detail corresponds to

$$\begin{array}{c} H \\ | \\ H-C-C \\ | \\ H \end{array} \begin{array}{c} O \\ \diagup \\ \diagdown \\ O-H \end{array} \longrightarrow \begin{array}{c} H \\ | \\ H-C-C \\ | \\ H \end{array} \begin{array}{c} O \\ \diagup \\ \diagdown \\ O^- \end{array} + H^+$$

Although acetic acid contains four hydrogen atoms, only the one attached to the oxygen atom comes off easily or is, in other words, an *acidic hydrogen atom*. This is a general feature of organic compounds: hydrogen atoms bonded to carbon atoms are seldom noticeably acidic in aqueous solution. The rare exceptions involve molecules in which the carbon atom has a much higher positive charge than usual.

There are many acids, both organic and inorganic, that contain more than one ionizable proton, including H_2SO_4, H_3PO_4, H_2CO_3, H_2SO_3 (sulfurous acid), and $HOOC-COOH$ (oxalic acid). In each of these *polyprotic* acids (sometimes, somewhat confusingly, called polybasic acids), the ionizable hydrogen atoms are attached to oxygen atoms, so the formulas might better be written $(HO)_2SO_2$, $(HO)_3PO$, and so on.

The various acidic hydrogen atoms in a polyprotic acid ionize to different extents. For example, sulfuric acid, H_2SO_4, is a strong acid insofar as the dissociation to give one hydrogen ion is concerned:

$$H_2SO_4 \longrightarrow H^+ + HSO_4^-$$

Except in very dilute solution (see Chap. 12), the ionization of the second hydrogen atom

$$HSO_4^- \longrightarrow H^+ + SO_4^{2-}$$

is not complete. HSO_4^- is a weak acid.

Some polyprotic acids do not contain oxygen; an example is H_2S (hydrogen sulfide), which is a very weak acid in aqueous solution. Conversely, there are many substances, both organic and inorganic, that contain the $-OH$ group but are not acids in aqueous solution. Examples include ethanol (CH_3CH_2OH) and other alcohols, a class of compounds containing a hydroxyl group attached to a carbon atom that is in turn linked to hydrogen or other carbon atoms. Typical alcohols do not ionize in aqueous solution, and they show neither acidic nor basic properties.

Strong and Weak Bases   It has perhaps occurred to the reader that there is another class of compounds containing O-H bonds that does not show acidic properties in aqueous solutions: the metallic hydroxides such as $NaOH$ and $Ca(OH)_2$. These compounds give OH^- ions in aqueous solution and thus are typical bases. The corresponding pure solid hydroxides are ionic, containing Na^+ and OH^- ions in one case and Ca^{2+} and OH^- in the other. Many weaker bases, as well as many acids, have the covalent structure X-O-H, where X represents the rest of the molecule. In some of these compounds the O-H bond is cleaved in solution to form H^+, whereas in others the X-O bond breaks to form OH^-. In still others neither bond breaks and no ionization at all occurs.

Typical strong bases include $NaOH$, KOH, and the hydroxides of the other alkali metals (Li, Cs, and Rb), as well as the hydroxides of the alkaline-earth elements other than beryllium (Be)—for example, $Mg(OH)_2$, $Ca(OH)_2$, and $Ba(OH)_2$. These alkaline-earth hydroxides are not very soluble. Only one of the two hydroxide ions separates completely from the positive ion, so solutions of the alkaline-earth hydroxides contain significant concentrations of MOH^+ ions (such as $MgOH^+$, $CaOH^+$, and $BaOH^+$) in addition to M^{2+} ions and OH^- ions.

The commonest weak base is ammonia, NH_3, which reacts to a small extent with water to form ammonium ions and hydroxide ions:

$$H_2O + NH_3 \longrightarrow OH^- + NH_4^+ \tag{9-5}$$

The fraction of ammonia molecules that undergo this reaction is small, and thus ammonia is considered a weak base. The reaction is not strictly a dissociation, but is rather a *proton-transfer* or *proton-exchange* process. As discussed in the next section, all acid-base reactions in aqueous solution can be viewed as proton-exchange reactions. Although for many years chemists considered that ammonium hydroxide, NH_4OH, was the predominant species responsible for the properties of a solution of ammonia in water, it has been shown that little or no ammonium hydroxide is present in such solutions or, indeed, exists under any circumstances.[2] In aqueous solution NH_3 forms hydrogen-bonded aggregates with water molecules. Both O-H---N and N-H---O bonds are present, as in

$$\tag{9-6}$$

When ammonia reacts with an acid, a proton is transferred to the NH_3 molecule, forming the ammonium ion, NH_4^+, as in (9-5). The unshared pair of electrons on NH_3 is responsible for its basicity:

$$\tag{9-7}$$

The ammonium ion is hydrogen bonded to water molecules through N-H---O bonds in a manner similar to that shown for NH_3 in (9-6).

Water as Acid and Base The neutralization reaction between H^+ and OH^-, Equation (9-3), goes nearly to completion, leaving only a very small concentration of unreacted H^+ ions and OH^- ions. Correspondingly, pure water is dissociated into H^+ and OH^- only to a very small extent:

$$H_2O \longrightarrow H^+ + OH^- \tag{9-8}$$

Water is thus a very weak base as well as an equally weak acid. A precise quantitative relation exists between the concentrations of the hydrogen and hydroxide ions as a result of (9-8); it is considered later [Equations (9-17) and (9-18)].

---

[2] If the species NH_4OH were present, the products in (9-5) would be formed by its dissociation. Undoubtedly the authority of Arrhenius' dissociation theory provided the original rationale for the belief in the species NH_4OH. Evidence against its existence has been provided only by modern physical methods, including various types of spectroscopy and diffraction.

9-2 The Hydronium Ion and Proton-Transfer Reactions in Aqueous Solutions

The Hydronium Ion The true hydrogen ion is a hydrogen atom stripped of its electron. In other words, it is the bare nucleus of a hydrogen atom—a proton. This is a very small particle compared to the size of a hydrogen atom—about 100,000 times smaller in diameter. It is also very small compared to the smallest other positive ions, such as Li^+ or Be^{2+}, which are comparable in size to atoms. The electrons surrounding another atom can therefore approach the positive charge of a bare hydrogen ion much more closely than that of any positive ion that has not been stripped of all its electrons. For this reason the bare hydrogen ion exists in reasonable concentrations only in hydrogen gas that has been subjected to an electric discharge or is at very high temperatures. No significant concentration of free H^+ exists in aqueous solutions because H^+ immediately combines with a water molecule to form the *hydronium ion,* H_3O^+:

$$\left[:O \overset{H}{\underset{H}{-}} H \right]^+$$

Recent evidence suggests that in aqueous solutions the hydronium ion is tied by hydrogen bonds to three more water molecules,

$$\text{(9-9)}$$

corresponding to the formula $H_9O_4^+$. As illustrated in Fig. 6-3, the water molecule is dipolar, with each hydrogen atom carrying a residual positive charge and the oxygen atom a residual negative charge. The positive charge on the $H_9O_4^+$ ion therefore attracts the oxygen atoms of additional water molecules, so that the complex is almost certainly even larger. The size of this complex is unknown, and it is in fact difficult to decide just where the influence of the original H_3O^+ ion on the surrounding water molecules ends. We may call this entire complex the *hydrated hydronium ion* and use the symbols H_3O^+ (*aq*) or just H_3O^+. However, in general we shall simply speak of the hydrogen ion and use the symbol H^+, the understanding always being that in an aqueous medium it represents a hydrated hydronium ion. Only when the situation requires it do we specifically refer to the hydronium or H_3O^+ ion.

Hydrogen ions may be transferred from one molecule to the next. Reactions such as

$$H_2O + H_3O^+ \longrightarrow H_3O^+ + H_2O$$

and

$$OH^- + H_2O \longrightarrow H_2O + OH^-$$

are among the fastest chemical reactions known. Indeed, most proton-transfer reactions are very fast.

The Brønsted Definition of Acids and Bases The fact that in aqueous solutions the hydrogen ion exists only as aquated hydronium ion means that the acid dissociation reactions discussed earlier are actually proton-transfer reactions, a new bond being formed as well as one being broken. Thus, representing an acid generally as HA, the reaction of "dissociation" (9-4) is really[3]

$$HA + O\overset{H}{\underset{H}{\diagup}} \longrightarrow A^- + \left[O\overset{H}{\underset{H}{-H}} \right]^+ \qquad (9\text{-}10)$$

Comparison of (9-10) with (9-5) shows that they involve the same fundamental process, a transfer of a proton from one species to another. This process is indeed the common feature of all acid-base reactions in aqueous media, a feature that led the Danish chemist Brønsted to propose in 1923 that an *acid be defined as a molecule or ion capable of losing a proton—a proton donor—and a base be defined as a proton acceptor.* Thus, in reaction (9-5) water behaves as an acid, donating a proton to the acceptor ammonia. In reaction (9-10), water acts as a base to which the acid HA donates a proton.

This way of viewing acids and bases leads to a natural definition of a *conjugate* acid-base pair (Latin: *conjugatus,* joined together). The conjugate base of any acid is the species obtained by removing a proton from the acid; thus Cl^- is the conjugate base of HCl, OAc^- is the conjugate base of HOAc, OH^- is the conjugate base of H_2O, and NH_3 is the conjugate base of the ammonium ion, NH_4^+. Conversely, the conjugate acid of any base is the acid obtained by adding a proton to the base; for example, H_3O^+ is the conjugate acid of the base H_2O, HCl is the conjugate acid of Cl^-, and so on. Some representative conjugate acid-base pairs are listed in Table 9-1. As these examples show, and as the definition implies, the charge on the conjugate acid is always more positive by one unit than that on the base.

A general formulation of the relation between the members of a conjugate acid-base pair is

$$HA \longrightarrow H^+ + A^- \qquad (9\text{-}11)$$

---

[3] In (9-10) and similar equations, the unshared electron pairs (on H_2O, H_3O^+, and other species) are omitted for typographical convenience.

Table 9-1
Typical Conjugate Acid-Base Pairs

Acid	Base	Acid	Base
HOAc	OAc^-	H_2CO_3	HCO_3^-
HCl	Cl^-	HCO_3^-	CO_3^{2-}
H_3O^+	H_2O	H_3PO_4	$H_2PO_4^-$
H_2O	OH^-	$H_2PO_4^-$	HPO_4^{2-}
NH_4^+	NH_3	HPO_4^{2-}	PO_4^{3-}
NH_3	NH_2^-		

Here HA represents the acid regardless of its actual charge or the charge on its conjugate base. If the emphasis is on the basic member of the pair, it is convenient to use the similar relation.

$$HB^+ \longrightarrow H^+ + B \qquad (9\text{-}12)$$

Here the base is symbolized generally by B, again regardless of what the actual charges might be.

With this definition, the general reaction between an acid and a base is one in which the accompanying proton-transfer results in the formation of a new acid and a new base, the conjugates of those present originally:

$$\underset{\text{Acid 1}}{HA} + \underset{\text{Base 2}}{B} \longrightarrow \underset{\text{Base 1}}{A^-} + \underset{\text{Acid 2}}{HB^+} \qquad (9\text{-}13)$$

Equations (9-5) and (9-10) are seen to represent reactions that may be formulated in just this way.

A number of species in Table 9-1 (for example, HCO_3^-) may act either as an acid or a base, losing or gaining a proton. This behavior is termed *amphiprotic* (Greek: *amphi*, both). An important example of amphiprotic behavior is that shown by water; one molecule reacts with another, the first acting as base and the second as acid:

$$H_2O + H_2O \longrightarrow H_3O^+ + OH^- \qquad (9\text{-}14)$$

This reaction is fundamentally the same reaction as the dissociation reaction (9-8) of water,

$$H_2O \longrightarrow H^+ + OH^- \qquad (9\text{-}15)$$

Comparison of (9-14) with (9-15) shows that (9-15) is properly considered to be a proton-transfer reaction of the general acid-base type (9-13). The same is true of the neutralization reaction (9-3), which is the reverse of the reaction just considered.

9-3 pH and Solution Stoichiometry

The pH The hydrogen-ion concentration of an aqueous solution is one of its important properties. Since this concentration may range over many powers of 10, and since these powers are mainly negative when the concentration is expressed in moles of hydrogen ion per liter of solution, as is common, it proves convenient to work with another quantity, the pH, which is the negative logarithm to the base 10 of the hydrogen-ion concentration. Symbolizing this concentration in moles per liter (or M) by $[H^+]$, we have[4]

$$pH = -\log[H^+] = -\log[H_3O^+] \qquad (9\text{-}16)$$

Note that the larger the H^+-ion concentration, the lower the pH. When $[H^+] = 1.00\ M$ the pH is zero, and for H^+-ion concentrations larger than this the pH becomes negative. For pure water at 24°C the pH is 7.00, corresponding to

---

[4]Strictly speaking $pH = -\log([H^+]/M)$, but the units moles per liter or M from $[H^+]$ are usually dropped, resulting in expressions (9-16). This is often done when the logarithm of a physical quantity such as a length or a pressure is taken (see also Appendix A).

$[H^+] = 1.00 \times 10^{-7} M$. Since hydroxide and hydronium ions are produced in equal numbers in pure water by reaction (9-14), the concentration of hydroxide ions in pure water is always equal to that of hydrogen ions. Thus, at 24°C, the concentration of OH^- ions is $[OH^-] = 1.00 \times 10^{-7} M$ and pOH (meaning $-\log[OH^-]$) is 7.00. Thus pH + pOH = 14.00 for water at 24°C.

In acidic and basic aqueous solutions, the concentrations of hydrogen and hydroxide ions are no longer equal, since acidic solutions contain excess H^+ and basic (or alkaline) solutions contain excess OH^-. However, as discussed at length in Chap. 12, in *any* aqueous solution at room temperature it is approximately true that

$$[H^+][OH^-] = 1.0 \times 10^{-14} M^2 \qquad (9\text{-}17)$$

or, expressed alternatively,

$$pH + pOH = 14.00 \qquad (9\text{-}18)$$

The number of significant figures with which a pH is expressed must be appropriately related[5] to the number with which the corresponding $[H^+]$ is given so that each can be calculated from the other without loss of precision.

Example 9-1

☐ **pH of Acidic Solutions** Give the pH of solutions with the following $[H^+]$ values in moles per liter: (*a*) 0.047; (*b*) 1.0; (*c*) 5.0; and (*d*) 3×10^{-6}.

Solution (*a*) We have $[H^+] = 0.047 M$ so that pH $= -\log 0.047 = \underline{1.33}$. [To obtain the answer without an electronic calculator we would use the procedure $-\log 0.047 = -\log (4.7 \times 10^{-2}) = -\log 4.7 - \log 10^{-2} = -0.67 - (-2.00) = \underline{1.33}$]. The number of significant figures is most easily established by finding the negative logarithm of a number that differs from the original number by one unit in the last place: $-\log 0.048 = 1.32$. The new pH value differs from the original only in the last place, implying that the answer, 1.33, is given to the proper number of significant figures. (*b*) $-\log 1.0 = \underline{0.00}$ $(-\log 1.1 = -0.04)$;   (*c*) $-\log 5.0 = \underline{0.70}$ $(-\log 5.1 = -0.71)$; (*d*) $-\log 3 \times 10^{-6} = \underline{5.5}$ $(-\log 4 \times 10^{-6} = 5.4)$. ∎

Exercise 9-1

☐ Give the pH of solutions with the following $[H^+]$ values in moles per liter: (*a*) 0.85; (*b*) 1.5×10^{-4}; (*c*) 3.5; (*d*) 5.5×10^{-6}. ∎

Example 9-2

☐ **H^+-Ion Concentration from the pH** What is $[H^+]$ for a solution in which (*a*) pH = 2.0; (*b*) pH = 3.75?

Solution

(*a*) $[H^+] = 10^{-2.0} M = \underline{1.0 \times 10^{-2} M}$
(*b*) $[H^+] = 10^{-3.75} M = \underline{1.8 \times 10^{-4} M}$

(To obtain the answer without an electronic calculator, we would write $10^{-3.75} = 10^{-4.00+0.25} = 10^{-4.00} \times 10^{0.25}$. Since the antilogarithm of 0.25 is 1.8, we have $[H^+] = 10^{-4} \times 1.8 M$.) ∎

---

[5]The *characteristic* of a logarithm (the number preceding the decimal point) merely denotes the power of 10 involved and is thus not a significant figure in the normal sense. Significant figures apply only to the remaining portion of the logarithm, the *mantissa*.

□ What is $[H^+]$ for a solution in which (*a*) pH $= -0.44$; (*b*) pH $= 4.68$? ■ **Exercise 9-2**

Solution Stoichiometry A convenient method for determining the concentration of a substance in solution is to measure the volume of one solution that reacts completely with a given volume of another. This technique is called *titration*. The experimental procedure consists in taking a known volume of a solution containing solute 1 at an unknown concentration and carefully adding a second solution containing a known concentration of solute 2 until all of 1 has reacted. From the reaction stoichiometry and the volume of 2, the concentration of 1 can be calculated. Two examples of such calculations follow; titration is discussed in detail in Chap. 12.

□ **Acid-Base Solution Reaction** A 20.00-ml sample of a 0.400 M solution of HCl was found to react completely with 25.00 ml of an NaOH solution of unknown concentration. What was the concentration of the base? **Example 9-3**

Solution The number of moles of HCl in the 20.00-ml sample is

$$\text{mol HCl} = 20.00 \text{ ml} \times \frac{1 \text{ liter}}{1000 \text{ ml}} \times 0.400 \text{ mol liter}^{-1}$$

$$= 0.00800 \text{ mol}$$

Because each molecule of HCl can furnish one H^+ and because one OH^- reacts with one H^+, there must be 0.00800 mol NaOH in the sample that reacts with the acid. If we let Y be the unknown molar concentration of the NaOH solution, then

$$25.00 \text{ ml} \times \frac{1 \text{ liter}}{1000 \text{ ml}} \times Y = 0.00800 \text{ mol}$$

or $\qquad\qquad Y = \dfrac{0.00800 \text{ mol}}{0.0250 \text{ liter}} = \underline{0.320 \ M}$ ■

□ A 22.00-ml sample of a 0.150 M solution of NaOH was found to react completely with 30.00 ml of an HNO_3 solution of unknown concentration. What was the concentration of the acid? ■ **Exercise 9-3**

Acids that are only slightly soluble in water usually can be dissolved easily in strong base, provided that the salt they form with the base is soluble,

$$\text{HA} + \text{NaOH} \longrightarrow \text{A}^- + \text{Na}^+ + \text{H}_2\text{O} \qquad\qquad (9\text{-}19)$$

A method to determine the amount of such an acid is to dissolve it in a known quantity of base, present in excess, and then to titrate the unreacted base with a strong acid. Similarly, slightly soluble bases can often be dissolved in a known quantity of a strong acid, present in excess, and the unreacted acid can be titrated with strong base. Such procedures are called *back titrations*.

□ **Back Titration** Benzoic acid is monoprotic and only slightly soluble in water. A 0.300-g sample of this acid was dissolved in 50.00 ml of 0.200 M NaOH. The resulting solution was then neutralized by the addition of 18.85 ml 0.400 M HCl. What is the molecular weight (MW) of benzoic acid? **Example 9-4**

Solution Since the final solution is neutral, the total moles of H^+ (from benzoic acid + HCl) that react must be equal to the moles of OH^- originally present.

$$\text{mol } H^+ \text{ from HCl} = 18.85 \text{ ml} \times \frac{1 \text{ liter}}{1000 \text{ ml}} \times \frac{1 \text{ mol } H^+}{\text{mol HCl}} \times \frac{0.400 \text{ mol HCl}}{\text{liter}}$$

$$= 0.00754 \text{ mol } H^+$$

mol H^+ from benzoic acid

$$= \frac{1 \text{ mol } H^+}{\text{mol benzoic acid}} \times 0.300 \text{ g} \times \frac{1 \text{ mol benzoic acid}}{\text{MW g}}$$

$$\text{mol } OH^- = 50.00 \text{ ml} \times \frac{1 \text{ liter}}{1000 \text{ ml}} \times \frac{1 \text{ mol } OH^-}{\text{mol NaOH}} \times \frac{0.200 \text{ mol NaOH}}{\text{liter}}$$

$$= 0.0100 \text{ mol } OH^-$$

Equating the total moles H^+ to moles OH^- gives

$$0.00754 + \frac{0.300}{MW} = 0.0100$$

$$\frac{0.300}{MW} = 0.0100 - 0.0075 = 0.0025$$

$$MW = \frac{0.300}{0.0025} = \underline{1.2 \times 10^2} \quad \blacksquare$$

Exercise 9-4

□ Adrenaline, $C_9H_{13}O_3N$ (MW = 183.2), is only slightly soluble in water. It dissolves readily in acid, by accepting a proton and being converted to the $C_9H_{14}O_3N^+$ ion. An impure 800-mg sample of adrenaline was dissolved in 25.00 ml of 0.200 *M* HCl. The resulting solution was then neutralized by the addition of 7.65 ml of 0.100 *M* NaOH. What is the percentage of adrenaline contained in the sample? (Assume that the impurities in the sample do not react with HCl.) ■

9-4   Acidic, Basic, and Amphoteric Properties of Oxides

Although it was believed in Lavoisier's day that substances were acidic because they contained oxygen, it was later seen that many oxygen compounds, such as the oxides of the alkali and alkaline-earth metals, are basic in character. It is possible and instructive to classify oxides on the basis of their acidic or basic properties. A water-soluble oxide is considered acidic if its aqueous solution is acidic. Examples include NO_2, SO_2, SO_3, $P_2O_3(P_4O_6)$, and $P_2O_5(P_4O_{10})$. Oxides that do not readily dissolve in water are said to be acidic if their solubility increases as the solution is made more basic—that is, with increasing $[OH^-]$—and decreases as the solution is made more acidic. Examples of water-insoluble acidic oxides are SiO_2 and SnO_2, the dioxides of silicon and tin.

Oxides are called basic if their aqueous solutions are alkaline or if their solubility increases with increasing H^+ concentration (decreasing pH) and decreases with increasing OH^- concentration (increasing pH). Among the insoluble basic oxides are those of many metals, for example, FeO, Fe_2O_3, CoO, and NiO.

The solubility of some oxides is increased both by an increase in the concentration of hydrogen ions and by an increase in the concentration of hydroxide ions. These oxides therefore show both acidic and basic character. Such behavior is termed *amphoteric*. The solubility of amphoteric oxides in pure water is usually limited. Examples are BeO and Al_2O_3, with the reactions

$$BeO + 2H^+ \longrightarrow Be^{2+} + H_2O \tag{9-20}$$

$$BeO + 2OH^- + H_2O \longrightarrow Be(OH)_4{}^{2-} \tag{9-21}$$

$$Al_2O_3 + 6H^+ \longrightarrow 2Al^{3+} + 3H_2O \tag{9-22}$$

$$Al_2O_3 + 2OH^- + 3H_2O \longrightarrow 2Al(OH)_4{}^- \tag{9-23}$$

In each of these reactions, it is assumed that suitable oppositely charged ions are present—for example, Cl^- or $NO_3{}^-$ in (9-20) and (9-22), and K^+ or Na^+ in (9-21) and (9-23). Their source could be solutions of HCl or HNO_3 for (9-20) and (9-22) or NaOH or KOH for the other two reactions. The ions $Be(OH)_4{}^{2-}$ and $Al(OH)_4{}^-$ are examples of complex ions, which are discussed in Sec. 9-6. Their -2 and -1 charges correspond to the sum of the charge on the related metal cation (Be^{2+}, Al^{3+}) and the four -1 charges from the OH^- that are part of each of these complex ions.

Other amphoteric oxides are ZnO, PbO, and SnO, which react with OH^- in H_2O to form the ions $Zn(OH)_4{}^{2-}$, $Pb(OH)_4{}^{2-}$, and $Sn(OH)_4{}^{2-}$ and react with H^+ in H_2O to form the (hydrated) cations Zn^{2+}, Pb^{2+}, and Sn^{2+}.

9-5 Aqueous Solutions of Salts

In an aqueous solution many salts are completely dissociated into ions. This is particularly true of salts resulting from reactions of strong acids with strong bases, salts such as $NaNO_3$ or KBr. Aqueous solutions and also melts of these salts contain no $NaNO_3$ or KBr molecules, only the ions Na^+ and $NO_3{}^-$, or K^+ and Br^-.

□ **Mixture of Electrolyte Solutions** The following solutions are mixed: 50 ml of 0.200 *M* barium nitrate, 50 ml of 0.100 *M* sodium nitrate, and 100 ml of 0.100 *M* barium chloride. Enough water is added to make the total volume 500 ml. Calculate the final concentrations of Ba^{2+}, Na^+, Cl^-, and $NO_3{}^-$.

Example 9-5

Solution All three salts, $Ba(NO_3)_2$, $NaNO_3$, and $BaCl_2$ are strong electrolytes. Each of the solutions mixed can therefore be thought of as the source of a specified number of moles of a positive and a negative ion. The 0.050 liter (50 ml) of $NaNO_3$ solution provides

$$0.050 \text{ liter} \times 0.100 \text{ mol Na}^+ \text{ liter}^{-1} = 5.0 \times 10^{-3} \text{ mol Na}^+$$

and 5.0×10^{-3} mol $NO_3{}^-$ as well because the dissociation of each $NaNO_3$ yields one $NO_3{}^-$ ion for each Na^+ ion. The barium nitrate solution contributes

$$0.050 \text{ liter} \times 0.200 \text{ mol Ba}^{2+} \text{ liter}^{-1} = 1.00 \times 10^{-2} \text{ mol Ba}^{2+}$$

and $2 \times 1.00 \times 10^{-2}$ mol $NO_3{}^-$ because each $Ba(NO_3)_2$ contains two $NO_3{}^-$ for each Ba^{2+}. Similarly the $BaCl_2$ solution contributes

$$0.100 \text{ liter} \times 0.100 \text{ mol Ba}^{2+} \text{ liter}^{-1} = 1.00 \times 10^{-2} \text{ mol Ba}^{2+}$$

and $2 \times 1.00 \times 10^{-2}$ mol Cl^-.

The total numbers of moles of ions in the resulting solution are the sums of the individual contributions:

Na^+: 5.0×10^{-3} mol

NO_3^-: $5.0 \times 10^{-3} + 2.00 \times 10^{-2} = 2.50 \times 10^{-2}$ mol

Ba^{2+}: $1.00 \times 10^{-2} + 1.00 \times 10^{-2} = 2.00 \times 10^{-2}$ mol

Cl^-: 2.00×10^{-2} mol

The concentration of each ion is found by dividing the number of moles by the *final* volume, 0.500 liter:

$$[Na^+] = \frac{5.0 \times 10^{-3} \text{ mol}}{0.500 \text{ liter}} = \underline{1.00 \times 10^{-2} \, M}$$

$$[NO_3^-] = \frac{2.50 \times 10^{-2} \text{ mol}}{0.500 \text{ liter}} = \underline{5.00 \times 10^{-2} \, M}$$

$$[Ba^{2+}] = \frac{2.00 \times 10^{-2} \text{ mol}}{0.500 \text{ liter}} = \underline{4.00 \times 10^{-2} \, M}$$

$$[Cl^-] = \frac{2.00 \times 10^{-2} \text{ mol}}{0.500 \text{ liter}} = \underline{4.00 \times 10^{-2} \, M}$$

One check on the calculations is to be sure that the answers are consistent with electrical neutrality: the concentration of positive charges must equal the concentration of negative charges.

Concentrations of positive charges		Concentrations of negative charges	
On Na^+:	$1.00 \times 10^{-2} \, M$	On NO_3^-:	$5.00 \times 10^{-2} \, M$
On Ba^{2+}:	$8.00 \times 10^{-2} \, M$	On Cl^-:	$4.00 \times 10^{-2} \, M$
TOTAL	$9.00 \times 10^{-2} \, M$	TOTAL	$9.00 \times 10^{-2} \, M$

■

Exercise 9-5

□ The following solutions are mixed: 20 ml of $0.200 \, M$ NH_4Cl, 30 ml $0.100 \, M$ $AlCl_3$, and 50 ml of $0.050 \, M$ $(NH_4)_2SO_4$. Enough water is added to make the total volume 200 ml. Calculate the final concentrations of NH_4^+, Al^{3+}, Cl^-, and SO_4^{2-}, assuming the three salts to be completely ionized. ■

When solutions of electrolytes are mixed, reactions may take place with resulting changes in the quantities of substances in solution. For example, acid-base reactions may remove some of the acids and bases, or insoluble solids may precipitate, removing some ions from the solution. In calculating the concentrations of the different species present in the final solution, one must take into account any such reactions and note as well which substances present are strong electrolytes and are therefore essentially completely dissociated into ions. Such reactions are discussed in Chaps. 12 and 13.

Hydrolysis Reactions Salts formed from weak acids or weak bases also dissociate completely into ions in aqueous solutions. Examples of such salts are sodium acetate, NaOAc, the salt of the weak acid HOAc and the strong base NaOH, and ammonium chloride, NH_4Cl, the salt of the weak base NH_3 and the strong acid HCl. Unlike solutions of salts formed from strong acids and strong bases, however, solutions of these salts contain some undissociated molecules of weak electrolytes formed by what is termed a *hydrolysis* reaction.

For example, a solution of NaOAc contains essentially no undissociated NaOAc molecules, but rather the ions Na^+ and OAc^-, which also exist in crystalline sodium acetate. In addition the solution contains a small concentration of HOAc molecules, produced by the reaction of OAc^- ions with water to form HOAc and OH^-:

$$H_2O + OAc^- \longrightarrow HOAc + OH^- \qquad (9\text{-}24)$$

This is a typical hydrolysis reaction, the reaction of a salt ion with water; the solution becomes basic because of the OH^- formed. Reaction (9-24) occurs to a small, but not a negligible, extent. Comparison with the general acid-base reaction (9-13) shows that (9-24) is simply another proton-transfer reaction. Water is the acid and OAc^- the base. A small fraction of the water molecules transfer protons to the acetate ions present, producing small concentrations of the conjugate base OH^- and the conjugate acid HOAc. In essence, OAc^- is trying to pull a proton from a water molecule, that is, it competes with OH^- for the proton. In this competition acetate ions finish a poor second, but they do succeed in removing some protons and thereby form some HOAc molecules.

A solution of ammonium chloride contains no undissociated NH_4Cl molecules, but rather chiefly NH_4^+ and Cl^- ions. However, it also contains a very small concentration of free ammonia formed by hydrolysis of NH_4^+:

$$NH_4^+ + H_2O \longrightarrow H_3O^+ + NH_3 \qquad (9\text{-}25)$$

This hydrolysis reaction produces a slightly acidic solution instead of the slightly alkaline one formed by sodium acetate. The essential difference here is that NH_4Cl contains the conjugate ions of a weak base (NH_3) and a strong acid (HCl), rather than those of a weak acid and a strong base. In each case, the hydrolysis reaction produces a small quantity of the weak electrolyte from which the salt was formed and a corresponding quantity of either H_3O^+ or OH^-.

Reaction (9-25) is another proton-exchange process, NH_4^+ being the acid (proton donor) and water the base (proton acceptor) on the left side. Water is a far weaker base than ammonia so that most of the protons remain attached to ammonia molecules in the form of ammonium ions, NH_4^+. All acid-base reactions in water may be viewed as a competition between bases.

Some salts (e.g., $HgCl_2$) dissociate only partially, but there are no simple rules about which salts are completely ionized in solution and which are not. We shall assume in general that dissociation of salts is complete.

Hydration of Ions We have already discussed the nature of the hydrogen ion in aqueous solutions (Sec. 9-2). Most other cations are also hydrated, that is, water molecules are bound to them more or less firmly. The strength of attachment of the water molecules increases with increasing positive charge and decreasing size of the ion, which suggests that electrostatic attraction of the ion for the negative end of the water dipole (Fig. 6-3) plays a significant role. Structural studies of many solids containing hydrated cations show that the oxygen atoms of the water molecules are indeed oriented toward the cation. This arrangement also permits formation of covalent bonds involving the unshared electron pairs on the oxygen atom of the water molecule.

The exact number of water molecules associated with any given cation in solution is difficult to specify. For example, there is evidence that Mg^{2+} is closely associated with six H_2O molecules, arranged octahedrally around it, corresponding to a formula $Mg(H_2O)_6^{2+}$. There is an additional layer of less strongly held water molecules. Should the magnesium ion in water be looked upon as a bare Mg^{2+}, as $Mg(H_2O)_6^{2+}$, or perhaps even as an ion containing more water molecules? The same question arises for practically all metal ions in aqueous solution. The present consensus is to assume the existence of definite species, such as $Mg(H_2O)_6^{2+}$, $Be(H_2O)_4^{2+}$ (tetrahedral), and $Fe(H_2O)_6^{3+}$; in each case the number of water molecules is assumed to be that in best agreement with available evidence, including arguments based on analogy with related compounds. Less firmly bound water of hydration is not counted.

When referring to ionic species in aqueous solutions, we adopt the convention already chosen for H^+: symbols such as Mg^{2+}, Be^{2+}, and Fe^{3+} are abbreviations that represent whatever species actually exist in solution. More explicit formulas such as $Mg(H_2O)_6^{2+}$ are used when this is particularly desirable—for example, in discussion of reactions that specifically involve the water molecules.

Anions such as SO_4^{2-} and Cl^- are also hydrated to some extent, but less is known about this than about the hydration of cations. However, both the number of water molecules involved and their strength of attachment are probably considerably less for anions than for cations. We use symbols such as Cl^- and NO_3^- to represent anions as they actually exist in aqueous solution, with the understanding that water molecules may be loosely attached.

Many uncharged molecules in aqueous solution are hydrogen bonded to water molecules—for example, NH_3 as shown in (9-6)—but the water molecules are rarely shown in formulas. Solutes in nonaqueous solvents may also be associated with solvent molecules; this phenomenon is called *solvation*.

Mg^{2+}

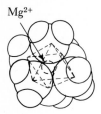

$Mg(H_2O)_6^{2+}$

Be^{2+}

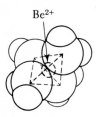

$Be(H_2O)_4^{2+}$

9-6 Complex Ions

Some complex ions have already been described in the preceding paragraphs and in Figs. 7-7 and 7-8. The term *complex ion* is used generally to refer to a polyatomic ion that consists of a central atom, usually of a metallic element, to which one or more atoms or groups of atoms are bonded. The attached atoms or groups of atoms are called *ligands* (Latin: *ligandus,* to be bound). Typical complex ions are listed in Table 9-2; the central atom is listed as a positive ion and the ligands as anions or neutral molecules. It is often useful to think of a complex ion as made up from components like those in Table 9-2, but this should not be taken to imply that the ligands and the central ion are linked by ionic bonds; usually the bonding is largely covalent (Chap. 27). The charge assigned to the central atom in Table 9-2 is determined by considering the charges assigned to the ligands and the overall charge on the complex ion. For historical reasons, compounds containing

Table 9-2
Typical Complex Ions

Complex ion	Central ion	Ligands	Coordination number
$CoCl_4^{2-}$	Co^{2+}	Cl^-	4
$Ag(NH_3)_2^+$	Ag^+	NH_3	2
$Mg(H_2O)_6^{2+}$	Mg^{2+}	H_2O	6
$Al(H_2O)_5OH^{2+}$	Al^{3+}	H_2O, OH^-	6
AlH_4^-	Al^{3+}	H^-	4
$Fe(CN)_6^{4-}$	Fe^{2+}	CN^-	6
$Zn(OH)_4^{2-}$	Zn^{2+}	OH^-	4
$Pt(NH_3)_4Cl_2^{2+}$	Pt^{4+}	NH_3, Cl^-	6
$PtCl_4^{2-}$	Pt^{2+}	Cl^-	4

complex ions are called *coordination compounds*. The ligands are said to be coordinated to the central atom; the number of atoms directly bonded to the central atom is referred to as the *coordination number*. The most common coordination numbers are 4 and 6.

The definition of a complex ion is not rigid or precise and there are minor variations in usage, but this causes no difficulties. Ions containing oxygen and only one other element—such as permanganate, MnO_4^-, sulfate, SO_4^{2-}, and nitryl, NO_2^+, which might be thought of as containing the oxide ion O^{2-} as the ligand, are not usually regarded as complex ions. These ions are so stable that substitution of ligands is much rarer than in more typical "complex ions" and there is no advantage in considering them as consisting of a central atom and ligands.

Hydrated metal ions may be viewed as complex ions in which the ligands are water molecules; consequently, the formation of other complex ions in solution usually involves displacing the ligand H_2O. For example, a solution of cupric sulfate, $CuSO_4$, in water has a light blue color that is due to hydrated Cu^{2+} ions, chiefly $Cu(H_2O)_4^{2+}$. When ammonia is added to the solution, an intense deep blue color appears, due chiefly to $Cu(NH_3)_4^{2+}$, the tetrammine cupric ion. Since, by convention, water of hydration attached to ions in aqueous solution is often not shown in formulas, this reaction is sometimes written as

$$Cu^{2+} + 4NH_3 \longrightarrow Cu(NH_3)_4^{2+} \qquad (9\text{-}26)$$

However, it might preferably be written as

$$Cu(H_2O)_4^{2+} + 4NH_3 \longrightarrow Cu(NH_3)_4^{2+} + 4H_2O \qquad (9\text{-}27)$$

to emphasize the fact that it is a *ligand-exchange reaction*.

Some hydrated cations, especially those with a high positive charge, behave as weak acids in solution. The reaction involves the transfer of a proton from a coordinated water molecule to a free water molecule—for example,

$$Fe(H_2O)_6^{3+} + H_2O \longrightarrow Fe(H_2O)_5OH^{2+} + H_3O^+ \qquad (9\text{-}28)$$

The hydrated ferric ion is in fact a slightly stronger acid than acetic acid [that is, it dissociates, by (9-28), slightly more than does acetic acid at a comparable concentration].

Most compounds of the transition metals are colored because of selective absorption of certain wavelengths of visible light. The colors of the complexes of a given transition metal depend upon the ligands that are attached. For example, $Fe(H_2O)_6^{3+}$ is pale violet. If a proton is removed, as in (9-28), the resulting $Fe(H_2O)_5OH^{2+}$ is yellow; if one of the water molecules is replaced by thiocyanate to form $Fe(H_2O)_5SCN^{2+}$, a deep red color appears. Similarly, the color of Cu^{2+} compounds is strongly dependent upon the ligands present. As already mentioned, $Cu(H_2O)_4^{2+}$ is blue and $Cu(NH_3)_4^{2+}$ is an intense deep blue. Crystalline $CuCl_2$ (Fig. 7-3), which may be made by direct combination of copper and chlorine, is yellow.

Some ligands contain more than one atom capable of coordinating a central atom. An example is ethylenediamine,[6]

$$H_2N\text{—}CH_2\text{—}CH_2\text{—}NH_2 \qquad (9\text{-}29)$$

often abbreviated by the letters *en*. The nitrogen atoms of its two amine groups

---

[6]Note that the —NH_2 group is called the amine group, with one *m*, while NH_3, when occurring in complexes such as in (9-26), is called ammine, with two *m*'s.

are capable of coordinating a metal ion in very much the same way as does the nitrogen atom of ammonia. It forms complexes such as $[Co(NH_3)_4(en)]^{3+}$, $[Co(NH_3)_2(en)_2]^{3+}$, and $[Co(en)_3]^{3+}$, in which each *en* is attached to the central cobalt atom by its two nitrogen atoms, taking the place of two NH_3. These ions all have octahedral structures, the first one, for example, being represented by the diagram

$$\left[\begin{array}{c} \text{CH}_2 \\ \text{CH}_2 \\ \text{NH}_2 \quad \text{NH}_2 \\ \text{H}_3\text{N} - \text{Co} - \text{NH}_3 \\ \text{H}_3\text{N} \\ \text{NH}_3 \end{array}\right]^{3+} \qquad (9\text{-}30)$$

Ligands such as ethylenediamine are called *multidentate* (many-toothed), and a complex in which a multidentate ligand is attached in several places to the same central atom is called a *chelate* (Greek: *chele*, claw).

Finally, complexes exist with more than one metal atom. These are called *polynuclear* complexes. An example is

$$\left[\begin{array}{c} \text{H} \\ \text{O} \\ (\text{NH}_3)_4\text{Co} \qquad \text{Co}(\text{NH}_3)_4 \\ \text{O} \\ \text{H} \end{array}\right]^{4+} \qquad (9\text{-}31)$$

The formation of polynuclear complexes sometimes leads to very large aggregates, of essentially infinite size, which may be insoluble. Thus, in the precipitation of $Zn(OH)_2$ by reaction of Zn^{2+} and OH^-, a possible schematic sequence is

$$\text{Zn}^{2+} \xrightarrow{4\text{OH}^-} \left[\begin{array}{c} \text{HO} \quad \text{OH} \\ \text{Zn} \\ \text{HO} \quad \text{OH} \end{array}\right]^{2-} \xrightarrow[2\text{OH}^-]{\text{Zn}^{2+}} \left[\begin{array}{c} \text{H} \\ \text{HO} \quad \text{O} \quad \text{OH} \\ \text{Zn} \quad \text{Zn} \\ \text{HO} \quad \text{O} \quad \text{OH} \\ \text{H} \end{array}\right]^{2-} \xrightarrow[2\text{OH}^-]{\text{Zn}^{2+}}$$

$$\left[\begin{array}{c} \text{H} \quad \text{H} \\ \text{HO} \quad \text{O} \quad \text{O} \quad \text{OH} \\ \text{Zn} \quad \text{Zn} \quad \text{Zn} \\ \text{HO} \quad \text{O} \quad \text{O} \quad \text{OH} \\ \text{H} \quad \text{H} \end{array}\right]^{2-} \longrightarrow \text{etc.} \quad (9\text{-}32)$$

This aggregation does not necessarily lead to crystalline structures. Hydroxides often precipitate in a gelatinous form that usually incorporates (hydrogen-bonded) H_2O molecules also, in amounts that depend on the conditions of precipitation.

Summary

Strong electrolytes are substances that in solution are almost completely dissociated into ions. In contrast, weak electrolytes are dissociated only to a small extent. The strength of an electrolyte should not be confused with its solubility or concentration.

Arrhenius defined an acid as a substance that yields H^+ ions when it is dissolved in water and defined a base as a substance that yields OH^- ions when it is dissolved in water. Neutralization, the characteristic reaction between such acids and bases, causes the distinctive properties of both the acid and base to disappear, water and salts being formed. Although the symbol H^+ is commonly used, protons in aqueous solutions are hydrated and no significant concentration of free H^+ exists.

According to Brønsted, an acid can be more generally defined as a molecule or ion capable of losing a proton—a proton donor. A base is a proton acceptor. A pair of related acid and base species, HA and A^- (for example, HCl and Cl^-) or HB^+ and B (for example, NH_4^+ and NH_3), is termed a conjugate acid-base pair. All acid-base reactions are viewed as proton transfers by which a new acid and new base are formed, the conjugates of those present originally:

$$HA + B \longrightarrow A^- + HB^+$$

Species that can act either as acids or bases are termed amphiprotic. An important example is water:

$$H_2O + H_2O \longrightarrow H_3O^+ + OH^-$$

A significant characteristic of a solution is its hydrogen-ion concentration. This is often specified by the pH, defined as $-\log[H^+]$, with $[H^+]$ the molar concentration of H^+. Similarly, $pOH = -\log[OH^-]$. A neutral aqueous solution is one in which $[H^+] = [OH^-]$. At $24°C$, a neutral solution has $pH = 7.00$; in any aqueous solution at $24°C$, $pH + pOH = 14.00$.

Oxides can be classified as acidic, basic, or amphoteric; amphoteric oxides exhibit both acidic and basic character. Many salts are completely dissociated into ions in aqueous solution, some only partially so. Moreover, some ions are partially hydrolyzed in aqueous solution; that is, some of the cations or anions or both react with water to form the respective conjugate base or acid species.

Ions in aqueous solution are generally hydrated. Water molecules attached to a small, highly charged cation, as in $Fe(H_2O)_6^{3+}$, may lose protons, thus displaying acidic character. Hydrated ions are examples of complex ions, a term used generally to describe a charged species in which ions or molecules (ligands) are attached to a central atom, usually a metal atom.

Terms and Concepts

Problems and Questions

9-1   pH Values and Concentrations (a) Convert the following pH values into pOH and H^+ concentrations: 4.7, 8.4, 11.22, 0.52, -0.30. (b) Convert the following molar H^+ concentrations into pH values: 6.0, 1.5×10^{-6}, 7×10^{-10}, 3×10^{-2}, 5×10^{-14}.

9-2   pH and pOH Convert the following pH values into pOH and into H^+ concentrations: 0.08; 4.7; 8.27; -0.3.

9-3   pH and pOH Convert the following molar H^+ concentrations into pH values and calculate the corresponding pOH: 6; 1.5×10^{-6}; 7×10^{-10}; 3×10^{-2}.

9-4   Solutions of Electrolytes (a) Distinguish clearly between the terms *strong* (as applied to electrolytes) and *concentrated* and between the terms *weak* and *dilute*. (b) Which of the following substances are strong electrolytes when dissolved in water, which are weak, and which are nonelectrolytes: KOH, sugar, NH_3, $Ca(NO_3)_2$, HCl, methyl alcohol, $LiBr$, Na_2SO_4, $HOAc$ (acetic acid), $NaOAc$ (sodium acetate), NH_4Cl?

9-5   Composition of Solutions (a) How many moles of magnesium ions and how many moles of nitrate ions are present in 150 ml of 2.0 M magnesium nitrate? Assume that $Mg(NO_3)_2$ is a strong electrolyte, but indicate how your answers would differ (if at all) if it were not. (b) How many grams of HNO_3 are needed to prepare 2.5 liter of 0.120 M nitric acid?

9-6   Dilution of HCl The "concentrated HCl" obtained from suppliers is usually an aqueous solution having a density of 1.198 g ml^{-1} that contains 40 percent HCl by weight. What volume of "concentrated HCl" must be added to water to prepare 1 liter of a 0.100 M solution?

9-7   Brønsted Acids and Bases Define the terms *acid* and *base* in the Brønsted sense. Then identify the acid and the base on each side of the following equations:

(a) $HOAc + CN^- \longrightarrow HCN + OAc^-$
(b) $HSO_4^- + HPO_4^{2-} \longrightarrow SO_4^{2-} + H_2PO_4^-$
(c) $NH_3 + O^{2-} \longrightarrow NH_2^- + OH^-$

9-8   Conjugate Acid-Base Pairs (a) Define the term *conjugate acid-base pair*, and illustrate it

with two examples not included in Table 9-1. (b) Like water, ammonia can act as both a base and an acid, although in aqueous solutions it normally behaves only as a base. Indicate the appropriate conjugate form when it acts as an acid and when it acts as a base. (c) Give the conjugate base of each of the following acids: H_3O^+, HS^-, HSO_4^-, $CH_3NH_3^+$, HF, H_3AsO_3. (d) Give the conjugate acid of each of the following substances acting as a base: OH^-, HSO_4^-, NO_3^-, $HOAc$, CO_3^{2-}.

9-9   Maximum Volume of Solution Given 1 liter of 0.200 M Na_2SO_4 and 1 liter water, what is the maximum volume of 0.00500 M Na_2SO_4 solution that can be made? Assume volumes additive.

9-10   Acid-Base Titration When 0.400 g of an unknown organic base is dissolved in 25.0 ml 0.200 M hydrochloric acid, the solution is acidic, but it can be neutralized by addition of 7.50 ml of 0.400 M potassium hydroxide. (a) If one molecule of the unknown base reacts with one proton, what is the molecular weight of the base? (b) If one molecule of base reacts with two protons, what is the molecular weight of the base?

9-11   Titration of Oxalic Acid One molecule of oxalic acid, $C_2H_2O_4$, can furnish two protons in reactions with strong bases. Suppose that a certain solution of oxalic acid reacted in this way when a 15.00-ml portion of the solution was neutralized by titration with 25.0 ml 0.600 M KOH. What was the molarity of the oxalic acid solution?

9-12   Titration of an Ammonia Solution Suppose that 22.5 ml of 0.420 M hydrochloric acid was needed to neutralize 18.0 ml of a certain ammonia solution. What was the molarity of the ammonia solution?

9-13   Reaction of a Hydrazine Solution Hydrazine, N_2H_4, is a base that can react with either one or two protons per molecule of base, depending on the reaction conditions. What are the weights of hydrazine that can react with 1 mol HCl?

9-14   Reaction of Citric Acid Solution Citric acid can be symbolized as H_3X, that is, it has three potentially acidic hydrogen atoms, one or more of which can react, depending on the reaction conditions. The molecular weight of citric acid is 192. What are the weights of citric acid that can react with 0.200 mol NaOH?

9-15 Reaction of Arsenic Acid Arsenic acid, H_3AsO_4, can be titrated with base in such a way that one, two, or three protons react per molecule of acid. When a 1.07-g sample of the acid was dissolved in 100 ml water, 30.0 ml of 0.500 M KOH was needed to react with the acid. How many protons were neutralized per molecule of acid?

9-16 Molecular Weight of an Organic Base Suppose that 30.0 ml of 0.224 M HCl is needed to neutralize 0.538 g of an unknown organic base. Give two possible values of the molecular weight of the unknown base and explain what they imply about the number of protons with which each molecule of base reacts.

9-17 Molecular Weight of an Unknown Acid When 2.00 g of an unknown organic acid A is dissolved in 20.0 ml of 0.400 M NaOH, the resulting solution is basic, but the solution can be neutralized by the addition of 10.0 ml of 0.200 M HCl. (*a*) Calculate the molecular weight of A on the assumption that only one proton of A reacts with base, and calculate another value on the assumption that n protons of A react with base. (*b*) The acid A is known to contain 5.9 percent cobalt (Co). What does this imply about the minimum value for n, defined as in (*a*)?

9-18 Analysis of a Sodium Carbonate Sample Sodium carbonate exists in various crystalline forms with different amounts of water of crystallization, including Na_2CO_3, $Na_2CO_3 \cdot 10H_2O$, and others. The water of crystallization can be driven off by heating; the amount of water removed depends on the temperature and duration of heating.

A sample of $Na_2CO_3 \cdot 10H_2O$ had been heated inadvertently and it was not known how much water had been removed. A 0.200-g sample of the solid that remained after the heating was dissolved in water, 30.0 ml of 0.100 M HCl was added, and the CO_2 formed was removed. The solution was acidic; 6.4 ml of 0.200 M NaOH was needed to neutralize the excess acid. What fraction of the water had been driven from the $Na_2CO_3 \cdot 10H_2O$?

9-19 Solution Composition (*a*) How many moles of barium ion and how many moles of chloride ion are there in 400 ml of a 0.30 M solution of $BaCl_2$, which may be assumed to dissociate completely into ions? (*b*) Suppose that 200 ml of 0.50 M

$NaNO_3$ is added to the solution in (*a*). Assume the volumes are additive and that no reactions occur. Calculate the concentrations of the four ions in the resulting solution.

9-20 Properties of Oxides (*a*) Explain what is meant by the term "acidic oxide" and illustrate your answer with equations for the reaction of SO_2 with H_2O and with a solution of NaOH. (*b*) Explain what is meant by the term "basic oxide" and illustrate your answer with equations for the reaction of CaO with H_2O and of NiO with hydrochloric acid. (*c*) Explain what is meant by the term "amphoteric oxide" and illustrate your answer with equations for the reaction of ZnO with a solution of NaOH and with a solution of HCl.

9-21 Solution of Sodium Carbonate An aqueous solution of sodium carbonate, Na_2CO_3, is titrated with strong acid to a point at which two protons have reacted with each carbonate ion. (*a*) If 20.0 ml of the carbonate solution reacts with just 40.0 ml of 0.50 M acid, what is the molarity of the carbonate solution? (*b*) If the solution contains 5.0 wt percent sodium carbonate, what is the density of the solution? (*c*) Suppose that you wanted to prepare a liter of an identical solution by starting with crystalline sodium carbonate decahydrate, $Na_2CO_3 \cdot 10H_2O$, rather than with solid Na_2CO_3 itself. How much of this substance would you need?

9-22 Solution of Hydrogen Iodide (*a*) What volume of gaseous HI at 25°C and 750 torr would be needed to form 250 ml of 6.0 M hydriodic acid? (*b*) What weight of magnesium carbonate would this solution react with if all the carbonate were converted to CO_2? What if all carbonate were converted to bicarbonate?

9-23 Analysis of a Vinegar Sample Suppose you are an analytical chemist for a large food concern and have obtained the following data pertaining to the analysis of vinegar for acetic acid content (acetic acid is CH_3COOH). When 17.0 ml of 1.00 M sodium hydroxide was added to a 15.0-g sample of vinegar, the solution became basic and 24.0 ml of 0.104 M sulfuric acid was needed to neutralize it. What is the percentage of acetic acid in the vinegar? Assume that vinegar contains no other acidic compounds.

9-24 Titration of Diphenylamine in Acetic Acid Weak organic bases are sometimes dissolved in pure acetic acid so that they may be more readily titrated. What volume of 0.0100 M solution of perchloric acid ($HClO_4$) in acetic acid would be needed to neutralize 10.0 ml of a 1.00 wt percent solution of diphenylamine, $C_{12}H_{11}N$, in acetic acid? Each diphenylamine molecule reacts with one proton in this reaction. Use the following densities, measured at 25°C, if needed:

	Density/(g ml^{-1})
Pure diphenylamine (solid)	1.159
Pure acetic acid (liquid)	1.049
0.0100 M $HClO_4$ in acetic acid	1.051
1.00 percent diphenylamine in acetic acid	1.050

9-25 Properties of an Unknown Acid When 1.50 g of an unknown acid A is dissolved in 25.0 g naphthalene ($C_{10}H_8$), the melting point of the naphthalene is depressed 2.55°C below its usual value. The molal freezing-point-depression constant for naphthalene is 6.8 K kg mol^{-1}. (a) What is the approximate molecular weight of A? (b) A 0.500-g sample of A is just neutralized by 25.0 ml of 0.500 M NaOH. How many protons reacted for each molecule of A in this reaction?

9-26 Neutralization (a) What volume of 0.400 M HNO_3 must be used to neutralize 240 ml of 0.300 M NaOH? (b) What are the concentrations of all ions in solution after the neutralization?

9-27 Mixing Strong Acids and Bases A solution is prepared by dissolving 7.00 g KOH in 500 ml water. Half of this solution is mixed with 250 ml of 0.240 M HCl. What is the pH of the resulting mixture?

9-28 Mixing Solutions of Electrolytes Calculate the concentrations of the principal ions present in a solution prepared by mixing 150 ml of 0.300 M NH_4NO_3, 200 ml of 0.240 M $NaNO_3$, 100 ml of 0.600 M $CaBr_2$, and enough water to bring the total volume to 800 ml. Assume that all the salts present are strong electrolytes.

9-29 Mixing Solutions of Electrolytes Calculate the concentrations of the principal ions present in a solution prepared by mixing 150 ml of 0.200 M $AgNO_3$ with 350 ml of 0.100 M $BaCl_2$. As-

sume that all substances present are strong electrolytes, that AgCl is insoluble, and that volumes are additive.

9-30 Mixing Solutions of Electrolytes Suppose that 150 ml of 0.200 M K_2CO_3 and 100 ml of 0.400 M $Ca(NO_3)_2$ are mixed together. Assume that the volumes are additive, that $CaCO_3$ is completely insoluble, and that all other substances that might be formed are soluble. Calculate the weight of $CaCO_3$ precipitated, and calculate the concentrations in the final solution of the four ions that were present initially.

9-31 Concentrations of Ions in a Mixture after Reaction Suppose that 15.0 liter of gaseous SO_3 at 0.40 atm and 20°C is passed into 500 ml of an aqueous solution that is 0.48 M in NH_3 and 0.40 M in $Ba(OH)_2$. Assume that the volume of the solution does not change and that barium sulfate is insoluble. Calculate the final concentrations of the principal ionic and molecular species present (other than water) and the weight of barium sulfate precipitated (if any).

9-32 Percentage of Ammonia in a Mixture A 5.0-liter flask contains a mixture of ammonia and nitrogen at 27°C and a total pressure of 3.00 atm. The sample of gas is allowed to flow from the flask until the pressure in the flask has fallen to 1.00 atm. The gas that escapes is passed through 1.50 liter of 0.200 M acetic acid. All the ammonia in the gas that escapes is absorbed by the solution and turns out to be just sufficient to neutralize the acetic acid present. The volume of the solution does not change significantly. (a) Will the electrical conductivity of the aqueous solution change significantly as the gas is absorbed? Give equations for any reactions, and calculate the final concentrations of the principal ions present (if any) at the end. (b) Calculate the percentage by weight of ammonia in the flask initially.

9-33 Analysis of a Gas Mixture A 3.00-liter flask contains a mixture of helium and gaseous HCl at a total pressure of 1.00 atm and 27°C. When the entire contents of the flask are passed into 500 ml of 0.200 M NaOH, the solution remains basic. However, addition of 100 ml of 0.100 M aqueous HCl to this resulting solution neutralizes it. (a) Calculate the total number of moles of gas in the flask. (b) Calculate the number of moles of HCl in the flask initially. (c) Calculate the partial pressures of HCl and of He in the flask initially.

9-34 Analysis of a Mixture of KCl and HBr It was desired to neutralize a certain solution X that had been prepared by mixing solutions of potassium chloride and hydrobromic acid. Titration of 10.0 ml X with 0.100 M silver nitrate required 50.0 ml of the latter. The resulting precipitate, containing a mixture of AgCl and AgBr, was dried and found to weigh 0.762 g. How much 0.100 M sodium hydroxide should be used to neutralize 10.0 ml solution X?

9-35 Analysis of Chloride Samples Suppose that you are determining chloride in many solid samples by titration of a weighed amount of sample with 0.2000 M silver nitrate. To simplify calculations you decide to weigh out an amount of sample each time such that the volume of silver nitrate used, in milliliters, will just equal the percentage of chloride in the sample. What weight of sample should you take?

9-36 Complex Ions (a) Define the terms complex ion, ligand, coordination number, chelate, polynuclear complex. (b) Give examples of formulas of hydrated ions with coordination numbers 4 and 6. Why are these formulas important when the formation and reactions of complex ions in aqueous solution are being considered?

Oxidation and Reduction

"Accordingly, for ions and their derivatives, oxidation in the larger sense . . . consists in the gain of positive charges . . . or in the equivalent loss of negative charges. Reduction conversely means loss of positive or gain of negative charges."
W. OSTWALD, 1899[1]

Oxidation and reduction reactions are common in every area of chemistry—biological processes that supply energy to organisms, corrosion of metals, flames, electric batteries, and many chemical syntheses, for example. These reactions differ from acid-base reactions in many ways, but there are also some important analogies and similarities. The present chapter is concerned with general features of oxidation and reduction, and with the balancing of equations for such reactions. Chapter 21 deals with the application of oxidation and reduction reactions in electrochemical systems.

10-1 Oxidation and Reduction Reactions

Originally *oxidation* meant combination with oxygen, as in the reactions

$$Cu + \tfrac{1}{2}O_2 \longrightarrow CuO \tag{10-1}$$

$$Fe + \tfrac{1}{2}O_2 \longrightarrow FeO \tag{10-2}$$

$$2Fe + \tfrac{3}{2}O_2 \longrightarrow Fe_2O_3 \tag{10-3}$$

Similarly, the term *reduction* was first used in the context of reduction of a metallic oxide to the metal. This may be achieved, for example, by heating the oxide with carbon, with hydrogen, or with other metals:

$$2FeO + C \longrightarrow 2Fe + CO_2 \tag{10-4}$$

$$CuO + H_2 \longrightarrow Cu + H_2O \tag{10-5}$$

$$Fe_2O_3 + 2Al \longrightarrow 2Fe + Al_2O_3 \tag{10-6}$$

Substances that combine readily with oxygen, such as carbon, hydrogen, and aluminum, are called reducing agents. They are oxidized to CO_2, H_2O, and Al_2O_3 in the reactions given.

It was only natural to extend the terms oxidation and reduction to reactions similar to those described but not involving oxygen. For example, metallic

---

[1] Ostwald wrote this before recognition of the fundamental part played in oxidation and reduction by electrons—the existence of which had been established only two years earlier.

lithium reacts vigorously not only with oxygen but also with fluorine and chlorine:

$$2Li + \tfrac{1}{2}O_2 \longrightarrow Li_2O \qquad (10\text{-}7)$$

$$Li + \tfrac{1}{2}F_2 \longrightarrow LiF \qquad (10\text{-}8)$$

$$Li + \tfrac{1}{2}Cl_2 \longrightarrow LiCl \qquad (10\text{-}9)$$

and there is little that is special about the reaction with oxygen. All three of these reactions are therefore called oxidations of lithium. In the resulting crystals of Li_2O, LiF, and LiCl, the lithium exists as the Li^+ ion, so that as far as lithium is concerned all three reactions may be described by the equation

$$Li \longrightarrow Li^+ + e^- \qquad (10\text{-}10)$$

that is, the lithium atom loses an electron. By the same token the other reactants, O_2, F_2, and Cl_2, gain electrons:

$$\tfrac{1}{2}O_2 + 2e^- \longrightarrow O^{2-} \qquad (10\text{-}11)$$

$$\tfrac{1}{2}F_2 + e^- \longrightarrow F^- \qquad (10\text{-}12)$$

$$\tfrac{1}{2}Cl_2 + e^- \longrightarrow Cl^- \qquad (10\text{-}13)$$

These similarities lead to the following generalizations:

Oxidation is defined as a *loss of electrons.*

Reduction is defined as a *gain of electrons.*

An *oxidizing agent* or *oxidant* ("ox") is an ion or molecule that *takes electrons away from* another ion or molecule, thereby oxidizing it. A *reducing agent* or *reductant* ("red") is an ion or molecule that *supplies electrons to* another ion or molecule, thereby reducing it. In these reactions, the oxidizing agent is reduced:

$$ox + \mathbf{n}e^- \longrightarrow red \qquad \text{(reduction)} \qquad (10\text{-}14)$$

The oxidizing agent is thus transformed into a reduced form that is potentially capable of losing electrons and could therefore be a reducing agent. Similarly, the reducing agent is oxidized, losing electrons and being transformed to a form capable of acting as an oxiding agent:

$$red \longrightarrow ox + \mathbf{n}e^- \qquad \text{(oxidation)} \qquad (10\text{-}15)$$

Two species related as in (10-14) or (10-15) are called a *redox couple*. In referring to redox couples in this text, the format *oxidized form/reduced form* will be used—for example, Na^+/Na, O_2/O^{2-}, or F_2/F^-.

A reaction that relates the oxidized and reduced forms of a redox couple is called a *half-reaction*. It may involve atomic or molecular species and ions, as in (10-10) to (10-13), or just ions, as for the couple Fe^{3+}/Fe^{2+}:

$$Fe^{3+} + e^- \longrightarrow Fe^{2+} \qquad (10\text{-}16)$$

Species other than oxidizing and reducing agents may also participate in redox reactions, as they do in the couples MnO_4^-/Mn^{2+} and $Cr_2O_7^{2-}/Cr^{3+}$:

$$MnO_4^- + 8H^+ + 5e^- \longrightarrow Mn^{2+} + 4H_2O \qquad (10\text{-}17)$$

$$Cr_2O_7^{2-} + 14H^+ + 6e^- \longrightarrow 2Cr^{3+} + 7H_2O \qquad (10\text{-}18)$$

Note that charges as well as atoms are balanced.

If oxidation occurred on a macroscopic scale without simultaneous reduction, electrons released during the oxidation would accumulate in large numbers, and the resulting large charge separations would be energetically very unfavorable. Consequently any macroscopic oxidation is always accompanied by a simultaneous reduction. The two half-reactions involved may be coupled electrically, as in a battery or in electrolysis (Chap. 21), each half-reaction occurring separately at an electrode that is able to furnish or take up the electrons being exchanged. A separate path is provided (e.g., a wire) to permit the electrons to flow readily from one electrode to the other. On the other hand, the oxidizing and reducing agents may exchange electrons directly, with no intervening electrical circuits. The general *redox reaction,* as an oxidation-reduction reaction is often called, is an electron-transfer reaction involving two redox couples, ox_1/red_1 and ox_2/red_2:

$$ox_1 + red_2 \longrightarrow red_1 + ox_2 \tag{10-19}$$

Note that by our generalizations oxygen is reduced in reactions (10-1) to (10-3). Similarly the production of oxygen by the electrode reaction

$$2H_2O \longrightarrow O_2(g) + 4H^+ + 4e^- \tag{10-20}$$

is called an oxidation of water since each water molecule loses two electrons.

Comparison of (9-11) with (10-15) and of (9-13) with (10-19) suggests some useful analogies between acid-base and redox reactions. An acid furnishes protons, while a reducing agent furnishes electrons; conversely a base accepts protons, while an oxidizing agent accepts electrons. A redox couple is analogous to a conjugate acid-base pair.

Oxidation States   The oxidation equation

$$Co^{2+} \longrightarrow Co^{3+} + e^- \tag{10-21}$$

for the reaction in aqueous solution is shorthand for

$$Co(H_2O)_6^{2+} \longrightarrow Co(H_2O)_6^{3+} + e^- \tag{10-22}$$

Similar reactions relate the ammonia complexes $Co(NH_3)_6^{2+}$ and $Co(NH_3)_6^{3+}$:

$$Co(NH_3)_6^{2+} \longrightarrow Co(NH_3)_6^{3+} + e^- \tag{10-23}$$

and the cyanide complexes $Co(CN)_6^{4-}$ and $Co(CN)_6^{3-}$:

$$Co(CN)_6^{4-} \longrightarrow Co(CN)_6^{3-} + e^- \tag{10-24}$$

While in all three cases the oxidation concerns the entire complex ion, it is convenient to consider that it is the Co atom that is oxidized, because all three reactants are related to Co^{2+} and the products to Co^{3+}. Thus the ion $Co(CN)_6^{4-}$ may be formed by the reaction of $Co(H_2O)_6^{2+}$ with $6CN^-$:

$$Co(H_2O)_6^{2+} + 6CN^- \longrightarrow Co(CN)_6^{4-} + 6H_2O \tag{10-25}$$

and $Co(CN)_6^{3-}$ is related to $Co(H_2O)_6^{3+}$ in a similar way. We say that in all three reactions cobalt is oxidized from the $+2$ to the $+3$ state[2] and write

$$Co(2) \longrightarrow Co(3) + e^- \tag{10-26}$$

---

[2] Although we use arabic numerals to indicate oxidation states here, in the nomenclature of inorganic chemistry it is customary to use roman numerals for oxidation state (see Appendix B).

Similarly, in reactions (10-7) to (10-9) we say that in each reaction lithium is oxidized from oxidation state zero to the $+1$ state, as expressed in (10-10) or by

$$\text{Li}(0) \longrightarrow \text{Li}(1) + e^- \qquad (10\text{-}27)$$

In the same fashion, oxygen in reactions (10-1) to (10-3) is said to be reduced from the zero state to the -2 state. By convention, oxygen is assigned oxidation state -2 in all its compounds except peroxides and others containing O-O bonds, and certain compounds with fluorine. This leads to the oxidation state $+1$ for hydrogen in water and in OH^-, and by convention this oxidation state is assigned to hydrogen in all compounds except metallic hydrides such as NaH. By similar rules it is possible to assign oxidation states to the atoms of other elements in a molecule or ion.

Oxidation states are used in the precise specification of many compounds by name (see Appendix B). Moreover, they make it possible to define the oxidation or reduction of atoms within molecules or ions, so that *oxidation is an increase in oxidation state and reduction is a decrease*. Finally, oxidation states are useful in balancing equations for oxidation-reductions, as described below. Some rules for the assignment of oxidation states are given in the next section.

10-2 Oxidation States and the Balancing of Redox Equations

Some Rules for Assigning Oxidation States

Rule 1. The sum of the oxidation states of all the atoms in a molecule or ion, must equal the charge on that molecule or ion. Thus the oxidation state of a monatomic ion, such as S^{2-} or Al^{3+}, must be the charge on the ion.

All the other rules are somewhat arbitrary.

Rule 2. Fluorine in its compounds has oxidation state -1.

Rule 3. Hydrogen in its compounds is assigned oxidation state $+1$ except in metallic hydrides (such as NaH), where it is given oxidation state -1.

Rule 4. Oxygen in its compounds is assigned oxidation state -2 except

(a) in compounds containing O-O bonds: peroxides (such as H_2O_2) and superoxides (such as KO_2), with oxidation states -1 and $-\frac{1}{2}$ for oxygen, respectively;

(b) in compounds with fluorine, where rule 2 takes precedence. Thus in OF_2 the oxidation state of oxygen is $+2$.

We have already used rules 2 through 4 without stating them, e.g., in the discussions of Equations (9-21) and (9-23) relating to $Be(OH)_4^{2-}$ and $Al(OH)_4^-$ and of the complex ions of Table 9-2. They also lead to the assignment of oxidation states $+6$ for sulfur in SO_4^{2-}, $+7$ for manganese in MnO_4^-, -3 for nitrogen in NH_3, $+4$ for silicon in SiO_2, and so on. In some situations oxidation states can be assigned by analogy. For example, application of rule 2 to PF_3 gives the phosphorus atom in this compound an oxidation state of $+3$. What then of P in PCl_3, PBr_3, and PI_3? The many similarities in the chemistry of the halogens (F, Cl, Br, and I) suggest that in these phosphorus trihalides it would be reasonable to assign oxidation state -1 to each halogen atom and $+3$ to phosphorus. In the complex ion CuF_4^{2-}, we assign oxidation state $+2$ to Cu, by rule 2. The same assignment seems reasonable for Cu in the similar ion $CuCl_4^{2-}$; note that oxidation state $+2$ holds also for Cu in $Cu(H_2O)_4^{2+}$, by application of rules 3 and 4.

Almost all compounds of the alkali and alkaline-earth metals (columns IA and IIA) are ionic, and it is usually reasonable to assume that the oxidation states of these elements are, respectively, $+1$ and $+2$ in their compounds. This assumption was made, for example, in the assignment of oxidation state -1 to H in NaH.

Since the rules by which oxidation states are assigned are in some respects arbitrary, oxidation states in any molecule or ion containing more than one atom are also to some degree arbitrary. They should not be taken to be the actual charges on the atoms concerned. They are, nonetheless, useful for a number of purposes, including balancing redox equations.

Example 10-1

☐ **Oxidation States** Assign oxidation states to the indicated atoms in the following species by means of the rules just given.
(a) Cr in Cr^{2+}, CrF_3, CrO_3, $HCrO_4^-$

Solution By rule 1, $\underline{+2}$; by rule 2, $\underline{+3}$; by rule 4, $\underline{+6}$; by rules 3 and 4, taking the charge into account, $\underline{+6}$.

(b) Si in SiH_4, SiO_3^{2-}, $SiCl_4$

Solution By rule 3, $\underline{-4}$; by rule 4, taking the charge into account, $\underline{+4}$; no rule specifically covers this situation, but by analogy with F, we assign -1 to Cl and thereby $\underline{+4}$ to Si in this compound. ■

Exercise 10-1

☐ Assign oxidation states to the indicated atoms in the following compounds: (a) I in IF_3, KI, HIO, I_2, KIO_4, IO_3^-, I_2O_7, $Na_2H_3IO_6$; (b) Fe in FeO_3^{2-}, $BaFeO_4$, $FeCl_3$, FeO, Fe_2O_3, Fe_3O_4. ■

Balancing Equations with the Aid of Oxidation States The equation for a redox reaction may readily be balanced by the method of oxidation states. When ions are involved, the reaction is assumed to take place in a medium favoring their existence and this medium is usually understood to be an aqueous solution. The reaction may then be defined solely by giving the oxidized and reduced forms of the couples present. *No other redox couples may be introduced in the balancing procedure,* although H^+, OH^-, and H_2O may participate in the reaction in the aqueous medium and thus may be used in balancing the equation. It is customary to use just the species H^+ and H_2O for acidic solutions, and just OH^- and H_2O for alkaline solutions. Either alternative is acceptable for neutral solutions.

Example 10-2

☐ **Oxidation of Ferrous Ion by Chlorine** When ferrous ion (Fe^{2+}) reacts with chlorine in an aqueous solution, the products are ferric ion (Fe^{3+}) and chloride ion. Write a balanced equation for this reaction.

Solution

Step 1. Assignment of oxidation states: The oxidation states of the various atoms present in the reactants and products are

	Start	End	Change in oxidation state per atom
Fe	$+2$ in Fe^{2+}	$+3$ in Fe^{3+}	$+1$
Cl	0 in Cl_2	-1 in Cl^-	-1

Step 2. Balancing changes in oxidation states: Since a chlorine molecule contains two chlorine atoms, each of which gains an electron, two electrons are gained by each Cl_2. These two electrons must come from two Fe^{2+}, because each Fe^{2+} loses one electron. The coefficients for the reactants and products thus are as follows:

$$2\,Fe^{2+} + Cl_2 \longrightarrow 2Fe^{3+} + 2\,Cl^-$$

The equation is balanced as written, with respect to both atoms and charges. ■

□ When iodide ion reacts with stannic ion (Sn^{4+}) in aqueous solution, the products are stannous ion (Sn^{2+}) and iodine (I_2). Write a balanced equation for this reaction. ■

Exercise 10-2

In most reactions in aqueous solution, more steps are needed for balancing than are shown in the preceding example, because oxygen in oxidation state -2, hydrogen in oxidation state $+1$, or both, are present in some of the reactants or products. These steps are illustrated in the following examples.

□ **Oxidation of Nitrite by Bromine in Water** When nitrite ion (NO_2^-) reacts with a dilute solution of bromine in water, the products are bromide ion and nitrate ion (NO_3^-). Write a balanced equation for this reaction, using the species OH^- and H_2O where required.

Example 10-3

Solution

Step 1. Assignment of oxidation states: The oxidation states of the reagents are

	Start	End	Change in oxidation state per atom
Br	0 in Br_2	-1 in Br^-	-1
N	$+3$ in NO_2^-	$+5$ in NO_3^-	$+2$
O	-2 in NO_2^-	-2 in NO_3^-	0

Note that only Br and N change oxidation state.

Step 2. Balancing of changes in oxidation states: The changes in oxidation state correspond to a gain of $2e^-$ by each Br_2 molecule and a loss of $2e^-$ by each nitrite ion. Since the number of electrons lost by one redox couple must be the same as that gained by the other, the coefficients for these reactants must be in the ratio 1:1.

$$NO_2^- + Br_2 \longrightarrow NO_3^- + 2Br^- \qquad (10\text{-}28)$$

This expression is not yet balanced with regard to all atoms nor with regard to charges.

Step 3. Balancing of charges: We add OH^- ions as needed to balance the charges. The charge on the left of (10-28) is -1; that on the right side is -3. Thus we add $2OH^-$ to the left side:

$$NO_2^- + Br_2 + 2OH^- \longrightarrow NO_3^- + 2Br^- \qquad (10\text{-}29)$$

This expression is still not completely balanced.

Step 4. Balancing of atoms: Since the charges and the atoms whose oxidation states change are now balanced, it remains only to balance the other atoms present, H and O, without any change in charge or oxidation state. The electrically neutral species H_2O may be added as needed. We note that there are $2 + 2 \times 1 = 4$ oxygen atoms on the left and 3 oxygen atoms on the right; thus to balance these atoms we add $4 - 3 = 1$ water molecules on the right

$$NO_2^- + Br_2 + 2OH^- \longrightarrow NO_3^- + 2Br^- + H_2O \qquad (10\text{-}30)$$

There are now $2 \times 1 = 2$ hydrogen atoms on the left and 2 hydrogen atoms on the right. Thus these atoms are also balanced, and (10-30) is completely balanced. It is always true that *the last kind of atom considered is necessarily found to be balanced if no errors have been made.* This serves as a check that all steps have been properly followed.

A final check on the balancing of all atoms and charges should be made after the complete equation has been written. ■

Exercise 10-3

☐ Oxygen reacts in an acidic solution with iodide ion to produce iodine, I_2, and water. Write a balanced equation for this reaction. ■

Example 10-4

☐ **Oxidation of Methanol by Dichromate** The products of the reaction in acid solution between dichromate ion, $Cr_2O_7^{2-}$, and methanol, CH_3OH, are Cr^{3+} ions and formaldehyde, H_2CO. Write a balanced equation for this reaction.

Solution

Step 1: The oxidation states of the reagents are

	Start	End
Cr	$+6$ in $Cr_2O_7^{2-}$	$+3$ in Cr^{3+}
C	-2 in CH_3OH	0 in H_2CO

Thus the changes in oxidation states are

Cr: -3 per Cr atom -6 per $Cr_2O_7^{2-}$ ion
C: $+2$ per C atom $+2$ per CH_3OH molecule

These changes correspond to a gain of $6e^-$ by each $Cr_2O_7^{2-}$ ion and a loss of $2e^-$ by each methanol molecule. The coefficients must therefore be in the ratio $1:3$,

$$Cr_2O_7^{2-} + 3CH_3OH \longrightarrow 2Cr^{3+} + 3H_2CO \qquad (10\text{-}31)$$

Step 2: Since the solution is acidic, we use H^+ to balance the charge. The total charge on the left of (10-31) is -2; that on the right is $+6$. Thus $8H^+$ must be added to the left:

$$8H^+ + Cr_2O_7^{2-} + 3CH_3OH \longrightarrow 2Cr^{3+} + 3H_2CO \qquad (10\text{-}32)$$

Step 3: We note that there are $7 + 3 = 10$ oxygen atoms on the left of (10-32) and 3 oxygen atoms on the right of this same expression. Thus we add $10 - 3 = 7$ water molecules on the right:

$$8H^+ + Cr_2O_7{}^{2-} + 3CH_3OH \longrightarrow 7H_2O + 2Cr^{3+} + 3H_2CO \quad (10\text{-}33)$$

There are now seen to be $8 + 3 \times 4 = 20$ hydrogen atoms on the left and $7 \times 2 + 3 \times 2 = 20$ hydrogen atoms on the right. Thus these atoms have automatically been balanced, and (10-33) is completely balanced. ■

□ The products of the reaction in acid solution between permanganate ion, $MnO_4{}^-$, and formaldehyde, H_2CO, are Mn^{2+} ions and formic acid, $HCOOH$ (not a peroxide). Write a balanced equation for this reaction. ■

Exercise 10-4

One step must be added to the procedure in Examples 10-1 to 10-4 when there are atoms or groups other than O and H that do not change oxidation state during the reaction. These atoms or groups should be balanced as soon as the oxidation states have been balanced, that is, before the charges are balanced (as in step 2 above). This step is usually a simple one.

When a group of atoms remains together throughout a reaction, it may be useful in balancing equations to assign an oxidation state to the entire group rather than to deal with individual atoms. This approach is used for the CN group in the following example.

□ **Balancing by Assigning Oxidation States to Groups** Write a balanced equation for the reaction

$$SCN^- + MnO_4{}^- \longrightarrow HCN + SO_4{}^{2-} + Mn^{2+} \quad (10\text{-}34)$$

occurring in an acidic solution.

Example 10-5

Solution The thiocyanate ion is the source of both HCN and $SO_4{}^{2-}$. The CN group in HCN is assigned the oxidation state -1, and for convenience we assign the same state to the CN in SCN^-, which implies then that S in SCN^- has oxidation state zero. Sulfur ends up at $+6$ in $SO_4{}^{2-}$, an overall loss of $6e^-$. The Mn originally in $MnO_4{}^-$ goes from $+7$ to $+2$, a gain of $5e^-$. The coefficients for balancing the changes in oxidation states are thus 5 and 6:

$$5SCN^- \text{ go to } 5SO_4{}^{2-}, \text{ with loss of } 30e^-$$
$$6MnO_4{}^- \text{ go to } 6Mn^{2+}, \text{ with gain of } 30e^-$$

Although the CN group does not change oxidation state this group comes only from the SCN^- and thus the HCN that appears as a product in (10-34) must be equal in number to the SCN^- on the left. At this point we have

$$5SCN^- + 6MnO_4{}^- \longrightarrow 5SO_4{}^{2-} + 6Mn^{2+} + 5HCN$$

The remaining steps are similar to those used in the previous example. We add $13H^+$ on the left to balance charges and then $4H_2O$ on the right to balance oxygen atoms. Hydrogen atoms then balance, confirming the correctness of the earlier steps:

$$13H^+ + 5SCN^- + 6MnO_4{}^- \longrightarrow 4H_2O + 5SO_4{}^{2-} + 6Mn^{2+} + 5HCN \quad ■$$

☐ Write a balanced equation for the reaction

$$H^+ + C_6H_5CH_3 + CrO_2Cl_2 \longrightarrow C_6H_5CHO + Cr^{3+} + Cl^- + H_2O$$

by assigning an (arbitrary) oxidation state to the group C_6H_5, which remains together in the reaction. ∎

Balancing Redox Equations with the Aid of Half-Reactions Equations for oxidation-reduction reactions may also be balanced by a quite different method utilizing the half-reactions of each redox couple involved in the overall reaction. The choice between this method and that of oxidation states is to a considerable extent a matter of personal preference. Half-reactions may be easier to use in very complex situations, and many of them are found in tabulations of standard electrode potentials for redox couples (Chap. 21). Their use in balancing redox equations is best illustrated by examples. Note that the sequence of steps differs from that used in the oxidation-state method. No knowledge of or assumptions about oxidation states are needed in applying the half-reaction method.

Example 10-6

☐ **Balancing by Half-Reactions** Balance the reaction in which ferric ion is reduced to ferrous ion by sulfurous acid, which in turn is oxidized to sulfate ion:

$$Fe^{3+} + H_2SO_3 \longrightarrow Fe^{2+} + SO_4^{2-}$$

Solution

Step 1. Separation into half-reactions: We first look for the two redox couples involved in this reaction. They are, as implied by the wording of the problem, Fe^{3+}/Fe^{2+} and SO_4^{2-}/H_2SO_3, with the first going as a reduction, the second as an oxidation.

Step 2a. Balancing of Fe^{3+}/Fe^{2+} couple: Since only Fe atoms are involved we need only add electrons on the appropriate side to produce a balanced half-reaction:

$$Fe^{3+} + e^- \longrightarrow Fe^{2+} \qquad (10\text{-}35)$$

Step 2b. Balancing of SO_4^{2-}/H_2SO_3 couple: Here atoms as well as electrons are needed to achieve a balanced equation. Always balance atoms first. The presence of H_2SO_3 makes the solution weakly acidic, so that H^+ and H_2O are available. We start with

$$H_2SO_3 \longrightarrow SO_4^{2-}$$

and add H_2O on the left to balance oxygen. Then H^+ will be needed on the right to balance hydrogen

$$H_2O + H_2SO_3 \longrightarrow H^+ + SO_4^{2-}$$

The coefficients still need attention. To balance the oxygen (maintaining the existing balance for sulfur) we need only one H_2O on the left; a factor of 4 with the H^+ then balances the hydrogen:

$$H_2O + H_2SO_3 \longrightarrow 4H^+ + SO_4^{2-}$$

Only the charge remains to be balanced. This is achieved by adding two electrons on the right:

$$H_2O + H_2SO_3 \longrightarrow 4H^+ + SO_4^{2-} + 2e^- \qquad (10\text{-}36)$$

Step 3. Combination of half-reactions: The two balanced half-reactions are combined in such a manner that the electrons cancel. Hence (10-35) must be multiplied by 2 and added to (10-36):

$$2Fe^{3+} + H_2SO_3 + H_2O \longrightarrow 2Fe^{2+} + 4H^+ + SO_4^{2-} \quad \blacksquare$$

☐ Balance the reaction

$$H^+ + IO_3^- + Sn^{2+} \longrightarrow Sn^{4+} + I^- + H_2O$$

by the method of half-reactions. ■

☐ **Balancing by Half-Reactions** Balance the reaction between *n*-propyl alcohol (C_3H_7OH) and permanganate in an acid solution to yield propanoic acid (C_2H_5COOH) and Mn^{2+}:

$$MnO_4^- + C_3H_7OH \longrightarrow Mn^{2+} + C_2H_5COOH$$

Example 10-7

Solution

Step 1. Separation into half-reactions: Two couples are involved in the reaction: MnO_4^-/Mn^{2+} and C_2H_5COOH/C_3H_7OH.

Step 2. Balancing of atoms and of charges: Since the solution is acidic we have H^+ and H_2O available. For the couple MnO_4^-/Mn^{2+}, the four oxygen atoms on the left must appear on the right, as $4H_2O$. Thus $8H^+$ are required on the left to furnish the hydrogen. With charges as yet unbalanced, we have

$$MnO_4^- + 8H^+ \longrightarrow Mn^{2+} + 4H_2O$$

The charges are now balanced by inserting the appropriate number of electrons, leading to the balanced half-reaction

$$5e^- + MnO_4^- + 8H^+ \longrightarrow Mn^{2+} + 4H_2O$$

This is a familiar reduction reaction; it is consistent with our earlier use of oxidation number +7 for MnO_4^- and +2 for Mn^{2+} but was derived with no consideration of oxidation numbers.

With the C_2H_5COOH/C_3H_7OH couple, one oxygen is needed on the C_3H_7OH side (the left) to balance oxygen atoms; it is supplied as H_2O. This gives an excess of four hydrogen atoms on the left, which must then be balanced on the right by $4H^+$ (since this reaction is in an acidic solution):

$$C_3H_7OH + H_2O \longrightarrow C_2H_5COOH + 4H^+$$

To balance the charges, $4e^-$ are required on the right:

$$C_3H_7OH + H_2O \longrightarrow C_2H_5COOH + 4H^+ + 4e^-$$

Step 3. Combination of half-reactions: For the electrons to cancel in the overall equation, the half-reactions must be multiplied by the coefficients 4 and 5 before they are combined:

$$32H^+ + 4MnO_4^- + 5C_3H_7OH + 5H_2O \longrightarrow$$
$$4Mn^{2+} + 16H_2O + 5C_2H_5COOH + 20H^+$$

The final equation is obtained by canceling the superfluous H^+ and H_2O:

$$12H^+ + 4MnO_4^- + 5C_3H_7OH \longrightarrow 11H_2O + 4Mn^{2+} + 5C_2H_5COOH \quad \blacksquare$$

Exercise 10-7

☐ Use the method of half-reactions to balance the equation for the reaction between dichromate ion, $Cr_2O_7^{2-}$, and methanol, CH_3OH, in acid to form CH_2O and Cr^{3+} (already balanced in Example 10-4 by the method of oxidation states). ∎

10-3 Redox Stoichiometry

Oxidation-reduction reactions in solutions are often rapid and well-suited for applications in chemical analysis. The principles involved parallel those pertinent to acid-base titrations and are illustrated in the following example and exercise.

Example 10-8

☐ **Redox Titration** Molybdenum forms a green $+3$ ion that is a strong reducing agent, being itself oxidized to molybdate, MoO_4^{2-}. A $0.100\ M$ solution of Mo^{3+} was used to reduce an unknown substance Y; exactly 20.0 ml of the solution was needed to reduce a 0.360-g sample of Y. If each molecule of Y accepted one electron, what is the molecular weight of Y? If each molecule of Y accepted n electrons, what would be the molecular weight of Y?

Solution The oxidation number of Mo at the start is $+3$ and at the end (in MoO_4^{2-}) is $+6$; thus each Mo loses three electrons.

$$\text{Number of moles of } Mo^{3+} \text{ initially} = 0.0200 \text{ liter} \times 0.100\ M$$
$$= 0.00200 \text{ mol}$$
$$\text{Number of moles of electrons lost} = 3 \times 0.00200 \text{ mol}$$
$$= 0.00600 \text{ mol}$$

Since this same number of electrons must have been gained by Y, we know that if each Y gained one electron, there were 0.00600 mol Y present. Hence the weight per mole of Y = 0.360 g/0.00600 mol = 60 g mol^{-1}. If each Y gained n electrons, then the number of moles of Y present must have been $0.00600/n$, and the weight per mole of Y would be

$$\frac{0.360 \text{ g}}{0.00600 \text{ mol}/n} = 60n \text{ g mol}^{-1} \quad \blacksquare$$

Exercise 10-8

☐ The blue chromous ion, Cr^{2+}, is an excellent reducing agent. Suppose it was used to titrate a solution of Fe^{3+}, the products being Fe^{2+} and Cr^{3+}. If 20.0 ml of $0.450\ M$ Fe^{3+} was reduced by 25.0 ml of the Cr^{2+} solution, what was the concentration of Cr^{2+}? ∎

Some substances can act as both a reducing agent and an oxidizing agent in the same reaction. To do so they must necessarily be capable of existing in at least three oxidation states. Consider copper, which can exist in the states 0, +1, and +2. The ion Cu^+ is unstable in aqueous solution because of the reaction

$$2Cu^+ \longrightarrow Cu(s) + Cu^{2+} \tag{10-37}$$

in which half of the Cu^+ is reduced to $Cu(s)$ as it oxidizes the other half to Cu^{2+}. Such a reaction in which a substance is both oxidized and reduced is called a *disproportionation.*

While in the example just given the ion in the intermediate oxidation state is unstable with respect to disproportionation, it does not follow that all ions in intermediate oxidation states will disproportionate. Ferrous ion, Fe^{2+}, for example, is stable in aqueous solution and does not disproportionate to $Fe(s)$ and Fe^{3+}:

$$3Fe^{2+} \xrightarrow{\quad//\quad} Fe(s) + 2Fe^{3+} \tag{10-38}$$

In fact, the reverse reaction occurs if $Fe(s)$ is placed in a solution containing Fe^{3+}.

Disproportionation of molecules can also occur. White phosphorus, P_4, disproportionates when it reacts with a hot solution of $Ba(OH)_2$:

$$P_4 + 3H_2O + 3\,OH^- \longrightarrow PH_3 + 3H_2PO_2^- \tag{10-39}$$

Here phosphorus goes from oxidation state 0 in P_4 to -3 in phosphine, PH_3, as well as to $+1$ in the hypophosphite ion, $H_2PO_2^-$. These two products must, therefore, be formed in the ratio 1:3.

The conditions that govern whether or not disproportionation will occur in a given situation are considered in Sec. 23-3.

Summary

Oxidation is defined as a loss of electrons and reduction as a gain of electrons. An oxidizing agent is a species capable of accepting electrons: $ox + ne^- \rightarrow red$; a reducing agent is a species capable of supplying electrons: $red \rightarrow ox + ne^-$. Two related species, ox and red, are called a redox couple; examples are Fe^{3+}/Fe^{2+} and MnO_4^-/Mn^{2+}. The equation that relates the two species in a redox couple is known as a half-reaction. Other molecules and ions, for example, H_2O and H^+, may also be involved in the half-reaction for a redox couple. A redox reaction is an electron-transfer reaction involving two redox couples: $ox_1 + red_2 \rightarrow red_1 + ox_2$.

When an entire molecule or ion is involved in a redox reaction, it is usually possible to ascribe the electron transfer to one atom within the species. In such cases the oxidation or reduction of the atom is described as an increase or decrease in its oxidation state. Oxidation states can usually be assigned by the use of simple rules.

One method for balancing equations for oxidation-reduction reactions is based on the balancing of changes in oxidation states. In another procedure, half-reactions are first balanced and then combined.

Some substances can act as both a reducing agent and an oxidizing agent in the same reaction. Such a reaction in which a substance is both oxidized and reduced is called a disproportionation.

Terms and Concepts

Problems and Questions

10-1 Reactions in Acidic Solutions Balance equations for the following reactions, supplying missing ingredients (if any) on each side, on the assumption that the reactions occur in acidic aqueous solutions.

(a) $MnO_4^- + Cl^- \rightarrow Mn^{2+} + OCl^-$
(b) $H_2O_2 + CH_3OH \rightarrow CO_2 + H_2O$
(c) $Ti^{2+} + NO_3^- \rightarrow N_2H_5^+ + TiO^{2+}$
(d) $Fe^{2+} + I_2 \rightarrow Fe^{3+} + I^-$
(e) $V^{2+} + H_2O_2 \rightarrow V(OH)_4^+$
(f) $Sn^{2+} + O_2 \rightarrow Sn^{4+}$
(g) $PH_4^+ + Cr_2O_7^{2-} \rightarrow P_4 + Cr^{3+}$
(h) $Fe^{3+} + I^- \rightarrow Fe^{2+} + I_3^-$

10-2 Reactions in Alkaline Solutions Balance equations for the following reactions, supplying missing ingredients (if any) on each side, on the assumption that the reactions occur in basic aqueous solutions.

(a) $MnO_4^- + Fe_3O_4 \rightarrow Fe_2O_3 + MnO_2$
(b) $CuO + NH_3 \rightarrow Cu + N_2$
(c) $As_2S_3 + NO_3^- \rightarrow HAsO_4^{2-} + S + NO_2$
(d) $CrO_4^{2-} + CN^- \rightarrow CNO^- + Cr(OH)_3$
(e) $Au + CN^- + O_2 \rightarrow Au(CN)_2^-$
(f) $AsO_2^- + O_2 \rightarrow AsO_4^{3-}$

10-3 Conventional Oxidation Numbers Listed below are a number of compounds. For each give the oxidation number obtained by rules 1, 2, 3, and 4 for the element indicated.

(a) P in H_3PO_3, P_4O_6, P_4O_{10}, P_4, H_3PO_4, H_3PO_2, HPO_4^{2-}, HPO_3, PH_4^+
(b) S in H_2S, SO_2, HSO_3^-, $S_2O_3^{2-}$, $S_4O_6^{2-}$, S_8, S_2^{2-}
(c) Mn in $Mn(OH)_2$, Mn^{3+}, Mn_3O_4, MnO_4^-, Mn_2O_7, MnO_4^{2-}, MnO_2
(d) Cl in Cl^-, $HClO$, ClO_2^-, $HClO_3$, Cl_2O, ClF
(e) N in NO, NO_2, HNO_3, NO_2^-, NH_3, N_2H_4, H_2NOH

10-4 Balancing of Reaction Equations Balance the following equations, adding, if necessary, water and hydrogen ions.

(a) $Cu^{2+} + CN^- \rightarrow Cu(CN)_4^{3-} + (CN)_2$
(b) $ClO_3^- + H_2C_2O_4 \rightarrow ClO_2 + CO_2$
(c) $NO_3^- + I_2 \rightarrow IO_3^- + NO_2$
(d) $ClO_3^- + H_2S \rightarrow SO_4^{2-} + Cl^-$
(e) $CuO + NH_3 \rightarrow Cu + N_2$
Identify the oxidizing agent in (a), (c), and (e), and the reducing agent in (b) and (d). For each give the number of electrons exchanged per formula unit.

10-5 Balancing of Reaction Equations Balance the following equations, adding, if necessary, water and hydrogen ions.

(a) $IO_3^- + N_2H_4 \rightarrow N_2 + I^-$
(b) $HBiO_3 + Mn^{2+} \rightarrow MnO_4^- + BiO^+$
(c) $BrO_2^- + NO_2^- \rightarrow Br_2 + NO_3^-$
(d) $NH_4^+ + MnO_4^- \rightarrow NO_3^- + Mn^{2+}$
Indicate the oxidizing agent in (b) and (d), and the reducing agent in (a) and (c). For each give the number of electrons exchanged per formula unit.

10-6 Redox Stoichiometry How many moles of Fe^{2+} could be oxidized to Fe^{3+} by 20.0 ml of $0.250\,M$ $Cr_2O_7^{2-}$ solution, if the chromium is reduced entirely to Cr^{3+}? Write a balanced equation for the reaction.

10-7 Redox Stoichiometry Stannous ion, Sn^{2+}, reacts with mercuric ion, Hg^{2+}, in the presence of Cl^- to form mercurous chloride, Hg_2Cl_2, and a complex of Sn^{4+}, $SnCl_6^{2-}$. Write a balanced equation for the reaction and then calculate the number of moles of mercuric ion that could be reduced by 25.0 ml of $0.600\,M$ Sn^{2+} solution (and an excess of chloride). How many moles of Hg_2Cl_2 would be formed?

10-8 Redox Titration (a) How many moles of Fe^{2+} could be oxidized to Fe^{3+} by 1 mol of $KMnO_4$ if the MnO_4^- was reduced to Mn^{2+}? (b) Suppose that all the Fe^{2+} in an impure sample of $FeSO_4$ was oxidized to Fe^{3+} by 20.0 ml of $0.350\,M$ $KMnO_4$, the latter being reduced to Mn^{2+}. If the sample weighed 9.0 g and contained no other reducing agent, what percentage of it was $FeSO_4$?

10-9 Redox Titration Vanadic ion, V^{3+}, forms green salts and is a good reducing agent, being itself changed in neutral solutions to the nearly colorless ion $V(OH)_4^+$. Suppose that 15.0 ml of a 0.200 M solution of vanadic sulfate, $V_2(SO_4)_3$, was needed to reduce completely a 0.540-g sample of an unknown substance X. If each molecule of X accepted just one electron, what is the molecular weight of X? Suppose that each molecule of X accepted three electrons; what would be the molecular weight of X then?

10-10 Molecular Weight of an Unknown Compound A new antibiotic, A, which is an acid, can readily be oxidized by hot aqueous permanganate; the latter is reduced to manganous ion, Mn^{2+}. The following experiments have been carried out with A: (*a*) 0.293 g A consumes just 18.3 ml of 0.080 M $KMnO_4$; (*b*) 0.385 g A is just neutralized by 15.7 ml of 0.490 M NaOH. What can you conclude about the molecular weight of A from (*a*), from (*b*), and from both considered together?

Introduction to Chemical Equilibrium

"When a system is in chemical equilibrium, a change in one of the parameters of the equilibrium produces a shift in such a direction that, were no other factors involved in this shift, it would lead to a change of opposite sign in the parameter considered."

HENRI-LOUIS Le CHÂTELIER, 1888

This chapter and the two that follow deal with the quantitative aspects of a fundamental concept in chemistry, chemical equilibrium. In the present chapter we consider general features of systems at equilibrium, including the quantitative relationships embodied in the equilibrium constant. Typical ionic equilibria and equilibria involving gases are used as illustrations. Chapter 12 is concerned with equilibria in acid-base systems, and Chap. 13 deals with some heterogeneous equilibria (equilibria involving different phases).

11-1 Background

The concept of dynamic equilibrium, involving a balance of two opposing processes that occur at equal rates, has been discussed in several different contexts in earlier chapters: the equilibrium between liquid and solid phases at the melting point of a crystalline solid (Sec. 5-2); the equilibrium between a gas and the corresponding condensed phases, leading to a definite vapor pressure of a liquid or a solid (Sec. 5-2); and the equilibrium between dissolved solute and excess solute in a saturated solution (Sec. 8-1). Furthermore, the reactions of aqueous acids and bases (Chap. 9) can be viewed in terms of chemical equilibria in which two (or more) bases compete for the available protons, as explained in Sec. 9-2 and in detail in Chap. 12.

Importance of Equilibrium All systems proceed spontaneously toward a position of equilibrium; that is, they tend to move toward equilibrium (although often only very slowly) in the absence of outside disturbances. They move away from a position of equilibrium only if they are perturbed from the outside, for example, by absorption of energy or by a sudden change of pressure. This is a generalization from many observations about the nature of our universe.

The phenomenon of chemical equilibrium is a very general one with applications to the entire range of chemistry. Consequently a firm grasp of the comparatively few general principles needed to understand equilibria can give one a broad command of many different aspects of chemistry, as illustrated throughout the remainder of this book. Many equilibria are established so slowly, however, that a large fraction of chemical problems involve nonequilibrium systems. For example, many of the most important chemical reactions in energy sources such as flames and in living organisms never reach equilibrium, and in many industrial processes the distribution of products depends on the relative rates of the reactions involved (as examined in Chap. 22) rather than on achieving equilibrium.

An understanding of nonequilibrium systems (which always tend toward equilibrium) is greatly helped by an appreciation of equilibrium systems. Hence the next few sections and chapters are important not only because of the significance of chemical equilibrium itself but also for the background they provide for situations in which equilibrium has not been achieved.

Dynamic Nature of Equilibrium In a system at equilibrium, there is no perceptible change in the macroscopic properties of the system with time. However, it is the very essence of chemical equilibria that they are dynamic. There is no net change because species lost in a chemical reaction are regenerated at an equal rate by the reverse reaction. This is quite in contrast to certain static equilibria of interest in physics and engineering.[1] For example, the dynamic nature of the equilibrium between liquid and vapor in a closed flask can readily be demonstrated: if some "labeled" molecules—e.g., molecules containing some radioactive or otherwise easily identifiable isotope of one of the atoms present—are introduced into one phase, they are very quickly found to be uniformly distributed between the two phases present. Similarly, an equilibrium between different molecules is dynamic. For example, two molecules of NO_2, a red-brown gas, readily combine to form molecules of N_2O_4, a colorless gas, which in turn can dissociate to form NO_2. In a flask containing NO_2 and N_2O_4 in equilibrium with each other at a given temperature and pressure, the numbers of moles of NO_2 and of N_2O_4 in the flask remain constant, but there is a continuous and very rapid conversion of NO_2 into N_2O_4 and vice versa.

A fundamental characteristic of the equilibria that are of interest to chemists is that very large numbers of particles—atoms, molecules, or ions—are involved. In other words, we are dealing with macroscopic samples. The statement that the concentrations and partial pressures of the components of an equilibrium mixture remain constant with time does not mean that there are no fluctuations at all in these quantities, but rather that these fluctuations, measured relative to the actual concentrations or pressures, are imperceptible because the numbers of molecules involved are so large (see Sec. 4-4).

An equilibrium of the type we are discussing may or may not involve any chemical reaction in the sense that different chemical species are formed, or even in the wider sense that chemical bonds are broken or formed. When solid CO_2 (dry ice) sublimes, for example, no chemical bonds are broken; the van der Waals interactions among the CO_2 molecules merely become less important as the molecules move apart in the gaseous phase. On the other hand, when CO_2 is dissolved in water, there *is* chemical interaction. Most of the dissolved CO_2 becomes "hydrated" in a way that is not understood in detail, with a small fraction of it forming carbonic acid, H_2CO_3. In both examples, the sublimation of solid

---

[1] For example, the balanced arch of a bridge does not collapse, because it is in static equilibrium: nothing happens (excepting vibrations of the bridge). Similarly, a ball that comes to rest at the bottom of a hole is in static equilibrium, at a position corresponding to a minimum in its gravitational potential energy. These static equilibrium situations are quite different from the dynamic ones of interest in chemistry.

Figure 11-1 (*a*) **Equilibrium;** (*b*) **Steady State**
In (*a*) liquid water and its vapor are at equilibrium because the rate of evaporation of water molecules is the same as their rate of condensation. In (*b*), a vessel receives water at a constant rate from a faucet. The water can leave the vessel through an opening at the bottom, at a rate that increases as the water level in the vessel rises. Unless the faucet is open too much or too little, the water level reaches a position where the rate of inflow just matches the rate of outflow; a steady state exists and the water level in the vessel remains constant.

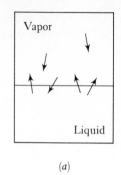
(*a*)

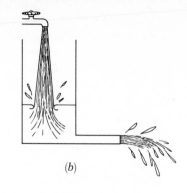
(*b*)

CO_2 and the dissolving of CO_2 in water, equilibrium situations may be established, characterized by constant partial pressures of $CO_2(g)$ (at a given temperature).

Contrast between Equilibrium and Steady State   It is possible to have dynamic situations, called *steady states,* in which even though no equilibrium exists, there is no apparent change in the properties of the system with time. *Dynamic equilibrium requires processes that move in opposite directions,* one process being in essence the reverse of the other, as in Fig. 11-1*a*. In a steady state, the processes are not so opposed and a continuing supply of new material (or energy) is required. A macroscopic example of a steady state is illustrated in Fig. 11-1*b*. Although the level in the tank remains constant, the flow out of the tank is not the reverse of the flow from the faucet—water must be steadily supplied. A steady state may exist without supply to the system from outside; it might occur through the slow reaction of a substance that is present in a large and therefore essentially constant amount.

11-2   Qualitative Considerations

In the brief discussion of the equilibria involved in the sublimation of CO_2 and the chemical reaction of CO_2 with water in the previous section, we pointed out that in each equilibrium the partial pressure of CO_2 was constant at a given temperature. Both equilibria are sensitive to temperature. An equilibrium may also be affected by pressure: the bubbles that appear in a bottle of soda, beer, or champagne when the cap is removed are evidence that the equilibrium solubility of CO_2 in water decreases when the pressure is lowered.

A qualitative principle of great practical value in thinking about the effects on equilibria of changes in the conditions affecting the equilibria was discovered independently and almost simultaneously during the last century by H. L. Le Châtelier and F. Braun. For some reason, however, Braun has received little credit for his insight, and this useful generalization is commonly identified only with the name of Le Châtelier.

Le Châtelier's Principle   Le Châtelier's principle is concerned with changes in systems initially at equilibrium. It predicts that a shift in the equilibrium will usually occur when some property of the system is changed, for example, if its volume is changed or if it is heated or cooled. The shift in equilibrium is in such a direction as to offset part of the effect of the change. Because of the shift, the change in pressure resulting from any change in volume is smaller than it would have been if the equilibrium had not shifted. Similarly, the change in temperature that occurs because of heating or cooling the system is smaller than it would have been if the equilibrium had not shifted. Examples of the application of Le Châtelier's principle are given in the following paragraphs.

Temperature Dependence of the Solubility When a substance dissolves, heat is usually evolved or absorbed. If heat is evolved, the process is said to be *exothermic* (Greek: *exo,* out of); if heat is absorbed, the process is termed *endothermic* (Greek: *endo,* within). Normally, the heat that is evolved or absorbed causes either an increase or a decrease in the temperature of the solution itself. The only way to avoid this would be to carry out the dissolving of the solute very slowly in a constant-temperature bath.

It is found experimentally that an *increase in temperature* leads to an *increase in solubility* if the process of dissolving is *endothermic* and to a *decrease in solubility* if the process of dissolving is *exothermic.* To see how these facts are in accord with Le Châtelier's principle, consider a solution that is at equilibrium with undissolved solute at a particular temperature. Equilibrium implies that no additional solute dissolves, so that the solution is saturated at the given temperature. Suppose the solution is heated. By Le Châtelier's principle the equilibrium must shift—that is, the concentration of solute in the saturated solution must change—in such a way that part of the energy introduced is used up. The actual change in the temperature of the solution is then smaller than it would have been if there had been no shift in the equilibrium.

Assume first that dissolving the solute is endothermic. If additional solute dissolves, some of the energy introduced will be used up because energy is absorbed in the process

$$\text{Solute} + \text{solvent} \longrightarrow \text{solution}$$

The remaining energy is available to increase the temperature of the solution, but the temperature increase will be smaller than it would have been if no additional solute dissolved. In the end the solubility has increased and the temperature is higher (but not as high as it would have been if the solubility had not changed). Thus Le Châtelier's principle leads to a conclusion in accord with the experimental facts: if dissolving the solute is an endothermic process, the solubility increases with increasing temperature.

Conversely, if dissolving the solute is exothermic, some of the energy introduced is used up by the reverse of dissolving, that is, by precipitation of part of the dissolved solute

$$\text{Solution} \longrightarrow \text{solute} + \text{solvent}$$

and the solubility decreases with increasing temperature. Finally, if no heat is evolved or absorbed when the solute dissolves, its solubility is not affected by a change in temperature.

The same result applies to any temperature-dependent equilibrium. As the temperature is increased, the products of any exothermic reaction will be less favored at equilibrium and the products of any endothermic reaction will be more favored. If a reaction is neither exo- nor endothermic, the equilibrium is not affected by a change in temperature.

Effect of Pressure on the Melting Point of Ice   The melting point of ice—the temperature at which liquid water and ice are in equilibrium—is 0°C at a pressure of 1 atm, but decreases as the pressure is increased (Fig. 5-7). The effect of pressure on the melting point can be understood in terms of Le Châtelier's principle when it is realized that there is a decrease in volume as ice changes to water. At 0°C the molar volume of ice is about 8 percent larger than that of water (the density of ice is about 0.92 g cm^{-3} whereas that of water is 1.00 g cm^{-3}).

Suppose the volume of a piece of ice at 0°C is forcibly decreased by squeezing the ice very hard. The decrease in the volume of the ice results in an increase of the pressure in the ice (just as a decrease in the volume of a gas causes an increase in pressure, although a much smaller increase than for ice for the same relative change in volume). Melting of the ice would then relieve some of the increase in pressure because it would permit the same total volume decrease at a lower pressure; that is, some of the volume decrease would result from the change of the ice to water. Thus a volume decrease, and therefore a pressure increase, favors liquid water over ice at 0°C. The equilibrium between water and ice at higher pressures occurs at lower temperatures (Fig. 5-7). This is, incidentally, not the usual behavior. For most substances an increase in pressure causes an increase in the temperature at which there is equilibrium between the solid and liquid states, the density of the solid phase being higher than that of the liquid phase.

These same results apply to any equilibrium. If a shift in the equilibrium can result in a change in the volume of the system, then the shift of equilibrium with increasing pressure will be in such a direction that the volume decreases as a result of the shift.

Le Châtelier's principle may be applied to many different equilibria of chemical interest, and we shall cite it again. The qualitative predictions from Le Châtelier's principle can be understood quantitatively in terms of equilibrium constants, considered in this chapter and the two that follow, and the change of equilibrium constants with temperature (Chap. 20).

Criteria for Equilibrium   One very important feature of any chemical equilibrium is that, for a given set of conditions, the same equilibrium situation is achieved whether it is approached from one side or the other. For example, at a given temperature and total pressure, the same concentrations of NO_2 and N_2O_4 will be obtained at equilibrium whether one starts with pure NO_2, or an equivalent amount of pure N_2O_4, or any mixture of the two. Similarly, the pressure of a vapor in equilibrium with a liquid at a given temperature will be the same whether one starts with pure liquid and lets it evaporate until equilibrium is reached or starts with pure vapor and allows it to condense.

It is sometimes difficult in practice to distinguish a true equilibrium, in which the concentrations remain constant with time, from a nonequilibrium situation in which the concentrations also do not change perceptibly because the rate of any possible reaction is simply too small to observe. For example, at ordinary temperatures and pressures a mixture of gaseous hydrogen and oxygen in the mole ratio 2:1 would be, at equilibrium, essentially completely converted into water, but such a mixture can be kept for many years apparently unchanged because the reaction is normally so slow.

To establish experimentally whether equilibrium does in fact exist, either of the following two criteria may be applied. (1) One or more of the variables (e.g., total pressure, temperature, or concentration of some component) that might affect a possibly existing equilibrium can be changed. If changes in the concentrations or partial pressures of the (other) substances involved then occur, and if these concentrations and partial pressures return to their initial values when the original values of the variables are restored, an equilibrium exists. (2) A possible equilibrium may be approached both from the side of the reactants and from the side of the products, in separate experiments carried out at the same temperature and pressure and with amounts of reactants in one experiment and of products in the other that are equivalent to one another. If the final concentrations and partial pressures are the same in each experiment, a true equilibrium exists.

Equilibrium and nonequilibrium situations may coexist in the same system. Thus, hydrogen peroxide is less stable than its possible decomposition products, O_2 and H_2O,

$$2H_2O_2(l) \longrightarrow 2H_2O(l) + O_2(g) \tag{11-1}$$

but the rate of this reaction and therefore establishment of the equilibrium of (11-1) is very slow in the absence of catalysts. Another reaction, the dissociation into H^+ and HO_2^-,

$$H_2O_2(l) \longrightarrow H^+ + HO_2^-$$

reaches equilibrium rapidly and can be examined by experiment in spite of the instability of the H_2O_2. As long as the rate of (11-1) is slow enough to be negligible, the H_2O_2 is said to be *metastable*, a term implying that a substance may persist (and thus *seem* stable) for a relatively long period, although it eventually changes (or may change) to a more stable form (see Sec. 5-2). Many commonly known substances are actually metastable. This is true, for example, of most organic compounds in the presence of oxygen, since they are readily combustible.

Position of Equilibrium It is frequently convenient to refer to the *position of equilibrium* in a chemical system, even though it is difficult or impossible to define this term in such a general way as to apply to all possible equilibria. Often the position of equilibrium is a measure of the *relative* amounts of reactants and products present at equilibrium. For example, the position of equilibrium is said to be far to the left when the proportion of products (written on the right in a chemical equation) in the equilibrium mixture is small. Conversely, the position of equilibrium is said to be far to the right when the proportion of reactants remaining at equilibrium is small. In some cases, however, the position of equilibrium is not represented by the relative amounts of some of the reacting species. Thus, the position of equilibrium between a liquid and its vapor is characterized by the vapor pressure of the liquid and is completely independent of the amount of liquid present (as long as there is *some* liquid).

In many systems the position of equilibrium can be shifted by changing the concentrations. For example, the degree of dissociation of a weak acid can be altered by changing its concentration. This is in marked contrast to the *equilibrium constant,* to be defined in Sec. 11-3, which is a constant (at a given temperature), independent of the concentrations of the components of the equilibrium mixture.

11-3   The Equilibrium Law

A simple quantitative relationship can be formulated among the concentrations of the reactants and products in any chemical system at equilibrium. For a reaction symbolized by the balanced equation[2]

$$a\text{A} + b\text{B} + \cdots \rightleftharpoons d\text{D} + e\text{E} + \cdots \tag{11-2}$$

with capital letters denoting chemical species and lowercase letters signifying their coefficients in the balanced equation, we define the *reaction quotient*

$$Q \equiv \frac{[\text{D}]^d[\text{E}]^e \cdots}{[\text{A}]^a[\text{B}]^b \cdots} \tag{11-3}$$

Both experiment and theory show that at any particular temperature (and total pressure), Q as defined by (11-3) is a constant at equilibrium. The value of Q for a system at equilibrium is called the *equilibrium constant* for the reaction at the temperature (and pressure) in question. This constant is almost invariably symbolized by K, often with a subscript to indicate the reaction involved or the units used:

$$\frac{[\text{D}]^d[\text{E}]^e \cdots}{[\text{A}]^a[\text{B}]^b \cdots} = K \qquad \text{(equilibrium)} \tag{11-4}$$

What this all-important equilibrium law—sometimes called the law of mass action—implies is that, *if a system is at equilibrium,* the ratio of the product of the concentrations of the reaction products, each raised to a power equal to its coefficient in a balanced chemical equation (11-2) for the reaction, to the product of the concentrations of the reactants, each in turn raised to a power equal to its coefficient in the chemical equation, is always the same at a given temperature.[3] Although the concentrations of individual reactants or products may differ considerably from one experiment to the next, they will always mutually adjust themselves by a shift in the position of equilibrium (provided that equilibrium is achieved) in such a way that (11-4) is satisfied.

In later chapters devoted to thermodynamics and reaction rates, we show why the particular quotient in (11-4) is a constant at a given temperature and pressure. For the moment we merely *assert* that (11-4) is true for any reaction (11-2) at equilibrium. The reaction quotient Q can, of course, be evaluated whether the reaction is at equilibrium or not, and its value is sometimes of interest even when equilibrium has not been reached. The quotient Q is *equal* to K when equilibrium has been established.

Conventions in Formulating Reaction Quotients   There are several arbitrary rules or conventions followed in the formulation of reaction quotients that must be learned by anyone working with them:

---

[2]The double arrow $\rightleftharpoons$ is the customary symbol for chemical equilibrium.

[3]The "constant" K may also depend on the total pressure of the system, but this effect is negligible unless pressures of hundreds of atmospheres are involved (see also Sec. 11-5, item 8). In any event, K is independent of the pressure for reactions among ideal gases.

I. The concentrations of the *products* of the reaction appear in the *numerator;* those of the *reactants* are in the *denominator.* This convention implies that if a reaction is reversed, the roles of the reactants and the products thus being interchanged, the new reaction quotient (call it Q') is the inverse of the initial one (Q); that is, $Q' = 1/Q$. Since at equilibrium $Q = K$, the equilibrium constant for the reverse process, K', is similarly related to K; that is, $K' = 1/K$.

II. Unless otherwise specified, solute concentrations are always expressed in molarities, symbolized by [   ]. (In some applications, mole fractions are used as concentration units instead, but this usage is then explicitly mentioned.)

IIIa. In a gaseous reaction mixture the concentrations of the gases in the reaction quotient are usually replaced by their partial pressures.

IIIb. In reactions of gases with substances in the solid or liquid states (including substances in solutions), the concentrations of the gases are also usually replaced by their partial pressures.

IV. Water that participates in a reaction occurring in a dilute aqueous solution is omitted in the formulation of the reaction quotient.

V. The rules given so far do not indicate how to deal with a *pure* solid or *pure* liquid that occurs as a reactant or a product, for the concept of concentration applies normally to solutions, not to pure phases. The convention for dealing with pure solids and pure liquids is a very simple one: they are *omitted* entirely from the reaction quotient.

Conventions I and II are self-explanatory, but the others require justification and further explanation. These five conventions establish a one-to-one relationship between a chemical equation and the reaction quotient associated with it (and thus also with the appropriate value of K).

Discussion of Conventions IIIa and IV For a gaseous reaction mixture, the gas concentrations appearing in Q are usually replaced by partial pressures (convention IIIa). This is permissible as long as the gases are at least nearly ideal, for if the gases participating in the reaction are specified by numbers $i = 1, 2, 3, \ldots$, the molar concentration for gas i is $[i] \equiv n_i/V = P_i/(RT)$. In other words, the number of moles per liter of gas i is proportional to its partial pressure at a given temperature. The proportionality factor is constant at the given temperature and can be made part of the equilibrium constant. We examine this relationship further in Sec. 11-4 (where examples are given) and in some of the comments below. If the units of pressure are not explicitly stated, they should be assumed to be atmospheres.

Convention IV states that when water participates in a reaction in a dilute aqueous solution, the concentration of H_2O is omitted in formulating Q. This is always done when $[H_2O]$ in the solution does not deviate by more than 10 percent or so from the value it has for pure water. This value, the concentration of H_2O in pure water, is $(998 \text{ g liter}^{-1})/(18.0 \text{ g mol}^{-1})$ or about 55 mol liter^{-1} (55 M). Convention IV will be adhered to unless explicitly stated otherwise.

Thus in the very important reaction of water (Sec. 9-2),

$$H_2O + H_2O \rightleftharpoons H_3O^+ + OH^- \tag{11-5}$$

the reaction quotient at equilibrium might be written as[4]

$$K_5' = \frac{[H_3O^+][OH^-]}{[H_2O]^2}$$

However, since $[H_2O]$ is always about 55 M in any dilute solution, this value is by convention combined with K_5', leading to

$$[H_3O^+][OH^-] = K_5'[H_2O]^2 \equiv K_5 \equiv K_w \tag{11-5a}$$

Finally, since $[H_3O^+]$ and $[H^+]$ mean the same thing (Sec. 9-2), we simply write

$$[H^+][OH^-] = K_w \tag{11-5b}$$

Thus, in water or any dilute aqueous solution, the product of the concentrations of H_3O^+ (or H^+) and OH^- is a constant at a given temperature. This constant, the *ion-product constant* of water, symbolized K_w, is equal to $1.008 \times 10^{-14} M^2$ at $25°C$. Thus, in pure water, $[H^+] = [OH^-] = 1.00 \times 10^{-7} M$ at this temperature.

The following two important examples are further illustrations of convention IV:

$$NH_3 + H_2O \rightleftharpoons NH_4^+ + OH^- \tag{11-6}$$

$$\frac{[NH_4^+][OH^-]}{[NH_3]} = K_6 \tag{11-6a}$$

$$H_2O + HOAc \rightleftharpoons H_3O^+ + OAc^- \tag{11-7}$$

$$\frac{[H_3O^+][OAc^-]}{[HOAc]} \equiv \frac{[H^+][OAc^-]}{[HOAc]} = K_7 \tag{11-7a}$$

Again the concentrations of all hydrated proton species are usually represented by $[H^+]$—no attempt is made to distinguish among the various formulas.

Conventions IIIb and V are discussed later (Sec. 11-5).

General Comments and Examples Many aspects of the chemical equilibrium law are difficult for those just learning to work with it, but the details that prove puzzling often vary from person to person. The following comments emphasize points worthy of special attention.

1. *Units of equilibrium constants:* The units of K often differ for different reactions. They always follow from and are identical with the units of the corresponding reaction quotient Q. Concentrations are always expressed in moles per liter (M) and partial pressures in atmospheres, unless explicitly stated otherwise. Thus the units of K_w [Equation (11-5a)] are (moles per liter)^2 or M^2, while those of K_6 and K_7 [Equations (11-6a) and (11-7a)] are moles per liter or M. The gas reaction

$$PCl_5(g) \rightleftharpoons PCl_3(g) + Cl_2(g) \tag{11-8}$$

is associated with the equilibrium condition

$$\frac{P_{PCl_3}P_{Cl_2}}{P_{PCl_5}} = K_8 \tag{11-8a}$$

---

[4]The subscript of K' used here refers to the equation number describing the pertinent chemical equilibrium.

and the units of K_8 are atmospheres (unless the partial pressures concerned are expressed in other units). The dissociation of gaseous HI into H_2 and gaseous I_2,

$$2HI(g) \rightleftharpoons H_2(g) + I_2(g) \tag{11-9}$$

is governed at equilibrium by the equation

$$\frac{P_{H_2} P_{I_2}}{P_{HI}^2} = K_9 \tag{11-9a}$$

with K_9 dimensionless because the units of the terms in the numerator cancel the units in the denominator.

Note that *the numerical value of an equilibrium constant depends upon the units in which it is expressed,* unless the sum of the exponents of the concentration terms is the same in the numerator as in the denominator of the reaction quotient and unless the same relation holds for the pressure terms. To illustrate, because there is one more pressure term in the numerator than in the denominator in (11-8a), K_8 has the dimensions of pressure and its numerical value depends upon the pressure units chosen. Thus, at 200°C, $K_8 = 2.2 \times 10^{-4}$ atm; if partial pressures are expressed in torr rather than atmospheres, however, the value of K_8 is $(2.2 \times 10^{-4}\text{ atm})(760\text{ torr/atm}) = 0.17$ torr. On the other hand, K_9 is dimensionless and its value is thus independent of the units used.

☐ **Equilibrium Constant from Experimental Data**   Nitric oxide, NO, reacts with oxygen to give nitrogen dioxide, NO_2

Example 11-1

$$2NO + O_2 \rightleftharpoons 2NO_2 \tag{11-10}$$

A sample of NO is burned in oxygen and the resulting mixture is allowed to come to equilibrium at 400°C. The partial pressures are found to be $P_{NO} = 1.9 \times 10^{-2}$ atm, $P_{O_2} = 2.8 \times 10^{-3}$ atm, and $P_{NO_2} = 1.0$ atm. Write the equilibrium constant for (11-10) and calculate its value at 400°C.

Solution   The conventions for writing the equilibrium constant specify that the partial pressure of each reactant and product be raised to the power equal to its stoichiometric coefficient

$$K_{10} = P_{NO_2}^2 / (P_{NO}^2 P_{O_2})$$

Inserting the values of the partial pressures gives

$$K_{10} = \frac{(1.0\text{ atm})^2}{(1.9 \times 10^{-2}\text{ atm})^2 \times 2.8 \times 10^{-3}\text{ atm}} = \underline{1.0 \times 10^6 \text{ atm}^{-1}} \ \blacksquare$$

☐ At 1 atm and 630°C $SO_3(g)$ is partly dissociated into $SO_2(g)$ and $O_2(g)$

Exercise 11-1

$$SO_3(g) \rightleftharpoons SO_2(g) + \tfrac{1}{2} O_2(g)$$

and the following partial pressures are found:

$$P_{SO_3} = 0.564\text{ atm}, \ P_{SO_2} = 0.290\text{ atm}, \ P_{O_2} = 0.145\text{ atm}$$

Write the equilibrium constant for the above equation and calculate its value at 630°C. ∎

2. *Equilibrium shift by addition or partial removal of a reacting species:* The law of chemical equilibrium embodied in (11-4) is a quantitative expression of the fact that when a reaction mixture is at equilibrium and an additional quantity of one of the reactants or products is supplied from the outside without significant change in volume, the equilibrium will shift so as to offset part of the increase in concentration of the added species.

Consider the dissociation of acetic acid, given by (11-7), with the corresponding equilibrium expression, (11-7*a*):

$$\frac{[H^+][OAc^-]}{[HOAc]} = K_7 \tag{11-7a}$$

The equilibrium constant K_7 is frequently called the *dissociation constant* or *acid constant* of acetic acid.[5] If a strong acid like HCl is added to a solution containing acetic acid, acetate ions, and hydrogen ions at equilibrium, the added hydrogen ions increase $[H^+]$ so that, at least momentarily, the quotient in (11-7*a*) no longer equals the dissociation constant of acetic acid. The equilibrium represented by (11-7) is, however, established extremely rapidly (in much less than a microsecond, 10^{-6} s, if the solution is adequately mixed). Some of the hydrogen ions react with acetate ions to form acetic acid, thereby reducing $[OAc^-]$ and $[H^+]$ and increasing [HOAc] until equilibrium again exists and (11-7*a*) is again satisfied. Thus, adding hydrogen ions to the original solution shifts the equilibrium of reaction (11-7) to the left; the addition of strong acid is said to repress the dissociation of acetic acid (or of any weak acid). Conversely, removal of H^+ from the original solution would cause the equilibrium to shift to the right.

Example 11-2

☐ **Shift in Position of Equilibrium** In a solution prepared by dissolving 0.10 mol acetic acid in enough water to make 1.0 liter, the concentration of acetate ion is $[OAc^-] = 1.3 \times 10^{-3} M$.

(*a*) What fraction of the original acetic acid dissociated and what is the concentration of H^+?

Solution From Equation (11-7) we note that the dissociation of each HOAc produces one OAc^-. Thus the fraction of the original HOAc that dissociated is $(1.3 \times 10^{-3})/(0.10) = \underline{1.3 \times 10^{-2}}$. Since by (11-7) one H^+ is formed for every OAc^-, $[H^+] = [OAc^-] = \underline{1.3 \times 10^{-3} M}$.

(*b*) The addition of a small amount of concentrated HCl to the solution increases the H^+-concentration to $0.10 M$. Calculate the fraction of the original HOAc that is now dissociated. The equilibrium constant for (11-7) is $1.8 \times 10^{-5} M$.

Solution The equilibrium expression (11-7a) is applicable and can be rearranged to give

$$\frac{[OAc^-]}{[HOAc]} = \frac{K_7}{[H^+]}$$

---

[5] Acid constants are often symbolized as K_a; the corresponding constants for bases, a typical one of which is given above as K_6 [Equation 11-6*a*], are often symbolized as K_b. Acid and base equilibria are discussed in detail in Chap. 12.

Inserting the values of K_7 and $[H^+]$ leads to

$$\frac{[OAc^-]}{[HOAc]} = \frac{1.8 \times 10^{-5}}{0.10} = 1.8 \times 10^{-4}$$

This is the ratio of $[OAc^-]$ to $[HOAc]$ at equilibrium, and the question calls for the ratio of $[OAc^-]$ to the *original* concentration of acetic acid, $[HOAc]_0$. Note, however, that within the precision of the data the equilibrium and initial concentrations of acetic acid are identical in part (*a*):

$$[HOAc] = [HOAc]_0 - [OAc^-] \approx [HOAc]_0$$

This relation must also be valid after the addition of HCl, hence

$$\frac{[OAc^-]}{[HOAc]_0} \approx \frac{[OAc^-]}{[HOAc]} = \underline{1.8 \times 10^{-4}}$$

Thus the addition of HCl has indeed decreased greatly the fraction of acetic acid dissociated. ■

□ In a solution prepared by diluting 0.20 mol HNO_2 (nitrous acid) suffi-   **Exercise 11-2**
ciently to make 1.0 liter, the concentration of the nitrite ion is $[NO_2^-] = 0.0093\ M$. (*a*) What fraction of the original HNO_2 dissociated and what is $[H^+]$? (*b*) The addition of a small amount of HCl increases $[H^+]$ to 0.15 M. What is the fraction of the original HNO_2 that is now dissociated? $K_a = 4.5 \times 10^{-4}$ for HNO_2. ■

3. *All sources of a reacting species must be considered:* It is important to realize that the concentration or partial pressure terms in the reaction quotient of any reaction include contributions from *all available sources,* even though some of these sources may not be related to the reaction in question.

To illustrate, in the example just discussed the term $[H^+]$ includes not only the hydrogen ions created by the dissociation of HOAc but also those coming from the added HCl. There is a further contribution from the dissociation of water, but it is completely negligible in this example and indeed is usually negligible in most of the applications of equilibria to weak acids and weak bases that we shall consider, other than pure water itself.

4. *Equilibrium constant and degree of dissociation:* Beginners often confuse the equilibrium constant with the degree of dissociation. The two are quite distinct. The *equilibrium constant* is defined by (11-4); the *degree of dissociation* of a compound X is defined as the number of moles of X that have dissociated divided by the number of moles of X that would be present if none had dissociated.

Consider the dissociation of N_2O_4:

$$N_2O_4(g) \rightleftharpoons 2NO_2(g) \qquad (11\text{-}11)$$

The degree of dissociation, α, is the number of molecules (or moles) of N_2O_4 that have dissociated divided by the number of N_2O_4 molecules (or moles) that would have been present if no dissociation had occurred.

Dissociation of N_2O_4 increases the number of molecules in the gas; for each N_2O_4 molecule that dissociates, two NO_2 molecules appear. If pressure and temperature are kept constant, this means, by Avogadro's law, that dissociation increases the gas volume. Since dissociation does not change the total mass of a given quantity of gas, it necessarily decreases the density (at a fixed P and T), which is consistent with the average molecular weight decreasing, as it must when dissociation occurs.

Example 11-3

☐ **Dissociation and Gas Density** At $25°C$ and 1.000 atm, the degree of dissociation of N_2O_4 is 0.184. What is the density of the gas mixture?

Solution If there were no dissociation, 92.0 g would constitute 1.000 mol gas. Of this, 0.184 mol is dissociated, leaving $(1.000 - 0.184)$ mol undissociated N_2O_4. By the dissociation equation (11-11) it is seen that each N_2O_4 molecule yields two molecules of NO_2. The dissociation of 0.184 mol N_2O_4 thus yields 2×0.184 mol NO_2. The total number of moles in the final gas mixture is $(1.000 - 0.184) + 2 \times 0.184 = 1.184$. Thus

	Before dissociation	After dissociation
Moles N_2O_4	1.000	$1.000 - 0.184$
Moles NO_2	0.000	2×0.184
Total moles	1.000	1.184

The volume taken up by 1.184 mol gas at $25°C$ and 1.000 atm is $V = nRT/P = (1.184 \times 0.0821 \times 298/1.000)$ liter $= 29.0$ liter. The density is therefore 92.0 g/29.0 liter $= \underline{3.17 \text{ g liter}^{-1}}$. ∎

Exercise 11-3

☐ At $102°C$ and 840 torr the degree of dissociation of SO_2Cl_2, according to the reaction

$$SO_2Cl_2(g) \rightleftharpoons SO_2(g) + Cl_2(g)$$

is 0.840. What is the density of the gas mixture? ∎

A fictional problem is discussed next as an illustration of the reasoning that is characteristic in dissociation problems.

Example 11-4

☐ **Cracking Walnuts** All the snowbound skier had was 10 dozen walnuts. He set about cracking them carefully, separating them neatly into half shells and intact kernels. After a while the light failed and he decided to count the results of his labors—the uncracked nuts left, the half shells, and the kernels. He dared not remove his mittens in the bitter cold and therefore could not tell what his fingers were touching. All he knew was that in the end he had 194 pieces. What was the degree of dissociation of the walnuts?

Solution If the kernels are symbolically denoted by K, the half shells by S, and the nuts therefore by KS_2, the cracking of nuts may be described by

$$KS_2 \longrightarrow K + 2S \tag{11-12}$$
$$n_0 - x \quad\quad x \quad 2x$$

where $n_0 = 120$ is the original number of nuts and x is the number of cracked nuts. In the end there were x kernels, $2x$ half shells, and $(n_0 - x)$ whole nuts. This is the situation described by the symbols just beneath (11-12). The total number of pieces is therefore $n = x + 2x + (n_0 - x) =$

$n_0 + 2x$, and this the skier found to be equal to 194: $n_0 + 2x = 120 + 2x = 194$. Thus, $x = \frac{74}{2} = 37$, so that 37 walnuts were cracked. To find the degree of dissociation we have to refer to the state in which all nuts were whole. Thus, $\alpha = x/n_0 = \frac{37}{120} = \underline{0.31}$. ■

☐ Some round logs of firewood were split into halves. Originally there were 24 logs and at the end there were 37 pieces of wood. What is the degree of dissociation of the logs? ■

Exercise 11-4

5. *Implications of the magnitude of K:* In reactions for which K differs considerably from 1, equilibrium normally favors the side of the reactants when K is much smaller than 1 and the side of the products when K is much larger than 1. Intermediate situations are characterized by values of K of the order of unity.[6] In qualitative comparisons of different equilibria involving similar reactions—for example, acid dissociations—what matters is not so much the exact numerical values of the equilibrium constants but rather the powers of 10 that go with them. In other words, differences in equilibrium constants by a factor of 2 or 3 seldom lead to marked chemical effects, for which differences of at least a few powers of 10 are required.

6. *Driving a reaction in a desired direction:* If the equilibrium constant for a reaction is very small, so that the reactants are normally favored over the products, the reaction may still be driven to the right, as is implied by comment 2, either by removing one or more of the products as it is formed or by adding a large excess of one or more of the reactants. The first of these possibilities is usually easier to accomplish. For example, if one of the products is a gas, it may be removed by reducing the pressure or by increasing the temperature. Thus, it is possible to produce HCl by the reaction of nearly water-free H_2SO_4 with NaCl:

$$H_2SO_4 + NaCl \rightleftharpoons HCl + NaHSO_4$$

Because HCl is appreciably more volatile than any other component of this reaction mixture, it can readily be removed by heating and the reaction is thereby driven to the right.

Often a reaction appears to be driven (even though no external action is required) because a precipitate appears among the products and is effectively removed because of its small solubility. For example, H_2S is a very weak acid in aqueous solution, and the concentration of S^{2-} in a saturated solution of H_2S is less than 10^{-14} M. Yet H_2S effectively dissociates completely in the presence of certain metal ions, such as Ag^+ or Hg^{2+}, that form very insoluble sulfides:

$$H_2S(aq) + 2Ag^+ \rightleftharpoons Ag_2S(s) + 2H^+ \qquad (11\text{-}13)$$

The right side of this equilibrium is favored because of the extreme insolubility of Ag_2S (Chap. 13). Similarly, one of the reaction products in an aqueous solution

---

[6]Unless K is dimensionless, its value depends on the units used, as stated in comment 1. This must be taken into account when considering the meaning of the numerical value of K. As discussed later, when the sum of the exponents in the numerator of the reaction quotient is different from that in the denominator, the position of equilibrium can be altered by changes in total concentration or pressure. Thus, although K_a for acetic acid is very small so that this acid is less than 1 percent dissociated in a 1 M solution, it is more than half dissociated at a concentration of 10^{-5} M, so that in this very dilute solution, the products are actually favored at equilibrium.

may be a slightly ionized substance, such as H_2O itself, which again causes the side of the products to be strongly favored. Thus the equilibrium position of the reaction

$$HCN + OH^- \rightleftharpoons CN^- + H_2O$$

in aqueous solution is strongly in favor of the products, at least partly because H_2O is formed.

Another method of achieving almost complete conversion of reactants into products is recycling. For example, in the Haber process by which N_2 and H_2 are combined to form ammonia,

$$N_2(g) + 3H_2(g) \rightleftharpoons 2NH_3(g) \tag{11-14}$$

the equilibrium constant is very small at the temperatures of $500°C$ and above at which this reaction is carried out industrially, and only a small fraction of the nitrogen and hydrogen is converted into ammonia at a given time. However, the ammonia formed is removed by being cooled until it liquefies. Not only does the equilibrium constant change in favor of NH_3 as the temperature is lowered, but the speed with which NH_3 decomposes into N_2 and H_2 at ordinary temperatures is essentially zero. The unreacted N_2 and H_2, still gaseous, are returned to the reaction chamber and combined with fresh supplies of these gases in a continuous process.

We return to additional general comments and examples in Sec. 11-5.

11-4   Homogeneous Equilibria Involving Gases

Our considerations in this section concern homogeneous reactions (the only phase present being that containing the gases). We assume that the reader is familiar with Chap. 4, especially Sec. 4-3. Since all gases are miscible with one another, all reactions involving *only* gases are homogeneous reactions—that is, they occur in a single phase. Some such reactions come to equilibrium quickly, others slowly. We begin by considering examples of some common situations, the computation of an equilibrium constant from some experimental data and the calculation of the distribution of reactants and products from given equilibrium constants. Other examples of gas-phase equilibria are illustrated in the problems at the end of the chapter. Some heterogeneous equilibria involving gases and solids are considered in Sec. 13-4.

Example 11-5

☐ **Equilibrium Constant from Total Pressure and Percentage Dissociation**   At $25°C$ and 2.000 atm total pressure, N_2O_4 is 13.3 percent dissociated into NO_2. What is the value of the equilibrium constant at this temperature?

Solution   Consider exactly 1 mol initially undissociated N_2O_4. After it comes to equilibrium with NO_2, 0.133 mol N_2O_4 has dissociated and twice this many moles of NO_2 have been formed, as indicated by the numbers beneath the symbols in (11-15):

$$N_2O_4 \rightleftharpoons 2NO_2 \tag{11-15}$$
$$\underset{1.000 - 0.133}{} \qquad \underset{2 \times 0.133}{}$$

The resulting gas mixture contains a total of $0.867 + 0.266 = 1.133$ mol gas for each mole of N_2O_4 initially present. Since we know the total pressure and can calculate the mole fraction of each of the two components, we can calculate their partial pressures by using Dalton's law (Sec. 4-3). The mole fraction of N_2O_4 is $\frac{0.867}{1.133} = 0.765$ and that of NO_2 is $\frac{0.266}{1.133} = 0.235$. Since the total pressure is 2.000 atm, the partial pressures, each equal to the mole fraction of the corresponding component times the total pressure, are 1.530 and 0.470 atm, respectively. The equilibrium constant is

$$K_{15} = \frac{P_{NO_2}^2}{P_{N_2O_4}} = \frac{(0.470 \text{ atm})^2}{1.530 \text{ atm}} = \underline{0.144 \text{ atm}} \quad \blacksquare$$

☐ At 3730 K and a total pressure of 2.00 atm molecular H_2 is just $\frac{1}{3}$ dissociated into atomic H. What is the value of the equilibrium constant at that temperature? ■

Exercise 11-5

☐ **Degree of Dissociation of Gaseous HI** The equilibrium constant (11-9a) for reaction (11-9) (restated below) is 0.0144 at 600 K. What is the degree of dissociation of gaseous HI at equilibrium at this temperature?

Example 11-6

Solution For convenience, we assume that if none of the HI had dissociated there would be 1.000 mol of it present. Let the degree of dissociation be x. Then from our original sample of HI, x mol will have dissociated and $1.000 - x$ will remain undissociated. Since the chemical equation (11-9) indicates that one mole of H_2 and one mole of I_2 are formed for each *two* moles of HI dissociated, the amounts of H_2 and I_2 formed will each be $x/2$ mol:

$$2HI(g) \rightleftharpoons H_2(g) + I_2(g) \qquad (11\text{-}9)$$
$$1.000 - x \qquad x/2 \quad x/2$$

The total number of moles is 1.000, regardless of the extent to which (11-9) has proceeded. Thus, the mole fraction of each component is, in this example, equal to the number of moles of that component. Hence, if the total pressure is P, the respective partial pressures are

$$P_{HI} = P(1.000 - x)$$
$$P_{H_2} = P_{I_2} = P(x/2)$$

and the reaction quotient, at equilibrium, given as 0.0144, is

$$\frac{P_{H_2}P_{I_2}}{P_{HI}^2} = \frac{P^2(x/2)^2}{P^2(1.000 - x)^2}$$

$$= \frac{x^2}{4(1.000 - x)^2} = 0.0144 \qquad (11\text{-}16)$$

The total pressure P cancels, so that the degree of dissociation x is independent of P. This is always true in a gas reaction when the number of gas molecules on each side of the chemical equation is the same. Equation (11-16) is readily solved by taking the square root of each side:

$$\frac{x}{2(1.000 - x)} = 0.120$$

or

$$x = 0.240 - 0.240x$$

and finally

$$x = \frac{0.240}{1.240} = \underline{0.194} \quad \blacksquare$$

Exercise 11-6

□ The equilibrium constant for the reaction

$$2NO \rightleftharpoons N_2 + O_2$$

is 11.4 at 4000 K. What is the degree of dissociation of NO at equilibrium at this temperature? ■

In the preceding example, the amounts of H_2 and I_2 present were equivalent by (11-9) because it was assumed that we started with pure HI and thus formed the decomposition products in equivalent quantities. However, this need not be so; we might be interested in the pressure of each constituent when we start with a mixture containing H_2 and I_2 in some proportion other than 1:1, which is the proportion in which they would be present if originally only HI were present.

The following example illustrates this more complex situation for another reaction, the decomposition of PCl_5 into PCl_3 and Cl_2. In this example, the products are *not* present in the proportion in which they are formed in the reaction.

Example 11-7

□ **Equilibrium between PCl_5, PCl_3, and Cl_2** The equilibrium constant for the reaction

$$PCl_5(g) \rightleftharpoons PCl_3(g) + Cl_2(g) \tag{11-8}$$

is 3.60 atm at 540 K. What are the equilibrium partial pressures of the constituents of a mixture that originally consists of 0.200 mol PCl_5 and 3.00 mol PCl_3 and is heated to 540 K at a total pressure of 1 atm?

Solution Suppose at equilibrium x mol PCl_5 has dissociated by (11-8). This leads to x mol PCl_3 and x mol Cl_2 leaving $(0.200 - x)$ mol PCl_5 mixed with $(3.00 + x)$ mol PCl_3 and x mol Cl_2. The total number of moles of gas is $(0.200 - x) + (3.00 + x) + x = 3.20 + x$, and each partial pressure is equal to the product of the total pressure of 1.00 atm and the respective mole fraction:

$$P_{PCl_5} = \frac{0.200 - x}{3.20 + x} (1.00 \text{ atm})$$

$$P_{PCl_3} = \frac{3.00 + x}{3.20 + x} (1.00 \text{ atm}) \tag{11-17}$$

$$P_{Cl_2} = \frac{x}{3.20 + x} (1.00 \text{ atm})$$

The equilibrium condition (11-8a) is

$$\frac{P_{PCl_3} P_{Cl_2}}{P_{PCl_5}} = K_8 = 3.60 \text{ atm}$$

Inserting values for the pressures, we get

$$\frac{(3.00 + x)(x)(1.00 \text{ atm})}{(0.200 - x)(3.20 + x)} = 3.60 \text{ atm}$$

To solve this equation we first multiply out the terms

$$3.00x + x^2 = 3.60 (0.640 - 3.00x - x^2)$$
$$= 2.304 - 10.80x - 3.60x^2$$

and then rearrange

$$4.60x^2 + 13.80x - 2.304 = 0$$

This equation can be solved by using the formula for the exact solution of a quadratic expression,

$$ax^2 + bx + c = 0$$

which is

$$x = \frac{-b \pm \sqrt{b^2 - 4ac}}{2a} \qquad (11\text{-}18)$$

with $a = 4.60$, $b = 13.80$, and $c = -2.304$ in this example. This approach leads to

$$x = \frac{-13.80 \pm [(13.80)^2 + 4 \times 4.60 \times 2.304]^{1/2}}{2 \times 4.60}$$

$$= \frac{-13.80 \pm \sqrt{232.83}}{9.20} = \frac{-13.80 \pm 15.26}{9.20}$$

$$= \frac{1.46}{9.20} = 0.159$$

The positive root has been taken since it is the only physically meaningful solution (the number of moles of Cl_2 cannot be negative). When this result is substituted into Equations (11-17) we obtain

$$P_{PCl_5} = \frac{0.200 - 0.159}{3.20 + 0.159}(1.00 \text{ atm}) = \underline{0.012 \text{ atm}}$$

$$P_{PCl_3} = \frac{3.00 + 0.159}{3.20 + 0.159}(1.00 \text{ atm}) = \underline{0.94 \text{ atm}}$$

$$P_{Cl_2} = \frac{0.159}{3.20 + 0.159}(1.00 \text{ atm}) = \underline{0.047 \text{ atm}}$$

As a check, we use these partial pressures to evaluate the equilibrium constant

$$K_8 = \frac{(0.94)(0.047)}{0.012} = 3.7 \text{ atm}$$

The result agrees with the input value of K_8 within the precision of our calculation. As a further check, we note that the sum of the three partial pressures is indeed 1.00 atm. ∎

Exercise 11-7

☐ The equilibrium constant for the reaction

$$SO_2Cl_2(g) \rightleftharpoons SO_2(g) + Cl_2(g)$$

is 2.65 atm at 375 K. What are the partial pressures of the constituents of a mixture that originally consists of 0.500 mol SO_2Cl_2 and 5.00 mol SO_2 and is equilibrated at 375 K and a total pressure of 1.00 atm? ∎

If the number of molecules on one side of an equation describing a gaseous equilibrium differs from the number of molecules on the other, a decrease in the volume of the system (a change corresponding to an increase in pressure) shifts the position of equilibrium toward the side with fewer molecules. For example, in the reaction of nitrogen with hydrogen to form ammonia [Equation (11-14)], increasing pressure favors the production of ammonia, an effect used in the commercial production of ammonia. This effect can be understood qualitatively in terms of Le Châtelier's principle.

11-5 Additional General Comments and Examples

We now return to important general features of the chemical equilibrium law, beginning with a further discussion of conventions V and IIIb for setting up the reaction quotient Q.

Conventions V and IIIb We start with convention V, according to which the "concentration" of any *pure* solid or liquid that is actually present at equilibrium is omitted from the quotient Q. We shall use the reaction described by the equation

$$AgCl(s) \rightleftharpoons Ag^+(aq) + Cl^-(aq) \qquad (11\text{-}19)$$

as an example. In this reaction solid AgCl dissolves to form the silver and chloride ions in solution. The resulting equilibrium might be written in the form[7]

$$AgCl(s) \rightleftharpoons AgCl(aq) \rightleftharpoons Ag^+ + Cl^- \qquad (11\text{-}20)$$

The concentration of molecular AgCl in the aqueous solution, though known experimentally to be extremely small, is constant as long as some solid AgCl is present and there is equilibrium. Thus we might write, considering the right-hand equilibrium in (11-20),

$$\frac{[Ag^+][Cl^-]}{[AgCl(aq)]} = K_{20} \qquad (11\text{-}20a)$$

but since $[AgCl(aq)]$ is constant at a given temperature, its value may be combined with K_{20} giving

$$[Ag^+][Cl^-] = K_{20}[AgCl(aq)] \equiv K_{19} \equiv K_{sp,AgCl} \qquad (11\text{-}19a)$$

The expression on the left side of this equation is often called the *ion product*. When it is evaluated with equilibrium concentrations, as here, its value is said to be equal to the *solubility-product* constant, symbolized K_{sp}, for the salt in question (Chap. 13). There are many other situations in which *pure* solids or liquids participate in a reaction and are present at equilibrium. In each of these cases the quotient Q contains no term representing pure solid or liquid.

We return finally to convention IIIb, by which the concentration $[i]$ of a gas that participates in a reaction is replaced by its partial pressure P_i. Again this convention applies to reactions in aqueous solutions. The justification for this procedure lies in Henry's law (Sec. 8-3).

---

[7]The symbol $AgCl(aq)$ emphasizes that the species in question is molecular AgCl, as it exists in aqueous solution, surrounded by H_2O molecules, as distinguished from solid silver chloride, represented as $AgCl(s)$. Strictly speaking the symbols Ag^+ and Cl^- should be replaced by $Ag^+(aq)$ and $Cl^-(aq)$, as they were in (11-19), because these species are hydrated also, but this is always assumed to be understood for *ions* in aqueous solutions.

The equilibrium expression for the following reaction occurring in aqueous solution,

$$2I_2(s) + 2H_2O \rightleftharpoons O_2(g) + 4I^- + 4H^+ \qquad (11\text{-}21)$$

illustrates all five conventions given at the start of Sec. 11-3:

$$P_{O_2}[I^-]^4[H^+]^4 = K_{21} \qquad (11\text{-}21a)$$

In setting up the quotient Q it is important to note what is on the left and what is on the right in the chemical equation and what the coefficients are (see, for example, comment 7 below). Moreover, symbols such as s and g must be used to specify phases completely whenever an ambiguity might exist. For example, the symbol $I_2(s)$ implies an excess of *solid* iodine at equilibrium and therefore the absence of any term referring to I_2 in the reaction quotient Q. If only I_2 vapor is present at equilibrium, the combination $I_2(g)$ must be used in the chemical equation to imply that the equilibrium quotient Q contains the term P_{I_2} (with the appropriate exponent). In contrast, no g is *needed* in combination with H_2, which is gaseous under all normal circumstances.

Further Comments on the Chemical Equilibrium Law We present now some further general remarks about equilibria, continuing from the comments in Sec. 11-3.

7. *Combination of reactions:* When two or more reaction equations are combined in an overall equation, the corresponding equilibrium constants may be combined to give the equilibrium constant for the overall reaction by applying the following three rules:

(a) When two or more chemical equations are added, the equilibrium constant for the resulting equation is the product of the constants of the original equations.

(b) When a chemical equation is multiplied by a factor q, the original equilibrium constant K becomes K^q. The factor q may be an integer or a fraction.

(c) Subtracting one reaction from a second is the same as reversing the first one and adding them. We know already, as mentioned in convention I (Sec. 11-3), that reversing a reaction implies changing K to $1/K$. Thus if the equation for reaction 1, with equilibrium constant K_1, is subtracted from that for reaction 2, with constant K_2, the constant for the resulting equation is $K = K_2/K_1$.

These rules are a direct consequence of the way in which the reaction quotient (11-3) depends upon the coefficients of the chemical species in the related reaction equation. The following two examples illustrate the application of the rules.

□ **Changing the Coefficients of a Balanced Equation** The equilibrium constant for the reaction

Example 11-8

$$N_2(g) + 3H_2(g) \rightleftharpoons 2NH_3(g) \qquad (11\text{-}22)$$

is $K_{22} = 3.3 \times 10^{-8}$ atm^{-2} at 1200 K. Suppose that the reaction is represented by the equation

$$\tfrac{1}{2}N_2(g) + \tfrac{3}{2}H_2(g) \rightleftharpoons NH_3(g) \qquad (11\text{-}23)$$

What is the value of the corresponding equilibrium constant K_{23}?

Solution If we write the reaction quotients for the two reactions,

$$\frac{P_{NH_3}^2}{P_{N_2}P_{H_2}^3} = K_{22} \quad \text{and} \quad \frac{P_{NH_3}}{P_{N_2}^{1/2}P_{H_2}^{3/2}} = K_{23}$$

we see that K_{22} is just the square of K_{23}. This is in accord with rule (*b*) above: Equation (11-23) is obtained from (11-22) by multiplying through by $\frac{1}{2}$, and hence $K_{23} = K_{22}^{1/2} = \underline{1.8 \times 10^{-4} \text{ atm}^{-1}}$. (Note also the change in units.) ∎

Exercise 11-8

☐ The equilibrium constant for the reaction

$$\tfrac{1}{2}N_2 + \tfrac{1}{2}O_2 \rightleftharpoons NO$$

is 0.202 at 3500 K. What is the equilibrium constant for the reaction

$$2NO \rightleftharpoons N_2 + O_2$$

at the same temperature? ∎

Example 11-9

☐ **Combination of Reactions** The two reactions

$$16H^+ + 2MnO_4^- + 10Cl^- \rightleftharpoons 8H_2O + 2Mn^{2+} + 5Cl_2(g) \quad (11\text{-}24)$$

and
$$Cl_2(g) + 2Fe^{2+} \rightleftharpoons 2Cl^- + 2Fe^{3+} \quad (11\text{-}25)$$

have the respective equilibrium constants K_{24} and K_{25}. What is the equilibrium constant K_{26} for the reaction

$$8H^+ + MnO_4^- + 5Fe^{2+} \rightleftharpoons 4H_2O + Mn^{2+} + 5Fe^{3+} \quad (11\text{-}26)$$

Solution Equation (11-26) may be obtained by combining the two preceding equations as follows: $\frac{1}{2}$(11-24) + $\frac{5}{2}$(11-25). The equilibrium constant for (11-26) is thus, by rules (*a*) and (*b*),

$$K_{26} = \sqrt{K_{24}K_{25}^5}$$

This result can be checked by writing out explicitly the expressions for the three quotients concerned. ∎

Exercise 11-9

☐ The two gas reactions

$$2H_2 + O_2 \rightleftharpoons 2H_2O$$

and
$$2CO + O_2 \rightleftharpoons 2CO_2$$

have the respective equilibrium constants 26.4 atm^{-1} and 0.376 atm^{-1} at 3500 K. What is the equilibrium constant at the same temperature for the reaction

$$CO_2 + H_2 \rightleftharpoons CO + H_2O? \quad ∎$$

A useful corollary of these rules is that in the treatment of an equilibrium situation, only those reactions need be considered that are *independent* of each other, that is, reactions with equations that are not combinations of others given. In other words, no reaction that can be represented as a combination of those explicitly considered needs to be included in examining the equilibrium situation, because the appropriate equilibrium conditions for it are automatically satisfied.

Thus, if any two of the three reactions (11-24) to (11-26) are known to be at equilibrium, the third one necessarily is also at equilibrium.

8. *Concentrated solutions and high partial pressures:* Although we have stated that at a particular temperature the equilibrium constant *is* a constant (as its name implies), this is strictly true only in the limit of infinite dilution. In this sense the equilibrium relation is like the ideal-gas law—it is a very good approximation for low concentrations and becomes less exact at higher concentrations. The reason for the change in the value of K is the influence of molecular and ionic interactions on the equilibrium, which increase as the concentrations of the reacting species become higher.

In highly accurate work the reaction quotient must be formulated in terms of effective concentrations called *activities*. For equilibria involving gases, effective pressures called *fugacities* must be introduced. If these substitutions are made, the reaction quotient is truly constant at equilibrium (at a given temperature and total pressure). Tables of activities and fugacities of common species in different situations are available, as are formulas for their calculation.

Some numerical values will give an idea of the orders of magnitude involved. For gas reactions at pressures of 1 atm or less, use of partial pressures rather than fugacities leads to errors of no more than about 1 percent and thus such a correction is negligible for our purposes. For solutions of simple *ionic* substances such as NaCl, the difference between concentration and activity exceeds 5 percent at a molarity of about 0.002; for *uncharged* species such as ethanol or acetone, the same order of discrepancy is reached only at concentrations of about 1 M. At intermediate concentrations typical activity corrections in aqueous solutions of ions are of the order of ± 20 percent; at high concentrations the ratio of the activity to the concentration (called the *activity coefficient*) may become very large.

9. *Adequacy of approximations*: It is seldom worthwhile to make calculations for equilibria in solution to better than 1 percent, and we shall only infrequently make them to better than a few percent. Furthermore, systematic errors will often make the accuracy even poorer. There are three principal reasons for the low precision and accuracy[8] of the results of such calculations:

(a) The use of concentrations in place of activities can lead to significant errors, as discussed in comment 8.

(b) The values of the equilibrium constants used are usually precise to at best a few percent, and their accuracy is often considerably lower than their precision. For example, one of the most commonly used constants is that pertaining to the product of the concentrations of Ag^+ and Cl^- in a solution saturated with silver chloride, the *solubility product* of AgCl; yet the values for this constant quoted in various recent sources vary by almost a factor of 2, from $1.6 \times 10^{-10}\ M^2$ to $2.8 \times 10^{-10}\ M^2$ at 25°C.

(c) Frequently a number of equilibria occur simultaneously but some of them are ignored in the calculations, either because they are believed to be of relatively minor importance and ignoring them simplifies the calculations appreciably or because one is unaware of them. Such ignored equilibria

[8]Precision and accuracy are defined in Sec. 3-1.

might often have had significant effects on the results of the calculations if they had been included.

Nevertheless, the relatively low precision and accuracy of even approximate equilibrium calculations are adequate for most purposes. The examples in Chaps. 12 and 13 show that the concentrations of species involved in equilibria in solution can vary over many powers of 10 (many *orders of magnitude*). Thus a calculation of these concentrations within 10 or 20 percent, or even within a factor of 2, almost always provides a quite meaningful picture of the actual equilibrium situation.

10. *Uniqueness of equilibrium concentrations*: An important and useful fact about chemical equilibria is that the concentrations of the different species involved, which satisfy the equilibrium equations and other conditions imposed upon the system, are *unique*. This means that there is only one set of concentration (and partial pressure) values that satisfies all the conditions that define a chemical system. If we wish to calculate these concentrations by solving the simultaneous equations describing the system, we may use any method that seems convenient, including guessing. The only criterion is that the values of the concentrations eventually arrived at satisfy all the equations concerned; if they do, they must necessarily be correct (provided that all the relevant equations have been included), regardless of how they were obtained.

It is for this reason that the intelligent use of approximations can be very helpful in analyzing equilibrium situations. If the initial approximations, made on the basis of an analysis of the problem and gradually accumulated experience, are reasonable, it is usually possible to simplify a complicated set of simultaneous equations so that they can be readily solved. Often the first answers so obtained will be sufficiently good—that is, the concentrations obtained will be consistent with all the equations, within the desired precision. If they are not, the *method of successive approximations* can be applied by substituting answers from the first approximation back into the equations and repeating the calculations to obtain a second approximation, and so on, until a set of values consistent within the desired precision is reached. We illustrate this procedure in examples in Chaps. 12 and 13. Often the first cycle is sufficient.

11. *Effect of temperature:* The way in which equilibrium constants depend upon temperature is related to whether the corresponding reaction is exothermic or endothermic, as discussed in the consideration of Le Châtelier's principle (Sec. 11-2). The position of equilibrium of an exothermic reaction is shifted to the left, toward the reactants, when energy is added externally and the temperature is thus increased. Correspondingly, the position of equilibrium of an endothermic reaction is shifted to the right if energy is added. This means that in the first of these two cases the equilibrium constant decreases with increasing temperature and that in the second it increases. If energy is withdrawn from the system and the temperature is lowered, the effects are reversed. The equilibrium constant of a reaction that neither evolves nor absorbs energy is not affected by temperature.

For example, the reaction between nitrogen and oxygen to form nitric oxide,

$$N_2 + O_2 \rightleftharpoons 2NO \tag{11-27}$$

is endothermic and the equilibrium constant at $25°C$ is 4×10^{-31}, so that the left side is very heavily favored. An increase in temperature causes the equilibrium constant to become very much larger, however, and at 2000 K it is equal to 4×10^{-4}. It is possible, therefore, to produce NO by "burning" nitrogen at this temperature, and this is done commercially in an electric arc.[9] The yield of NO is only of the order of a few percent because the equilibrium constant still favors the left side of (11-27), but the NO formed is removed, thus driving the reaction to the right. The NO must be quickly chilled to room temperature to prevent it from decomposing again to N_2 and O_2. At room temperature the rate at which equilibrium is established is so slow that the NO can be kept without decomposing. It is easily converted to NO_2 and then to nitric acid (Sec. 23-5).

12. *Names of constants:* Many different names are attached to equilibrium constants; usually they relate to the nature of the reaction involved, such as dissociation constants and association constants. Beginners are sometimes overwhelmed by these different names and get the notion that somehow the various constants differ in fundamental ways. In fact, they do not and the reaction quotient for any reaction can be formulated uniquely by following the five conventions given earlier. The names associated with different constants are primarily a matter of convenience in identifying the nature of the reaction or whatever other feature may be of interest.

13. *A requirement—equilibrium must exist:* Finally, there is a rule that may seem most obvious: do not attempt to apply an equilibrium constant to a system unless all the substances appearing in the corresponding chemical equation *are present* and the system *is* at equilibrium. For example, the solubility product of AgCl, Equation (11-19a), applies *only* if some solid AgCl is in equilibrium with an aqueous solution containing Ag^+ and Cl^-, that is, only when the solution is saturated with AgCl. It is of course possible to have a solution containing Ag^+ and Cl^- that is not saturated with AgCl—for example, the ocean. In such a solution, the product of $[Ag^+]$ and $[Cl^-]$ will be less than $K_{sp,\text{AgCl}}$. Similar considerations apply to any equilibrium involving two or more phases—some of each phase must be present.

With a single-phase (homogeneous) equilibrium, the only question is whether the equilibrium is established with sufficient rapidity. If so, all species in the equilibrium equation will necessarily be present. We shall assume that the equilibria for which calculations are made in this chapter and the two that follow are established so rapidly that rate effects are unimportant.

---

[9]The nitrogen oxides found in the exhaust of automobiles and jet engines and involved in the production of smog are formed by a similar process. The air drawn into the cylinders of the typical auto engine is subjected to high temperatures during part of the combustion cycle.

Summary

Chemical equilibrium is a dynamic situation in which there is a balance of processes that move in opposite directions. When a system at equilibrium is disturbed, there is usually a shift in the equilibrium in such a direction as to offset partially the effect of the disturbance. This is known as Le Châtelier's principle.

The macroscopic properties of a system at equilibrium do not change with time. Constancy of properties is not proof of equilibrium, however. Two criteria can be used to establish experimentally whether equilibrium does exist: (1) a system that has been disturbed will return to its initial condition when the disturbance is removed; (2) an equilibrium may be approached both from the side of the reactants and from the side of the products.

For a reaction symbolized by the balanced equation

$$a A + b B + \cdots \rightleftharpoons d D + e E + \cdots$$

the reaction quotient is defined as

$$Q \equiv \frac{[D]^d [E]^e \cdots}{[A]^a [B]^b \cdots}$$

At any particular temperature, Q is a constant at equilibrium. The value of Q for a system at equilibrium is called the equilibrium constant, usually symbolized by K. There are several rules governing the form of the reaction quotient that must be learned.

The position of equilibrium is a measure of the relative amounts of reactants and products present at equilibrium. It can be shifted by the addition or partial removal of products or reactants. Such a shift does not alter the equilibrium constant.

The values (and units) of equilibrium constants depend on how the chemical reaction is written: which substances are the reactants and which the products, the stoichiometric coefficients used, and whether a solid or liquid participant in the equilibrium is present as a pure substance or not. When two or more reaction equations are added or subtracted to give an overall reaction, the corresponding equilibrium constants may be multiplied or divided to give the equilibrium constant for the overall reaction.

It is only at infinite dilution that the equilibrium constant formulated with concentrations and partial pressures is strictly constant. In highly accurate work the reaction quotient must be formulated in terms of effective concentrations called activities and effective pressures called fugacities. It is seldom worthwhile to make calculations for equilibria in solution to better than a few percent.

Terms and Concepts

Problems and Questions

11-1 Le Châtelier's Principle Suppose that each of the following systems is at equilibrium. Predict the effect on each equilibrium of (i) increasing the volume at constant temperature, and (ii) cooling the system at constant pressure:

(a) $H_2(g) + Br_2(g) \rightleftharpoons 2HBr(g)$ (exothermic)
(b) $2N_2(g) + O_2(g) \rightleftharpoons 2N_2O(g)$ (endothermic)
(c) $2NO(g) + Cl_2(g) \rightleftharpoons 2NOCl(g)$ (exothermic)

11-2 Reaction Quotients Set up the reaction quotients Q that correspond to the following overall chemical equations (occurring in aqueous solution):

(a) $MnO_4^- + 5Cl^- + 8H^+ \rightleftharpoons$
$$Mn^{2+} + \tfrac{5}{2}Cl_2(g) + 4H_2O$$
(b) $2MnO_4^- + 10Cl^- + 16H^+ \rightleftharpoons$
$$2Mn^{2+} + 5Cl_2(g) + 8H_2O$$

(c) $MnO_2(s) + 2Cl^- + 4H^+ \rightleftharpoons$
$$Mn^{2+} + Cl_2(g) + 2H_2O$$
(d) $2MnO_4^- + 3Mn^{2+} + 2H_2O \rightleftharpoons$
$$5MnO_2(s) + 4H^+$$
Note any relationships between the different Q's.

11-3 Reaction Quotients and Le Châtelier's Principle Consider the following reactions:
  (I) $2C_2H_6(g) + 7O_2(g) \rightleftharpoons 4CO_2(g) + 6H_2O(g)$
 (II) $COCl_2(g) \rightleftharpoons CO(g) + Cl_2(g)$
(III) $FeC_2O_4(s) \rightleftharpoons FeO(s) + CO(g) + CO_2(g)$
(a) Write the expression for the reaction quotient for each reaction. (b) Reaction II is endothermic. Suppose that an equilibrium mixture of the three gases is heated. How will the degree of dissociation of $COCl_2$ change? (c) Predict the effect on the amount of CO of a decrease in volume for reaction II and of a decrease in the amount of FeO(s) in reaction III.

11-4 Calculation of an Equilibrium Constant Sulfur trioxide, SO_3, can be formed by the oxidation of SO_2

$$2SO_2(g) + O_2(g) \rightleftharpoons 2SO_3(g)$$

When a mixture of SO_2 and O_2 is allowed to equilibrate at a high temperature, the following equilibrium pressures are observed:

$$P_{SO_2} = 0.340 \text{ atm} \qquad P_{O_2} = 0.250 \text{ atm}$$
$$P_{SO_3} = 0.317 \text{ atm}$$

(a) Calculate the equilibrium constant for the reaction at this temperature. (b) What is the equilibrium constant for the reaction $SO_2(g) + \frac{1}{2}O_2(g) \rightleftharpoons SO_3(g)$?

11-5 Equilibrium Constant from Experimental Data Nitric oxide, NO, reacts with oxygen to give nitrogen dioxide, NO_2

$$2NO + O_2 \rightleftharpoons 2NO_2$$

Nitric oxide is burned in oxygen and the mixture is allowed to come to equilibrium at 400°C. The partial pressures are found to be $P_{NO} = 1.9 \times 10^{-2}$ atm, $P_{O_2} = 2.8 \times 10^{-3}$ atm, and $P_{NO_2} = 1.0$ atm. Write the expression for the reaction quotient and calculate its value at 400°C.

11-6 Gas Reaction At 1000 K the equilibrium constant of the exothermic reaction

$$H_2O(g) + CO(g) \rightleftharpoons CO_2(g) + H_2(g)$$

is 1.37. In which direction does the equilibrium position shift if (a) the pressure is raised at 1000 K and (b) the temperature is raised to 1100 K at constant pressure? (c) What is the composition of the equilibrium mixture at 1000 K if originally 1.00 mol H_2O and 1.00 mol CO are permitted to react and the total pressure is maintained at 1.00 atm? Would your answer be different if the total pressure were different?

11-7 Dissociation of a Gas Nitrosyl bromide, NOBr, is a gaseous compound that can be formed by the reaction

$$2NO + Br_2 \rightleftharpoons 2NOBr$$

At 25°C, the equilibrium constant for this reaction is 116 atm^{-1} when all substances are in the gaseous state. The reaction is exothermic (at constant volume as well as at constant pressure). (a) Suppose that these compounds are introduced into a reaction chamber so that, if no reaction occurred, the partial pressures would be NOBr, 0.80 atm; NO, 0.40 atm; Br_2, 0.20 atm. Will any reaction take place? If so, will there be net formation or decomposition of NOBr? Explain. (b) Suppose that the temperature is raised to 100°C after the equilibrium is established at 25°C. Will there be net formation or decomposition of NOBr as a result of the temperature change? Explain. (c) Suppose that helium (which does not react with any of the ingredients of the mixture) is pumped into the flask when it is at equilibrium at 25°C until the total pressure in the flask is doubled. What will be the effect on the amount of NOBr present? Think carefully before answering.

11-8 Gaseous Reaction and Equilibrium At a certain temperature, with all substances present as gases, the equilibrium constant for the formation of SO_2Cl_2 from SO_2 and Cl_2 is 1.0×10^{-2} atm^{-1}. (a) Suppose that at this temperature enough of each gas is introduced into a flask so that, if no reaction occurred, the partial pressures would be:

$$P_{SO_2} = P_{Cl_2} = 0.50 \text{ atm} \qquad P_{SO_2Cl_2} = 0.020 \text{ atm}$$

Will there be any change in the amount of SO_2Cl_2 present? If so, will the amount increase or decrease? (b) What will be the pressure of each of the three gases at equilibrium?

11-9 Oxygen-Ozone Equilibrium The formation of ozone from oxygen,

$$3O_2(g) \rightleftharpoons 2O_3(g)$$

is a highly endothermic reaction, with $K = 1 \times 10^{-50}$ atm^{-1} at 25°C. (*a*) What would be the pressure of ozone in equilibrium with the oxygen in ordinary air at 25°C ($P_{O_2} = 0.2$ atm) if this equilibrium were established? (*b*) How would the proportion of ozone change as the temperature increased?

11-10 Hydrogen-Bromine Equilibrium At 600 K the equilibrium constant for the reaction $H_2(g) + Br_2(g) \rightleftharpoons 2HBr(g)$ is 3×10^{14}. What will be the partial pressures of H_2 and Br_2 in equilibrium with HBr at 1 atm at this temperature? What fraction of the HBr will have decomposed?

11-11 Gas Equilibrium It is known from experiment that phosphorus pentachloride vapor dissociates as follows:

$$PCl_5(g) \rightleftharpoons PCl_3(g) + Cl_2(g)$$

When a flask that initially contained only pure PCl_5 is heated to 250°C, the PCl_5 is 69 percent dissociated at a total pressure of 2.00 atm. Calculate the equilibrium constant for the dissociation reaction at 250°C.

11-12 Association of Nitrogen Dioxide A 46.0-g sample of NO_2 occupies 15.7 liter at 20°C and 700 torr. (*a*) What is the number of moles of gas as computed from the ideal-gas equation? (*b*) Give a quantitative explanation for the result of (*a*) based on the reaction $2NO_2(g) \rightleftharpoons N_2O_4(g)$.

11-13 Dissociation of Sulfur Trioxide At 627°C and 1 atm, SO_3 is partly dissociated into SO_2 and O_2:

$$SO_3(g) \rightleftharpoons SO_2(g) + \tfrac{1}{2}O_2(g)$$

The density of the equilibrium mixture is 0.925 g liter^{-1}. What is the degree of dissociation of SO_3 under these circumstances?

11-14 Dissociation of O_2 At 4000 K and 1.000 atm, O_2 is 60.9 percent dissociated to atomic O. What is the percentage dissociation at 4000 K and 0.500 atm?

11-15 Gas Equilibrium Constant To determine the equilibrium constant at 445°C of the reaction

$$H_2(g) + I_2(g) \rightleftharpoons 2HI(g)$$

a mixture of 0.915 g I_2 and 0.0197 g H_2 was heated to and held at 445°C until equilibrium was established. It was found that 0.0128 g H_2 was left unreacted. Compute the equilibrium constant K for this reaction, including appropriate units.

11-16 Dissociation of a Gas The gas X_2Z is stable at ordinary temperatures but dissociates partially to gaseous X_2 and gaseous Z when it is heated. Suppose a sample of X_2Z, initially at 1.00 atm and 27°C, is confined in a flask of fixed volume and is heated to 627°C. If the pressure is then found to be 4.20 atm, what is the degree of dissociation?

11-17 Relation between Equilibrium Constants Expressed in Pressure Units and in Concentration Units Although reaction quotients for reactions involving gases are usually formulated in terms of pressures, they are occasionally expressed in terms of molar concentrations. The respective equilibrium constants are often symbolized as K_p and K_c, the subscripts indicating whether pressure units or concentration units have been used. The relation between these two constants is particularly simple when the gases can be assumed to be ideal. Let Δn be the change in the number of molecules of gas during the reaction; that is, Δn is the number of molecules of gaseous products minus the number of molecules of gaseous reactants. Show then that $K_p = (RT)^{\Delta n} K_c$. (*Hint:* Consider the relation between the number of moles per liter of a gas and its pressure.)

11-18 Dissociation of Phosphorus Vapor Between about 800 and 1200°C phosphorus vapor consists of a mixture of P_4 and P_2. At 1000°C and a pressure of 0.200 atm, the density of phosphorus vapor is found to be about 0.178 g liter^{-1}. What is the degree of dissociation of P_4 into P_2 under these conditions?

11-19 Combination of Equilibria The equilibria

$$NH_3 + H_2O \rightleftharpoons NH_4^+ + OH^-$$
$$HOAc \rightleftharpoons H^+ + OAc^-$$

happen to have equilibrium constants of equal values, $1.8 \times 10^{-5}\,M$, while the ion product $[H^+][OH^-]$ has the value $1.0 \times 10^{-14}\,M^2$, all at $25°C$. What is the value, at $25°C$, of the equilibrium constant for the reaction

$$NH_4^+ + OAc^- \rightleftharpoons NH_3 + HOAc$$

11-20 Relation between Equilibrium Constants
Express the equilibrium constant for the reaction

$$NH_3 + HNO_2 \rightleftharpoons NH_4^+ + NO_2^-$$

in terms of the constants for the following reactions:

$$HNO_2 \rightleftharpoons H^+ + NO_2^- \qquad K_1 = 4.5 \times 10^{-4}$$
$$NH_3 + H_2O \rightleftharpoons NH_4^+ + OH^- \quad K_2 = 1.8 \times 10^{-5}$$
$$H_2O \rightleftharpoons H^+ + OH^- \qquad K_w = 1.0 \times 10^{-14}$$

Ionic Equilibria in Aqueous Solution: Acids and Bases

12

". . . Acids and bases are substances that are capable of splitting off or taking up hydrogen jons, respectively."

J. N. BRØNSTED, 1923

12-1 Introduction

The early emergence of the concepts of acid, base, and salt, discussed in Chap. 9, was one of the first stages in the development of modern chemistry: the classification of compounds into types with characteristic properties and reactions. The implications of these concepts have evolved over the years, but an understanding of the effects of acids and bases is still essential for all those who deal with systems in which chemical reactions occur. Chemists frequently need to control pH precisely in order to ensure that a particular reaction mixture will yield the desired set of products rather than some quite different set. Those working in many other areas of science, technology, and medicine must understand acids and bases also—for example, engineers concerned with the properties of materials in various environments in which corrosion or other deterioration may occur; doctors and nurses who must recognize abnormalities in body fluids such as blood and urine, many of which are related to pH changes; geologists trying to understand the action of groundwater in the formation of limestone caves; biologists concerned with the regulation of metabolic processes in animals and plants. The list is almost endless.

Many of the properties of acids and bases have been described in detail in Chap. 9, and the reader should review that chapter carefully, especially Secs. 9-1 to 9-5, before studying the present chapter. A familiarity with Chap. 11 is also essential. The conventions for reaction quotients (Sec. 11-3) and many of the general comments and examples are especially relevant.

12-2 The Strengths of Acids and Bases

Degree of Dissociation The dissociation of an acid or a base involves a proton transfer to or from H_2O rather than a simple dissociation alone. Nonetheless, the concept of degree of dissociation (Sec. 11-3) is commonly used for acids and bases. For example, acid HA may be said to be 1.4 percent dissociated in a 2.000 M

and Bases

aqueous solution; this means that 98.6 percent of the HA, or 1.972 mol liter^{-1}, remains undissociated while 1.4 percent, or 0.028 mol liter^{-1}, has formed H_3O^+ and A^-.

The degree of dissociation of an acid or base in a given solution can be estimated from measurements of any of several properties of the solution and comparison of the results with those to be expected for various degrees of dissociation. For example, one might measure the pH, the electrical conductivity, or some colligative property such as the freezing point relative to that of the pure solvent (see Sec. 8-3). Solutions of a strong and a weak acid of comparable concentration, or solutions of a strong and a weak base, will differ markedly with regard to each of these properties because essentially all molecules of strong electrolytes dissociate into ions whereas comparatively few molecules of weak electrolytes dissociate. As a result, the total concentration of solute species present is higher for strong electrolytes than for weak electrolytes in solutions of similar molarity. Colligative properties depend on the total concentration of *all* solute species, as demonstrated in the following examples.

□ **Degree of Dissociation**   Estimate the degree of dissociation of the solute in each of the following solutions:

(a) A 1.00 *M* solution of an unknown acid, HX, that has a pH of 1.6.

(b) A 0.010 *M* solution of sodium hydroxide in water, which freezes at $-0.037°C$. The molal freezing-point-depression constant for water is $k_f = 1.86$ K kg mol^{-1} (Sec. 8-3).

Example 12-1

Solution   If you are uncertain about the precise significance of *degree of dissociation,* review comment 4 of Sec. 11-3 and Examples 11-3 and 11-4.

(a) The dissociation of HX in water occurs by transfer of a proton to a water molecule,

$$HX + H_2O \rightleftharpoons X^- + H_3O^+ \tag{12-1}$$

As is customary (Chap. 9), we represent H_3O^+ as H^+. Since $pH \equiv -\log_{10}[H^+]$, pH of 1.6 implies that

$$[H^+] = 10^{-1.6} = 10^{0.4} \times 10^{-2.0} = 2.5 \times 10^{-2}\ M$$

One hydrogen ion is produced from each molecule of HX that dissociates; hence the number of moles of HX dissociated per liter must be 2.5×10^{-2}. The solution is 1.00 *M*, and thus there would be 1.00 mol HX liter $^{-1}$ if none dissociated. Consequently the degree of dissociation is $2.5 \times 10^{-2}/1.00$, or 2.5 percent. HX is a weak acid in water.

(b) The molality of the solution, *m*, can be calculated from the freezing-point depression by Equation (8-11):

$$m = \frac{-\Delta T}{k_f} = \frac{0.037}{1.86} = 0.020\ \text{mol kg}^{-1}$$

This means that there is 0.020 mol of solute particles in each kilogram of water, and since this aqueous solution is dilute, 1 liter weighs 1 kg (within the precision of the data). Thus a 0.010 *M* solution of NaOH contains about 0.020 mol of solute species per liter. The conclusion is that the NaOH has dissociated essentially completely, two ions being formed for every formula unit of the compound. (Measurements of colligative properties can give information about dissociation but *not* about whether the dissociation products are charged. Other means, such as tests of electrical conductivity, are needed for that. When we state here that two *ions* are formed from every formula unit of the compound, we are assuming that such other tests have been made.) ■

Exercise 12-1

☐ Estimate the degree of dissociation of the solute in each of the following solutions: (*a*) A 0.100 *M* solution of an unknown acid, HX, that has a pH of 2.5. (*b*) A 0.030 *M* solution of another unknown acid. The solution melts at $-0.058°C$. The molal freezing-point-depression constant for water is $k_f = 1.86$ K kg mol^{-1}. ■

Dissociation Constants Comparisons of acids and bases in terms of their degree of dissociation are of limited value because the degree of dissociation of a weak electrolyte varies with concentration. More precise statements about the relative strengths of various acids and bases can be made by comparison of the equilibrium constants for their dissociation, which are independent of concentration. The dissociation constant for an acid in aqueous solution (the acid constant) is symbolized by K_a and that for a base by K_b. For a monoprotic acid HA with only one acidic hydrogen atom, the reaction in water is

$$\text{HA} + \text{H}_2\text{O} \rightleftharpoons \text{A}^- + \text{H}_3\text{O}^+ \tag{12-2}$$

with
$$K_a = \frac{[\text{H}_3\text{O}^+][\text{A}^-]}{[\text{HA}]} = \frac{[\text{H}^+][\text{A}^-]}{[\text{HA}]} \tag{12-3}$$

For polyprotic acids, such as H_2SO_4 and H_3PO_4, the protons are furnished in successive steps, for each of which there is a corresponding equilibrium constant. These successive constants are usually distinguished by subscripts. For example, for phosphoric acid the successive equilibria are described by the following chemical equations, shown with the corresponding equilibrium constants:

$$\text{H}_3\text{PO}_4 + \text{H}_2\text{O} \rightleftharpoons \text{H}_2\text{PO}_4^- + \text{H}_3\text{O}^+$$
$$K_1 = \frac{[\text{H}^+][\text{H}_2\text{PO}_4^-]}{[\text{H}_3\text{PO}_4]} = 7.1 \times 10^{-3} \tag{12-4}$$

$$\text{H}_2\text{PO}_4^- + \text{H}_2\text{O} \rightleftharpoons \text{HPO}_4^{2-} + \text{H}_3\text{O}^+$$
$$K_2 = \frac{[\text{H}^+][\text{HPO}_4^{2-}]}{[\text{H}_2\text{PO}_4^-]} = 6.2 \times 10^{-8} \tag{12-5}$$

$$\text{HPO}_4^{2-} + \text{H}_2\text{O} \rightleftharpoons \text{PO}_4^{3-} + \text{H}_3\text{O}^+$$
$$K_3 = \frac{[\text{H}^+][\text{PO}_4^{3-}]}{[\text{HPO}_4^{2-}]} = 4.4 \times 10^{-13} \tag{12-6}$$

These constants for phosphoric acid, K_1, K_2, and K_3, are usually referred to as the first, second, and third dissociation constants of phosphoric acid; however, K_2 and K_3 can equally well be regarded as the first dissociation constants of the acids $H_2PO_4^-$ and HPO_4^{2-}, respectively. The dihydrogenphosphate ion, $H_2PO_4^-$, behaves as a base in the equilibrium (12-4), accepting a proton, and as an acid in the equilibrium (12-5), donating a proton. Similarly, HPO_4^{2-} behaves as a base in (12-5) and as an acid in (12-6).

Any base B may react with water to form the conjugate acid BH^+ and hydroxide ions,

$$B + H_2O \rightleftharpoons BH^+ + OH^- \qquad (12\text{-}7)$$

with the corresponding equilibrium constant, K_b, for B given by the expression

$$K_b = \frac{[BH^+][OH^-]}{[B]} \qquad (12\text{-}8)$$

An extremely useful relation may be derived by considering the product of the acid constant for any acid HA, given by (12-3), and the base constant for the corresponding conjugate base A^-. By analogy with (12-7), the reaction of A^- as a base is

$$A^- + H_2O \rightleftharpoons HA + OH^- \qquad (12\text{-}9)$$

with

$$K_{b,A^-} = \frac{[HA][OH^-]}{[A^-]} \qquad (12\text{-}10)$$

The product of $K_{a,HA}$ and K_{b,A^-} is then

$$K_{a,HA}K_{b,A^-} = \frac{[H^+][A^-]}{[HA]}\frac{[HA][OH^-]}{[A^-]} = [H^+][OH^-] = K_w \qquad (12\text{-}11)$$

which has the value 1.0×10^{-14} at $25°C$. *Thus, the product of the acid constant of an acid and the base constant of its conjugate base is the ion product of water.* This completely general relationship for reactions in aqueous solutions makes it easy to calculate K_b of a base when K_a for the conjugate acid is known and vice versa. It implies that if an acid is strong—for example, if K_a is greater than 1—then the conjugate base must be very weak, with K_b less than 10^{-14}. In terms of the reactions involved, this implies that if HA is so strong an acid that it dissociates almost completely in water to form H_3O^+ and A^-, then A^- is so weak a base that it has essentially no tendency to take a proton away from the hydronium ion, H_3O^+. Conversely, if a base is so strong that when dissolved in water it tends to be converted almost completely into the conjugate acid and hydroxide ions, by (12-7) or (12-9)—that is, if the base is much stronger than OH^- in a competition for protons—then the conjugate acid of this base must be a significantly weaker acid than H_2O.

□ **K_a and K_b for Conjugate Pairs** (*a*) Estimate the acid constant for NH_4^+ in water, given that K_b is 1.8×10^{-5}. (*b*) Estimate the base constant for cyanide ion (CN^-) in water, given that K_a for HCN is 2×10^{-9}.

Example 12-2

Solution With (12-11) for any conjugate pair in water, we have

(*a*) $K_{a,NH_4^+} = \dfrac{K_w}{K_{b,NH_3}} = \dfrac{1.0 \times 10^{-14}}{1.8 \times 10^{-5}} = \underline{5.5 \times 10^{-10}}$

and

(*b*) $K_{b,CN^-} = \dfrac{K_w}{K_{a,HCN}} = \dfrac{1.0 \times 10^{-14}}{2 \times 10^{-9}} = \underline{5 \times 10^{-6}}$

Thus ammonium ion is a very weak acid in water, and cyanide ion is a moderately weak base (see Table 12-1), not very much weaker than ammonia. ∎

Table 12-1
Classification of the
Strengths of Acids and Their
Conjugate Bases* in Water

| | DISSOCIATION CONSTANTS | | |
Strength of acid	Acid	Conjugate base	Strength of base
Strong	$K_a \geq 1$	$K_b \leq 10^{-14}$	Ineffective
Moderately weak	$1 > K_a \geq 10^{-7}$	$10^{-14} < K_b < 10^{-7}$	Very weak
Very weak	$10^{-7} > K_a > 10^{-14}$	$10^{-7} \leq K_b < 1$	Moderately weak
Ineffective	$K_a \leq 10^{-14}$	$K_b \geq 1$	Strong

*The categories used are based on suggestions by N. D. Cheronis.

Exercise 12-2

☐ The base hydroxylamine, NH_2OH, may be thought of as derived from ammonia by replacement of one H of NH_3 by an OH. It reacts with water according to the equation

$$NH_2OH + H_2O \rightleftharpoons NH_3OH^+ + OH^-$$

Its base constant, K_b, is 6.6×10^{-9}. (a) Estimate K_a of NH_3OH^+. (b) Estimate K_b for hypochlorite ion (ClO^-) in water, given that K_a for hypochlorous acid (HOCl) is 1.1×10^{-8}. ■

The Leveling Effect of Water on Strong Acids and Bases Since the equilibrium of the dissociation reaction for any strong acid lies, by definition, overwhelmingly to the right, there are essentially no undissociated HA molecules in the solution. All the HA that might be present is converted to H_3O^+ and the (ineffective) conjugate base A^-. This is true, for example, for HI, HBr, HCl (but not HF), $HClO_4$, H_2SO_4 (but not HSO_4^-), and HNO_3. Aqueous solutions of any strong acid are thus essentially solutions of the same acid, H_3O^+, and different ineffective bases, the anions of the corresponding acids. Any difference in the strengths of strong acids in water is thus masked by the fact that no acid appreciably stronger than H_3O^+ can exist in dilute aqueous solution in other than very small concentrations. Water is said to have a *leveling effect* on strong acids. To illustrate this effect, consider 0.10 M solutions of HNO_3 and HCl. The respective values of K_a have been estimated to be about 24 and about 10^5. This difference seems large, but it means little in terms of completeness of dissociation. In 0.10 M solutions, HNO_3 is about 99.6 percent dissociated and HCl is about 99.9999 percent dissociated; hence both solutions contain H_3O^+ at a concentration of essentially 0.10 M.

Differences between strong acids become apparent when the acids are dissolved in a nonaqueous solvent that can accept a proton and thereby form a stronger acid than H_3O^+—that is, a solvent that is a weaker base than water. For example, the acetic acid molecule, HOAc, can accept a second proton[1] and thus form the acid H_2OAc^+, which is a considerably stronger acid than H_3O^+. When HCl is dissolved in anhydrous acetic acid, the equilibrium of the reaction

$$HCl + HOAc \rightleftharpoons Cl^- + H_2OAc^+ \tag{12-12}$$

is no longer completely in favor of the right side; there are substantial numbers of HCl molecules in this solution. This is true for some other strong acids as well. By comparisons in different solvents, it is possible to assess the relative strengths of acids that are almost equally strong in water. In this way it has been shown that $HClO_4 > HCl > H_2SO_4 > H_3O^+$ (see also Table 12-2).

---

[1]The second proton is attached to the oxygen atom that is not already part of a hydroxyl group in the CH_3COOH molecule; H_2OAc^+ is thus $CH_3C(OH)_2^+$.

Conjugate pair of acid and base		Acid constant	
$HClO_4$	ClO_4^-	$\sim 10^9$	
HI	I^-	$\sim 10^7$	
HBr	Br^-	$\sim 10^6$	
HCl	Cl^-	$\sim 10^5$	
H_2SO_4	HSO_4^-	$\sim 10^3$	
$HClO_3$	ClO_3^-	$\sim 10^3$	
H_3O^+	H_2O	55 †	
HNO_3	NO_3^-	24	
H_2CrO_4	$HCrO_4^-$	1.2	
HSO_4^-	SO_4^{2-}	1.2×10^{-2}	
H_2SO_3	HSO_3^-	1.0×10^{-2}‡	
$HClO_2$	ClO_2^-	1.0×10^{-2}	
H_3PO_4	$H_2PO_4^-$	7.1×10^{-3}	
HF	F^-	6.7×10^{-4}	
HNO_2	NO_2^-	4.5×10^{-4}	
$HCOOH$	$HCOO^-$	1.7×10^{-4}	
$HOAc$	OAc^-	1.8×10^{-5}	$(K_b = 5.5 \times 10^{-10})$
H_2CO_3	HCO_3^-	4.4×10^{-7}§	
$HCrO_4^-$	CrO_4^{2-}	3.2×10^{-7}	
H_2S	HS^-	9.1×10^{-8}	
$H_2PO_4^-$	HPO_4^{2-}	6.2×10^{-8}	
HSO_3^-	SO_3^{2-}	5.0×10^{-8}	
$HClO$	ClO^-	1.1×10^{-8}	
HCN	CN^-	2.0×10^{-9}	$(K_b = 5.0 \times 10^{-6})$
H_3BO_3	$B(OH)_4^-$	6.4×10^{-10}	
NH_4^+	NH_3	5.5×10^{-10}	$(K_b = 1.8 \times 10^{-5})$
HCO_3^-	CO_3^{2-}	4.8×10^{-11}	
H_2O_2	HO_2^-	2.4×10^{-12}	
HPO_4^{2-}	PO_4^{3-}	4.4×10^{-13}	
HS^-	S^{2-}	1.2×10^{-15}	
H_2O	HO^-	1.8×10^{-16}¶	
NH_3	NH_2^-	$\sim 10^{-23}$	
HO^-	O^{2-}	$\sim 10^{-35}$	

*Values of equilibrium constants often differ widely in different reference sources, depending in part on whether they were extrapolated to zero concentration or not. Estimated values mainly by G. Schwarzenbach.

†By (12-2) and (12-3) with HA and A^- being, respectively, H_3O^+ and H_2O, $K_a = [H_3O^+][H_2O]/[H_3O^+] = [H_2O] = 55$.

‡Acid constant for which the term $[H_2SO_3]$ is the total concentration of both actual H_2SO_3 and dissolved (hydrated) SO_2 that has not reacted with H_2O to form H_2SO_3. The acid constant referred to just the species H_2SO_3 has the value 4.0×10^{-2}.

§Acid constant for which the term $[H_2CO_3]$ is the total concentration of both actual H_2CO_3 and dissolved (hydrated) CO_2 that has not reacted with H_2O to form H_2CO_3. The acid constant referred to just the species H_2CO_3 has the value 1.7×10^{-4}.

¶$K_a = [HO^-][H^+]/[H_2O]$. ($[H_2O]$ is retained in the denominator by analogy with other acid constants.)

Just as the strongest acid that can exist in significant quantity in an aqueous solution is H_3O^+, the strongest base that can be present in water in appreciable concentrations is the hydroxide ion, OH^-. Any base that is stronger than OH^- will take protons away from water molecules and form OH^- and the ineffective conjugate acid of the base. Water thus has a leveling effect on strong bases, just as it does on strong acids. For example, the amide ion, NH_2^-, is a stronger base than OH^-. When the crystalline ionic compound sodium amide, $NaNH_2$, is dissolved in water, the amide ion reacts vigorously and almost completely to form its conjugate acid, NH_3, and OH^-:

$$NH_2^- + H_2O \rightleftharpoons NH_3 + OH^- \qquad (12\text{-}13)$$

Hydroxide ion is so much stronger a base than NH_3 that the NH_3 has little effect on the pH of the resulting solution. The pH is essentially the same as that of a solution of sodium hydroxide of the same concentration.

Values of K_a and K_b for many of the more common acids and bases are included in Table 12-2 in order of decreasing acid strength (and in the alphabetized reference Table D-3). Few base constants are listed explicitly in Table 12-2, but any missing value of K_b can readily be derived from the corresponding value of K_a with the help of (12-11). Important moderately weak acids include the bisulfate ion, HSO_4^-, $K_a = 1.2 \times 10^{-2}$; phosphoric acid, H_3PO_4, $K_a = 7.1 \times 10^{-3}$; formic acid, $HCOOH$, $K_a = 1.7 \times 10^{-4}$; and acetic acid, $HOAc$, $K_a = 1.8 \times 10^{-5}$. The conjugate base of each of these moderately weak acids (respectively, SO_4^{2-}, $H_2PO_4^-$, $HCOO^-$, and OAc^-) is a very weak base. Typical very weak acids include hydrocyanic acid, HCN, $K_a = 2.0 \times 10^{-9}$; boric acid, H_3BO_3, which is effectively monoprotic,[2] $K_a = 6.4 \times 10^{-10}$; ammonium ion, NH_4^+, $K_a = 5.5 \times 10^{-10}$; and bicarbonate ion, HCO_3^-, $K_a = 4.8 \times 10^{-11}$. The conjugate base of each of these very weak acids [respectively, CN^-, $B(OH)_4^-$, NH_3, and CO_3^{2-}] is a moderately weak base and thus is a stronger base than the conjugate form is an acid. It is an interesting coincidence that perhaps the most common weak acid and weak base, acetic acid and ammonia, have identical dissociation constants—that is, K_a for $HOAc$ and K_b for NH_3 both equal 1.8×10^{-5}.

As Table 12-2 indicates, an enormous range of strengths of acids and bases has been measured, at least approximately. The values of K_a and K_b have been extended beyond the limits measurable in water by comparisons in other solvents. Rules for estimating values of acid constants for many inorganic oxyacids are given in Sec. 24-1.

12-3 Equilibrium Calculations: Dissociation and Hydrolysis

pH, pOH, and pK As discussed in Sec. 9-3, the use of the symbol pH for the negative base-10 logarithm of the concentration of H^+ has been generalized, so that pOH means $-\log[OH^-]$, and the same symbol is used for equilibrium constants in similar logarithmic form, with $pK \equiv -\log K$.

Since

$$[H^+][OH^-] = K_w$$

it follows that

$$pH + pOH = pK_w$$

At $25°C$, $pK_w = 14.00$. When the concentrations of H^+ and OH^- are equal, as they

---

[2] H_3BO_3, or $B(OH)_3$, is unusual because it does not lose a proton when acting as an acid in water; it accepts a hydroxide ion instead:

$$B(OH)_3 + 2H_2O \rightleftharpoons B(OH)_4^- + H_3O^+$$

are in pure water, one speaks of a neutral medium, for which

$$[H^+] = [OH^-] = \sqrt{K_w}$$

and thus

$$pH = pOH = \frac{pK_w}{2} \qquad (12\text{-}14)$$

which is 7.00 at 25°C. Conversely, when the pH of any solution at 25°C is 7.00, it follows from (12-14) that $[H^+]$ and $[OH^-]$ are equal, so that the solution is neutral. Since K_w varies with temperature, it is only approximately true that a solution with pH = 7.00 is neutral at temperatures other than 25°C. For example, at 0°C, $K_w = 1.14 \times 10^{-15}$, so that $pK_w = 14.94$ and the pH of a neutral solution is 7.47. At 60°C, $K_w = 9.6 \times 10^{-14}$, corresponding to $pK_w = 13.02$, so that the pH of a neutral solution is 6.51. At 37°C (body temperature), K_w is 2.6×10^{-14} and the pH of a neutral solution is 6.79.

As mentioned in Chap. 11, we shall not hesitate to use approximations in equilibrium calculations provided that no errors larger than 10 or 20 percent result. We shall therefore assume the pH of a neutral solution to be 7.0 and pK_w to be 14.0, at temperatures not far from 25°C. Acid and base constants also change with temperature, but we shall neglect these changes also. They are usually smaller than the change of K_w.

The pH scale in dilute solutions ranges from about 0 to 14. In more concentrated solutions, this range is exceeded at each end. On the acid side, the pH becomes negative for concentrated solutions. In concentrated H_2SO_4, the lowest existing pH has been estimated to be about -10; in concentrated solutions of NaOH and KOH, the pH is believed to reach 18 or 19. In such concentrated solutions, a definition of pH in terms of hydrogen-ion activity (rather than concentration) is required.

Strong Acids and Bases In an aqueous solution of a strong acid, essentially all molecules of the strong acid have transferred their protons to water molecules, thereby furnishing an equal quantity of hydrogen ions and of the base that is conjugate to the strong acid. The hydrogen-ion concentration and the concentration of conjugate base are thus equal to the concentration that the acid would have if it did not dissociate. We neglect the contribution to the hydrogen-ion concentration from the dissociation of water, which is almost always possible in such solutions, as discussed in the following paragraph. The concentration of the strong-acid species itself is essentially zero because it has completely dissociated. Thus, in a 1.0 M solution of HCl, for example, $[H^+] = [Cl^-] = 1.0\,M$, so that the pH = 0.0, while $[HCl] \approx 0$.

The effect of the dissociation of water can be understood by noting that this reaction produces H^+ and OH^- in equal numbers. In a solution of a strong acid, water is the only source of hydroxide ions, so that $[OH^-]$ is just equal to the contribution to the total hydrogen-ion concentration that comes from the dissociation of H_2O. In an acid solution, $[OH^-]$ is necessarily less than $10^{-7}\,M$ and the contribution to $[H^+]$ from the dissociation of water is thus less than $10^{-7}\,M$. For example, suppose that calculation of $[H^+]$ with complete neglect of the dissociation of water gives $[H^+] = 10^{-6}\,M$, or pH = 6. This implies that $[OH^-] = 10^{-8}\,M$, so that the contribution to $[H^+]$ from the dissociation of water is $10^{-8}\,M$, only 1 percent of the value obtained when this contribution is neglected. Any error made by ignoring the contribution from the dissociation of

H_2O is thus negligible unless the pH is close to 7. Thus for strong acids the dissociation of water can be ignored unless the concentration of the acid is below $10^{-6} M$, which is so low a concentration that it has little importance in practical situations: it corresponds to about one drop of $1 M$ acid added to more than 40 liters of water.

In a solution of a strong base the situation is similar to that for a strong acid. For example, in $0.1 M$ NaOH, $[OH^-] = [Na^+] = 0.1 M$, so that pOH $= 1$ and pH $= 14 - 1 = 13$, while $[NaOH] \approx 0$. The contribution to $[OH^-]$ from the dissociation of water is negligible. In either an acidic or a basic solution, the effect of the dissociation of water can be ignored unless the pH is within less than one unit of 7.

Weak Acids A weak monoprotic acid is not completely dissociated in an aqueous solution; that is, proton transfer from the acid to H_2O is incomplete. The interrelation between the degree of dissociation and the acid constant K_a is illustrated in the following example.

Example 12-3

☐ **Calculation of K_a from Degree of Dissociation** A weak acid HA is 2 percent dissociated in a $1.00 M$ solution. What is the dissociation constant of HA?

Solution   The reaction of interest is

$$H_2O + HA \rightleftharpoons H_3O^+ + A^-$$

The concentration of HA that would be present *if there were no dissociation* is $1.00 M$. However, 2 percent of the HA present has dissociated, so its actual concentration is 98 percent of $1.00 M$ (that is, $0.98 M$) and by its dissociation it forms H_3O^+ and A^-, each at a concentration of $0.02 M$. Summarizing:

	[HA]	[H_3O^+]	[A^-]
If no dissociation	1.00	~ 0	0
At equilibrium, after dissociation	0.98	0.02	0.02

Thus, when we evaluate the equilibrium constant by inserting the *equilibrium* concentrations into the reaction quotient, we get

$$K_a = \frac{[H^+][A^-]}{[HA]} = \frac{(0.02)(0.02)}{0.98} M = \underline{4 \times 10^{-4} M}$$

The degree of dissociation is, as given, 2 percent; that is,

$$\text{degree of dissociation} = \frac{[A^-]}{[HA] + [A^-]} = \frac{0.02}{0.02 + 0.98}$$
$$= \tfrac{0.02}{1.00} = 0.02$$

Note that the denominator in the expression for the degree of dissociation of a weak electrolyte in solution is always just the number of moles per liter that would be present if none had dissociated. ∎

Exercise 12-3

☐ A weak acid HA is 5 percent dissociated in a $0.100 M$ solution. Estimate the dissociation constant of HA. ∎

To calculate the pH of a solution of a weak monoprotic acid, we proceed as in Example 12-4.

□ **The pH of Dilute Acetic Acid** What is the pH of 0.20 M acetic acid?

Example 12-4

Solution We start with 0.20 mol HOAc molecules liter^{-1} and assume that at equilibrium x mol liter^{-1} have lost their protons, leaving $(0.20 - x)$ mol HOAc liter^{-1}. The concentrations of the different species are then as shown on the line below the reaction equation:

$$HOAc \rightleftharpoons H^+ + OAc^-$$
$$0.20 - x \qquad x \qquad x$$

This neglects the contribution made to [H$^+$] by the dissociation of water, which causes an error of less than 1 percent if [H$^+$] turns out to be larger than 10^{-6} M and [OH$^-$] smaller than 10^{-8} M, as discussed earlier. Insertion of the concentrations in terms of x into the equilibrium expression results in the equation

$$\frac{[H^+][OAc^-]}{[HOAc]} = \frac{x^2}{0.20 - x} = K_a = 1.8 \times 10^{-5} \qquad (12\text{-}15)$$

This is a quadratic equation, and such an equation can always be solved by an exact method, but that is not necessary here. The use of approximations is frequently faster and simpler; one need only check the answer obtained to be sure that it is consistent with the assumptions made in the initial approximations.

Inspection of (12-15) shows that the right side is very small, 1.8×10^{-5}. The left side can only be small if x is small relative to 0.20, for if x is comparable to 0.20, the x^2 term in the numerator cannot be very small and the term $0.20 - x$ in the denominator might itself be small, which would only make the whole expression on the left larger.

Thus we suspect that a good approximation is that x is small compared to 0.20, that is, that

$$0.20 - x \approx 0.20 \qquad (12\text{-}16)$$

within the desired precision. Substitution of the approximation (12-16) into the denominator of the quadratic expression in (12-15) gives

$$x^2 \approx 0.20 \times 1.8 \times 10^{-5} = 3.6 \times 10^{-6}$$

or
$$x = [H^+] \approx 1.9 \times 10^{-3} \, M$$

and
$$pH \approx 3.0 - \log 1.9 = \underline{2.7}$$

Since x is less than 1 percent of 0.20, our approximation (12-16) was justified. Furthermore, because the pH is so low, we know also that the contribution to [H$^+$] from the dissociation of water, measured by the [OH$^-$] in the solution, is negligible. ■

□ What is the pH of 0.100 M HA, if the value of K_a for HA is 4.5×10^{-6}? ■

As emphasized in Sec. 12-2, and especially in the discussion accompanying Example 12-2, the conjugate form of a moderately weak base is itself a very weak acid. Thus, for example, salts containing ammonium ion can have an acidic reaction by the equilibrium $NH_4^+ + H_2O \rightleftharpoons NH_3 + H_3O^+$, provided that the other ion present does not have an even stronger opposing effect. Salts formed from other weak bases behave similarly.

Example 12-5

□ **pH of a Solution of a Salt of a Weak Base and a Strong Acid** Ethylamine, $C_2H_5NH_2$, is a weak base very similar to ammonia. It reacts with acids to form salts containing the ethylammonium ion, $C_2H_5NH_3^+$, and the conjugate base of the acid. For example, ethylamine reacts with HCl to form ethylammonium chloride, $C_2H_5NH_3Cl$, a strong electrolyte that dissociates completely in solution into $C_2H_5NH_3^+$ and Cl^-. Since Cl^- is the conjugate base of a strong acid, HCl, it shows no significant tendency to react with water. The ethylammonium ion is, however, a very weak acid, just as is NH_4^+. If K_b for ethylamine is 4.7×10^{-4}, what is the pH of a 0.10 M solution of ethylammonium chloride?

Solution The solution is acidic because of the reaction

$$C_2H_5NH_3^+ + H_2O \rightleftharpoons C_2H_5NH_2 + H_3O^+ \qquad (12\text{-}17)$$

and the acid constant for $C_2H_5NH_3^+$ can be derived from K_b for its conjugate base by the usual relation analogous to (12-11):

$$K_{a,\mathrm{BH}^+} K_{b,\mathrm{B}} = K_w = 1.0 \times 10^{-14} \qquad (12\text{-}18)$$

Thus, here

$$K_a = \frac{1.0 \times 10^{-14}}{4.7 \times 10^{-4}} = 2.1 \times 10^{-11} \qquad (12\text{-}19)$$

If we symbolize the ethylammonium ion by BH^+ and ethylamine by B and let the unknown $[H^+]$ be x, we have

$$[B] = [H^+] = x \qquad \text{and} \qquad [BH^+] = 0.10 - x$$

and

$$K_a = \frac{[B][H^+]}{[BH^+]} = \frac{x^2}{0.10 - x} = 2.1 \times 10^{-11} \qquad (12\text{-}20)$$

It is quite evident from the smallness of K_a that the value of x must itself be very small and hence that the approximation $0.10 - x \approx 0.10$ is plausible. Thus we get

$$x^2 \approx 0.10 \, (2.1 \times 10^{-11})$$
$$[H^+] = x \approx 1.5 \times 10^{-6} \, M$$

Checking the validity of our approximation, we find it justified. Furthermore, $[H^+]$ is slightly greater than 10^{-6} M so that ignoring the contribution from the dissociation of water is also reasonable since it causes an error of less than 1 percent. Finally, the pH of the solution is

$$\mathrm{pH} = -\log(1.5 \times 10^{-6}) = -\log 1.5 + 6.0 = \underline{5.8} \quad \blacksquare$$

Exercise 12-5

□ Hydrazine, H_2NNH_2, can be thought of as being derived from ammonia by replacement of an H of NH_3 by NH_2. Its base constant is $K_b = 8.5 \times 10^{-7}$. What is the pH of a 0.2 M solution of hydrazinium chloride, NH_2NH_3Cl? ■

We now consider an example in which the approximation $[H^+] \ll c$, where c is the molarity of the acid, is not valid.

□ **pH of a Solution of HClO$_2$** What is the pH of a $0.10\,M$ solution of chlorous acid, $HClO_2$, for which K_a is 1.0×10^{-2}?

Example 12-6

Solution Proceeding as in Examples 12-4 and 12-5, we obtain the equation for the hydrogen-ion concentration x:

$$\frac{x^2}{0.10 - x} = 1.0 \times 10^{-2} \tag{12-21}$$

Since K_a is not very small, x is not negligible with respect to 0.10 and thus our earlier approximation, $c - x \approx c$, is no longer valid. We can solve (12-21) either by the method of successive approximations or exactly with the formula for the solution of a quadratic equation. We illustrate both approaches.

In applying the *method of successive approximations,* we proceed as in Examples 12-4 and 12-5, using the approximation $0.10 - x \approx 0.10$ on the left side of (12-21) even though we believe that this may lead to a significant error. The result is

$$x \approx (0.10 \times 1.0 \times 10^{-2})^{1/2} = 0.032 \tag{12-22}$$

which shows that, indeed, the approximation $0.10 - x \approx 0.10$ is not a good one. However, we can consider (12-22) to be a good *first approximation* to the correct value of x and substitute it for x in the expression $0.10 - x$ in (12-21). This leads to

$$x^2 \approx (0.10 - 0.032)\, 1.0 \times 10^{-2}$$

and hence

$$x \approx (0.068 \times 10^{-2})^{1/2} = 0.026 \tag{12-23}$$

This is a *second approximation* to x. Repeating the procedure, we obtain

$$x^2 \approx (0.10 - 0.026)1.0 \times 10^{-2}$$

and

$$x \approx (0.074 \times 10^{-2})^{1/2} = 0.027$$

This is within 4 percent of the preceding value and therefore as close to the correct value as we may expect, unless activities are taken into account. We therefore conclude that $[H^+] = [ClO_2^-] = 0.027\,M$ and $[HClO_2] = 0.10 - 0.027 = 0.07\,M$. As a check we note that if these concentrations are inserted into the reaction quotient the correct value of K_a is obtained:

$$\frac{[H^+][ClO_2^-]}{[HClO_2]} = \frac{(2.7 \times 10^{-2})^2}{0.07} = \frac{7.3 \times 10^{-4}}{7 \times 10^{-2}} = 1.0 \times 10^{-2}$$

One characteristic of the method of successive approximations is illustrated by the application here: the successive approximate values sometimes straddle the correct answer, being alternatively on one side of it or the other, greater or smaller. If convergence is slow, that is, if successive values differ markedly, it can sometimes be accelerated by averaging two successive values, which will normally give an approximation better than either value.

Ionic Equilibria in Aqueous
Solution: Acids and Bases

When convergence is rapid, as in this example, no such averaging is needed.

We can also solve (12-21) by using the formula for the exact solution of a quadratic expression, Equation (11-18), with $a = 1$, $b = 0.010$, and $c = -1.0 \times 10^{-3}$. Thus

$$x = \frac{-0.010 \pm (1.0 \times 10^{-4} + 4.0 \times 10^{-3})^{1/2}}{2}$$

$$= \frac{-0.010 \pm 0.064}{2}$$

Only the plus sign gives a physically meaningful value for x, the hydrogen-ion concentration, which cannot be negative. This leads to

$$[H^+] = x = \frac{0.054}{2} = 0.027\ M$$

which is the same result as that obtained in the third approximation above. The pH of the solution is $2.00 - \log 2.7 = \underline{1.57}$, with the final digit very uncertain. ∎

Exercise 12-6

☐ What is the pH of a $0.20\ M$ solution of an acid for which K_a is 2.5×10^{-2}? ∎

Which method, successive approximations or exact solution, should be used to solve problems like this one? The techniques lead to the same answer—the choice is a matter of personal taste.

Weak Bases Calculations for weak bases are in all respects analogous to those for weak acids. We shall consider as examples both an uncharged weak base, whose reaction may be represented by

$$B + H_2O \rightleftharpoons HB^+ + OH^- \tag{12-24}$$

and the salt of a weak acid, such as sodium acetate, which acts as a very weak base in water because acetate ion is the conjugate base of a moderately weak acid. This reaction can be represented by

$$A^- + H_2O \rightleftharpoons HA + OH^- \tag{12-25}$$

which differs from (12-24) only in the charges on the basic and the acidic species.

Example 12-7

☐ **The pH of a Solution of Ammonia** What is the pH of $0.50\ M$ ammonia, for which $K_b = 1.8 \times 10^{-5}$?

Solution We start with 0.50 mol NH_3 liter^{-1} and assume that at equilibrium x mol liter^{-1} has reacted by the equation given below. The final concentrations are as indicated:

$$\underset{0.50-x}{NH_3} + H_2O \rightleftharpoons \underset{x}{NH_4^+} + \underset{x}{OH^-}$$

The contribution of OH^- from the dissociation of H_2O has been neglected. The equilibrium condition yields

$$\frac{[NH_4^+][OH^-]}{[NH_3]} = \frac{x^2}{0.50 - x} = K_b = 1.8 \times 10^{-5}$$

If we assume that x is small compared to 0.50, then $0.50 - x \approx 0.50$ and we find $x^2 \approx 0.50 \times 1.8 \times 10^{-5}$ and $x = [OH^-] \approx 3.0 \times 10^{-3} M$. All assumptions are seen to be justified. Therefore, $pOH = 3.00 - \log 3.0 = 2.5$ and $pH = 14.0 - pOH = \underline{11.5}$. ∎

Exercise 12-7

☐ Hydroxylamine, H_2NOH, is a weak base for which $K_b = 1.2 \times 10^{-8}$. What is the pH of a 0.25 M solution of hydroxylamine in water? ∎

Example 12-8

☐ **Estimation of K_a for an Acid from the pH of a Solution of the Sodium Salt of the Acid**   The pH of a 0.10 M solution of the sodium salt of a certain acid HX is 9.5. What is K_a for the acid?

Solution   The reaction involved is (12-25). Since the pH is 9.5, we know that $pOH = 14 - pH = 4.5$, so that $[OH^-] = 10^{-4.5} \approx 3 \times 10^{-5} M$. This must also be the concentration of HX, inasmuch as HX and OH^- are formed in equal numbers in reaction (12-25). Consequently the concentration of unreacted X^- is $0.10 - 3 \times 10^{-5} = 0.10$. Substitution of these equilibrium concentrations in the reaction quotient for the reaction of X^- as a base gives

$$K_{b,X^-} = \frac{[OH^-][HX]}{[X^-]} = \frac{(10^{-4.5})^2}{0.10} = 1.0 \times 10^{-8}$$

Hence

$$K_{a,HX} = \frac{K_w}{K_{b,X^-}} = \frac{1.0 \times 10^{-14}}{1.0 \times 10^{-8}} = \underline{1.0 \times 10^{-6}}$$

Note that in this example we have ignored the effect of Na^+ in the solution. The fact that NaOH is a very strong base in water implies that the sodium ion has essentially no tendency to act as an acid, even though the hydrated species, $Na(H_2O)_n{}^+$, could in principle lose a proton. Thus, the position of equilibrium in the potential reaction

$$Na(H_2O)_n{}^+ + H_2O \rightleftharpoons NaOH(H_2O)_{n-1} + H_3O^+ \qquad (12\text{-}26)$$

lies almost completely to the left, with no detectable amount of undissociated NaOH present. The species $Na(H_2O)_n{}^+$ also has no tendency to accept a proton. Similar arguments support our neglect of the effect of Cl^- in the example dealing with the pH of ethylammonium chloride (Example 12-5). ∎

Exercise 12-8

☐ The pH of a 0.25 M solution of the sodium salt of a certain acid HX is 8.2. What is K_a for the acid? ∎

Hydrolysis of Salts   Salts are ionic compounds that often show little covalent character; they are usually strong electrolytes. However, the anion A^- of a strong electrolyte may be conjugate to a weak acid HA. In this case some of the A^- in an aqueous solution of the electrolyte reacts with H_2O to form HA and OH^-. Similarly, if the cations HB^+ of such a salt are conjugate to a weak base, some of them transfer their protons to water, forming the base B and H_3O^+. These are, of course, the very reactions illustrated in Examples 12-5 and 12-8. The term *hydrolysis* has long been applied to reactions of this general type, as emphasized in Sec. 9-5. We avoided using the term *hydrolysis* in presenting Examples 12-5 and

12-8 in order to stress that, in fact, hydrolytic reactions are in no way different from other acid-base reactions and may be handled identically. It is important to know what the term *hydrolysis* implies, since it is still commonly used, but it is equally important to recognize that it is not some new and distinct phenomenon for which new principles and techniques must be learned.

No hydrolysis occurs in solutions of salts that may be considered to have been formed by reaction of strong acids with strong bases (for example, NaCl, KNO_3, $LiClO_4$), since neither the anion nor the cation tends to react with water. Thus such solutions are neutral, with pH = 7.

Aqueous solutions of salts that might have been formed from weak bases and strong acids (for example, NH_4NO_3) are acidic because of reactions analogous to (12-17) in Example 12-5. Similarly, solutions of salts of strong bases and weak acids (for example, NaOAc) are alkaline because of reactions analogous to (12-25). If both the acid and the base from which the salt might be formed are weak, hydrolysis of each ion occurs, but the reaction that produces the weaker conjugate form predominates. Thus, when HA is a weak acid and B a weak base, solutions of the salt BH^+A^- will be basic if K_a for HA is smaller than K_b for B and will be acidic if the converse is true. In the unusual event that K_a for HA is equal to K_b for B, solutions of the salt will be neutral, with pH = 7, even though extensive hydrolysis of each ion occurs. Such is the case for ammonium acetate NH_4OAc (see Prob. 12-38 at the end of the chapter).

12-4 Buffer Solutions

Many chemical reactions are sensitive to changes in pH; that is, the yield of products and even the nature of the products may be altered appreciably if the pH changes significantly during the course of the reaction. This is especially true of biochemical reactions important to the proper metabolism and functioning of animals and plants, but it is by no means limited to them. Since many chemical processes either produce or consume protons or other acids, large pH changes normally accompany such reactions unless there is a mechanism for preventing such changes. All organisms are equipped with naturally occurring chemical systems that minimize the effects of the addition or removal of acid, and many similar systems have been devised for use with reactions that do not occur in organisms. Any chemical system that can provide a resistance to pH change is termed a *buffer*, in analogy with the common use of this word to denote anything that can absorb the effect of a shock or otherwise minimize the effects of some drastic change.

It is characteristic of chemical buffer systems that *they contain substantial amounts of both a weak acid and its corresponding weak conjugate base*. Typical components of buffer systems include HOAc and OAc^-, NH_4^+ and NH_3, $H_2PO_4^-$ and HPO_4^{2-}, and HCO_3^- and CO_3^{2-}.

Buffer solutions resist pH changes because added H^+ or OH^- does not remain as such in the solution but is removed by reaction with one of the forms of the conjugate acid-base pair that constitutes the buffer system. For example, if H^+ is added to an acetic acid–acetate buffer, most of it reacts with some of the acetate, converting it to acetic acid; conversely, if OH^- is added to the same buffer, most of it is removed by reaction with acetic acid to form acetate ion. There is always *some* pH change when acid or base is added to a buffer, but the change is far less than it would be if the solution were not buffered. This is illustrated shortly in Example 12-9.

Equilibria of acids and bases in aqueous solutions are established so quickly that for practical purposes we may assume that acid-base systems, including buffers, are at equilibrium as soon as the components are uniformly mixed. Thus if we have an aqueous solution containing an acid HX, with dissociation constant K_a, we may always assume that

$$\frac{[H^+][X^-]}{[HX]} = K_a$$

or

$$\frac{[X^-]}{[HX]} = \frac{K_a}{[H^+]} \tag{12-27}$$

Hence, if we know the pH of any aqueous solution and the acid constants of any acids present, we also know the ratio of the molar concentrations of the basic and acidic forms of each conjugate pair present. Conversely, if we know the ratio of $[X^-]$ to $[HX]$, we know $K_a/[H^+]$ and can, for example, calculate K_a for an acid if the pH is known or calculate the pH if K_a is known. These relations hold for *any* solution of an acid or a base. What makes them particularly relevant and useful for buffer solutions is that the ratio of the molar concentrations of the members of a conjugate acid-base pair in a buffer (i.e., the left-hand side of (12-27)) can almost always be calculated easily with the help of simple approximations:

1. If the two conjugate forms have been mixed directly, with no other acid or base added, this ratio is very nearly the same as the ratio of the initial concentrations of the conjugate acid and base in the solution, c_a/c_b, ignoring any possible dissociation or "hydrolysis". Thus

$$[H^+] = K_a \frac{[HX]}{[X^-]} \approx K_a \frac{c_a}{c_b} \tag{12-28}$$

as long as *both* members of the conjugate pair are present in significant concentration, as must be true in a buffer.

2. If the required significant concentration of one member of the conjugate pair is produced by reaction of the other member with strong acid or strong base, Equation (12-28) still applies, but now c_a and c_b are to be interpreted as the concentrations of the acid and base forms of the conjugate pair *after* allowance for the reaction with the added base or acid. For example, a buffer containing equal concentrations of NH_4^+ and NH_3 might be made by adding 0.5 mol HCl to a solution containing 1 mol NH_3. The added H^+ reacts almost completely with the NH_3, thus converting half of the latter to NH_4^+ and leaving the other half unchanged; this reaction occurs because it is impossible to have a high concentration of H^+ and a high concentration of a base present simultaneously. (Think about the definition of a base.) The same buffer might have been prepared by starting with a given amount of NH_4^+ and adding half as many moles of OH^-, the OH^- converting half of the NH_4^+ to NH_3.

☐ **Two Buffer Solutions** What is the pH of each of the following solutions? (*a*) A solution prepared by adding 0.50 mol acetic acid and 0.25 mol

Example 12-9

245

potassium acetate to enough water so that the final volume is 5.0 liter. (b) A solution prepared by adding 0.050 mol KOH to just 1.00 liter of 0.150 M acetic acid.

Solution (a) Equation (12-28) may be applied directly, with

$$c_a = \frac{0.50 \text{ mol}}{5.0 \text{ liter}} = 0.100 \ M$$

$$c_b = \frac{0.25 \text{ mol}}{5.0 \text{ liter}} = 0.050 \ M$$

Since K_a for acetic acid is 1.8×10^{-5},

$$[H^+] \approx 1.8 \times 10^{-5} \times \frac{0.100}{0.050}$$

$$= 3.6 \times 10^{-5} \ M$$

and pH $\approx$ 4.44. (b) Equation (12-28) may still be applied, but first we must take into account the fact that the added strong base (OH⁻) will react with the acid present (acetic acid) to produce the conjugate base (acetate ion). As long as the acid is present in significant excess this reaction may be assumed to go to completion within the precision of any calculations involving equilibria in solution. We have initially (0.150 mol liter⁻¹) (1.00 liter) = 0.150 mol acetic acid, and add to it 0.050 mol OH⁻, which will convert 0.050 mol of the acid into 0.050 mol acetate ion and leave 0.100 mol acetic acid unreacted. The concentrations after reaction are then

$$c_a = \frac{0.100 \text{ mol}}{1.00 \text{ liter}} = 0.100 \ M$$

$$c_b = \frac{0.050 \text{ mol}}{1.00 \text{ liter}} = 0.050 \ M$$

These are the same concentrations found in (a), so the pH is the same, 4.44. ∎

Exercise 12-9

□ What is the pH of each of the following solutions? (a) A solution prepared by adding 0.25 mol ammonia and 0.75 mol ammonium chloride to enough water so that the final volume is 2.5 liter. (b) A solution prepared by adding 0.20 mol HCl to just 2.0 liter 0.40 M ammonia. ∎

The reason that the approximation in (12-28) is a good one is easy to understand qualitatively. Imagine a solution containing both acetic acid and sodium acetate, each 0.10 M before any allowance for possible dissociation of the acid or reaction of acetate ion as a base. These reactions in water are:

$$HOAc + H_2O \rightleftharpoons OAc^- + H_3O^+ \qquad K_a = 1.8 \times 10^{-5} \qquad (12\text{-}29)$$

$$OAc^- + H_2O \rightleftharpoons HOAc + OH^- \qquad K_b = 5.5 \times 10^{-10} \qquad (12\text{-}30)$$

From the earlier discussion and examples (see especially Example 12-4), we know that in a 0.1 M solution a weak acid with $K_a \approx 10^{-5}$ is only about 1 percent dissociated, even in the absence of a large added quantity of its conjugate base. Thus in such a solution the initial concentration of the acid is essentially the same as its concentration after equilibrium is established. If an appreciable quantity of the conjugate base (here acetate ion) is added to the solution, it suppresses the dissociation indicated by (12-29) for reasons understandable in terms of the equilibrium law or of Le Châtelier's principle; hence the approximation in (12-28) becomes even better. In the same way, the hydrolysis of acetate ion represented by (12-30) is less than 0.01 percent in a 0.1 M acetate solution even in the absence of added acetic acid; in a buffer containing added HOAc an even smaller fraction of the OAc$^-$ reacts. Here again the approximation that the equilibrium concentration of the acetate is the same as its initial concentration is an excellent one. These same arguments apply to most buffers, since usually the concentrations of X$^-$ and HX are each appreciably greater than the concentration of OH$^-$ and H$^+$.

□ **Comparison of an Acidic Buffer with a Solution of a Strong Acid at the Same pH** Suppose that you have a large volume of each of the following solutions:

Example 12-10

1. Water brought to pH $= 5.0$ by addition of strong acid (e.g., to 10 liter of water, add 0.10 ml of 1.0 M HCl).

2. A buffer solution prepared by adding 1.0 mol HX ($K_a = 1.0 \times 10^{-5}$) and 1.0 mol NaX to several liters of water and diluting to 10 liter. Thus this buffer is 0.10 M in HX and 0.10 M in X$^-$, and

$$[H^+] = K_a \frac{[HA]}{[A^-]} \approx 1.0 \times 10^{-5} \left(\frac{0.10}{0.10} \right) = 1.0 \times 10^{-5} \, M$$

Compare the effects of adding to separate 1.0-liter portions of each of these solutions
(a) 50 ml of 1.0 M HCl (that is, 0.05 mol H$^+$)
(b) 50 ml of 1.0 M NaOH (that is, 0.05 mol OH$^-$)

Solution

1. *Unbuffered solution:* At the start the strong-acid solution at pH $= 5.0$ contains 1.0×10^{-5} mol H$^+$ liter^{-1} and no other acid.
(a) Addition of the 0.05 mol H$^+$ in 50 ml of 1.0 M HCl to the 1.0 liter of solution gives a total of 0.05 mol H$^+$ in 1.05 liter so that $[H^+] = 5 \times 10^{-2} \, M$ and pH $= 2.0 - 0.7 = 1.3$, a change of 3.7 pH units.
(b) Addition of the 0.05 mol OH$^-$ in 50 ml of 1.0 M NaOH neutralizes the minute quantity of H$^+$ initially present (10^{-5} mol) and since there is no other acid available leaves an excess of essentially all the OH$^-$. Hence $[OH^-] = \frac{0.05}{1.05} = 5 \times 10^{-2} \, M$, pOH $= 1.3$, and pH $= 12.7$, a change of 7.7 units.

2. *Buffered solution:* Addition of H$^+$ to the buffer results in the conversion of X$^-$ into HX because X$^-$ is a sufficiently strong base (HX a sufficiently weak acid) that no large concentrations of both H$^+$ and X$^-$ can coexist in water. Similarly, addition of OH$^-$ to the buffer causes conversion of HX to X$^-$ because no appreciable concentrations of both HX and OH$^-$ can coexist. As long as large quantities of both X$^-$ and HX are present in the buffer, almost all the added H$^+$ or OH$^-$ is used up in this way.

(a) When 0.05 mol H^+ is added, it changes 0.05 mol X^- into HX, so we have

$$\text{mol HX in solution} \approx 0.10 + 0.05 = 0.15$$
$$\text{mol } X^- \text{ in solution} \approx 0.10 - 0.05 = 0.05$$

The total volume has increased slightly, by 50 ml (0.05 liter). This has no significant effect on the resulting pH, however, since it is the *ratio* of the concentrations of X^- and HX that is important [see (12-28)] and in the calculation of this ratio from the corresponding numbers of moles of each member of the conjugate pair, the volume occurs in both the numerator and the denominator. Thus

$$\frac{[X^-]}{[HX]} \approx \frac{\frac{0.05}{1.05}}{\frac{0.15}{1.05}} = \frac{0.05}{0.15} = \frac{1}{3} \tag{12-31}$$

and by (12-27),

$$\frac{K_a}{[H^+]} = \frac{[X^-]}{[HX]} \approx \frac{1}{3} \tag{12-32}$$

or

$$[H^+] = 3K_a$$

and

$$pH = pK_a - \log 3 = 5.0 - 0.5 = 4.5$$

The addition of acid has indeed lowered the pH, as is reasonable, but it has lowered it only 0.5 pH units instead of 3.7, as it did for the unbuffered solution.

(b) The addition of 0.05 mol OH^- to a portion of the original buffer at pH = 5.0 may be analyzed in a parallel fashion. It results in the conversion of 0.05 mol HX into 0.05 mol X^-. Thus, after the addition we have

$$\text{mol } X^- \text{ in solution} \approx 0.10 + 0.05 = 0.15$$
$$\text{mol HX in solution} \approx 0.10 - 0.05 = 0.05$$

and the ratio of the molar concentration of X^- to that of HX is 3. Hence,

$$\frac{K_a}{[H^+]} = \frac{[X^-]}{[HX]} \approx 3 \tag{12-33}$$

Consequently pH = pK + log 3 = 5.0 + 0.5 = 5.5. The addition of 0.05 mol OH^- has, as expected, made the solution more basic, but the *change* is only 0.5 pH units as contrasted with the 7.7-unit change for the unbuffered solution.

This example shows that even a relatively dilute buffer resists pH change considerably more than does a solution of a strong acid at the same pH. ■

Exercise 12-10

☐ Suppose that you have a large volume of each of the following solutions:

1. Water brought to pH = 6.5 by addition of strong acid.
2. A buffer solution that contains 1.0 mol HY ($K_a = 3.2 \times 10^{-7}$) and 1.0 mol NaY in 10 liter of water, so that $[H^+] = 3.2 \times 10^{-7}$ and pH = 6.5.

Compare the effects of adding to separate 1.0-liter portions of each of these solutions: (a) 25 ml of 1.0 M HCl; (b) 25 ml of 1.0 M NaOH. ■

The two fundamental characteristics of a buffer are its pH and its capacity to resist pH change. A solution can be buffered at almost any pH in the range from about 3 to about 11 by appropriate choice of the conjugate pair to be used and of the ratio of their concentrations. The pair selected must be one with a pK_a near the desired pH, preferably within 0.5 units. The ratio of the concentrations of the two forms of the conjugate pair is chosen in accord with (12-28) to give the desired pH.

The capacity of a buffer to resist pH change depends solely on the amounts of the two conjugate forms present in the solution. The larger these amounts, the smaller is the change in their ratio and hence in the ratio of their concentrations, for a given addition of acid or base, and by (12-28) the smaller is the change in [H$^+$] and thus in pH.

☐ **Selection of Components of a Buffer** Suppose that you need to prepare a buffer at pH = 9.0. Select several different acid-base pairs from Table 12-2 that would be appropriate and describe the composition of one such buffer.

Example 12-11

Solution From (12-28) we see that, since we want the concentrations of both X$^-$ and HX to be appreciable, the most appropriate acid to choose must have pK within about 0.5 units of 9.0; that is, K_a should fall in the range between 3×10^{-9} and 3×10^{-10}. Three acids in Table 12-2 do fall in this range, H_3BO_3 ($K_a = 6.4 \times 10^{-10}$), NH_4^+ ($K_a = 5.5 \times 10^{-10}$), and HCN ($K_a = 2 \times 10^{-9}$). Any of them could be used.

If an ammonia–ammonium ion buffer were used, the ratio of the concentrations of the conjugate forms should be

$$\frac{[NH_3]}{[NH_4^+]} = \frac{K_a}{[H^+]} = \frac{5.5 \times 10^{-10}}{1.0 \times 10^{-9}} = \frac{5}{9}$$

Any solution in which the ratio of the concentration of ammonium ion to ammonia was 9:5 would suffice—for example, 0.18 M ammonium chloride and 0.10 M ammonia, or 0.9 M ammonium nitrate and 0.5 M ammonia. The latter would have a higher capacity for a given volume.

Boric acid and one of its salts, or even HCN and NaCN, could also be used, with different relative proportions, chosen in accord with (12-28). ■

☐ Select from Table 12-2 an acid-base pair suitable for preparing a buffer at pH = 4.3 and describe the composition of the buffer. ■

Exercise 12-11

The following additional points about buffers are worth remembering:

1. As long as the concentrations of the conjugate acid and base are much greater than [OH$^-$] or [H$^+$], the pH of the buffer can be calculated from the ratios of the numbers of moles of acid and base present in the solution, without need to calculate the concentrations explicitly. Thus, if we let V be the volume of the buffer solution, we can multiply *both* the numerator and the denominator of the ratio c_a/c_b by V without affecting the value of this ratio:

$$\frac{K_a}{[H^+]} = \frac{[X^-]}{[HX]} \approx \frac{c_b}{c_a} = \frac{Vc_b}{Vc_a} = \frac{\text{mol } X^- \text{ present}}{\text{mol HX present}} \qquad (12\text{-}34)$$

2. The pH of a buffer is to a first approximation independent of its volume, at least for volume changes that are not excessive. Thus dilution will not markedly change the pH of a buffer unless the volume changes by several orders of magnitude.

3. A buffer that may be used for absorption of *either* acid or base is most effective when $c_a = c_b$, which implies $pH = pK_a$. Under these conditions the maximum amount of either H^+ or OH^- can be absorbed for a given change in the ratio $[X^-]/[HX]$ and thus for a given pH change.

12-5 Titration of Acids and Bases

Titration has been mentioned briefly in Sec. 9-3. It is a method for determining the amount or the concentration of a substance in solution by quantitative reaction with a solution of some second substance, called the titrant, that is gradually added to the first. The solutions used are usually liquid but may be gaseous; sometimes the titrant is generated in the first solution rather than being added to it externally. The titrant is added gradually until some signal from the reaction mixture indicates that the amount of reagent added is stoichiometrically equivalent (by the equation for the reaction) to the amount of the substance initially present. The signal may be a change in color of some added compound (called an *indicator*) so chosen that it changes color at or near the desired point in the reaction. Alternatively, the signal might be the appearance of a precipitate, an electrical indication of some kind, or any other event that occurs at a point in the reaction sufficiently well defined to be suitable for quantitative purposes.

Our present concern is with the titration in aqueous solutions of acids with bases, or vice versa, but some of the results are more widely applicable, both to acid-base reactions in nonaqueous media and to other kinds of reactions as well—e.g., oxidation-reduction reactions or the formation of a slightly soluble precipitate, as in the reaction of a solution of a silver salt with a soluble chloride.

During the titration of an acid with a base, or a base with an acid, there is a continual change of the acid into its conjugate base, or vice versa, and there is a simultaneous change in the pH of the solution. Before analyzing the general problem of the variation of pH as a function of the amount of base (or acid) added during a titration, we consider the variation in the proportions of the two members of a conjugate acid-base pair at different pH values.

Distribution of Species for a Monoprotic Acid It is useful to write expressions for the fraction of the total concentration of a given conjugate acid-base species in solution that is present in each separate form, HA and A^-. This is readily done with the help of the equilibrium expression for the acid-base pair in solution in the form

$$\frac{[A^-]}{[HA]} = \frac{K_a}{[H^+]} = 10^{pH} 10^{-pK_a} = 10^{pH-pK_a} \tag{12-35}$$

We then have, for the fraction in the form HA,

$$\frac{[HA]}{[HA] + [A^-]} = \frac{[HA]}{[HA](1 + [A^-]/[HA])} = \frac{1}{1 + [A^-]/[HA]} = \frac{1}{1 + 10^{pH-pK_a}} \tag{12-36}$$

and similarly, for the fraction in the form A^-,

$$\frac{[A^-]}{[HA] + [A^-]} = \frac{1}{[HA]/[A^-] + 1} = \frac{1}{1 + 10^{pK_a-pH}} \tag{12-37}$$

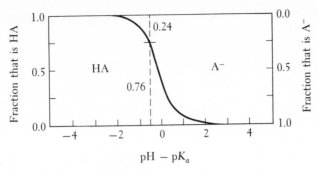

Figure 12-1 Fractions of HA and of A⁻ as a Function of pH − pK_a

The curve represents the fraction of HA + A⁻ that is HA if the left-hand scale is used and the fraction that is A⁻ if the right-hand scale is used. The curve is symmetric about its midpoint, which corresponds to pH = pK_a and to equal concentrations of HA and A⁻, so that each fraction is 0.5. The change from about 99 percent HA to about 1 percent HA takes place within a pH range of four units, and the largest part of this change, from 90 to 10 percent, occurs within about two pH units.

To obtain the fractions of HA and A⁻ at a given pH, a vertical line is drawn with the pH − pK_a value as the abscissa. The portion of the line that is in the field labeled "HA" represents the fraction that is HA; the portion in the field labeled "A⁻" represents the fraction that is A⁻. Thus the dashed line indicates a distribution of 76 percent HA and 24 percent A⁻ at a pH − pK_a of −0.50.

The sum of the fractions (12-36) and (12-37) is unity, and these fractions can thus be represented by a single curve as shown in Fig. 12-1. Comparison of (12-36) and (12-37) shows that the two expressions are symmetric about the point pH = pK_a, or pH − pK_a = 0, at which point each fraction is 0.5 because the concentrations of A⁻ and HA are equal. The fractions (12-36) and (12-37) depend only on the difference between the pH and the pK_a for the acid of interest and are completely general; that is, they apply to any monoprotic acid. The corresponding curves for different acids are therefore all identical when plotted as a function of pH − pK_a. When the fractions are expressed as a function of pH, rather than pH − pK_a, each curve is centered at the pK_a of the corresponding acid. A set of such curves is shown in Fig. 12-2. Curves of the general shape illustrated in Figs. 12-1 and 12-2 are said to be sigmoidal because of their resemblance to the letter S (Greek: *sigma*, S).

Figure 12-2 Distribution of Species for Several Conjugate Acids and Bases

The figure represents a superposition of distribution curves for several separate conjugate acid-base pairs. The curves are of the kind shown in Fig. 12-1, each moved to the proper pK_a value. The abscissa here is the actual pH, while in Fig. 12-1 it is the pH relative to pK_a. The curves at the extreme left and extreme right apply to water itself, that is, to the pairs H_3O^+/H_2O and H_2O/OH^-.

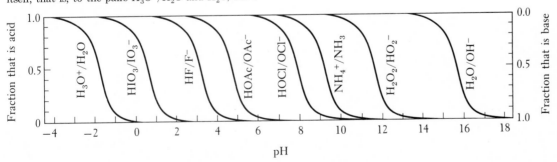

Titration Curves: The Change in pH During Titration   A knowledge of the pH at different points during a titration is valuable because it makes possible the choice of an appropriate indicator for signaling the conclusion of the titration, provides insight into the precision with which the titration can be done, and helps to explain a simple method for measuring K_a (or K_b) for a given weak acid–weak base conjugate pair. In this section, we illustrate the calculation of the pH at representative points during a titration and outline the shapes of typical titration curves, i.e., graphs of pH against the fraction of added reagent needed to reach equivalence.

For simplicity we assume in all the following calculations that there is no volume change during titration. The titrant may, for example, be assumed to be generated directly in the titration vessel with no volume change, and this is in fact sometimes possible. However, even when titrant is added from a buret, the volume change during titration is almost always sufficiently small that the resulting dilution does not change the characteristic features of the titration curves. At most it only shifts the pH by a few tenths of a unit over part of the curve but does not change the shape of the curve significantly.

First, suppose that we titrate a monoprotic acid HA with a base such as NaOH. The acid may be weak or strong; the base is assumed to be strong, because this is an experimental advantage, as will be seen. During the course of the titration, the added base changes HA into A^- (unless the acid is so strong that all HA is dissociated to begin with) and H_3O^+ changes into H_2O. We assume that the original solution contains only HA and A^-; the total of HA + A^- remains constant and represents the number of moles of acid present initially. For convenience we define the degree of titration τ as

$$\tau = \frac{\text{moles of base added}}{\text{moles of acid present initially}} \qquad (12\text{-}38)$$

At the beginning of the titration, τ is 0. At the point at which τ is exactly 1, the amount of base added is equivalent to the total of HA + A^-. This is called the *equivalence point,* and the titration should be stopped as close to this value as possible. We will calculate the pH at the start (before any base is added), at the equivalence point, and at several other points and plot the pH as a function of τ (Fig. 12-3).

Figure 12-3   Titration Curve for a Weak Acid (K_a = 10^{-5}) with a Strong Base

The initial concentration of the acid is 0.10 M; volume changes during titration are ignored. The eight points calculated in Example 12-12 lie on this curve. The titration starts at $\tau = 0$, with no base added as yet. As base is added, τ increases. At $\tau = 1$ the equivalence point has been reached, and values of τ larger than 1 correspond to an excess of base.

The rate of pH change is greatest in the region of the equivalence point; it changes by four units between $\tau = 0.99$ and $\tau = 1.01$. The broad, rather flat region of the curve between $\tau \approx 0.1$ and $\tau \approx 0.9$ is called the *buffer region* because in this region of composition, characteristic of buffer solutions, the pH changes only slowly as strong base is added. The pH at the midpoint of this region ($\tau = 0.5$) is equal to pK_a of the acid.

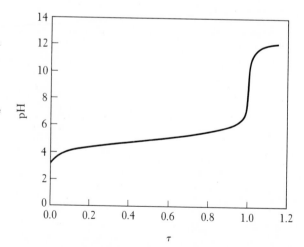

Example 12-12

☐ **Titration Curve for a Weak Acid and a Strong Base**   Consider the titration of a 0.10 M solution of a weak acid HA, $K_a = 1.0 \times 10^{-5}$, with a strong base. Assume no volume changes and calculate the pH at the following values of τ: 0.00, 0.09, 0.50, 0.91, 0.99, 1.00, 1.01, 1.10.

Solution

$\tau = 0.00$.   This corresponds to calculation of the pH of 0.10 M HA. We have, as in Example 12-4,

$$\frac{[H^+]^2}{0.10 - [H^+]} = 1.0 \times 10^{-5}$$

which leads to $[H^+] = 1.0 \times 10^{-3}\, M$ and pH = 3.0.

$\tau = 0.09$.   At this value of τ, about 9 percent of HA has been converted to A^- and about 91 percent remains as HA. The solution is thus a buffer, with

$$\frac{K_a}{[H^+]} = \frac{[A^-]}{[HA]} \approx \frac{9}{91} = 0.10$$

whence $\log[H^+] = \log 10 + \log K_a$, or $-\log[H^+] = -1.0 - \log K_a$. Thus pH $\approx -1.0 + pK_a = 4.0$.

$\tau = 0.50$.   Here $[A^-] \approx [HA]$, so pH $\approx pK_a = 5.0$.

$\tau = 0.91$.   At this point in the titration,

$$\frac{[A^-]}{[HA]} \approx \frac{91}{9} = 10$$

so $[H^+] \approx 0.10\, K_a$ and pH $\approx 1.0 + pK_a = 6.0$.

$\tau = 0.99$.   Here it is approximately true that

$$\frac{[A^-]}{[HA]} = \frac{99}{1} \simeq 10^2$$

so pH $\approx pK_a + 2.0 = 7.0$.

$\tau = 1.00$.   This is the equivalence point, and a little reflection about the nature of the species present at this juncture simplifies the calculation of the pH. In the titration of 0.10 M HA with NaOH, the solution at the equivalence point is indistinguishable from a solution of NaA at an equivalent concentration, 0.10 M, if we ignore volume changes. In such a solution the very weak base A^- reacts with water to form equal numbers of OH^- ions and HA molecules:

$$A^- + H_2O \rightleftharpoons HA + OH^- \qquad (12\text{-}39)$$

$$0.10 - x \qquad\qquad x \qquad x$$

If the equilibrium represented by (12-39) is approached from the opposite direction by mixing OH^- and HA in equivalent amounts, the position of equilibrium is necessarily identical. Thus the reverse of (12-39), the titration reaction, does not go quite to completion even at the equivalence point—a little OH^- and HA remain in the solution—precisely because A^- is so weak a base that it hydrolyzes perceptibly in water.

Calculation of the pH at the equivalence point is now simply a matter of calculating the pH of a solution of a weak base A^- at the appropriate concentration and with $K_b = K_w/K_a = 1.0 \times 10^{-14}/(1.0 \times 10^{-5}) = 1.0 \times 10^{-9}$. The procedure is identical to that illustrated in Example 12-7. From (12-39) we have $[OH^-] = [HA] = x$ and $[A^-] = 0.10 - x$. Hence,

$$\frac{x^2}{0.10 - x} = 1.0 \times 10^{-9}$$

which leads to $x^2 \approx 1.0 \times 10^{-10}$ and $x \approx 1.0 \times 10^{-5}$. Thus $pOH \approx 5.0$ and $pH \approx \underline{9.0.}$

$\tau = 1.01.$ The solution now contains $0.10\,M$ A^- and a 1 percent excess of OH^-, which means that $[OH^-] \approx 1.0 \times 10^{-3}$. Thus $pOH \approx 3.0$ and $pH \approx \underline{11.0,}$

$\tau = 1.10.$ Here there is a 10 percent excess of OH^- in a $0.10\,M$ solution of A^-. Thus $[OH^-] \approx 1.0 \times 10^{-2}$, $pOH \approx 2.0$, and $pH \approx \underline{12.0.}$ ■

Exercise 12-12

□ Consider the titration of a $0.050\,M$ solution of a weak acid HA, $K_a = 3.2 \times 10^{-7}$, with a strong base. Assume no volume changes and calculate the pH at the following values of τ, the ratio of (moles base added) to (moles of acid originally present): 0.00, 0.09, 0.50, 0.91, 0.99, 1.00, 1.01, 1.10. ■

The curve found in Example 12-12 is plotted in Fig. 12-3. Titration curves for acids of different strengths, as well as for bases, can be derived in a similar manner. The precision with which the equivalence point can be found is directly related to the slope of the curve at the equivalence point ($\tau = 1.0$), that is, to the magnitude of the change of pH (with volume of titrant added) near this point.

It is worth noting that when one speaks of "neutralizing" an acid with a base, it usually means adding a quantity of the base just equivalent to that of the acid. The pH of the resulting solution will *not* be neutral—that is, the pH will *not* be 7—unless both the acid and the base are strong (or unless they are both weak and it happens that K_a for the acid is just equal to K_b for the base).

12-6 Polyprotic Acids

A number of common and important acids have more than one dissociable proton. These include carbonic acid, H_2CO_3, phosphoric acid, H_3PO_4, sulfuric acid, H_2SO_4, and oxalic acid, $(COOH)_2$. Some highly charged cations—for example, $Al(H_2O)_6^{3+}$ and $Fe(H_2O)_6^{3+}$—also behave as polyprotic acids in water. The successive reactions and corresponding equilibrium expressions for the three-stage dissociation of phosphoric acid were given in (12-4) to (12-6). The acid constants for the successive dissociation steps of a number of polyprotic acids are listed in Tables 12-2 and D-3.

No new principles need be learned to understand the equilibria and reactions of polyprotic acids in water. The equilibria seem more complicated and formidable than for a monoprotic acid because the successive stages of dissociation lead to more species (for example, H_3A, H_2A^-, HA^{2-}, A^{3-}) and, of course, all equilibria must be satisfied simultaneously. In fact, however, the consideration of solutions of many polyprotic acids can be greatly simplified because at most only two of the successive species can be present in appreciable concentration in any solution. Dealing with these is then not much more difficult than analyzing a situation involving a monoprotic acid.

We first illustrate the distribution of species as a function of pH for an important diprotic acid, H_2CO_3, and discuss pH calculations for polyprotic acids. Then we examine a titration curve for a common polyprotic acid, H_3PO_4.

Distribution of Carbonate Species as a Function of pH The relative concentrations of H_2CO_3, HCO_3^-, and CO_3^{2-} as the pH is changed in a solution containing any of these species can be calculated in a manner similar to that used for the monoprotic acid in Sec. 12-5. The results are plotted in Fig. 12-4, where each of the three species is represented by a separate curve. At no pH are more than two of the forms present in appreciable concentrations, because K_1 and K_2 differ so greatly (by about 10^4). At low pH, almost all the carbonate is present as H_2CO_3; at high pH (above about 12), almost all is present as CO_3^{2-}; and at pH values between 8 and 9, most is present as HCO_3^-. When the pH is not far from pK_1 the two important species present are the undissociated acid, H_2CO_3, and its conjugate base, HCO_3^-. When the pH is not far from pK_2, the two predominant species present are the acid and base corresponding to the second dissociation, HCO_3^- and CO_3^{2-}. Curves similar to those in Fig. 12-4 can be calculated for any polyprotic acid.

Figure 12-4 Fractions of H_2CO_3, HCO_3^-, and CO_3^{2-} as a Function of pH
Each species is represented by a distinct curve, as indicated, in contrast to the graphs for monoprotic acids in Figs. 12-1 and 12-2. The sigmoidal curve on the left represents the fraction of H_2CO_3, and that on the right the fraction of CO_3^{2-}. The central bell-shaped curve represents the fraction of HCO_3^-, which rises to nearly 1.0 in the pH range about 8 to 9 and diminishes at higher and at lower pH. The crossing points correspond respectively to pK_1 (on the left) when $[H_2CO_3] = [HCO_3^-]$ and to pK_2 (on the right) when $[HCO_3^-] = [CO_3^{2-}]$; the maximum of $[HCO_3^-]$ is at $(pK_1 + pK_2)/2$.

The graph shows that neither HCO_3^- nor CO_3^{2-} can coexist with H^+ at appreciable concentrations; even if $[H^+]$ is as low as 10^{-4} (pH = 4), essentially all carbonate exists as H_2CO_3. Similarly, neither HCO_3^- nor H_2CO_3 can coexist with OH^- in significant concentrations; when $[OH^-]$ is 10^{-2} or higher, essentially all carbonate species are in the form of CO_3^{2-}.

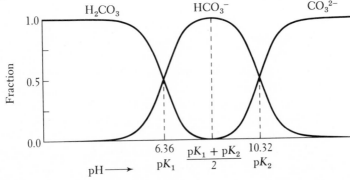

pH Calculations It is relatively straightforward to calculate the pH of any solution containing species related to a polyprotic acid with dissociation constants that differ by 10^3 or more. We illustrate this for an acid H_2A.

1. A solution containing appreciable quantities only of H_2A itself is almost entirely governed by the reaction $H_2A \rightleftharpoons HA^- + H^+$, with the equilibrium determined by K_1, unless the acid is exceptionally strong. Even for 0.1 M sulfuric acid, for which K_1 is about 10^3, the contribution of the further dissociation of HSO_4^- is only about 10 percent; for acids with smaller values of K_1 (or for more concentrated sulfuric acid solutions), the contribution of the second dissociation is negligible. The calculation can then be handled exactly as for a monoprotic acid.

2. A solution containing only the salt Na_2A (or any salt of A^{2-} with a non-hydrolyzing cation) has a pH dominated by the reaction of A^{2-} as a base with water:

$$A^{2-} + H_2O \rightleftharpoons HA^- + OH^- \tag{12-40}$$

with $K_b = K_w/K_2$. Further reaction of HA^- with water contributes negligibly.

3. Any solution that contains appreciable concentrations of *two* of the neighboring forms (H_2A and HA^-, or HA^- and A^{2-}) will be a buffer if both the acidic form and its conjugate base are weak.

4. The only remaining possibility is a solution prepared from or corresponding to a salt of HA^-, for example, $NaHA$. The calculation for such a solution is somewhat lengthy, but the result is simple: $[H^+] \approx \sqrt{K_1K_2}$, or pH $\approx$ 0.5 $(pK_1 + pK_2)$, a result that is independent of the concentration of the solution provided only that $[HA^-] \gg K_1$ and $K_2 [HA^-] \gg K_w$.

This analysis of solutions prepared from a diprotic acid H_2A can be extended to solutions of acids with three or more protons, provided that the consecutive acid constants differ by three or more powers of 10. For example, the pH of a NaH_2PO_4 solution is 0.5 $(pK_1 + pK_2)$, where K_1 and K_2 are the first two acid constants of H_3PO_4, and that of a Na_2HPO_4 solution is $0.5(pK_2 + pK_3)$. These pH values are also those of the two equivalence points in a titration of H_3PO_4 or of PO_4^{3-}.

Example 12-13

□ **Predominant Species at Different pH Values** Use the K_1 and K_2 values for H_2S to predict the predominant species in a sulfide solution at pH = 5, 7, and 11.

Solution From Table 12-2, for H_2S $K_1 = 9.1 \times 10^{-8}$ and $K_2 = 1.2 \times 10^{-15}$; thus $pK_1 = 7.04$ and $pK_2 = 14.9$. At pH = 5, two pH units more acidic than pK_1, essentially everything will be in the most acidic form (compare Fig. 12-4), which for this system is H_2S. More specifically, $[HS^-]/[H_2S] = K_1/[H^+] = 9 \times 10^{-3}$, so there is less than 1 percent as much HS^- as H_2S. On the other hand, at pH = 7, both HS^- and H_2S are present in significant amounts, their concentrations being about equal since pH $\approx pK_1$. At pH = 11, almost everything is in the form of HS^-; this pH is about equal to 0.5 $(pK_1 + pK_2)$ and thus at this pH the fraction present as HS^- reaches its maximum (as shown for HCO_3^- in Fig. 12-4). The concentrations of H_2S and S^{2-} at pH = 11 are only about 10^{-4} that of HS^-, since pK_1 and pK_2 each differ by almost 4 units from 11. ■

☐ Use the values for K_1 and K_2 for phosphorous acid (Table D-3) to predict the predominant species present if a solution of phosphorous acid is brought successively to pH = 0, 2, 4, and 8. ■

Example 12-14

☐ **pH of a Carbonate Solution** (a) What is the pH of a solution prepared by dissolving 0.20 mol Na_2CO_3 and 0.10 mol $NaHCO_3$ in 2.0 liter water? (b) What will be the pH after 0.12 mol HCl is added to this solution?

Solution (a) In this buffer system, the base is CO_3^{2-} and the acid is HCO_3^-, with $K_a = 4.8 \times 10^{-11}$ (Table 12-2). Initially

$$c_b = [CO_3^{2-}] = \frac{0.20 \text{ mol}}{2.0 \text{ liter}} = 0.10 \; M$$

and

$$c_a = [HCO_3^-] = \frac{0.10 \text{ mol}}{2.0 \text{ liter}} = 0.05 \; M$$

Thus

$$[H^+] \approx K_a \frac{c_a}{c_b} = 4.8 \times 10^{-11} \frac{0.05}{0.10} = 2.4 \times 10^{-11} \; M \qquad \text{and } \underline{pH = 10.6}$$

(b) When HCl, a strong acid, is added to the solution, the added H^+ reacts with the strongest base present, CO_3^{2-}, converting it into its conjugate acid, HCO_3^-. The addition of 0.12 mol H^+ converts 0.12 mol CO_3^{2-} into HCO_3^-, so now

$$c_b = [CO_3^{2-}] = \frac{0.20 - 0.12}{2.0} = 0.04 \; M$$

and

$$c_a = [HCO_3^-] = \frac{0.10 + 0.12}{2.0} = 0.11 \; M$$

Thus

$$[H^+] \approx 4.8 \times 10^{-11} \times \frac{0.11}{0.04} = 1.3 \times 10^{-10} M \qquad \text{and } \underline{pH = 9.9} \; ■$$

☐ (a) What is the pH of a solution prepared by dissolving 0.50 mol $KHSO_3$ and 0.20 mol Li_2SO_3 in 2.5 liter of water? (b) What will be the pH after 0.20 mol NaOH is added to this solution? ■

Buffer solutions involving polyprotic acids and their salts are common, particularly in biological systems. The following example illustrates the preparation of a phosphate buffer and the calculation of its pH.

Example 12-15

☐ **pH of a Phosphate Solution** What is the pH of a solution made by adding 0.30 mol NaOH, 0.25 mol Na_2HPO_4, and 0.25 mol H_3PO_4 to water and diluting to 1.000 liter?

Solution Remembering that at most two neighboring members of the series OH^-, PO_4^{3-}, HPO_4^{2-}, $H_2PO_4^-$, H_3PO_4, H^+ can coexist in significant concentrations, we begin by mentally reacting members at extreme ends of this sequence. Thus we use up as much of the OH^- as is needed to convert all H_3PO_4 to $H_2PO_4^-$. The resulting concentrations are $[H_3PO_4] \approx 0$, $[OH^-] = 0.05 \; M$, and $[H_2PO_4^-] = 0.25 \; M$. ($[Na^+]$ remains at

$0.30 + 2(0.25) = 0.80$ M throughout.) Next the remaining OH^- is imagined to react with the $H_2PO_4^-$, the result being $[OH^-] \approx 0$, $[H_2PO_4^-] = 0.20$ M, and $[HPO_4^{2-}] = 0.05 + 0.25 = 0.30$ M. These, then, are the major components. They lead to the hydrogen-ion concentration

$$[H^+] = K_2 \frac{[H_2PO_4^-]}{[HPO_4^{2-}]} = K_2 \left(\frac{0.20}{0.30}\right) = 0.67 \times 6.2 \times 10^{-8}$$
$$= 4.1 \times 10^{-8} \, M \quad \text{and pH} = \underline{7.4}$$

The other concentrations in the solution are

$$[OH^-] = \frac{1.0 \times 10^{-14}}{4.1 \times 10^{-8}} = 2.4 \times 10^{-7} \, M$$

$$[H_3PO_4] = \frac{[H^+][H_2PO_4^-]}{K_1} = \frac{0.20 \times 4.1 \times 10^{-8}}{7.1 \times 10^{-3}} = 1.2 \times 10^{-6} \, M$$

$$[PO_4^{3-}] = K_3 \frac{[HPO_4^{2-}]}{[H^+]} = \frac{0.30 \times 4.4 \times 10^{-13}}{4.1 \times 10^{-8}} = 3.2 \times 10^{-6} \, M$$

All these concentrations are seen to be small compared to those of $H_2PO_4^-$ and HPO_4^{2-}. ■

Exercise 12-15

☐ What is the pH of a solution made by mixing 0.10 mol NaOH, 0.15 mol Na_2HPO_4, and 0.15 mol NaH_2PO_4 and diluting to 1 liter? ■

Titration of Phosphoric Acid We now consider the change in pH during the titration of a typical polyprotic acid with a strong base. Figure 12-5 shows a titration curve—the change in pH as a function of the ratio of the volume of base added to the volume of acid originally present—for 0.10 M H_3PO_4 titrated with 0.10 M NaOH.

Suppose that the original volume of 0.10 M H_3PO_4 is 100 ml. When the volume of 0.10 M NaOH added has reached 100 ml, practically all the phosphate present exists as $H_2PO_4^-$. The titration up to this point is therefore described by the equation

$$H_3PO_4 + OH^- \longrightarrow H_2PO_4^- + H_2O \tag{12-41}$$

As more base is added the chief reaction occurring is

$$H_2PO_4^- + OH^- \longrightarrow HPO_4^{2-} + H_2O \tag{12-42}$$

This reaction is essentially complete when the volume of base added has reached 200 ml, because practically all the phosphate present exists as HPO_4^{2-} at this point. The titration curve shows steep portions at both situations just described, and titrations to both points are feasible. However, as even more base is added no further steep portion appears on the titration curve. When the volume of base added has reached 300 ml only about two-thirds of the phosphate is in the form of PO_4^{3-}, one-third still being present as HPO_4^{2-}. The reason is that PO_4^{3-} is a strong enough base to compete effectively with OH^- for protons. In other words, although the equilibria for reactions (12-41) and (12-42) lie far to the right, this is not true for the reaction

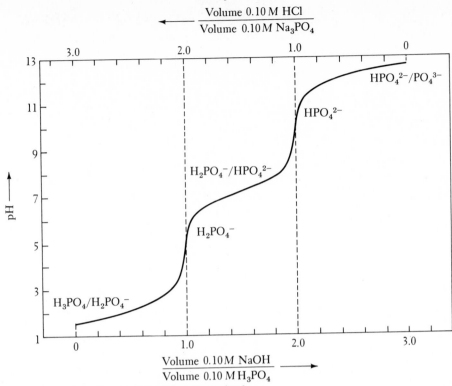

Figure 12-5 Titration Curve for Phosphoric Acid
The curve applies to 0.10 M solutions and describes the pH change during a titration of H_3PO_4 with strong base (bottom scale) as well as that during a titration of PO_4^{3-} with strong acid (top scale). Two equivalence points are shown, in steep portions of the pH curve, the principal species being those indicated.

$$HPO_4^{2-} + OH^- \rightleftharpoons PO_4^{3-} + H_2O \tag{12-43}$$

A titration of one or two of the hydrogens of H_3PO_4 is possible, but not one that involves all three hydrogens.

Titration curves for other polyprotic acids can be discussed in a similar fashion, provided that the successive acid constants differ by several powers of 10. The same considerations apply equally to mixtures of two or three acids, as long as the acid constants are suitably different. When the ratio of acid constants becomes smaller than 10^3, titration of separate acids in the same solution (or of different acid protons of the same acid) becomes difficult.

Hydrated Species Hydrated metal ions (Sec. 9-5) sometimes act as acids in water, and their conjugate bases—metal ions associated with one or more hydroxide ions, as well as with some water molecules perhaps—often show basic properties. The higher the charge and the smaller the size of the ion, the greater its acid strength when hydrated. For example, $Al(H_2O)_6^{3+}$, with $K_1 \approx 10^{-5}$, is about as strong an acid as acetic acid. This is primarily a consequence of the high electric field of the cation, which polarizes the attached water molecules sufficiently to make dissociation of protons from them far easier than from ordinary water molecules. The hydrated Al^{3+} ion is, in fact, a polyprotic acid, losing one proton from each water molecule successively to form a series of products:[3]

$$
\begin{array}{cc}
\text{Acid} & \text{Base} \\
[\text{Al}(\text{H}_2\text{O})_6]^{3+} \rightleftharpoons & [\text{AlOH}(\text{H}_2\text{O})_5]^{2+} + \text{H}^+ \\
[\text{AlOH}(\text{H}_2\text{O})_5]^{2+} \rightleftharpoons & [\text{Al}(\text{OH})_2(\text{H}_2\text{O})_4]^+ + \text{H}^+ \\
[\text{Al}(\text{OH})_2(\text{H}_2\text{O})_4]^+ \rightleftharpoons & \text{Al}(\text{OH})_3(\text{H}_2\text{O})_3 + \text{H}^+ \\
\text{Al}(\text{OH})_3(\text{H}_2\text{O})_3 \rightleftharpoons & [\text{Al}(\text{OH})_4(\text{H}_2\text{O})_2]^- + \text{H}^+
\end{array}
\qquad (12\text{-}44)
$$

Note that $\text{Al}(\text{OH})_3(\text{H}_2\text{O})_3$ is the only uncharged species in the series. It is rather insoluble, as is often true of uncharged coordination species containing OH groups. When acid is added to $\text{Al}(\text{OH})_3(\text{H}_2\text{O})_3$, the aluminum hydroxide dissolves with the formation of $\text{Al}(\text{OH})_2(\text{H}_2\text{O})_4{}^+$ and the other positively charged species in (12-44); one, two, or all three of the OH groups in the neutral species are converted to H_2O. It also dissolves when base is added, forming $\text{Al}(\text{OH})_4(\text{H}_2\text{O})_2{}^-$ and perhaps other negatively charged species. This is the amphoteric behavior of aluminum hydroxide mentioned in Sec. 9-4.

Many other metal ions form series of hydrates and hydroxides similar to those of aluminum. Some of the uncharged hydroxides are amphoteric, while others are not. For example, the species $\text{Ca}(\text{OH})_2(\text{H}_2\text{O})_4$ is not amphoteric and forms a series with $[\text{CaOH}(\text{H}_2\text{O})_5]^+$ and $[\text{Ca}(\text{H}_2\text{O})_6]^{2+}$, for which the usual symbols are $\text{Ca}(\text{OH})_2$, $[\text{CaOH}]^+$, and Ca^{2+}. Again the hydrated $\text{Ca}(\text{OH})_2$ is only slightly soluble in water. It readily dissolves upon addition of acid, but not of base, because species like $[\text{Ca}(\text{OH})_3(\text{H}_2\text{O})_3]^-$ do not appear to form. The tendency of $\text{Ca}(\text{H}_2\text{O})_6{}^{2+}$ to lose protons is so much weaker than that of $\text{Al}(\text{H}_2\text{O})_6{}^{3+}$ that Ca^{2+} shows essentially no acidic reaction in water; in other words, $\text{CaOH}(\text{H}_2\text{O})_5{}^+$ is a relatively strong base, far stronger than $\text{AlOH}(\text{H}_2\text{O})_5{}^{2+}$.

---

[3] As mentioned in Chap. 9, the water molecules are frequently omitted from the formulas for hydrated species; thus $\text{Al}(\text{H}_2\text{O})_6{}^{3+}$ is abbreviated as Al^{3+} and $\text{Al}(\text{OH})_4(\text{H}_2\text{O})_2{}^-$ is written as $\text{Al}(\text{OH})_4{}^-$. In the present context of conjugate acid-base pairs, the water molecules are of course crucial, being the source of the protons.

Summary

A measure of the strength of an acid or a base is its degree of dissociation. The strength of an acid HA can be more precisely defined in terms of its acid constant, $K_a = [\text{H}^+][\text{A}^-]/[\text{HA}]$, and the strength of a base B in terms of its base constant, $K_b = [\text{HB}^+][\text{OH}^-]/[\text{B}]$. The product of the acid constant for any acid HA and the base constant for the corresponding conjugate base, A^-, is the ion product of water, K_w. No substantial quantity of an acid stronger than H_3O^+ or a base stronger than OH^- can remain unreacted in dilute aqueous solution. This phenomenon, known as the leveling effect, masks any differences in strengths of strong acids (or of strong bases) in water.

Estimates of the degree of dissociation of an acid or base in a given solution can be derived from measurements of a colligative property of the solution or of its pH.

The pH of dilute solutions of strong or weak acids or bases and their salts can usually be calculated by simple approximations. A solution of a salt of a strong acid and a weak base is acidic because of hydrolysis, the reaction in which undissociated base and H_3O^+ are formed. Similarly, a solution of a salt of a strong base and a weak acid is basic, the result of hydrolysis in which undissociated acid and OH^- are produced. If both the acid and the base from which the salt might be formed are weak, hydrolysis of each ion occurs, and the solution is basic if K_a is smaller than K_b and acidic if the converse is true.

The pH of a solution can be stabilized against the influence of added acids or bases by buffering. A buffer solution contains substantial amounts of both a weak acid and its corresponding weak conjugate base, for example, HOAc and OAc$^-$. The concentration of H$^+$ in a buffered solution is given by the relation $[H^+] \approx K_a c_a / c_b$, where K_a is the dissociation constant of the acid and c_a and c_b are, respectively, the concentrations of the conjugate acid and base in the solution after any reactions have reached equilibrium. Effective buffering requires that the amounts of the two conjugate forms be large compared to the amounts of H$^+$ and OH$^-$ that are added to or generated within the solution.

The amount or concentration of a substance in solution can be determined quantitatively by titration. Graphs of pH against the fraction of added reagent, which are known as titration curves, have a characteristic sigmoidal shape for the titration of a monoprotic acid or base.

The equilibrium constants for the successive stages of dissociation in water of many polyprotic inorganic acids differ greatly in magnitude. As a result, no more than two of the species produced in the dissociation are present in appreciable concentration in solution at any pH. Calculations of the pH for polyprotic acids therefore parallel those for monoprotic acids.

Hydrated metal ions sometimes act as acids in water. The greater the charge and the smaller the size of the central ion, the greater the acid strength of the hydrated species.

Terms and Concepts

Problems and Questions

12-1 Freezing Point of Solutions Arrange the following solutions in order of decreasing freezing point, placing that with the highest freezing point first: (a) 0.20 M KOH; (b) 0.10 M sugar; (c) 0.25 M NH$_3$; (d) 0.04 M BaCl$_2$; (e) 0.04 M HNO$_3$.

12-2 Acid Constants (a) The acid HX is 4.5 percent dissociated in a 0.10 M solution. What is its acid constant? (b) The degree of dissociation of the acid HY in a 0.50 M solution is 0.15. What is its acid constant?

12-3 Very Weak Acid The pH of a 0.10 M solution of the very weak acid HA is 5.0. (a) What is the acid constant for HA? (b) What is the degree of dissociation of HA in its 0.10 M solution? (c) What would be the concentration of a solution of HA in which the degree of dissociation was twice that in a 0.10 M solution?

12-4 Weak Base A certain weak base B has a base constant of 6×10^{-4} M, corresponding to the reaction B + H$_2$O $\rightleftharpoons$ BH$^+$ + OH$^-$. Calculate the pH of a 0.20 M aqueous solution of B.

12-5 Base Constant A 0.10 M solution of a certain base has a pH of 11.50. What is the base constant K_b of this base?

12-6 Degree of Dissociation of a Base Suppose that a 0.40 M solution of a weak base has the same hydroxide-ion concentration as a 0.020 M solution of sodium hydroxide. What is the degree of dissociation of the weak base in its solution? Explain.

12-7 pH of Solutions Calculate the pH of the following solutions: (a) 0.100 M HF; (b) 0.050 M HCl; (c) 0.025 M NaOH; (d) 0.200 M NaF; (e) 0.100 M NH_2OH (hydroxylamine, $K_b = 1.2 \times 10^{-8} M$).

12-8 Concentration and pH A certain solution of acetic acid ($K_a = 1.8 \times 10^{-5} M$) has a pH of 3.30. What is its molarity?

12-9 Solution of a Strong Base A solution of a certain strong base has pH = a, where a may be assumed to be some number between 10.0 and 14.0, although its exact value is unimportant. If the solution is diluted with water to five times its original volume, what will be the pH of the new solution formed?

12-10 Solution of an Acid The pH of a certain solution of an acid HX is 3.50. The dissociation constant of the acid is 2.0×10^{-6}. (a) Is HX a strong acid or a weak acid? (b) What is the concentration of X^- in the solution? (c) What is the concentration of HX in the solution? (d) What is the degree of dissociation of the acid?

12-11 Relative Strength of Bases Arrange the following bases in the order of their tendency to combine with protons, putting that with the greatest proton affinity first: NH_3, Cl^-, OH^-, acetate ion, H_2O.

12-12 Dissociation of HCN What is the percentage dissociation of HCN in a 0.50 M solution ($K_a = 2 \times 10^{-9} M$)? What will be the percentage dissociation if 90 ml of water is added to 10 ml of the 0.50 M solution?

12-13 Dissociation of Acetic Acid Although acetic acid is normally regarded as a weak acid, it is about 34 percent dissociated in a $10^{-4} M$ solution at 25°C. It is less than 1 percent dissociated in 1 M solution. Discuss this variation in degree of dissociation with dilution in terms of Le Châtelier's principle, and explain how it is consistent with the supposed constancy of equilibrium constants.

12-14 Ammonia Solution A 10.0-ml sample of aqueous ammonia weighs 9.60 g and has a pH of 12.0. What is the percentage by weight of ammonia in this solution? ($K_b = 1.8 \times 10^{-5} M$)

12-15 Acid Constant of a Weak Acid Suppose that a 0.10 M aqueous solution of a monoprotic acid HX has just 11 times the conductivity of a 0.0010 M aqueous solution of HX. What is the approximate dissociation constant of HX? (*Hint:* In thinking about this problem, consider what the ratio of the conductivities would be if HX were a strong acid and if HX were extremely weak, as limiting cases.)

12-16 Properties of an Unknown Acid A 0.100 M solution of a certain compound C in water is acidic, conducts electricity, and freezes at -0.25°C. When 20.0 ml of the solution is titrated with base, 16.0 ml 0.250 M NaOH is required to reach an end point. Calculate the number of acidic hydrogen atoms on each molecule of C in reactions with NaOH, and calculate the degree of dissociation of the acid in 0.100 M aqueous solution, assuming that one molecule of C dissociates to form only two ions in this solution.

12-17 pH of a Nitrite Solution The dissociation constant of nitrous acid, HNO_2, is about 5×10^{-4}. What would be the pH of a 0.10 M solution of KNO_2?

12-18 Aqueous Solutions at 0°C At 0°C, $K_w = 1.14 \times 10^{-15}$. Is a solution at 0°C with pH = 7.0 neutral, acidic, or basic? What is the ratio of $[H^+]$ to $[OH^-]$?

12-19 Ammonium Ion Discuss the justification for this statement: "Although one does not normally regard NH_4^+ as an acid, it is actually only slightly weaker as an acid than hydrocyanic acid, HCN, in aqueous solution."

12-20 Hydrolysis State clearly what information you would need in order to predict whether a given salt BX would give a neutral solution, an acidic solution, or a basic solution when dissolved in water.

12-21 Solution of NaCN Calculate the concentrations of Na^+, CN^-, HCN, H^+, and OH^- in a 0.20 M solution of sodium cyanide. Calculate also the degree of hydrolysis of CN^- in this solution. (K_a for HCN = $2 \times 10^{-9} M$.)

12-22 Sodium Hypochlorite Solution Sodium hypochlorite, $NaClO$, is a salt of hypochlorous acid, $HClO$. What is the pH of a $0.100\,M$ $NaClO$ solution?

12-23 Hydrolysis In a $0.10\,M$ solution of the salt KA the anion A^- is 8.0 percent hydrolyzed to HA. What is the acid constant K_a of the acid HA?

12-24 Solution Containing Acid and Salt What is the pH of a solution that is $0.1\,M$ in HNO_2 and $0.2\,M$ in $Ca(NO_2)_2$?

12-25 Buffer Solution Find the approximate pH that results when 0.12 mol of solid $NaOAc$ is added to 1 liter of $0.10\,M$ $HOAc$.

12-26 Comparison of Three Solutions Three flasks, labeled A, B, and C, contained aqueous solutions of the same pH. It was known that one of the solutions was $1.0 \times 10^{-3}\,M$ in nitric acid, one was $6 \times 10^{-3}\,M$ in formic acid, and one was $4 \times 10^{-2}\,M$ in the salt formed by the weak organic base aniline with hydrochloric acid ($C_6H_5NH_3Cl$). (Formic acid is monoprotic.) (a) Describe a procedure for identifying the solutions. (b) Compare qualitatively (on the basis of the preceding information) the strengths of nitric and formic acids with each other and with the acid strength of the anilinium ion, $C_6H_5NH_3{}^+$. (c) Show how the information given may be used to derive values for K_a for formic acid and K_b for aniline. Derive these values.

12-27 Solution of Novocain Novocain, the commonly used local anaesthetic, is a weak base with $K_b = 7 \times 10^{-6}\,M$. (a) If one had a $0.020\,M$ solution of Novocain in water, what would be the approximate concentration of OH^- and the pH? (b) Suppose that you wanted to determine the concentration of Novocain in a solution that is about $0.020\,M$ by titration with $0.020\,M$ HCl. Calculate the expected pH at the equivalence point.

12-28 Solution of Potassium Benzoate Benzoic acid is a weak acid with K_a about $6 \times 10^{-5}\,M$. Explain with the aid of Le Châtelier's principle the effects of the following changes in conditions on the degree of hydrolysis of potassium benzoate in a dilute solution. Assume that activity effects are negligible. (a) Addition of KNO_3 to the solution; ignore any volume change. (b) Addition of $NaOH$ to the solution; ignore any volume change. (c) Dilution of the solution with water.

12-29 Solution of Sodium Pivalate A $0.100\,M$ solution of pivalic acid has a pH of 3.0. What would be the pH of $0.10\,M$ sodium pivalate?

12-30 Acetic Acid–Acetate Buffer What volume of $1.0\,M$ KOH must be added to 10 ml of $1.0\,M$ $HOAc$ to give a pH of 5.0? (K_a for $HOAc = 1.8 \times 10^{-5}\,M$.)

12-31 Dimethylamine and Its Solutions Dimethylamine, $(CH_3)_2NH$, is very soluble in water, forming a weakly basic solution similar to that formed by ammonia. (a) Write a reaction, analogous to that of ammonia with water, that can account for the basicity of a solution of dimethylamine in water. If the OH^- concentration of a $0.50\,M$ solution of this compound is about $5 \times 10^{-3}\,M$, what is K_b for dimethylamine? (b) Suppose that 40.0 ml of a $0.40\,M$ solution of dimethylamine in water is neutralized by 32.0 ml of a solution of a strong acid. What can you conclude about the molarity of the acid if it is monoprotic? What would be the molarity of the acid if it had two acidic hydrogen atoms per molecule in its reaction with dimethylamine? (c) [Assume answer from part (a).] What will be the pH of a $1.0\,M$ solution of dimethylamine that has been 90 percent neutralized with a strong acid?

12-32 Formic Acid and Its Solutions The dissociation constant of formic acid, which is monoprotic, is $1.7 \times 10^{-4}\,M$. (a) What is the approximate degree of dissociation of formic acid in a $0.10\,M$ solution? (b) What is the approximate pH of $0.10\,M$ potassium formate? (c) What is the approximate pH of a solution prepared by adding 10 ml of $1.0\,M$ HCl to 30 ml of $1.0\,M$ sodium formate?

12-33 Buffer Capacity: Buffers of Different Concentration Suppose you have a buffer at pH 5.0 that is $1.00\,M$ in HX and $1.00\,M$ in X^-, where HX is a weak acid with $K_a = 1.0 \times 10^{-5}\,M$. Calculate the effects of adding 50 ml of $1.0\,M$ HCl to a 1-liter portion of this buffer and 50 ml of $1.0\,M$ $NaOH$ to a separate 1-liter portion. Compare the results with those in Example 12-10 for the similar buffer with $0.1\,M$ concentrations.

12-34 Preparation of a Buffer Suppose that you want to prepare a buffer solution with pH $= 4.0$ and have available 2 liter of $1.0\,M$ $NaOH$ and 2 liter of $1.0\,M$ phenylacetic acid, for which K_a is $5 \times 10^{-5}\,M$. Describe and explain how you could prepare at least 2 liter of the desired buffer, using only the solutions given. By how much would the pH of 1 liter of your buffer change if $\frac{1}{3}$ liter of 1.0 M KOH were added to it?

12-35 Ammonia–Ammonium Ion Buffer (a) Describe how you would prepare 1 liter of a buffer at pH = 9.0, using 1.0 mol NH_3 and as much of any strong acid as needed. (b) A 40-ml sample of a 0.10 M solution of nitric acid is added to 20 ml of 0.30 M aqueous ammonia. What is the pH of the resulting solution?

12-36 Arsenate Buffer Arsenic acid, H_3AsO_4, is very similar to phosphoric acid. It is a triprotic acid with $K_1 \approx 6 \times 10^{-3}\,M$, $K_2 \approx 2 \times 10^{-7}\,M$, and $K_3 \approx 3 \times 10^{-12}\,M$. What is the ratio of $HAsO_4^{2-}$ to $H_2AsO_4^-$ in a solution at pH = 7.0? What volume of 0.10 M NaOH should be added to 100 ml of 0.10 M H_3AsO_4 to give a solution with pH = 7.0?

12-37 Preparation of a Buffer Imagine that you want to do physiological experiments at a pH of 6.0 and the organism with which you are working is sensitive to most available materials other than a certain weak acid, H_2Z, and its sodium salts. K_1 and K_2 for H_2Z are $3 \times 10^{-2}\,M$ and $5 \times 10^{-7}\,M$. You have available 1.0 M aqueous H_2Z and 1.0 M NaOH. How much of the NaOH solution should be added to 1.0 liter of the acid solution to give a buffer at pH = 6.0?

12-38 Ammonium Acetate Solution Show that NH_4OAc is about 0.6 percent hydrolyzed in aqueous solution, that to a good approximation this degree of hydrolysis is independent of the concen-tration of the salt, and that the pH of the solution is close to 7.0. (*Hint:* Since OAc^- is a much stronger base than water, and NH_4^+ a much stronger acid than water, the only important reaction in the solution is the direct reaction $NH_4^+ + OAc^- \rightleftharpoons NH_3 + HOAc$.)

12-39 Exact Neutralization (a) How many milliliters of 0.1000 M NaOH must be added to 100.0 ml of 0.1000 M HClO to reach pH = 7.0? $K_a = 1.1 \times 10^{-8}\,M$. (b) Repeat the calculation for 0.1000 M solutions of acids with $K_a = 1.0 \times 10^{-6}\,M$ and $1.0 \times 10^{-4}\,M$.

12-40 Degree of Dissociation For many weak-acid solutions, it is sometimes said to be a good approximation that the degree of dissociation α is related to the acid constant K_a and the molarity of the acid, c, by $\alpha = \sqrt{K/c}$. Show that in fact this relation is valid (within 5 percent) only if $K/c \leq 10^{-2}$, and that if K/c is greater than about 10^2, then $\alpha \approx 1 - c/K$.

12-41 Titration Curve for Pyridine Pyridine is a very weak base with $K_b = 2 \times 10^{-9}\,M$. Like ammonia, it forms salts with strong acids. (a) Calculate the pH of a 0.20 M aqueous solution of pyridine. (b) Calculate the pH of a 0.10 M solution of the salt formed by the reaction of pyridine with HCl. (c) Using the results of (a) and (b), as well as calculations for at least two other points, sketch the titration curve for the titration of 0.20 M pyridine with 0.20 M HCl. Suggest why in practice the titration would not be easy to perform very precisely.

The Solubility Product and Heterogeneous Equilibria

"The solubility of a salt decreases . . . in the presence of a second one with a common ion."
W. NERNST, 1893

13

13-1   Introduction

In this chapter we continue the discussion of important ionic equilibria in aqueous solutions begun in Chaps. 11 and 12, focusing here on heterogeneous equilibria, that is, equilibria involving more than one phase. The emphasis on ionic aqueous solutions is a direct consequence of our environment. Many natural processes on this planet's surface, as well as many of the reactions that have been discovered and applied in the laboratory and in industry, involve ionic aqueous solutions.

Equilibria between ionic solids and their solutions, examined in the first section of the chapter, are important in the chemistry of both the ocean and "fresh" water. The waters of the earth contain dissolved salts picked up as water passes over rocks and clays. Ionic solubility equilibria also underlie several practical methods of analysis, a few of which are illustrated in examples below.

In a later section we examine the effect of changes in pH on the solubility of salts of weak acids. We conclude with a discussion of some representative chemical equilibria between gases and solids.

13-2   Equilibria between Ionic Solids and Aqueous Solutions: The Solubility-Product Principle

In writing the reaction quotient for the equilibrium between a pure ionic solid and its aqueous solution, the solid is conventionally omitted (Sec. 11-3, convention V, and especially Sec. 11-5). The corresponding equilibrium constant is referred to as K_{sp}, the solubility product or the solubility-product constant for the ionic compound being considered.

Although the dissolving of *any* ionic compound may be described in terms of a solubility equilibrium and a corresponding solubility-product constant, the usual application of solubility products is to substances of low solubility. This is true in part because these applications are often the most interesting ones and in part because substances of high solubility lead to concentrations at which activities differ greatly from concentrations, so that there are large deviations from the solubility-product relationship formulated with concentrations (see Sec. 11-5, comment 8).

The form of the solubility-product relation depends on the coefficients in the chemical equation that describes the dissolving of the substance. This is illustrated in the following example.

Example 13-1

□ **The Form of the Ion Product** Formulate the ion products for barium sulfate, $BaSO_4$, and calcium phosphate, $Ca_3(PO_4)_2$.

Solution The dissolving of $BaSO_4$ is described by

$$BaSO_4(s) \rightleftharpoons Ba^{2+} + SO_4^{2-} \tag{13-1}$$

The ion product is therefore given by the solubility-product relationship

$$[Ba^{2+}][SO_4^{2-}] \leqslant K_{sp} \qquad \text{for } BaSO_4 \tag{13-2}$$

The "smaller than" sign in (13-2) applies in the absence of solid; the "equal" sign applies in the presence of the solid at equilibrium. For the second salt, the corresponding relationships are

$$Ca_3(PO_4)_2(s) \rightleftharpoons 3Ca^{2+} + 2PO_4^{3-} \tag{13-3}$$

and

$$[Ca^{2+}]^3[PO_4^{3-}]^2 \leqslant K_{sp} \qquad \text{for } Ca_3(PO_4)_2 \tag{13-4}$$

∎

Exercise 13-1

□ Formulate the ion products for calcium fluoride, CaF_2, and lanthanum carbonate octahydrate, $La_2(CO_3)_3 \cdot 8H_2O$. ∎

The value of K_{sp} for a given substance must be determined experimentally. This may be done by measuring the solubility, as illustrated in the following example, unless the solubility is so small that its measurement presents difficulties. Other methods are available that do not suffer from this limitation; one of them is discussed in Chap. 21, which deals with electrochemistry.

Example 13-2

□ **K_{sp} from Solubility Measurement** In reference works, solubilities are frequently expressed in units of grams per 100 ml. For Ag_2CrO_4 the solubility in water at 25°C is 0.0027 g per 100 ml. Estimate K_{sp} for this substance.

Solution The formula weight of Ag_2CrO_4 is 332. Thus the molar solubility is

$$\frac{0.0027 \text{ g}}{100 \text{ ml}} \times \frac{1 \text{ mol}}{332 \text{ g}} \times \frac{1000 \text{ ml}}{\text{liter}} = 8.1 \times 10^{-5} \, M$$

The equation for the process of dissolving Ag_2CrO_4 (dissolution) is

$$Ag_2CrO_4(s) \rightleftharpoons 2Ag^+ + CrO_4^{2-}$$

This means that in the saturated solution the concentration of CrO_4^{2-} is $8.1 \times 10^{-5} \, M$ and the concentration of Ag^+ is twice as great, $1.6 \times 10^{-4} \, M$. Therefore,

$$K_{sp} = [Ag^+]^2[CrO_4^{2-}] = (1.6 \times 10^{-4} \, M)^2(8 \times 10^{-5} \, M)$$
$$= 2.1 \times 10^{-12} \, M^3 \quad ∎$$

Exercise 13-2

□ The solubility of CaF_2 (FW = 78.1) at 18°C is reported to be 1.6 mg in 100 ml water. What is K_{sp} at that temperature? ∎

Solutions in Pure Water   In the foregoing example we assumed that the ionic substance was dissolving in pure water rather than in water containing other ions, and we also ignored any possible hydrolysis. The effects of hydrolysis can be estimated with methods of Chap. 12, but to simplify the presentation we shall neglect them. In most cases the degree of hydrolysis is sufficiently small that its effects are not significant. They can occasionally be appreciable, however, especially because the concentrations of ions produced in saturated solutions of slightly soluble compounds are very low and the degree of hydrolysis increases with dilution (by Le Châtelier's principle).

The next examples illustrate how the solubility of a slightly soluble substance in water can be calculated from its solubility-product constant.

□ **Solubility Estimation from K_{sp}**   Estimate the solubility at $25°C$ of FeS, for which $K_{sp} = 5 \times 10^{-18} M^2$. Give the answer in moles per liter and in grams per 100 ml.

Example 13-3

Solution   Let the solubility be y mol liter^{-1}, so that $[Fe^{2+}] = y$ and $[S^{2-}] = y$, because each formula unit of the salt yields one Fe^{2+} and one S^{2-}

$$FeS(s) \rightleftharpoons Fe^{2+} + S^{2-} \qquad (13\text{-}5)$$

Therefore

$$[Fe^{2+}][S^{2-}] = (y)(y) = y^2 = 5 \times 10^{-18} M^2$$

Taking the square root of both sides gives $y = 2.2 \times 10^{-9}$ mol liter^{-1}, or $y = 2.2 \times 10^{-9} M$. To calculate the solubility in grams per 100 ml, we need to know the formula weight of FeS, which is 87.9. Hence for 100 ml the solubility is

$$(2.2 \times 10^{-9} \text{ mol liter}^{-1} \times 87.9 \text{ g mol}^{-1})\left(\frac{1 \text{ liter}}{10 \times 100 \text{ ml}}\right) = \frac{1.9 \times 10^{-8} \text{ g}}{100 \text{ ml}} \quad ■$$

□   Estimate the solubility, in moles per liter and in milligrams per 100 ml, of $AgIO_3$ (silver iodate, FW = 283), for which $K_{sp} = 3.0 \times 10^{-8} M^2$. ■

Exercise 13-3

□ **Solubility of Lanthanum Iodate, $La(IO_3)_3$**   Estimate the solubility of $La(IO_3)_3$ and calculate the concentration of IO_3^- in a solution in equilibrium with solid $La(IO_3)_3$, for which $K_{sp} = 6.2 \times 10^{-12} M^4$ at $25°C$.

Example 13-4

Solution   The equation for the dissolution of $La(IO_3)_3$ is

$$La(IO_3)_3(s) \rightleftharpoons La^{3+} + 3IO_3^- \qquad (13\text{-}6)$$

Thus, if the solubility of $La(IO_3)_3$ is represented by y mol liter^{-1}, the concentrations of the ions are given by $[La^{3+}] = y$ and $[IO_3^-] = 3y$, because each formula unit of the salt yields one La^{3+} and three IO_3^- ions. Therefore

$$[La^{3+}][IO_3^-]^3 = y(3y)^3 = 27y^4 = 6.2 \times 10^{-12} M^4$$

whence

$$y^4 = \frac{6.2}{27} \times 10^{-12} M^4 = 0.23 \times 10^{-12} M^4$$

$$y = 6.9 \times 10^{-4} \text{ mol liter}^{-1}$$

The solubility of $La(IO_3)_3$ is 6.9×10^{-4} mol liter^{-1}. The concentration of iodate ion is

$$[IO_3^-] = 3y = 2.1 \times 10^{-3}\, M \quad \blacksquare$$

Exercise 13-4

☐ Estimate the solubility of $Ba(IO_3)_2$ (barium iodate, FW $= 487$) and calculate the concentration of IO_3^- in equilibrium with solid $Ba(IO_3)_2$ ($K_{sp} = 6.0 \times 10^{-10}\, M^3$). ∎

Note that each 3 in the term $(3y)^3$ in Example 13-4 has its origin in the coefficient of IO_3^- in the formula of $La(IO_3)_3$ and thus in the equation for its dissolution, (13-6). In the solubility product, the *exponent* 3 arises from the rules for formulating Q's; the *coefficient* 3 arises because we have chosen to express the concentration of IO_3^- in terms of the molar solubility of the salt and there are three IO_3^- in each formula unit.

Be very careful in studying problems like Example 13-4 not to imagine, through inattention to definitions, that the concentration of iodate has been *tripled* as well as raised to the third power. It has not been; $[IO_3^-]$ has been multiplied by no factor. The concentration of IO_3^- is three times the molar solubility, and thus, since there is no other source of either IO_3^- or La^{3+}, $[IO_3^-] = 3[La^{3+}]$. If we chose to represent the iodate concentration by z, as we might, then $[IO_3^-] = z$, $[La^{3+}] = z/3$, and $K_{sp} = (z/3)\, z^3 = z^4/3$. This would not imply that we had multiplied the La^{3+} concentration by one-third, but only that we had defined the molar solubility as $z/3$. Remember: *no ion concentration in a solubility-product expression, or in any other reaction quotient, is ever multiplied by any factor.*

Remember too that the solubility-product principle can be applied only to saturated solutions, i.e., only when equilibrium exists (see Sec. 11-5, comment 13). In addition, be sure to distinguish carefully between the *solubility product,* which is an equilibrium constant, and the *solubility,* which is a measure of the concentration of some species in a saturated solution in equilibrium with another phase, in this case a solid phase.

The magnitudes of the solubilities of two substances need not parallel the magnitudes of their solubility products. For example, the solubility of $La(IO_3)_3$ ($7 \times 10^{-4}\, M$) is somewhat greater than that of $SrSO_4$ ($5.3 \times 10^{-4}\, M$), despite the fact that the numerical value of K_{sp} for the second compound (2.8×10^{-7}) is greater by nearly 10^5 than that of the first (6.2×10^{-12}). The point is that comparison of the values of the two solubility products is meaningless because they have different units, those of K_{sp} for $SrSO_4$ being M^2 and those of K_{sp} for $La(IO_3)_3$ being M^4.

Table D-2 lists K_{sp} values for many common substances. The units are not given in the table but follow at once from the stoichiometry of the compounds and the convention that all concentrations are molarities.

Solubility of an Ionic Substance in a Solution Containing One of Its Ions: The Common-Ion Effect The solubility of a sparingly soluble ionic substance is sharply decreased by the presence of another ionic substance when the two have an ion in common. The following example demonstrates this *common-ion effect.*

Example 13-5

☐ **Solubility of $SrSO_4$ in 0.10 M Na_2SO_4** What is the solubility in mol liter^{-1} of $SrSO_4$ ($K_{sp} = 2.8 \times 10^{-7}$) in 0.10 M Na_2SO_4?

Solution Because Na_2SO_4 is a strong electrolyte, the sulfate concentration is 0.10 M before any $SrSO_4$ dissolves. Let the unknown solubility be z. Then

$$[Sr^{2+}] = z$$
$$[SO_4^{2-}] = 0.10 + z$$

Substituting these expressions into the solubility product expression gives

$$K_{sp} = [Sr^{2+}][SO_4^{2-}] = z(0.10 + z) = 2.8 \times 10^{-7}$$

a quadratic equation in z. Instead of applying the usual formalism for solving quadratic equations, we note that z is probably very small compared to 0.10, so that z can be neglected in the combination $0.10 + z$. This gives

$$z(0.10 + z) \approx z(0.10) = 2.8 \times 10^{-7}$$
$$z = 2.8 \times 10^{-6}$$

This result is indeed small relative to 0.10, so the approximation is justified and the solubility is $\underline{2.8 \times 10^{-6} \, M}$. ∎

The solubility of $SrSO_4$ in pure water is $\sqrt{K_{sp}} = 5.3 \times 10^{-4} \, M$, so the presence of 0.10 M Na_2SO_4 (or of sulfate ion at this concentration from any other source) has decreased the solubility nearly 200-fold. Similarly, the presence of any soluble strontium salt, such as $SrCl_2$ or $Sr(NO_3)_2$, would provide Sr^{2+} ions and hence decrease the solubility of $SrSO_4$. Figure 13-1 shows how the solubility of $SrSO_4$ in solutions of Na_2SO_4 depends on the concentration of Na_2SO_4. The same figure applies to the solubility of $SrSO_4$ in solutions of $SrCl_2$.

☐ What is the solubility in mol liter^{-1} of $AgIO_3$ ($K_{sp} = 3.0 \times 10^{-8}$) in 0.20 M KIO_3? ∎

Exercise 13-5

The common-ion effect is often important in quantitative analysis. Suppose, for example, that the amount of SO_4^{2-} in a solution is to be determined by precipitating the sulfate as $BaSO_4$ ($K_{sp} = 1.1 \times 10^{-10}$) and weighing the dried precipitate. It is necessary that essentially all the SO_4^{2-} be removed from the solution. This can be accomplished by using an excess of barium ions in precipitating the sulfate, so that the concentration of SO_4^{2-} remaining is appreciably smaller than it would be if Ba^{2+} and SO_4^{2-} were present in equal numbers.

☐ **Precipitation of Fe(OH)$_2$** The solubility of ferrous hydroxide, $Fe(OH)_2$, in water is about 0.6 mg liter^{-1}. Calculate its K_{sp} and then calculate the

Example 13-6

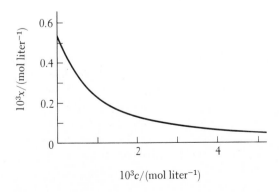

Figure 13-1 The Common-Ion Effect
The solubility x of $SrSO_4$ in a solution of Na_2SO_4 is plotted as a function of the concentration c of Na_2SO_4. Since $[Sr^{2+}] = x$ and $[SO_4^{2-}] = c + x$, we obtain $x(c + x) = K_{sp} = 2.8 \times 10^{-7}$. This equation is quadratic in x and linear in c. For plotting x against c it is therefore easiest to assume values for x and solve for c. Note the decrease in solubility caused by the increase in the concentration of the SO_4^{2-} ion, the ion common to both $SrSO_4$ and Na_2SO_4. There is a similar decrease of the solubility when $SrSO_4$ is dissolved in a $SrCl_2$ solution, the common ion now being Sr^{2+}. The solubility of $SrSO_4$ is measured by $[SO_4^{2-}]$ in this new situation so that now $x = [SO_4^{2-}]$ while $[Sr^{2+}] = c + x$. Thus again, $x(c + x) = 2.8 \times 10^{-7}$.

minimum pH needed to precipitate ferrous ion so completely that no more than 1.0 μg (microgram, or 10^{-6} g) of the ion remains in solution per liter.

Solution The formula weight of $Fe(OH)_2$ is 90. Thus the solubility is

$$\frac{0.6\ \text{mg}}{\text{liter}} \times \frac{1\ \text{g}}{1000\ \text{mg}} \times \frac{1\ \text{mol}}{90\ \text{g}} = 7 \times 10^{-6}\ M$$

The value of K_{sp} is then

$$K_{sp} = [Fe^{2+}][OH^-]^2 = (7 \times 10^{-6}\ M)(2 \times 7 \times 10^{-6}\ M)^2$$
$$= \underline{1.4 \times 10^{-15}\ M^3}$$

The maximum concentration of Fe^{2+} must be no greater than 1.0×10^{-6} g liter^{-1} or

$$[Fe^{2+}]_{max} = 1.0 \times 10^{-6}\ \text{g liter}^{-1} \times \frac{1\ \text{mol}\ Fe^{2+}}{56\ \text{g}}$$
$$= 1.8 \times 10^{-8}\ M$$

The minimum concentration of OH^- can be deduced from the solubility-product principle. When the solution is saturated,

$$[OH^-]^2 = \frac{K_{sp}}{[Fe^{2+}]}$$

Hence, when the Fe^{2+} concentration is $[Fe^{2+}]_{max}$

$$[OH^-]^2 = \frac{K_{sp}}{[Fe^{2+}]_{max}} = \frac{1.4 \times 10^{-15}\ M^3}{1.8 \times 10^{-8}\ M} = 8 \times 10^{-8}\ M^2$$
$$[OH^-] = 2.8 \times 10^{-4}\ M$$
$$pOH = -\log(2.8 \times 10^{-4}) = 3.6$$
$$pH = 14.0 - 3.6 = 10.4$$

If $[OH^-]$ is less than $2.8 \times 10^{-4}\ M$, then, in a saturated solution, $[Fe^{2+}]$ would have to be greater than $[Fe^{2+}]_{max}$ by the solubility-product principle. Thus the concentration of Fe^{2+} will be less than 1.0 μg liter^{-1} when

$$[OH^-] \geqslant \underline{2.8 \times 10^{-4}}$$

or equivalently when pOH $\leqslant 3.6$ and therefore $\underline{pH \geqslant 10.4}$. ■

Exercise 13-6

☐ The solubility of magnesium hydroxide, $Mg(OH)_2$ (FW = 58.3), is about 9 mg liter^{-1}. Calculate its K_{sp} and then calculate the minimum pH needed to precipitate Mg^{2+} so completely that no more than 1.0 μg of the Mg^{2+} remains in solution per liter. ■

Solutions Containing Several Sparingly Soluble Salts The solubilities of sparingly soluble salts that are in equilibrium with the same solution but that have no ions in common are independent of each other. If the salts do have ions in common, their solubility-product equations must be satisfied simultaneously. The following example illustrates this point.

Example 13-7

☐ **Simultaneous Solubilities of $PbSO_4$ and $SrSO_4$** The solubility-product constants of $PbSO_4$ and $SrSO_4$ are, respectively, 1.7×10^{-8} and 2.8×10^{-7}.

What are the values of $[SO_4^{2-}]$, $[Pb^{2+}]$, and $[Sr^{2+}]$ in a solution at equilibrium with both substances?

Solution Since the dissolving of each formula unit of $PbSO_4$ gives one Pb^{2+} ion and one SO_4^{2-} ion, the concentration of sulfate ions originating from this source is given by $[Pb^{2+}]$. A similar argument applies to the SO_4^{2-} ions arising from the dissolving of $SrSO_4$, so that the total concentration of SO_4^{2-} is

$$[SO_4^{2-}] = [Pb^{2+}] + [Sr^{2+}] \qquad (13\text{-}7)$$

The solubility-product relationships yield the equations $[Pb^{2+}] = 1.7 \times 10^{-8}/[SO_4^{2-}]$ and $[Sr^{2+}] = 2.8 \times 10^{-7}/[SO_4^{2-}]$. When these are inserted into (13-7), we get

$$[SO_4^{2-}] = \frac{1.7 \times 10^{-8} + 2.8 \times 10^{-7}}{[SO_4^{2-}]} \qquad (13\text{-}8)$$

and thus

$$[SO_4^{2-}]^2 = (0.17 + 2.8) \times 10^{-7} = 3.0 \times 10^{-7}$$
$$[SO_4^{2-}] = \underline{5.5 \times 10^{-4}\,M}$$

From this result we obtain

$$[Pb^{2+}] = \frac{1.7 \times 10^{-8}}{5.5 \times 10^{-4}} = \underline{3.1 \times 10^{-5}\,M}$$

and
$$[Sr^{2+}] = \frac{2.8 \times 10^{-7}}{5.5 \times 10^{-4}} = \underline{5.1 \times 10^{-4}\,M}$$

As a check, we note that, within their precision, the three concentrations found satisfy (13-7). ∎

□ The solubility-product constants of $BaCrO_4$ and $BaSO_4$ are, respectively, 3×10^{-10} and 1.1×10^{-10}. What are the values of $[Ba^{2+}]$, $[SO_4^{2-}]$, and $[CrO_4^{2-}]$ in a solution at equilibrium with both substances? ∎

Exercise 13-7

Limitations of Solubility-Product Calculations As noted earlier, concentrations arrived at in solubility-product calculations sometimes deviate appreciably from those found experimentally. In part, this low accuracy arises from failure to correct for activity effects; however, the chief factor is usually that other equilibria simultaneously involving the ions of interest are disregarded. We have mentioned hydrolysis as one possible complication; others include the formation of undissociated neutral molecules and of complex ions. These latter effects are well illustrated by solutions of AgCl in the presence of excess Ag^+ and especially in the presence of excess Cl^-. First, when AgCl(*s*) dissolves, it yields not only Ag^+ and Cl^- ions, but also a very small but not always negligible concentration of undissociated AgCl *molecules* in solution. A second effect, the formation of the complex ion $AgCl_2^-$ and of others containing even more chloride, enters when $[Cl^-]$ begins to exceed about $10^{-3}\,M$. This effect substantially increases the solubility of AgCl(*s*). Thus, when silver ions are precipitated for purposes of analysis by adding a reagent containing Cl^- ions, the solubility of the resulting AgCl(*s*) is decreased in the presence of a small excess of Cl^- but is increased substantially by a large excess.

We have considered several different situations in which a number of equilibria are operative simultaneously and compete with each other for at least one of the ions or molecules present. These include all acid-base equilibria, in which two bases (one of them water) compete for a proton. Example 13-7 concerns a heterogeneous competing equilibrium in which Pb^{2+} and Sr^{2+} are simultaneously in equilibrium with their corresponding sulfates and thus compete for SO_4^{2-}. In this section we examine another class of heterogeneous competing equilibria of some importance, chiefly by means of illustrative examples.

There are many almost insoluble salts of weak acids. It is a general rule that the solubility of these salts increases with increasing acidity of a solution (decreasing pH), provided that the weak acid is itself soluble. The examples that follow are concerned with the quantitative aspects of this phenomenon; qualitatively it is easy to understand. Imagine a salt MX formed by the metal ion M^+ and the anion X^-, with X^- being the conjugate base of a weak acid HX. In a saturated solution of MX(s) we have

$$MX(s) \rightleftharpoons M^+ + X^- \tag{13-9}$$

with the concentrations of the ions satisfying the solubility-product expression for MX(s),

$$[M^+][X^-] = K_{sp} \tag{13-10}$$

If the pH is so high that no appreciable quantity of X^- picks up a proton in solution, the concentrations of M^+ and X^- will be equal, and each will equal $\sqrt{K_{sp}}$, which is then a measure of the solubility of MX. As the pH is lowered, however, a significant number of the X^- ions will pick up protons and be converted into HX, because HX is a weak acid. When $pH = pK_a$, $[X^-] = [HX]$ and thus 50 percent of the X^- will have been changed to HX; when $pH = (pK_a - 1)$, 91 percent will have been converted. As the acidity increases, more and more of the X^- that goes into solution is transformed into HX. Consequently it is no longer true that $[M^+] = [X^-]$. However, it is true that $[M^+] = [X^-] + [HX]$, so that $[M^+] > [X^-]$, the disparity becoming larger as the pH decreases. Thus, to satisfy (13-10), more MX dissolves than in a solution in which essentially all the X^- remains as X^-; the solubility, measured by $[M^+]$ or by $([X^-] + [HX])$, becomes greater.

The argument is in principle the same regardless of the charges on M and X or of their proportions in the salt; for example, the salt might be $(M^{2+})_3(X^{3-})_2$. It applies to salts of many different soluble weak acids—phosphates, carbonates, acetates, even sulfates at high enough acidity, sulfides, and others. It applies as well to the solubility of hydroxides and oxides, which may in this context be regarded as salts of the very weak acid H_2O.

Example 13-8

☐ **Solubility of Silver Acetate at pH = 3.0**   What is the solubility of AgOAc in a solution buffered at pH = 3.0? K_{sp} for AgOAc = 2.5×10^{-3} and K_a for HOAc = 1.8×10^{-5}.

Solution   As AgOAc dissolves, acetate ions are produced, some of which combine with H^+ to form HOAc. This permits more AgOAc to dissolve until the concentrations of Ag^+ and OAc^- reach the values permitted by the

13-3 Effect of Changes in pH
on the Solubility of Salts
of Weak Acids

solubility-product constant. If we let the solubility of AgOAc in this solution be S mol liter^{-1}, then

$$[Ag^+] = S = [OAc^-] + [HOAc]$$

since the total concentration of acetate in all forms must equal the silver-ion concentration inasmuch as Ag^+ and OAc^- are present in 1:1 proportion in AgOAc. However, we know that

$$[HOAc] = [OAc^-]\frac{[H^+]}{K_a} \tag{13-11}$$

so that

$$S = [OAc^-] + [HOAc] = [OAc^-] + [OAc^-]\frac{[H^+]}{K_a}$$

$$= [OAc^-]\left(1 + \frac{[H^+]}{K_a}\right) \tag{13-12}$$

Multiplying by $[Ag^+]$ gives

$$[Ag^+]S = [Ag^+][OAc^-]\left(1 + \frac{[H^+]}{K_a}\right) \tag{13-13}$$

However, we know that $[Ag^+] = S$, so the left side is just S^2, and if undissolved AgOAc is present at equilibrium, the right-hand side is a function of pH only:

$$S^2 = K_{sp}\left(1 + \frac{[H^+]}{K_a}\right) \tag{13-14}$$

In this particular example, with $[H^+] = 1.0 \times 10^{-3}\ M$ and $K_a = 1.8 \times 10^{-5}$, we obtain

$$S^2 = K_{sp}\left(1 + \frac{1.0 \times 10^{-3}}{1.8 \times 10^{-5}}\right) = K_{sp}(1 + 56)$$

and

$$S = (57K_{sp})^{1/2} = 7.5K_{sp}^{1/2} = 7.5 \times 5.0 \times 10^{-2}$$
$$= \underline{0.38\ M}$$

At high pH, where essentially no HOAc is formed, the solubility of AgOAc is $K_{sp}^{1/2} = 5.0 \times 10^{-2}\ M$. Thus, it has been increased $7\frac{1}{2}$-fold by the increased acidity. ∎

☐ What is the solubility of LiF in a solution buffered at pH 2.5? K_{sp} for LiF $= 5.0 \times 10^{-3}$; K_a for HF $= 6.7 \times 10^{-4}$. ∎ **Exercise 13-8**

Equation (13-14) is perfectly general for a salt M^+X^-. When the pH is well above pK_a, then $[H^+]/K_a$ is very small and $S \approx K_{sp}^{1/2}$. In solutions sufficiently acidic that $[H^+]/K_a$ is large, the solubility is increased by approximately the factor $([H^+]/K_a)^{1/2}$; that is, $S \approx K_{sp}^{1/2}([H^+]/K_a)^{1/2}$.

☐ **Prevention of Precipitation of Mg(OH)$_2$** How many moles of NH_4Cl must be added to 100 ml of 0.10 M NH_3 so that, when this buffer mixture is added to 100 ml of 0.020 M $MgCl_2$, no precipitate of $Mg(OH)_2$ will form? K_{sp} **Example 13-9** for $Mg(OH)_2$ is 1.8×10^{-11} and K_b for NH_3 is 1.8×10^{-5}. Ignore any possible complexes with NH_3.

Solution Addition of NH_4Cl, which is a weak acid, reduces the OH^- concentration in the NH_3 solution and creates a buffer. We must calculate the maximum concentration of OH^- that can be present without the formation of $Mg(OH)_2$ and then the minimum concentration of NH_4^+ needed in a solution with $[NH_3] = 0.10\ M$ to produce at most this concentration of OH^-.

Since the $MgCl_2$ solution is $0.020\ M$ and is to be diluted twofold by addition of the buffer, the final $[Mg^{2+}] = 0.010$. Thus, to prevent precipitation,

$$[Mg^{2+}][OH^-]^2 < 1.8 \times 10^{-11}$$

or

$$[OH^-]^2 < \frac{1.8 \times 10^{-11}}{0.010} = 1.8 \times 10^{-9}$$

so that $[OH^-] = \sqrt{18} \times 10^{-5}\ M = 4.2 \times 10^{-5}\ M$ is the upper limit of permissible concentrations.

The ammonia buffer is governed by the relationship

$$\frac{[NH_4^+]}{[NH_3]} = \frac{K_b}{[OH^-]} = \frac{1.8 \times 10^{-5}}{[OH^-]}$$

Replacement of $[OH^-]$ by its upper permissible limit just calculated yields a lower limit for the ratio $[NH_4^+]/[NH_3]$, because $[OH^-]$ is in the denominator.

$$\frac{1.8 \times 10^{-5}}{\sqrt{18} \times 10^{-5}} = 0.42 \leq \frac{[NH_4^+]}{[NH_3]}$$

Hence the concentration of ammonium ion must be above 0.42 times that of NH_3. In 100 ml of $0.10\ M\ NH_3$, this means about $0.05\ M\ NH_4^+$, or about 5×10^{-3} mol of NH_4Cl, should be added to the 100-ml sample. ∎

Exercise 13-9

☐ A buffer is prepared by adding 0.020 mol of sodium acetate, NaOAc, to acetic acid, HOAc, and then adjusting the volume to 100 ml. What is the minimum number of moles of acetic acid required in the buffer to prevent $Al(OH)_3$ from precipitating when the buffer is added to 100 ml of a $0.010\ M\ Al_2(SO_4)_3$ solution? $K_{sp} = 2.0 \times 10^{-33}\ M^4$ for $Al(OH)_3$, and $K_a = 1.8 \times 10^{-5}\ M$ for acetic acid. ∎

A few salts of common weak bases are not very soluble. Their solubility increases with increasing pH for the same reason that the solubility of the salt of a weak acid increases with decreasing pH, and they can be treated in an exactly analogous fashion.

13-4 Heterogeneous Equilibria Resulting from Reactions of Gases with Solids

Many heterogeneous equilibria in which gases play a role are interesting and important. We have already discussed in some detail vapor-pressure equilibria for both liquids and solids (Chap. 5). Before giving illustrations that concern chemical equilibria we return to conventions III*b* and V of Sec. 11-3 for setting up the reaction quotient Q. According to convention IIIb, the concentrations of gases are replaced by their partial pressures even when the reaction mixture does not consist entirely of gases; by convention V, the "concentrations" of *pure* solids or liquids are deleted from Q. Two illustrative chemical equations and the related equilibrium expressions are

$$PCl_5(s) \rightleftharpoons PCl_3(g) + Cl_2(g) \tag{13-15}$$
$$P_{PCl_3}P_{Cl_2} = K_{15} \tag{13-15a}$$

$$2HgO(s) \rightleftharpoons 2Hg(l) + O_2(g) \tag{13-16}$$
$$P_{O_2} = K_{16} \tag{13-16a}$$

The symbols (s) and (l) imply that the substances concerned are *pure* solids and *pure* liquids, respectively, and also that these substances are *actually present* as pure solids or liquids under the equilibrium situations of interest. The symbol (g) implies purity only if just one gas is present, because all gases are miscible with one another.

How can we justify the rule that pure solids and pure liquids are omitted entirely from reaction quotients? The reason that solid PCl_5 need not be mentioned in (13-15a) is a direct consequence of the fact that, at whatever temperature (13-15a) is applicable, the vapor pressure of solid PCl_5 is a constant as long as the solid PCl_5 is pure. Let us call this vapor pressure $P^*_{PCl_5}$. If the system is at equilibrium, the gas phase is saturated with PCl_5 vapor and we may rewrite (13-15) in an alternative form:

$$PCl_5(s) \rightleftharpoons PCl_5(g \text{ at } P^*_{PCl_5}) \rightleftharpoons PCl_3(g) + Cl_2(g) \tag{13-17}$$

In other words, since the vapor of PCl_5 (at a pressure $P^*_{PCl_5}$) is in equilibrium with the solid, which is in turn in equilibrium with gaseous PCl_3 and Cl_2, the PCl_5 vapor is itself in equilibrium with the two decomposition products. We might then write for the right-hand equation in (13-17):

$$\frac{P_{PCl_3}P_{Cl_2}}{P^*_{PCl_5}} = K_{17} \tag{13-17a}$$

However, since $P^*_{PCl_5}$ is a constant at this temperature, it is most convenient to combine it with K_{17} thus leading to

$$P_{PCl_3}P_{Cl_2} = K_{17}P^*_{PCl_5}$$

Above, in (13-15a), we have merely used the symbol K_{15} for this product $K_{17}P^*_{PCl_5}$.

It is important to note the distinction between the equilibrium (13-15) and the corresponding one in which no solid PCl_5 is present.

$$PCl_5(g) \rightleftharpoons PCl_3(g) + Cl_2(g) \tag{13-18}$$

for which we have

$$\frac{P_{PCl_3}P_{Cl_2}}{P_{PCl_5}} = K_{18} \tag{13-18a}$$

Because no solid PCl_5 is present, the pressure of gaseous PCl_5 can vary widely, from zero up to the vapor pressure, and hence must be included explicitly in the reaction quotient[1] for (13-18).

---

[1] Equation (13-17a) is a special case of (13-18a), and K_{18} and K_{17} are identical. On the other hand, the value and even the units of K_{15} and K_{18} are different (atm² and atm, respectively, if partial pressures are expressed in atmospheres).

The justification for the omission of any pure liquid from the reaction quotient is based on identical reasoning. As long as the liquid is pure, its vapor pressure is constant at a given temperature. For example, in (13-16)—the equilibrium between solid mercuric oxide (HgO), liquid mercury, and gaseous oxygen—the only term that appears in the reaction quotient is the pressure of oxygen, which must be constant $(=K_{16})$ once equilibrium is established.

If, however, the mercury is diluted with some other material—e.g., by dissolving another metal in it to make a solution known as an amalgam—then the vapor pressure of the mercury will decrease in accord (at least approximately) with Raoult's law. Explicit account must then be taken in the reaction quotient of the fact that the liquid phase is no longer pure. Similarly, if any solid is present as a solid solution, its decrease in concentration must be included explicitly in formulating the reaction quotient. This is why it must be emphasized that only *pure* solids and liquids can be omitted from these quotients.

We now turn to a more detailed examination of some typical equilibria arising as a result of chemical reactions between gases and solids. Because intimate contact between different phases is difficult to establish, heterogeneous equilibria are often established slowly. For example, many common metals tend to react with some normal components of the earth's atmosphere, chiefly O_2, H_2O, and acidic oxides such as CO_2, NO_2, and SO_2. This corrosion, as it is called, is usually slow at ordinary temperatures, in part because some of the reactions involved are intrinsically slow and in part because a complex series of transformations usually takes place during corrosion. Many other heterogeneous gas-solid reactions also take time to reach equilibrium, but for some comparatively simple reactions equilibrium is achieved fairly rapidly. To illustrate the principles involved, we choose the reaction of CO_2 with a basic oxide to form a carbonate.

Carbon dioxide has been present in the earth's atmosphere at least since the dawn of life on this planet 2 billion or so years ago, and it plays a central role in most known forms of life today. It also figures prominently in many geological processes, because it is a widespread acidic oxide, reacting with water to form the weak acid H_2CO_3 (carbonic acid). This acid, and CO_2 itself, react with bases to form carbonates, which are important constituents of many common rocks and minerals. The most abundant carbonate is $CaCO_3$, which occurs in an enormous variety of forms; it is produced by many marine organisms, being the major ingredient of coral, pearl, and the shells of most mollusks and crustaceans. The shells of dead marine animals accumulate in thick deposits on the floor of the sea and are slowly transformed by geological processes into such common rocks as limestone and marble.

When $CaCO_3$ is heated sufficiently, it decomposes to form calcium oxide (CaO) and CO_2:

$$CaCO_3(s) \rightleftharpoons CaO(s) + CO_2(g) \qquad (13\text{-}19)$$

The equilibrium condition for this reaction at any temperature is simply

$$P_{CO_2} = K_{19} \qquad (13\text{-}19a)$$

In other words, if at a particular temperature there is to be equilibrium between $CaCO_3(s)$ and $CaO(s)$, it can occur only at a specific pressure of CO_2 in this system, given by the value of K_{19} at that temperature. This equilibrium pressure of CO_2 is sometimes called the dissociation pressure of $CaCO_3$; it increases rapidly with increasing temperature (Fig. 13-2).

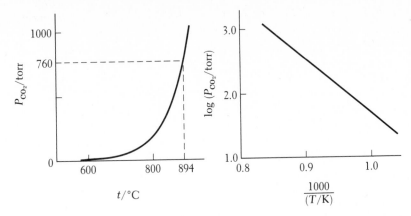

It is important to recognize that even though neither solid $CaCO_3$ nor solid CaO appears explicitly in the equilibrium expression (13-19a), both these substances must be present if the equilibrium (13-19) is to exist. Their absence from (13-19a) implies only that the position of equilibrium is independent of the amount of either solid *as long as some of each is present.* If only CaO(s) and CO_2 are present, with no $CaCO_3$, the pressure of the CO_2 can have any value between zero and the equilibrium pressure (K_{19}) at the temperature of the system. If only $CaCO_3(s)$ and CO_2 are present, the pressure of the CO_2 must be greater than or equal to the equilibrium pressure. These conditions are analogous to those that govern vapor-liquid equilibria for pure substances. Recall (Sec. 5-2) that liquid and vapor coexist in equilibrium at a particular temperature only at a specific pressure, the vapor pressure. A single vapor phase exists at pressures below the vapor pressure, and a single liquid phase exists at higher pressures. The phase relations in the $CaCO_3$–CaO–CO_2 system are illustrated in Fig. 13-3, in which the pressure of CO_2 over a mixture of CaO(s) and $CaCO_3(s)$ at 800°C is plotted as a function of the weight fraction y of $CaCO_3$. At this temperature, the dissociation pressure of $CaCO_3$ is 180 torr.

What can be said about the temperature at which some form of $CaCO_3(s)$ such as limestone[2] will decompose in air? To answer this question we need only the data of Fig. 13-2 and an estimate of the partial pressure of CO_2 in air. This quantity varies from place to place on the earth's surface. It depends on the density of plant life, of industrial activity, and of other consumers and producers of CO_2, but an average value for P_{CO_2} in the earth's atmosphere is about 0.2 torr. Thus log (P_{CO_2}/torr) is about -0.7. Extrapolation of the right-hand graph in Fig. 13-2 indicates that this pressure corresponds approximately to a value of 1.3 for $1000\,(T/\text{K})^{-1}$. Thus for atmospheric CO_2 there is equilibrium between $CaCO_3(s)$ and CaO(s) when T is around $\frac{1000}{1.3}$ K, that is, somewhat below 800 K or around 500°C. This means that at any temperature below 500°C, the pressure of CO_2 in the atmosphere exceeds the dissociation pressure and $CaCO_3(s)$ is stable whereas CaO(s) (quicklime) is unstable and reacts with CO_2 to form $CaCO_3$. The latter reaction is one of several important ones that occur during the hardening of

---

[2] Actually, each different form of $CaCO_3(s)$ has its own dissociation pressure. The pressures vary according to the relative stabilities of the forms, being lowest for the most stable form, but this variation is very small for the different common forms, such as limestone, marble, and chalk.

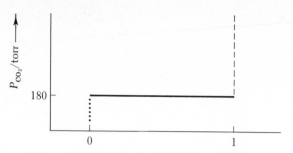

Fraction of solid that is $CaCO_3(s)$, y

Figure 13-3 **Phase Relations in the $CaCO_3$–CaO–CO_2 System at 800°C**

When y, the weight fraction of $CaCO_3(s)$ in the solid phase, is between 0 and 1, both $CaO(s)$ and $CaCO_3(s)$ are present, and P_{CO_2} must be equal to 180 torr at 800°C (solid line). If the pressure of the CO_2 were larger than 180 torr, the $CaO(s)$ would react with the CO_2 until either all the CaO had been used up or P_{CO_2} had decreased to 180 torr. Likewise, if P_{CO_2} were smaller than 180 torr in the presence of both solids, some $CaCO_3(s)$ would decompose, yielding CO_2, until either P_{CO_2} had increased to 180 torr or all the $CaCO_3(s)$ had been used up.

When $y = 0$ and thus no $CaCO_3(s)$ is present, P_{CO_2} (dotted line) may be smaller than 180 torr (and remain smaller) because no source of CO_2 exists to increase its pressure. Likewise, when $y = 1$ and no $CaO(s)$ is present, P_{CO_2} (dashed line) may be larger than 180 torr (and remain larger) because the system contains no substance that can react with the excess CO_2 and thereby reduce its pressure. Only the horizontal line at 180 torr corresponds to an equilibrium situation, with all three substances present.

cement. Above 500°C, $CaCO_3(s)$ is unstable relative to atmospheric CO_2 and decomposes into $CaO(s)$ and CO_2.

With a substance such as $CaCO_3(s)$ that can decompose to form a gaseous product, there is an important distinction between heating it in a closed system and in an open one such as the open air. In a closed system, the gas that is released accumulates and its concentration in the gas phase therefore rises, which causes an increase in its partial pressure (over and above that caused by any increase in temperature). Eventually the partial pressure of the accumulated gas reaches the dissociation pressure of the decomposing solid (if the sample of solid has not completely decomposed before this point) and equilibrium is established. On the other hand, in an open system, where the gas produced during the decomposition is swept away or diffuses away, the partial pressure of this gas does not increase significantly around the sample. Thus, if this pressure was initially below the dissociation pressure at the temperature of heating, it will remain so and the sample will eventually decompose completely.

Equilibria between other carbonates and the related oxides and CO_2 can be formulated in a fashion parallel to (13-19) and (13-19a). For those oxides appreciably more basic than CaO, such as Na_2O and the oxides of the other alkali metals, the dissociation pressure of the corresponding carbonate is much lower at a given temperature than it is for $CaCO_3$. Conversely, with carbonates such as Ag_2CO_3 and $PbCO_3$, formed by the reaction of less basic oxides with CO_2 (Ag_2O and PbO are appreciably less basic than CaO), the dissociation pressures are considerably higher at a given temperature and these substances are easier to decompose by heating. For many less basic oxides, the dissociation pressures are so high even at ordinary temperatures that the corresponding carbonates are not stable in the normal atmosphere and thus do not normally exist.

Another group of important heterogeneous equilibria involves substances capable of liberating oxygen when heated. Among these are the oxides of some of the less reactive metals such as mercury [Equation (13-16)] and silver and the peroxides of some of the alkaline-earth metals:

$$2BaO_2(s) \rightleftharpoons 2BaO(s) + O_2(g) \qquad (13\text{-}20)$$

$$2Ag_2O(s) \rightleftharpoons 4Ag(s) + O_2(g) \qquad (13\text{-}21)$$

The same principles apply to equilibria between water vapor and different hydrates of a substance (including the anhydrous form)—such as $CaSO_4$, $CaSO_4 \cdot \frac{1}{2}H_2O$, $CaSO_4 \cdot 2H_2O$—and also the saturated solution. Only *neighboring* forms (for example, $CaSO_4$ and $CaSO_4 \cdot \frac{1}{2}H_2O$) in this sequence of solid and liquid phases can coexist at equilibrium with water vapor, and for each pair of neighboring phases and a given temperature the equilibrium partial pressure of H_2O is fixed. If two phases that are not neighbors ($CaSO_4$ and $CaSO_4 \cdot 2H_2O$, for example) are present in a system, one will lose water or the other will gain water until only two neighboring phases exist, together with H_2O at the appropriate equilibrium pressure, or until only a single solid phase of composition intermediate between the original phases is present. When only one solid phase is present, the partial pressure of H_2O may vary within certain bounds.

□ **Gas-Solid Equilibrium** Ammonium chloride is a white crystalline solid that is volatile at elevated temperatures, decomposing completely in the vapor phase to form gaseous ammonia and HCl.

Example 13-10

$$NH_4Cl(s) \rightleftharpoons NH_3(g) + HCl(g)$$

When an excess of solid NH_4Cl is heated in a previously evacuated flask at 340°C until equilibrium is established, the total pressure in the flask is 1.00 atm. (a) Formulate the reaction quotient for the reaction occurring, and give the numerical value of the corresponding equilibrium constant at 340°C. (b) Suppose that an excess of solid NH_4Cl is introduced into a flask in which the partial pressure of ammonia at 340°C is 1.00 atm. What will be the pressure of HCl and the total pressure when equilibrium is established?

Solution (a) Because NH_4Cl is a solid in the equation as written, it does not appear in the reaction quotient,

$$Q = P_{NH_3}P_{HCl}$$

At equilibrium at 340°C the *total* pressure is 1.00 atm. The only two gases present, NH_3 and HCl, are present in equimolar amounts; by Avogadro's law, they must have equal pressures. Therefore,

$$P_{NH_3} = P_{HCl} = 0.50 \text{ atm}$$

$$K = (0.50 \text{ atm})^2 = \underline{0.25 \text{ atm}^2}$$

(b) Let the equilibrium pressure of HCl be x atm. Then the pressure of NH_3 produced (with the HCl) by decomposition of NH_4Cl will also be x and the total pressure of NH_3 will be $(1.00 + x)$. Since the system is at equilibrium,

$$(1.00 + x)x = 0.25 \text{ atm}^2$$

This gives

$$x = \underline{0.21 \text{ atm}} = P_{HCl}$$

The pressure of NH_3 is then 1.21 atm, and the total pressure is $\underline{1.42 \text{ atm.}}$ ■

Exercise 13-10

□ Phosphonium bromide, PH_4Br, is a volatile crystalline solid that is completely decomposed in the vapor phase into gaseous phosphine, PH_3, and gaseous hydrogen bromide, HBr,

$$PH_4Br(s) \rightleftharpoons PH_3(g) + HBr(g)$$

The total pressure above an excess of PH_4Br in a previously evacuated flask is 291 torr at 23°C. (*a*) What is the equilibrium constant for the above reaction at 23°C? (*b*) Suppose that extra HBr has been introduced into the flask so that the total pressure at equilibrium at 23°C is 500 torr. What are P_{PH_3} and P_{HBr}? ■

Summary

Equilibria between ionic solids and aqueous solutions can be described by the solubility-product constant, K_{sp}. The form of the solubility product for a substance depends on the coefficients in the chemical equation that describes the dissolving of the substance. The value of K_{sp} can be obtained from solubility data or other suitable measurements.

Examples are given that demonstrate how the solubility of a slightly soluble substance in water can be calculated from its K_{sp}. The solubility of a sparingly soluble ionic substance is decreased by the presence of another ionic substance when the two have an ion in common. This phenomenon, called the common-ion effect, is a direct result of the requirement that the solubility product be constant at equilibrium. The common-ion effect can be used to control precipitation in quantitative analysis procedures.

The solubilities of sparingly soluble salts that are in equilibrium with the same solution but that have no ions in common are independent of each other. If the salts do have ions in common, their solubility-product equations must be satisfied simultaneously.

It is a general rule that the solubility of a salt of a weak acid increases with decreasing pH (provided that the weak acid is itself soluble). Similarly, the solubilities of salts of weak bases increase with increasing pH. These effects can be understood quantitatively as the result of simultaneous equilibria in which ions in equilibrium with a solid react in solution to form undissociated acids or bases.

If an equilibrium involves two pure condensed phases and a gas, the partial pressure of the gas must be constant at a given temperature. This principle underlies the phase relationships in the $CaCO_3$–CaO–CO_2 system and, in particular, explains the stability of $CaCO_3$ exposed to the earth's atmosphere at temperatures below 500°C.

Terms and Concepts

Solubility-product principle	p. 265
Solubility of salts in pure water	p. 267
Common-ion effect	p. 268
Simultaneous solubility-product equilibria	p. 270
pH and the solubility of a salt of a weak acid	p. 272
Heterogeneous equilibria involving gases and condensed phases	p. 274

Problems and Questions

13-1 Form of the Solubility Product Let S equal the solubility in moles per liter for each of the following salts. Express the solubility product of each salt in terms of S: (a) $AgBr$; (b) PbF_2; (c) Ag_2CrO_4; (d) Ag_3AsO_4; (e) $BaSO_4$.

13-2 Solubilities from K_{sp} Values Calculate the solubility in water of each of the following substances from its K_{sp} value. Express the solubility in moles per liter and in grams per 100 ml.
(a) $BaCrO_4$ $K_{sp} = 3 \times 10^{-10} M^2$
(b) Ag_2SO_4 $K_{sp} = 1.6 \times 10^{-5} M^3$
(c) $Mg(OH)_2$ $K_{sp} = 1.8 \times 10^{-11} M^3$
(d) $PbSO_4$ $K_{sp} = 1.7 \times 10^{-8} M^2$
(e) $Fe(OH)_3$ $K_{sp} = 10^{-36} M^4$

13-3 K_{sp} from Solubility Data The solubilities of the following ionic compounds are given in grams per 100 ml of saturated solution in water. Calculate the solubility-product constant for each substance.
(a) SrF_2 1.1×10^{-2}
(b) Ag_2S 9×10^{-16}
(c) CuI 4×10^{-5}
(d) Ag_3PO_4 6.7×10^{-4}
(e) $PbCrO_4$ 4.5×10^{-6}

13-4 Common-Ion Effect The solubility product of $ZnCO_3$ is $3 \times 10^{-8} M^2$. Calculate the solubility of $ZnCO_3$ in water, in $0.04 M$ $Zn(NO_3)_2$, and in $0.04 M$ K_2CO_3.

13-5 Common-Ion Effect The solubility product of $Mg(OH)_2$ is $1.8 \times 10^{-11} M^3$. Calculate the solubility of $Mg(OH)_2$ (a) in $0.050 M$ $MgBr_2$, and (b) in $0.050 M$ $NaOH$.

13-6 Solubility of Silver Carbonate The solubility product of silver carbonate is $5 \times 10^{-12} M^3$ at 25°C. (a) Calculate the approximate solubility (in moles per liter) of silver carbonate in water and its solubility in milligrams per 100 ml. (b) Calculate the solubility of silver carbonate in $0.1 M$ $AgNO_3$ and in $0.1 M$ Na_2CO_3.

13-7 Solutions of Lead Fluoride The solubility product of PbF_2 is about $4 \times 10^{-8} M^3$ at 25°C. (a) Calculate the solubility of PbF_2 in water in milligrams per 100 ml. (b) Calculate the solubility in mol liter^{-1} of PbF_2 in $0.20 M$ $Pb(NO_3)_2$. (c) The solubility of PbF_2 in nitric acid is appreciably greater than its solubility in pure water. Explain.

13-8 Test for Chloride Ion Suppose that a 100-ml sample of a solution is to be tested for chloride ion by addition of 1.0 ml of $0.20 M$ $AgNO_3$. What is the minimum number of grams of Cl^- that must be present in order for some $AgCl$ to be formed? (K_{sp} for $AgCl$ is $1.8 \times 10^{-10} M^2$.)

13-9 Insoluble Carbonates The solubility of $CaCO_3$ in water is about 7 mg liter^{-1}. Show how one can calculate the solubility product of $BaCO_3$ from this information and from the fact that when sodium carbonate solution is added slowly to a solution containing equimolar concentrations of Ca^{2+} and Ba^{2+}, no $CaCO_3$ is formed until about 90 percent of the Ba^{2+} has been precipitated as $BaCO_3$.

13-10 Precipitation of $PbBr_2$ The solubility product of $PbBr_2$ is $7 \times 10^{-5} M^3$. If 10 ml of a $0.02 M$ $NaBr$ solution is added to 10 ml of $0.2 M$ $Pb(NO_3)_2$ solution, will any solid $PbBr_2$ be present at equilibrium?

13-11 Solubility of Carbonates in Acids It is sometimes asserted that carbonates are soluble in strong acids because a gas is formed that escapes (CO_2). Suppose that CO_2 were extremely soluble in water (as, for example, ammonia is) so that it did not leave the site of the reaction, but that otherwise its chemistry was unchanged. Would calcium carbonate be soluble in strong acids? Explain.

13-12 Solubility of CaF_2 in an Acidic Solution The solubility of CaF_2 at pH = 1.0 is about $5.4 \times 10^{-3} M$. Use this information, together with the solubility product of CaF_2 ($K_{sp} = 3.4 \times 10^{-11} M^3$), to estimate the dissociation constant of HF.

13-13 Strontium Carbonate and Strontium Fluoride The solubility-product constant for $SrCO_3$ is $1.1 \times 10^{-10} M^2$ and that for SrF_2 is $2.9 \times 10^{-9} M^3$. Calculate the molar solubility of strontium carbonate in a solution in which the fluoride-ion concentration is held constant at $0.10 M$ by calculating the equilibrium concentration of CO_3^{2-} in such a solution. Is $[Sr^{2+}]$ equal to $[CO_3^{2-}]$ at equilibrium? If not, why not?

13-14 Separation of Fe(III) and Ni(II) The solubility products of $Fe(OH)_3$ and $Ni(OH)_2$ are about $10^{-36} M^4$ and $6 \times 10^{-18} M^3$, respectively. Find the approximate pH range suitable for the separation of Fe^{3+} and Ni^{2+} by precipitation of $Fe(OH)_3$ from a solution initially 0.01 M in each ion, as follows: (a) calculate the lowest pH at which all but 0.1 percent of the Fe^{3+} will be precipitated as $Fe(OH)_3$; (b) calculate the highest pH possible without precipitation of $Ni(OH)_2$.

13-15 Simultaneous Precipitation of AgCl and Ag_2CrO_4 The solubility products of AgCl and Ag_2CrO_4 are $1.8 \times 10^{-10} M^2$ and $2.0 \times 10^{-12} M^3$, respectively. Suppose that a very dilute $AgNO_3$ solution is added dropwise to a solution that is 0.0010 M in Cl^- and 0.010 M in CrO_4^{2-}. Ignore volume changes. Will AgCl or Ag_2CrO_4 precipitate first? Approximately what fraction of the anion first precipitated will have been removed when the second one starts to precipitate? (These results are made use of in the *Mohr titration* of Cl^- solutions with a solution of $AgNO_3$ of known concentration; a small amount of K_2CrO_4 is added to the solution. Since Ag_2CrO_4 is red-orange in color, the color of the precipitating solid changes from white to a reddish hue as the end point is reached, that is, when almost all the Cl^- has been precipitated.)

13-16 Heterogeneous Equilibrium Consider the reaction of Example 13-10

$$NH_4Cl(s) \rightleftharpoons NH_3(g) + HCl(g)$$

for which the equilibrium constant is 0.25 atm^2 at 340°C. Suppose the partial pressure of ammonia at 340°C in equilibrium with solid NH_4Cl is 0.75 atm. What will be the pressure of HCl?

13-17 Heterogeneous Equilibrium (a) Formulate the reaction quotient for the endothermic reaction

$$AgCl \cdot NH_3(s) \rightleftharpoons AgCl(s) + NH_3(g)$$

What is the effect on P_{NH_3} at equilibrium if (b) additional AgCl(s) is added or (c) additional $NH_3(g)$ is pumped into or out of the system, provided that neither of the two solid phases shown in the chemical equation is completely used up? (d) What is the effect on P_{NH_3} of lowering the temperature?

13-18 Thermal Decomposition of Barium Peroxide A 20.0-g sample of BaO_2 is heated to 794°C in a closed evacuated vessel of volume 5.00 liter. How many grams of the peroxide are converted to BaO(s)? Neglect the volume of the solid.

$$2BaO_2(s) \rightleftharpoons 2BaO(s) + O_2(g) \qquad K = 0.50 \text{ atm}$$

13-19 Dissociation of Calcium Carbonate One formula weight $CaCO_3(s)$ is heated to 850°C in a closed evacuated vessel of volume 10.0 liter. What fraction of the carbonate is converted to the oxide? Neglect the volume of the solids.

$$CaCO_3(s) \rightleftharpoons CO_2(g) + CaO(s)$$
$$\text{at } 850°C, \ K = 0.49 \text{ atm}$$

13-20 Solid and Three Gases Solid ammonium carbonate decomposes according to the equation

$$(NH_4)_2CO_3(s) \rightleftharpoons 2NH_3(g) + CO_2(g) + H_2O(g)$$

At a certain elevated temperature the total pressure of the gases NH_3, CO_2, and H_2O generated by the decomposition of, and at equilibrium with, pure solid ammonium carbonate is 0.400 atm. Calculate the equilibrium constant for the reaction considered. What would happen to P_{NH_3} and P_{CO_2} if P_{H_2O} were adjusted by external means to be 0.200 atm without changing the relative amounts of $NH_3(g)$ and $CO_2(g)$ and with $(NH_4)_2CO_3(s)$ still being present?

13-21 Calcium Chloride as a Drying Agent Calcium chloride is an effective drying agent. It reacts with water vapor to form a series of hydrates, the simplest of which is the monohydrate:

$$CaCl_2(s) + H_2O(g) \rightleftharpoons CaCl_2 \cdot H_2O(s)$$

The equilibrium constant for this reaction as written is 25 torr^{-1} at 25°C. (a) Formulate the expression for the equilibrium constant of this reaction, and give the equilibrium pressure of water over a mixture of anhydrous calcium chloride and its monohydrate at 25°C. (b) The equilibrium pressure rises as the temperature increases; i.e., calcium chloride is a less effective drying agent at higher temperatures. Is the reaction above exothermic or endothermic? Explain. (c) Suppose that a stream of air at 25°C and 1 atm, containing water vapor at a pressure of 0.025 atm, is passed over anhydrous $CaCl_2$ at a rate of 2.0 liter min^{-1}. If there is initially 1.00 mol $CaCl_2$ present, how long will it remain effective as a drying agent if the reaction above is assumed to be that which determines its effectiveness?

13-22 Decomposition of Solid PCl_5 When an excess of solid PCl_5 is heated in a previously evacuated flask at a certain temperature, the total pressure in the flask is 0.250 atm. The vapor pressure of PCl_5 at this temperature is 0.050 atm. (*a*) Give the numerical value of the equilibrium constant, defined by Equation (13-15a). (*b*) Suppose that at this temperature an excess of solid PCl_5 is introduced into a flask in which the partial pressure of PCl_3 is 0.200 atm. What will be the pressure of Cl_2 and the total pressure when equilibrium is established?

Particles, Waves, and Quantization

14

"There have probably been few events in the history of science that have had such extraordinary consequences within the brief span of one generation as Planck's discovery of the elementary quantum of action. This discovery not only has formed, to an ever increasing degree, the basis for bringing order into our knowledge of atomic events . . . , but at the same time it has caused a complete change in our fundamental ways of describing natural phenomena. . . . This new knowledge has shattered the conceptual foundations not only of classical physics but also of our ordinary modes of thinking. It is to just the liberation gained in this way that we owe the marvelous progress that has been made in our insight into natural phenomena during the past generation."
NIELS BOHR, 1929[1]

The foundations for an understanding of atomic structure, the periodic classification of the elements, and the binding of atoms together to form molecules are provided by the concepts and methods of quantum theory, which have revolutionized physics and chemistry during this century. This chapter introduces some of the fundamental ideas needed to understand the applications of quantum theory to problems of atomic and molecular structure, which are discussed in the following chapters.

14-1 Properties of Particles and Waves

During the first quarter of this century it was discovered that electromagnetic radiation such as light behaves in many circumstances like a stream of particles and that particles such as electrons and protons behave in many circumstances like waves. Both concepts—that of particles and that of waves—have their roots in classical physics, a description of the familiar macroscopic world. Their application to phenomena in the microscopic realm, phenomena involving individual atoms and molecules and their interaction with radiation, is a considerable extrapolation. The inadequacy of this extrapolation is presumably the reason that neither the wave nor the particle concept alone is sufficient to explain events on an atomic scale, where the laws of classical physics must be replaced by those of quantum mechanics. Before explaining what is implied by these statements, more needs to be said about particles and waves.

Particles Particles can be *counted* and follow well-defined paths (*trajectories*) when in motion. They have attributes such as *mass* and (often) *charge* and, if

---

[1] *Naturwissenschaften,* vol. 17, pp. 483 and 486, 1929.

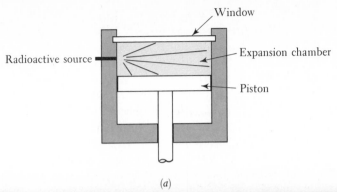

Window

Radioactive source

Expansion chamber

Piston

(a)

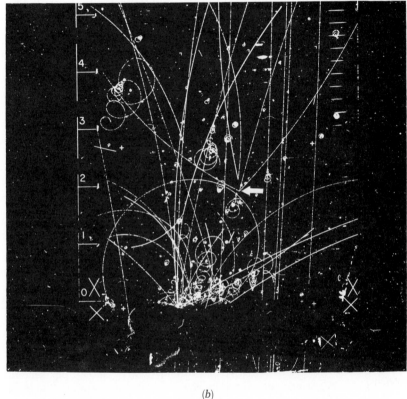

(b)

Figure 14-1   (a) Schematic Diagram of a Wilson Cloud Chamber; (b) Bubble Chamber Tracks

In (a) the expansion chamber contains air saturated with the vapor of water or ethanol. When the piston is moved downward quickly, the air expands and its temperature falls. The vapor becomes supersaturated and tends to condense in tiny droplets on the ions created when charged particles of high energy traverse the cloud chamber. The droplets make the particle trajectories visible.

In (b) in a bubble chamber tiny vapor bubbles form along the paths of charged particles passing through a liquid, such as liquid H_2, the pressure of which has just been reduced to slightly below the boiling pressure. A magnetic field causes the paths to be curved, permitting conclusions to be drawn about the momentum (the product of mass and velocity) of each particle. The arrow indicates the appearance of tracks produced by charged particles that originate from the decay of a neutral species.

moving, *kinetic energy* and *momentum*. Sometimes the observation of these properties is difficult and may require ingenious experimentation. For example, elaborate pieces of equipment such as a Wilson cloud chamber or a bubble chamber (Fig. 14-1) may be needed to observe the trajectories of subatomic particles.

The tracks of some kinds of particles have never been observed, yet these particles are still believed to exist. Sometimes detailed analyses of events that involve particles with visible tracks show discrepancies in the expected balances of kinetic energy and momentum (Study Guide), which can be explained by assuming the existence of a particle with an invisible track. To postulate the existence of a particle on such evidence is not as outlandish as it seems. Remember that it is in fact impossible to *prove* the existence of *any* particle (or of

285

anything else). The postulation of the existence of *anything* is always done in order to explain a set of observed events in a natural and unlabored way. No way has been found to make visible the trajectories of neutral particles such as neutrons and photons. However, the initial and final points of such trajectories can often be located because the particles that create a neutral particle upon collision or by radioactive decay may leave visible tracks. Moreover, when new particles are created from a neutral particle they also may have detectable tracks (Fig. 14-1*b*). In summary, then, we speak of particles when we observe events that can be described in terms of at least some typical attributes of particles.

Waves Some of the measurable attributes of waves are *frequency v* ("nu"), the number of vibrations per second; *amplitude A_0*; and *wavelength* λ ("lambda") (Fig. 14-2). The *velocity* of the wave, *u*, is equal to the distance traveled by the wave in one second. This distance must contain λ exactly *v* times; hence,

$$u = \lambda v \qquad\qquad (14\text{-}1)$$

Example 14-1

☐ **Properties of Waves** In the musical scale used internationally the frequency corresponding to the note A in the fourth octave on a piano is arbitrarily chosen as 440 s^{-1}. If the speed of sound in air is 330 m s^{-1}, what is the wavelength of sound in air corresponding to this A?

Solution Rearrangement of (14-1) gives

$$\lambda = u/v$$
$$\lambda = \frac{330\text{ m s}^{-1}}{440\text{ s}^{-1}} = \underline{0.750\text{ m}} \quad\blacksquare$$

Exercise 14-1

☐ The string in a piano for the note A in the fourth octave is 450 mm long, which is just one-half of the wavelength of the vibration that is excited when that string is struck. The frequency of this note is 440 s^{-1}. What is the speed with which this excitation travels on the struck string? (Note that the speeds with which a vibration travels may be quite different in different media.) ■

Another quantity that can be used to characterize a wave is the wave number $\tilde{v}$, the reciprocal of the wavelength,[2] $\tilde{v} = \lambda^{-1}$. The *intensity* (energy per second) associated with a wave is proportional to the square of its amplitude. The *phase* γ of the wave specifies the position of the crest with respect to a fixed point. As the wave travels, its phase changes, and since the choice of fixed point and time of measurement is arbitrary, the concept of phase itself may seem arbitrary. However, the *difference in phase* of two waves of the same wavelength, *traveling with the same speed*, is *not* arbitrary; it is independent both of the time of measure-

---

[2] Note that here the tilde (∼) does not indicate a molar quantity as it usually does in this text.

Figure 14-2 A Wave

The wavelength λ of the wave shown is the distance between neighboring crests or troughs or other corresponding points of the wave train. The phase γ is the position of a crest relative to a fixed point such as the origin. It may be specified as a fraction of λ, γ/λ being about 0.12 in this example.

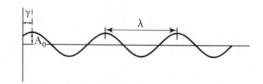

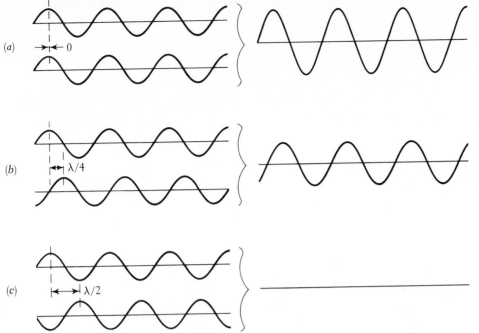

Figure 14-3 Interference of Two Waves

Three examples are shown of what occurs when two waves of the same wavelength and equal amplitude add. The two separate waves are shown on the left and their sum, or resultant wave, on the right. The three examples are characterized, respectively, by (a) a phase difference zero— total reinforcement; (b) a phase difference $\lambda/4$—partial reinforcement; (c) a phase difference $\lambda/2$—complete destruction. The resultant wave has the same wavelength λ [except in case (c), where the term *wavelength* has no meaning].

ment and of the choice of reference point for measuring the phase of either wave separately. Phase differences are extremely important in the interaction of waves. Various phase differences between two waves are illustrated in Fig. 14-3.

In certain experiments, waves show the phenomenon of *interference*. This means that the amplitudes of waves of the same wavelength, which may come from different directions, may add or subtract when they meet (Fig. 14-3). When the crest of one wave coincides with the crest of the other (which means that troughs also coincide), there is reinforcement or addition of the amplitudes: the two waves are *in phase* and undergo *constructive interference*. When the crest of one wave coincides with the trough of the other, the two waves cancel. They are *out of phase*; there is *destructive interference*.

14-2   Light and Electromagnetic Radiation

The Electromagnetic Spectrum   In 1672 Newton published his first scientific paper, an account of experiments in which sunlight was passed through a glass prism and was spread out into a rainbow-colored band (Fig. 14-4a). Further experiments convinced him that white light was a mixture of light of all colors of the rainbow, and he suggested that each color was associated with different-wavelength vibration of a fluid called the "ether", which filled all space but could not be perceived.

The wave nature of light was not established until the early years of the nineteenth century when Thomas Young, A. J. Fresnel, and others showed that it offered an explanation for all the properties of light that had been observed, particularly interference phenomena.

In 1864, Maxwell showed on theoretical grounds that the properties of light were those expected of electromagnetic waves. Associated with such waves are a

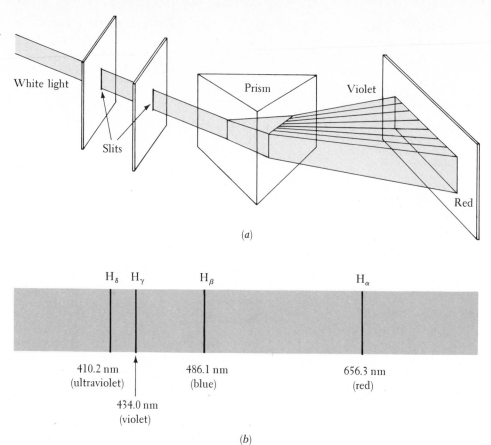

(a)

(b)

Figure 14-4 Schematic Spectrograph and a Typical Atomic Emission Spectrum
(a) In the optical range, where glass prisms, lenses, and photographic plates can be used, the
spectrum may be examined after the light has passed through a fine slit and been spread out by
a prism or grating. (b) Atomic spectra are characterized by spectral lines, each of which consists
of radiation of a particular wavelength, or frequency, also called *monochromatic radiation*. Each
kind of atom has its own characteristic sets of spectral lines in absorption and in emission. The
spectrum of H is typical. The diagram shows both the wavelengths and the customary symbols
for the four most common lines in the emission spectrum of H (the *Balmer* lines).

magnetic field and an electric field that oscillate in a plane perpendicular to the
direction of travel. All electromagnetic waves have the same speed in empty
space, called the speed of light and symbolized by c. The whole wavelength range
of electromagnetic waves constitutes the *electromagnetic spectrum*. Since, as
discussed below, visible light is part of the electromagnetic spectrum, the term
light is often used to designate electromagnetic radiation in general.

Planck's Radiation Law Matter may absorb or emit electromagnetic radiation.
The intensity of the radiation emitted by an object at each wavelength (its
emission spectrum) or the amount of radiation absorbed at each wavelength (its
absorption spectrum) can be measured by instruments related to the simple
prism used by Newton (Fig. 14-4).

Absorption and emission spectra were first intensively investigated during the
last decades of the nineteenth century. Absorption was recognized as representing a gain in energy, emission a loss. Although the energy received by absorption

at certain wavelengths may lead to specific phenomena, such as the emission of radiation at a new wavelength, the general result of absorption is a temperature increase of the body absorbing the radiation. Emission of radiation correspondingly causes a temperature decrease. In such a situation there is always the possibility of equilibrium—the radiation intensity may be such that just as much radiation (and thus energy) is emitted by an object at each wavelength as is absorbed by that same object at that wavelength. It is found that when there is equilibrium between radiation and matter at a given temperature, the relative intensities of light of each wavelength do not depend on the chemical or physical nature of the object (Fig. 14-5). Radiation that has this particular spectral composition is called *blackbody radiation.*

As shown in Fig. 14-5, when the temperature increases, the maximum in the spectral distribution of blackbody radiation shifts to shorter wavelengths, and the total intensity increases rapidly. A related effect is familiar to everyone. When a piece of metal (or anything else) that does not burn is heated sufficiently it first begins to glow with a dull reddish color (red has the longest visible wavelength; see Fig. 14-8). As the object is heated further the amount of light emitted increases greatly and the color becomes more orange-yellow (a shorter wavelength than red). If the object is heated to a high enough temperature it becomes brilliantly "white hot".

Many unsuccessful efforts were made to derive from theory the experimentally well-established spectral distribution of blackbody radiation. Finally, in 1900, Planck succeeded in finding by trial a mathematical formula to fit the curves, and a few weeks later he derived this formula—now known as Planck's radiation law—by theoretical arguments. Planck showed that one revolutionary assumption was needed: his radiation law could be derived only if it was assumed that absorption and emission of light by matter involve discrete units of energy $h\nu$, where h is called Planck's constant. He called these packages of energy *quanta* (the plural of *quantum*). Their size is proportional to ν, the frequency of the radiation.

Planck was cautious and said only that the absorption and emission of light were quantized but not that the light itself must therefore exist as particles. The quantization of energy that he postulated was a new phenomenon and was accepted only hesitantly by many; even the quantization of matter, in the form of discrete atoms, was not accepted by all the leading physicists and chemists of

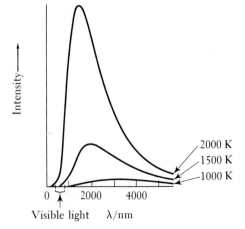

Figure 14-5 The Spectral Distribution of Blackbody Radiation at Three Temperatures

The intensity (energy per second) of blackbody radiation is shown as a function of the wavelength at three different temperatures. Only a small part of the total wavelength range shown constitutes visible light. Note that (1) the total intensity (total area under a curve) increases rapidly as the temperature is increased; (2) the maximum of the intensity curve shifts toward shorter wavelengths as the temperature is increased; (3) only a small fraction of the total intensity falls into the visible range, even at 2000 K. Most of the radiation is in the infra-red region. We can feel it because it causes heating as it is absorbed by our bodies even at temperatures below those at which a dull reddish glow can be seen. The formula that describes these curves was derived by Planck.

289

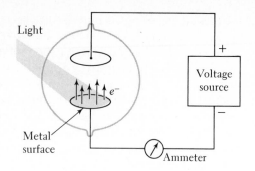

Figure 14-6 The Photoelectric Effect
When a metal surface is irradiated by light of sufficiently short wavelength (for some metals ultraviolet radiation may be required), electrons are emitted by the metal. The apparatus schematically shown here provides a way to measure the number of electrons emitted in a given time (the photocurrent).

that day, and hence it is not surprising that Planck's quantum hypothesis met with some skepticism. However, the real test of any hypothesis is that it can be used to explain more than one independent set of facts, more than those it was tailored to fit. Within a very few years the quantum hypothesis had passed this test magnificently: Einstein had applied it to radiation and Bohr to atomic structure, and even the skeptics were convinced.

The Photoelectric Effect When light shines on a metal surface, the surface may emit electrons, a phenomenon called the *photoelectric effect* (Fig. 14-6). In 1905 Einstein, by an analysis of this effect, showed that under certain conditions electromagnetic radiation has the characteristics of particles. It had been determined that the maximum kinetic energy of the electrons emitted by the metal depends not upon the *intensity* of the light—that is, not upon the *total energy* striking the surface per second—but only upon the *frequency* of the light (Fig. 14-7). This is in contradiction to the supposed wave nature of light. If light behaved like ordinary waves, the electrons emitted would be expected to have less and less kinetic energy as the intensity of the light is decreased. What happens

Figure 14-7 The Photoelectric Effect
In this schematic representation photons of energy hν are shown as dark spheres and electrons as light spheres. (*a*) When a photon of sufficiently high energy strikes the surface of a metal an electron is emitted. The kinetic energy of the electron is the difference between the energy of the photon and the energy with which the electron is bound in the metal. (*b*) The intensity of the light beam corresponds to the number of photons per second. If the light intensity is increased the number of electrons emitted per second increases proportionally, but their kinetic energy is unchanged. (*c*) If the light is of higher frequency, more energy is given to the electrons—their kinetic energy is increased. A minimum frequency ν_{min} is required for emission because the photon energy must at least equal the binding energy of the electrons.

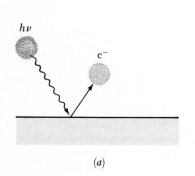

(*a*)

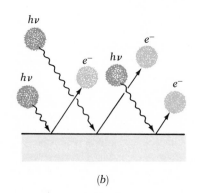

(*b*)

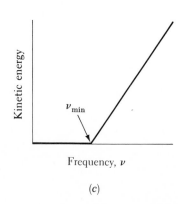

(*c*)

instead is that if the intensity of the light is cut in half, only half as many electrons are emitted but the kinetic energy of each electron remains at the original value provided that the color of the light—that is, its frequency—is not changed. As the intensity is decreased further, fewer and fewer electrons emerge, but at *any* level of intensity some electrons are emitted, and their maximum kinetic energy does not depend on the level of illumination.

Einstein saw that the *one* assumption that explained this effect was that in some way light had particle nature and that the energy of these light quanta or *photons* was constant for electromagnetic radiation of a given frequency. A decrease in light intensity would thus result in a decrease in the number of photons involved, but not in a change of their energy. At low intensity levels the number of photons might be so small that just *one* photon would strike the metal surface every few seconds or hours, but these photons would have the same energy and each would cause the emission of one electron. As long as there was *any* light, it would exist as photons and cause electron emission. Furthermore, the energy of the electrons emitted increases as the frequency of the light used for irradiation increases, and the energy of each photon is found to be just that demanded by Planck's earlier considerations.

Physical Characteristics of Photons   The *energy* of a photon is found to be

$$\epsilon = h\nu = \frac{hc}{\lambda} \tag{14-2}$$

where ν and λ are the frequency and wavelength of the radiation and c is the velocity of light, with both λ and c pertaining to radiation in a vacuum. The quantity h is Planck's constant, $h = 6.6262 \times 10^{-34}$ J s in SI units. The mass of a photon of a given frequency is obtained by dividing (14-2) by c^2, using Einstein's relationship[3] $\epsilon = mc^2$ (Study Guide):

$$m = \frac{\epsilon}{c^2} = \frac{h\nu}{c^2} = \frac{h}{c\lambda} \tag{14-3}$$

The momentum of a photon is obtained by multiplying its mass by its velocity:

$$\text{Momentum of photon} = \frac{h\nu}{c^2} c = \frac{h\nu}{c} = \frac{h}{\lambda} \tag{14-4}$$

These expressions for photon mass and momentum have also been established experimentally.

The statement that light may behave like particles as well as waves means that light may exhibit some of the attributes of particles as well as those of waves.

---

[3] It should be noted that a photon has mass only by virtue of the fact that its velocity is the velocity of light, c. By the special relativity theory, if a particle has the mass m_0 when at rest, it has the mass

$$m = \frac{m_0}{\sqrt{1 - v^2/c^2}} \tag{1}$$

when it is moving with velocity v. At the limit where v approaches the velocity of light c, the denominator approaches zero and hence the mass m approaches infinity unless the rest mass m_0 is also zero. Two conclusions follow: (1) only particles of rest mass zero can move at the speed of light; (2) when such particles do move at the speed of light, Equation (1) yields the indeterminate expression $\frac{0}{0}$ for their mass. The mass can be found if the energy is known, as shown for photons with Equation (14-3).

Which attributes are exhibited depends on the experimental situation. When light passes through slits, interference phenomena occur that are characteristic of waves; when light interacts with matter, phenomena such as energy and momentum transfer bring out its particle nature. In interactions with matter, the particle aspects of electromagnetic radiation become more important as the frequency becomes larger, because the mass of the photon increases with frequency. Thus, particle aspects of radio waves, which are of relatively low frequency (Table 14-1), are almost never observed.

Energy Relationships Although the energy of electromagnetic radiation is quantized, the amounts of energy in two "packages", or quanta, of different frequency are not the same, because by (14-2) these amounts are directly proportional to the frequency (and thus inversely proportional to the wavelength). It is very helpful to have a feeling for the energy of the individual photons associated with radiation of a given frequency or wavelength. Consider, for example, blue-green light of wavelength 5.0×10^2 nm (or 5.0×10^{-7} m), which corresponds to the wave number[4] $\tilde{\nu} = 1/\lambda = 2.0 \times 10^4$ cm^{-1} and the frequency $\nu = c/\lambda = c\tilde{\nu} = (2.0 \times 10^4 \text{ cm}^{-1})(3.0 \times 10^8 \text{ m s}^{-1})(100 \text{ cm/m}) = 6.0 \times 10^{14} \text{ s}^{-1}$. A photon of this frequency has the energy $h\nu = (6.63 \times 10^{-34} \text{ J s})(6.0 \times 10^{14} \text{ s}^{-1}) \approx 4.0 \times 10^{-19}$ J. It is often advantageous to be able to express energies in different units so that comparisons of different kinds of energy quantities can be made; the conversion factors in Tables A-4 and A-5 may be used. Two of the most common energy units in chemistry and chemical physics are (1) the *electronvolt*, eV, which is the kinetic energy of an electron that has been accelerated, for example, by being placed between two parallel metal plates, with an electrical potential difference of exactly 1 V; and (2) the *kilojoule per mole*, kJ mol^{-1}, the energy in kilojoules of 1 mol of whatever is being considered.

The conversion from joules to electronvolts is made by using the relationship $1 \text{ eV} = 1.6022 \times 10^{-19}$ J (Table A-5), which gives the photon energy considered in the previous paragraph, 4.0×10^{-19} J, a value in electronvolts of about $\frac{4.0}{1.6} = 2.5$ eV. The energy units electronvolts permit easy comparison with the energy stored in familiar chemical sources of electrical energy, such as storage batteries and flashlight batteries. A 2.5-V battery can provide electrons with a kinetic energy about equal to the energy of a 500-nm photon. In this sense such a battery and blue-green light are energetically equivalent. It is easy to remember that the energies of ordinary chemical reactions correspond to a few electronvolts because common batteries, which operate by means of chemical reactions, furnish potentials of a few volts per cell and can thus supply electrons with energies of a few electronvolts.

To find the energy in kilojoules of a *mole* of photons of wavelength 500 nm, we multiply by N_A. The energy of N_A photons of this blue-green light is thus

$$(6.0 \times 10^{23} \text{ photons/mol})(4.0 \times 10^{-19} \text{ J/photon}) = 24 \times 10^4 \text{ J mol}^{-1}$$
$$= 2.4 \times 10^2 \text{ kJ mol}^{-1}$$

This energy (or, more precisely, 239 kJ mol^{-1}) is just below the bond energy of Cl_2, the energy required to convert 1 mol Cl_2 molecules into atoms, which is

---

[4]The customary units for $\tilde{\nu}$ are cm^{-1}.

242 kJ mol^{-1}. Thus, a photon of slightly shorter wavelength than 500 nm (for example, 490 nm) has sufficient energy to split one Cl-Cl bond.

It is important to recognize the special nature of the unit kilojoules per mole, and in particular the difference between it and the kilojoule. While the kilojoule is a plain energy unit, the unit kilojoules per mole implies that the energy quantity considered pertains to exactly 1 mol of whatever is being discussed, each of the N_A individuals in the mole having on the average $1/N_A$ of the total energy. For example, the statement that the dissociation energy of the Cl-Cl bond is 242 kJ mol^{-1} means that to break the bonds of N_A *molecules* of Cl_2 requires 242 kJ. The statement that the dissociation energy is 2.5 eV means that it takes 2.5 eV to break *one* Cl-Cl bond. It also takes $242/(6.02 \times 10^{23}) = 4.02 \times 10^{-22}$ kJ to break this one bond.[5]

□ **Thermal Energy at Room Temperature** What is the average translational energy for a gas at 25°C in joules per mole and electronvolts per molecule?

Example 14-2

Solution As shown in Sec. 4-4, this energy is $\frac{3}{2}RT$ for 1 mol gas, or

$$1.5 \times 8.314 \text{ J K}^{-1} \text{ mol}^{-1} \times 298 \text{ K} = 3716 \text{ J mol}^{-1}$$
$$= \underline{3.72 \text{ kJ mol}^{-1}}$$

This is equivalent (see Table A-5) per molecule to

$$3.72 \text{ kJ mol}^{-1} \times 0.01036 \frac{\text{eV}}{\text{kJ mol}^{-1}} = \underline{0.0385 \text{ eV}} \quad \blacksquare$$

□ The atmosphere of the planet Venus has been reported to reach temperatures as high as 750 K. What is the average translational energy of a gas corresponding to this temperature, in joules per mole and in electronvolts per molecule? ■

Exercise 14-2

The quantitative relation between the energy of a photon and the frequency of radiation, Equation (14-2), can be combined with the principle of conservation of energy to determine energy relations in processes involving the emission or absorption of light. In the photoelectric effect, for example, energy conservation requires that the energy of a photon absorbed by the metal be equal to the energy imparted to the electron emitted:

Photon energy = binding energy of electron + electron kinetic energy

There are two terms on the right because the electron is bound in the metal and energy must be expended to remove it. If the photon energy is less than the binding energy, an electron will not be emitted; energy in excess of the binding energy appears as the kinetic energy of the free electron. The photon energy can be expressed by (14-2), and the kinetic energy of the electron can be written in terms of its mass and speed

$$h\nu = \text{binding energy} + \tfrac{1}{2}mv^2 \qquad (14\text{-}5)$$

---

[5] It is sometimes customary to make "per molecule" part of the units of a quantity (e.g., as in 2.5 eV per molecule). This practice is followed in Example 14-2, but we shall not do it generally.

Example 14-3

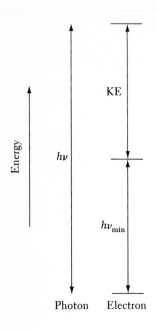

Energy

$h\nu$

$h\nu_{\text{min}}$

KE

Photon   Electron

□ **Photoelectric Effect** When light of wavelength 4.00×10^{-7} m strikes the surface of a calcium electrode, electrons with a kinetic energy of 6.3×10^{-20} J are emitted. Calculate the binding energy in electronvolts of electrons in calcium and the minimum frequency of light required for observation of the photoelectric effect.

Solution By combining (14-2) and (14-5) we can write

$$h\nu = hc/\lambda = \text{binding energy} + \text{kinetic energy}$$

Binding energy $= hc/\lambda - \text{kinetic energy}$

$$= \frac{6.63 \times 10^{-34}\,\text{J s} \times 3.00 \times 10^{8}\,\text{m s}^{-1}}{4.00 \times 10^{-7}\,\text{m}} - 6.3 \times 10^{-20}\,\text{J}$$

$$= 4.97 \times 10^{-19}\,\text{J} - 6.3 \times 10^{-20}\,\text{J} = 4.34 \times 10^{-19}\,\text{J}$$

$$= 4.34 \times 10^{-19}\,\text{J} \times \frac{1\,\text{eV}}{1.602 \times 10^{-19}\,\text{J}} = \underline{2.71\,\text{eV}}$$

The minimum frequency required for observation of the photoelectric effect is the frequency for which the electron is emitted with no kinetic energy—at any lower frequency the energy imparted to an electron will be less than the binding energy. Thus

$$h\nu_{\text{min}} = \text{binding energy}$$

$$\nu_{\text{min}} = \frac{\text{binding energy}}{h}$$

$$= \frac{4.34 \times 10^{-19}\,\text{J}}{6.63 \times 10^{-34}\,\text{J s}} = \underline{6.55 \times 10^{14}\,\text{s}^{-1}} \blacksquare$$

Exercise 14-3

□ The binding energy of an electron in strontium is 2.74 eV. Calculate the kinetic energy of an electron emitted from strontium when it is irradiated with light of wavelength 4.20×10^{-7} m. ∎

Regions of the Electromagnetic Spectrum The electromagnetic spectrum may be divided into different regions, depending on the modes of production and detection of the radiation, as well as on other criteria. Table 14-1 illustrates one such scheme; the optical and "visible" regions are shown in Fig. 14-8. There are, of course, no sharp boundaries; adjacent regions blend into one another smoothly.

Radio waves, the region of the spectrum with wavelengths above 0.1 m, had few chemical applications until recently because the corresponding energies are

Table 14-1
The Electromagnetic Spectrum

RANGE					
$\bar{\nu}/\text{cm}^{-1}$	ν/s^{-1}	λ/m	Energy/eV	Designation	Mode of generation
(0)–10^{-1}	(0)–3×10^{9}	(∞)–10^{-1}	0–10^{-5}	Radio waves	Electrical oscillations
10^{-1}–10^{6}	3×10^{9}–3×10^{16}	10^{-1}–10^{-8}	10^{-5}–10^{2}	Optical radia-tion	Atomic, molecular, and ionic events
10^{6}–10^{9}	3×10^{16}–3×10^{19}	10^{-8}–10^{-11}	10^{2}–10^{5}	X rays	Transitions of inner atomic electrons
10^{9}–(∞)	3×10^{19}–(∞)	10^{-11}–(0)	10^{5}–(∞)	γ rays	Nuclear phenomena

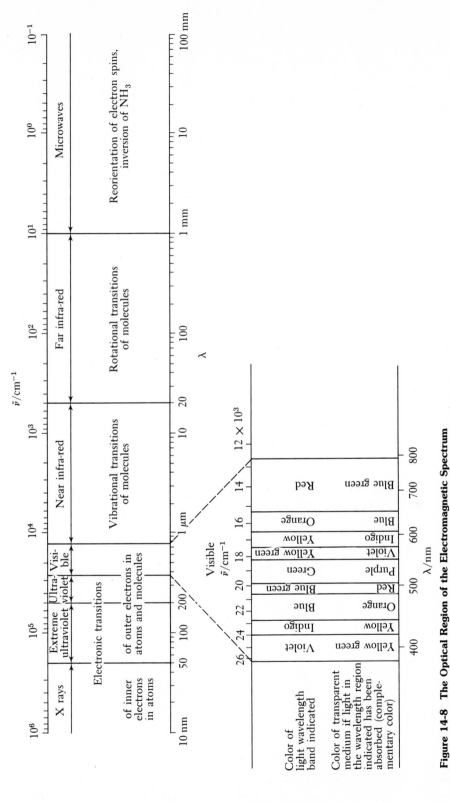

Figure 14-8 The Optical Region of the Electromagnetic Spectrum

Note the complementary colors indicated at the bottom of the enlarged visible portion of the spectrum.

very low. Radio waves have become important chemically during the last three decades because of the development of nuclear magnetic resonance (nmr), a powerful tool that involves radio-frequency spectroscopy and is used for studying problems of molecular structure (Sec. 23-1).

The portion with wavelengths from about 10^{-8} to 10^{-1} m is called *optical radiation*. Visible light (400 to 750 nm) represents only a minor part of the region, but the term *optical* is applied to this entire region of the spectrum because the radiation can be manipulated by the same types of instruments that are used for light, such as diffraction gratings, interferometers, lenses, and prisms.

Radiation in the visible and nearby ultraviolet and infra-red regions of the spectrum is emitted by atoms, molecules, and ions that have been excited thermally or in some other way, as by an electric discharge. Such radiation can also be emitted by species formed in a high-energy state by a chemical reaction. Conversely, many important reactions, known generally as photochemical reactions, are initiated by the absorption of photons in these regions of the spectrum. As indicated in the upper portion of Fig. 14-8, radiation in different parts of the optical region can be used to study changes in different kinds of molecular energy. In a very rough way, one can generalize that radiation in the far infra-red region corresponds to changes in rotational energy of molecules, that in the near infra-red to changes in vibrations of atoms relative to one another, and that in the visible and near ultraviolet to changes in the arrangement of the outer electrons in atoms and molecules.

X rays are generated by energy changes involving the inner electrons of atoms, and thus study of these changes yields information about the inner electronic structure of the atom. X rays are also produced when electrons or other particles are stopped rapidly. There are no practical lenses available for X rays because the wavelengths are of the order of atomic dimensions, so that arrays of atoms act like gratings and diffract the radiation rather than simply altering its speed. This behavior can be turned to advantage, and X-ray diffraction is an important tool in the determination of molecular structure.

Gamma rays are produced by changes (transitions) from high-nuclear-energy states to lower ones, states that involve different "arrangements" of the components of nuclei in the same sense that photons in the visible and ultraviolet

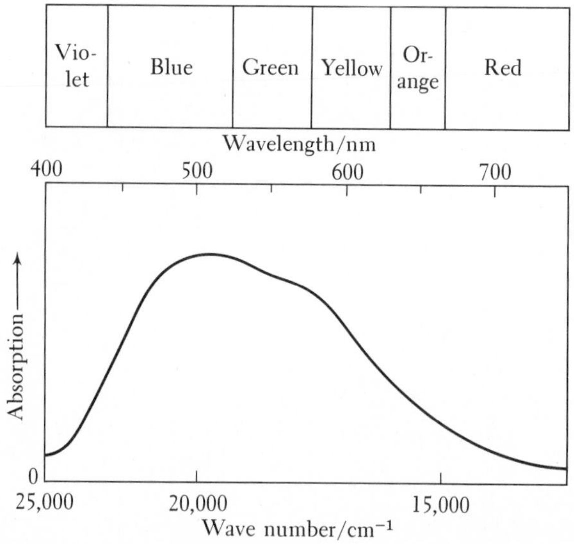

Figure 14-9 A Continuous Absorption Spectrum
This graph represents the absorption spectrum of a solution of $Ti(H_2O)_6{}^{3+}$ in water. The relative intensity of absorption of radiation is plotted as a function of the frequency. The sample shows maximum absorption in the blue-green region of the spectrum, and consequently the transmitted light is primarily in the violet and orange-red regions. Thus the solution appears purple when viewed by transmitted light.

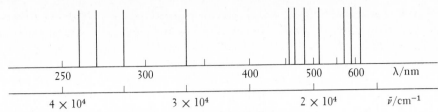

Figure 14-10 Line Emission Spectrum of Atomic Sodium
The lines represent the wavelengths at which atomic sodium emits radiation when in an electric discharge or a flame. The different lines occur with very different intensity, the most prominent feature being the two lines at 589.6 nm and 589.0 nm (not separated in the figure). This radiation is the cause of the intense yellow-orange color of sodium lamps or of flames containing sodium. The intensities are not represented here because of the difficulty of varying the shading in a line drawing. Each line is an image of the slit system of the spectrometer (see Fig. 14-4a).

regions of the spectrum correspond to transitions between states involving different arrangements of the outer electrons in atoms. The fact that γ rays have energies some 10^6 times those of photons in the visible region implies that the energy changes accompanying the rearrangement of nuclear components (nuclear reactions) are of the order of a million times those of ordinary chemical reactions, which involve the rearrangement of the outer electrons of atoms. The enormous potential of nuclear reactions as energy sources—for peaceful uses such as electric power generation and propulsion of ships and space vehicles as well as for destructive purposes—is a direct consequence of this fact.

The absorption and emission spectra of any atomic or molecular species (Fig. 14-4) are very characteristic and are often used for identification and quantitative analysis. These spectra depend greatly on the conditions under which they are obtained, such as state of aggregation, temperature, and solvent present (if any). A spectrum that consists of a continuous range of frequencies (and thus wavelengths) is called a *continuous spectrum* (Fig. 14-9), whereas in a *line spectrum* (Fig. 14-10) only certain very narrow regions of frequencies, called spectral lines, are obtained.

14-3 The Wave Nature of Particles

The de Broglie Wavelength The wave nature of particles was discovered some two decades later than the particle nature of light. It was in 1923 that Louis de Broglie first published a satisfactory argument (later amplified in his doctoral thesis) that there might be a wave nature to particles just as there is a particle nature to electromagnetic waves. De Broglie showed that if particles had wave properties, the wavelength associated with them should be

$$\lambda = \frac{h}{p} = \frac{h}{mv} \qquad (14\text{-}6)$$

This is called the *de Broglie wavelength* of a particle of mass m, speed v, and momentum p. The relationship given in (14-6) is the same as that which applies to photons, Equation (14-4).

The first particle for which a wave nature was demonstrated experimentally was the electron. For nearly three decades after their discovery, electrons had been regarded only as particles. They had been recognized as the carriers of electricity in metals, their mass and charge had been measured, their tracks had been seen in cloud chambers, and they had been identified as the once mysterious "beta rays" from radioactive substances. It did not occur to anyone to perform experiments that might demonstrate that particles had wavelike properties until de Broglie's thesis appeared. However, soon afterward it was shown experimentally by G. P. Thomson that the passage of electrons through thin films of crystalline material was accompanied by diffraction effects similar to those exhibited by X rays. C. Davisson and L. H. Germer showed independently that reflection of electrons by crystalline surfaces also produced diffraction effects. Other particles, including entire atoms and molecules, have also been shown to have wave character. Indeed, the diffraction of electrons and neutrons has been developed into a powerful method for investigating the atomic structure of matter.

Example 14-4

□ **De Broglie Wavelength** What is the de Broglie wavelength of an electron of speed 100 km s^{-1}?

Solution The momentum is

$$p = mv = (9.1 \times 10^{-31} \text{ kg})(100 \times 10^3 \text{ m s}^{-1})$$
$$= 9.1 \times 10^{-26} \text{ kg m s}^{-1}$$

Inserting this value into de Broglie's formula for the wavelength, we obtain

$$\lambda = \frac{6.6 \times 10^{-34} \text{ kg m}^2 \text{ s}^{-1}}{9.1 \times 10^{-26} \text{ kg m s}^{-1}}$$

$$= 0.73 \times 10^{-8} \text{ m} = 0.73 \times 10^{-8} \times \frac{1 \text{ nm}}{10^{-9} \text{ m}}$$

$$= \underline{7.3 \text{ nm}} \quad \blacksquare$$

Exercise 14-4

□ What is the de Broglie wavelength of a proton of speed 50 km s^{-1}? ∎

The Uncertainty Principle An important aspect of the wave-particle duality of matter and radiation is the uncertainty relation, discovered first by W. Heisenberg in 1927 through a careful analysis of the procedures required to determine the location and the momentum of a particle.

Consider, for example, a particle whose momentum is precisely known. Suppose that we wish to determine its position at a given time. To do this we must *observe* the particle, which requires a beam of light. The most precise optical measurements of distance give results in terms of numbers of wavelengths of the light beam used for observation. Thus, as the wavelength of the light used is shortened, the precision of the measurement of the position improves because the comparison units on the measurement scale (wavelengths of light) become smaller. However, light consists of photons, each with the momentum h/λ, and the observation of the particle requires that it collide with at least one photon and deflect this photon toward the observer. This collision necessarily transfers momentum to the particle observed, so that the actual momentum of the particle is different from that known precisely before collision, and the measurement is uncertain by this amount. The amount of momentum transferred to the particle increases as light of shorter and shorter wavelength is used. Hence the more accurately the position of the particle is measured, the less precise is the final knowledge of its momentum. Conversely, if the disturbance of the particle momentum is diminished by choosing radiation of long wavelength, the precision with which the particle location is known decreases. Careful analysis of this situation leads to the *Heisenberg uncertainty relation*

$$\delta x \, \delta p_x \gtrsim h \qquad (14\text{-}7)$$

which relates δx, the uncertainty in position of the particle in a direction, and δp_x, the uncertainty in its momentum in the same direction. The same relationship emerges again and again upon analysis of different methods of measurement. It thus appears to be independent of the particular method employed to measure x and p_x, and Bohr and Heisenberg postulated that the relationship was part of the nature of matter and radiation. Similar relationships are obtained for the y and z directions. Perfect knowledge of the position of a particle thus precludes *any* knowledge of its momentum and thus its velocity. Conversely, perfect knowledge of the momentum and velocity of a particle precludes any knowledge of its position.

By analysis of the measurement of the energy of a system, Heisenberg and Bohr discovered that the uncertainty $\delta\epsilon$ in this measurement depends on the time δt taken for the measurement in such a way that[6]

$$\delta\epsilon \, \delta t \gtrsim h \qquad (14\text{-}8)$$

The variable pairs (x,p_x), (y,p_y), (z,p_z), and (t,ϵ) are called pairs of *conjugate variables,* and the uncertainty principle states that conjugate variables may not be known *simultaneously* with a precision that would violate (14-7) or (14-8). There are other pairs of conjugate variables for which analogous relationships exist.

The effects of the uncertainty principle can be seen in many different phenomena. For example, vibrational motion of atoms in solids persists even at the lowest possible temperatures. If it did not, the momentum of any atom would be zero (except for possible motion of the crystal), the atomic positions could be measured quite precisely by diffraction methods, and the product of the uncertainties in position and momentum would be far less than h. Similarly, the energies of states with very short lifetimes can never be measured precisely. They inevitably turn out to be "fuzzy", that is, repeated measurements give a spread of values of the energy rather than one sharply defined value. When the lifetime is very short, the uncertainty in it is correspondingly low and hence, by (14-8), the uncertainty in the energy must increase. We discuss a number of other manifestations of this most fundamental principle in later chapters; one of the most illuminating is its bearing on the sizes of atoms (Sec. 15-6).

□ **Uncertainty Principle** If the position of a hydrogen atom is known with an uncertainty of 1.0×10^{-11} m, what is the minimum uncertainty in its velocity?

Example 14-5

Solution The *minimum* uncertainty is fixed by the Heisenberg uncertainty relation (14-7). Given the uncertainty in position, the corresponding mini-

---

[6]Note that the dimensions of h are mass $\times$ length^2/time, which are the same as the dimensions of energy $\times$ time.

mum uncertainty in momentum is

$$\delta p = \frac{h}{\delta x} = \frac{6.63 \times 10^{-34}\,\text{J s}}{1.0 \times 10^{-11}\,\text{m}}$$

$$= 6.6 \times 10^{-23}\,\text{kg m s}^{-1}$$

Since the momentum is mv, the uncertainty in the momentum is $m\delta v$. Thus

$$\delta v = \frac{\delta p}{m} = \frac{6.6 \times 10^{-23}\,\text{kg m s}^{-1}}{1.67 \times 10^{-27}\,\text{kg}}$$

$$= \underline{4.0 \times 10^4\,\text{m s}^{-1}} \quad \blacksquare$$

Exercise 14-5   □ The velocity of a helium atom is known with an uncertainty of 2.0×10^4 m s^{-1}. What is the minimum uncertainty in its position? ■

Summary

In classical physics a clear distinction can be made between particles and waves. Particles can be counted and possess attributes such as mass, kinetic energy, momentum, and often charge. Typical attributes of waves are the wavelength λ, frequency ν, and speed of propagation u, which fulfill the relation $\lambda = u/\nu$. Other characteristics of waves are amplitude, phase, and intensity. Waves can show the phenomenon of interference. In constructive interference, waves in phase combine with a resultant increase in amplitude; destructive interference is the combination of waves out of phase to produce a wave of smaller amplitude.

Light and other forms of electromagnetic radiation exhibit the properties of both waves and particles. Light waves of frequency ν behave in many ways like photons, particles of energy $h\nu$ and momentum $h\nu/c$, where h is Planck's constant and c is the speed of light. The photoelectric effect is evidence for the particle nature of light. Particles such as electrons and protons (and even atoms and molecules) can display wave properties. The de Broglie relation $\lambda = h/mv$ gives the wavelength associated with a particle of momentum mv.

The whole wavelength range of electromagnetic waves constitutes the electromagnetic spectrum. The electromagnetic spectrum can be classified into different regions, including radio waves, optical radiation (subdivided into infra-red, visible, and ultraviolet regions), X rays and γ rays. Radio waves are of long wavelength (low frequency, low energy); γ rays and X rays are of short wavelength (high frequency, high energy).

Uncertainties δx and δp_x are associated, respectively, with the simultaneous measurement of the position x and momentum p_x of a particle in a given direction. The Heisenberg uncertainty principle, $\delta x\,\delta p_x \gtrsim h$, sets a fundamental limit on the precision of such measurements and therefore represents a limit on the information that can be obtained in an experiment. A similar relationship holds for certain other pairs of variables, such as energy and time. Thus $\delta \epsilon\,\delta t \gtrsim h$, where $\delta \epsilon$ is the uncertainty in a measurement of the energy of a system ϵ and δt is the uncertainty in the time interval t during which the measurement is made.

Terms and Concepts

Problems and Questions

14-1 Characteristic Properties of Waves and Particles Describe as precisely as possible some of the characteristics of (macroscopic) waves and of (macroscopic) particles.

14-2 Photoelectric Effect Explain what is meant by the *photoelectric effect*. Why did it play an important role in the development of the particle-wave interpretation of the nature of electromagnetic radiation?

14-3 Frequency and Wavelength (a) Calculate the frequency in s^{-1} of electromagnetic radiation of wavelength 300 m, 1 cm, 500 nm, 0.1 nm. Indicate in what region of the spectrum each of these wavelengths would be found. (b) Calculate the wavelength of a photon of frequency 10^{12} s^{-1}, 10^{15} s^{-1}, 10^{19} s^{-1}. Indicate in what region of the spectrum each of these wavelengths would be found.

14-4 Photoelectric Effect Photons of wavelength 273 nm or less are needed to cause ejection of electrons from a surface of electrically neutral tungsten. (a) What is the energy barrier, in joules and in electronvolts, that electrons must overcome to leave an uncharged piece of tungsten? (b) What is T such that kT is equal to this energy, with k the Boltzmann constant? (c) What is the maximum kinetic energy, in joules and in electronvolts, of electrons ejected from tungsten by photons of wavelength 150 nm?

14-5 Photoelectric Effect A minimum of 2.26 eV is required to knock electrons from the surface of an uncharged potassium electrode. What maximum kinetic energy do such electrons have when the wavelength of the incident light is 300 nm?

14-6 Photodissociation In the initial stages of the chemical reactions that produce photochemical smog, an N—O bond in nitrogen dioxide is broken by the absorption of light:

$$NO_2 + h\nu \longrightarrow NO + O$$

This reaction requires light of wavelength shorter than 430 nm. What is the minimum energy required to break this N—O bond?

14-7 Planck Radiation Law The maximum in Planck's formula for the emission of blackbody radiation can be shown to occur at a wavelength $\lambda_{max} = 0.20 \, hc/kT$. The radiation from the surface of the sun approximates that of a blackbody with $\lambda_{max} = 4650$ Å. What is the approximate surface temperature of the sun?

14-8 Wave Nature of Particles Indicate what experimental evidence there is for the fact that some "particles" show wavelike properties.

14-9 Speed and de Broglie Wavelength A proton has a de Broglie wavelength of 100 nm. What is its speed?

14-10 De Broglie Wavelengths of Macroscopic Objects What is the de Broglie wavelength of a 1-mg weight moving at a speed of 1 cm s^{-1}? What is it for the earth (6.0×10^{24} kg) moving at an average orbital velocity of 30 km s^{-1}?

14-11 De Broglie Wavelengths of Neutrons and Atoms Thermal neutrons are neutrons with speeds characteristic of the translational energies of gases at room temperature. What is the wavelength of neutrons of a speed that corresponds to v_{rms} at 298 K? What are the wavelengths of Ar atoms and of Xe atoms under the same circumstances?

14-12 Photoelectric Effect and de Broglie Wavelength Photons of wavelength 3150 Å or less are needed to eject electrons from a surface of electrically neutral cadmium. (a) What is the energy barrier that electrons must overcome to leave an uncharged piece of cadmium? (b) What is the maximum kinetic energy of electrons ejected from a piece of cadmium by photons of wavelength 2000 Å? (c) Suppose the electrons described in (b) were used in a diffraction experiment. What would be their wavelength?

14-13 Uncertainty Principle State the uncertainty relations between the coordinates and the momentum components of a particle and between the energy and the lifetime of a state.

14-14 Uncertainty Principle (*a*) Assume that a 1.0-mg weight is positioned with an accuracy of 1.0×10^{-3} mm. What is the minimum momentum of the weight consistent with this positional uncertainty? What velocity and what kinetic energy correspond to this momentum? (*b*) Repeat the calculation for a hydrogen atom positioned with an accuracy of 1.0×10^{-3} Å.

14-15 Uncertainty Principle The excited state of an atom responsible for the emission of a photon typically has an average life of 10^{-10} s. What energy uncertainty corresponds to this value? What is the corresponding uncertainty in the frequency associated with the photon?

14-16 Wavelengths of a Neutron and an Electron The neutron has a mass 1839 times that of the electron. What is the relation between the wavelength of an electron and that of a neutron that has the same kinetic energy as the electron?

14-17 Light Pressure It has been suggested that spacecraft could be powered by the pressure exerted by sunlight striking a sail. The force exerted on a surface is the momentum mv transferred to the surface per second. Assume that photons of 6000 Å light strike the sail perpendicularly. How many must be reflected per second by 1 cm^2 of surface to produce a pressure of 10^{-6} atm?

Atomic Structure

"I am reading your paper in the way a curious child eagerly listens to the solution of a riddle with which he has struggled for a long time, and I rejoice over the beauties that my eye discovers, but which I must study in much greater detail to comprehend fully."
MAX PLANCK, IN LETTER TO ERWIN SCHRÖDINGER, APRIL 2, 1926

"But everything resolves itself with unbelievable simplicity and unbelievable beauty, everything turns out exactly as one would wish, in a perfectly straightforward manner, all by itself and without forcing."
ERWIN SCHRÖDINGER, IN LETTER TO MAX PLANCK, JUNE 11,1926

15

Before we consider how the particle-wave duality of matter and radiation has influenced current views of the structure of matter, it is instructive to examine earlier theories of atomic structure, particularly that developed by Niels Bohr in 1913. This chapter begins with a brief description of the classical picture of Rutherford's nuclear atom, the difficulties with that picture, and Bohr's resolution of these difficulties by his bold and imaginative application of quantum theory. It then continues with a discussion of the more modern quantum theory (or *wave mechanics*), developed during the 1920s, and its application to the structure of atoms.

15-1 Rutherford's Atom Model

As discussed in Sec. 1-5, atoms consist of a tiny, positively charged nucleus with a diameter of the order of 10^{-4} Å and an appropriate number of electrons surrounding the nucleus. Rutherford, basing his ideas largely on the results of his scattering experiments with alpha particles, had conceived of the atom as a planetary system with the nucleus as central sun and the electrons as orbiting planets.

By the laws of classical electrodynamics, however, an electron moving in a circular orbit would emit electromagnetic radiation of a frequency corresponding to its number of revolutions per second. Since radiation is a form of energy, the energy of the system would decrease. This lowering of the energy would mean a decrease in the radius of the orbit of the electron and a simultaneous increase in its speed. The electron would thus be expected to spiral ever faster and ever closer to the nucleus, until the radius of its orbit reached the value zero and the atom collapsed or some other catastrophic event occurred. This train of events is quite in contradiction with experience. First, atoms are stable; their collapse is not observed. Second, this model also requires that the frequency of the orbital motion should change as the radius of the orbit changed. The radiation emitted by a collection of many atoms should therefore cover a whole range of frequencies; that is, according to this model, atomic spectra would be expected to be *continuous.* In actuality, however, atoms have *discrete spectra* or *line spectra,* and for each kind of atom the observed radiation shows only certain specific frequencies, as illustrated for atomic hydrogen in Fig. 14-4 and for sodium in Fig. 14-10.

Regularities in these spectra had been sought and found during the last two decades of the nineteenth century, principally by J. H. Balmer, a Swiss mathematician, and J. R. Rydberg, a Swedish physicist. It was determined empirically, for example, that the wave number of any line in the spectrum of hydrogen can be described with high accuracy by the formula

$$\tilde{\nu} = 1.09678 \times 10^5 \left(\frac{1}{n_1^2} - \frac{1}{n_2^2} \right) \text{cm}^{-1} \tag{15-1}$$

where n_1 and n_2 are integers, n_2 always being greater than n_1. For $n_1 = 2$ and $n_2 = 3, 4, 5,$ and 6 the wave numbers obtained from (15-1) are those of the four Balmer lines in the H spectrum (Fig. 14-4).

15-2 Bohr's Model of the Hydrogen Atom

Bohr's Postulates   As we have just seen, the predictions of classical theory applied to the Rutherford atom model are in direct contradiction to experimental facts. Bohr succeeded in saving the model by boldly adding two novel postulates to the rules of classical physics, chosen simply because they gave correct results for the spectrum of the hydrogen atom. The postulates were:

1.   The only allowed orbits are those for which the radius is given by[1]

$$r = \frac{nh}{2\pi m v} \tag{15-2}$$

where m is the mass of the electron, v is its speed, and h is Planck's constant. The integer n is a characteristic of the orbit called its *quantum number.* For these orbits the laws of classical electrodynamics are suspended, so that a circling electron is forbidden to emit electromagnetic radiation. The system therefore loses no radiant energy and the orbits are circles and not spirals. The atom is said to be in a *stationary state.*

2.   When an electron changes orbits, the energy difference $\Delta\epsilon$ between the two orbits shows up as electromagnetic radiation. If the new orbit is of lower energy, radiation is emitted by the system; if it is of higher energy, radiation is absorbed. The frequency ν_rad of the radiation involved has nothing to do with the frequency of the orbiting electron but obeys the fundamental relationship

$$|\Delta\epsilon| = h\nu_\text{rad} \tag{15-3}$$

This is an extension of Planck's relationship [Equation (14-2)], applied now to the transitions[2] between stationary states of atoms. An atomic transition causes the emission, or is caused by the absorption, of a photon with energy $h\nu_\text{rad} = |\Delta\epsilon|$.

---

[1] Equation (15-2) follows from Bohr's postulate that the angular momentum of the electron, *mvr,* is an integral multiple of $h/2\pi$.

[2] The term *transition* is used to refer to a change from one definite energy state to another.

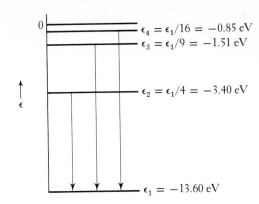

$\epsilon_4 = \epsilon_1/16 = -0.85$ eV
$\epsilon_3 = \epsilon_1/9 = -1.51$ eV

$\epsilon_2 = \epsilon_1/4 = -3.40$ eV

$\epsilon_1 = -13.60$ eV

Figure 15-1 Four Energy Levels of the Hydrogen Atom
The energy zero corresponds to complete separation of electron and nucleus of the hydrogen atom. In other words $|\epsilon_1|$ is just the *ionization energy,* the minimum energy required to separate the proton and electron of a hydrogen atom in the ground state. When the electron of a hydrogen atom is in the excited state with $n = 2$, it may drop into the ground state, with the energy difference $\epsilon_2 - \epsilon_1$ appearing as a photon. Conversely, a hydrogen atom in the ground state may absorb a photon of this same energy and be excited to the state with the electron in the level $n = 2$. The situation is similar for other excited states.

The energy ϵ_n of an electron in a hydrogen atom orbit of quantum number n can be derived by combining Bohr's postulates with the classical expressions for the potential energy and the kinetic energy of an electron moving in a circular orbit about a proton. These different energy values are called *energy levels;* the state of lowest energy is called the *ground state,* the others *excited states.* Since there is an attraction between the proton and the electron, the energy is lowest when the electron is closest to the nucleus; energy must be put into the atom to excite the electron and move it further from the nucleus. A few energy levels of the hydrogen atom are shown in Fig. 15-1. It is customary to assign the energy zero to the state in which electron and nucleus are completely separated and the separated particles possess no kinetic energy. This implies that negative energy values must be assigned to all other states shown in Fig. 15-1.

The energy change for the transition from state n_1 to state n_2 is

$$\Delta\epsilon = \epsilon_{n_2} - \epsilon_{n_1} \qquad (15\text{-}4)$$

If the level ϵ_{n_2} is lower than the level ϵ_{n_1}, $\Delta\epsilon$ is negative and the transition is associated with the emission of a photon of frequency $|\Delta\epsilon|/h$. Conversely, if ϵ_{n_2} is above ϵ_{n_1}, a photon of frequency $\Delta\epsilon/h$ must be available and be absorbed by the atom to make the transition occur.[3]

Bohr's theory successfully predicts the frequencies of the lines in the spectrum of atomic hydrogen. It leads to an equation identical in form to (15-1) in which the integers n_1 and n_2 are identified as the quantum numbers of the lower and upper levels, respectively. The theory also predicts the frequencies of the lines in the spectra of hydrogenlike ions, that is, of one-electron ions such as He^+, Li^{2+}, and Be^{3+}. These spectra are similar to that of hydrogen. An equation like (15-1) also applies to them, except that the proportionality constant varies directly as the square of the atomic number, Z^2, as Bohr's theory predicts it should.

☐ **Energy Corresponding to a Spectral Line** What is the energy in joules, in kilojoules per mole, and in electronvolts corresponding to the Balmer line H_α that arises from the transition between the states $n_2 = 3$ and $n_1 = 2$?

Example 15-1

Solution By (15-1), $\tilde{\nu} = 1.09678 \times 10^5 \left(\frac{1}{4} - \frac{1}{9}\right)$ cm^{-1} = 1.5233×10^4 cm^{-1}. From $\tilde{\nu} = \lambda^{-1}$ and $\nu = c/\lambda$, it follows that the energy equivalent to the wave

---

[3] Under appropriate conditions, the transition may also be brought about by supplying the same energy in another form—for example, by collision with another atom or molecule that has high kinetic or electronic energy.

number $\tilde{\nu}$ is $h\nu = hc\tilde{\nu}$. For the wave number just given, this energy is $(1.5233 \times 10^4 \text{ cm}^{-1})$ $(2.998 \times 10^8 \text{ m s}^{-1})$ (100 cm/m) $(6.626 \times 10^{-34} \text{ J s}) = \underline{3.026 \times 10^{-19} \text{ J}}$. (See inside back cover for the values of c and h.) From Table A-5 we find that $1 \text{ eV} = 1.6022 \times 10^{-22} \text{ kJ} = 96.48 \text{ kJ mol}^{-1}$. The energy is therefore also $\underline{1.889 \text{ eV}}$ and $\underline{182.22 \text{ kJ mol}^{-1}}$. ∎

Exercise 15-1

□ What is the energy, in joules, in kilojoules per mole, and in electronvolts corresponding to the spectral line H_β that arises from the H-atom transition between the states $n_2 = 4$ and $n_1 = 2$? ∎

Example 15-2

□ **Ionization Energy of the Hydrogen Atom** What is the ionization energy of the hydrogen atom?

Solution This is the energy needed to lift an electron from the ground state of the atom to the state in which the electron is at an "infinite" distance—that is, to the state representing the zero of the energy scale, a state associated with the quantum number $n_2 = \infty$ (Fig. 15-1). The energy desired is $-\epsilon_1$, which by (15-1) with $n_1 = 1$ and $n_2 = \infty$ corresponds to $\tilde{\nu} = 1.09678 \times 10^5 \text{ cm}^{-1}$. The energy relationships used in the preceding example lead to the following equivalent values: $\underline{1312 \text{ kJ mol}^{-1}}$, $\underline{13.60 \text{ eV}}$, and $\underline{2.179 \times 10^{-18} \text{ J}}$. ∎

Exercise 15-2

□ What is the ionization energy (kilojoules per mole, electronvolts, and joules) of the He^+ ion? Note that an equation like (15-1) applies to one-electron ions, with the proportionality constant multiplied by Z^2, as mentioned just preceding Example 15-1. ∎

The Old Quantum Theory During the decade after Bohr's first application of quantum theory to the atom, a set of rules was developed for treating many different systems, rules now referred to collectively as the *old quantum theory*. Other phenomena on the atomic level were treated by imposing upon the laws of classical theory rules of quantization similar to those used by Bohr. Results were usually in agreement with experiments; however, there were some notable failures of the old quantum theory—for example, its inability to explain certain features of the infra-red spectra of diatomic molecules such as HCl or to explain so fundamental a chemical fact as the formation of H_2 molecules. Furthermore the theory was unsatisfactory in that Bohr's atom model was two-dimensional, the orbits being planar. The theory was also objectionable on aesthetic grounds: it was a patchwork of arbitrary rules, justified only by the fact that they usually worked. There was no set of central postulates from which the entire theory was derived by a sequence of logical steps.

All this was changed around 1926 by the advent of three theories: Schrödinger's wave mechanics, Heisenberg's matrix mechanics, and a more abstract theory by P. A. M. Dirac. These three theories proved to be mathematically equivalent, even though they represented seemingly very different approaches. Bohr's atom model and other work had provided important guidelines in their creation. Only a descriptive treatment can be given here of this present-day quantum theory. The theory that lends itself best to such a description is that of Schrödinger's wave mechanics.

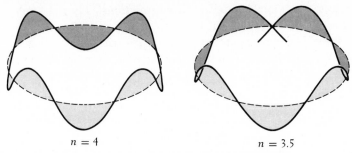

$$n = 4 \qquad\qquad n = 3.5$$

Figure 15-2 De Broglie Waves for Permitted and Prohibited Orbits
For the allowed orbit illustrated schematically here ($n = 4$), the de Broglie wave remains in
phase with itself upon successive revolutions and does not destroy itself by interference. In the
forbidden orbit ($n = 3.5$) there is a phase difference of $\lambda/2$ and therefore complete destruction by
interference after only one revolution. For other nonintegral values of n, destruction by interference may take more than one revolution but is still inevitable.

15-3 Matter Waves and Wave Functions

Some Consequences of the Existence of Matter Waves De Broglie's matter
waves of wavelength $\lambda = h/mv$ (Sec. 14-3) gave a partial clue as to why Bohr's
circular orbits worked so well. According to Bohr's postulate, the only allowed
orbits are those for which $mv = nh/2\pi r$ [Equation (15-2)]. This result can be used
to calculate λ_n, the wavelength of an electron in the nth orbit:

$$\lambda_n = \frac{h}{mv} = h\,\frac{2\pi r_n}{nh} = \frac{2\pi r_n}{n} \qquad (15\text{-}5)$$

The wavelength of the orbiting electron is thus a simple fraction of the circumference of the orbit, so that the circumference is an integral multiple of the
wavelength and in successive revolutions the waves are in phase and therefore
reinforced (Fig. 15-2). Orbits of any other circumference do not exist—are forbidden—because the electron wave would no longer be in phase with itself after
going around once and would thus destructively interfere with itself. The permitted orbits are those for which waves do not die out. Waves corresponding to
such orbits are called *standing waves* and are well known in physics. For example,
when a piano string is struck it vibrates in a standing-wave pattern.

The Schrödinger Equation While de Broglie's investigations provided a flash of
insight into the inner workings of the atom, they furnished only the wavelengths
associated with particles. It was several years before Schrödinger proposed his
celebrated equation that expresses the law to which matter waves are subject and
provides a systematic approach to problems of quantum theory. We present here
only a simple description of some of the solutions of the Schrödinger equation.
There will be no attempt to provide mathematical derivations. Most of the
features of wave mechanics will have to be accepted on faith. This is a traditional
cause of uneasiness for many students, but it must be lived with until the
mathematics needed for a more complete understanding is learned.

The Wave Function ψ The Schrödinger equation is of a general type often
referred to as wave equations because waves of all kinds—ocean waves, sound
waves, electromagnetic waves—obey equations of this general form. Solutions of

307

the Schrödinger equation are called *wave functions*. They describe, in a way explained below, possible states of the system.

A knowledge of the wave function for a system permits the calculation of the probability of finding that system in any particular state (for example, of finding an electron in a particular region of space). Many other quantities of interest can also be calculated if the wave function is known—for example, the permitted energy levels, the average values of the positions and momenta, and, for certain systems in excited states, the probability that the system will return to the most stable state and emit radiation within a given time.

One of the most important characteristics of the Schrödinger equation is that it has solutions *only* for a set of special values of ϵ, the possible energies of the system. The discrete energy levels of atoms and molecules are thus explained in a natural way, and no arbitrary quantization postulate such as that in the Bohr theory need be invoked. The concept of a wave function can be thought of as an extension of the de Broglie standing-wave picture shown in Fig. 15-2. Different states of a hydrogen atom are described by the different three-dimensional standing waves (the wave functions, ψ [Greek letter psi]) that are solutions of the

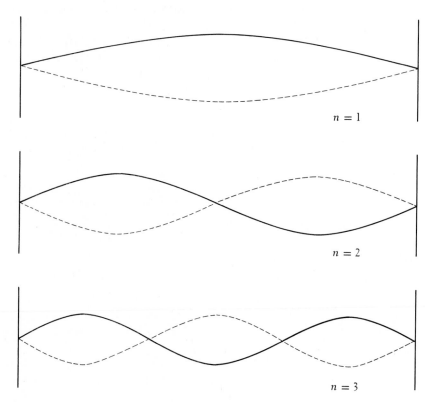

$n = 1$

$n = 2$

$n = 3$

Figure 15-3 Standing Waves on a Piano String

The string is clamped at its ends and therefore cannot vibrate at these points. For a string of length L each standing wave of the string must be of wavelength λ_n, given by

$$\lambda_n = \frac{2L}{n} \qquad n = 1, 2, 3, \cdots$$

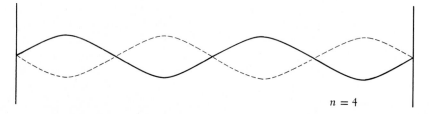

$n = 4$

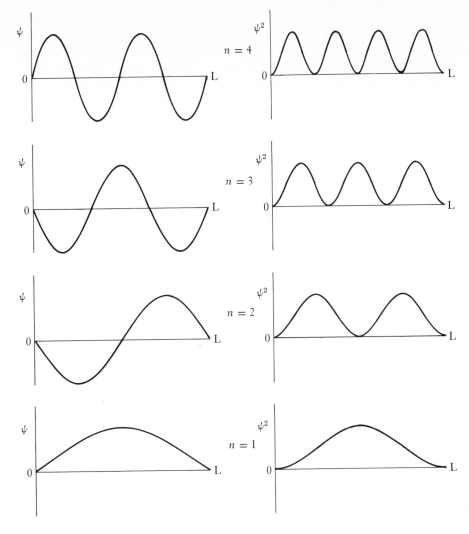

**Figure 15-4 Standing Waves in
a One-Dimensional Box**
The quantity plotted vertically is,
in (*a*), the wave function ψ and,
in (*b*), its square ψ^2. In all diagrams the x coordinate is plotted
horizontally.

Schrödinger equation. The value of ψ for each different possible state of the
hydrogen atom varies throughout space. The probability of finding the electron
of the hydrogen atom in any given region of space is proportional to the value of
ψ^2 in that region of space. Note that ψ^2 is always positive, as must be true for a
probability, while ψ can be positive or negative.

The solutions of the Schrödinger equation for a simple one-dimensional model
can demonstrate the essential features found in more complex cases. If the
Schrödinger equation is solved for an electron confined in a one-dimensional box,
the wave functions have the characteristics of a vibrating piano string (Fig. 15-3).
Standing waves are observed in such a string because the clamped ends of the
string cannot move. The wave functions (Fig. 15-4) are subject to the same
condition: the electron cannot move out of the box. Hence the probability of
finding it, ψ^2, must fall to zero at the box edge. As with the vibrating string,
standing waves of only certain wavelengths can exist. For a box of length L, the
allowed wavelengths are given by

$$\lambda_n = \frac{2L}{n} \qquad n = 1, 2, 3, \ldots \tag{15-6}$$

where n is called the *quantum number* of the electronic state represented by the standing wave. The quantum number is also used as a subscript for the different quantities associated with the state, such as λ_n.

The energy associated with each state can be found by substituting (15-6) into the de Broglie relation (14-6)

$$\lambda_n = \frac{h}{mv_n} = \frac{2L}{n} \tag{15-7}$$

and solving for v_n, the speed associated with an electron in state n

$$v_n = \frac{nh}{2mL} \tag{15-8}$$

The energy ϵ_n is simply the kinetic energy

$$\epsilon_n = \frac{1}{2} mv_n^2 = \frac{n^2h^2}{8mL^2} \tag{15-9}$$

Several features of this result, which is illustrated in Fig. 15-5, are of great importance. The electron can assume only certain prescribed energies; all other energies are forbidden. Neighboring energy levels ϵ_{n+1} and ϵ_n are separated by

$$\Delta\epsilon = \epsilon_{n+1} - \epsilon_n = \left(\frac{h^2}{8mL^2}\right)[(n+1)^2 - n^2]$$

$$= \frac{(2n+1)h^2}{8mL^2}$$

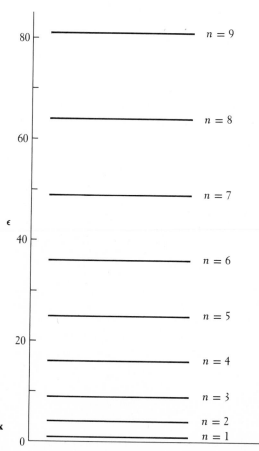

Figure 15-5 The Nine Lowest Allowed Energies for an Electron in a Box
The units of the vertical scale are $(h^2/8\,mL^2)$.

The energy separation $\Delta\epsilon$ is large only for particles of small mass, such as electrons or protons, and for lengths on the atomic scale. For such cases, therefore, there are discrete energy levels. The spacing between adjacent levels is so small for macroscopic systems that essentially all values of the energy are possible and there is no trace of quantization.

The probability of location of the electron is given by ψ^2. For $n = 1$, ψ^2 has its maximum value at the center of the box (Fig. 15-4). This is the place where the electron is most likely to be found, although there is significant probability that at any instant it might be found elsewhere. The plot of ψ^2 for $n = 2$ has two maxima, at $L/4$ and $3L/4$, which represent points of maximum probability. There is no chance that an electron in the state with $n = 2$ will be found at the center of the box—ψ^2 is identically zero there. Points at which ψ (and therefore ψ^2) is zero are called *nodes*. With the exception of the state with $n = 1$, all states for an electron in a box have nodes at positions other than at the ends of the box, as well as at the ends.

As will be seen in later sections, similar characteristics are found in the solutions of the Schrödinger equation for more complex systems.

15-4 Atomic Orbitals and Energy Levels

The Allowed Energies The Schrödinger equation for an atom consisting of a nucleus with charge Ze and one electron of mass m and charge $-e$ has solutions only for energy values[4]

$$\epsilon_n = -\frac{2\pi^2 mZ^2 e^4}{n^2 h^2 \alpha^2} \tag{15-10}$$

with the zero of potential energy chosen as the state with the electron completely removed. The quantum number n may have any positive integral value, 1, 2, 3, ... These are the energies that permit the existence of standing waves for the wave function ψ. For all but the lowest energy level ($n = 1$) there exists more than one possible wave function. Such energy levels are said to be *degenerate*, and quantum numbers in addition to n are needed to designate the different solutions ψ of the Schrödinger equation.

Nodes occur in the wave functions for atoms; these nodes are surfaces in three-dimensional space. Some of the nodal surfaces are spherical, centered at the atomic nucleus. Other possible nodes are planes that pass through the nucleus. One node always present is a sphere of infinite radius, because the wave functions of atoms go to zero at distances far from the nucleus. This corresponds to the fact that the probability of finding an electron, which depends on ψ^2, approaches zero at large distances from the nucleus.

The Quantum Numbers Wave functions for electrons in atoms are characterized by four quantum numbers. Only two of these are important in specifying electron energies in atoms and they are discussed in the following pages: the *principal* quantum number n and the *azimuthal* quantum number l.

The principal quantum number n is identical with the n used in (15-10). While n may assume the value of any positive integer, $n = 1, 2, 3, 4, . . .$, the quantum number l is restricted to the range $l = 0, 1, 2, . . . , n - 1$. Thus for the state with

[4]$\alpha = 1.113 \times 10^{-10}$ C^{2}J^{-1} m^{-1} (see Appendix A, Equation A-3).

$n = 1$, only the value $l = 0$ is possible, but for $n = 2$ there are two permissible values of l, 0 and 1. In general, the value of l indicates the number of nodal planes in the wave function.

For one-electron (hydrogenlike) atoms and ions (He^+, Li^{2+}, . . .), the energy is essentially independent of the value of l. This is not true when there is more than one electron, however, and hence for most atoms and ions the energy depends both on n and on l. As a result, two subscripts are needed to denote the energy of each level, ϵ_{nl}. This dependence on l removes part of the degeneracy of the energy levels of the hydrogen atom, but some degeneracy remains: for each combination of values of n and l, there exist $2l + 1$ solutions of the wave equation for the many-electron atom that correspond to the same energy ϵ_{nl}. When $l = 0$, there is only one energy state, but when $l = 1$, $2l + 1 = 3$, and there are thus three states identical in energy (a threefold degeneracy). When $l = 2$, there is a fivefold degeneracy.

It is customary to denote states with different values of the quantum number l by letters as follows:[5]

Value of l	Letter used
0	s
1	p
2	d
3	f

Values of l beyond 3 are used only rarely. Figure 15-6 shows the number of different states corresponding to different n,l pairs and also indicates, very schematically, the relative energies of the different states in many-electron atoms. When n is 1, the only permitted value of l is zero. This state is labeled $1s$ and is nondegenerate. When $n = 2$, l may be 0 or 1, corresponding to one $2s$ and three $2p$ states. For $n = 3$, l may be 0, 1, or 2, and there are nine states: one $3s$, three $3p$, and five $3d$. When $n = 4$, l may be 0, 1, 2, or 3, corresponding to one $4s$, three $4p$, five $4d$, and seven $4f$ states. The states corresponding to higher values of l for $n = 5$ and 6 (and more) are not shown. This important pattern of energy levels will be considered again in the following chapter. The $(2l + 1)$-fold degeneracy of the levels, and their relative positions, are of great importance for an understanding of the periodic system of the elements.

It must be emphasized that diagrams such as that in Fig. 15-6 are only schematic. The relative positions of levels that are very close, such as $3d$ and $4s$, or $4d$ and $5s$, are not the same in all atoms; sometimes one is slightly higher, sometimes the other. Furthermore, the vertical scale in this schematic diagram is greatly distorted. For example, the separation of the $1s$ and $2s$ levels is tens, hundreds, or even thousands of times as great as the separation of, say, the $4s$ and $5s$ levels, the relative difference increasing with increasing atomic number. Nevertheless, the diagram serves as a useful guide to the distribution of electrons among energy levels, those of lowest energy being filled first according to rules we discuss in Sec. 16-2. The principal rule is that no more than two electrons can occupy any of the individual states depicted in Fig. 15-6.

---

[5] The letters s, p, d, and f were once used by spectroscopists to characterize certain lines. They are abbreviations of *s*harp, *p*rincipal, *d*iffuse, and *f*ine.

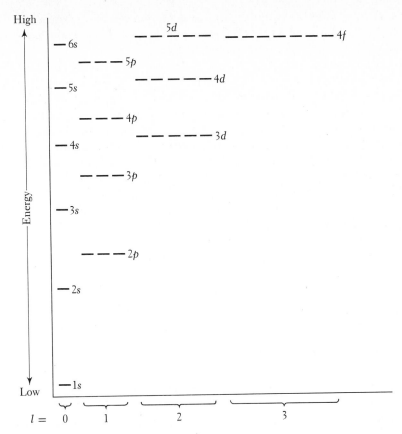

Figure 15-6 Schematic Energy-Level Diagram for a Many-Electron Atom
The degeneracies of the *s, p, d,* and *f* levels are seen to be 1, 3, 5, and 7, respectively, regardless of the value of *n*. The relative heights of levels that are very close together (such as 4s and 3d) may shift in different atoms for reasons discussed in the following pages. The energy scale in this schematic diagram is not linear, being especially compressed at the lower end (see text).

Description of the Different Atomic Orbitals   The solutions ψ of the Schrödinger equation for an atom are called *atomic orbitals* because they represent the wave-mechanical counterpart of the orbits of the Bohr model. They are the products of two functions. One of these functions, denoted here by $R(r)$, depends only on the distance r from the center of the atom and not on the direction; it is thus spherically symmetric. This function $R(r)$ is called the *radial* part of ψ. The second function, denoted by $\Phi(\theta,\varphi)$, depends only on the direction (that is, on θ and φ; see Fig. 15-7) and is independent of r; it is known as the *directional* or *angular* part of ψ. Thus

$$\psi(r,\theta,\varphi) = R(r)\Phi(\theta,\varphi)$$

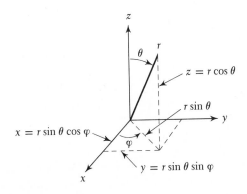

Figure 15-7 Spherical Polar Coordinates
The angles θ (colatitude) and φ (azimuth or longitude) are defined as illustrated. Consideration of the relation of x, y, and z to r, θ, and φ shows that

$$x = r \sin \theta \cos \varphi$$
$$y = r \sin \theta \sin \varphi$$
$$z = r \cos \theta$$

and also that

$$r^2 = x^2 + y^2 + z^2$$

313

The functional form of $R(r)$ and $\Phi(\theta,\varphi)$ for some of the commonest atomic orbitals[6] is shown in Table 15-1, and various graphical representations of these and other orbitals are given in Figs. 15-8 to 15-10 and 27-6.

Radial Part   The shape of the radial part of ψ, denoted $R(r)$, depends on the quantum numbers n and l. Examples of different radial parts of orbitals $R_{nl}(r)$ are shown in Table 15-1. They all fall off exponentially to zero for distances large compared to atomic dimensions. However, of greater interest than $R(r)$ itself is the probability of finding the electron at a specified distance r from the nucleus. This probability is proportional to r^2R^2. Important examples of radial probabilities are depicted in Fig. 15-8. The maximum in the curve for the 1s orbital is at 0.53 Å, the radius of the smallest orbit in the Bohr theory, r_B. Thus this radius retains a fundamental significance even in the wave-mechanical picture of the hydrogen atom, but instead of being the precise radius of the 1s orbit of the electron, it is the radius at which the probability of finding the electron is greatest. For example, the probability of finding the electron at about 0.20 Å or about 1.10 Å is half as great as that of finding it at about the Bohr radius.

The curves in Fig. 15-8 show that as n increases, the electron is on the average further from the nucleus, in agreement with the increase of potential energy with increasing n. A more subtle feature, which is of great importance in understanding the relative energies of the states in diagrams like Fig. 15-6 and thus in interpreting the periodic table, may be discerned by comparing the curves in Fig.

---

[6] The orbitals discussed here have been obtained by solving the wave equation for a one-electron atom and are thus strictly applicable only for atoms in which the potential energy is spherically symmetric and coulombic, that is, proportional to $-(1/r)$. However, the angular parts, which are most important for bonding considerations because of their directional character, are valid for all (isolated) atoms, in the approximation in which each electron may be considered to be in the average field of all the others. The radial parts shown here apply, at least qualitatively, after appropriate scaling.

Table 15-1
Radial and Angular Dependence[*][†] of Some Orbital Wave Functions, $\psi(r,\theta,\varphi) = R(r)\Phi(\theta,\varphi)$ **for One-Electron Atoms**

Orbital	$R(r)$	$\Phi(\theta,\varphi)$
1s	e^{-kr}	1
2s	$(2-kr)e^{-kr/2}$	1
$2p_x$	$kre^{-kr/2}$	$\sin\theta\cos\varphi = x/r$
$2p_y$	$kre^{-kr/2}$	$\sin\theta\sin\varphi = y/r$
$2p_z$	$kre^{-kr/2}$	$\cos\theta = z/r$
3s	$(27-18kr+2k^2r^2)e^{-kr/3}$	1
$3p_x$	$kr(6-kr)e^{-kr/3}$	$\sin\theta\cos\varphi = x/r$
$3p_y$	$kr(6-kr)e^{-kr/3}$	$\sin\theta\sin\varphi = y/r$
$3p_z$	$kr(6-kr)e^{-kr/3}$	$\cos\theta = z/r$
$3d_{xy}$	$k^2r^2e^{-kr/3}$	$\sin^2\theta\cos\varphi\sin\varphi = xy/r^2$
$3d_{yz}$	$k^2r^2e^{-kr/3}$	$\sin\theta\cos\theta\sin\varphi = yz/r^2$
$3d_{zx}$	$k^2r^2e^{-kr/3}$	$\sin\theta\cos\theta\cos\varphi = xz/r^2$
$3d_{x^2-y^2}$	$k^2r^2e^{-kr/3}$	$\sin^2\theta(\cos^2\varphi - \sin^2\varphi) = (x^2-y^2)/r^2$
$3d_{z^2}$	$k^2r^2e^{-kr/3}$	$3\cos^2\theta - 1 = (3z^2-r^2)/r^2$

[*] See Fig. 15-7 for the relation of r, θ, φ to x, y, z. The constant $k = Z/r_B$, where Z is the atomic number of the atom and r_B is the Bohr radius, 0.53 Å.

[†] Certain constant terms in each function have been omitted here since our present concern is the dependence of the functions on direction and distance from the nucleus. The relative magnitudes of these constants are reflected in the relative magnitudes of the different functions.

The angular functions $\Phi(\theta, \varphi)$ can be expressed quite simply in terms of the ratios $x/r, y/r$, and z/r, which depend on θ and φ only and not on r (see Fig. 15-7). These ratios appear on the right-hand side of the equations in col. 3 of the table. The subscripts of the different orbitals are simplified versions of these expressions.

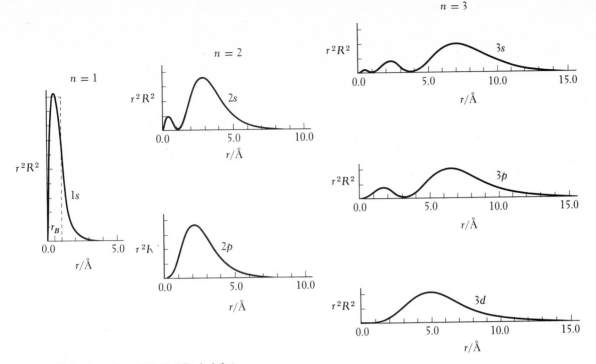

Figure 15-8 Examples of Radial Probabilities

The plots are drawn so that the area under any curve between any two values of r (say, r_1 and r_2) equals the probability that an electron occupying the corresponding orbital is at a distance from the nucleus between r_1 and r_2. The total area under any of the curves is unity, since there is unit probability (i.e., there is certainty) that the electron lies between $r = 0$ and $r = \infty$. The dotted rectangle in the figure illustrating the $1s$ curve is of unit area, and it is apparent that the area of this rectangle is comparable to the area under the curve. The horizontal scale applies for H-atom orbitals and is increasingly contracted as the atomic number Z increases.

15-8 for electrons with the same n but different l—for example, $2s$ and $2p$, or $3s$, $3p$, and $3d$. Although the average distance from the nucleus decreases slightly with increasing l for all the orbitals of a given n, it can be seen that an s electron is more likely to be found near the nucleus than is a p electron, and a p electron in turn more likely than a d electron. The relative energies of the s, p, and d electrons of a given n (Fig. 15-6) are consistent with this pattern of decreasing penetration toward the nucleus with increasing l.

Angular Part The angular part of the wave function, designated Φ, does not depend on the kind of atom or ion involved or on the principal quantum number n. It does, however, depend on the azimuthal quantum number l, and for a given l there are $2l + 1$ angular functions—e.g., one s function, three independent p functions, five independent d functions, and so forth. The shapes of these functions are important because they explain many aspects of molecular structure. Their squares, Φ^2, represent an angular probability distribution; that is, they contain information about the probability of finding an electron in a given direction. They are the counterpart of the nondirectional, spherically symmetric function R^2 that describes the radial dependence of ψ^2. As mentioned, the functions Φ are found to be identical for all atoms and monatomic ions, a statement that is not true for the functions R.

Example 15-3

☐ **Nodes of Orbitals** The positions at which the value of the wave function for an atomic electron becomes zero (i.e., the nodes of the wave function) are important in many applications because if $\psi = 0$ then $\psi^2 = 0$, and the electron density in that region thus vanishes. Show with the help of Table 15-1 that the 2s orbital for a hydrogen atom has a node at $r = 2r_B = 1.06$ Å and that the x, z plane is a nodal plane for a p_y orbital.

Solution The radial dependence $R(r)$ for the 2s orbital is given in Table 15-1 as

$$(2 - kr)e^{-kr/2}$$

This will become zero if either $(2 - kr)$ is zero or $e^{-kr/2}$ is zero. The latter factor approaches zero as r becomes infinite; however, $(2 - kr)$ becomes zero when $r = 2/k$, and since $k = Z/r_B$, this node occurs at $r = 2/(Z/r_B) = 2r_B/Z$. For a hydrogen atom, $Z = 1$, so the node is at $2r_B = 2(0.53$ Å$) = 1.06$ Å (see Fig. 15-8).

For a p_y orbital, the angular dependence (Table 15-1) is $\sin\theta \sin\phi$ or y/r. The latter form shows that the wave function vanishes when $y = 0$ (which is true for any point in the x, z plane). The same conclusion may be reached from the $\sin\theta \sin\phi$ dependence; this vanishes when either $\sin\theta$ or $\sin\phi$ becomes zero, that is, when $\theta = 0°$ or $180°$, or $\phi = 0°$ or $180°$ (see Fig. 15-7). These conditions define the x, z plane. This node for a p_y orbital is evident for $2p_y$ in Fig. 15-10a and especially 15-10b. ∎

Exercise 15-3

☐ Show with the help of Table 15-1 that any 3p orbital for a hydrogen atom has a radial node at about 3.18 Å (see also Fig. 15-8) and that any p_z orbital has the x, y plane as a nodal plane (see Fig. 15-10a). ∎

Complete Atomic Orbitals Let us now consider the complete atomic orbitals resulting from the multiplication of the radial parts R by appropriate angular parts Φ (Figs. 15-9 and 15-10).

s *orbitals* For all s states $\Phi(\theta, \phi) = 1$. Thus, the complete s orbitals, which are the product of Φ and R, are spherically symmetric—they depend only on r and not on direction. In other words, the probability of finding an s electron at some distance r from the nucleus is the same in any direction. Representations of this distribution of probability are shown in Fig. 15-9.

p *orbitals* For these orbitals, there are three independent angular functions (which correspond to the $2l + 1 = 3$ degenerate states for $l = 1$). The three p orbitals for any specific value of n are identical in shape but differ in orientation, being aligned along three mutually perpendicular directions. Those for $n = 2$ are illustrated in Fig. 15-10. The subscript x, y, or z refers to the coordinate axis along which the p orbital is oriented. For example, the maximum in the probability for a p_x orbital is along the x axis. The probability is also large in directions near this axis, but it falls to zero in directions that are in the y, z plane, perpendicular to the x axis.

d *orbitals* When $l = 2$ there are five independent orbitals. They are shown schematically in Fig. 27-6.

Orbits and Orbitals What is left of the orbits of the Bohr theory? For each there is a corresponding orbital or wave function, the square of which gives a probabil-

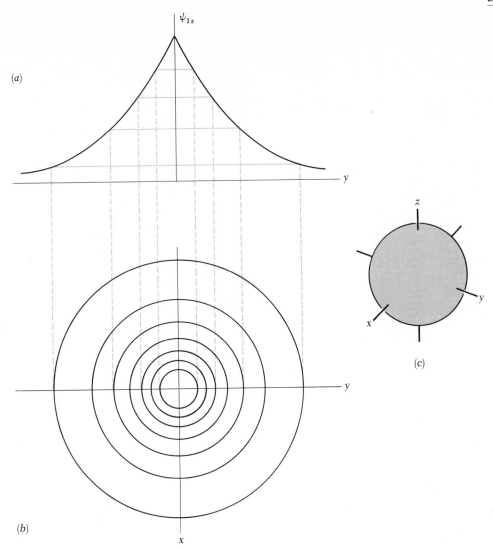

Figure 15-9 Representations of the Hydrogen-Atom 1s Function, ψ_{1s}

(a) The function ψ_{1s} along any line through the center of the atom, e.g., the y axis. (b) Each
circle represents a different constant value of ψ_{1s} on any plane through the center of the atom
(e.g., the x,y plane). As one goes toward the center, the 1s function increases by a constant
amount from one circle to the next, much as the altitude indicated on a contour map changes
by a constant amount from one contour to the next. The dashed vertical lines show the relation
of circles in (b), corresponding to given values of ψ_{1s}, to levels in (a) corresponding to those same
values of ψ_{1s}. (c) In three dimensions the surfaces of constant ψ_{1s} (and thus constant $\psi_{1s}{}^2$) are
spheres around the center of the atom. Any such sphere is also associated with a certain probability of finding inside that sphere the electron described by the 1s function. For example this
probability is 0.90 for a sphere of radius 1.40 Å about the atomic center. It is often useful to
represent a 1s orbital (or any s orbital for any atom) symbolically by such a sphere.

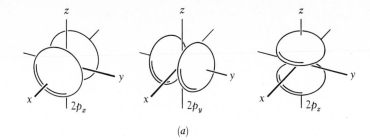

(a)

Figure 15-10 Representations of 2p Orbitals

(a) The surfaces of constant $|\psi|$ (and thus constant ψ^2) for the three 2p orbitals for an H atom. The two surfaces for each orbital nearly touch each other. (b) Contour diagram of a section through an H atom showing constant values of ψ for the $2p_y$ orbital.

The probability of finding an electron in the $2p_x$, $2p_y$, or $2p_z$ state inside the corresponding surfaces depicted in (a) is about 90 percent when the maximum extension of the surfaces along the respective coordinate axes is 4.2 Å. In (b), the change in ψ from one contour to the next is constant. Dashed contours represent negative values of ψ, and the innermost contour on each side represents the largest absolute value of ψ. The marks along the coordinate axes are at 1-Å intervals.

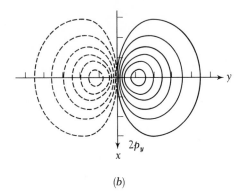

(b)

ity density for finding the electron, and associated with an electron in any given orbital is a certain energy. It is also possible to calculate for each orbital the probability that the momentum, and hence the velocity, of the electron lies in any particular range of values. While each orbital can thus be given particlelike attributes, it would be a futile proposition to attempt a detailed description of an electron *orbit*. All that can be said is that if we were to locate the electron of a hydrogen atom in a given state and were to repeat the experiment many times, we would find distributions of location, momentum, and certain other measurable properties as predicted by wave mechanics. What the electron does in the absence of the disturbances that are an inextricable part of making the observation is an undecidable question.

15-5   Electron Spin and the Pauli Exclusion Principle

The electron also has a quantized property called *spin*. In a simple picture, electron spin is associated with rotation of the electron. A rotating charge behaves like a magnet and is affected by external magnetic fields. The spin can have two, and only two, possible orientations relative to the direction of a magnetic field. These two directions are commonly described as *parallel* and *antiparallel* to the external field. Although this description is not quite accurate, it is unambiguous and therefore useful. Two spin states are available to an electron even in the absence of an external field, and it is convenient to describe these two states by terms such as "spin up" and "spin down". These two states of "opposite spins" have the same energy in the absence of an external field and somewhat different energies in the presence of one.

It is possible to understand many features of the periodic system of the elements by considering two factors: (1) the positions and degeneracies of the atomic energy levels and (2) the *Pauli exclusion principle*. A common statement of Pauli's principle for atomic electrons is that at most two electrons may occupy the same orbital and that these two electrons must have opposite spins. Such electrons are said to be *paired*. An electron that occupies an orbital singly is said to be *unpaired*, since there is no second electron of antiparallel spin in the same orbital. The exclusion principle is of great importance not only to the understanding of the periodic table, as we shall see in the next chapter, but also to a comprehension of the behavior of matter in general—it plays a fundamental role, for example, in the behavior of electrons in metals, in semiconductors, and in plasmas.

15-6 Wave Mechanics and the Uncertainty Principle

We return to one of the more general questions of quantum theory—the uncertainty relation

$$\delta x \, \delta p_x \gtrsim h \tag{15-11}$$

(where p_x again has the meaning of the x component of the momentum).

From the point of view of wave mechanics the uncertainty relation can be understood as follows. If a particle moving in the x direction has the exact momentum p_x, it is represented by a wave of the exact wavelength $\lambda = h/p_x$. However, by its very nature such a wave extends uniformly from minus infinity to plus infinity. The probability of finding the particle is thus the same everywhere; that is, there can be no information available as to the whereabouts of the particle. On the other hand, the description of a particle that is known to be in a certain region of space requires a wave function that is nonzero *only in this region*. It is a mathematical fact that such a highly localized wave can be constructed only by the superposition of many waves of *different wavelengths*, waves that represent a whole spread of λ's. But different λ's mean different values of p_x, so that an uncertainty δp_x exists. This is how δx and δp_x are inextricably interwoven.

One might ask whether the uncertainty relations are an attribute of the act of measuring or whether they represent a fundamental property of matter and radiation. If particles exist, is it that they do not *possess* precise location and precise momentum at the same time, or is it just impossible *to know* their values? These are questions that have no answer as yet. There are some scientists, for example, who believe that particles such as electrons exist, that their behavior is described by the solutions of the Schrödinger equation, and that any description other than one on a probability basis is impossible. Others think that the solutions of the Schrödinger equations describe not the behavior of particles themselves but only *the state of knowledge* that is accessible to physical measurements. There are other possible interpretations of the equations of wave mechanics. A choice among these interpretations cannot be made on any rational basis at this time. It belongs to the realm of philosophy or metaphysics and is a matter of personal preference. At least part of the problem appears to stem from the fact that both the particle concept and the wave concept are of macroscopic origin. It should not be surprising that difficulties arise when these concepts are applied in the atomic domain.

One illuminating application of the uncertainty principle is in providing an estimate of the minimum size of atoms consistent with the known binding energies of electrons. The diameter of the region within which an electron moves in a hydrogen atom is of the order of a few angstroms. Why can't it be much smaller? Suppose that it were 0.01 Å. This would imply that the uncertainty in the position of the electron could be no greater than 0.01 Å or 10^{-12} m, and hence, by (15-11), the uncertainty in its momentum could be no less than about 7×10^{-34} J s$/10^{-12}$ m $\approx 7 \times 10^{-22}$ kg m s^{-1}. Thus the momentum itself might be this large, which means that the kinetic energy of the electron might be as large as

$$\epsilon_k = \frac{p^2}{2m} = \frac{49 \times 10^{-44}}{18 \times 10^{-31}} \,\text{J} \approx 3 \times 10^{-13} \,\text{J}$$

which is about 10^6 eV. However, the binding energy (the potential energy) of an electron in a hydrogen atom is only 13 eV; hence, an electron with a kinetic energy of 10^6 eV could not be contained in a hydrogen atom. Noting that the kinetic energy varies as the square of the momentum and that the uncertainty in the momentum varies inversely with the uncertainty in the position, we see that if we *increase* the size of the region within which the electron is confined by a factor of 300, to about 3 Å, the uncertainty in the momentum is *decreased* by a factor of 300 and the maximum kinetic energy is then *decreased* by 10^5 to the order of magnitude of 10 eV. This energy is consistent with the known binding energy. Thus, the electron cannot be confined in a region of dimensions much smaller than 3 Å unless the binding energy is appreciably greater than a few electronvolts. This general inverse relation between the kinetic energy of a particle and the size of the region within which it is confined in a system of atomic dimensions is of wide applicability.

Summary

Bohr postulated that electrons move about the nucleus in circular orbits characterized by a quantum number n, and that when an electron changes orbits (undergoes a change in n) the energy difference $\Delta\epsilon$ between the two orbits appears as electromagnetic radiation of frequency $\nu = |\Delta\epsilon|/h$. The Bohr model of the atom permitted a successful calculation of the line spectrum associated with atomic hydrogen but failed to account in detail for the properties of other systems.

In modern quantum theory, the properties of atomic and molecular systems (e.g., the allowed energy levels for electrons in atoms, the positions and momenta of molecules in a fluid) can be described by the Schrödinger equation. Solutions of this equation are known as wave functions (ψ). The probability of finding particles, such as electrons, in different regions of space is given by the values of ψ^2 in those regions.

The energies of electrons in atoms depend on two quantum numbers: n, which can assume any positive integral value; and l, which can have values $l = 0, 1, 2, \ldots, (n-1)$. There are $(2l + 1)$ different wave functions associated with each pair of n and l values. These wave functions are also known as orbitals, a term applied too to the corresponding energy levels. A general energy-level diagram can be constructed for many-electron atoms that shows the relative energies of each orbital. The $(2l + 1)$ levels for each allowed value of l are degenerate, that is, they have the same energy.

Orbitals with $l = 0, 1, 2, 3$ are designated respectively by the letters s, p, d, f. Each wave function is the product of a radial function and an angular function. As the quantum number n increases, the electron is on the average farther from the nucleus. All s orbitals are spherically symmetric—the probability of finding an electron at some distance r from the nucleus is the same in any direction. Other orbitals are direction-dependent; for example, the three p orbitals (for each $n > 1$) are oriented along mutually perpendicular directions.

Electrons have a quantized property called spin, which can have two values, often called "spin up" and "spin down". The assignment of quantum numbers to electrons in atoms is limited by the Pauli exclusion principle: at most two electrons may occupy the same orbital and these two electrons must have opposite spins (antiparallel spins).

The position of a particle is represented in wave mechanics by the constructive interference of waves with a certain range of wavelengths and therefore of momenta. The larger the range of wavelengths used the sharper is the representation of the position; the quantitative expression of this complementary relationship is just the Heisenberg uncertainty principle.

Terms and Concepts

Problems and Questions

15-1 The Bohr Atom Discuss the difficulties with the Rutherford nuclear model of the atom according to classical physics, and state the postulates that Bohr made in deriving the formula for the spectrum of a hydrogen atom from this model.

15-2 Spectra of Atoms and Ions with Only One Electron (a) How are the frequencies ν, wave numbers $\tilde{\nu}$, and wavelengths λ of the spectra of the species He^+, Li^{2+}, and Be^{3+} related approximately to those in the spectrum of the hydrogen atom? (b) Calculate ν and λ for the spectral lines of H and He^+ when $n_1 = 1$ and $n_2 = 2, 3, 4$ and when $n_1 = 2$ and $n_2 = 3, 4, 5$. In which range of the spectrum do these lines lie? Give approximate colors if they are in the visible range. [Observed wavelengths for H are, in nanometers: 121.6 (1–2), 102.6 (1–3), 97.2 (1–4), 656 (2–3), 486 (2–4), 434 (2–5).

15-3 Ionization Energies of One-Electron Ions Calculate the energy required to remove (to infin-

ity) the electron in the lowest possible state (the ground state) of the ions He^+, Li^{2+}, and Be^{3+}.

15-4 Balmer Lines The Balmer lines in the spectrum of the hydrogen atom are those for which $n_1 = 2$ in Equation 15-1. Calculate the shortest and the longest possible wavelengths for the Balmer lines.

15-5 Visible Spectra Find the transitions in the hydrogen atom for which $n_1 = 1$, and those for which $n_1 = 2$, that correspond to lines in the visible region of the spectrum.

15-6 Consequences of the Uncertainty Principle Indicate the implications of the uncertainty principle with regard to the sharply defined orbits of the Bohr atom.

15-7 Dissociation Energy of a Molecule Photochemical dissociation of a molecule can occur if the molecule absorbs a photon that has sufficient energy to dissociate that molecule. The dissociation of a *mole* of iodine requires about 150 kJ. Would a photon of yellow light of wavelength 600 nm have sufficient energy to dissociate a *molecule* of iodine? What wavelength would be just sufficient?

15-8 Energy States The yellow-orange light of sodium lamps consists primarily of two wavelengths: 589.59 and 589.00 nm. What is the energy of each excited state involved relative to the Na ground state? What is the energy difference?

15-9 Radiation from an Atom The energy needed to ionize an atom of element Y when it is in its most stable state is 500 kJ mol^{-1}. However, if an atom of Y is in its lowest excited state, only 120 kJ mol^{-1} is needed to ionize it. What is the wavelength of the radiation emitted when an atom of Y undergoes a transition from the lowest excited state to the ground state?

15-10 Relationship between Wavelengths Suppose an atom in an excited state can return to the ground state in two steps. It first falls to an intermediate state, emitting radiation of wavelength λ_1, and then to the ground state, emitting radiation of wavelength λ_2. The same atom can also return to the ground state in one step, with the emission of radiation of wavelength λ. How are λ_1, λ_2, and λ related? How are the frequencies of the three radiations related?

15-11 Photoelectrons The minimum energy required to liberate an electron from the surface of lithium is 2.37 eV. What is the maximum kinetic energy of photoelectrons emitted by lithium when it is illuminated by the radiation associated with the transition of an electron in a hydrogen atom from the level with $n = 2$ to that with $n = 1$?

15-12 The Franck-Hertz Experiment When atoms in a gas are bombarded with electrons of known energy, they absorb no energy unless the electron energy is above a minimum value. As the atoms absorb energy from the electrons, they begin to emit photons. Explain. If the experiment is performed on mercury vapor, the threshold for absorption is 4.90 eV. What wavelength do you expect for the photons emitted by Hg atoms bombarded with electrons of this energy?

15-13 *p* States and *s* States (*a*) State clearly what is implied by *p orbital* in terms of quantum

numbers and "shape". (*b*) The energy difference between the most stable state of the Li atom and the lowest excited state, in which the outer electron is in a 2*p* orbital, is 1.85 eV, or 2.96×10^{-19} J. However, the energy difference between the 2*s* and 2*p* levels in the ion Li^{2+} is less than 0.0002 of this value. Explain.

15-14 Radial Portion of Atomic Orbitals Study Fig. 15-8 and deduce what graphs of r^2R^2 for $n = 4$, $l = 0, 1, 2, 3$ must look like. Sketch these graphs.

15-15 Radial Nodes Find the positions of the radial nodes in the 3*s* orbital for the hydrogen atom by finding the values of r for which $R(r)$ (Table 15-1) goes to zero. Check your results by comparison with Fig. 15-8, noting that $r_B = 0.53$ Å.

15-16 Atomic Orbitals Consider an *s* orbital and a *p* orbital of the same principal quantum number. What can you say about: (*a*) the principal quantum number; (*b*) the angular momentum quantum number for each orbital; (*c*) the angular variation of the electron distribution for the *s* orbital and (one of) the *p* orbitals; (*d*) the relative energy of an electron in the *s* orbital and in (one of) the *p* orbitals? Assume the atom is polyelectronic, and discuss the reason for any difference in energy in terms of the shapes of the orbitals.

15-17 Angular Nodes (*a*) Consider the definition of the angles θ and φ (Fig. 15-7) and the angular dependence of a p_x orbital wave function (Table 15-1). Show that this orbital has a nodal plane (i.e., the wave function goes to zero) at $x = 0$. (*b*) Similarly, show that a d_{zx} orbital has nodal planes at $z = 0$ and $x = 0$. Show also that a $d_{x^2-y^2}$ orbital has nodal planes at $x = y$ and $x = -y$, that is, at $\varphi = 45°$ and 135°.

15-18 Comparison of Energy-Level Spacings For the hydrogen atom, the transition from the 2*p* state to the 1*s* state is accompanied by the emission of a photon with an energy of 16.2×10^{-19} J. For an iron atom, the same transition (2*p* to 1*s*) is accompanied by the emission of X rays of wavelength 0.193 nm. What is the energy difference between these states in iron? Comment on the reason for the variation (if any) in the 2*p*-1*s* energy-level spacing for these two atoms.

The Relation between Periodic Properties and Electronic Structure

"The periodicity in the properties of the elements is connected with the continuing build up and completion of the various electron groups that takes place with increasing atomic number."
N. BOHR, 1923

"In an atom there never can be two or more . . . electrons for which . . . the values of all quantum numbers . . . are the same."
W. PAULI, 1925

16

The periodicity of the properties of the elements was considered in an introductory way in Secs. 2-3 and 2-4. We now return to a more detailed consideration of this topic in the light of the picture of the electronic structure of atoms discussed in Chap. 15.

16-1 General Relationships in the Periodic Table

It is helpful to get an overall view of the periodic system (Fig. 16-1) before studying the relation between periodicity and electronic structure. Several generalizations may be made on the basis of the region of the periodic table in which a given element is found. The most useful of these classifies elements as metals[1], nonmetals, and metalloids (those with intermediate properties).

About three-fourths of all known elements are metallic. The typical nonmetals occupy a triangular region on the right side of the table, with the long side of the triangle extending from the vicinity of boron (B, $Z = 5$) and carbon (C, $Z = 6$) to that of astatine (At, $Z = 85$) and radon (Rn, $Z = 86$). All elements to the right of this line are nonmetallic. The elements lying along a strip one or two elements wide and adjacent to this line show some metallic properties and some properties typical of nonmetals. They sometimes exist in several different crystalline forms, some of which may exhibit more metallic properties than others. These are the *metalloids;* they include boron (B), silicon (Si), germanium (Ge), arsenic (As), antimony (Sb), tellurium (Te), and polonium (Po), which lie in fields indicated by shading in Fig. 16-1. All the other elements, lying to the left of this strip, are metals.

Metallic character increases not only as one goes to the left in the usual form of the periodic table but also as one proceeds downward within any group. Thus the

---

[1]Characteristic properties of metals are described in Sec. 7-4 and Chap. 25.

		H 1		He 2			

IA	IIA	IIIA	IVA	VA	VIA	VIIA	0
Li 3	Be 4	B 5	C 6	N 7	O 8	F 9	Ne 10
Na 11	Mg 12	Al 13	Si 14	P 15	S 16	Cl 17	Ar 18

IA	IIA	IIIB	IVB	VB	VIB	VIIB	VIII			IB	IIB	IIIA	IVA	VA	VIA	VIIA	0
K 19	Ca 20	Sc 21	Ti 22	V 23	Cr 24	Mn 25	Fe 26	Co 27	Ni 28	Cu 29	Zn 30	Ga 31	Ge 32	As 33	Se 34	Br 35	Kr 36
Rb 37	Sr 38	Y 39	Zr 40	Nb 41	Mo 42	Tc 43	Ru 44	Rh 45	Pd 46	Ag 47	Cd 48	In 49	Sn 50	Sb 51	Te 52	I 53	Xe 54
Cs 55	Ba 56	La 57 *	Hf 72	Ta 73	W 74	Re 75	Os 76	Ir 77	Pt 78	Au 79	Hg 80	Tl 81	Pb 82	Bi 83	Po 84	At 85	Rn 86
Fr 87	Ra 88	Ac 89 †	(Rf) 104	(Ha) 105	106												

*Lanthanides	Ce 58	Pr 59	Nd 60	Pm 61	Sm 62	Eu 63	Gd 64	Tb 65	Dy 66	Ho 67	Er 68	Tm 69	Yb 70	Lu 71
†Actinides	Th 90	Pa 91	U 92	Np 93	Pu 94	Am 95	Cm 96	Bk 97	Cf 98	Es 99	Fm 100	Md 101	No 102	Lr 103

Figure 16-1   One Form of the Periodic Table of the Elements

The slanting lines connect groups of elements with similar properties and closely related outer electronic structure (*congeners*). This strong relationship is indicated by the letter A, added to the Roman numerals used as labels for the columns. There is a much weaker relationship between the elements of a column A and those in the column B with the same Roman numeral, consisting principally (but not exclusively) in these elements having certain oxidation states in common. Elements in shaded fields are called *metalloids*.

most metallic elements are cesium (Cs) and francium (Fr), which are in the lower left corner of the table; the least metallic is fluorine (F), in the upper right corner. One prominent characteristic of metals is the ease with which their atoms can form positive ions by loss of electrons, as symbolized by the equation

$$X(g) \longrightarrow X^+(g) + e^- \tag{16-1}$$

The energy required to make this reaction proceed is called the *first ionization energy* (first, because it is the first electron that is removed from a neutral atom). The metals Rb and Cs have the lowest ionization energy, the noble gases He and Ne the highest (Fig. 16-2). It is also difficult to remove electrons from the least metallic elements, F and Cl. More generally, a plot of ionization energy against atomic number (Fig. 16-2) shows a very marked periodicity, illustrating the regular recurrence of certain trends in the ionization energy as the atomic number increases. There are similar general correlations of many other properties with position in the periodic table.

A Brief Survey of Congeners   The term *congeners* implies a group of elements with closely related properties in a given column of the periodic table. The *alkali metals* (column IA: Li, Na, K, Rb, Cs, and the rare Fr) are all soft, low-melting metals of low density. One electron is easily removed from each atom, giving singly charged positive ions M^+. The metals react rapidly with atmospheric oxygen, the reactivity increasing from Li to Cs. The reaction of the alkali metals

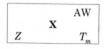

		AW
	X	
Z		T_m

X   Chemical symbol

Z   Atomic number

AW   Atomic weight (except in parenthesis, when mass number of most stable or best known isotope is given)

T_m Melting temperature

with water is violent, and the resulting alkali hydroxides give strongly basic solutions in water.

The *alkaline-earth metals* (column IIA: Be, Mg, Ca, Sr, Ba, and Ra) are harder and denser than the alkali metals. In suitable media, they easily form dipositive ions M^{2+}, but their reactivity is notably lower than that of the alkali metals. Their hydroxides, not very soluble in water, give basic solutions.

The *congeners of boron* (column IIIA: B, Al, Ga, In, Tl) tend to form ions M^{3+}, although the B^{3+} ion itself is not found, even hydrated, in ionic compounds or aqueous solution. The oxides of the elements in this group, for example, Al_2O_3, show acidic as well as basic character; they are said to be amphoteric (Sec. 9-4).

In the *carbon group* (column IVA: C, Si, Ge, Sn, Pb), there is a tendency to form four covalent bonds. Carbon is a nonmetal and shows only slight metallic characteristics in the form of graphite. Silicon and germanium are metalloids, and tin shows nonmetallic properties in one of its different forms or *allotropes* (Greek: *allo,* other; *trope,* way, manner)—gray tin. Lead is a metal. The oxides of the lighter elements in this column are weakly acidic; those of the heavier are amphoteric. These variations in properties exemplify the trend toward increasing metallic character with higher atomic number within a column in the system. Although, with the exception of the metalloid boron, all the elements in groups I to III are metals, the same trend in metallic character can be seen within these groups as an increase in reactivity when going down a column.

The *nitrogen group* (column VA: N, P, As, Sb, Bi) contains only one element with metallic properties: bismuth, the heaviest of the group. Nitrogen and phosphorus are nonmetals; arsenic and antimony are metalloids. The oxides of nitrogen, phosphorus, and arsenic are acidic; those of antimony and bismuth are amphoteric.

The *oxygen group* (column VIA) contains three nonmetals, O, S, and Se, and two metalloids, Te and Po. Sulfur and selenium form acidic oxides; those of tellurium are amphoteric. The elements of column VIA are sometimes called the *chalcogens.*

The *halogens* (column VIIA: F, Cl, Br, I, and the very rare At) are all strongly nonmetallic. The elements are all very reactive, F_2 the most so. Their compounds

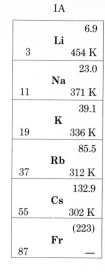

IA		
		6.9
	Li	
3		454 K
		23.0
	Na	
11		371 K
		39.1
	K	
19		336 K
		85.5
	Rb	
37		312 K
		132.9
	Cs	
55		302 K
		(223)
	Fr	
87		—

IIA		
		9.0
	Be	
4		1560 K
		24.3
	Mg	
12		922 K
		40.1
	Ca	
20		1112 K
		87.6
	Sr	
38		1043 K
		137.3
	Ba	
56		1002 K
		(226)
	Ra	
88		973 K

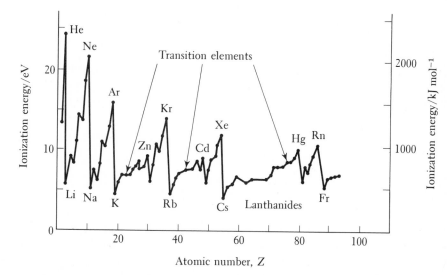

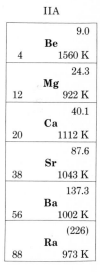

Figure 16-2 The First Ionization Energy of the Elements as a Function of the Atomic Number
The ionization energy in a given period is lowest for the alkali metals and increases in a general way as Z is increased. Deviations from this rule can be understood by considering the details of the electron configurations of the atoms involved, as explained in Sec. 16-3. In a given group of congeners, such as the noble gases or the alkali metals, the ionization energy usually decreases slowly as Z increases.

325

IIIA

	10.8
B	
5	2340 K

	27.0
Al	
13	933 K

	69.7
Ga	
31	303 K

	114.8
In	
49	430 K

	204.4
Tl	
81	577 K

IVA

	12.0
C	
6	~3820 K*

	28.1
Si	
14	1685 K

	72.6
Ge	
32	1210 K

	118.7
Sn	
50	505 K

	207.2
Pb	
82	600 K

VA

	14.0
N	
7	63 K

	31.0
P	
15	317 K

	74.9
As	
33	885 K†

	121.8
Sb	
51	904 K

	209.0
Bi	
83	545 K

with hydrogen (HF, HCl, HBr, and HI) are all acids. The oxides of Cl, Br, and I are also acidic.

The *noble gases* (column 0: He, Ne, Ar, Kr, Xe, and Rn) are almost completely inert chemically. It was only in 1962 that the first noble-gas compound was prepared, and compounds are known only for the noble gases of higher atomic number. The forces between noble-gas atoms are very small, exceptionally so for He. These gases are therefore difficult to condense, and their behavior at ordinary pressures and temperatures is closely described by the ideal-gas equation.

The *transition metals* form the B columns or B groups (excepting the post-transition metals in column IIB, Zn, Cd, and Hg, but including the three columns VIII, the lanthanides and the actinides). The relationships between corresponding A and B groups are rather tenuous. For example, the elements in column IB (Cu, Ag, and Au) resemble the alkali metals (column IA) insofar as they tend to form compounds in the $+I$ oxidation state[2], such as CuI, AgBr, and AuCl, but these elements are far less reactive than the alkali metals. Furthermore, Cu forms many stable compounds containing Cu^{2+} (for example, $CuSO_4$) and Au forms compounds containing Au^{3+} (for example, $AuCl_3$). The relationships of the other pairs of corresponding A and B groups are similarly weak. The ions of the IIB metals Zn, Cd, and Hg are doubly positive, but many compounds of mercury exist in which mercury has the oxidation state $+I$, and there are other exceptions. The IIIB metals Sc, Y, La, and Ac form chiefly $+3$ ions, as do all rare-earth metals or lanthanides. In group IVB (Ti, Zr, Hf, and Th) the $+IV$ state predominates, and this group is perhaps more closely related to the elements in the corresponding A group than is the case for any other A and B groups. In the succeeding groups VB, VIB, VIIB, and VIII, the situation becomes more complicated, although the highest oxidation state is $+V$ in group VB, $+VI$ in VIB, and $+VII$ in VIIB. It is characteristic of transition elements that they are all metallic, that they can exist in many oxidation states, and that many of them form colored salts.

Diagonal Relationships The interesting "diagonal" relationship expressed by the locations of the metalloids in the periodic table (Fig. 16-1) also exists among some elements of the first and second short periods. Thus Li is in some respects similar to Mg (e.g., both form carbonates, phosphates, and fluorides that are sparingly soluble in water, and both form nitrides with ease). Be shows similarity to Al (e.g., both show a strong tendency to form covalent compounds, and both form amphoteric oxides and volatile halides), and there is some relationship between B and Si (both form acidic oxides and volatile hydrides and halides). However, no pronounced relationship exists between C and P, between N and S, or between O and Cl.

$$\text{Li} \quad \text{Be} \quad \text{B} \quad \text{C} \quad \text{N} \quad \text{O} \quad \text{F}$$
$$\text{Na} \quad \text{Mg} \quad \text{Al} \quad \text{Si} \quad \text{P} \quad \text{S} \quad \text{Cl}$$

---

[2] Oxidation states are here designated by roman numerals, as is common (Appendix B).

* Sublimes at slightly lower temperature.

† Sublimation temperature at 1 atm; mp = 1090 K at 28 atm.

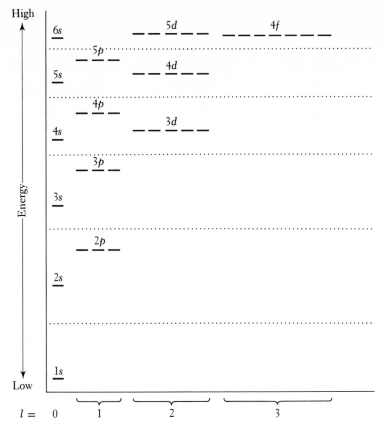

**Figure 16-3 Schematic Energy-Level Diagram
for a Many-Electron Atom**
The energy gaps (indicated by dotted lines) that
separate certain sets of levels are very pronounced
at low energies, less so at high energies. Electrons
occupying high-energy orbitals are the easiest to
remove from the atom.

16-2 Atomic Structure and the Periodicity of Chemical Properties

The Aufbau Principle The key to an understanding of the periodic system in
terms of atomic structure was furnished by the Pauli exclusion principle (Sec.
15-5), in combination with the system of atomic energy levels discussed in Chap.
15 and portrayed schematically in Fig. 15-6 and in Fig. 16-3. Pauli and Bohr were
the first to recognize this fact, and A. Sommerfeld pioneered in establishing the
details. The fundamental ideas are these:

1. The system of relative energy levels, reproduced schematically in Fig. 16-3,
   applies to all atoms. The *relative* energies of the orbitals remain approximately the same from element to element, with only rare interchanges of
   relative heights.
2. By the Pauli principle, each orbital may hold at most two electrons, which
   must have opposite spin.
3. The orbitals are filled in order with the electrons available, starting with the
   orbital of lowest energy (the $1s$ orbital, closest to the nucleus) and assigning
   at most two electrons to any orbital. This procedure is called, after Sommerfeld, the *Aufbau principle* (German: *Aufbau*, building up).

The First 10 Elements Starting with hydrogen ($Z = 1$) we see that the one
electron occupies the lowest available orbital, the $1s$ orbital. In helium ($Z = 2$)
both electrons are able to occupy the $1s$ orbital provided that their spins are

VIA		
		16.0
	O	
8		54 K
		32.1
	S	
16		388 K
		79.0
	Se	
34		494 K
		127.6
	Te	
52		723 K
		(210)
	Po	
84		527 K

opposed. It is customary to describe the electron configuration of these two atoms by the following notation:

$$H \qquad 1s$$
$$He \qquad 1s^2$$

The superscript indicates the number of electrons that occupy the orbital in question when that number is larger than 1.

The next element, lithium, has three electrons. By the Pauli principle the $1s$ orbital is now filled and the third electron must be placed in the $2s$ orbital. The fourth element, beryllium, has four electrons, which fill the $1s$ and $2s$ orbitals in pairs with antiparallel spins. The configurations of Li and Be are thus $1s^2\,2s$ and $1s^2\,2s^2$, respectively. In all subsequent elements the $1s$ and $2s$ orbitals are filled by two pairs of electrons.

The next orbitals are the three $2p$ orbitals, all of which have the same energy in an isolated atom. These orbitals can hold a total of six electrons. Together with the $1s$ and $2s$ orbitals, they are thus able to accommodate the electrons of the next six elements, B, C, N, O, F, and Ne. A total of four electrons always occupy the $1s$ and $2s$ orbitals. In boron the additional electron is in one of the $2p$ orbitals, leading to the configuration $1s^22s^22p$. An interesting phenomenon occurs in the next two elements: the extra electrons go into separate $2p$ orbitals, leading to the configuration $1s^22s^22p2p$ for carbon and $1s^22s^22p2p2p$ for nitrogen. The spins of the electrons in the separate $2p$ orbitals are all parallel, which is permitted by the exclusion principle and leads to two and three parallel spins for C and N, respectively (Table 16-1).

This pattern of filling the degenerate $2p$ energy levels depicted in Table 16-1 has been established experimentally by interpretation of atomic spectra. It follows a generalization known as *Hund's rule:* the order of filling degenerate orbitals in an atom (or molecule) is such that as many electrons remain unpaired as possible. The nitrogen atom, having three p electrons, has the maximum number of unpaired electrons in this period. For the next three elements there is a pairing of the electrons in the $2p$ orbitals. In oxygen, the configuration is $1s^22s^22p^22p2p$, so that two electrons with parallel spin remain. The electron configuration of fluorine is $1s^22s^22p^22p^22p$ and that of neon is $1s^22s^22p^22p^22p^2$. The symbols involving the $2p$ orbitals, or other sets of degenerate orbitals, are usually abbreviated by ignoring any differences between the different $2p$'s and writing, for example, $1s^22s^22p^3$ for N and $1s^22s^22p^5$ for F, as in Table 16-1.

‡At about 25 atm. Helium cannot be solidified at lower pressures.

VIIA		
	F	19.0
9		54 K
	Cl	35.5
17		172 K
	Br	79.9
35		266 K
	I	126.9
53		387 K
	At	(210)
85		—

0		
	He	4.0
2		1 K‡
	Ne	20.2
10		25 K
	Ar	39.9
18		84 K
	Kr	83.8
36		116 K
	Xe	131.3
54		161 K
	Rn	(222)
86		202 K

IB		
	Cu	63.5
29		1356 K
	Ag	107.9
47		1234 K
	Au	197.0
79		1336 K

IIB		
	Zn	65.4
30		693 K
	Cd	112.4
48		594 K
	Hg	200.6
80		234 K

IIIB		
	Sc	45.0
21		1812 K
	Y	88.9
39		1799 K
	La	138.9
57		1193 K
	Ac	(227)
89		1323 K

IVB		
	Ti	47.9
22		1943 K
	Zr	91.2
40		2125 K
	Hf	178.5
72		2500 K
	Rf	
104		

Atomic number	Element	Electron configuration	Occupancy of the $2p$ orbitals	
5	B	$1s^2 2s^2 2p$	↑ — —	
6	C	$1s^2 2s^2 2p^2$	↑ ↑ —	
7	N	$1s^2 2s^2 2p^3$	↑ ↑ ↑	
8	O	$1s^2 2s^2 2p^4$	↑↓ ↑ ↑	
9	F	$1s^2 2s^2 2p^5$	↑↓ ↑↓ ↑	
10	Ne	$1s^2 2s^2 2p^6$	↑↓ ↑↓ ↑↓	

**Table 16-1
Electron Configuration for the Elements B to Ne**

Some Definitions Before we consider the normal electronic structures of atoms of the remaining elements, the implications of a few terms that are used repeatedly and sometimes in slightly varying contexts must be examined.

The word *orbital* is used to refer to a particular solution of the Schrödinger equation for the electronic structure of an atom (or more generally a molecule). However, at times the word *orbital* is also used to allude to the *region of space* in which the corresponding value of ψ^2 has its largest values—that is, the region of space in which an electron occupying that orbital is likely to spend most of its time. This geometric implication of the word is reinforced in many people's minds by "orbital shapes", such as those shown in Figs. 15-9 and 15-10. Finally, the term *orbital* occasionally has energetic, as well as spatial, implications.

The word *subshell* is used to refer to a group of orbitals with a given n and l. Thus, for example, the three $2p$ orbitals are said to form a subshell, as are the three $4p$ orbitals, the seven $4f$ orbitals, and so forth.

Finally, the word *shell* is used in several different ways, of which two are most common: (1) the first usage is to refer to a group of orbitals with a given value of n (and varying values of l). Thus, for example, the $3s$, $3p$, and $3d$ orbitals are in this sense said to constitute one shell. (2) The second usage is less well defined. It designates groups of subshells of comparable energy as constituting a shell, regardless of whether they all have the same n or not. The orbitals lying between the dotted horizontal lines in Fig. 16-3 constitute shells in this sense, a sense that is usually more useful for chemists because the closeness of the energy levels involved markedly influences the relation of the chemical properties of the elements concerned.

Since the word *shell* normally has specific geometric implications, implying a spherical or ellipsoidal distribution that is thin relative to its radius, it is important to realize that its use in discussing atomic structure does *not* usually have this implication. The term *outer shell,* although it seems to imply something about shape and position, really refers to relative energy. An outer-shell electron (or more simply, an outer electron) is one that is relatively easy to remove from an atom because its ionization energy is comparatively low, whereas an inner-shell electron is much harder to remove. There is *some* implication of position as well, since the inner-shell electrons are harder to remove just because they are on the average much closer to the nucleus than the outer ones, but as we saw in Sec. 15-4, the shapes and relative penetrations of different orbitals vary considerably.

The Remaining Elements The electron configurations of all elements are summarized in Table 16-2. The configurations of the noble-gas atoms are particularly important because of the unusual stability of these elements (manifested by their high ionization energies, indicated by the successive peaks of the curve in Fig. 16-2). These successive noble-gas configurations correspond to filling all the

Table 16-2
The Electron Configurations of the Atoms of the Elements

Element	Atomic number	POPULATIONS OF SUBSHELLS											Element
		$1s$	$2s$	$2p$	$3s$	$3p$	$3d$	$4s$	$4p$	$4d$	$5s$	$5p$	
H	1	1											H
He	2	2											He
Li	3		1										Li
Be	4		2										Be
B	5		2	1									B
C	6	He	2	2									C
N	7	core	2	3									N
O	8		2	4									O
F	9		2	5									F
Ne	10	2	2	6									Ne
Na	11				1								Na
Mg	12				2								Mg
Al	13		Ne		2	1							Al
Si	14		core		2	2							Si
P	15				2	3							P
S	16				2	4							S
Cl	17				2	5							Cl
Ar	18	2	2	6	2	6							Ar
K	19							1					K
Ca	20							2					Ca
Sc	21						1	2					Sc
Ti	22						2	2					Ti
V	23						3	2					V
Cr	24						5	1					Cr
Mn	25						5	2					Mn
Fe	26			Ar			6	2					Fe
Co	27			core			7	2					Co
Ni	28						8	2					Ni
Cu	29						10	1					Cu
Zn	30						10	2					Zn
Ga	31						10	2	1				Ga
Ge	32						10	2	2				Ge
As	33						10	2	3				As
Se	34						10	2	4				Se
Br	35						10	2	5				Br
Kr	36	2	2	6	2	6	10	2	6				Kr
Rb	37										1		Rb
Sr	38										2		Sr
Y	39									1	2		Y
Zr	40									2	2		Zr
Nb	41									4	1		Nb
Mo	42					Kr				5	1		Mo
Tc	43					core				6	1		Tc
Ru	44									7	1		Ru
Rh	45									8	1		Rh
Pd	46									10			Pd
Ag	47									10	1		Ag
Cd	48									10	2		Cd
In	49									10	2	1	In
Sn	50									10	2	2	Sn

Table 16-2 (Continued)

Element	Atomic number	POPULATIONS OF SUBSHELLS										Element
		4d	5s	5p	4f	5d	6s	6p	5f	6d	7s	
Sb	51	Kr core 10	2	3								Sb
Te	52	10	2	4								Te
I	53	10	2	5								I
Xe	54	{Kr}* 10	2	6								Xe
Cs	55						1					Cs
Ba	56						2					Ba
La	57					1	2					La
Ce	58				2		2					Ce
Pr	59				3		2					Pr
Nd	60				4		2					Nd
Pm	61				5		2					Pm
Sm	62				6		2					Sm
Eu	63				7		2					Eu
Gd	64				7	1	2					Gd
Tb	65				9		2					Tb
Dy	66				10		2					Dy
Ho	67				11		2					Ho
Er	68				12		2					Er
Tm	69	Xe core			13		2					Tm
Yb	70				14		2					Yb
Lu	71				14	1	2					Lu
Hf	72				14	2	2					Hf
Ta	73				14	3	2					Ta
W	74				14	4	2					W
Re	75				14	5	2					Re
Os	76				14	6	2					Os
Ir	77				14	9						Ir
Pt	78				14	9	1					Pt
Au	79				14	10	1					Au
Hg	80				14	10	2					Hg
Tl	81				14	10	2	1				Tl
Pb	82				14	10	2	2				Pb
Bi	83				14	10	2	3				Bi
Po	84				14	10	2	4				Po
At	85				14	10	2	5				At
Rn	86	{Kr}* 10	2	6	14	10	2	6				Rn
Fr	87										1	Fr
Ra	88										2	Ra
Ac	89									1	2	Ac
Th	90									2	2	Th
Pa	91								2	1	2	Pa
U	92								3	1	2	U
Np	93	Rn core							5		2	Np
Pu	94								6		2	Pu
Am	95								7		2	Am
Cm	96								7	1	2	Cm
Bk	97								9		2	Bk
Cf	98								10		2	Cf
Es	99								11		2	Es
Fm	100								12		2	Fm
Md	101								13		2	Md

* {Kr} symbolizes the configuration of the krypton atom, $1s^2 2s^2 2p^6 3s^2 3p^6 3d^{10} 4s^2 4p^6$, which is also termed the *krypton core* in atoms of higher atomic number.

levels up to the successive horizontal dotted lines in Fig. 16-3. The periodicity of chemical properties reflects the fact that similar outer-electron arrangements recur at regular intervals, with successively larger values of the principal quantum numbers of the orbitals involved; inspection of Table 16-2 reveals these regular recurrences. For convenience the inner electrons are represented by reference to the electronic arrangement of the corresponding noble gas, designated by the terms "He core", "Ar core", and the like or by the symbols {He}, {Ar}, and so forth.

Table 16-2 shows that the characteristic outermost electron configuration of the noble-gas atoms (excepting that for He, which is $1s^2$) is ns^2np^6, with $n = 2, 3,$ 4, 5, and 6. The element immediately following each of the noble gases in the periodic table is an alkali metal, the atoms of which easily lose one electron and form a positive ion of unit charge. As implied by the dotted lines in Fig. 16-3, in each of the alkali-metal elements this outermost electron is in a substantially higher energy level than all the others. For similar reasons, atoms of the elements that are two places beyond each noble gas—the alkaline-earth metals, Be, Mg, Ca, Sr, Ba, and Ra—are prone to lose their two outermost electrons to form ions. All these ions—M^+ for the alkali metals and M^{2+} for the alkaline-earth metals—have the electron configuration of the noble-gas atoms just before them in the table. It is very hard to remove more electrons from these ions, not only because of the stability of the noble-gas configurations but also because of the coulomb attraction of the positive ion for the negatively charged electrons. Ions such as Na^{2+} and Ca^{3+} are formed only under extreme conditions, in high-energy electric discharges, stellar atmospheres, and the like.

Atoms that have three electrons in excess of a noble-gas configuration may form $+3$ ions, but removal of electrons becomes harder as the charge on the resulting ion increases. This effect becomes less important as the quantum number n increases, because the increased size of atoms and ions (and hence the larger electron-nucleus separation) makes removal of electrons easier. Thus Al^{3+} is common, but not B^{3+}.

Atoms of elements that *lack* just one electron of a filled shell may acquire an electron (if available) so readily that energy is liberated. This attests again to the special stability of the noble-gas configurations. Thus, gaseous hydrogen atoms and halogen atoms all liberate energy when combining with electrons to form -1 ions, each of which has a noble-gas configuration; this energy is a measure of the *electron affinity* of these atoms, a property considered in Sec. 16-3.

Table 16-3
Mnemonic Diagram of Approximate Order of Filling Orbitals*

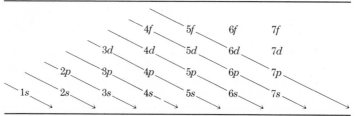

* The atomic orbitals are arranged in rows s, p, d, f, beginning with the s orbitals at the bottom and starting each higher row one place farther to the right than the preceding row, so that orbitals of the same principal quantum number are above each other. The approximate sequence in which orbitals are filled is indicated by the slanted arrows, proceeding from left to right. Minor deviations from this sequence are discussed in the text.

Study of Table 16-2 reveals that the order of filling the levels in successive atoms follows a quite regular pattern, with only occasional fluctuations. We shall discuss the reasons for these exceptions shortly. The approximate order in which the levels are filled can easily be remembered with a mnemonic scheme such as that depicted in Table 16-3.

At first glance the periodic table may seem formidable, because the systematization provided by considering it in terms of the electronic structures of the atoms is not apparent. However, when one is armed with a knowledge of the pattern of orbital energy levels and the principles governing the way in which these orbitals become filled with electrons, the systematic interrelationships among the 106 elements begin to emerge. We urge you to examine the general features of Table 16-2 and relate them to the periodic chart depicted in Fig. 16-1 and the schematic energy-level diagram of Fig. 16-3. Among the features to note are:

1. The A columns, and column 0, correspond to filling s and p subshells.
2. The B columns and column VIII, 10 columns in all, correspond to filling d subshells.
3. The lanthanides, or rare-earths, and the actinides correspond to filling f subshells, which explains why there are 14 of each.
4. The outer-electron configuration (ignoring the principal quantum number) is the same within each of the A columns and column 0; it is nearly the same within each of the columns containing a transition element,[3] as well as in column IIB. (There are a few variations among congeners in the transition elements; for example, Ni, Pd, and Pt are congeners and have many similarities in properties, but their most stable outer-electron configurations are, respectively, $3d^8 4s^2$, $4d^{10}$, and $5d^9 6s$. These variations are not easy to explain in simple terms and reflect the fact that any d level is almost identical in energy with the next higher s level so that very subtle effects can alter relative electron populations.)

Interpretation of Some Details Why should there be unusual stability for the noble-gas configurations? How can one rationalize the fluctuations of ionization energy within any period, illustrated in Fig. 16-2? How can one explain the general trend of ionization energy within any group of congeners? What is the significance of the twofold change in subshell population in going from V ($Z = 23$) to Cr ($Z = 24$), $d^3 s^2$ to $d^5 s$, or from Ni ($Z = 28$) to Cu ($Z = 29$), $d^8 s^2$ to $d^{10} s$? Some of the effects can be understood at least qualitatively. To do so, we must bear in mind many of the principles and results discussed in earlier chapters, most notably the coulomb attraction of unlike charges and repulsion of like

---

[3]There is no unanimity among chemists on a definition of the term *transition element,* but a common usage, which we have adopted, is to denote in this way any element that has a partly filled d or f subshell in any of its commonly occurring oxidation states, including the elemental state [oxidation state (0)]. The elements of column IIB (Zn, Cd, and Hg) have a filled d subshell (and either an empty or filled f subshell) in both the elemental state ($d^{10} s^2$) and the other common oxidation states, including that corresponding to the $+2$ ion (d^{10}); this is why they are usually considered not to be transition elements. Many of the characteristics that the transition elements have in common, including variable oxidation state and color of many of their compounds, are closely associated with the presence of incomplete d or f subshells (Chaps. 26 and 27). More than half of all known elements (59 out of 106) are transition elements.

Table 16-4
First Ionization Energies of the First 12 Elements*

Element	eV	kJ mol^{-1}	Element	eV	kJ mol^{-1}
H	13.6	1310	N	14.5	1400
He	24.6	2370	O	13.6	1310
Li	5.4	520	F	17.4	1680
Be	9.3	900	Ne	21.6	2080
B	8.3	800	Na	5.1	500
C	11.3	1090	Mg	7.6	740

* Rounded to the nearest 10 kJ mol^{-1}.

charges, the Pauli principle, and the varying shapes and extent of approach to the nucleus of different orbitals, depicted and discussed in Sec. 15-4.

We begin by seeking to explain the variation of ionization energy of the first dozen elements, for in so doing we shall consider most of the effects that provide the answers to our questions. These ionization energies are depicted graphically in Fig. 16-2 and are listed in Table 16-4.

The ionization energy of the hydrogen atom can be calculated exactly, and both the Bohr theory and present-day quantum theory give the experimental value, 13.6 eV. The spacing of the energy levels for one-electron atoms and ions varies directly with Z^2, so that the energy required to remove the *second* electron from a helium atom (that is, the ionization energy of He$^+$ when its single electron is in the $1s$ orbital) is $(2^2 \times 13.6)$ eV = 54.4 eV. Since both electrons of a helium atom are in the $1s$ orbital, one might at first imagine that the energy needed to remove the first would be the same as that for the second. It is actually far smaller, 24.6 eV, only about 1.8 times that of hydrogen instead of 4 times as great. Does this indicate a failure of the theory? Not at all, for a neutral helium atom is not a one-electron atom, and it is just because of the presence of two electrons in the $1s$ orbital that the first ionization energy is much smaller than the second. Each electron diminishes the effective nuclear charge felt by the other electron as long as both are present, because of the electron-electron repulsion resulting from their like charges; each electron is thus said to "screen" or "shield" the nucleus, in part, from the other. Because the electrons tend to avoid one another, this shielding is not complete; the effective nuclear charge is about 1.69, as compared with the actual charge of 2 and the value of 1 that would be appropriate if the shielding by the other electron were complete. Because the radius of the $1s$ orbital is small, the first ionization energy of helium is high—indeed, the highest for any neutral atom.

The first ionization energy of the next atom, lithium, involves removing an electron from the $2s$ rather than the $1s$ orbital, the third electron being excluded from the $1s$ orbital in accord with the Pauli principle. Not only is the radius of the $2s$ orbital (that is, the position of maximum probability density, illustrated in Fig. 15-8) significantly greater than that for a $1s$ orbital for the same atom, but the nucleus is very effectively screened by the two inner-shell electrons so that the effective nuclear charge for the $2s$ electron is nearer 1 than 3. Thus the ionization energy of lithium is far smaller than that of He and only about 40 percent that of H. The screening is not, of course, perfect. As Fig. 15-8 shows, the $2s$ electron has a significant, although small, probability of being at a distance comparable to the radius of the $1s$ orbital, where the effect of the nucleus is much greater and where, at the same time, the $1s$ electrons will tend to avoid being near the $2s$ electron because of coulomb repulsion.

The electron configuration of Be, $1s^2 2s^2$, is related to that of Li much as that of He is to that of H, and the relation of the ionization energies is similar, that of Be being a little more than 1.7 times that of Li. Inspection of Fig. 16-2 or Table 16-4 shows that the first ionization energy of the next atom, B, is somewhat smaller than that of Be; the added electron is in a $2p$ orbital, $2s$ being fully occupied. As Fig. 16-3 indicates, the $2p$ orbital has somewhat higher energy than the $2s$ and thus less energy is required to remove an electron from it. The reason that a $2s$ electron is more stable becomes apparent when one considers orbital shapes. A close study of Fig. 15-8 suggests that there is a higher likelihood for a $2s$ electron than for a $2p$ electron to be very near the nucleus and hence to be in a more stable state. This is borne out by a more detailed analysis. The resulting differ- ence in ionization energy between B and Be is small but perceptible, as noted in Fig. 16-2 and Table 16-4.

In the ground state of a carbon atom, the two $2p$ electrons have parallel spins and occupy different orbitals, in accord with Hund's rule. This configuration is more stable than one with the electrons paired in the same $2p$ orbital because electron-electron repulsions are minimized when the electrons occupy different $2p$ orbitals. As illustrated in Fig. 15-10, the different p orbitals are oriented at $90°$ to one another and thus electrons in them are in quite different regions of space. As long as the electrons in a degenerate set of orbitals have parallel spins, they must occupy different orbitals; hence the state with parallel spins ensures that the electron-electron repulsion will be minimized. The ionization energy of carbon is higher than that of boron because the electrons in the different $2p$ orbitals do not effectively shield the nucleus from one another and the nuclear charge is larger for carbon.

Hund's rule applies as well to the configurations of N and O atoms, as men- tioned earlier, for just the same reason—the minimization of the coulomb inter- electron repulsions. The ionization energy of N is higher than that of C by about the same amount by which the ionization energy of C exceeds that of B, and the interpretation of the difference is similar. However, as Fig. 16-2 and Table 16-4 show, the ionization energy of the next element, O, is below that of N. Can this be rationalized in terms of the principles we have been using? The nuclear charge has again increased but now, for the first time, one of the $2p$ orbitals contains two electrons. Removal of one of these electrons is made easier by their mutual repulsion. Evidently this increased repulsion is slightly more important than the increased attraction by the nucleus, and the removal of one electron from O is slightly easier than from N. With F and Ne, the electron being removed also comes from a $2p$ orbital already containing two electrons, so the principal difference from O is the increasing nuclear charge, and the ionization energy increases progressively.

The next atom, Na, with 11 electrons, must have its outer electron in a $3s$ orbital, the $2p$ subshell now holding six electrons, the maximum number permit- ted by the Pauli principle. Just as for Li, this outer s electron is quite effectively shielded from the nucleus by the completed inner shells, and, since the effective radius of the $3s$ orbital is greater than that of the $2s$, the ionization energy of Na is somewhat lower than that of its congener Li. Similarly, the ionization energy of Mg is somewhat less than that of Be. In general, ionization energy decreases within a group of congeners as the atomic number increases just because the outer-electron configurations of any group of congeners involve similarly shaped orbitals with gradually increasing radii.

Hund's rule applies to the filling of any degenerate level, such as d or f; furthermore, unpaired electrons in different orbitals will have parallel spins.

Example 16-1

□ **Hund's Rule** How many unpaired electrons (that is, electrons of a given spin that are not matched by electrons of opposite spin) are associated with the ground state of $_{45}$Rh? Use the information given in Table 16-2.

Solution The only occupied orbitals that are not completely filled with electrons are the five $4d$ and the one $5s$ orbitals, which have eight and one electrons, respectively. The scheme of these states, occupied by as many electrons of parallel spin as possible (Hund's rule), is

showing that there are three unpaired electrons. ■

Exercise 16-1

□ How many unpaired electrons are associated with the ground state of $_{65}$Tb? (See Table 16-2.) ■

The general arguments used above in analyzing the trends in ionization energy among the first dozen elements can be applied throughout the periodic table, although it is often difficult to predict the details of electronic structure. Several points are worthy of attention. The special stability of half-filled subshells, illustrated by the fact that N has a higher ionization energy than either of its neighbors, C and O, is noticeable many times. Similarly, the fact that filled subshells represent relatively stable configurations is often evident, whether it is the ns^2 characteristic of the alkaline-earth metals or, even more strikingly, the configurations $3d^{10}4s^2$, $4d^{10}5s^2$, and $5d^{10}6s^2$ characteristic of Zn, Cd, and Hg (column IIB). These effects are particularly noticeable in Fig. 16-2.

With very close energy levels, such as $3d$ and $4s$, or $4f$, $5d$, and $6s$, the exact order of filling and the relative stability may change from one atom or ion to another, as the effects of nuclear charge, interelectronic repulsion, and shielding of the nucleus vary. These variations cannot be simply rationalized. Note, for example, in Table 16-2 that one $5d$ orbital (as well as $6s$) is occupied in La ($Z = 57$) before the first $4f$ orbital receives an electron and that two $6d$ orbitals (as well as $7s$) in Th ($Z = 90$) contain electrons before the first $5f$ orbital is occupied. It is note-worthy too that when transition-metal (and column IIB) ions are formed, the s electrons are lost first rather than the d electrons, which have a lower principal quantum number. Thus, the d orbitals of these elements appear to gain slightly in relative stability when positive ions are formed. Typical configurations observed for some of these ions are shown in Table 16-5.

Table 16-5
Configuration of Outer Electrons of Some Ions of Transition and Related Elements

Ion	Configuration	Ion	Configuration
Ti^{3+}	d^1	Co^{3+}, Fe^{2+}	d^6
V^{3+}	d^2	Co^{2+}	d^7
Cr^{3+}, V^{2+}	d^3	Ni^{2+}	d^8
Mn^{3+}, Cr^{2+}	d^4	Cu^{2+}	d^9
Fe^{3+}, Mn^{2+}	d^5	Zn^{2+}, Cu^+	d^{10}

16-3 Orbital Energies and Chemical Properties

The arrangement of atomic energy levels described earlier not only provides a basis for understanding the general structure of the periodic table but explains in addition many chemical properties of individual elements. These two subjects are closely intertwined. We have already discussed briefly the interpretation of the general properties of the noble gases, alkali metals, halogens, and other groups. In this final section we consider a few aspects of the general behavior of *valence electrons,* as the electrons in the incompletely filled shell are called. The shell that is partially occupied by these electrons is termed the *valence shell.* We begin with two important physical quantities that characterize the electrons and orbitals of the valence shell, the ionization energy and the electron affinity, and then proceed to a brief consideration of some other significant atomic properties.

Ionization Energy The first ionization energy has been illustrated (Fig. 16-2) and discussed earlier. It is important to note that it is the energy needed to remove an electron from a neutral *atom* of the element in the *gas* phase:

$$X(g) \longrightarrow X^+(g) + e^- \tag{16-1}$$

Thus the ionization energy of hydrogen is the energy needed to ionize an H atom rather than an H_2 molecule, and that of sodium refers to monatomic sodium vapor, not to metallic sodium, a solid at ordinary temperatures. When ionization energy is defined in this specific way, the effects of interaction with other atoms in molecules or in the larger aggregates characteristic of condensed phases are removed and the striking periodicity displayed in Fig. 16-2 is revealed.

The energies needed to remove a second, then a third, and then still further electrons from an atom are called, respectively, the second, third, and higher ionization energies. Their variations with the atomic number and electron configuration of the atoms and ions involved can be analyzed in a fashion very similar to the one we have used for the first ionization energy, taking into account also the effect of the increased ionic charge as more and more electrons are removed. For example, a graph of second ionization energy against atomic number would resemble in many ways Fig. 16-2. However, the vertical changes would be larger because more energy is required to remove an electron from a positive ion than from a neutral atom, and the curve would be shifted one place to the right because a singly charged positive ion is isoelectronic[4] with the neutral atom of atomic number smaller by 1.

Electron Affinity The energy liberated when an electron is added to a gaseous atom, converting it into an ion with a single negative charge, is called the *electron affinity.* Thus in (16-2),

$$X(g) + e^- \longrightarrow X^-(g) + \epsilon_1 \tag{16-2}$$

the electron affinity corresponds to the energy ϵ_1. It is important to note that, for historical reasons, the sign convention for defining electron affinity is opposite to

---

[4]Two species with the same number of electrons are said to be *isoelectronic;* thus H^-, He, and Li^+ are isoelectronic, as are O^{2-}, F^-, Ne, Na^+, Mg^{2+}, and Al^{3+}, and even molecular species such as N_2, CO, and NO^+.

**Table 16-6
Approximate Electron
Affinities**

Element	eV	kJ mol^{-1}	Element	eV	kJ mol^{-1}	Element	eV	kJ mol^{-1}
H	0.754	73	C	1.27	122	F	3.40	328
Li	0.620	60	N	≤ 0	≤ 0	Cl	3.62	349
Na	0.546	53	P	0.74	72	Br	3.36	325
K	0.501	48	O	1.46	141	I	3.06	295
			S	2.08	200			

that used for defining ionization energy: a positive ionization energy signifies that energy is *required* to remove an electron, whereas a positive electron affinity indicates that energy is *liberated* when an electron is added. The electron affinity of an atom is equal to the ionization energy of the related anion, as indicated by (16-3), which is just the reverse of (16-2).

$$\epsilon_1 + X^-(g) \longrightarrow X(g) + e^- \tag{16-3}$$

Comparison of (16-3) with (16-1) shows that ϵ_1 is indeed the ionization energy of the anion X^-, as well as being [by (16-2)] the electron affinity of X.

Table 16-6 gives electron affinities for some elements; those shown here are positive, but electron affinities are by no means positive for all the elements. Precise values of electron affinities are usually hard to obtain, but the values given in Table 16-6 are consistent with our previous discussions: they are large for the halogens, whose -1 ions have the stable noble-gas configurations, are smaller for oxygen and sulfur, and rather small (but still positive) for hydrogen, corresponding to formation of the hydride ion, H^-, which is isoelectronic with He. The values for alkali metal atoms are positive, although even smaller than that for H; they decrease with increasing Z. Be and Mg have *negative* electron affinities, around -0.5 to -0.8 eV. All known *second* electron affinities of atoms [corresponding to adding an electron to a singly charged negative ion, a reaction such as $O^-(g) + e^- \longrightarrow O^{2-}(g)$] are also negative, as a consequence of the repulsion between the negative ion and the electron to be added to it.

Electronegativity *Electronegativity* is a measure of the tendency of a bonded atom to attract electrons from other atoms in the same molecule or ion; the greater the electronegativity, the larger this tendency. A large electronegativity difference between two bonded atoms results in significant partial ionic character of the bond. Because electronegativity is a property of an atom in a bonded state, it depends to a small extent on the nature of the orbitals that the atom uses in bonding (which may vary from one compound to the next) and on the nature and number of its bonded neighbors. Thus the electronegativity of a given atom may vary slightly from one compound to the next. Attempts to establish a quantitative scale of electronegativity involve correlating it with properties that can be measured precisely. Several different scales have been proposed. The two most common are that due to Pauling, who obtained electronegativities from a study of the energies needed to break covalent bonds, and that suggested by Mulliken, who equated the electronegativity of an atom with the average of its first ionization energy and its electron affinity. These scales give quite similar values when adjusted so that the electronegativity of fluorine is 4.0 on each. Representative values are shown in Fig. 16-4, in which the elements are arranged in the same rows as in the periodic table except that the horizontal scale is one of electronegativity.

Several features of Fig. 16-4 are worth remembering:

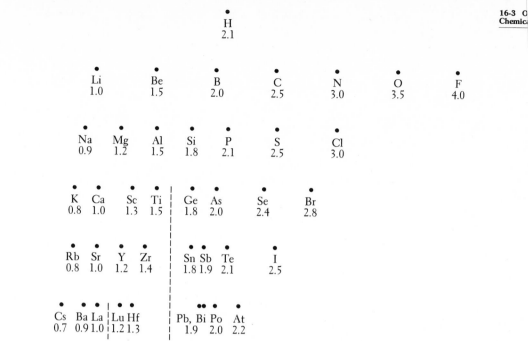

Figure 16-4 **Diagram of Electronegativities**

The diagram resembles a distorted periodic table, the elements of each A column being displaced toward the left with increasing atomic number. This displacement is strongest for the halogens, weakest for the alkali metals. There is little leftward displacement in the B columns (the transition metals, including the lanthanides and actinides); for these columns the electronegativities average about 1.7, varying from about 1.1 for most of the lanthanides to around 2.3 for the less reactive metals of higher atomic number, such as Pt and Au. Few of the transition elements are shown here; the vertical dashed lines represent average values of the electronegativities of the missing transition elements. (*From Linus Pauling and Peter Pauling, "Chemistry", W. H. Freeman and Co., San Francisco. Copyright © 1975.*)

1. The values of the electronegativities of the first-row elements, Li through F, are rather widely and equally spaced, ranging from 1.0 for Li to 4.0 for F. Hydrogen (2.1) is near the middle of the table, slightly less electronegative than carbon (2.5).

2. If one considers the usual A columns of the periodic table, the diagram of electronegativities has a characteristic slant, the bottom part being displaced toward the left relative to the top. In other words, there is a decrease in electronegativity as one proceeds down a given A column. This effect is much more pronounced for the halogens (F, 4.0; Cl, 3.0; Br, 2.8; I, 2.5; At, 2.2) than for the alkali metals (Li, 1.0; Na, 0.9; K, 0.8; Rb, 0.8; Cs, 0.7).

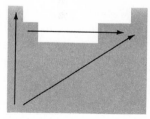

Electronegativity Trends

3. Fluorine, the most electronegative element, is significantly more electronegative (by 0.5 units on this scale) than oxygen, which is the second most electronegative. Nitrogen and chlorine, which have about equal electronegativities, come next, about 0.5 units less than oxygen. At the other end of the scale, the electro*positive* elements, the differences are much smaller. Cesium, the least electronegative element (other than the very rare Fr) is, for example, only 0.1 units less electronegative than rubidium and potassium.

4. The electronegativity of most metals is between 1.2 and 1.8, but the alkali metals lie significantly below this range. Gold and the "platinum metals", the elements in the last two rows of column VIII, have electronegativities between 2.2 and 2.4. Metalloids have electronegativities close to 2.0.

Pauling has suggested a quantitative correlation of the fraction of ionic character of a covalent bond between elements X and Y with the difference in the electronegativities of X and Y. The larger the difference, the greater the partial ionic character of the bond, 50 percent ionic character corresponding to an electronegativity difference of about 1.7 units on the scale of Fig. 16-4. For example, bonds between F or O and alkali metal atoms are predominantly ionic, while those between pairs of elements not widely separated horizontally in Fig. 16-4 are largely covalent.

Other Properties We have already alluded to the way in which metallic character varies with position in the periodic table (Sec. 16-1) and shall have occasion in later chapters to examine various other properties in a similar fashion. For example, the relative effective sizes of ions, as well as of neutral atoms, both in bonding and in nonbonding situations, can be understood quite well in terms of their positions in the periodic table and their electronic structures. Thus, in the isoelectronic series O^{2-}, F^-, Na^+, Mg^{2+}, Al^{3+}, the effective radius of the ion (Fig. 17-6) decreases progressively and significantly with increasing atomic number because of the increased attraction by the nucleus without corresponding change in electron configuration. Another important type of correlation is that of the acid-base character of the oxides of elements with position in the periodic table: in general, the oxides of the most electropositive elements are strongly basic, those of elements with high electronegativity are acidic, and those of intermediate elements lie between and may even be amphoteric. There is a clear correlation with oxidation state as well, not entirely independent of that with electronegativity inasmuch as electronegativity varies somewhat with oxidation state.

Summary

It is useful to classify the elements as either metals, nonmetals, or metalloids. Nonmetals are confined to a triangular region in the upper right of the periodic table. A strip one or two elements wide adjacent to the triangle comprises the metalloids: B, Si, Ge, As, Sb, Te, and Po. The metals constitute the remainder of the table. Metallic character increases as one goes to the left in the periodic table and, within a group, as one proceeds downward.

Keys to understanding the periodic system are the characteristic spacing of the energies of the different atomic orbitals and the Aufbau principle, the method of establishing electron configurations by filling each orbital with at most two electrons of opposite spin, starting with the orbital of lowest energy. A convenient notation for the electron configuration uses superscripts to indicate the electron population of each orbital, as in $1s^2 2s^2 2p$.

Orbitals with the same values of n and l comprise a subshell. The order of filling of subshells is such that as many electrons remain unpaired as possible (Hund's rule). The word *shell* refers to orbitals with the same value of n or to groups of subshells of comparable energy. Electrons in an incompletely filled shell are called valence electrons, and this shell is known as the valence shell (or, loosely, as the outer shell).

Many atomic properties exhibit "periodicity", that is, the relation of the properties of adjacent elements is repeated in an approximate way in each period as the atomic number is increased. Typical periodic properties include the first and higher ionization energies (the energies *required* to remove one or more electrons from a gaseous atom); the electron affinity (the energy *liberated* when a gaseous atom acquires an extra electron); and the electronegativity, a quantity that measures the tendency of a bonded atom to attract electrons from other atoms.

Terms and Concepts

Problems and Questions

16-1   Predicted Properties of a New Element   It has been predicted that a nucleus with 114 protons and 184 neutrons might be sufficiently stable to be isolated in appreciable quantities (with a half-life of 1000 years or more for its most rapid disintegration), if it could be made in some way. What would be the approximate atomic weight of this new element? What would be its probable outer-electron configuration (beyond the Rn core)? What element would it most closely resemble in properties?

16-2   Memorizing Part of the Periodic System   (*a*) Reproduce from memory the first two short periods of the periodic system (extending from He through Ar) and the columns containing the noble gases, the alkali and alkaline-earth metals, the halogens, and the chalcogens (congeners of oxygen). (*b*) Do the same for the elements K to Kr.

16-3   Noble Gases   Show the detailed electron configuration of all noble gases. Use notation such as $1s^2 2s^2 2p^6$, with the orbitals approximately in sequence of increasing energy.

16-4   Electron Configurations of Some Atoms   Without consulting any of the tables in this chapter, give plausible electron configurations ($1s^2 2s^2 2p^6 \cdots$) of the following atoms (atomic numbers are given in parentheses): C (6), N (7), P (15), Sc (21), Cr (24), Ni (28), Cu (29), Zn (30), Ga (31), As (33), Zr (40), Pd (46), Te (52), and Au (79). Do the same for these ions: V^{3+} (23), Cr^{3+} (24), Fe^{2+} (26), Fe^{3+} (26), Co^{2+} (27), Co^{3+} (27), and Ni^{2+} (28).

16-5   Excited State   Define the term *excited state* as applied to an atom or molecule. Illustrate by giving the electron configuration for an excited state of a sodium atom.

16-6   Electron Configurations   Give the electron configuration ($1s^2 2s^2 2p \cdots$) of (*a*) the ground state of Ga, I^-, Sr^+; (*b*) an excited state of Xe^+, Ca, O.

16-7   Isoelectronic Species   Separate the following into groups of isoelectronic species: (*a*) Li^+, NH_4^+, Ca^{2+}, Cl^-, CH_4, Ne, He; (*b*) Li^+, NH_3, H_3O^+, S^{2-}, Na^+, H^-, K^+.

16-8   Definitions and Distinctions   Define each of the following terms carefully, distinguishing it from those grouped with it: (*a*) orbital, subshell, shell; (*b*) ionization energy, electron affinity, electronegativity; (*c*) ground state, excited state, ionized state.

16-9   Halogens   What is the characteristic outer-shell configuration of the halogens? How is this correlated with their electron affinity?

16-10 Hund's Rule How many unpaired electrons are associated with the ground states of the atoms $_6$C, $_{24}$Cr, $_{64}$Gd, $_{67}$Ho, and $_{96}$Cm? Use the information in Table 16-2.

16-11 Ionization Energy According to Table 16-4, the first ionization energy of helium is 2370 kJ mol^{-1}, the highest for any element. (a) Define *ionization energy* and discuss why that for He should be so high. (b) Which element would you expect to have the highest *second* ionization energy? Why? (c) Suppose that you wished to ionize some helium by shining electromagnetic radiation on it. What is the maximum wavelength you could use?

16-12 Ionization Energies The energy needed to remove one electron from a gaseous potassium atom is only about two-thirds as much as that needed to remove one electron from a gaseous calcium atom, yet nearly three times as much energy is needed to remove one electron from K$^+$ as from Ca$^+$. What explanation can you give for this contrast? What do you expect to be the relation between the ionization energy of Ca$^+$ and that of neutral K?

16-13 Arrangement of Substances in Order Without consulting any tables, arrange the following substances in order and explain your choice of order: (a) Mg^{2+}, Ar, Br$^-$, Ca^{2+} in order of increasing radius; (b) Na, Na$^+$, O, Ne in order of increasing ionization energy; (c) H, F, Al, O in order of increasing electronegativity.

16-14 Differences between Cl and S Both the electron affinity and the ionization energy of chlorine are higher than the corresponding quantities for sulfur. Explain in terms of the electronic structure of the atoms.

16-15 Different Electronic States (a) Give the complete electron configuration ($1s^2 2s^2 2p \cdots$) of Al in the ground state. (b) The wavelength of the radiation emitted when the outermost electron of Al falls from the 4s state to the ground state is about 395 nm. Calculate the energy separation (in joules) between these two states in the Al atom. (c) When the outermost electron in Al falls from the 3d state to the ground state, the radiation emitted has a wavelength of about 310 nm. Draw an energy diagram of the states and transitions discussed here and in (b). Calculate the separation (in joules) between the 3d and 4s states in Al. Indicate clearly which has higher energy.

16-16 Relative Stabilities of Electrons in Different Orbitals What experimental evidence does the periodic table provide that an electron in a 5s orbital is slightly more stable than an electron in a 4d orbital for the elements with 37 and 38 electrons?

16-17 Comparison of Properties Among the elements with atomic numbers from 14 to 24, identify the element that has: (a) the highest first ionization energy; (b) the highest third ionization energy; (c) the highest electron affinity; (d) the lowest electronegativity; (e) the most basic oxide; (f) the most acidic oxide; (g) metalloid properties.

16-18 Ionic Character of Bonds Explain what is meant by the term *electronegativity*. Discuss briefly how it varies in the periodic table and how it may be used to assess qualitatively the ionic character of bonds.

16-19 Bond Character Suppose that a bond is formed between the elements in each of the following pairs. Arrange these bonds in order of increasing ionic character. (a) C,H; (b) C,O; (c) C,C; (d) Mg,O.

Chemical Bonds:
I. Experimental Facts
and Lewis Formulas

17

"I am inclined to think, that when our views are sufficiently extended, to enable us to reason with precision concerning the proportions of elementary atoms, we shall find the arithmetical relation alone will not be sufficient to explain their mutual action, and that we shall be obliged to acquire a geometrical conception of their relative arrangement in all three dimensions of solid extension."
W. H. WOLLASTON,[1] 1808

"Before leaving the subject of the atom it is desirable to refer to the new idea which is revolutionizing chemistry, namely that the different atoms have different exchangeable values or valencies. An atom is the seat of attractive forces, and the valency is the number of centers of such force."
ANONYMOUS ESSAY,[1] 1869

This chapter and the next one deal with the nature of the interactions between atoms that we characterize as *bonding*. Bonding is fundamental to all considerations of chemical structure and to all interpretations of chemical reactivity as well, since reactions necessarily change the way atoms are combined. A brief introduction to chemical bonding was presented in Chap. 6. This chapter begins with an examination of some of the experimental evidence that leads us to believe that the usual concept of a chemical bond—a strong and specific attractive interaction between atoms—is a useful one. This is followed by a discussion of Lewis formulas, considered briefly in Chap. 6. Theories of chemical bonding are discussed in Chap. 18.

17-1 Background

Although the *fact* that atoms are strongly attracted to one another in various combinations in different substances was self-evident from the earliest days of consideration of the atomic nature of matter, the interpretation of the *nature* of this attractive force developed only very slowly. Coulomb had firmly established experimentally the law of interaction of charged particles in the 1780s, only two decades before Dalton's atomic theory was published. It is thus not surprising that the earliest fruitful speculation about the nature of the attractive force between atoms was that it was essentially electrical in nature, resulting from the attraction of opposite charges. This view was developed first by Berzelius, a Swedish chemist, not long after Dalton's work had appeared, and it was very

---

[1] Quoted from W. G. Palmer, "A History of the Concept of Valency to 1930", pp. 27 and 89, Cambridge University Press, New York, 1965.

helpful in rationalizing the chemistry of many substances—the general class of compounds we now recognize as ionic. However, it could not explain the properties of many other substances, including most organic compounds, as Berzelius himself reluctantly recognized. It was only after more than a century had passed that it was possible to show, using the methods of quantum mechanics, that Berzelius' fundamental notion was correct.

As the concept of the combining power of atoms—valency—developed during the nineteenth century, the concept of the chemical bond developed simultaneously. The Russian chemist A. M. Butlerov first suggested in 1861 that a chemical formula should indicate clearly just how each atom is linked to other atoms in a molecule. He also suggested that all the properties of a compound were determined by its molecular structure and that it should be possible to deduce this formula by finding the different ways in which the molecule could be synthesized. Indeed, it was just through study of the chemical reactions by which molecules are formed, and of those in which molecules are fragmented in different ways, that all the detailed schemes of structural formulas involving bonds were gradually developed.

The use of symbols to represent bonds was a natural step. It provided at once a shorthand notation and a stimulus to thinking about the abstract concept of chemical bonds. The men who developed schemes of notation in the mid-nineteenth century—Couper, Crum Brown, Frankland, Kekulé, and others—were careful to point out that their early formulas had no geometric significance. However, a series of experimental and theoretical developments by Kekulé, Körner, Pasteur, van't Hoff, Le Bel, Werner, and others gradually led to explicit inferences about the shapes of molecules and the relative orientations of the different bonds formed by various atoms. Most of these inferences have been completely verified by modern physical methods of structure analysis developed during the present century, which have revealed the precise shapes of molecules and the detailed three-dimensional arrangements of atoms in many different substances. Methods for studying the energies involved in chemical bonds have also been developed, chiefly during the last 50 years. In the next section we examine some of the experimental facts about the interactions of atoms in molecules.

Before we proceed, however, a word of caution is needed. There is a tendency, among both beginning students and some experienced chemists, to regard chemical bonds in molecules as fixed entities with rather specific, well-defined properties. The fact that we have a conventional symbolic notation for bonds, draw schematic pictures of them involving orbital overlaps, present tables listing typical (equilibrium) bond distances and bond energies, and in other ways endow bonds with concrete properties should never obscure the fact that they are only useful abstractions for explaining the observed properties of certain substances. The concept of the chemical bond in nonionic molecules as a specific and rather strong attractive interaction between two atoms is limited.

17-2 Experimental Information about Atomic Interaction in Molecules

Bond Energies The dissociation energy of a diatomic molecule—that is, the energy to dissociate a diatomic molecule from its lowest energy state into its component atoms in their lowest energy states—can be considered the energy of

H—CH$_3$	425
H—CH$_2$	465
H—CH	417
H—C	335

Table 17-1
Some Speci[...]
Bond Dissoc[...]
Energies/kJ [...]

the chemical bond formed between the two atoms. Although the concept of the energy of a bond is quite natural for diatomic molecules, it cannot be extended to polyatomic molecules without some qualifications.

Consider the energy of the C-C bond in ethane. It might be taken to be the energy by which ethane is dissociated into two methyl radicals in the gas phase:

$$C_2H_6(g) \longrightarrow 2CH_3(g) \qquad (17\text{-}1)$$

However, this procedure for determining the bond energy involves the assumption that the dissociation does not change the energies attributable to the C-H bonds, i.e., that the energies of the C-H bonds are the same in the methyl radical, CH_3, as they are in ethane. If these energies do not remain the same, then some of the energy change in (17-1) will be caused by changes in the C-H bonds.

A feeling for the variation of the energies of nominally identical bonds can be gotten by examining the energy changes in reactions such as

$$CH_4(g) \longrightarrow CH_3(g) + H(g) \qquad (17\text{-}2)$$
$$CH_3(g) \longrightarrow CH_2(g) + H(g) \qquad (17\text{-}3)$$

If there were no effect of molecular structure on the energy of a bond, then the energy required to remove successive H's from CH_4 would be constant. The data in Table 17-1 show that in fact these energies vary by as much as 18 percent from the mean value, 411 kJ mol^{-1}.

Even though there is a spread in the values of the energy that can be associated with the C-H bond, it is often useful in approximate calculations to assume that all C-H bonds have the same energy, irrespective of the molecule in which they are found. This energy, called the *bond energy,* is an average value derived from thermochemical and spectroscopic data. Consistent sets of bond energies can be derived for bonds between many other different pairs of atoms in a similar way. Some representative values are given in Table 17-2.

Some elements, most notably C, N, O, and their nearby congeners, can form double and triple bonds. In fact, as discussed in Sec. 17-6 and in Chap. 18, bonds of character intermediate between single and double, and between double and triple, can also be considered to be present in many compounds. Bond energies for

	H	C	N	O	F	Cl	Br	I	Si
H	440	420	390	460	560	430	370	300	300
C		340	290	350	440	330	280	240	290
N			160	180	270	200			
O				140	210	210	220	240	430
F					160	250	250	280	590
Cl						240	220	210	400
Br							190	180	290
I								150	210
Si									190

Table 17-2
Representative Single-
Bond Energies* /kJ mol^{-1}

* Rounded to the nearest 10 kJ mol^{-1}. The bond energies given here are applicable only to reactions occurring in the gas phase. They can be used for liquids or solids when contributions from the energies of vaporization or sublimation are taken into account (Chap. 19).

various kinds of multiple bonds have been estimated by the same methods used for single bonds. A few representative values are listed in Table 17-3. There is a marked increase in bond energy with increasing multiplicity of the bond, which is in accord with the qualitative view that a multiple bond should be stronger than a single bond. The same view leads to the expectation that a multiple bond should be shorter than a single bond, a relationship that is confirmed experimentally.

Bond energies can be used to predict approximate energy changes in many reactions, as demonstrated in Example 17-1.

Example 17-1

☐ **Use of Bond Energies** Estimate the approximate energy changes accompanying the following reactions:

(a) $H_3CCH_3(g) + H_2(g) \longrightarrow 2CH_4(g)$

(b) $CH_4(g) + 2Cl_2(g) \longrightarrow CH_2Cl_2(g) + 2HCl(g)$

Solution In anticipation of the sign convention for energy changes in chemical reactions (Chap. 19), we will define the overall energy change as positive when energy is absorbed (that is, when more energy is required to break bonds than is released when new bonds are formed). Thus the energy change is defined as the sum of the energies of the bonds broken minus the sum of the energies of the bonds formed. In reaction (a), one C-C bond is broken and one H-H bond is broken; two C-H bonds are formed. Thus the approximate energy change should be

$$[(\text{C-C}) + (\text{H-H})] - 2(\text{C-H}) = [(340 + 440) - 2(420)]\,\text{kJ}$$
$$= -60\,\text{kJ per mole of ethane reacted (energy is released)}$$

For comparison, the experimentally observed change is -65 kJ per mole of ethane. A discrepancy of 10 or 20 kJ is not unusual when bond energies are used; nevertheless, the possibility of making estimates of energies of reaction within about this accuracy is valuable.

In reaction (b), two C-H bonds and two Cl-Cl bonds are broken and two C-Cl bonds and two H-Cl bonds are formed. Thus the energy change would be expected to be about

$$2[(\text{C-H}) + (\text{Cl-Cl}) - (\text{C-Cl}) - (\text{H-Cl})] = 2(420 + 240 - 330 - 430)\,\text{kJ}$$
$$= -200\,\text{kJ per mole of methane reacted}$$

The observed change is -202 kJ per mole of methane. Thus for this reaction, bond energies give a remarkably good prediction of the energy change. ∎

**Table 17-3
Representative Single-,
Double-, and Triple-Bond
Energies* /kJ mol⁻¹**

Elements	Single bond	Double bond	Triple bond
O—O	140	400	
N—N	160	420	950
C—C	340	620	810
C—O	350	720	
C—N	290	620	890

*Rounded to the nearest 10 kJ mol⁻¹.

☐ Estimate the approximate energy changes accompanying the following reactions:

(a) $H_3CCH_3(g) + H_2O(g) \longrightarrow H_3CCH_2OH(g) + H_2(g)$

(b) $H_2C{=}O(g) + H_2(g) \longrightarrow H_3COH(g)$ ∎

Bond Distances A characteristic bond distance can be associated with each kind of bond. Because, even at the lowest temperatures, atoms vibrate relative to one another at frequencies around 10^{13} s^{-1}, interatomic distances are continually changing in all molecules by 0.1 Å or more. Despite this fact, average or equilibrium distances between atomic nuclei in many molecules and ions, and in many unstable species as well, have been measured with considerable precision, often appreciably better than 0.01 Å (10^{-12} m). The principal methods used for such measurements are spectroscopy and the diffraction of X rays, neutrons, and electrons. Some average values for the lengths of different kinds of bonds are presented in Table 17-4.

The internuclear distance characteristic of a C-H bond varies over a range of only about 4 percent (about 0.04 Å) in different substances, and the variation is only around 1 percent if bonds in similar groups are compared (e.g., the C-H bonds in CH_3 groups in different molecules or the C-H bonds in benzene and molecules related to it). Comparable relative variations are found for other bonds that cannot have significant double- or triple-bond character (for example, O-H or C-halogen). On the other hand, some bonded interatomic distances, such as carbon-carbon and carbon-oxygen, vary by about 20 percent in different compounds. Most of this variation is attributable to multiple-bond formation, and when this is taken into account, only small variations remain, comparable with those found for C-H bonds. Covalent bonding radii for different elements have been derived from tables such as Table 17-4. They make it possible to predict bond distances quite accurately for almost any molecule whose structural formula is known.

Bond Angles If the average positions of the atoms in a polyatomic molecule are known accurately, it is possible to calculate the angle subtended at an atom by any two atoms bonded to it. For example, atoms X, Y, and Z may be arranged linearly or in a bent configuration. The angle of interest is the angle subtended at Y by X and Z. It is called the bond angle because it is normally assumed that a bond is coincident with the line between the nuclei, an assumption that is probably a very good approximation in most situations. In any event, the bond angle defined in this way is an experimentally accessible quantity whereas angles involving electron distributions in orbitals are not.

N—N	1.45	C—N	1.48	C—C	1.54	C—O	1.43	O—O	1.45
N=N	1.25	C=N	1.28	C=C	1.34	C=O	1.21	O=O	1.21
N≡N	1.09	C≡N	1.16	C≡C	1.20				
H—H	0.74	C—H	1.09			N—H	1.01	O—H	0.97
F—F	1.42	C—F	1.38	H—F	0.92				
Cl—Cl	1.99	C—Cl	1.77	H—Cl	1.27				
Br—Br	2.28	C—Br	1.94	H—Br	1.41				
I—I	2.67	C—I	2.14	H—I	1.61				

Table 17-4
Representative Bond Distances/Å

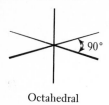

Octahedral

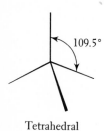

Tetrahedral

Trigonal planar

Experimentally observed values of bond angles range from around 60° (in three-membered rings) to 180°, but they tend to cluster around certain characteristic values. The illustrations in Chap. 7 show typical bonded configurations. Angles near 90° are common at atoms that are surrounded octahedrally—such as Cu in the $CuCl_2$ structure (Fig. 7-3), Cd in the $CdCl_2$ structure (Fig. 7-5), or the central atom in many complex ions [for example, $PtCl_6^{2-}$ or $Al(H_2O)_6^{3+}$]—and in ions and molecules with other configurations as well. Bond angles near 109.5° are extremely common because this is the angle subtended at the center of a regular tetrahedron by any two of the corners. Many atoms have four bonded neighbors arranged either exactly or approximately tetrahedrally. Such arrangements are found in the methane molecule (CH_4), the diamond crystal (Fig. 7-6), and the $Zn(OH)_4^{2-}$ ion (Fig. 7-7), in all of which the angle is 109.5° (called the *tetrahedral angle*). There are many other molecules of similar structure in which the four substituents on the central atom are not identical and in which the angles depart by a few degrees from the ideal tetrahedral value. Bond angles near 120° are also common, usually (but not always) occurring whenever three atoms are bound to a central atom in a planar arrangement. Thus the bond angle in the layers of the boron nitride and the graphite structures (Fig. 7-4) is 120°, as is the angle subtended at the carbon atom by any two of the oxygen atoms of the carbonate ion, CO_3^{2-}. Angles near 180°, giving rise to a linear or nearly linear bonded arrangement of three (or more) atoms, occur in a few situations, including some triatomic molecules such as CO_2 and $HgCl_2$, the simple organic molecule acetylene (HC≡CH), and the octahedral species just mentioned. It is also noteworthy that isoelectronic species show comparable angles. Thus CO_2 and NO_2^+ are both linear, and ozone (O_3) and nitrite ion (NO_2^-) are bent to about the same degree.

We shall see in Chap. 18 that a good empirical correlation of the configuration about a central atom can be made if bonded atoms and unshared pairs of electrons are taken into account.

Torsion Angles about Bonds Information accumulated during the last few decades about the three-dimensional architecture of molecules has made it much easier to understand and visualize the ways in which molecules interact and react with one another, since molecular interactions depend intimately on the exact

Figure 17-1 Torsion Angle
(*a*) A torsion angle φ can be defined for a sequence of four bonded atoms (W—X—Y—Z). It is the angle of twist, around the bond X—Y, of the plane of the three atoms X, Y, and Z relative to the plane of the three atoms W, X, and Y.
(*b*) The torsion angle about the O—O bond in hydrogen peroxide, H_2O_2.

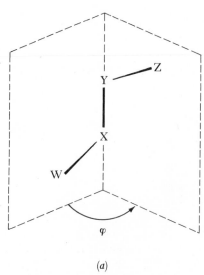

(*a*)

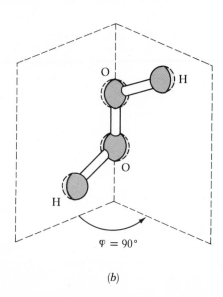

(*b*)

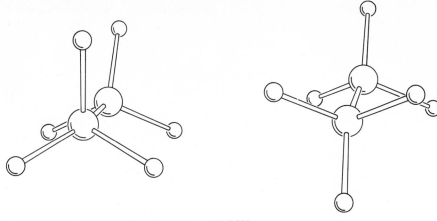

Figure 17-2   Two Conformations of Ethane, H_3CCH_3
The two CH_3 groups in ethane may be rotated relative to each other through any angle. The figure shows two of the many possible relative orientations or conformations. The molecule on the left is said to be in the *eclipsed* conformation because, viewed along the C—C bond, the hydrogen atoms in one CH_3 group cover, or eclipse, those in the other. The molecule on the right is in the *staggered* conformation. The torsion angle (Fig. 17-1) about the C—C bond is 0° in the eclipsed conformation and 60° in the staggered conformation.

spatial arrangements of atoms. Thus chemists have had to learn to think about and visualize the three-dimensional implications of molecular formulas. In order to specify three-dimensional molecular geometry unambiguously, it is necessary to specify more than all the bond distances and bond angles. This is because any two points, corresponding to a bond distance, define only a one-dimensional object (a line), and any three points not on the same line, corresponding to a bond angle different from 180°, define only a two-dimensional object (a plane). One convenient and illuminating way of providing additional information for specifying the three-dimensional arrangement of the atoms in a molecule is to give the torsion angle about each bond (Fig. 17-1). The torsion angle about a bond X-Y can be defined by the positions of the two bonded atoms and of one bonded neighbor of each (W and Z in Fig. 17-1), that is, by the positions of four atoms W, X, Y, Z bonded in sequence. As explained in the legend of Fig. 17-1 the torsion angle represents the twist of the plane of one group of three consecutive atoms (X-Y-Z) relative to that of the other group of three (W-X-Y).

Rotation about a single bond is usually almost unhindered and so a wide range of torsion angles is possible. However, it has been found experimentally that certain torsion angles are preferred, varying to some extent with the nature of the bonded atoms X and Y. Rotation about *double* bonds is not possible at normal temperatures. Furthermore, all atoms directly bonded to two atoms joined by a double bond lie essentially in one plane, together with the double-bonded atoms (for example, $H_2C=CHCl$), and thus torsion angles for double bonds are always near 0 or 180°. For *triple* bonds, the directly bonded neighbors lie on a line with the triple-bonded atoms (for example, $HC\equiv CH$ and $H—C\equiv N$) and the torsion-angle concept does not apply.

Structures differing only in angles of rotation about single bonds are referred to as different *conformations*. Two of the possible conformations for ethane are illustrated in Fig. 17-2. These two conformations, which represent the maximum and the minimum in the torsional potential energy, differ in energy by

349

12 kJ mol^{-1}. The energy of the eclipsed conformation (left) is higher than that of the staggered conformation (right). However, rotation about the C-C bond in ethane is essentially unhindered at room temperature because the energy difference is only about $5RT$. Conformations about C-C bonds in which the groups attached to each carbon atom are not all the same often have a larger difference in energy favoring the preferred conformation and rotation may be more difficult.

Conformational questions are of considerable importance in many areas of chemistry. They play an especially significant role in modern organic chemistry and biochemistry.

17-3 The Ionic Bond

The attraction between oppositely charged ions, which constitutes the ionic bond, is primarily the result of their electrostatic interaction [Equation (6-1)]. Repulsive forces caused by van der Waals interactions become appreciable as the ions approach, as indicated by the very steep portion on the short-distance side of the potential energy curve. The fact that this curve and the curves for the interactions typified as H—H and Ar~~~Ar in Fig. 6-7 are extremely steep implies that the model of an atom or ion as a hard (i.e., incompressible) sphere of definite radius is a good first approximation. This model has proved especially useful in consideration of ionic crystals.

Ionic Crystals Many crystals can usefully be regarded as composed of ions packed together in a systematic and efficient way, with the only significant attractive forces between the ions arising from their electrostatic interaction, covalent forces being negligible. All the evidence for the existence of ions in solids cannot be reviewed here in detail, but the more important facts include:

1. When many solids, including sodium chloride and other familiar salts, are melted (or vaporized), the resulting melt shows very high electrical conductivity, attributable to ions that are now highly mobile. Although the ions are in relatively fixed positions in a crystalline solid, they may migrate slowly, and this low mobility can also be observed.
2. The *lattice energy* of an ionic crystal of empirical formula MX is defined as the energy released in the reaction

$$M^+(g) + X^-(g) \longrightarrow MX(s) \tag{17-4}$$

that is, it is the energy released during the formation of the crystal from its components in their gaseous states. Lattice energies can be determined indirectly from experimental measurements. They can also be calculated from a knowledge of the structure of the crystal and the energy of interaction of the ions as a function of their distance apart—essentially the electrostatic attraction and the van der Waals repulsion.

When comparisons are made between experimental and calculated lattice energies, the agreement is remarkably good for crystals believed to be essentially ionic in nature. For example, typical theoretical and experimental values are respectively, in kJ mol^{-1}, 760 and 770 for NaCl, 670 and 670 for KCl, 610 and 570 for CsI. The agreement is significantly poorer for crystals in which the bonds between the ions presumed to be present are believed to have appreciable covalent character.

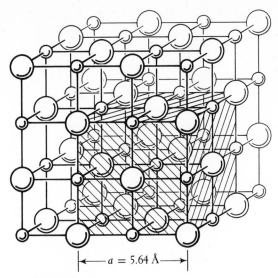

$\longleftarrow a = 5.64 \text{ Å} \longrightarrow$

Figure 17-3 The Sodium Chloride Structure
The entire structure of a crystal can be built up by periodic repetition by translation in three dimensions of a small portion called the *unit cell.* One unit cell of the sodium chloride structure shown here is indicated by diagonal hatching; it has an edge length $a = 5.64$ Å at room temperature. The larger spheres represent Cl^-, the smaller ones Na^+. To show geometric relationships more clearly, the spheres have been reduced in scale and connected by lines that must not be mistaken for covalent bonds.

This is a face-centered cubic (fcc) structure, so called because the ions of each type occupy the corners and the centers of the faces of a cube (of side a). Each Cl^- has 6 Na^+ ions as closest neighbors, at distances $L = a/2$. Further close neighbors are 12 Cl^- ions at $L\sqrt{2}$, 8 Na^+ ions at $L\sqrt{3}$, and many more too numerous to list here. The environment of each Na^+ is similar; the only difference is that the roles of Cl^- and Na^+ are interchanged. Many substances have this structure. Among them are the halides of Li, Na, and K; CaO, MgO, NiO, PbS, AgF, $AgCl$, and $AgBr$. Some of them (for example, PbS, $AgCl$, and $AgBr$) are considered to have substantial covalent character.

Ionic Coordination Geometry and Ionic Radii The structures of ionic crystals of compositions described by the formulas AB or AB_2 are usually quite simple. Some typical structures are shown in Figs. 17-3 and 17-4. Electrostatic forces are nondirectional—that is, the electric field around an ion is spherically symmetric. Hence, to the extent that only electrostatic forces play a role, the characteristic arrangement of the ions of opposite charge immediately surrounding a given ion (an arrangement referred to as the coordination geometry of the ion) is determined by the relative sizes of the ions involved. The most common coordination geometries found in ionic crystals, those of coordination numbers 4, 6, and 8, are illustrated in Fig. 17-5, which also shows typical geometries of coordination numbers 2 and 3. The coordinated ions often occupy the corners of a regular polyhedron (a tetrahedron, an octahedron, or a cube, respectively), with the oppositely charged ion at its center.

The structures of many simple inorganic salts can be classified in a few structural types, of which the NaCl structure is a representative example. Different structural types correspond to different ways of combining the characteristic coordination polyhedra, in accord with the stoichiometry of the compound. The structural type is to some extent determined by the ratio of the sizes

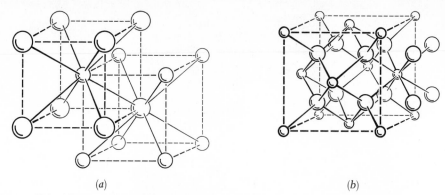

(a)   *(b)*

Figure 17-4 (a) The Cesium Chloride Structure; (b) the CaF₂ (Fluorite) Structure

In the cesium chloride structure each Cs^+ ion (smaller spheres) is surrounded by 8 Cl^- ions (larger spheres) at the corners of a cube, and each Cl^- ion is surrounded in the same way by 8 Cs^+ ions. Either of the two cubes shown in the figure can serve as the cubic unit cell of the structure. This structure is sometimes *erroneously* called body-centered cubic (bcc), but the ion in the center of either of the two cubes shown is of a different kind than the ions at the corners of the same cube. The structure is therefore *simple cubic,* an ion of a *given* kind occupying only the corners of a cube.

The calcium fluoride (fluorite) structure also has symmetry characteristic of a cube; the dashed lines indicate the cubic unit cell of the crystal. The large spheres represent the anions (F^-), each being surrounded tetrahedrally by four cations (Ca^{2+}), indicated by the smaller spheres. Each Ca^{2+} is, in turn, surrounded by 8 F^- ions at the corners of a cube.

Examples of crystals with the CsCl structure are the chlorides, bromides, and iodides of Cs and Tl. SrF_2, BaF_2, and CdF_2 are typical substances with the fluorite structure; Li_2O, Li_2S, and K_2Te have the so-called anti-fluorite structure, in which the roles of cations and anions are interchanged.

of the cation and the anion, as well as by the relative numbers of the different ions (the empirical formula of the compound).

The picture of ions as hard spheres with well-defined radii (Fig. 17-6) has proved very useful, particularly in interpreting the properties of ionic crystals. For example, it has provided the key to an understanding of the phenomenon of *isomorphism* (Greek: *isos,* equal; *morphe,* form) of ionic crystals, that is, the

Figure 17-5 Five Common Coordination Geometries

The names indicate the geometry of the coordination, that is, the geometric figure whose edges are represented by the dashed lines in each drawing and whose corners are occupied by the coordinated atoms (shown here by light spheres). The ion with which they are coordinated, shown dark, lies at the center of the geometric figure. The tetrahedron, octahedron, and cube, with four, six, and eight corners, respectively, are three of the most common coordination polyhedra in ionic crystals. The respective coordination numbers are 4, 6, and 8. Linear and planar triangular coordination (coordination numbers 2 and 3) exist in isolated molecules and ions, as do the other coordination geometries shown.

 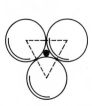 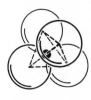 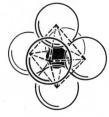 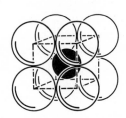

Linear   Planar triangular   Tetrahedral   Octahedral   Cubic

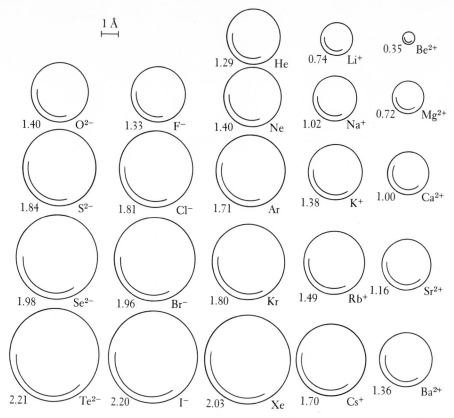

Figure 17-6 Approximate Sizes of Some Ions and Isoelectronic Noble Gas Atoms
The number below each ion or atom is its radius in angstroms. Note that the ions (and noble-gas atoms) in each horizontal row are isoelectronic with one another and that their radii decrease with increasing positive (or decreasing negative) charge, as expected. Each column contains a set of congeners, for which the radii increase as the atomic number increases. Whereas the radii of the spheres depicting the ions were obtained from crystal data, those for the noble-gas atoms were deduced from the detailed behavior of the gases, so that the two kinds of radii are not strictly comparable.

existence of different crystalline substances with very similar crystal shapes and chemical formulas. It has long been known that certain ions can often be substituted partially or completely in crystals for others of the same charge. However, this phenomenon long puzzled chemists because it seemed to bear no obvious relation to chemical similarity. For example, it is seldom possible to substitute K^+ for Na^+, or vice versa, in crystals, despite their close chemical similarity in many other ways. On the other hand, rubidium ion (Rb^+), ammonium ion (NH_4^+), and thallous ion (Tl^+), although chemically quite different, form many isomorphous crystals. These observations can be easily explained on the basis of the hard-sphere model for ions. Rb^+, NH_4^+, and Tl^+ have nearly identical radii, 1.49, 1.48, and 1.50 Å, respectively,[2] while the radii of Na^+ and K^+, 1.02 and 1.38 Å, respectively, are quite different from one another.

Another use of ionic radii is to provide an indication of the oxidation state of the ion concerned. The radii of different ions formed by a given element decrease markedly with increasing positive charge. For example, the radius of Tl^+, just cited, is 1.50 Å; that of Tl^{3+} is 0.88 Å. The radii of Cu^+ and Cu^{2+} are 0.96 and 0.73 Å, respectively. The effect is usually sufficiently pronounced, at least for lower oxidation states, that the effective radius of an ion in a crystal can be a fairly reliable indicator of the oxidation state.

One of the most striking applications of ionic radii is in predicting the arrangement of the nearest-neighbor anions around a given cation in a crystal, that is, the coordination polyhedron of a given cation-anion pair. The predictions are

---

[2] The NH_4^+ ion is not spherical and the value quoted is an "effective" radius.

based on the hard-sphere ion model and on the assumption that the distance between centers of oppositely charged ions will be as small as possible. It is also assumed that each cation will be surrounded by as many coordinated anions as possible, provided that each anion is in contact with the cation; this implies that the larger the relative size of the cation, the larger its coordination number. The important quantity is thus the ratio of the radii of the cation and the anion (see Probs. 17-4 and 17-5). The structures of many ionic crystals can be understood with the help of this model.

Ionic radii should not be taken too literally, despite their usefulness. No set of ionic radii is completely consistent in the sense that it can be used with unfailing reliability for the prediction of interatomic distances, the structures of ionic compounds, the existence of isomorphous crystals, the oxidation states of ions, and other properties that reflect effective ionic sizes. Furthermore, effective radii vary with coordination number. Ions are not, in reality, completely incompressible spheres.

17-4 The Octet Rule

The valence shells of the elements in each of the two short eight-element periods contain four orbitals and thus can accommodate a total of eight electrons with paired spins—the orientation of four of the spins being opposed to that of the other four. G. N. Lewis postulated that there is a tendency of these atoms to have their valence shells either empty or else filled by a full complement of eight electrons. This is part of a rule known as the *octet rule*. There are two ways in which atoms may achieve a completely filled or a completely empty valence shell. The first is the outright *gain* or *loss* of electrons, that is, the formation of ions. A second and equally important way of attaining a full octet of valence electrons is the *sharing* of electrons, the formation of a covalent bond.

The octet rule is only infrequently violated for the elements of the first short period (Li through F), and then only by the atom's having *fewer* than eight electrons in its valence shell. A second part of the octet rule is, then, that the elements of the first short period are *never found to have more than eight valence electrons.* On the other hand, some of the elements in the second short period (Na through Cl) do have more than eight valence electrons in some of their compounds, although in many they have just an octet.

The reasons why the octet rule is strictly obeyed in the first short period, but may be broken occasionally in the second, become apparent when the energies of the orbitals involved are considered (Fig. 16-3). In the first short period only the $2s$ and $2p$ orbitals are available for use by valence electrons, because the energies of the $3s$ and $3p$ orbitals are so much higher as to be prohibitive. The situation is less stringent for the second short period, because the $3d$ orbitals are much closer in energy to the $3s$ and $3p$ orbitals than the $3s$ and $3p$ orbitals are to $2s$ and $2p$ orbitals. It is thus understandable that the elements Na through Cl may occasionally use $3d$ orbitals for purposes of sharing electrons, particularly if this is favored energetically by the formation of strong bonds.[3]

---

[3] The reader may wonder whether $4s$ electrons, which have about the same energy as $3d$ electrons, are similarly used; the answer is that they are not, because their spherically symmetric distribution does not favor the formation of strongly directed bonds.

A corollary of the octet rule is that hydrogen and helium may never have more than two valence electrons. The reason is that the large energy gap between the $1s$ orbital and those in the next shell makes it unprofitable for H or He to use $2s$ or $2p$ orbitals.

The situation is more complex in the longer periods. Two general observations may be made. (1) In many compounds of elements in these periods the atoms contain—either by sharing or by outright transfer—a complement of electrons that corresponds to the next higher or lower noble gas. In particular this holds for elements near the beginning or near the end of the period, such as the alkali and alkaline-earth metals, the halogens, and the chalcogens. (2) As the principal quantum number n increases, the level spacings narrow. Thus the use of orbitals outside the valence shell proper becomes less and less prohibitive. This explains, for example, why noble gases like Kr, Xe, and Rn are able to form compounds although no similar compounds have been prepared for He, Ne, and Ar.

17-5 The Covalent Bond: Lewis Model

As indicated in Chap. 6, the nature of chemical bonding in nonionic compounds began to be understood only after the discovery of the electron and the development of the model of the nuclear atom, with extranuclear electrons, by Rutherford and Bohr. The idea that covalent bonds involve shared pairs of electrons was first proposed by Lewis in 1916, but it was not until the development of quantum mechanics in the mid-1920s that a clear understanding of covalent bonding became possible. While the Lewis picture of bonding is qualitative and cannot account for the structures of some molecules, it is still in common use because it is simple and, when used judiciously, is capable of predicting the structures and properties of many species. We shall use the Lewis model as an introduction to the detailed discussion of the covalent bond; the quantum-mechanical theory is developed in Chap. 18.

Lewis Formulas The application of the octet rule to bonding is greatly helped by a simple scheme devised by Lewis for representing electronic formulas. In these *Lewis formulas,* which were introduced in Sec. 6-4, shared pairs of electrons of opposite spin are represented by a line connecting the bonded atoms, unshared outer-shell electrons are represented by dots, and the nucleus and the inner-shell electrons are represented by the symbol for the element.

There is no necessary geometric significance to the relative positions of the atoms or the electron pairs in a Lewis formula; that is, the formula should in general be considered only schematic. However, when it is easily possible to display significant geometric features with such a formula, it is helpful to do so. Typical Lewis formulas (electron-dot formulas) are

$$
\text{H}-\overset{\displaystyle ..}{\underset{\displaystyle ..}{\text{I}}}: \qquad \text{H}-\overset{\displaystyle ..}{\underset{\displaystyle |}{\text{O}}}: \qquad \overset{\displaystyle \text{H}}{\underset{\displaystyle \text{H}}{\text{H}-\text{N}}}: \qquad \overset{\displaystyle \text{H}}{\underset{\displaystyle \text{H}}{\text{H}-\text{C}-\text{H}}} \qquad (17\text{-}5)
$$

Inspection shows that in each of these molecules the octet rule is satisfied; i.e., when the shared pairs are counted with *each* of the atoms that share them, the atoms achieve the electron configuration of the nearest noble gas. In each of the last three molecules the bond angle is not far from tetrahedral and an effort is sometimes made to display, in the formula, the bent shape of H_2O, the pyramidal shape of NH_3, and the tetrahedral shape of CH_4.

The Lewis formulas of charged molecules—that is, of ions containing several atoms—are no different from those of neutral molecules. It is convenient to place them within brackets so that the charge of the ion can properly be shown as an attribute of the whole group rather than of one of the constituent atoms. Examples of such ions are H_3O^+, BH_4^-, ClO^-, and SO_4^{2-}:

$$\left[\begin{array}{c} H \\ H-O: \\ H \end{array}\right]^{+} \quad \left[\begin{array}{c} H \\ H-B-H \\ H \end{array}\right]^{-} \quad \left[:\ddot{Cl}-\ddot{O}:\right]^{-} \quad \left[\begin{array}{c} :\ddot{O}: \\ :\ddot{O}-S-\ddot{O}: \\ :\ddot{O}: \end{array}\right]^{2-} \quad (17\text{-}6)$$

Isoelectronic species have similar Lewis formulas and, as mentioned earlier, similar geometric configurations. In the examples shown here, H_3O^+ is isoelectronic with NH_3, and BH_4^- is isoelectronic with CH_4 (and with NH_4^+). Both ClF and the isoelectronic ClO^- have Lewis formulas similar to those of any halogen molecule. In the sulfate ion, the four oxygen atoms are situated tetrahedrally around the sulfur atom, and the oxygens are all equivalent—i.e., there is nothing to distinguish any one of them from the others. It is difficult to represent this tetrahedral ion on a plane surface because of the numerous unshared electron pairs that must be portrayed.

Rules for Constructing Lewis Formulas A set of general rules can be given to serve as a guide in the construction of Lewis formulas.

1. The structural (or topological) formula of the molecule or ion concerned, that is, the way in which the atoms are connected to one another, must be known. One can sometimes make sensible guesses about which of two or more possible structural formulas might be more reasonable, but since isomers exist for many structures, it is generally essential to know which atoms are bonded to each other in the species of interest.

2. The Lewis formula must show all the outer-shell (or valence) electrons of the atoms involved, appropriately increased or diminished to take the overall charge into account when considering an ion. For example, the formula

$$\left[:\ddot{O}-H\right]^{-}$$

shows a total of eight electrons, corresponding to six valence electrons for oxygen, another for hydrogen, and one additional electron for the single negative charge of the ion. Similarly, in the Lewis formula for sulfate ion [Formula (17-6)] 32 electrons are shown: six valence electrons for each of the four oxygen atoms, six for the sulfur atom, and two additional electrons corresponding to the charge -2.

3. Each atom in the Lewis formula should have a completed valence shell (or possibly an expanded one for elements beyond Ne). This means two electrons for H, Li^+, and Be^{2+}, eight electrons for second-row elements, and eight (or sometimes 10 or 12) electrons for elements of higher atomic number. Expansion of the valence shell beyond 12 electrons (six pairs) is rare. Only

occasionally does the valence shell contain fewer than eight shared and unshared electrons (four pairs).

☐ **Lewis Formulas** Draw Lewis formulas for H_2S and PH_3.

Example 17-2

Solution The electron configurations of the valence shells of the atoms H, S, and P are

$$\text{H} \cdot \qquad : \overset{\cdot\cdot}{\underset{}{\text{S}}} \cdot \qquad : \overset{\cdot\cdot}{\underset{}{\text{P}}} \cdot$$

Pairing of the lone electrons of two H atoms with those of one S atom leads to the formula

$$\begin{array}{c} \text{H} \\ | \\ \text{H}-\overset{}{\underset{\cdot\cdot}{\text{S}}}: \end{array}$$

Similar pairing for three H atoms and one P atom yields

$$: \text{P}\!\!\begin{array}{c}\nearrow \text{H} \\ -\text{H} \\ \searrow \text{H}\end{array}$$

Note the analogy with the second and third formulas of (17-5). ■

☐ Draw Lewis formulas for Cl_2O and PBr_3. ■

☐ **Lewis Formulas for HSO_4^- and H_2SO_4** Draw Lewis formulas for the hydrogen sulfate ion and sulfuric acid by starting with the Lewis formula for the sulfate ion in (17-6) and adding one or two protons to it.

Example 17-3

Solution It is immaterial to which of the four oxygen atoms each of the protons is added, except that they should be attached to different oxygen atoms. The electronic formula of SO_4^{2-} remains almost unchanged, since each proton carries no electrons with it. For each added proton one of the unshared pairs on a given oxygen atom becomes shared. The resulting formulas are

$$\left[\begin{array}{c} :\overset{\cdot\cdot}{\underset{\cdot\cdot}{\text{O}}}: \\ | \\ \text{H}-\overset{\cdot\cdot}{\underset{\cdot\cdot}{\text{O}}}-\text{S}-\overset{\cdot\cdot}{\underset{\cdot\cdot}{\text{O}}}: \\ | \\ :\overset{}{\underset{\cdot\cdot}{\text{O}}}: \end{array} \right]^{-} \qquad \begin{array}{c} :\overset{\cdot\cdot}{\underset{}{\text{O}}}: \\ | \\ :\overset{\cdot\cdot}{\underset{\cdot\cdot}{\text{O}}}-\text{S}-\overset{\cdot\cdot}{\underset{\cdot\cdot}{\text{O}}}: \\ | \quad\quad \text{H} \\ :\overset{}{\underset{\cdot\cdot}{\text{O}}}: \\ | \\ \text{H} \end{array} \qquad (17\text{-}7)$$

In these species, the H-O-S bond angles are not far from tetrahedral, but it is awkward to represent them all as bent because of the crowding of the formula. Note that in all these formulas, the octet rule is satisfied for every atom but hydrogen, which must achieve the He configuration for stability. ■

☐ Draw Lewis formulas for the hydrogen phosphate ion, HPO_4^{2-}, the dihydrogen phosphate ion, $H_2PO_4^-$, and the phosphoric acid molecule, H_3PO_4. ■

Lewis formulas for species containing multiple bonds are straightforward. Each shared pair of electrons is symbolized by a line between the atoms, as illustrated by the formulas for ethylene (C_2H_4), formaldehyde (H_2CO), acetylene (C_2H_2), nitrogen, and cyanide ion:

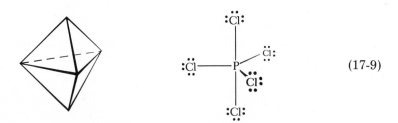

$$H_2C=CH_2 \qquad H_2C=\ddot{O}: \qquad H-C\equiv C-H \qquad :N\equiv N: \qquad [:C\equiv N:]^- \qquad (17\text{-}8)$$

Again note that each atom has a completed valence shell (an octet for each atom except H) and also that ethylene and formaldehyde are isoelectronic, as are acetylene, nitrogen, and cyanide ion.

As noted before, the octet rule is not entirely inviolable. Third-row elements (especially Si, P, S, and Cl) and those with higher atomic numbers, which have relatively low-lying d orbitals available, sometimes use more than four orbitals for bond formation and occupancy by unshared outer-electron pairs.

Phosphorus pentachloride (PCl_5) and sulfur hexafluoride (SF_6) both have structures that violate the octet rule as far as the P and S atoms are concerned. There are five bonds to the phosphorus atom in PCl_5, a molecule that has the shape of a *trigonal bipyramid*—that is, two triangular pyramids sharing an equilateral triangular face. A chlorine atom occupies each corner, with the phosphorus atom at the center of the polyhedron:

$$(17\text{-}9)$$

The total number of electrons displayed in this formula is 40 (20 pairs), corresponding to the seven valence electrons of each of five chlorine atoms and the five valence electrons of the phosphorus atom. The five P-Cl bonds in PCl_5 are not all equivalent; the three in the *equatorial* plane (the plane of the face shared by the two pyramids) are different from the two to the apices of the trigonal bipyramid (called *axial* bonds).

The SF_6 molecule is octahedral, with the S atom at the center of the octahedron and an F atom at each of the six corners. All six bonds are equivalent in this molecule. Its Lewis formula is

$$(17\text{-}10)$$

(The dashed lines are shown only to outline the octahedron whose six corners are occupied by the fluorine atoms.)

There are molecules and ions for which several possible Lewis formulas exist, each different, yet all about equally acceptable. As explained in Sec. 17-6, when this is true only a combination of all these Lewis formulas provides an appropriate description of the molecule.

In most of the molecules and ions for which Lewis formulas have been given so far, each of the atoms joined by a covalent bond has contributed one electron to the bond. However, there are some bonds for which both electrons may be regarded as supplied by only one of the atoms—for example, in the product of the reaction between ammonia and boron trifluoride (BF_3):

$$\text{(17-11)}$$

The product of this reaction is stable, whereas BF_3 is very reactive. Study of the Lewis formula given here for BF_3 shows that the B atom has only a sextet of electrons, not an octet, which helps to explain the reactivity of BF_3. In the product formed in (17-11), all atoms have completed valence shells.

The reaction between ammonia and a proton to form NH_4^+ is one in which the electrons in the fourth N-H bond are contributed by only one of the atoms:

$$\text{(17-12)}$$

Note, however, that once the NH_4^+ is formed, all four bonds in it are equivalent and indistinguishable.

Comparison of the reactions illustrated in (17-11) and (17-12) shows that there is a fundamental similarity. In each of them, the ammonia molecule donates its pair of electrons to another molecule or ion that needs additional electrons for greater stability. Since the reaction of ammonia with a proton (17-12) is a typical acid-base reaction, Lewis proposed that reactions like (17-11) also be regarded as acid-base reactions. In this general sense, a *Lewis acid* is any molecule or ion that will accept a pair of electrons and a *Lewis base* is any molecule or ion that will donate a pair of electrons. For example, H^+ and BF_3 are Lewis acids and OH^- and NH_3 are Lewis bases. Not all species with unshared pairs of electrons in the valence shell are good Lewis bases, although many are.

17-6 Resonance among Different Lewis Formulas

The Basic Concept   There are some molecules and ions for which several acceptable Lewis formulas can be drawn, no one of which alone is adequate to describe the electronic structure or explain the observed properties. For example, there are three acceptable formulas for the nitrate ion:[4]

$$\left\{ \left[\overset{..}{\underset{..}{O}}\overset{..}{\underset{..}{O}} N=\overset{..}{\underset{..}{O}} \right]^{-}, \left[\overset{..}{\underset{..}{O}}\overset{..}{\underset{..}{O}} N-\overset{..}{\underset{..}{O}}: \right]^{-}, \left[\overset{..}{\underset{..}{O}}\overset{..}{\underset{..}{O}} N-\overset{..}{\underset{..}{O}}: \right]^{-} \right\} \qquad (17\text{-}13)$$

$$(a) \qquad\qquad (b) \qquad\qquad (c)$$

Note that the formulas are indeed different; (b) and (c) should not be regarded simply as formula (a) turned 120° and 240°. The atomic nuclei (including inner shells) should be thought of as remaining fixed while the valence electrons are distributed in three different patterns. None of these formulas describes the nitrate ion satisfactorily. Each formula predicts two different kinds of nitrogen-oxygen bonds, but these bonds are known from X-ray studies of crystals containing NO_3^- to be of equal length and equivalent in every way. The three oxygen atoms are thus equivalent to one another; they lie at the corners of an equilateral triangle, the center of which is occupied by the nitrogen atom.

No single Lewis formula, employing integral numbers of shared pairs and unshared pairs, can be drawn that *does* reflect these properties of the nitrate ion. This ion may instead be regarded as a blend or superposition of the three formulas shown. We say that the nitrate ion is a *resonance hybrid* of the three formulas (17-13), and it is often symbolized, as in (17-13), by separating the contributing formulas by commas and enclosing them within braces. The N-O bond is also said to have one-third double-bond character, because each of the N-O bonds is double in one of the three equally contributing formulas and single in the other two. A shorthand way of indicating that a bond is intermediate between single and double is to write the partial bond as a dashed line. Thus NO_3^- can be represented as

$$\left[\overset{O}{\underset{O}{}} N\text{==}O \right]^{-} \qquad (17\text{-}14)$$

A formula like (17-14) expresses the equivalence of the oxygen atoms, and of the three bonds, but it cannot be considered exactly comparable to a Lewis formula since it does not show unshared pairs (of which there is an average of $\frac{8}{3}$ on each oxygen atom).

Because the term *resonance* is used in physics in a sense that implies an oscillation between two states, the incorrect notion sometimes arises that the phenomenon here described as resonance implies an oscillation back and forth among the various contributing structures. The nitrate ion does *not* rapidly change structure so as to correspond at one instant to one of the possible formulas in (17-13) and at a later instant to another. Rather, it corresponds at any instant to a hybrid of all three formulas, as implied by (17-14). An analogy due to J. D. Roberts is helpful. A man who had come across a rhinoceros described the animal as something between a dragon and a unicorn. With this

---

[4]The implication of the braces, {}, will be explained shortly.

description he did not imply that the animal alternated between being a dragon and a unicorn. Moreover he used in his description creatures that exist in the imagination only, just as formulas (*a*) to (*c*) of (17-13) relate to hypothetical species.

Thus resonance is not a phenomenon but rather a *description*. It serves to overcome a limitation of the description of molecules by Lewis formulas, which necessarily involve integral numbers of shared pairs of electrons. Whenever resonance is needed to describe the structure of a molecule, the molecule is found to have higher stability than would correspond to a single Lewis formula that might be used to describe it; in appropriate cases, it is found to have higher symmetry as well. Thus the nitrate ion itself is more stable than would be a hypothetical ion corresponding to any one of the three contributing formulas (17-13). The difference between the energy of one of these hypothetical ions and the actual energy of the nitrate ion is called the resonance energy. It is important to note that the apparently enhanced stability (the resonance energy) and other special properties are *not conferred* on the molecules *by resonance*. Rather, they are a consequence of the fact that *classical valence formulas were chosen as the starting point of the description*.

Benzene and Related Compounds The most frequently cited example of a resonance hybrid is benzene, C_6H_6, for which two equivalent Lewis formulas can be drawn:

$$(17\text{-}15)$$

These valence bond formulas were first proposed by A. Kekulé and are often referred to as the Kekulé formulas for (or the Kekulé forms of) benzene. However, neither of them represents the properties of benzene satisfactorily. For example, compounds containing carbon-carbon double bonds, $C=C$, usually react rapidly with bromine, as illustrated by the reaction

$$(17\text{-}16)$$

Similar reactions with benzene are extremely slow. Furthermore, it is known from spectroscopy and X-ray and electron diffraction that the carbon atoms of benzene form a *regular* planar hexagon, which would not be expected if (17-15*a*) or (17-15*b*) alone were appropriate. If benzene could be described correctly by one of the Kekulé formulas, isomerism of the type illustrated by formulas (*a*) and (*b*) of (17-17) would be observed:

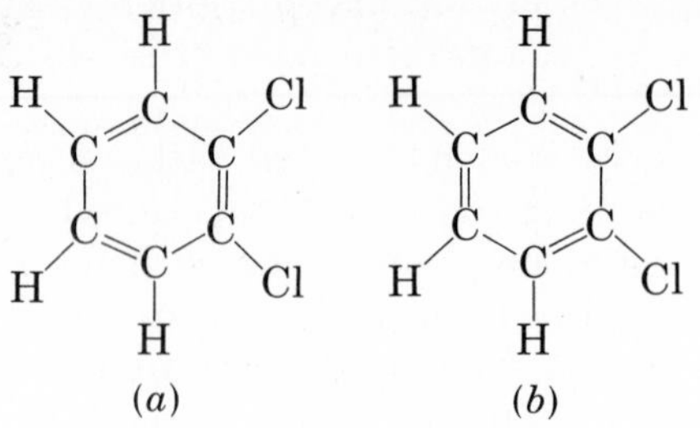

$$\text{(17-17)}$$

(a) (b)

In (a) the two carbon atoms that carry the chlorine atoms are connected by a double bond, in (b) by a single bond, so that the formulas correspond to different compounds. No evidence for such isomerism has ever been found.

A shorthand notation is often used for benzene and related compounds. It shows only the six-membered ring skeleton of single and double bonds, as well as atoms *other than* hydrogen attached to ring carbons, as exemplified by

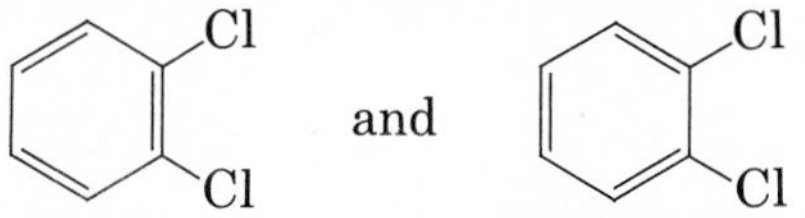

and

$$\text{(17-18)}$$

for the hypothetical isomers just discussed. In the same notation the resonance formulation for benzene is

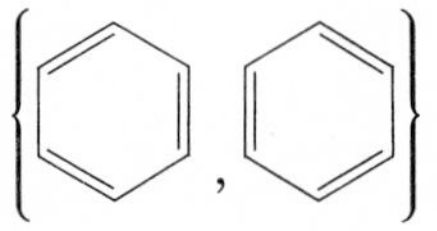

$$\text{(17-19)}$$

Alternative representations often used are

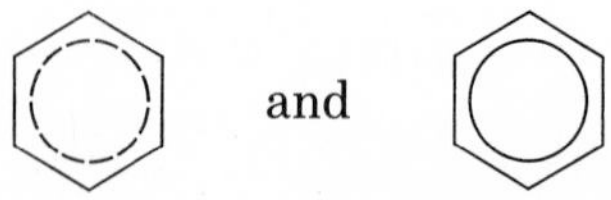

and

$$\text{(17-20)}$$

where the dashed or solid circles indicate, as before, additional partial bonding. In benzene each carbon-carbon bond has half double-bond character. As with the nitrate ion, the benzene molecule is significantly more stable than would be the hypothetical molecule with a structure like either one of the Kekulé forms of benzene.

Benzene is the simplest representative of a large number of organic compounds, termed *aromatic compounds,* many of which contain one or more six-membered rings like that in benzene. Examples of such compounds include the dichlorobenzene depicted in (17-18) and naphthalene, $C_{10}H_8$, for which the formula analogous to (17-20) is

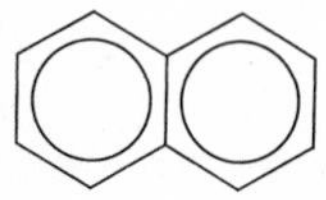

As before, the corners indicate carbon atoms and it is implied that a hydrogen atom is attached to each of the eight carbon atoms that is linked to only two other carbons.

The virtue of the resonance description is that it extends the scheme of classical valence theory and permits relating the properties of molecules that cannot be described by the scheme of single Lewis formulas to the properties of molecules that can be so described. However, other ways have also been developed to describe the properties of molecules that do not correspond to single Lewis formulas. The most successful of them employs molecular orbitals, described in Sec. 18-3.

□ **Resonance Formulas and Variation in Bond Distances** Write resonance formulas for nitric acid, HNO_3, and indicate how you would expect the different N-O bond lengths in this molecule to compare with each other and with that in NO_3^-. Recall that increasing double-bond character corresponds to a shorter distance.

Example 17-4

Solution There are two equivalent resonance formulas for nitric acid:

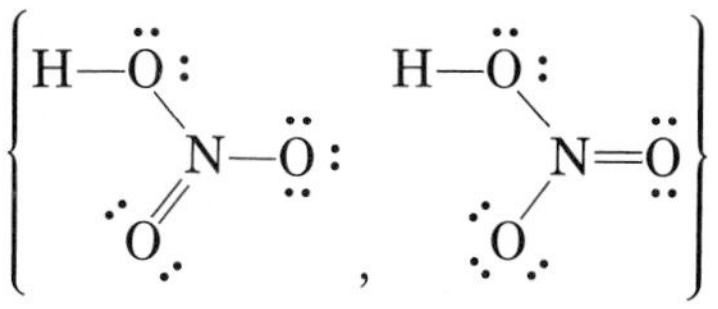

The bond from the N atom to the O atom of the OH group has no double-bond character in either formula; each of the other two bonds from N to O has one-half double-bond character. The latter two bonds should thus be equal in length and shorter than the N-OH bond; they should also be shorter than the N-O bonds in the nitrate ion. These expectations agree with experimental observations. ■

□ Compare the carbon-oxygen distances in the formic acid molecule,

Exercise 17-4

$$\underset{H-C-OH,}{\overset{\overset{\textstyle O}{\|}}{}}$$ with each other and with those in the formate ion, $\underset{H-C-O^-,}{\overset{\overset{\textstyle O}{\|}}{}}$ for which resonance formulas should be written. ■

Summary

The strengths of chemical bonds in molecules are estimated from bond energies, average values of the energies required to break bonds between specified pairs of atoms. Tables of bond energies can be used to predict approximately the energy changes accompanying chemical reactions. The three-dimensional structure of a molecule can be described in terms of: (1) the bond distances between neighboring atoms; (2) the bond angles defined by the relative positions of an atom and two neighboring atoms bonded to it; (3) the torsion angles about bonds, determined by the relative positions of four atoms bonded in sequence. Structures differing only in the angles of rotation about single bonds are referred to as different conformations.

The ionic bond is the consequence of electrostatic attraction between oppositely charged ions. A purely ionic bond is an idealization. Many crystals can be regarded as composed of spherical ions packed together in a systematic and efficient way. Since ionic radii differ little from one crystal to another, it is possible to predict the arrangement of nearest-neighbor anions around a given cation. Tetrahedral, octahedral, and cubic coordination geometries are commonly found in ionic crystals. Evidence for the existence of ions in solids includes

363

the electrical conductivity of melts and the agreement between calculated and experimental lattice energies.

The valence shells of the elements in the first two periods tend to be either empty or filled by a full complement of eight electrons. This generalization is known as the octet rule. The valence shell of the elements in the second period is occasionally expanded by the use of the 3*d* orbitals.

The covalent bond can be described as an attractive interaction in which a pair of electrons is shared between two atoms. In Lewis formulas for the representation of bonding, a covalent bond is indicated by a line between atoms and unshared electrons are shown by dots. There are some molecules and ions for which several Lewis formulas can be drawn. An adequate description of the bonding in such species by Lewis formulas requires a blend or superposition of several electronic structures, a situation termed resonance.

Terms and Concepts

Bond energies	p. 344
Bond distances and angles	p. 347
Torsion angles	p. 348
Rotation about single bonds	p. 349
Rotation about double bonds	p. 349
Coplanarity of atoms bonded to two double-bonded atoms	p. 349
Collinearity of atoms bonded to two triple-bonded atoms	p. 349
Conformations	p. 349
Ionic bond	p. 350
Ionic crystal (crystal structures of NaCl, CsCl, CaF_2)	p. 350
Lattice energy	p. 350
Coordination geometry	p. 351
Effective ionic radius	p. 352
Isomorphism	p. 352
Octet rule	p. 354
Sharing of electrons	p. 354
Covalent bond	p. 355
Lewis electron-dot formulas of molecules and ions	p. 355
Isoelectronic molecules	p. 356
Multiple bonds	p. 358
Violation of octet rule	p. 358
Trigonal bipyramidal structure; equatorial and axial bonds	p. 358
Octahedral structure	p. 358
Lewis acids and bases	p. 359
Rules for Lewis formulas	p. 356
Resonance hybrids	p. 360
Resonance and bond distances	p. 363

Problems and Questions

17-1 Bond Energies Use bond energies to compute approximate values for the energy changes associated with the following gaseous reactions:

(*a*) $H_2 + Cl_2 \longrightarrow 2HCl$
(*b*) $3Cl_2 + N_2 \longrightarrow 2NCl_3$
(*c*) $C_2H_4 + H_2 \longrightarrow C_2H_6$
(*d*) $2H_2 + N_2 \longrightarrow N_2H_4$

17-2 Bond Energies (*a*) Use bond energies to predict the approximate changes in energy that accompany the following reactions, where X is Cl, Br, or I:

$$CH_4 + X_2 \longrightarrow CH_3X + HX$$

(*b*) Would your prediction apply equally to hydrocarbons other than CH_4, such as C_2H_6 or C_8H_{18}?

17-3 Torsion Angles In 1,2-dichloroethane (ClH_2CCH_2Cl), one hydrogen atom on each carbon of ethane (Fig. 17-2) has been replaced by a chlorine atom. The two stable conformations of this molecule have torsion angles of 60 and 180°. Sketch each of these conformations, with drawings similar to those of Fig. 17-2.

17-4 Packing of Unequal Spheres (*a*) The centers of three touching spheres (see Fig. 17-5) of radius *a* form the corners of an equilateral triangle (of edge 2*a*) while a fourth sphere of radius *b*, touching the other three, is centered in the middle of the triangle. What is the ratio *b*/*a*? (*b*) Repeat these calculations for a planar square with four spheres of radius *a* at the corners and one of radius *b* at the center. (*c*) Repeat for six spheres of radius a and one of radius b, with an octahedral configuration.

17-5 Tetrahedral Angles A set of four alternate corners A, B, C, and D of a cube (Fig. 17-7) define a regular tetrahedron. Make a sketch of the cube and the tetrahedron; let the tetrahedral edge (AB) be d, and let P be the center of the tetrahedron (and the cube). (a) By solid geometry, vector algebra, or any other means calculate the distance AP in terms of d. (b) Show that the angle APB is 109.5°. (c) Suppose four equal spheres of radius a are arranged at the corners of a tetrahedron. Find the radius b of the sphere located at the center of the tetrahedron that will just touch each of the four spheres at the corners; express the answer in terms of the ratio b/a, as in Prob. 17-4.

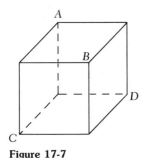

Figure 17-7

17-6 Radii of Isoelectronic Ions What generalization can you make about the relation between the ionic radii of a pair of isoelectronic monatomic ions? Explain it with reference to some particular pair of isoelectronic ions of your own choosing.

17-7 Environment of the Ions in CsCl The arrangement of the ions in a crystal of CsCl is shown in Fig. 17-4a. The edge of the cubic unit cell is 4.12 Å. (a) Consider a Cl^- ion and give the number of closest Cs^+ ions, their distance from the Cl^- ion, and the polyhedron these Cs^+ ions define. (b) Give the same kind of information for the Cl^- ions nearest the Cl^- ion considered. (c) Describe the environment of a Cs^+ ion in the same way.

17-8 Solid Phosphorus Pentachloride Although PCl_5 vapor has the structure shown in Fig. 17-8, the crystalline solid formed by condensation of the vapor is found to consist of a mixture of equal numbers of two ions, PCl_4^+ and PCl_6^-. Write Lewis formulas for these species and suggest geometries by analogy with known species.

Figure 17-8

17-9 Environment of the Ions in CaF$_2$ The arrangement of the ions in a crystal of CaF_2 is shown in Fig. 17-4b, the edge of the unit cube being 5.46 Å. Describe (a) the arrangement of nearest F^- ions about each Ca^{2+} ion, as in Prob. 17-7, and (b) the arrangement of nearest Ca^{2+} and F^- ions about each F^- ion.

17-10 Covalent and Ionic Compounds Diagram reasonable electronic structures for the following species. Use a line to represent a shared pair of electrons and indicate outer unshared electrons by dots. If the compound is predominantly ionic, indicate the ions by enclosing them in brackets, with the appropriate charge shown, and give the structures for the individual ions.

HBr, I_2, CCl_4, CH_2F_2, HCN, $SiBr_4$, Na_3PO_4, LiF, NCl_3, H_2Se, NH_4BH_4, SCl_2, PH_3, Li_2S, $MgBr_2$

17-11 Unstable Species Many molecules that are unstable under normal terrestrial conditions have been observed spectroscopically in comets, together with some familiar species as well. These include CN, C_2, OH, and CH, together with CH_4, NH_3, and CO. The pressure in the central portion of a comet has been estimated to be no higher than 10^{-10} torr, and it is presumably far lower in the tail of the comet. Suggest Lewis electronic structures for the various unstable species mentioned. What feature of these structures might explain why the species are unstable under normal conditions on earth? Why do they apparently persist in comets?

17-12 Lewis Acid Aluminum chloride, $AlCl_3$, is a strong Lewis acid. Explain what the term *Lewis acid* implies and how this property of aluminum chloride is interpreted in terms of its electronic structure.

17-13 Lewis Acid (a) The reaction of BF_3 with ammonia to form F_3BNH_3 is usually cited as the classic example of reaction of a Lewis acid (which is seeking an unshared electron pair) with a base. Is it reasonable to consider this also as an oxidation-reduction reaction? If so, why? If not, in what essential way does it differ? How does it differ from the reaction of a Brønsted acid with a base? (b) Contrast the electronic structures of the following two distinctly different compounds: NH_3BF_3 and NH_4BF_4.

17-14 The Resonance Concept Explain why the concept of resonance was introduced into structural chemistry. Illustrate your explanation with its application to the structure of the nitrite ion, NO_2^-.

17-15 Resonance and Bond Distances The carbonate ion, CO_3^{2-}, has a planar structure with three equal bond angles and three equal carbon-oxygen distances. Draw three equivalent Lewis formulas for this ion. Draw two structures for the bicarbonate ion, $HOCO_2^-$, and predict which of the carbon-oxygen bonds in these two ions would be the shortest and which the longest. Would you expect the bonds in CO_2 to be similar in length to any of the carbon-oxygen bonds in these ions, or shorter, or longer? Explain.

17-16 Lewis Formulas Draw Lewis formulas for the following sets of molecules or ions. Show important resonance formulas where appropriate. (a) $Be(OH)_4^{2-}$, SO_2, H_3PO_2 (in which two H's, an O, and an OH group are linked to the phosphorus), NNO, PH_4^+
(b) SO_3, ClO_4^-, O_2^{2-}, NCS^-, H_2NNH_2, C_5H_5N (containing a six-membered ring with five C's, each attached to one H, and one N), ClO_2^-

17-17 Resonance and Bond Distances The B-F distance in BF_3 is 1.30 Å; that in BF_4^- is 1.40 Å. Explain.

Chemical Bonds:
II. Theories of
Bonding and
Molecular Structure

"It is not unfair to say that in . . . practically the whole of theoretical chemistry, the form in which the mathematics is cast is suggested, almost inevitably, by experimental results. This is not surprising when we recognize how impossible is any exact solution of the wave equation for a molecule. Our approximations to an exact solution ought to reflect the ideas, intuitions, and conclusions of the experimental chemist."
C. A. COULSON,[1] 1952

"Consequently it is expected that of two orbitals in an atom the one that can overlap more with an orbital of another atom will form the stronger bond with that atom, and, moreover, the bond formed by a given orbital will tend to lie in that direction in which the orbital is concentrated."
L. PAULING,[2] 1939

18

This chapter is a continuation of the preceding one. The first section describes quantum-chemical theories of the covalent bond. Next comes a discussion of the most common types of combinations of orbitals on a given atom (hybrid orbitals) that are used in explaining the geometry of molecules. An empirical scheme for correlating molecular shape with electronic structure is also presented. The chapter concludes with an introduction to molecular orbital theory.

18-1 The Covalent Bond: Quantum Theory

General Remarks about Quantum-Chemical Theories of Bonding An accurate treatment of the binding together of atoms in molecules requires quantum mechanics. The systems involved may be represented by wave functions that are solutions of the Schrödinger wave equation appropriate to the molecule considered, just as for an atom. However, highly accurate solutions have been obtained to date only for the simplest diatomic ions and molecules, such as H_2^+ (the hydrogen molecule-ion), H_2, and HLi. For more complex species—even for such simple molecules as O_2 and H_2O—the exact solution of the Schrödinger equation, even if it were possible, would yield a wave function of such detail and complexity as to make it impossible even to record. What is of interest to the chemist is a

---

[1]C. A. Coulson, "Valence", 2d ed., pp. 113–114, Oxford University Press, New York, 1961.
[2]Linus Pauling, "The Nature of the Chemical Bond", 3d ed., p. 108, Cornell University Press, Ithaca, N.Y., 1960. Copyright 1939 and 1940, 3d ed. © 1960, by Cornell University.

method of constructing and improving simplified approximate wave functions
that can be used for the calculation of energies, geometries, and other properties
of molecular systems. Many different approaches to the problem of constructing
suitable wave functions and thus obtaining approximate solutions to the
Schrödinger equation for molecular systems have been proposed. We shall discuss
qualitatively two of the simplest of them.

The justification of these quantum-chemical approximations is that, like any
theory, they permit the explanation and categorization of a large variety of
phenomena. Each of the approaches is based on wave-mechanical principles but
includes additional postulates, different for each approach. These postulates are
compatible with wave mechanics but are chosen so that they greatly simplify the
process of obtaining solutions. In each of these theories, results obtained from
application of wave mechanics to simple situations are assumed to be general and
are applied to situations too complicated for rigorous treatment.

The most widespread of the current theories of molecular structure are the
valence bond and the *molecular orbital* theories. An essential feature of both is
that chemical bonding arises as a direct consequence of the overlap of orbitals on
nearby atoms. The theories differ in their postulates and in details of treatment,
and they are in part tailored to fit different types of molecules and ions. Thus in
a given situation one or the other of them may be most easily applied and most
revealing in its results. They would merge into one and the same definitive theory
of molecular structure if it were possible to refine sufficiently the approximations
used.

Molecular Orbital Treatment of the Hydrogen Molecule-Ion and the Hydrogen Molecule

The principal idea of the molecular orbital method is to start with all
the nuclei and inner shells of the atoms making up the molecule to be treated and
formulate suitable approximate wave functions that extend over the entire
molecular domain and are hence called *molecular orbitals*. These orbitals are
then filled with the valence electrons of the original atoms, in accordance with
the Pauli exclusion principle (Sec. 15-5) and (for the ground state of the molecule) in the order of increasing energy of the orbitals. In other words, an Aufbau
principle is involved, similar to that used in the description of the electronic
structure of atoms.

In the simplest version of the theory the molecular orbitals are formed by
combining the atomic orbitals of the valence shells of the various atoms constituting the molecule. According to the principles of quantum chemistry, such
combining of orbitals always yields the same number of resulting orbitals as there
were individual orbitals at the start—there is, so to speak, an "orbital conservation principle".[3]

Consider a system of two hydrogen atoms. Each atom has a $1s$ orbital available
in its valence shell, which (like all s orbitals) is spherically symmetric (Fig. 15-9).
Since we begin with two orbitals, there must be two ways to combine them so as
to produce two molecular orbitals. By the rules of quantum chemistry the first
combination is their sum, $\psi_+ = \psi_A + \psi_B$ (Fig. 18-1), and the second is their
difference, $\psi_- = \psi_A - \psi_B$ (Fig. 18-2). When the atoms are far apart (Fig. 18-1a;

---

[3]This principle, applied here to the combination of atomic orbitals on separate atoms, holds also for
the combination of orbitals on the *same* atom, in the formation of what are termed *hybrid orbitals*
(discussed in Sec. 18-2).

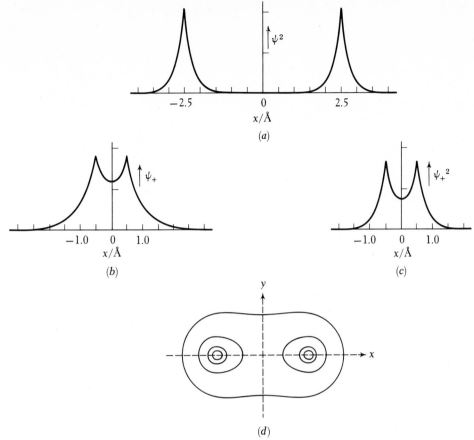

Figure 18-1 Representations of ψ and ψ^2 for the Sum of Two Hydrogen-Atom 1s Orbitals.

(a) Cross section of the electron probability for one electron and two hydrogen nuclei (protons) separated by 5.0 Å, along the line through the center of each proton. Interaction is essentially zero, and diagrams of ψ_+^2 and ψ_-^2 are identical; i.e., the term $\pm\psi_A\psi_B$ is negligible.

(b) Cross section of the sum of two hydrogen 1s orbitals, $\psi_+ = \psi_A + \psi_B$, when the protons are 1.06 Å apart, the internuclear separation observed in H_2^+.

(c) Cross section of ψ_+^2, the electron probability, along the same line as in (a) and (b). The probability of finding the electron is highest near the protons, but there is also a high probability of finding the electron in the region between the protons.

(d) Contours of the electron probability density ψ_+^2 with a distance of 1.06 Å between the hydrogen nuclei. The view in (c) is a section of (d) along the x axis. The contours in (d) are at a constant increment of electron probability density. [The scale along x is expanded by a factor of about 3 relative to that in (a), (b), and (c).]

even 5 Å is far apart for two hydrogen atoms), they hardly interact because neither of the two 1s wave functions has a value significantly different from zero near the position of the other atom. At smaller separations, however, the orbitals overlap noticeably—i.e., both wave functions have values significantly different from zero in the same region of space, called the region of overlap (Figs. 18-1b and 18-2a). The electron probability density ψ^2 is then significantly different from the sum of the probability densities of the two separate atoms. That is, if ψ_A and ψ_B are the orbitals of the separate atoms A and B, the square of their sum is $(\psi_+)^2 = (\psi_A + \psi_B)^2 = \psi_A^2 + \psi_B^2 + 2\psi_A\psi_B$. This function thus differs from

$\psi_A{}^2 + \psi_B{}^2$, the sum of the probability distributions for the two separate atoms, when the product $2\psi_A\psi_B$ is significantly different from zero. This occurs when the wave functions of the two atoms overlap, that is, in a region in which both ψ_A and ψ_B are significantly different from zero. Similar considerations apply to the difference $\psi_A - \psi_B$, because $(\psi_-)^2 = (\psi_A - \psi_B)^2 = \psi_A{}^2 + \psi_B{}^2 - 2\psi_A\psi_B$.

Examination of Fig. 18-1c and d shows that when the orbitals (i.e., the wave functions) are added to each other there is enhancement of the electron density in the region between the nuclei and thus a relatively high probability of finding electrons in this region. This preponderance of negative charge between the (positively charged) nuclei tends to pull the nuclei together. At small separations, however, the mutual electrostatic repulsion between nuclei is sufficiently great to exceed the attraction caused by the electron density between them. At an intermediate distance there is a balance between attraction and repulsion and a stable configuration results.

Figure 18-2 shows the result of combining the atomic orbitals by taking their difference. At large distances (not shown), there is again no interaction, but as the atoms approach one another, the molecular wave function (the difference of the 1s functions of the two atoms) remains equal to zero halfway between the nuclei. Consequently the square of the wave function, the probability density, also falls to zero in this region. The electron density is somewhat enhanced on the side of each nucleus away from the other nucleus, but in contrast to the situation when the *sum* of the atomic wave functions is taken, there is no buildup of negative charge between the nuclei to offset the internuclear electrostatic repulsion.

Figure 18-2 Representations of ψ and ψ^2 for the Difference of Two Hydrogen-Atom 1s Orbitals
(a) Cross section, along the line through the centers of the protons, of the difference of two hydrogen 1s orbitals, $\psi_- = \psi_A - \psi_B$, when the protons are 1.06 Å apart. The scale is the same as that for Fig. 18-1b.

(b) Cross section of the electron probability density $\psi_-{}^2$ along the same line as in (a). The density is highest near the protons, as it was for $\psi_+{}^2$, but it falls to zero in the region between them.

(c) Contours of the electron probability density $\psi_-{}^2$ with a distance of 1.06 Å between the protons. Note the enhancement of the density on the side of each proton away from the other proton. Diagram (b) is a cross section of (c) along the x axis. The contours are the same as in Fig. 18-1d except that the lowest contour is the nodal line $x = 0$ representing the nodal y, z plane, for which ψ_- and $\psi_-{}^2$ are zero.

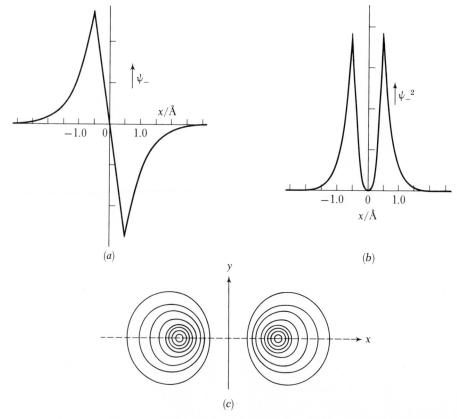

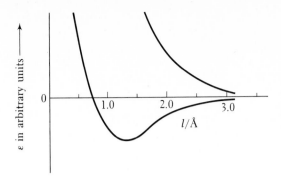

Figure 18-3   Energy of Two Hydrogen Nuclei and One Electron as a Function of the Internuclear Distance
The curves have been calculated by the molecular orbital method. The lower curve has a minimum that corresponds to the formation of the stable molecule-ion H_2^+. The upper curve represents repulsion between the hydrogen nuclei at all distances and no formation of a molecule.

The contrast between the effects of adding the atomic orbitals and of subtracting them (Figs. 18-1 and 18-2) is all-important to an understanding of bonding. In the first case, the molecular orbital that is formed is said to be a *bonding orbital;* that is, when electrons occupy it, the atoms are held together in the stable combination that we refer to as a molecule. In the second case, the molecular orbital formed is referred to as an *antibonding orbital,* implying that if it is occupied by electrons, their density in the region between the nuclei is actually diminished relative to what it would be in the absence of any interaction, so that there is a repulsion of the nuclei instead of an attraction.

Let us consider in more detail what happens for the molecules H_2^+ and H_2. In H_2^+ there is only one valence electron available, which accordingly can occupy either the bonding or the antibonding molecular orbital. The two resulting energy curves are shown in Fig. 18-3.

In the H_2 molecule there are two valence electrons available, one from each hydrogen atom. If both electrons occupy the bonding orbital, they must, by the Pauli principle, have antiparallel spins. This is the molecular orbital description of the H_2 molecule in its ground state. It is also possible to place one of the two electrons in the bonding orbital and the other in the antibonding orbital, in which case the spins of the two electrons may be parallel or antiparallel. Detailed calculations show that for antiparallel spins there is still some attraction, so that this represents an excited state of the H_2 molecule. When the spins are parallel, there is repulsion at all distances and therefore no formation of a stable (but excited) molecular state. Finally, both electrons may be placed in the antibonding orbital, in which case their spins must be antiparallel. This again corresponds to repulsion at all internuclear distances and therefore does not represent a molecular state. The energy diagram of Fig. 18-3 applies only qualitatively to H_2 because, among other things, it does not take the electrostatic repulsion between the two electrons into account. However, elaborate quantum-mechanical calculations that include this and other possible interactions have yielded excellent agreement between observed and calculated properties of both H_2^+ and H_2, such as the internuclear distances (1.06 and 0.74 Å, respectively) and the bond energies (256 and 433 kJ mol^{-1}, respectively).

The Valence Bond Method   When the molecular orbital description of bonding is applied to diatomic molecules such as H_2, the bonding orbitals correspond closely to the conventional chemical picture in which a bond is a line connecting two atoms. This correspondence is more difficult to retain in polyatomic molecules, and a different theoretical approach, which is more closely related to chemical intuition, is often useful. This procedure is called the *valence bond method*. In its simplest version, one starts with the atoms in the molecule and describes the bonding as an interchange of electrons between each pair of atoms that are bonded to each other.

In the valence bond treatment, the bonding in H_2 is represented as arising from the overlap of the two 1s orbitals of the two hydrogen atoms. The Li_2 molecule may be considered to be similar, except that 2s rather than 1s orbitals are used for it:

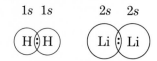

Applications of the valence bond method to more complex species are described in the following sections.

18-2   Hybridization

We consider now the nature of the atomic orbitals that are used to accommodate the shared and unshared electron pairs characteristic of Lewis formulas. Many of these same atomic orbitals also play an important role in the construction of molecular orbitals.

General Remarks   A valence bond procedure based on the ordinary atomic orbitals (for example, s, p) is satisfactory for many diatomic molecules. There are difficulties, however, in applying it to most larger molecules, for example, molecules of tetrahedral structure such as CH_4 (methane), SiH_4 (silane), and CCl_4 (carbon tetrachloride). As discussed in Sec. 17-2, the four outer atoms in each of these molecules are situated at the corners of a regular tetrahedron. They are linked to the central C or Si atom by four covalent bonds that are equivalent to one another by symmetry. This equivalence of the bonds is at first sight difficult to understand, because the orbitals of the central atom available for bonding are the s and p orbitals in the valence shell (a 2s and three 2p orbitals for carbon; a 3s and three 3p orbitals for silicon), which are quite different in their directional properties and thus are by no means equivalent. How can these orbitals be used to form four equivalent bonds? The answer to this question lies in the description of the bonding by a formalism called *hybridization,* in which an s orbital and three p orbitals on the central atom are combined to create four new orbitals that are equivalent to each other and point toward the corners of a regular tetrahedron. Many other combinations of different orbitals on a given atom can also be formed, with other directional properties.

The reason for the importance of hybridization is that suitable hybrid orbitals have highly directional character, leading to better overlap with orbitals of other atoms to which the atom of interest is bonded. In particular, the nondirectional character of s orbitals (which are spherically symmetric) makes them poorly suited for bonding unless they are combined in hybrids, and they are used by themselves only when no other orbitals of comparable energy are available, as in hydrogen.

Hybridization Combination of orbitals centered on the same atom produces a composite orbital called a *hybrid*. The signs associated with wave functions must be taken into account when a hybrid is formed. Consider, for example, the combination of an *s* and a *p* orbital. In representing a *p* orbital schematically, we indicate that the two lobes have opposite signs by

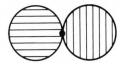

where the vertical hatching indicates the region in which the wave function is positive and the horizontal hatching denotes the negative region; the dot shows the position of the nucleus.[4] In this representation an *s* orbital is

The addition of *s* and *p* orbitals on the same atom can be indicated schematically by superimposing the diagrams

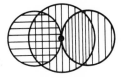

In regions in which the hatching is the same, the two orbitals have the same sign and therefore reinforce. In crosshatched regions there is some cancellation because the orbitals are opposite in sign. The resultant hybrid therefore looks like this

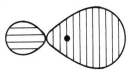

The new orbital is more strongly dependent on direction than either of its "parents", a characteristic that is important in bonding. Hybrid orbitals can also be formed from more than two orbitals on the same atom. The orbitals combined should not differ too much in energy. This means, for example, that no satisfactory hybrid would result from combining orbitals such as 2*s* and 3*p*, although *s* and *p* orbitals of the same principal quantum number can be hybridized effectively.

We now turn to an examination of the geometric aspects of some important hybrid orbitals. Note that in each of the examples, the number of hybrid orbitals formed is just the same as the number of separate orbitals at the start.

---

[4] The diagram is highly schematic and does not show that for *np* orbitals with $n \geq 3$ the signs of inner and outer portions may be different because of the radial nodes (see Exercise 15-3). The same is true for *ns* orbitals with $n \geq 2$ (Example 15-3).

Hybridization of *s* and *p* Orbitals   Three important kinds of hybrids can be formed by combining *s* and *p* orbitals. One *s* and one *p* orbital can be hybridized to form what are known as *sp hybrids*. Two such hybrids can be formed. They point in opposite directions from the center of the atom and thus are at 180° to each other. A simple molecule that may be described by *sp* bonding is the hypothetical BeH_2 molecule,

$$H—Be—H$$

in which the bonds are formed by the overlap of the *sp* orbitals of the Be atom with the 1*s* orbitals of the hydrogen atoms, as portrayed schematically in Fig. 18-4. The expected geometry of the BeH_2 molecule is *linear,* in accord with the 180° angle between two *sp* hybrids. Experimental verification of the structure of this molecule has not so far been possible, because only a polymeric form, represented as $(BeH_2)_n$, is known. A molecule for which *sp* hybrids are believed to be used is $HgCl_2$, which is linear; the two Hg-Cl bonds have appreciable covalent character. Presumably *sp* hybrids formed from the 6*s* and 6*p* orbitals of the Hg atom are involved in the bonding. An important application of *sp* hybridization is in one description of triple bonding (Fig. 18-12).

When one *s* and two *p* orbitals are hybridized in appropriate proportions, three equivalent hybrid orbitals result. These three hybrids, termed *sp*² *hybrids,* enclose angles of 120°. The plane is that characteristic of the two *p* orbitals; thus the remaining *p* orbital, not used in forming the hybrids, is perpendicular to this plane.

An example of a molecule with *sp*² hybridization is the unstable molecule BH_3,

which is formed when B_2H_6 is decomposed by bombardment with high-energy electrons. The BH_3 molecule is expected (but not as yet known) to be planar, with 120° bond angles (Fig. 18-4). The methyl cation, CH_3^+, a short-lived reaction intermediate that is isoelectronic with BH_3, is expected to have this same *planar trigonal* configuration, a conjecture that has been verified experimentally. The planar trigonal configuration is also known to exist in trimethylboron, $B(CH_3)_3$, as well as in BCl_3 and other boron trihalides. Other instances of *sp*² hybridization are considered later, including its use in a description of double bonds (Fig. 18-12).

When one *s* and three *p* orbitals are appropriately combined, four equivalent hybrids directed toward the corners of a tetrahedron may be formed. They are termed *sp*³ *hybrids* and make angles of 109°28′ with each other. This is how hybridization makes it possible to describe the bonding in CH_4, CCl_4, SiH_4, and other tetrahedral molecules; *sp*³ orbitals on the central atom, together with suitable overlapping orbitals of the peripheral atoms, hold the shared electron pairs (Fig. 18-4).

It is also possible to combine *s* and *p* orbitals to form hybrids that are not equivalent to one another. For example, one *s* and three *p* orbitals can be hybridized in such a way that the proportions of *s* and *p* character in the hybrid orbitals formed differ from $\frac{1}{4}s$ and $\frac{3}{4}p$, which is the blend that makes the usual *sp*³ hybrids exactly equivalent and tetrahedrally directed. If the proportions of *s* and

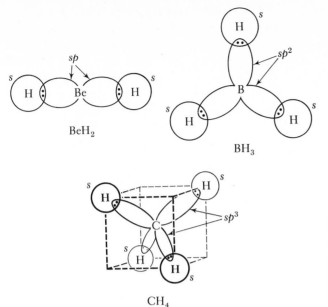

Figure 18-4 Bonding with sp, sp^2, and sp^3 Hybridization
The sp, sp^2, and sp^3 hybrids of the central atoms are shown in stylized form. BeH$_2$ is a hypothetical molecule that has not yet been observed. BH$_3$ has been observed, but its structure has not yet been proved. Nonetheless these two species serve as illustrative examples of sp and sp^2 hybridization of the orbitals on the central atom. CH$_4$ exemplifies sp^3 hybridization.

p character differ in the four hybrid orbitals formed, the hybrids point toward the corners of a *distorted* tetrahedron rather than a regular one.

Atoms with Unshared Pairs There are many molecules and ions in which not all electron pairs of the valence-shell octet of an atom are shared. In spite of this it is often reasonable to describe the bonding orbitals by sp^3 hybridization, with the unshared pairs also occupying sp^3 orbitals. The oxygen atom of the water molecule may serve as an example. Its valence shell contains two unshared pairs of electrons, and two shared pairs that constitute the covalent bonds to the hydrogen atoms. In one extreme description the shared and unshared pairs may be considered to use four equivalent sp^3 hybrids; in the other extreme the unshared pairs occupy the $2s$ and one of the three $2p$ orbitals while the shared pairs use the remaining two $2p$ orbitals. The first of these situations would lead to an expected H-O-H angle equal to the tetrahedral angle of $109.5°$, the second to an H-O-H angle of $90°$, corresponding to the angle between two p orbitals. The bond angle in water is actually $104.5°$, which suggests that the situation is closer to sp^3 hybridization.

Another molecule containing an unshared pair on the central atom is NH$_3$. This molecule is pyramidal in shape, with the nitrogen atom at the apex of the pyramid and the three hydrogen atoms forming its base. Again we may consider that four equivalent sp^3 orbitals of the nitrogen atom are used, both for the shared pairs of electrons that bond the hydrogen atoms and for the unshared pair. An alternative is that the bonding pairs use the three $2p$ orbitals, with the unshared pair in the $2s$ orbital. The actual bond angle of $107.3°$ suggests hybridization close to equivalent sp^3 orbitals. The unshared pair then occupies an sp^3 orbital extending from the nitrogen atom in a direction away from the base of the pyramid.

How reasonable is it to expect sp^3 hybridization in these molecules? These hybrids provide greater bond strength than would pure p orbitals, since they are more highly directed, i.e., more concentrated in particular directions, and also extend further out from the nucleus. Both these factors contribute to more effective overlap with bonded neighbors. On the other hand, to make a $2s$ orbital available for sharing electrons, one of the electrons in it must be promoted (raised) to a $2p$ state. Since the difference between the $2s$ and $2p$ states of oxygen is about 840 kJ mol^{-1} and the corresponding difference for nitrogen is about 630 kJ mol^{-1}, a high price must be paid for promotion. Other important effects are the mutual repulsion that the two hydrogen atoms in H_2O or the three hydrogen atoms in NH_3 incur if they are too close to each other. There is also interaction between the hydrogen atoms and the unshared pairs (or pair) on the central atom. The balance of all these effects is hard to assess and only qualitative conclusions can be drawn, such as that the hybridization is somewhere between the two extremes described.

It is interesting that the bond angles in the hydrides of the congeners of N and O are quite close to 90° (91 to 94°). This may reflect in part the larger size of the central atoms involved, which increases the distances between hydrogen atoms and thus decreases their mutual repulsion, but whatever the cause, there is presumably little or no hybridization of s and p orbitals in these molecules. Tetrasubstituted congeners of carbon, as well as the corresponding ions in neighboring columns (for example, PH_4^+ and $GaBr_4^-$), are tetrahedral in shape and presumably use sp^3 orbitals on the central atom.

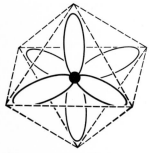

Figure 18-5   Geometry of sp^3d^2 Hybrids

The hybrid orbitals of the central atom are shown in stylized form. The six sp^3d^2 orbitals are equivalent by symmetry.

Hybridization Involving d Orbitals   A number of different hybrids can be formed by combining d orbitals with s and p orbitals. Only the most important of these will be discussed, sp^3d^2 hybrids.

The six equivalent sp^3d^2 *hybrids* point toward the six corners of an octahedron (Fig. 18-5) and are often called octahedral orbitals. The d orbitals used in these hybrids may either have the same principal quantum number as the s and p orbitals, in which case they are known as *outer d orbitals* (for example, $4d$ orbitals hybridized with $4s$ and $4p$), or a principal quantum number one less, in which case they are called *inner d orbitals* (for example, $3d$ orbitals hybridized with $4s$ and $4p$). For the elements in columns IIIA to VIIA of the periodic table (the congeners of B, C, N, O, and F) and for the noble gases, the inner d orbitals are either completely filled and therefore do not participate in hybridization or (when n is 3 or less) there are no inner d orbitals. Thus, for these elements only outer d orbitals can be involved in d orbital hybrids. Since these outer d orbitals are of higher energy than the ns and np orbitals, promotion energy is required. In the first short period, when n is less than 3, the promotion energy is prohibitive; the $3d$ orbitals would have to be hybridized with the $2s$ and $2p$ orbitals. This is why the valence shells of the atoms Li through F never contain more than eight unshared or shared electrons, in accord with the octet rule.

The situation is different with the transition elements, for which at least some of the *inner d* orbitals are usually available for hybridization. Thus when some inner d orbitals are unfilled, all electrons in the $(n-1)d$, ns, and np levels of the atom may be counted as valence electrons. For example, the elements Ti, V, ..., Ni have 4, 5, ..., 10 valence electrons, respectively (see Table 16-2). Transition-metal compounds often exhibit geometries in accord with sp^3d^2 and other d-

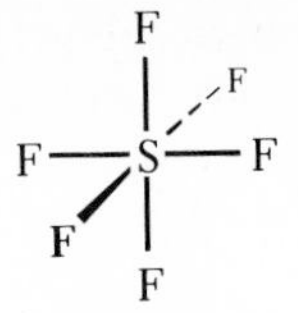

F
| /F
F—S—F
/ |
F F

Octahedral
SF_6, SeF_6, TeF_6
PF_6^-, PCl_6^-, SbF_6^-
$GeCl_6^{2-}$, $SnCl_6^{2-}$, $SnBr_6^{2-}$
$PbCl_6^{2-}$, $ZrCl_6^{2-}$, $CdCl_6^{4-}$

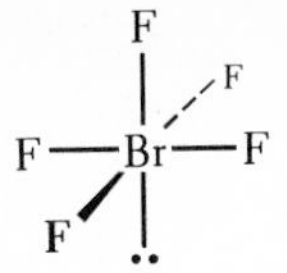

F
| /F
F—Br—F
/ |
F ..

Square pyramidal
BrF_5, IF_5, SbF_5^{2-}, $XeOF_4$

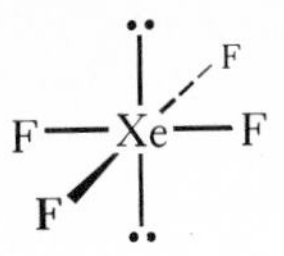

..
| /F
F—Xe—F
/ |
F ..

Square planar
XeF_4, ICl_4^-, BrF_4^-

Figure 18-6 Representative Structures Based on Octahedral (sp^3d^2) Hybridization Involving Outer d Orbitals
One or two of the atoms or groups bonded to the central atom may be replaced by unshared electron pairs, the overall octahedral configuration (including, presumably, the unshared pairs) being at least approximately maintained. Note the noble-gas compounds $XeOF_4$ and XeF_4.

orbital hybridization. However, this valence bond description has certain short-comings, particularly in its failure to account for the spectra and reactivities of transition-metal complexes. A more satisfactory approach to bonding in transition-metal compounds is considered in Chap. 27.

Examples of Structures Involving Outer d Orbitals The valence shell of S includes the $3s$ and $3p$ orbitals and can be expanded by the inclusion of the $3d$ orbitals. In SF_6 there are 12 electrons in the valence shell of the sulfur atom, six from the S atom itself and one from each of the six fluorine atoms to which it is bonded. These 12 electrons are used to form the six covalent bonds in the molecule, so there are no unshared pairs on the sulfur atom. The sp^3d^2 hybridization in this molecule is also found in many other molecules and ions, a few of which are illustrated in Fig. 18-6. In a number of the examples one or two of the bonded atoms have been replaced by unshared pairs.

The concept of hybridization is an aid to the understanding of the geometric configurations of many different molecules and ions. It must always be remembered, however, that hybridization itself is not a physical phenomenon, any more than is resonance. It is simply a useful way of describing the orbitals used in bonding in terms of particular sets of "atomic" orbitals (s, p, d, . . .) that are suitable for describing the properties of isolated atoms and ions but are far less suitable in descriptions of bonding.

□ **Hybrid Orbitals** Indicate the specific individual orbitals that might be used by the central atom in forming possible hybrid orbitals in the following molecules and ions. Indicate also the kind of hybrid expected to be formed and the approximate bond angle if that kind of hybrid is formed.

Example 18-1

$$CF_4, \ NI_3, \ BI_3, \ CuCl_2^-, \ AsF_6^-, \ AlH_4^-, \ NH_2^-, \ Cl_2CH_2$$

Solution We begin by considering the orbitals available in the valence shell of the central atom, and then we consider how the shared and unshared electron pairs on the central atom in the Lewis formula for the species can most plausibly be assigned to hybrid orbitals constructed from the orbitals available:

Species	Available orbitals on central atom	Lewis formula*	No. of valence electron pairs on central atom	Expected hybrid	Likely bond angle
CF_4	$2s$, three $2p$	$:\ddot{F}:$ $:\ddot{F}:\underset{\ddot{F}:}{\overset{..}{C}}:\ddot{F}:$	4	sp^3	109.5°
NI_3	$2s$, three $2p$	$:\ddot{I}:\underset{:\ddot{I}:}{\overset{..}{N}}:\ddot{I}:$	4	sp^3	~109°
BI_3	$2s$, three $2p$	$\overset{:\ddot{I}\quad :\ddot{I}.}{\underset{:\ddot{I}:}{B}}$	3	sp^2	120°
$CuCl_2^-$	$4s$, three $4p$	$[:\ddot{Cl}:Cu:\ddot{Cl}:]^-$	2	sp	180°
AsF_6^-	$4s$, three $4p$, five $4d$	(like PCl_6^-)	6	sp^3d^2	90°
AlH_4^-	$3s$, three $3p$	$\left[H:\underset{H}{\overset{H}{Al}}:H\right]^-$	4	sp^3	109.5°
NH_2^-	$2s$, three $2p$	$\left[H:\underset{H}{\ddot{N}}:\right]^-$	4	$\sim sp^3$	~105–109°
Cl_2CH_2	$2s$, three $2p$	$H:\underset{:\ddot{Cl}:}{\overset{H}{C}}:\ddot{Cl}:$	4	$\sim sp^3$	~109–110°

*Dots are used here for bonding electrons as well as for unshared electrons.

Note that AlH_4^- is isoelectronic with SiH_4 and that NH_2^- is isoelectronic with H_2O; hence a bond angle of 105° in NH_2^-, comparable to that in H_2O, might be expected. The different bond angles in Cl_2CH_2 would not all be expected to be the same, but they will be not far from 109 to 110°, the average tetrahedral angle, and the hybridization will approximate sp^3. ∎

As indicated in this example, the hybridization about a central atom is usually predictable from a consideration of the number of electron pairs around the central atom and the orbitals they occupy.

Exercise 18-1

☐ Indicate the specific individual orbitals that might be used by the central atom in forming possible hybrid orbitals in the following molecules and ions. Indicate also the kind of hybrid expected to be formed and the approximate bond angle if that kind of hybrid is formed.

$$PF_3, \quad BF_3, \quad GeCl_4, \quad SF_2 \quad ∎$$

Rules for Predicting Geometric Configuration An empirical set of rules for the prediction of the geometric configuration of the bonded neighbors about a central atom can be summarized rather concisely. These rules can be rationalized and refined by considering that electron pairs in the valence shell of the central atom, whether unshared or in bonds, tend to remain as far apart as possible, and that repulsions between unshared pairs are somewhat greater than those between unshared pairs and bonded pairs, which are in turn somewhat greater than those between bonded pairs. An appreciation of hybridization also helps in understanding the rules, which are as follows:

1. Formulate the Lewis structure(s) for the molecule or ion.
2. Count the number of bonded atoms attached to the atom considered and the number of unshared electron pairs on that atom. (Do not include unshared pairs that occupy *inner d* orbitals.)
3. Add the number of bonded neighbors to the number of unshared pairs. The geometric configuration about the central atom, including both the attached bonded atoms *and* the (presumed positions of the) unshared pairs, can then be predicted from Table 18-1. Since the positions of the atoms themselves, but not of the unshared pairs, are all that can be observed experimentally, the atomic arrangement in molecules with basically similar electronic structures may be described in quite different words (for example, CH_4, tetrahedral; NH_3, pyramidal; H_2O, bent), as in the right-hand column of Table 18-1.
4. The following points are worth noting:
   (*a*) It is the number of bonded neighbors, not the number of bonds, that is counted. Hence double or triple bonds, or possible resonance involving different numbers of bonds to the central atom, have no bearing on this correlation scheme.
   (*b*) The configuration will not be completely symmetric unless all the atoms bonded to the central atom are the same, but it will usually approximate the regular figure—an equilateral triangle, a regular tetrahedron, and so on.
   (*c*) Remember that any diatomic molecule is necessarily linear and any three atoms necessarily lie in the same plane. Thus, applying the term *linear* to a diatomic molecule or *planar* to a triatomic one is redundant.
   (*d*) If the total of bonded neighbors plus unshared pairs is four and sp^3 orbitals are available, the configuration will be approximately tetrahedral. For some transition-metal complexes with one *d*, one *s*, and two *p* orbitals available, a square planar configuration about the central atom sometimes results, e.g., in some Ni(II) and Pt(II) complex ions. We do not consider these here, nor do we consider totals of bonded neighbors plus unshared pairs larger than six, although the scheme can be extended to cover these cases.

□ **Molecular Geometry** Explain the square pyramidal molecular geometry of BrF_5 by the rules just given.

Example 18-2

Solution The number of electrons in the valence shell of the Br atom is $7 + 5 = 12$; there are five shared pairs plus one unshared pair. The basic geometry is that of an octahedron, one corner of which is occupied by the unshared pair and the other five by the five F atoms. ■

□ Predict the molecular geometry of the ion IF_4^- by the rules just given. ■

Exercise 18-2

Table 18-1
Correlation of Geometric Configuration and Electronic Structure

No. of bonded neighbors plus unshared pairs	Possible suitable hybrids*	Expected configuration of bonded neighbors and unshared pairs		Observed examples of configurations of bonded neighbors
2	sp (linear)		Linear	CO_2, NO_2^+, NNO, OCS; $HgCl_2$, Hg_2Cl_2(Cl—Hg—Hg—Cl), $Ag(NH_3)_2^+$; H—C≡C—H, H—C≡N
3	sp^2 (trigonal planar)		Trigonal planar	BCl_3, NO_3^-, CO_3^{2-}, SO_3, GaI_3
			Bent	SO_2, NO_2^-
4	sp^3 (tetrahedral)		Tetrahedral	CH_4, $CHCl_3$; NH_4^+, PBr_4^+, SO_4^{2-}; $Ni(CO)_4$
			Trigonal pyramidal	NH_3, H_3O^+, PCl_3, SO_3^{2-}, XeO_3
			Bent	H_2O, SCl_2, $O(CH_3)_2$
5	sp^3d (trigonal bipyramidal†)		Trigonal bipyramidal	PCl_5, PF_5, PF_3Cl_2
			Trigonal bipyramidal with one corner unoccupied	SF_4, $TeCl_4$, $[IO_2F_2]^-$
			T-shaped	BrF_3, ClF_3
			Linear	I_3^-, ICl_2^-, XeF_2
6	sp^3d^2 (octahedral‡)		Octahedral	SF_6, $[Fe(CN)_6]^{3-}$, $[Fe(CN)_6]^{4-}$
			Square pyramidal	BrF_5, $[SbF_5]^{2-}$
			Square planar	$[ICl_4]^-$, XeF_4

* The orbitals in all but the trigonal bipyramidal set are equivalent by symmetry (provided the peripheral atoms are all of the same kind). In this set the three hybrids in the x,y plane and at 120° from each other are equivalent by symmetry, as are, separately, the two hybrids in the $\pm z$ direction. See also the discussion of Formula (17-9).

† Unshared pairs always occupy positions in the central ("equatorial") plane.

‡ See Fig. 18-6 for additional examples. When two unshared pairs are present they are opposite each other, leading to a square planar arrangement.

18-3   Molecular Orbitals

Introduction   The simple qualitative view of molecular orbital (MO) theory presented here is designed to give some notion of the way in which chemists attempt to construct orbitals characteristic of the molecule as a whole and also to give some insight into the usefulness of this approach.

The usual procedure for formulating a molecular wave function (a molecular orbital) is by linear combination of atomic orbitals, one atomic orbital being chosen for each atom included. The number of MOs formed is the same as the number of atomic orbitals taken initially. Each MO corresponds to a particular energy; sometimes several MOs correspond to the same energy and thus are said to be degenerate. Each orbital can hold at most two electrons, and if two electrons do occupy an orbital, their spins must be opposed, in accord with the Pauli principle. The MOs are filled with electrons in the order of increasing energy, just as atomic orbitals are. When a molecule is excited electronically, an electron is moved from a stable orbital (usually the highest occupied one) to a higher empty or half-filled orbital. We begin our discussion by considering MOs for some diatomic molecules and ions.

Diatomic Molecules: General Features   The simplest diatomic molecule is H_2, the molecular orbitals for which are made up by combining the $1s$ orbital for each of the two H atoms, as discussed in Sec. 18-1 and illustrated in Figs. 18-1 and 18-2. It will be recalled that two molecular orbitals are formed, one corresponding to a greater stability and one to a lesser stability than for the separated atoms. The

Figure 18-7   Overlap of *s* Orbitals and the Corresponding Atomic and Molecular Energy Levels
When s orbitals are added, the resulting molecular orbital ψ_+ is denoted σ_s (sigma s); the implications of the symbol σ are given in the text. Similarly, when s orbitals are subtracted, the antibonding orbital ψ_- formed is denoted by σ_s^*. The schematic energy-level diagram shows the energies of an electron in an orbital on each of the separated atoms and in the bonding orbital σ_s and the antibonding orbital σ_s^*. The s orbitals may be $1s$, $2s$, or higher s orbitals, as appropriate for the valence electrons involved. The representation of the orbitals in this and the next few figures is highly schematic.

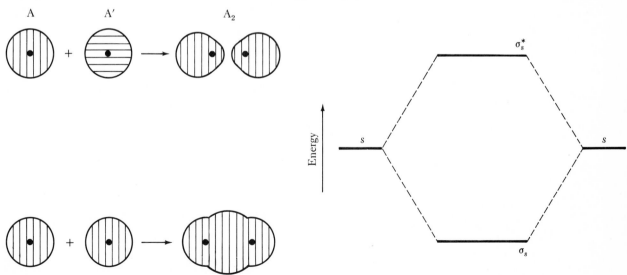

more stable is called a *bonding* orbital and the less stable an *antibonding* orbital. The relationships between these orbitals are illustrated schematically in Fig. 18-7, which may be taken to apply more generally to the overlap of s orbitals of any pair of identical atoms (here designated A and A′) in a molecule A_2, rather than just to hydrogen. The difference between the energies of the two molecular orbitals and that of the isolated atomic orbitals depends upon the degree of orbital overlap and thus varies with both the nature of the orbitals and the identity of the atoms.

Two cases of overlap of *directed* orbitals (such as p orbitals) must be considered: (1) when they are pointed along the molecular axis and (2) when they are pointed at right angles to the molecular axis. The overlap of two p orbitals directed toward each other along the internuclear axis (here designated the x axis) is depicted schematically in Fig. 18-8. At the left is shown the summation of the angle-dependent parts of the two overlapping p_x orbitals of atoms A and A′, corresponding to the formation of the bonding MO. The antibonding MO is formed by taking the difference of the orbitals, which is equivalent to reversing the signs of the lobes of one of them and adding. The energy levels are illustrated on the right. The separation of the levels is greater than for the overlap of s

Figure 18-8 Overlap of p_x Orbitals and the Corresponding Atomic and Molecular Energy Levels
The orbitals point along the molecular axis, here the x axis; the horizontal and vertical lines denote differing signs of the wave function. Because overlap of the p_x orbitals is greater than for the s orbitals (Fig. 18-7), the energy separation of the bonding and antibonding MOs is greater than for the σ_s orbitals. The MOs formed by the combination of p orbitals directed along the internuclear axis are designated σ_p and σ_p^*.

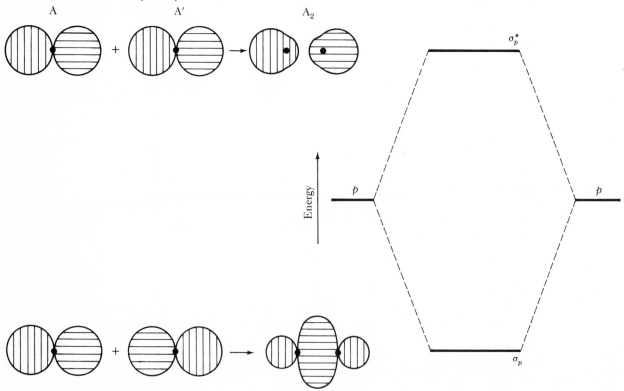

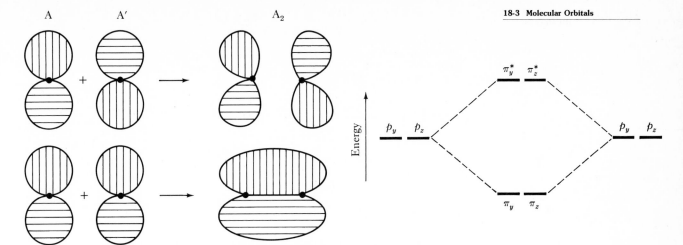

Figure 18-9 Overlap of p_y or p_z Orbitals and the Corresponding Atomic and Molecular Energy Levels
The overlap is smaller than for s orbitals or for p orbitals directed along the internuclear axis, and hence the separation of the
MOs is smaller. The diagrams on the left refer to the combination of two p_y orbitals or two p_z orbitals of neighboring atoms A
and A′. The resulting bonding orbitals (center, lower) are designated π_y or π_z, the antibonding orbitals (center, upper) π_y^* or π_z^*.
These bonding and antibonding orbitals are sometimes abbreviated π_p and π_p^*. The symbol π is explained in the text.

orbitals, because the directional character of the p orbitals makes their overlap
more effective when they are directed toward each other.

The situation for orbitals directed at right angles to the internuclear axis (for
example, p_y or p_z) is shown schematically in Fig. 18-9. The overlap is smaller than
in either of the two cases already discussed (s–s and p_x–p_x), and thus the separation of the energies of the corresponding MOs is also smaller.

As indicated in the legends to Figs. 18-7 to 18-9, the various MOs are designated
by symbols that include the Greek letters σ and π. Orbitals labeled π have a nodal
plane (a plane in which ψ, and thus ψ^2, is 0), with the internuclear axis lying in
this plane. The signs of these orbitals differ on opposite sides of this plane, a
property that they share with p orbitals. Sigma orbitals have no nodal plane. The
subscripts (s, p, x, and so on) of σ and π indicate the types of atomic orbitals
used to construct the MOs; such subscripts are sometimes omitted when there is
no ambiguity or when they are not of importance. The superscript * is used to
denote an antibonding orbital.

Figure 18-10 represents a composite of the energy levels of the three cases
discussed and illustrated in Figs. 18-7 to 18-9. One s level and three p levels for the
two A atoms of an A_2 molecule are shown on the left and right. The levels
corresponding to the MOs formed from these atomic orbitals are depicted in the
center. The exact spacing and the relative order of the energy levels of a molecule
can be determined only by experiment, but the general energy-level scheme
shown in Fig. 18-10 is applicable to most diatomic molecules.

The Different A_2 Molecules The electronic structures of various A_2 molecules
can be deduced with the help of the energy level diagram of Fig. 18-10. The
procedure is to construct molecular orbitals, find their relative energies, and fill

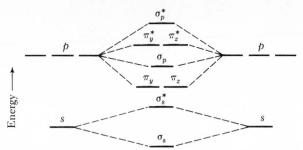

Figure 18-10   Atomic and Molecular Energy Levels Involving s and p Orbitals
On the left and the right of the diagram are shown the energies of the s and the p orbitals of
the atoms A and A'. In the center are the levels of the corresponding bonding and antibonding
molecular orbitals. It is believed that in most A_2 molecules the σ_p level is slightly above the π_y
and π_z levels, as indicated, but the positions of these levels may sometimes be reversed.

the orbitals with the available valence electrons, starting at the lowest level. It is
this last step that we now consider for different ions and molecules.

For hydrogen and helium, the only accessible atomic orbitals are $1s$ orbitals,
so that only the σ_s and σ_s^* of Fig. 18-10 are relevant. We begin with H_2^+; the one
electron available is placed in the σ_s orbital (see also the earlier discussion in Sec.
18-1). In H_2 the two valence electrons are both placed in the σ_s orbital, with
antiparallel spins. The separation between bonding and antibonding levels should
be somewhat different from that in H_2^+ because the internuclear distance is
different and because there is energy of repulsion between the two electrons. The
bond energy of H_2 is found to be about 1.7 times that of H_2^+ (Table 18-2).

In He_2^+ three electrons are available; two go into the σ_s orbital (as in H_2), while
for the third only the relatively high-energy σ_s^* orbital is available. The occupancy of σ_s^* by one electron offsets in part the effect of the two electrons in the
bonding orbital and thus has a destabilizing effect, which is why σ_s^* is called an
antibonding orbital. The spacing of the levels and the energy of interaction of the
electrons are different from those in H_2^+ and H_2, but the stability of He_2^+ is
comparable to that of H_2^+ (Table 18-2).

For the hypothetical molecule He_2, both σ_s and σ_s^* would be filled by
electron pairs. The bond energy would be essentially zero because the effect of
bonding and antibonding electrons would balance. This is the explanation in MO
theory for the nonexistence of the He_2 molecule.

We consider now the molecular orbital picture of A_2 molecules in which the
element A ranges from Li through Ne. In all such A_2 molecules the $1s$ electrons of
the separate atoms would completely fill the σ_s and σ_s^* orbitals formed by the
atomic $1s$ orbitals, with zero total stabilization energy, as for He_2. It is thus
considered that the $1s$ electrons of the separate atoms do not interact and that
only the valence electrons need be considered. The same reasoning applies to all
filled atomic shells.

**Table 18-2
Simplest Diatomic
Molecules and Ions**

Species	Electron configuration	Bond distance/Å	Bond energy/kJ mol^{-1}
H_2^+	σ_s	1.06	256
H_2	σ_s^2	0.74	433
He_2^+	$\sigma_s^2 \sigma_s^*$	1.08	244
He_2	$\sigma_s^2 \sigma_s^{*2}$		

Table 18-3
Homonuclear Diatomic
Molecules for Elements
Li through Ne

Species	σ_s	σ_s^*	π_y	π_z	σ_p	π_y^*	π_z^*	σ_p^*	Bond distance/Å	Bond energy/kJ mol^{-1}
Li$_2$	2								2.67	105
Be$_2$	2	2								
B$_2$	2	2	1	1					1.59	288
C$_2$	2	2	2	2					1.31	627
N$_2^+$	2	2	2	2	1				1.12	610
N$_2$	2	2	2	2	2				1.09	950
O$_2^+$	2	2	2	2	2	1			1.12	622
O$_2$	2	2	2	2	2	1	1		1.21	492
O$_2^-$	2	2	2	2	2	2	1		1.26	
F$_2$	2	2	2	2	2	2	2		1.42	160
Ne$_2$	2	2	2	2	2	2	2	2		

The molecular orbitals constructed from the $2s$ and $2p$ orbitals of the atoms considered will be assumed to lead to the energy level system shown in Fig. 18-10, and it remains to fill the levels with the valence electrons available and to compare the results with the known experimental facts about these species. The configurations predicted on this basis are summarized in Table 18-3, where observed internuclear distances and bond energies are also shown. (A diatomic molecule A$_2$, with atoms of the same atomic number, is called *homonuclear;* a molecule AB is called *heteronuclear.*)

At room temperature lithium is a metallic solid; the Li$_2$ molecule exists only in the vapor phase. The bonding in Li$_2$ is similar to that in H$_2$, but the bond energy of Li$_2$ is considerably smaller and the interatomic distance larger than in H$_2$. This is attributed to mutual repulsion of the $1s$ electrons of the two Li atoms.

In Be$_2$ the bonding by the electron pair in the σ_s orbital is nominally balanced by the pair in the σ_s^* orbitals, with a net result of zero bonding. The experimental fact is that at room temperature beryllium is a metallic solid. The normal boiling point of Be is 2970°C and the vapor is monatomic.

Boron is also a solid at room temperature, the boron atoms being linked into a three-dimensional network. The boiling point is 2550°C and the vapor contains B$_2$ molecules. Their electron configuration is shown in the diagram on the left of Fig. 18-11. The two electrons of highest energy are seen to be in the two degenerate levels π_y, π_z. Since two orbitals of equal energy are available, the spins of these electrons need not be opposed, and in analogy to the situation in atoms (Hund's rule) it is expected that parallel spins are energetically favored. This is found to be so: the B$_2$ molecule has two unpaired electrons.

The number of bonds in B$_2$ can be said to be one, because the electrons in the σ_s^* orbital largely offset the bonding effects of those in the σ_s orbital, so that only

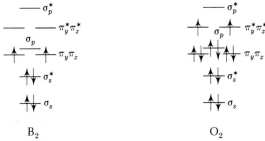

Figure 18-11   The Electron Configuration of B$_2$ and O$_2$

In both molecules the highest levels filled are doubly degenerate and contain two electrons. The spins of these electrons are parallel, analogous to the parallel spins encountered for degenerate atomic levels (Hund's rule).

the electrons in the π_y and π_z orbitals contribute to the net bonding. In this sense the number of effective "bonds" corresponding to occupied molecular orbitals is half the difference between the number of electrons in bonding orbitals and the number of electrons in antibonding orbitals:

Nominal no. of "bonds" $= \frac{1}{2}$(bonding electrons − antibonding electrons) (18-1)

This nominal number of bonds is sometimes called the *bond order*.

The room-temperature modifications of carbon are diamond and graphite, both of them solid. However, C_2 molecules exist at high temperatures in the vapor phase. This molecule has two bonds, as shown by applying (18-1) to the data in Table 18-3. The bond energy is 627 kJ mol^{-1} in C_2 and 288 kJ mol^{-1} in B_2, while the bond distance is 1.31 Å in C_2 and 1.59 Å in B_2. Thus all these experimental data are in accord with the MO picture.

In N_2 the bonding corresponds to a triple bond. It is interesting to compare the increase of bonding energy and the decrease of the internuclear distance in the sequence B_2, C_2, and N_2 with the increasing number of bonds. N_2 is diamagnetic[5] and thus has no unpaired electrons, as expected.

The fact that the O_2 molecule is observed to be paramagnetic, to a degree corresponding to two unpaired electrons, is inconsistent with the obvious Lewis formula that one can write for this species,

$$:\ddot{O}\!=\!\ddot{O}:$$

and so this formula was early recognized as inadequate. One of the first successes of molecular orbital theory was its explanation of the paramagnetism of oxygen. The level scheme in Fig. 18-10 leads to the prediction that the two highest-energy electrons occupy the degenerate π_y^* and π_z^* orbitals and thus remain unpaired (Fig. 18-11). The nominal number of bonds in the MO view of O_2 is two, as it is in the Lewis formula, but the interpretation is different.

In F_2 the nominal number of bonds is again one. In Ne_2 it is zero, corresponding to the nonexistence of Ne_2 and the chemical inertness of Ne.

Three ions of the type A_2^+ or A_2^- are also listed in Table 18-3. For two of these, N_2^+ and O_2^+, the relation (18-1) gives the number of bonds as 2.5. The bond energies and interatomic distances are seen to be similar in the two ions. It is particularly instructive to compare N_2 with N_2^+ and O_2 with O_2^+. When an electron is removed from N_2, the interatomic distance increases and the bond energy decreases, whereas removal of an electron from O_2 has just the opposite effects. The MO picture provides a ready explanation of this difference, since with N_2 the electron is removed from a bonding orbital, thus producing a destabilization, whereas with O_2 the electron is removed from an antibonding orbital, giving an increase in stability. The interatomic distance in O_2^- (its bond energy is not known) is also consistent with the MO view, for this distance is greater than that in O_2, in accord with the addition of an electron to the antibonding π_y^*, π_z^* system.

This discussion can easily be extended to the higher periods of the periodic table. The filling of molecular orbitals with electrons according to an Aufbau

---

[5] Most substances have only weak inherent magnetic properties. These show up when the substance is placed in a nonuniform magnetic field. If the substance is pulled into the field, it is said to be *paramagnetic;* if it is pushed away from the field, it is termed *diamagnetic*. Paramagnetism is related to the presence of unpaired electrons in a substance; in diamagnetic substances all the electrons are paired.

principle is a convenient procedure, but remember that the relative positions of the energy levels may depend on the number of electrons present, that is, the levels may not always be in the same sequence as in Fig. 18-10. The procedure is nonetheless very helpful as an accounting of filled and unfilled orbitals in the final molecule.

□ **MO Treatment of Br$_2$ and S$_2$**   Give the configuration of the valence electrons and the number of effective bonds for the molecules Br$_2$ and S$_2$.

Example 18-3

Solution   In Br the valence shell consists of the $4s$ and $4p$ orbitals, the Ar shell and the $3d$ subshell being complete. Molecular orbitals can be constructed from the $4s$ and $4p$ orbitals with approximate energy levels as shown in Fig. 18-10. Filling these with the 14 valence electrons of the two bromine atoms, we obtain the configuration

$$\sigma_s^2(\sigma_s^*)^2(\pi_{y,z})^4\sigma_p^2(\pi_{y,z}^*)^4 \qquad (18\text{-}2)$$

for the ground state of Br$_2$. The number of electrons in bonding orbitals is 8 and that in antibonding orbitals is 6, so the number of bonds between the two Br atoms is 1. The molecules F$_2$, Cl$_2$, Br$_2$, I$_2$, and At$_2$ all have the same configuration of valence electrons.

In sulfur the valence shell consists of the $3s$ and the $3p$ orbitals, and again molecular orbitals with energies as shown in the center of Fig. 18-10 can be formed. The total of valence electrons for the two sulfur atoms is 12, and they fill the available orbitals as for O$_2$ (Fig. 18-11) with the π_y^* and π_z^* levels filled singly by electrons of parallel spin. The electron configuration is

$$\sigma_s^2(\sigma_s^*)^2(\pi_{y,z})^4\sigma_p^2\pi_y^*\pi_z^* \qquad (18\text{-}3)$$

The last two symbols, π_y^* and π_z^*, are written individually to signify that the orbitals are singly occupied. The electron configuration of S$_2$ parallels that for O$_2$; S$_2$ is a paramagnetic molecule and has two bonds. ■

□ Give the configuration of the valence electrons and the number of effective bonds for the molecules P$_2$ and Te$_2$, both of which have been observed at elevated temperatures. ■

Exercise 18-3

Bonding in C$_2$H$_4$ and C$_2$H$_2$ The bonding in ethylene (C$_2$H$_4$) and acetylene (C$_2$H$_2$) can be described in several ways, each consistent with the classical picture of a double bond in ethylene and a triple bond in acetylene. One important description involves the use of sp^2 hybrid orbitals for the carbon atoms in ethylene and sp hybrids for those in acetylene. We form sp^2 hybrids of the $2s$, $2p_x$, and $2p_y$ orbitals of each carbon atom and use them, with the $1s$ orbitals of the hydrogen atoms, to form a framework of five bonds that tie together the carbon and hydrogen atoms, using 10 of the 12 ($= 2 \times 4 + 4 \times 1$) valence electrons of the C$_2$H$_4$ molecule (Fig. 18-12). These bonds are called *sigma* (σ) *bonds* because, like s orbitals, they have no nodal planes. They are cylindrically symmetric about each of the corresponding internuclear lines.

The $2p_z$ orbital on each carbon atom, which is not used in forming the sp^2 hybrid, extends perpendicular to the plane of the sp^2 hybrid. When the planes formed by the two CH$_2$ groups coincide—that is, when all six atoms of the molecule lie in the same plane, as they normally do in the unexcited ethylene

387

molecule—these two p orbitals overlap to the maximum extent possible and form what is called a π *molecular orbital* (Fig. 18-12). The final two valence electrons of the ethylene molecule occupy this orbital, and this electron pair constitutes a *pi (π) bond,* the second bond of the double bond in ethylene. In the carbon-carbon double bond, the overlap of the p_z orbitals constituting the π bond is less effective than is that of the sp^2 hybrids composing the σ bond. This accounts for the fact that the energy of the carbon-carbon double bond (620 kJ mol^{-1}) is less than twice that for the C-C single bond (340 kJ mol^{-1}).

If the planes of the two CH$_2$ groups in ethylene are rotated relative to each other, the overlap of the p_z orbitals (Fig. 18-12b, upper part) decreases. When the angle between the planes becomes 90°, the overlap is zero, which corresponds to breaking the π bond. Thus it is the π bond that keeps the molecule planar and keeps the two CH$_2$ groups from rotating around the σ bond that joins them. In fact, any six atoms joined in the configuration

$$\begin{array}{ccc} A & & E \\ & B{=}D & \\ C & & F \end{array}$$

(18-4)

lie in the same plane, or very nearly so. The molecular rigidity resulting from the presence of the π bond makes possible the existence of geometrical isomers in double-bonded molecules when A differs from C and E differs from F (Sec. 29-1).

The triple bond can be described in a similar way. In acetylene (C$_2$H$_2$), for example, we begin with a framework of σ bonds that use the 1s orbitals of the hydrogen atoms and sp hybrids on the carbon atoms (Fig. 18-12). This leads to a linear H—C—C—H sequence. The remaining 2p orbitals of the carbon atoms, a

Figure 18-12 Bonding in Ethylene and Acetylene
The two diagrams (a) are schematic cross sections through the s orbitals of the hydrogen atoms and the hybridized orbitals of the carbon atoms (sp^2 for ethylene and sp for acetylene), which are used by shared electron pairs to form frameworks of σ bonds. Diagrams (b) show the remaining p orbitals of the carbon atoms, which are joined as symbolized in diagrams (c) to be used by π electrons to form one π bond for ethylene and two for acetylene. The σ bonds are indicated by heavy lines (dashed if concealed) in diagrams (b) and (c).

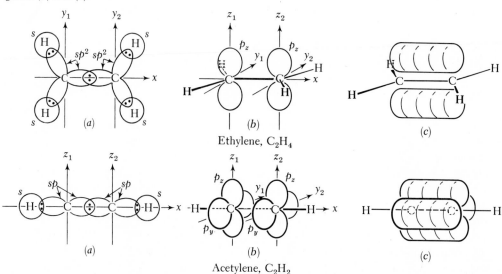

Ethylene, C$_2$H$_4$

Acetylene, C$_2$H$_2$

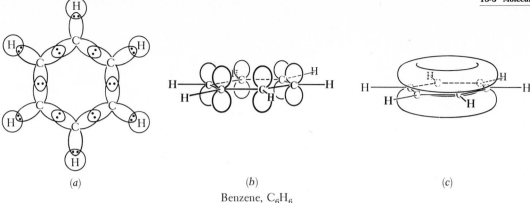

(a) (b) (c)

Benzene, C_6H_6

Figure 18-13 Bonding in Benzene
In benzene, with a total of 30 ($= 6 \times 4 + 6$) valence electrons, the framework, consisting of 6 carbon sp^2 and 6 hydrogen s orbitals, requires 24 σ electrons (a). The remaining 6 electrons are distributed over the remaining 6 unused p_z orbitals (b), which can be combined in various ways. The lowest-energy orbital of benzene is symbolized by the two doughnut-shaped arrangements (c).

pair of them on each atom, perpendicular to the H—C—C—H line, provide for two π bonds. A triple bond is thus pictured as the sum of a σ and two π bonds. This view not only explains the observed linear structure of any sequence of four atoms bonded thus:

$$A—B{\equiv}C—D \tag{18-5}$$

but also explains why quadruple bonds cannot be formed (unless there is expansion of the valence shell), because there are no further overlapping orbitals by which atoms B and C could share an additional electron pair.

Delocalization of Electrons An area of chemistry in which the picture of σ and π bonds has been particularly fruitful is that of delocalization of electrons, illustrated here by the example of benzene, C_6H_6.

The bonding of the atoms in benzene may be represented as in Fig. 18-13. The six p_z orbitals on adjacent atoms overlap to form an extended π *system,* a set of six molecular orbitals extending over all six carbon atoms. The six π electrons occupy the three lowest-energy orbitals of this set, the most stable of which is the one illustrated in Fig. 18-13. The electrons occupying these orbitals are said to be *delocalized*. It is generally true that when the description of a molecule requires a set of resonance formulas, the molecular orbital description will include delocalized π electrons.

Other molecules and ions for which we have earlier used the resonance description (for example, NO_3^-, Sec. 17-6) can equally well be described in terms of a σ-bond framework and delocalized π electrons. One important class of molecules with delocalized electrons is that which can be described by a sequence of alternating single and double bonds. Double bonds so arranged are said to be *conjugated;* examples of molecules with conjugated double bonds are 1,3-butadiene, $CH_2{=}CHCH{=}CH_2$, and benzene. When the array of conjugated double bonds over which the delocalized electrons are able to range is large, there is often strong absorption in the visible region of the spectrum. In other words, the molecule is strongly colored. An example is the dye indigo, known since ancient times:

$$\text{(18-6)}$$

In conclusion, it should always be remembered that the theories and formalisms expounded in this and the preceding chapters are all *models*—extremely useful models of considerable predictive power, but models nonetheless. To what degree such models reflect features of the real world is a metaphysical question, as is true for any theory. The feeling of satisfaction conveyed by a "good" theory is largely tied to inherent aesthetic beauty, to compactness and economy of basic postulates, to logical structure, and to consistency and uniqueness of predictions.

Summary

The bonding in molecules can be described approximately by the valence bond and molecular orbital theories. In the valence bond theory, bonding is pictured as arising from the overlap of atomic orbitals. A feature of the theory is the hybridization of $s, p, d, \ldots$ orbitals to form new orbitals that have strongly directional properties. The number of hybrid orbitals formed is always the same as the number of separate orbitals that are combined. Important examples of hybrid orbitals are:

1. *sp hybrids,* two equivalent orbitals formed from one s and one p orbital. There is a 180° angle between the sp hybrids.
2. *sp^2 hybrids,* three equivalent orbitals formed from one s and two p orbitals. These three orbitals lie in a plane and enclose angles of 120°.
3. *sp^3 hybrids,* four equivalent orbitals directed toward the corners of a tetrahedron. They are formed from one s and three p orbitals.
4. *sp^3d^2 hybrids.* These six orbitals, which are formed from one s, three p and two d orbitals, are directed toward the vertices of an octahedron.

The geometric configuration of the bonded neighbors of an atom can usually be predicted from a knowledge of the sum of the number of bonded neighbors and the number of unshared electron pairs.

In the molecular orbital theory of bonding the molecule is considered as a whole. Orbitals that extend over most or all of the atoms are constructed by combining one atomic orbital for each atom that is included. Each molecular orbital can contain at most two electrons, which must be of opposite spin. When electrons are in bonding orbitals the electron density in the region between the nuclei is relatively high, and there is a stabilization of the species. The electron density between the nuclei is diminished when electrons are in antibonding orbitals, and there is a repulsion of the nuclei. Electronic structures of molecules and ions can be explained by an Aufbau procedure in which electrons are placed into molecular orbitals in order of increasing energy. The effective number of bonds in a diatomic molecule is half the difference between the number of electrons in bonding orbitals and the number in antibonding orbitals.

Bonding in molecules with multiple (double or triple) bonds may be represented by a framework of σ bonds supplemented by π bonds. In double bonds, one π bond is formed by the overlap of two p orbitals; the two π bonds in a triple bond arise from two mutually perpendicular pairs of p orbitals. The six electrons in benzene that are not involved in the σ-bond framework occupy three of the six orbitals in a π system that extends over all six carbon atoms; these electrons are said to be delocalized.

When resonance formulas are required for the Lewis picture of a molecule, the molecular orbital description will include delocalized π electrons.

Terms and Concepts

Problems and Questions

18-1 Wave Functions for H_2 The wave function for the $1s$ orbital of a hydrogen atom can be written $\psi = e^{-r/r_0}$, where r is the distance from the nucleus and r_0 is the Bohr radius, 0.53 Å. (a) Calculate ψ at the following values of the relative distance from the nucleus, r/r_0: 0, 0.2, 0.4, 0.6, 0.8, 1.0, 1.2, 1.4, 1.6, 1.8. (b) Assume that the distance between two protons is $2r_0$. The $1s$ wave function for one of the H atoms is denoted by ψ_A and that for the other H atom as ψ_B. Use the table you prepared in (a) to calculate $\psi_+ = \psi_A + \psi_B$ and $\psi_- = \psi_A - \psi_B$ at a point midway between the protons (i.e., at a distance r_0 from each nucleus). (c) Calculate ψ_+, ψ_+^2, ψ_-, and ψ_-^2 at a number of other distances from the nuclei and plot the results. Compare the graphs you obtain with those in Figs. 18-1 and 18-2.

18-2 Hybrid Orbitals (a) Make a list of hybrid orbitals that include only s and p functions. Indicate their directional character. (b) Extend the list to include at least one hybrid that also includes one or more d functions.

18-3 Bonding in Molecules Consider the molecules H_2Te, AsH_3, SiH_4, and P_2. Specify the atomic orbitals of the central atom used for bonding, as well as the hybrids formed. Predict bond angles for the first three molecules.

18-4 Definitions and Distinctions Define each of the following terms, distinguishing it carefully from those terms grouped with it: (a) atomic orbital, molecular orbital, hybrid orbital; (b) bonding orbital, antibonding orbital; (c) inner d orbital, outer d orbital; (d) tetrahedral configuration, octahedral configuration; (e) square pyramidal configuration, square planar configuration.

18-5 Paramagnetic Molecules Molecules and ions in which the number of electrons with spins of one orientation is different from the number with spins of opposite orientation, so that the spins do not add to zero, are paramagnetic, and a permanent magnetic moment is associated with them. Which molecules would you expect to be paramagnetic: NO_2, $NOCl$, N_2O_4, S_2, ClO_2, Cl_2O?

18-6 Electron Configuration and Molecular Shape Give plausible electron configurations for the following ions and molecules. Indicate by an appropriate word or phrase the expected shape. (a) PF_3, PBr_5, HNO_3, AlH_4^-; (b) ClO_3^-, PCl_6^-, NNN^-, D_2Te (D = deuterium)

18-7 Geometry of Molecules Consider the molecules and ions listed below and describe the geometry of each by using one or two of the following terms: linear, bent, planar, tetrahedral, pyramidal.

(a) H_2S, SO_4^{2-}, BF_3, C_2H_4, NCS^-
(b) O_3, CS_2, ClO_2, $HNCS$, SbH_3
(c) SO_2, NCl_3, OCS, CH_2Cl_2, $PbCl_4$
(d) Cl_2CO, SO_3^{2-}, CO_3^{2-}, $ClNO$, HCN

18-8 Dipole Moment and Molecular Geometry Consider the following molecules and ions and predict which of them have dipole moments (Fig. 6-3) and which do not: $CHCl_3$, CH_2F_2, $H_3C—O—CH_3$, $N(CH_3)_3$, $N(CH_3)_4^+$, $Cl_3Si—SiCl_3$, $Cl_3Si—O—SiCl_3$.

18-9 Dipole Moment and Molecular Geometry On the basis of the structures of the molecules and the electronegativity differences between the bonded atoms, make qualitative comparisons of the dipole moments (see Fig. 6-3) of the following pairs of molecules by using the symbols > (greater than), = (equal to), or < (smaller than):

HF	HCl	F_2O	CO_2
CCl_4	CH_4	PH_3	NH_3
H_2S	H_2O	BF_3	NF_3

18-10 Comparison of Congeners Both PCl_3 and NCl_3 are known compounds. However, although PCl_5 is a known substance, NCl_5 has never been prepared; what explanation can you give for this fact? Similarly, discuss why OF_6 is not known, although SF_6 is a well-characterized substance.

18-11 Concepts of Molecular Structure Theory Explain what is meant by the following terms as used in current discussions of atomic and molecular structure: electron affinity, hybrid atomic orbital, degenerate energy level, subshell, Pauli exclusion principle, outer-shell electron, antibonding molecular orbital, polar molecule.

18-12 Shapes and Signs Associated with Orbitals (a) Give an example, including a sketch of the "shape" (indicating clearly what is being plotted), of a spherically symmetric orbital and of an orbital with one angular node (show where this node is and indicate what the term *angular node* implies). (b) Why is the sign of each component orbital wave function in different regions of space important in the formation of molecular orbitals and hybrid orbitals? Illustrate with a sketch.

18-13 Double Bond and Isomers Draw three possible isomeric structures consistent with the formula C_2H_2FCl.

18-14 Orbital Overlap (a) Explain what is meant by "orbital overlap". (b) Why is a sigma bond formed from p orbitals significantly stronger than a pi bond formed from p orbitals on the same atoms at the same distance? (c) When two hydrogen atoms are 1 Å apart there is appreciable overlap of the $1s$ orbitals, resulting in both bonding and antibonding orbitals. Since the same atomic orbitals, at the same distance, are used to form each of these molecular orbitals, why is it that the energies of these molecular orbitals are so different?

18-15 Bonding in Carbon Dioxide The CO_2 molecule is linear, the carbon atom being at the center. Describe the bonding in terms of σ and π bonds, including resonance among possible valence bond (Lewis) formulas.

18-16 Acetylides Acetylene, C_2H_2, forms ionic compounds such as CaC_2 and Na_2C_2. These compounds contain the ion C_2^{2-}. Predict the electron configuration of this ion and predict whether its bond distance and its bond energy will be about the same as, greater than, or less than those of C_2.

18-17 Molecular Orbitals and Diatomic Molecules and Ions (a) Give the electron configurations in terms of filled molecular orbitals [such as $\sigma_s^2(\sigma_s^*)^2(\pi_{x,y})^4\sigma_p^2(\pi_{x,y}^*)^4(\sigma_p^*)^2$] for the following diatomic species: B_2, C_2, N_2, N_2^+, O_2^{2-}, O_2^-, O_2, and O_2^+. (b) The interatomic distances in the last four species are, in sequence, 1.49, 1.26, 1.21, and 1.12 Å. How are these values related to the bonding? (c) The interatomic distances in N_2 and N_2^+ are 1.09 and 1.12 Å; in Cl_2 and Cl_2^+ they are 1.99 and 1.89 Å. Explain why the values in each pair vary as they do. (d) Explain the existence of the species H_2^+ and He_2^+.

18-18 MO Treatment for Diatomic Species Give the configuration of the valence electrons and the number of effective bonds for Na_2, which is found in sodium vapor, and Ar_2^+, which can be detected in a mass spectrometer.

18-19 Molecular Orbitals for NO (*a*) Draw a molecular orbital diagram for NO as follows: on the left draw the $2s$ and $2p$ states for N; on the right those for O. The energies of the oxygen atomic states are somewhat below the corresponding nitrogen states, because O is more electronegative than N. Combine the states into σ, σ^*, π, and π^* molecular states. Label the states and show how they are filled with the valence electrons of N and O. (*b*) Predict the electron configurations of NO, NO^+, and NO^- in terms of occupancies of molecular orbitals. On the basis of these predictions, arrange the three species in order of decreasing bond energy and in order of decreasing bond distance. The known distances are NO, 1.15 Å; NO^+, 1.06 Å.

The First Law of Thermodynamics: Energy and Enthalpy

19

"A theory is the more impressive the greater the simplicity of its premises, the more varied the subjects it ties together, and the wider the domain of its applicability. Thus the deep impression that classical thermodynamics made upon me. It is the only physical theory of general content about which I have the conviction that it will never be overturned within the bounds of applicability of its fundamental concepts (for the special attention of inveterate sceptics)."

ALBERT EINSTEIN, 1946[1]

The chief topic of this chapter is energy in chemical systems that contain large numbers of atoms or molecules. Our approach is macroscopic, through the postulates of thermodynamics. We begin with a brief exposition of the nature and the purpose of thermodynamics and continue with a development of the first law of thermodynamics (the principle of conservation of energy), starting with basic concepts of work, internal energy, and heat. Some chemical applications follow. A review of the discussion of temperature and temperature scales given in Sec. 4-2 is recommended.

19-1 The Nature of Thermodynamics

Thermodynamics is the science of heat and temperature and, in particular, of the laws governing the conversion of heat into mechanical, electrical, or other macroscopic forms of energy. It is a central branch of science with important applications in chemistry and other physical sciences, in biology, and in engineering.

Thermodynamics is a macroscopic theory, concerning quantities such as pressure, temperature, and volume. It is both the strength and the weakness of thermodynamics that the relationships based upon it do not depend on any microscopic explanation of physical phenomena. The *strength* is that thermodynamic relationships are not affected by the changes in microscopic explanations that continue to occur as the theories of atomic and molecular interactions and structure are modified and improved. On the contrary, the conclusions of atomic and molecular theories *must not* contradict those of thermodynamics, so that thermodynamics can be used as a guide in the development of microscopic theories. Much of thermodynamics was, in fact, developed at a time when many

---

[1] P. A. Schilpp (ed.), "Albert Einstein, Philosopher-Scientist", Tudor Publishing Co., New York, 1951.

scientists did not believe in atoms and molecules, long before detailed atomic theories were available.

The *weakness* of thermodynamics is that it does not provide the deep insight into chemical and physical phenomena that can be obtained from microscopic models and theories. Although thermodynamics is a completely self-contained macroscopic theory, it is nevertheless possible to find a microscopic interpretation of it in what is called statistical mechanics, which provides considerable insight and is of great value for a full understanding of thermodynamics. On the other hand, much more mathematics is needed to understand statistical mechanics than to understand thermodynamics.

Several postulates, referred to as *laws,* are basic to thermodynamics. Although these laws are consistent with the results of all known experiments, this great mass of observations serves only as support, not as proof. The "laws" of thermo-dynamics are postulates, or axioms, as are all "laws" of nature.

19-2   The First Law of Thermodynamics

System and Surroundings   As a prelude to thermodynamic reasoning, the physical universe is divided conceptually into the *system,* the part of particular interest, and the *surroundings*. The system may be complex and contain all kinds of substances, machinery, and equipment, or it may be very simple and consist only of a homogeneous sample of matter, such as a certain volume of air. What is included in the system is a matter of convenience that depends on the question being considered.

Energy Changes in Chemical Systems   The concepts of work and energy are most familiar in mechanical systems. For example, work must be performed to lift an object from one height to another. The change in the object's height is a tangible result of the work performed. In winding a grandfather clock (Fig. 19-1), we do work on a weight and the potential energy of the weight is increased by the amount of work done. As the clock ticks away, the weight moves downward and the energy that drives the clock is obtained from the gradual decrease in the potential energy of the weight.

Work can also be performed on *chemical* systems. The condition of any system is called its *state* and is described by the values of macroscopic properties such as pressure, temperature, and density that need to be specified to duplicate the system precisely. Since the properties of a substance may depend on each other, usually only a few need to be specified. For example, the volume and density of a quantity of gas are determined by the pressure and temperature. The number of variables that need to be specified to define the state of any particular system, in addition to P and T, depends in part on the nature of the system and the focus of one's interest. If unpaired electrons are present, for example, the strength of any magnetic field present must be specified for a complete description; with a two-phase system containing a solid and a gas, it may also be necessary to specify the surface area in contact with the gas if adsorption is important. However, we shall normally ignore such specialized effects.

When work is performed on a system, the state of the system changes. For example, when work in the form of vigorous stirring is performed on a liter of water, the temperature of the water rises, say from 25 to 30°C, and its density

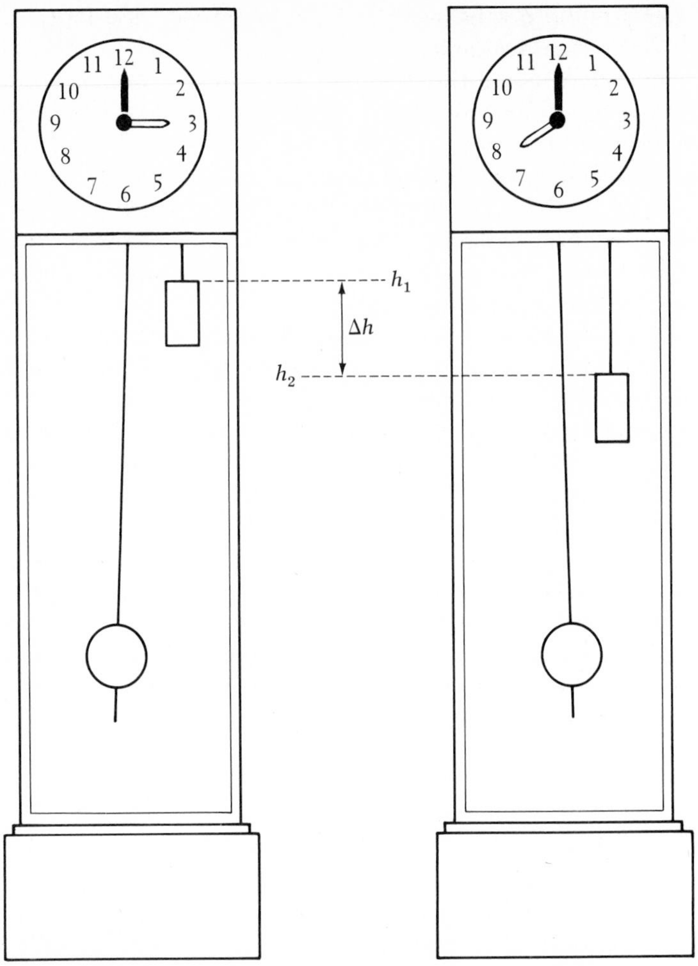

Figure 19-1 Work and Energy in a Mechanical System

Work is performed when the weight is raised in winding the clock. The potential energy of the weight is thereby increased. The energy required to operate the clock comes from the loss in potential energy when the weight falls. The change in height $\Delta h = h_2 - h_1$ is directly proportional to the change in potential energy, $mg\Delta h$, with m the mass and g the acceleration due to gravity.

decreases. The water has been brought to a different state, characterized by particular physical properties.

This change in the state can also be accomplished without performing work. If water at 25°C is brought into contact with surroundings that are at a higher temperature, its temperature can rise to 30°C. Energy is transferred from the surroundings to the system. The energy transferred solely as a result of temperature differences is called *heat.*

The First Law of Thermodynamics It is possible to assign an energy to every state of a system. Changes in this energy, which is called the *internal energy,* are therefore related to changes in state. If the internal energy of a system in some initial state is symbolized by U_1 and its internal energy in some final state by U_2, then it is observed that the change in internal energy, ΔU, is given by

$$\Delta U = U_2 - U_1 = q + w \tag{19-1}$$

where q is the amount of energy transferred to the system in the form of heat and w is the work performed on the system. Equation (19-1) is called the *first law of thermodynamics.*

We use the convention that work is positive when it is performed by the surroundings on the system. Work performed by the system on the surroundings is therefore negative. The same sign convention is used for heat. Heat that is absorbed by the system, that is, energy transferred as heat from the surroundings to the system, is by convention positive, and energy transferred as heat from the system to the surroundings is negative.

Equation (19-1) is a definition of the *change* in internal energy, ΔU, and not of the internal energy itself. The *absolute internal energy* of a system remains undefined in thermodynamics, and an arbitrary constant may be added to any set of U values. Since we deal only with differences of internal energy, such a constant always drops out. Microscopically, the internal energy of a substance is the entire energy of its atoms or molecules, kinetic and potential. For thermodynamic considerations, however, analysis of the many interactions on the molecular level is not needed.

The internal energy is an *extensive* property, that is, its value is proportional to the amount of substance in the system. If the internal energy of 2 mol of a substance is U, then the internal energy of 1 mol is $\frac{1}{2}U$ and of 4 mol is $2U$. Some properties, for example temperature and pressure, do not depend on the amount of substance present. Such properties are said to be *intensive*.

The symbol Δ in (19-1) has a special meaning. It is reserved for the change in quantities called *state functions*. These are properties that depend only on the state of the system and not on its history (i.e., how it was prepared). In the mechanical example of the grandfather clock, the height h of the weight is a state function (Fig. 19-1). Typical thermodynamic state functions are volume, temperature, pressure, density, and internal energy.

When the state of a system is altered, the change of any state function depends, by definition, only on the initial and final states (Fig. 19-2). It must therefore be

Figure 19-2 Thermodynamic Paths and State Functions

A system in an initial state defined by the values of the pressure and temperature P_1 and T_1 is to be brought to a final state specified by P_2 and T_2. The path along which the change occurs is defined by the succession of states through which the system travels between the initial and final states. One such path, indicated by dark dashed lines, consists of two steps, a change in temperature from T_1 to T_2 at constant pressure P_1 followed by a pressure change from P_1 to P_2 at constant temperature T_2. The pressure and temperature could also be varied simultaneously as shown by the path indicated by the solid line.

The change in any state function X is the same for any path. For state functions the change depends only on the values of the function in the initial and final states and not on the path: $\Delta X = X(P_2, T_2) - X(P_1, T_1)$.

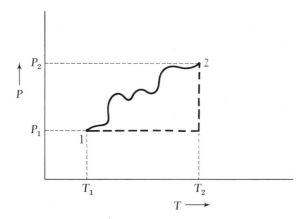

independent of the path along which the change in state has taken place. Note that the Δ symbol has not been used with either w or q, because they are not state functions. It is only the sum of q and w, ΔU, that represents a change in a state function. As we have pointed out for the two methods of changing the temperature of water, the same change in the internal energy can be produced by a process in which $w = 0$ or one in which $q = 0$. This change can also be accomplished along many other paths for which neither q nor w is zero but for which the sum of q and w is always the same: $q + w = \Delta U$.

It is important to note that heat is defined only in terms of processes involving energy transfer. It is as meaningless to say a system contains heat as to say it contains work. In this context a system may only contain energy. To give an example from mechanics, a tightly coiled spring contains potential energy that may be used to perform work, but the spring does not therefore contain work. A hot body may transfer energy to a colder body; this transferred energy is called heat but neither body can properly be said to contain heat. Phrases such as "heat transfer" and "absorption or liberation of heat" are commonly used but they do not mean that heat is a special form of stored energy.

Heat is sometimes confused with temperature as, for example, in the term "hot", which implies a high temperature and has nothing directly to do with energy being transferred as heat. Similarly, to "heat" an object means to increase its temperature. While this is often done by energy transfer in the form of heat, it can equally well be done by energy transfer in the form of work, as when a bearing becomes hot through friction.

Adiabatic and Isolated Systems The rate at which energy is transferred as heat between a system and its surroundings can be decreased by thermal insulation. Plastic foam, such as that used in picnic coolers, and silvered glass walls enclosing a vacuum, as in thermos bottles, are common thermal insulators. With careful design it is possible to construct containers so well insulated that the transfer of energy as heat, i.e., solely because of temperature differences, is negligible during the time it takes to carry out some experiment or process of interest. Systems enclosed in such containers are referred to as *adiabatic* systems (Greek: *a*, not; *diabatos,* passable). The term *adiabatic* is applied in general to any system or process for which $q = 0$.

The change in internal energy of an adiabatic system is necessarily

$$\Delta U = w \qquad \text{(adiabatic)} \tag{19-1a}$$

because $q = 0$. It is possible to change the internal energy of an adiabatic system by performing work; although q must be zero, w need not be.

If a system is allowed no interchange of energy with its surroundings—either as work or heat—it is said to be *isolated*. There can be no change in the internal energy of an isolated system, since both q and w are zero,

$$\Delta U = 0 \qquad \text{(isolated)} \tag{19-1b}$$

In other words, internal energy cannot be created in an isolated system, nor can it be lost—it is *conserved*. Furthermore, if a system is not isolated, it is always possible to imagine it combined with its surroundings into a "supersystem" that *is* isolated and for which, consequently, $\Delta U_{tot} = 0$. Hence $\Delta U_{sys} + \Delta U_{surr} = \Delta U_{tot} = 0$, or $\Delta U_{surr} = -\Delta U_{sys}$. *Any change in the internal energy of a system*

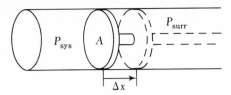

Figure 19-3 P, V Work
When the gas expands and moves the piston of cross-sectional area A through the distance Δx, it performs work $F\Delta x$, where $F = AP_{surr}$. The volume increase of the gas is $\Delta V = A\Delta x$ and the work is thus $P_{surr}A\Delta x = P_{surr}\Delta V$. This is equal to $-w$, because work is performed *on* the surroundings when ΔV is positive; positive work is, by convention, assumed to be performed *by* the surroundings on the system.

must always be accompanied by an equal and opposite change in the internal energy of the surroundings.

The term *isothermal* (Greek: *isos,* equal; *thermos,* hot) is applied to systems held at constant temperature, for example by immersing them in a large water bath whose temperature is thermostatically controlled.

The internal energy of a sample of an *ideal gas* depends only on the temperature of the gas; it is independent of the pressure or the volume. There are no forces between the molecules of an ideal gas and thus no potential energy of interaction that might vary with the distance between molecules. Hence $\Delta U = 0$ for any *isothermal* expansion, contraction, or mixing of *ideal gases* (although it will not in general be zero if a reaction occurs, or for a process involving the condensation of a nearly ideal real gas).

P,V Work   An important type of work is *pressure-volume work* (or *P,V* work for short). This is work performed when a system expands or contracts against an external pressure P_{surr}.

As a special case, consider the system to be a substance contained in a cylinder of cross section A with a movable (frictionless) piston (Fig. 19-3). The piston is to be moved a distance Δx against an external pressure P_{surr}. By definition, pressure is force divided by area. The work performed by the system is $F\Delta x = -w$, and since $F = P_{surr}A$ is the force exerted by the piston on the system, $w = -P_{surr}A\Delta x$. The important minus sign arises from the fact that for a positive Δx (that is, an expansion) the system performs work on the surroundings, which by our convention represents negative work. In general, work of expansion is negative and work of contraction, performed by the surroundings on the system, is positive. Since $A\Delta x$ is the volume increase ΔV of the system,

$$w = -P_{surr}\Delta V = -P_{surr}(V_2 - V_1) \qquad (19\text{-}2)$$

where V_1 and V_2 are the initial and final volumes.

☐ **Change in Internal Energy of a Gas on Expansion**   A 1-liter sample of gas is kept under adiabatic conditions, that is, no energy can be transferred into or out of the gas as heat. The gas is allowed to expand to 10.0 liter against a constant pressure of 1 atm. The expansion is slow enough that only *P,V* work is performed and any work expended to overcome friction or to accelerate the piston is negligible. Calculate (*a*) the work, (*b*) the change in the internal energy of the gas, and (*c*) the change in the internal energy of the surroundings.

Example 19-1

Solution   (a) Since the pressure in the surroundings is constant, the work w is

$$w = -P_{surr}(V_2 - V_1) = -1.00 \text{ atm} \times (10.0 - 1.0) \text{ liter}$$
$$= -9.0 \text{ liter atm} = -9.0 \text{ liter atm} \times \frac{101 \text{ J}}{\text{liter atm}}$$
$$= \underline{-9.1 \times 10^2 \text{ J}}$$

(See Table A-4 for the conversion from liter atmospheres to joules.)

(b) For an adiabatic process, $q = 0$, so

$$\Delta U = q + w = 0 - 9.1 \times 10^2 \text{ J} = \underline{-9.1 \times 10^2 \text{ J}}$$

(c) Since energy is conserved, any loss of internal energy by the system results in the gain of an equal amount of energy by the surroundings. Thus

$$\Delta U_{surr} = -\Delta U_{sys} = \underline{9.1 \times 10^2 \text{ J}} \quad \blacksquare$$

Exercise 19-1

☐ A 3.0-liter sample of gas is expanded adiabatically against an outside constant pressure of 15.0 atm until the volume has increased to 7.0 liter. The expansion is slow enough that only P,V work is involved. Calculate (a) the work, (b) the change in the internal energy of the gas, and (c) the change in the internal energy of the surroundings. ∎

Whenever a system open to the atmosphere undergoes a change in volume, P,V work is done. For example, when 1 liter of water is heated from 25 to 30°C, its volume increases by 1.4 cm^3. If the system is open to the air at 1 atm pressure, the work performed is 1.4×10^{-3} liter atm or 0.14 J. The statement on page 396 that this change in state can be accomplished without performing work is therefore imprecise. A more accurate statement is that the work performed is negligible compared to the 21 kJ of energy that must be transferred to raise the temperature of the water (see Prob. 19-7).

Reversible and Irreversible Paths   Changes in the state of a system may be brought about along reversible or irreversible paths. A *reversible path* is one that may be followed in all its details in either direction. At any point along the path the direction of change may be reversed by a very small alteration in a variable such as the temperature or the pressure.

Consider, for example, the following change in state: An ideal gas, initially at a pressure $P_i = 1$ atm and a volume $V_i = 1$ liter, is brought at constant temperature to the state in which the pressure and volume are $P_f = \frac{1}{2}$ atm and $V_f = 2$ liter. There are many paths along which this change of state can take place. Some of these are illustrated in Fig. 19-4.

The gas may be thought of as being in an apparatus with a tightly fitting piston, similar to that shown in Fig. 19-3, so that the expansion can proceed only very slowly. A catch prevents the piston from moving prematurely from its initial position. Path (a) consists of a two-step process. First the external pressure acting on the piston is kept at $\frac{2}{3}$ atm. When the catch is released the gas pushes the piston back and expands until its pressure equals the applied pressure and the piston stops. At this point the volume is $\frac{3}{2}$ liter. Now the external pressure is dropped to $P = \frac{1}{2}$ atm, and the gas expands to the final volume, 2 liter. At this point the pressure of the gas is again equal to the external pressure. The work

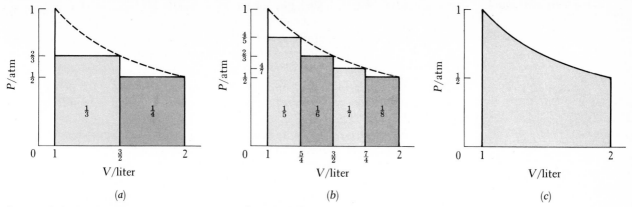

Figure 19-4 Constant-Temperature Expansion of an Ideal Gas

(a) Two-step expansion. (b) Four-step expansion. (c) Expansion by N steps in the limit that N is infinite. The dashed line in (a) and (b) is the same curve as that in (c); it corresponds to the equation $PV = 1$ liter atm. The number on each rectangle in (a) and (b) is its area, $P\Delta V$ for the step. The work performed by the gas is the negative of the sum of the areas.

performed by the gas in this two-step expansion is $w = -P_1\Delta V_1 - P_2\Delta V_2 = -\frac{2}{3}(\frac{3}{2} - 1) - \frac{1}{2}(2 - \frac{3}{2}) = -\frac{1}{3} - \frac{1}{4} = -0.583$ liter atm.

If the gas is allowed to expand until the piston stops against successive pressures of $\frac{4}{5}, \frac{4}{6}, \frac{4}{7}$, and $\frac{4}{8}$ atm (process (b)), the volume changes will each be $\frac{1}{4}$ liter and the work performed is -0.634 liter atm. An eight-step process gives $w = -0.663$ liter atm.

As the number of steps is increased the changes in pressure between each of the steps become smaller. The changes in volume also decrease and with each decrease in step size the total work delivered to the surroundings becomes larger. By the methods of calculus (Appendix C) it is possible to calculate the work for an infinite number of steps [Fig. 19-4(c)]

$$w = -nRT\ln\frac{V_f}{V_i} = -PV\ln\frac{V_f}{V_i} \qquad (19\text{-}3)$$

In this example,

$$w = -(1.000 \text{ liter atm}) \times \ln 2 = -0.693 \text{ liter atm}$$

Equation (19-3) gives the maximum work that can be obtained in an isothermal expansion of an ideal gas.

What are the characteristics of this maximum-work path? At each point during the expansion the pressure of the gas nearly equals the pressure exerted by the piston—there is almost no force acting on the piston. Thus, an expansion can be changed into a compression by an exceedingly small increase of the outside pressure. Similarly, compression can be changed to expansion by an exceedingly small decrease of the outside pressure. The process is said to be *reversible* because the direction along the path can be reversed by very small changes in the conditions.

Strictly speaking, the word "reversible" applies only to the limiting case of an infinite number of infinitesimally small steps with P_{sys} negligibly different from P_{surr} at all points along the path from the initial state to the final state. Such a

401

process would be infinitely slow so that reversible expansion of a gas is an idealized concept. In practice one can carry out expansion or compression of a gas, and other changes in state, in such a way as to approach reversible conditions very closely. For example, for the process illustrated in Fig. 19-4, with as few as 16 steps, the work performed by the system, $-w$, is 2.2 percent below the value given for true reversibility; with 256 steps the difference is only 0.3 percent and with 1024 steps only 0.04 percent. Under such nearly reversible conditions it is justifiable to take $P_{sys} = P_{surr}$.

More work must be done in an irreversible isothermal compression than in a reversible isothermal compression between the same initial and final states. What becomes of the extra work done in the irreversible compression? The corresponding energy must appear in the surroundings. It may, for example, speed up the gas molecules momentarily and then be transferred as heat to the thermostat that maintains the temperature of the gas.

Example 19-2

□ **Isothermal Reversible Expansion of an Ideal Gas** What is the work performed on the surroundings by the reversible expansion of 1.00 liter of an ideal gas at a constant temperature of 300 K, from 10.0 to 1.00 atm?

Solution The final volume of the gas is, by Boyle's law, 10.0 liter. By (19-3) the work performed on the surroundings is $-w = nRT \ln (V_2/V_1)$. Here the unknown number of moles of gas, n, is related to P_1 and V_1, the initial pressure and volume of the gas, by $n = P_1V_1/RT$. Hence

$$-w = P_1 V_1 \ln \frac{V_2}{V_1} = (10.0 \text{ atm})(1.00 \text{ liter})(\ln \tfrac{10.0}{1.00})$$

$$= (10.0)(\ln 10.0) \text{ liter atm} = 23.03 \text{ liter atm}$$

$$= 23.03 \text{ liter atm} \times 101.3 \text{ J/liter atm}$$

$$= \underline{23.3 \times 10^2 \text{ J}}$$

An alternative approach is to calculate n explicitly:

$$n = \frac{10.0 \text{ liter atm}}{(0.0821 \text{ liter atm mol}^{-1} \text{ K}^{-1})(300 \text{ K})}$$

$$= 0.406 \text{ mol}$$

Hence
$$-w = nRT \ln 10.0$$

$$= (0.406 \text{ mol})(8.31 \text{ J mol}^{-1} \text{ K}^{-1})(300 \text{ K})(2.30)$$

$$= \underline{23.3 \times 10^2 \text{ J}}$$

Comparison with the result of Example 19-1, the irreversible expansion of 1.00 liter of gas against a constant pressure of 1.00 atm, shows that the work performed by the gas in the reversible expansion is $\frac{23.3}{9.1} = 2.6$ times larger. ■

Exercise 19-2

□ What is the work performed by the surroundings when 2.50 liter of an ideal gas is reversibly compressed at a constant temperature of 400 K from 2.00 atm to 3.00 atm? ■

Heat Capacity The heat required to raise the temperature of a system by one degree is called the *heat capacity C* of the system. The heat capacity per mole of a substance is called the *molar heat capacity.* The heat capacity for one gram of a substance is called the *specific heat;* the molar heat capacity is the specific heat multiplied by the weight (in grams) per mole. Molar quantities are indicated by a tilde, as for the molar volume $\widetilde{V}$; hence the molar heat capacity is given the symbol $\widetilde{C}$.

The energy required to change the temperature of a substance at constant volume from T to $T + \Delta T$ is given by the product of the heat capacity at constant volume, C_V, and the temperature change ΔT. If this change is brought about solely by absorption of heat and there is no accompanying reaction or phase change, then

$$q_V = C_V \Delta T = n\widetilde{C}_V \Delta T \qquad (19\text{-}4a)$$

where the subscript V has been used to specify the constant-volume path along which the temperature change occurs. If only P, V work is considered, as is often useful, $w = 0$ for a constant-volume process and hence

$$\Delta U = w + q = 0 + q_V = C_V \Delta T$$

The relation

$$\Delta U = C_V \Delta T \qquad (19\text{-}4b)$$

is general for any temperature change at constant volume, whether the change is caused by a transfer of energy as heat or as work or both.

Another common mode of heating is at constant pressure, and the corresponding heat capacity is denoted by C_P. Thus, if by heat transfer the temperature of a substance is to be changed from T_1 to T_2 at constant pressure (with no phase change or reaction), the amount of heat involved is

$$q_P = C_P \Delta T = n\widetilde{C}_P \Delta T \qquad (19\text{-}4c)$$

An equation that parallels (19-4b) and relates $C_P \Delta T$ to the change of a function of state is given later (Equation 19-17). The heat capacity at constant pressure is (for all substances) never smaller than that at constant volume; for gases, it is distinctly larger because the gas expands appreciably when heated at constant pressure so a significant amount of energy is expended in P,V work. For an *ideal* gas, the difference between $\widetilde{C}_P$ and $\widetilde{C}_V$ is

$$\widetilde{C}_P - \widetilde{C}_V = R$$

The heat capacity of a substance is temperature-dependent. For our purposes, however, the change of $\widetilde{C}_P$ and $\widetilde{C}_V$ with temperature is usually sufficiently small to be neglected. $\widetilde{C}_P$ and $\widetilde{C}_V$ are always positive; thus if ΔT is positive, heat must be supplied to the substance. If ΔT is negative, heat is liberated from the substance.

□ **Temperature Change and Heat Transfer** Find the heat transferred to or from the surroundings when the temperatures of the following systems are changed: (*a*) a copper penny at 1.0 atm from 20.0 to 25.0°C; (*b*) 150 g water at 1.0 atm from 25.0 to 10.0°C; (*c*) 0.500 mol methanol at 1.0 atm from 15.0 to 25.0°C; and (*d*) the same system and temperature change as in (*c*) but at constant volume. At constant pressure the heat capacity of the penny is $C_P = 1.21$ J K^{-1} and the specific heat of water is 4.18 J g^{-1} K^{-1}. The pertinent molar heat capacities of methanol are $\widetilde{C}_P = 81.6$ J mol^{-1} K^{-1} and

Example 19-3

$\tilde{C}_V = 64.0 \text{ J mol}^{-1} \text{ K}^{-1}$. These heat capacities remain constant in the temperature ranges specified within the precision of the values given.

Solution Applying (19-4a) and (19-4c) as needed, we have:

(a) $q_P = (1.21 \text{ J K}^{-1})(5.0 \text{ K}) = \underline{6.0 \text{ J}}$

(b) $q_P = (150 \text{ g})(4.18 \text{ J g}^{-1} \text{ K}^{-1})(-15.0 \text{ K}) = \underline{-9.4 \text{ kJ}}$

(c) $q_P = (0.500 \text{ mol})(81.6 \text{ J mol}^{-1} \text{ K}^{-1})(10.0 \text{ K}) = \underline{408 \text{ J}}$

(d) $q_V = (0.500 \text{ mol})(64.0 \text{ J mol}^{-1} \text{ K}^{-1})(10.0 \text{ K}) = \underline{320 \text{ J}}$

In (b) heat is transferred to the surroundings (that is, q is negative); in the other three cases heat is transferred to the system (q is positive). ∎

Exercise 19-3

☐ Find the heat transferred to or from the surroundings when the temperatures of the following systems are changed: (a) a 5.0-g piece of Ag from 15.0 to 30.0°C at a constant pressure of 1.0 atm; (b) the same system and temperature change but at constant volume; (c) 0.200 mol of gaseous ethane, C_2H_6, from 20.0 to 10.0°C at a constant pressure of 1.0 atm; (d) the same system and temperature change but at constant volume. The pertinent values of the molar heat capacities are: Ag, $\tilde{C}_P = 25.2 \text{ J mol}^{-1} \text{ K}^{-1}$, $\tilde{C}_V = 24.1 \text{ J mol}^{-1} \text{ K}^{-1}$; C_2H_6, $\tilde{C}_P = 48.0 \text{ J mol}^{-1} \text{ K}^{-1}$, $\tilde{C}_V = 39.7 \text{ J mol}^{-1} \text{ K}^{-1}$. ∎

19-3 Energy Changes in Chemical Reactions

The application of thermodynamics to chemical reactions is a straightforward extension of the principles that have already been discussed. The system consists of a quantity of some substance or mixture of substances that, in the initial state, is termed the reactants. In the final state these substances have been converted into new substances called the products. The change in internal energy as a result of a chemical reaction

$$aA + bB + \cdots \longrightarrow cC + dD + \cdots$$

is then given by

$$\Delta U = U_f - U_i = \Sigma U_{prod} (P_f, T_f, \ldots) - \Sigma U_{react} (P_i, T_i, \ldots) \qquad (19\text{-}5)$$

where the quantity ΣU_{prod} is the sum of the internal energies of all products. If c mol of product C and d mol of product D are formed, then

$$\Sigma U_{prod} = c\tilde{U}_C + d\tilde{U}_D$$

ΣU_{react} is similarly defined for all reactants. The number of moles used in the sums must be those in the balanced equation if ΔU is to be meaningful. Note that it is necessary to specify the initial and final conditions, such as pressure and temperature, in order to characterize the initial and final states.

Enthalpy When a reaction does not take place at constant volume, P,V work must be taken into account. Although for some processes the P,V work done may be of major interest, in chemistry it is usually just incidental. Chemical reactions are commonly performed in vessels open to the atmosphere, so that the pressure during a reaction remains essentially constant. If gases are evolved or used up, or if the volume of the system changes in any other way during the reaction, the resulting P,V work done on the system is, by (19-2), $-P\Delta V$. Frequently the work done on or by the system is of no particular interest but it must nevertheless be taken into account.

By introduction of a new state function, the enthalpy, this $-P\Delta V$ term can be taken into account automatically. The *enthalpy H* (Greek: *enthalpein,* to warm), an energy quantity, is defined by the equation

$$H = U + PV \tag{19-6}$$

Because U, P, and V are functions of state, H is also. For a system undergoing a change of state along a constant-pressure path so that $\Delta P = 0$, with no work other than P,V work, we have

$$\begin{aligned} \Delta H = \Delta U + \Delta(PV) &= \Delta U + PV_2 - PV_1 \\ &= \Delta U + P(V_2 - V_1) = \Delta U + P\Delta V \end{aligned} \tag{19-7}$$

and also

$$w = -P\Delta V$$

From the first law, $\Delta U = q + w$, we obtain, for a constant-pressure path with only P,V work,

$$\Delta U = q_P - P\Delta V \tag{19-8}$$

which leads, with (19-7), to

$$q_P = \Delta U + P\Delta V = \Delta H \tag{19-9}$$

Thus the heat absorbed during a change in state at constant pressure[2] when only P,V work is done is equal to the change in a state function, the enthalpy.

The quantity called the *heat of reaction* is usually defined as the heat transferred *to the surroundings*. Thus[3] it is $-q$. For an exothermic reaction at constant pressure, heat is liberated by the reaction; $-q_p = -\Delta H$ is positive and ΔH is negative. For an endothermic reaction, heat is absorbed and thus if the reaction is carried out at constant pressure, ΔH is positive. For example, the "bond energy", as used in Sec. 17-2, is the change in *enthalpy* associated with the *breaking* of a bond.

Phase transitions (e.g., melting) of pure substances at fixed temperature are necessarily constant-pressure processes (Chap. 5). When a solid substance melts, the increase in its enthalpy is called the *heat of fusion* or *enthalpy of fusion*. Similarly the increase in enthalpy that accompanies the vaporization of a liquid or of a solid is called the *heat* or *enthalpy of vaporization* or *sublimation*. Table 19-1 gives some examples of enthalpy changes for phase transitions as well as for some common chemical reactions. Reversing a reaction implies a change in sign of

---

[2] We have implicitly assumed in deriving (19-9) that the pressure in the *system,* which is what P stands for in (19-6) and (19-7), is the same as the pressure in the *surroundings,* which is the meaning of P in $w = -P\Delta V$. This is commonly true for constant-pressure processes.

[3] In some books the heat of reaction $-q_p$ is indicated in combination with the chemical equation, as shown by the following example:

$$H_2(g) + \tfrac{1}{2}O_2(g) \longrightarrow H_2O(l) + 285.8 \text{ kJ}$$

This symbolism is intended to imply that 285.8 kJ of energy in the form of heat is liberated—is a "product" of the reaction. However, it is important to realize that the 285.8 kJ represents $-\Delta H$ of the reaction.

Table 19-1
Enthalpy Changes for Some Phase Changes and Chemical Reactions*

Phase change or reaction	ΔH/kJ	t/°C
(1) $H_2O(s) \longrightarrow H_2O(l)$	6.02	0
(2) $H_2O(l) \longrightarrow H_2O(g)$	44.0	25
	40.6	100
(3) $Br_2(l) \longrightarrow Br_2(g)$	31.0	59
(4) $I_2(s) \longrightarrow I_2(g)$	62.3	25
(5) C (graphite) $\longrightarrow$ C(g)	716.7	25
(6) C (graphite) $+ O_2(g) \longrightarrow CO_2(g)$	-393.5	25
(7) $CO(g) + \frac{1}{2}O_2(g) \longrightarrow CO_2(g)$	-283.0	25
(8) $H_2(g) + \frac{1}{2}O_2(g) \longrightarrow H_2O(l)$	-285.8	25

*The values refer to as many moles of reactants and products as are shown by the coefficients of the chemical equations, the initial states being reactants (or initial phase) only and the final states products (or final phase) only. The pressure is 1 atm.

the value of ΔH. The usual SI units of ΔH are *kilo*joules, but many compilations are in terms of kilocalories; ΔH refers, as it always will in this book, to as many moles of reactants and products as are indicated by the coefficients of the overall chemical equation given.

For any *isothermal* expansion, contraction, or mixing of *ideal gases,* $\Delta H = 0$. This follows from the fact that for such processes $\Delta U = 0$ and the fact that for an ideal gas at a constant temperature $\Delta(PV) = 0$ (Boyle's law).

Combining Enthalpies of Reaction Since enthalpy is a state function, the enthalpy increase for any change of state is independent of the path. This has important implications for the enthalpy changes of chemical reactions: when the chemical equations describing such reactions are added or otherwise combined into an equation describing a new reaction, the corresponding ΔH values, combined in the same way, yield the ΔH for the new reaction. This is because the ΔH does not depend on the actual course of the reaction. This important result is called *Hess' law*. Its application is illustrated by the following three examples.

Example 19-4

□ **Combustion of Graphite to Carbon Monoxide** Find the enthalpy change that accompanies the combustion of graphite to carbon monoxide at 25°C by considering the reactions in Table 19-1.

Solution The equation of the reaction may be obtained by subtracting reaction (7) of Table 19-1 from reaction (6):

(6)	C (graphite) $+ O_2(g)$	$\longrightarrow$	$CO_2(g)$	$\Delta H_6 = -393.5$ kJ
(7)	$CO(g) + \frac{1}{2}O_2(g)$	$\longrightarrow$	$CO_2(g)$	$\Delta H_7 = -283.0$ kJ
	C (graphite) $+ \frac{1}{2}O_2(g)$	$\longrightarrow$	$CO(g)$	$\Delta H = ?$

The enthalpy change for the new reaction may therefore be obtained by subtracting ΔH of (7) from that of (6):

$$\Delta H = \Delta H_6 - \Delta H_7 = [-393.5 - (-283.0)]\,\text{kJ} = \underline{-110.5\,\text{kJ}}$$

It is possible to combine graphite and oxygen to form CO not only directly, or by a sequence of the reactions represented by (6) and (7), but along many other paths involving any number of steps. For all these paths the total of the ΔH values is -110.5 kJ. ■

☐ At 25°C the following reactions are accompanied by the enthalpy changes indicated:

$$Si(s) + O_2(g) \longrightarrow SiO_2(s) \qquad \Delta H = -910.9 \text{ kJ mol}^{-1}$$
$$2 Si(s) + 3 H_2(g) \longrightarrow Si_2H_6(g) \qquad \Delta H = 80.3 \text{ kJ mol}^{-1}$$
$$H_2(g) + \tfrac{1}{2} O_2(g) \longrightarrow H_2O(l) \qquad \Delta H = -285.8 \text{ kJ mol}^{-1}$$

What is the enthalpy change at 25°C for the reaction

$$Si_2H_6(g) + \tfrac{7}{2}O_2(g) \longrightarrow 2SiO_2(s) + 3H_2O(l) \quad \blacksquare$$

Many reaction enthalpies have to be measured indirectly by following a roundabout path leading from the reactants to the products or from the products to the reactants. An experimental procedure that is often followed is to burn separately, in oxygen, all reactants and products that are combustible, to measure the corresponding individual *enthalpies of combustion,* and to combine the results.

Example 19-5

☐ **Use of Enthalpies of Combustion** Calculate the enthalpy change ΔH for the reaction

$$2C \text{ (graphite)} + 2H_2(g) + H_2O(l) \longrightarrow C_2H_5OH(l) \qquad (19\text{-}10)$$

from the measured enthalpies or heats of combustion of ethanol, graphite, and hydrogen at 25°C:

$$C_2H_5OH(l) + 3O_2(g) \longrightarrow 2CO_2(g) + 3H_2O(l) \qquad (19\text{-}11)$$
$$\Delta H_{11} = -1366.7 \text{ kJ}$$

$$C \text{ (graphite)} + O_2(g) \longrightarrow CO_2(g) \qquad (19\text{-}12)$$
$$\Delta H_{12} = -393.5 \text{ kJ}$$

$$H_2(g) + \tfrac{1}{2}O_2(g) \longrightarrow H_2O(l) \qquad (19\text{-}13)$$
$$\Delta H_{13} = -285.8 \text{ kJ}$$

Solution The chemical equation (19-10) can be obtained by combining the other equations in the way indicated symbolically by $2[(19\text{-}12)+(19\text{-}13)]$ $-(19\text{-}11)$. Thus

$$\Delta H = 2(\Delta H_{12} + \Delta H_{13}) - \Delta H_{11}$$
$$= 2(-393.5 - 285.8) \text{ kJ} - (-1366.7) \text{ kJ} = \underline{8.1 \text{ kJ}} \quad \blacksquare$$

☐ Calculate ΔH for the reaction

$$12 C \text{ (graphite)} + 11 H_2O(l) \longrightarrow C_{12}H_{22}O_{11} \text{ (sucrose, } s)$$

from the heat of combustion of sucrose, $\Delta H = -5648 \text{ kJ mol}^{-1}$,

$$C_{12}H_{22}O_{11}(s) + 12 O_2(g) \longrightarrow 12 CO_2(g) + 11 H_2O(l)$$

and from data given in the preceding example. ∎

Example 19-6

□ **Reaction Enthalpy and Specification of Initial and Final States**
Calculate ΔH for a reaction like (19-10) but with $H_2O(l)$ on the left replaced by $H_2O(g)$. Use data from Table 19-1 and Example 19-5.

Solution The chemical equation of the desired reaction

$$2C \text{ (graphite)} + 2H_2(g) + H_2O(g) \longrightarrow C_2H_5OH(l)$$

results by adding the chemical equation

$$H_2O(g) \longrightarrow H_2O(l)$$

with $\Delta H = -44.0$ kJ to Equation (19-10), for which ΔH is 8.1 kJ. The desired enthalpy change is therefore the sum of the values quoted: $\Delta H = (-44.0 + 8.1)$ kJ = $\underline{-35.9 \text{ kJ}}$. Note the difference of the result from that in Example 19-5 and therefore the importance of carefully specifying initial and final states. ■

Exercise 19-6

□ The enthalpy change associated with the reaction

$$C \text{ (graphite)} \longrightarrow C \text{ (diamond)}$$

is 1.9 kJ mol^{-1}. Use this information and the answer to Exercise 19-5 to calculate ΔH for the reaction

$$12C \text{ (diamond)} + 11H_2O(l) \longrightarrow C_{12}H_{22}O_{11} \text{ (sucrose, } s) \quad ■$$

Standard Enthalpies Within the framework of thermodynamics, only relative values of $H = U + PV$ are defined because only relative values for U are defined. It is, however, possible to assign zero enthalpy to suitable reference states *by convention*. This is of great value because it permits the assignment and tabulation of a standard enthalpy for each substance. These standard enthalpies are useful in many computations, as will be seen. The conventions chosen fulfill the experimental requirement that a reference state can be reached easily and reproducibly.

These conventions may be separated into two parts, the definition of standard states for all substances and the definition of reference states on which tabulated enthalpy values are based.

1. A substance is in its *standard state* when under a pressure of 1 atm and at a standard temperature, usually chosen to be 25 °C or 298 K (more accurately 298.15 K, but the difference of 0.15 K is normally insignificant and will be neglected). For dissolved molecular or ionic species a concentration (or more accurately, an "effective" concentration or activity) of 1 mol liter^{-1} is implied. Quantities referring to standard states are designated by a superscript ⊖ (such as $X^{\ominus}$ for a quantity X);[4] if the standard state is chosen to be at a temperature T that differs from 298 K, as is sometimes useful, this is indicated by showing the value of T in parentheses, $X^{\ominus}(T)$. Thus $X^{\ominus}$ specifies a standard state of 1 atm, 1 M for dissolved species, and 298 K, whereas $X^{\ominus}(1000)$ refers to the same pressure and concentration but at 1000 K. In other words, the temperature of a standard state is flexible and may need

---

[4] The plimsoll ⊖ is a load-line mark on merchant vessels to prevent their overloading; it is named after Samuel Plimsoll, whose efforts resulted in its legal adoption by the English Parliament in 1876.

special mention, whereas pressure and concentration are fixed.

The standard states discussed here must not be confused with the standard conditions customarily abbreviated by STP (standard temperature and pressure); STP conditions are often used when specifying quantities of a gas and imply 1 atm and $0°C$ (Sec. 4-3).

2. The *reference states* for tabulated enthalpies are the *chemical elements in their most stable forms and in their standard states.* Under these conditions *zero enthalpy* is assigned to each element. To all pure substances are then assigned *standard molar enthalpies of formation* $\tilde{H}_f^{\ominus}$ that represent the enthalpy change, positive or negative, when 1 mol of the substance is formed at the standard pressure and temperature from the elements in their most stable forms. In some texts this quantity is also given, among others, the symbol $\Delta\tilde{H}_f^{\ominus}$, $\Delta\tilde{H}_f^{\circ}$, or $\Delta\tilde{H}_f^{\circ}$. Often we shall call it more briefly the *standard molar enthalpy.*

The conventions about zero enthalpy are arbitrary, in the same way as is the choice of the zero level for measurement of altitudes on earth. They are chosen for convenience.

□ **Standard Molar Enthalpy** What is the standard enthalpy (of formation) $\tilde{H}_f^{\ominus}$ of liquid water?

Example 19-7

Solution For the reaction

$$H_2(g) + \tfrac{1}{2}O_2(g) \longrightarrow H_2O(l) \qquad (19\text{-}14)$$

$$\Delta H^{\ominus} = H_f^{\ominus}\{H_2O(l)\} - H_f^{\ominus}\{H_2(g)\} - \tfrac{1}{2}\tilde{H}_f^{\ominus}\{O_2(g)\}$$

The enthalpy change at standard conditions is -285.8 kJ (Table 19-1). The standard molar enthalpy $\tilde{H}_f^{\ominus}$ of liquid water is thus $\underline{-285.8 \text{ kJ mol}^{-1}}$ since $H_2(g)$ and $O_2(g)$ are elements in their standard states and, by definition of the enthalpy of formation of an element, $\tilde{H}_f^{\ominus}$ for an element in its standard state is zero. ■

□ What is the standard molar enthalpy of $H_2O(g)$ (see Table 19-1)? ■

Table D-4 contains the standard molar enthalpies $\tilde{H}_f^{\ominus}$ for a number of elements, compounds, atoms, and aqueous ions, some in forms that are not those most stable at standard conditions.

When the values of $\tilde{H}_f^{\ominus}$ for the reactants and products of a reaction are known, it is possible to compute the standard enthalpy change $\Delta H^{\ominus}$ for the reaction from the relation

$$\Delta H^{\ominus} = \Sigma H_{f\,\text{prod}}^{\ominus} - \Sigma H_{f\,\text{react}}^{\ominus}$$

This extremely useful procedure is illustrated in the following examples, with $\tilde{H}_f^{\ominus}$ values shown in kilojoules per mole beneath the chemical symbols.

□ **Combination of Standard Molar Enthalpies** Water gas, an equimolar mixture of CO and H_2, is made from carbon and water by the reaction

Example 19-8

$$H_2O(g) + C \text{ (graphite)} \longrightarrow CO(g) + H_2(g) \qquad (19\text{-}15)$$

$$\quad\;\; -241.8 \qquad\qquad 0 \qquad\qquad\; -110.5 \qquad\quad 0$$

What is $\Delta H^{\ominus}$ for this reaction?

Solution The change in standard enthalpies is

$$\Delta H^{\ominus} = (-110.5 + 241.8) \text{ kJ} = \underline{131.3 \text{ kJ}}$$

The reaction is actually carried out at around 600°C by passing steam over hot coal. The $\Delta H^{\ominus}$ just given applies to 25°C and shows the reaction to be endothermic at that temperature. It is also endothermic at 600°C, and to maintain this temperature while making water gas the steam is turned off every few minutes and replaced by a brief blast of hot air. ■

Exercise 19-8

☐ What is $\Delta H^{\ominus}$ for the reaction

$$4NH_3 + 5\,O_2 \longrightarrow 4NO + 6H_2O(g) \quad ■$$
$${-46.2} \qquad 0 \qquad\quad 90.4 \quad\; -241.8$$

The enthalpy change when forming molecules from atoms in the gas phase is frequently of interest.

Example 19-9

☐ **Combination of Atoms into Molecules** What would be the enthalpy change if benzene vapor were made from carbon and hydrogen atoms?

Solution

$$6C(g) + 6H(g) \longrightarrow C_6H_6(g) \qquad\qquad (19\text{-}16)$$
$$\;716.7 \qquad 217.9 \qquad\quad 82.9$$

The enthalpy change under standard conditions is

$$\Delta H^{\ominus} = (82.9 - 6 \times 716.7 - 6 \times 217.9) \text{ kJ} = \underline{-5525 \text{ kJ}} \quad ■$$

Exercise 19-9

☐ What would be the enthalpy change if $CO_2(g)$ were made from carbon and oxygen atoms (see Table D-4)? ■

The energy and enthalpy of melting are nearly equal, but for vaporization there is a significant difference between ΔU and ΔH.

Example 19-10

☐ **Energy of Vaporization and of Melting** (*a*) What is $\Delta \tilde{U}$ for the vaporization of water at 25°C? (*b*) What is $\Delta \tilde{U}$ for the melting of ice at 0°C? Use needed data from Table 19-1. At 0°C the density of ice is 0.915 g ml^{-1} and that of liquid water is 0.99987 g ml^{-1}.

Solution (*a*) From Table 19-1 the value of $\Delta \tilde{H}$ at 25°C is 44.0 kJ mol^{-1}. (Note the temperature-dependence of $\Delta \tilde{H}$, the value at 100°C being 40.6 kJ mol^{-1}.) For any constant-pressure process $\Delta \tilde{U} = \Delta \tilde{H} - P\Delta \tilde{V}$. Neglecting the volume of the liquid water compared to that of the vapor and assuming ideal gas behavior, we get

$$-P\Delta \tilde{V} = -P(\tilde{V}_g - \tilde{V}_l) \approx -P\tilde{V}_g \approx -P\frac{RT}{P} = -RT$$

Thus

$$\Delta \tilde{U} = 44.0 \text{ kJ mol}^{-1} - 8.31 \text{ J K}^{-1}\text{mol}^{-1} \times 298 \text{ K} \times \frac{1 \text{ kJ}}{1000 \text{ J}}$$

$$= (44.0 - 2.48) \text{ kJ mol}^{-1} = \underline{41.5 \text{ kJ mol}^{-1}}$$

(b) For the melting of 1 mol ice,

$$\Delta \tilde{V} = 18.0 \text{ g mol}^{-1} \left(\frac{1 \text{ ml}}{1.000 \text{ g}} - \frac{1 \text{ ml}}{0.915 \text{ g}} \right)$$

$$= 18.0 \, (1.000 - 1.093) \text{ ml mol}^{-1} = -1.67 \text{ ml mol}^{-1}$$

Thus

$$-P\Delta\tilde{V} = -1.00 \text{ atm} \times (-1.67 \text{ ml mol}^{-1}) \times \frac{1 \text{ liter}}{1000 \text{ ml}}$$

$$= 1.67 \times 10^{-3} \text{ liter atm mol}^{-1}$$

$$= 1.67 \times 10^{-3} \text{ liter atm mol}^{-1} \times \frac{1 \text{ J}}{0.00987 \text{ liter atm}}$$

$$= 0.169 \text{ J mol}^{-1} = 1.7 \times 10^{-4} \text{ kJ mol}^{-1}$$

In contrast to part (a), $P\Delta\tilde{V}$ is so small as to be negligible when ice is melted at 0°C and 1 atm and $\Delta\tilde{U}$ is almost exactly equal to $\Delta\tilde{H}$. Hence by Table 19-1,

$$\Delta\tilde{U} \approx \Delta\tilde{H} = \underline{6.02 \text{ kJ mol}^{-1}}. \quad \blacksquare$$

□ What is $\Delta\tilde{U}$ for the vaporization of bromine at 59°C? What is $\Delta\tilde{U}$ of sublimation of iodine at 25°C? (Use needed data from Table 19-1.) Relevant densities are: $Br_2(l)$ at 59°C, 3.1 g ml^{-1}; $I_2(s)$ at 25°C, 4.93 g ml^{-1}. ■   **Exercise 19-10**

Enthalpy and Heat Capacity The heat capacity at constant pressure, C_P, is closely related to H. To increase the temperature of any substance from T to $T + \Delta T$ at constant pressure by a process involving only P,V work, a quantity of heat

$$q_P = \Delta H = C_P \, \Delta T \tag{19-17}$$

must be transferred to the system. In this equation it is assumed that C_P does not depend on temperature; hence the equation is strictly valid only for small values of ΔT.

□ **Mixing of Snow and Water** In a thermos bottle supplied with a heat-insulating stopper, 100.0 ml of liquid water at 25°C and 10.0 g of snow at −10.0°C are mixed under adiabatic conditions. The pressure is 1.00 atm. What is the final temperature? The enthalpy of fusion of ice is 334 J g^{-1} and the constant-pressure specific heats (assumed to be temperature-independent) are 2.09 J g^{-1} K^{-1} for ice and 4.18 J g^{-1} K^{-1} for liquid water.   **Example 19-11**

Solution The enthalpy change of the contents of the thermos bottle is zero for the prescribed adiabatic, constant-pressure process. Consideration of the relative amounts and temperatures of the snow and the water at the start indicates that the final temperature will not be below 0°C. As the temperature of the liquid water decreases, its loss of enthalpy is balanced by the enthalpy increase of the snow being warmed to 0°C and the snow being melted. If there is sufficient snow to cool all the original liquid to 0°C, the final temperature will be 0°C; otherwise it will be somewhat higher, with all the snow melted. We begin by assuming the latter condition; we note below how to check whether this assumption is correct. If we denote the part of the system that was originally water as 1 and the part originally snow as 2, we can write

$$\Delta H_1 + \Delta H_2 = 0$$

Let the final temperature be t_f and replace the unit K^{-1} by the unit $°C^{-1}$ (the degrees being equal in magnitude). Then by (19-17)

$$\Delta H_1 = C_P \Delta T = 100.0 \text{ g} \times 4.18 \text{ J g}^{-1} \, °C^{-1} \times (t_f - 25.0°C)$$

ΔH_2 is the sum of three terms: the enthalpy change when 10 g of snow is heated from $-10°C$ to the melting point, $0.0°C$; the enthalpy of fusion of the snow at $0.0°C$; and the enthalpy change when the 10.0 g of water formed is heated from $0.0°C$ to t_f.

$$\Delta H_2 = 10.0 \text{ g} \times 2.09 \text{ J g}^{-1} \, °C^{-1} \times [0.0 - (-10)]°C + 10.0 \text{ g} \times 334 \text{ J g}^{-1}$$
$$+ 10.0 \text{ g} \times 4.18 \text{ J g}^{-1} \, °C^{-1} \times (t_f - 0.0°C)$$

Combining ΔH_1 and ΔH_2 and solving for t_f gives

$$t_f = \frac{6.90 \times 10^3 \text{ J}}{460 \text{ J} \, °C^{-1}} = \underline{15.0°C}$$

and our assumption that the final temperature was above $0°C$ was justified. If, however, the amount of snow at low temperature had been sufficient to cool all the water to $0°C$ without melting all the snow, our equations would have yielded a negative value for t_f, contrary to our assumption. An alternative relation could then be used to calculate the fraction of the snow that would have to melt in order to cool all the liquid to $0°C$. ∎

Exercise 19-11

☐ What is the result of mixing snow and water as described in the example if 100.0 g snow at $0.0°C$, rather than 10.0 g at $-10.0°C$, is used? ∎

Summary

The state of a system is described by the values of macroscopic properties such as pressure, temperature, and density that need to be specified to duplicate the system precisely. An energy, called the internal energy, can be assigned to every state of a system. The first law of thermodynamics, $\Delta U = q + w$, relates the change in the internal energy of a system, ΔU, to q, the energy transferred to the system in the form of heat, and w, the work performed on the system. The symbol Δ is used for changes in quantities called state functions whose values depend only on the state of a system and not on its history. Changes in other quantities, such as q and w, depend on the path along which the change in state has taken place.

Any process or system for which $q = 0$ is said to be adiabatic. Isolated systems have no interchange of energy with their surroundings and therefore have constant internal energy. The term isother-

mal is applied to changes of state that occur at constant temperature. For an ideal gas, $\Delta U = 0$ for any isothermal process. A reversible change in state occurs along a path that may be followed in all its details in either direction. Such changes may be reversed by a very small alteration in a variable such as pressure or temperature and must necessarily be carried out slowly.

The work performed on a system when its volume expands or contracts an amount ΔV against a constant external pressure P_{surr} is

$$w = -P_{surr} \Delta V$$

For a reversible isothermal process in which n mol of an ideal gas undergoes a change in volume from V_1 to V_2 the work is

$$w = -nRT \ln \frac{V_2}{V_1}$$

The energy required to change the temperature of a substance is given by the product of the temperature change ΔT and the relevant heat capacity, C_P for constant-pressure paths and C_V for processes that occur at constant volume. If only P, V work is considered, at constant volume $\Delta U = q_V = C_V \Delta T$. For an ideal gas,

$$\widetilde{C}_P - \widetilde{C}_V = R$$

A state function called the enthalpy is defined by

$$H = U + PV$$

For changes in state at constant pressure in which only P,V work is done, $\Delta H = q_p$. The enthalpies of ideal gases are unaffected by any isothermal expansion, contraction, or mixing. Enthalpy changes as a result of chemical reactions are given by

$$\Delta H = \Sigma H_{\text{prod}} - \Sigma H_{\text{react}}$$

When chemical equations are added or otherwise combined into an equation describing a new reaction, the corresponding ΔH values, combined in the same way, yield the ΔH for the new reaction (Hess' law).

The standard molar enthalpy of formation $\widetilde{H}_f^{\circ}$ is the enthalpy change when one mole of a substance is formed at the standard pressure (1 atm) and temperature (usually 298 K) from the elements in their most stable forms. Hence, by definition, $H_f^{\circ} = 0$ for any element in its most stable form. Tables of standard enthalpies can be used to calculate enthalpy changes accompanying chemical reactions.

Terms and Concepts

Problems and Questions

19-1 Concepts and Definitions Distinguish clearly: (*a*) heat and molecular kinetic energy; (*b*) an isothermal process and an adiabatic process; (*c*) intensive and extensive properties; (*d*) a change in state that occurs adiabatically and a change in state that occurs in an isolated system.

19-2 Cooling a Gas at Constant Pressure A 2.00-mol sample of an ideal gas initially at 350 K and 1.50 atm was cooled at constant pressure until its volume was 35.0 liter. During this process the gas lost 1256 J as heat. Assume the gas to be the system and calculate (*a*) the initial volume, (*b*) the final temperature, (*c*) the work done, (*d*) the change in internal energy, (*e*) the change in enthalpy, and (*f*) $\widetilde{C}_P$ for the gas.

19-3 Isothermal Expansion of a Gas Exactly 3 mol of an ideal gas is expanded isothermally at 300 K so that the volume is increased by a factor of exactly 4 and work against a constant outside pressure of 2.00 atm is performed. Find w, q, ΔU, and ΔH. The initial pressure is 8.00 atm.

19-4 Isothermal Expansion of a Gas Two moles of an ideal gas is expanded at 0°C from a volume of 10.0 liter to 22.4 liter against a constant external pressure of 2.00 atm. (*a*) Calculate w and q. (*b*) If the temperature of the gas is kept at 0°C by means of an ice-water bath, will there be any change in the bath as the expansion takes place? If so, how many grams of ice will melt or how many grams of water will freeze? (Enthalpy of fusion of ice, 334 J g^{-1}.)

19-5 Reversible Expansion of a Gas Exactly 5 mol of an ideal gas is expanded reversibly from 50.0 to 150.0 liter at a constant temperature of 150°C. Calculate w in liter atmospheres and in joules.

19-6 Reversible Isothermal Compression of Helium Two moles of helium is compressed isothermally and reversibly at -73°C. The work done is 5.0 kJ. If the initial pressure was 1.0 atm, what is the final pressure?

19-7 Heating Water How much heat is required to change the temperature of exactly 1 liter of water from 25.0°C to 30.0°C at 1 atm? C_P of liquid water is 4.18 J K^{-1} g^{-1}; the density of liquid water is 1.00 g ml^{-1}.

19-8 Heat Capacities How much heat is required to change the temperature of 100 ml benzene from 20 to 50°C (a) at constant pressure, (b) at constant volume? $\widetilde{C}_P = 133$ J mol^{-1} K^{-1}; $\widetilde{C}_V = 92$ J mol^{-1} K^{-1}; MW = 78; and density = 0.88 g ml^{-1} at 20°C.

19-9 Stirring Water A 10.0-kg weight falls through a height of 3.00 m, causing a paddle wheel to stir 500 g water initially at 25.00°C. What is the final temperature of the water? (C_P (H_2O) = 4.18 J K^{-1} g^{-1}, $g = 9.80$ m s^{-2})

19-10 Temperature of a Mixture A 300-g silver bar at 80°C is placed in an ice-water bath at 0.0°C. The ice-water bath contains 10.0 g ice and 100 g H_2O, and is well-insulated. Calculate the final temperature of the system. (Specific heats: Ag, 0.234 J K^{-1} g^{-1}; water, 4.18 J K^{-1} g^{-1}; ice, 2.09 J K^{-1} g^{-1}; enthalpy of fusion of ice, 334 J g^{-1}).

19-11 Iced Drink Ice cubes at 0.0°C are used to cool 250 ml water from 30.0 to 0.0°C. How much of the ice melts, assuming no heat transfer between the water and its container or other parts of its surroundings? The enthalpy of fusion of ice is 6.02 kJ mol^{-1}, $\widetilde{C}_P$ of liquid water is 75.3 J mol^{-1} K^{-1}, and the density of liquid water is 1.000 g ml^{-1}.

19-12 Mixing of Gases A box with walls of a thermally nonconducting material is divided into two compartments by a partition of the same material. One compartment contains 0.40 mol He at 20°C and 1.00 atm; the other contains 0.60 mol N_2 at 100°C and 2.00 atm. The partition is removed so that the two gases mix. What is the final temperature and pressure? The molar heat capacities of He and of N_2 at constant volume are 12.5 and 20.7 J mol^{-1} K^{-1}, respectively. The gases may be assumed to be ideal gases and to behave as such on mixing. (*Hint:* Imagine the gases first to come to the final temperature while at their original volumes; then allow them to mix.)

19-13 Changing Ice to Steam Suppose that 1.00 mol ice at -20°C is heated at a constant pressure of 1.00 atm until it has been transformed into 1.00 mol steam at 127°C. Calculate q, w, ΔH, and ΔU for this process. Use relevant data from Table 19-1 as needed. The densities of ice and water are 0.92 and 1.00 g ml^{-1} and may be assumed to be independent of temperature. The heat capacities of ice, water, and steam at 1 atm are, respectively, 38, 75, and 36 J mol^{-1} K^{-1} and may be assumed to be independent of temperature. Assume that steam behaves like an ideal gas.

19-14 Implications of the First Law Indicate whether each of the following statements is true or false as it stands. If the statement is not true, indicate in what way it is false and whether it could be made into a true statement by a slight change in wording. If the statement is true but unnecessarily restricted, indicate what qualifying words or phrases can be omitted.

(a) The work done by the system on the surroundings during a change in state is never greater than the decrease in the energy of the system.

(b) The enthalpy of a system cannot change during an adiabatic process.

(c) When a system undergoes a given isothermal change in state, its change in enthalpy does not depend upon the process involved.

(d) When a change in state occurs, the increase in the enthalpy of the system must equal the decrease in the enthalpy of the surroundings.

(e) The equation $\Delta U = q + w$ is applicable to any macroscopic process, provided no electrical work is performed by the system on the surroundings.

(*f*) No change in state occurring within an isolated system can cause a change in its energy or its enthalpy.

(*g*) For any constant-pressure process, the increase in enthalpy equals the heat absorbed whether or not electrical work is done during the process.

(*h*) A reversible process is one in which the amount of energy lost by the system is just sufficient to restore the system to its initial state.

(*i*) When an imperfect gas expands into a vacuum, it does work because the molecules of the gas have been separated from one another against an attractive (van der Waals) force.

19-15 Enthalpy of Solution When 40.0 g of NaCl is dissolved in 200 ml water at 20°C, the temperature decreases by 1.5 K; a similar experiment using KCl produces a temperature drop of 9.1 K. When 50 g of a mixture of NaCl and KCl is dissolved in 200 ml water a temperature lowering of 7.9 K is observed. What is the weight percentage NaCl in the mixture?

19-16 Enthalpy of Reaction The standard enthalpy of formation of gaseous methylamine, CH_3NH_2, is -28.0 kJ mol^{-1}. Calculate, with the help of Table D-4, the change in enthalpy, and say whether heat is evolved or absorbed if the following reaction is carried out at 25°C and 1 atm:

$$CH_3NH_2(g) + H_2(g) \longrightarrow CH_4(g) + NH_3(g)$$

19-17 Energy of a Reaction From the values of $\widetilde{H}_f^{\circ}$ in Table D-4 find ΔH at 25°C for the reaction

$$CH_4(g) + 2\,O_2(g) \longrightarrow CO_2(g) + 2H_2O(l)$$

Estimate ΔU for the same process.

19-18 Enthalpy of Reaction Write a balanced equation for the reaction of 1 mol $CH_4(g)$ with Cl_2 to form $HCl(g)$ and $CHCl_3(l)$, and then, with the help of Table D-4, calculate the enthalpy change for this reaction.

19-19 Enthalpy of Combustion The standard enthalpy of formation of liquid cyclohexane, C_6H_{12}, is $-153.6\text{ kJ mol}^{-1}$. (*a*) Calculate the enthalpy of combustion of 2 mol liquid cyclohexane to CO_2 and $H_2O(l)$ at 25°C. (*b*) What is the heat of combustion for the same reaction at constant volume?

19-20 Heating Water Exactly 1 kg of water originally at 25°C is heated in a beaker by burning under it 1 liter of propane (C_3H_8) gas (measured at 25°C and at 1 atm pressure). Assume that 60 percent of the energy evolved is transferred to the water, while the remainder is lost in vaporizing the water formed in combustion, in heating gases, and in radiation. What is the final temperature of the water?

19-21 Hess' Law of Heat Summation Given the enthalpy changes at a certain temperature for reactions (*a*) and (*b*), calculate that for (*c*). Is (*c*) endothermic or exothermic at this temperature?

(*a*) $3H_2(g) + N_2(g) \longrightarrow 2NH_3(g)$
$$\Delta H = -92.4\text{ kJ}$$

(*b*) $2H_2(g) + O_2(g) \longrightarrow 2H_2O(g)$
$$\Delta H = -483.7\text{ kJ}$$

(*c*) $4NH_3(g) + 3O_2(g) \longrightarrow 2N_2(g) + 6H_2O(g)$

19-22 Enthalpy of Formation (*a*) The enthalpy of combustion of benzoic acid, C_6H_5COOH, at 25°C is -3227 kJ mol^{-1}. What is the enthalpy of formation? Use values of standard enthalpies of formation listed in Appendix D as needed. (*b*) The enthalpy of combustion of succinic acid, $HOOCCH_2CH_2COOH$, is -1494 kJ mol^{-1}. What is its enthalpy of formation?

19-23 Combustion of Nitromethane Nitromethane, CH_3NO_2, is a good fuel. It is a liquid at ordinary temperatures. When the liquid is burned, the reaction involved is chiefly

$$2CH_3NO_2(l) + \tfrac{3}{2}O_2(g) \longrightarrow$$
$$2CO_2(g) + N_2(g) + 3H_2O(g)$$

The standard enthalpy of formation of liquid nitromethane at 25°C is -112 kJ mol^{-1}; other relevant values may be found in Appendix D. (*a*) Calculate the enthalpy change in the burning of 1 mol liquid nitromethane to form gaseous products at 25°C. State explicitly whether the reaction is endothermic or exothermic. (*b*) Would more or less heat be evolved if *gaseous* nitromethane were burned under the same conditions? Indicate what additional information (if any) you would need to calculate the exact amount of heat, and show just how you would use this information.

19-24 Enthalpy of Reaction (*a*) With the help of data in Table D-4, calculate the heat evolved or absorbed (say which) when the following reaction is carried out at 25°C at constant pressure:

$$CCl_4(l) + H_2(g) \longrightarrow CHCl_3(l) + HCl(g)$$

(*b*) Calculate also the enthalpy of vaporization of CCl_4 at 25°C from data in Table D-4, and determine the enthalpy of the above reaction if gaseous CCl_4 instead of liquid CCl_4 were used as the reactant.

19-25 Energy and Enthalpy of Combustion The combustion of 1 mol gaseous acetylene, C_2H_2, to liquid water and gaseous CO_2 yields 1303 kJ at constant volume and at 20°C. What is the enthalpy of combustion of acetylene under the same conditions except that the pressure is kept constant at 1 atm? Consider all gases involved in the reaction to be ideal and neglect the volume of the liquid water produced.

19-26 Determination of CO Concentration Dry air containing a small amount of CO was passed through a tube containing a catalyst for the oxidation of CO to CO_2. Because of the heat evolved in this oxidation the temperature of the air increased by 3.2 K. Calculate the weight percentage CO in the sample of air. Assume that C_P for air is 1.01 J K^{-1} g^{-1}.

19-27 Combustion of Isobutene When 1 mol isobutene, a gas with formula C_4H_8, is burned at 25°C and 1 atm to form CO_2 and gaseous water, the enthalpy change is -2528 kJ. (*a*) Calculate, with the aid of any information needed from Table D-4, Appendix D, the standard enthalpy of formation of isobutene. (*b*) Suppose that 0.50 mol isobutene is burned adiabatically at constant pressure in the presence of an excess of oxygen, with 5.0 mol oxygen left at the end of the reaction. The heat capacity of the reaction vessel is 700 J K^{-1} and pertinent molar heat capacities (in joules per kelvin per mole) are $CO_2(g)$, 37; $H_2O(g)$, 34; $O_2(g)$, 29. What is the approximate final temperature of this system (including the reaction vessel)?

19-28 Maximum Temperature Find the maximum possible temperature that may be reached when 0.5 mol Ca(OH)$_2$ (*s*) is allowed to react with 1 liter of a 1 *M* HCl solution, both initially at 25°C. Assume that the final volume of the solution is 1 liter and that the C_P of the solution is constant and equal to that of water, 4.18 J K^{-1} g^{-1}.

The Second Law of Thermodynamics: Entropy and Free Energy

20

"Suppose we think of . . . a body of a higher and one of a lower temperature . . . ; through inter-action between the bodies this state is changed; according to the second law this change must always occur so that the total entropy of all bodies increases; according to our present interpreta-tion this means nothing else than that the probability of the entire state of these bodies is always increasing; the system of bodies always proceeds from a less probable to a more probable state."
L. BOLTZMANN, 1877

"The second law of thermodynamics has the same degree of truth as the statement that if you throw a tumblerful of water into the sea, you cannot get the same tumblerful of water out again."
J. C. MAXWELL, 1870

20-1 Introduction

It is a central fact of experience that there is a natural direction for all kinds of changes in the physical universe. At ordinary pressures, ice always melts at temperatures above $0°C$; water at such temperatures and pressures never freezes. Similarly, if a piece of ice is dropped into some boiling water, the net result is the formation of water at some intermediate temperature. The reverse never happens: a sample of water at, say, $40°C$ never spontaneously turns into a mixture of ice and boiling water. To give still another example, a gas at a given temperature expands spontaneously to fill all the volume accessible to it; it never shrinks in volume spontaneously at the same temperature. For any of the changes just described, the process can occur in either direction in accord with the first law of thermodynamics; yet in fact each happens in only one direction.

Everyone is aware that there is a natural direction for most processes. Suppose that a movie of an egg being broken is run backward. Everyone watching it immediately recognizes that what seems to be is impossible, that time has apparently been reversed by the trick of reversing the sequence in which the individual frames are projected. When we speak of a natural direction of reactions, we are implicitly assuming a direction for time. Processes that occur in this "natural direction" are referred to as *spontaneous*. The fact that a process is spontaneous implies nothing about its rate. Many spontaneous processes proceed at an imperceptibly slow rate for reasons that we consider in Chap. 22.

What factors determine the natural direction of a process? Consider a stone held at some distance above the floor. When the stone is released it falls. In the process of falling its potential energy is converted to kinetic energy and the kinetic energy is in turn dissipated when the stone collides with the floor. When it has come to rest on the floor it has reached a position of minimum gravitational energy. The decrease in gravitational energy governs the direction of this process. The inverse process, the rise of the stone from the floor, requires an investment of energy and never occurs spontaneously.

Many spontaneous chemical changes are also accompanied by a loss of energy. For example, water vapor at a temperature of 99°C and a pressure of 1 atm will spontaneously condense to a liquid with the release of the enthalpy of vaporization. On the other hand, many spontaneous processes occur with an absorption, rather than a release, of energy. Liquid water at 101°C and a pressure of 1 atm will vaporize spontaneously even though the process is endothermic. Clearly some factor in addition to energy must be at work in determining the natural direction of processes.

What is this factor? If there were *no* forces between molecules, a substance would never condense to a liquid. It would always fill its container uniformly. A given molecule would be just as likely to be in one part of the container as in any other—the molecules would be randomly distributed throughout the container. We would never observe a spontaneous process in which the molecules condense and occupy only a fraction of the container. Of course, there *are* attractive forces between all molecules, but there is a natural tendency of molecules to fill the available volume, and it is this tendency that must be considered, together with the forces between the molecules, in determining the natural direction of an evaporation or condensation process.

So long as the energy of intermolecular attraction is weak compared to the average molecular kinetic energy, as for example in helium at room temperature, the tendency for the random distribution of the molecules is the dominant factor. Helium cannot be condensed to a liquid at room temperature. The tendency for randomization exists even for substances that are liquid or solid at room temperature, but it does not dominate.

The principle that emerges is that the direction of change is a result of a balance between the tendency for the *potential energy* to be minimized (corresponding to condensation) and the tendency of the molecules to fill the available space uniformly. In this balance the second factor becomes increasingly important with increasing temperature because, as the *kinetic energy* of the molecules rises, it becomes sufficient to overcome the intermolecular attraction.[1]

Thus, in the example of water at 99°C and 1 atm the dominant factor is the loss in potential energy when the average distance between molecules decreases, and condensation occurs spontaneously. At 101°C and 1 atm, the tendency to fill the available volume, a tendency that increases with temperature, has become so large that it outweighs the energy factor. As a result the spontaneous change occurs in the opposite direction.

The direction of spontaneous change in any other chemical system can be similarly analyzed as the result of competition between the tendency of the potential energy to be minimized and a second factor that is weighted increasingly as the temperature rises. We have described this factor in vague terms but it is possible to arrive at a precise definition of it in terms of thermodynamics. The

---

[1]There are many more ways of distributing a given number of molecules over a large volume than over a small one and so, *if* these different distributions are equally favorable energetically, it is much more likely that the molecules will be spread over the large volume. However, in the presence of relatively strong intermolecular attractions, distributions in which the molecules are concentrated in a small volume *are* favored energetically. The molecules may then remain concentrated, i.e., in a liquid or solid phase.

Table 20-1
Examples of Reversible and
Irreversible Processes

Reversible process	Parallel irreversible process
Expansion of a gas when $P_{sys} \approx P_{surr}$. A slight increase in P_{surr} will cause the gas to contract.	Expansion of a gas into a vacuum. No slight change in pressure can stop the expansion.
Heating of a system immersed in a large bath, with T_{sys} almost equal to T_{bath}. A slight decrease in T_{bath} will cause heat to flow out of the system into the bath.	Heating of a system by immersing it in a bath at a temperature much higher than T_{sys}. The direction of heat flow, from the bath into the system, cannot be altered by small changes in T_{bath}.
Ice in water at 0°C. A slight heat flow into the ice-water mixture will melt some ice; a slight withdrawal of heat from the mixture will freeze some water.	Ice in water at 20°C. The ice will melt even if the water is cooled several degrees (to any temperature above 0°C).
A chemical reaction at equilibrium, e.g., $$H_2O + HOAc \rightleftharpoons H_3O^+ + OAc^-$$ A slight increase in the concentration of HOAc will produce more ions; a slight increase in the concentration of either ion will produce more HOAc.	A chemical reaction far from equilibrium, e.g., a piece of zinc dropped into acid, $$Zn + 2H^+ \longrightarrow Zn^{2+} + H_2$$ No small (or even large) increase in the concentration of Zn^{2+} or in the pressure of H_2 can reverse the direction of the reaction.

factor may be expressed as the product of the absolute temperature of the system and the change in a function of the system called its *entropy* (Sec. 20-2). It is also possible to find a function of any chemical system that plays for chemical processes the same role that potential energy does for ordinary mechanics of macroscopic objects. In other words, this function always decreases as equilibrium is approached, reaching a minimum at the position of equilibrium. It is termed the *free energy* (Sec. 20-4).

Reversibility and Irreversibility   The concept of reversibility is essential to an understanding of entropy and its applications. We considered in Sec. 19-2 the reversible expansion of an ideal gas and emphasized that a reversible change is one whose direction can be reversed at any point along its path by a very small change in the conditions. Table 20-1 gives some examples of various kinds of reversible processes and parallel irreversible processes.

20-2   Entropy

We begin this section with an *assertion* about the properties of the state function known as entropy, symbolized by S. A definition of entropy follows shortly.

For any process that can actually occur, the accompanying entropy change always satisfies this condition:

$$\Delta S_{tot} = (\Delta S_{sys} + \Delta S_{surr}) \geq 0 \qquad (20\text{-}1)$$

Here ΔS_{tot} is the entropy change of the entire "universe", composed of that portion identified as the system and the remainder characterized as surroundings. The statement (20-1) provides a criterion for finding the natural direction of any process. That direction must be such that the total entropy increases; the condition $\Delta S_{tot} = 0$ corresponds to equilibrium conditions or a change carried out reversibly, so that there is no natural direction. For an irreversible change, that is, a system moving toward the equilibrium condition (although not necessarily

419

reaching it), the *total* entropy always increases. This does not mean that the entropy of *either* the system *or* the surroundings cannot decrease during some process, but only that if either of these entropies does decrease, the other must increase at least as much so that the sum of the entropy changes is either zero or positive.

The Second Law of Thermodynamics Equation (20-1) is one of a number of ways of expressing the second law of thermodynamics. The implications of the second law and its applications are far from being as intuitively evident as those of the first law. However, intuitively reasonable consequences of the second law can be demonstrated by specific examples once a way of evaluating entropy changes has been defined. We turn now to this question.

Entropy Changes for Isothermal Processes Consider first a special case. For an isothermal process, the change in the state function S is given by

$$\Delta S = \frac{q_{rev}}{T} \tag{20-2}$$

that is, the change in entropy is equal to the heat transferred to the system along a reversible path divided by the absolute temperature T of the system. While there is an important restriction on the path, that of reversibility, the path is otherwise arbitrary provided that it leads from the given initial state to the desired final state. The restriction to reversibility applies only to (20-2), whereas ΔS itself is entirely independent of the path relating the states considered, because S is a state function. In effect, (20-2) provides a way to *calculate* ΔS for a specified change of state, and the result applies whether or not the change is *actually* effected along the reversible path used in the computation, along any other reversible path, or along any irreversible path. As stressed in Sec. 19-2, along different paths both q and w usually vary, although their sum [ΔU, by (19-1)] is independent of the path. It turns out that q_{rev} is always greater than q_{irrev}; thus $\Delta S > q_{irrev}/T$.

Application to Changes of State Let us apply (20-2) to a reversible change of state, such as the vaporization of a liquid at its boiling temperature T_b, that is, at the temperature at which the outside pressure on the liquid is equal to the vapor pressure of the liquid (Sec. 5-2). Pressure and temperature stay constant for this process. We know from Sec. 19-3 and in particular Equation (19-9) that as long as only P, V work is done, $q_{rev} = \Delta H_{vap}$ at T_b (for the pressure considered). Therefore

$$\Delta S_{vap} = \frac{\Delta H_{vap}}{T_b} \tag{20-3}$$

that is, the entropy of vaporization at the boiling temperature at the pressure considered is the enthalpy of vaporization divided by the (absolute) boiling temperature.

Example 20-1

☐ **Molar Entropy of Vaporization of Water** What is the entropy of vaporization per mole of water at 100°C and 1 atm? The molar enthalpy of vaporization under the conditions stated, for which there is equilibrium between liquid water and water vapor, is 40.6 kJ mol^{-1}.

$$\Delta\widetilde{S}_{vap} = \frac{4.06 \times 10^4 \, \text{J mol}^{-1}}{373 \, \text{K}} = \underline{109 \, \text{J mol}^{-1}\text{K}^{-1}} \quad \blacksquare$$

☐ What is the molar entropy of vaporization of Br_2 at its boiling point of 59°C at 1 atm? (See Table 19-1.) ■

Equation (20-3) is an example of a more general expression for a phase transformation,

$$\Delta S_{tr} = \frac{\Delta H_{tr}}{T_{tr}} \tag{20-4}$$

relating the entropy of transformation ΔS_{tr} to the enthalpy of transformation ΔH_{tr} and the transformation temperature T_{tr}, all quantities being associated with phase equilibrium at the temperature and pressure considered. Note that equilibrium implies reversibility because any phase change at equilibrium can be made to proceed in either direction by very slight changes in the conditions (for example, in temperature or pressure). An example of such a transformation might be from one solid phase to another, as from orthorhombic to monoclinic sulfur or from graphite to diamond. Other common phase transformations are fusion (melting) and vaporization.

☐ **The Entropy of Melting of Bromine** Bromine melts at -7°C, and the molar enthalpy of melting is 10.8 kJ mol^{-1}. What is the molar entropy of melting?

Example 20-2

Solution The melting temperature in kelvins is $(273 - 7) \, \text{K} = 266 \, \text{K}$, so that

$$\Delta\widetilde{S}_{melt} = \frac{\Delta\widetilde{H}_{melt}}{T_{melt}} = \frac{10.8 \times 10^3 \, \text{J mol}^{-1}}{266 \, \text{K}}$$
$$= \underline{40.6 \, \text{J mol}^{-1}\text{K}^{-1}} \quad \blacksquare$$

☐ Solid sulfur can exist in several crystalline forms with different molecular arrangements. Two common forms are "rhombic", stable at room temperature, and "monoclinic", stable at temperatures above 96°C. The two forms are at equilibrium with each other at 96°C and 1 atm. The molar enthalpy of monoclinic S is 0.30 kJ mol^{-1} higher than that of rhombic S. What is the molar entropy of the transformation from rhombic to monoclinic S at 96°C and 1 atm? ■

Isothermal Expansion of an Ideal Gas Consider the entropy change accompanying the isothermal expansion of an ideal gas, a process treated in Sec. 19-2. For a fixed amount of an ideal gas U is a function of T only, and so $\Delta U = 0$ for any isothermal change. Hence $w = -q$ for such changes. For a reversible expansion of n mol of gas from the volume V_1 to the volume V_2, it follows from (19-3) that

$$w_{rev} = -nRT \ln \frac{V_2}{V_1} = -q_{rev}$$

Combining the second portion of this equation with (20-2) we get

$$\Delta S = \frac{q_{rev}}{T} = nR \ln \frac{V_2}{V_1} \tag{20-5}$$

This equation holds for *any* isothermal expansion of an ideal gas, reversible or not, because S is a function of state. Note that this entropy change is positive, since V_2/V_1 is greater than 1 for an expansion and thus $\ln (V_2/V_1)$ is positive.

Example 20-3

☐ **Change in State for an Ideal Gas** The pressure on 1.00 mol helium is reduced from 1.00 to 0.100 atm at 50°C. Assume the gas to be ideal. What are ΔU, ΔH, and ΔS?

Solution For an ideal gas U and H depend only on the temperature, and therefore $\Delta U = \Delta H = \underline{0.}$ When n and T are constant, $V_2/V_1 = P_1/P_2$ for an ideal gas so (20-5) can be written

$$\Delta S = nR \ln \frac{P_1}{P_2} = 1 \text{ mol} \times 8.31 \text{ J K}^{-1} \text{ mol}^{-1} \times \ln \frac{1.00}{0.100}$$

$$= 8.31 \times 2.30 \text{ J K}^{-1} = \underline{19.1 \text{ J K}^{-1}} \blacksquare$$

Exercise 20-3

☐ The pressure on 0.50 mol of neon is increased from 1.00 to 5.00 atm at 200°C. Assume the gas to be ideal. What are ΔU, ΔH, and ΔS? ■

Entropy Change on Heating We have considered so far entropy changes for isothermal processes only. To evaluate the change in entropy when the temperature changes, we imagine a series of reversible processes at temperatures differing so little from one another that each process may be assumed to occur essentially at a constant temperature. The overall change in state is then the sum of the changes in these individual steps. We apply (20-2) to each step and sum the results to get the total change. Consider a process in which a system is heated from a temperature T_1 to T_f in the absence of a phase transition or chemical reaction. Assume that step 1 occurs at T_1 (the change in temperature, δT_1, during the process being very small relative to T_1), and let the entropy change in step 1 be δS_1 and the corresponding heat absorbed be $\delta q_{rev,1}$. The heat capacity appropriate to the path (which might be a constant-volume path, a constant-pressure path, or some other) is C. Then, for step 1,

$$\delta S_1 = \frac{\delta q_{rev,1}}{T_1} = \frac{C \, \delta T_1}{T_1} \tag{20-6}$$

Parallel equations can be written for each succeeding step, of which we imagine a large number, until the final temperature, T_f, has been reached. Then for the overall change

$$\Delta S = \text{sum over all steps of } \delta S_i = \text{sum of } \frac{C \, \delta T_i}{T_i} \tag{20-7}$$

The sums in (20-7) can be evaluated by the methods of integral calculus (Appendix C); they are very similar in form to those involved in calculating the work done in a reversible expansion [Sec. 19-2, Equation (19-3)]. The result is

$$\Delta S = C \ln \frac{T_f}{T_1} = n\tilde{C} \ln \frac{T_f}{T_1} \tag{20-8}$$

provided it can be assumed that the heat capacity is independent of the temperature, usually a good approximation if the temperature change is not too large. Note that (20-8) gives the change in entropy whether or not the heating is actually *done* reversibly, because entropy is a state function.

If a phase change does occur as energy is being added to or taken from the system, then the overall difference in entropy is given by the sum of the entropy change accompanying the phase change [Equation (20-4)] and that accompanying the change in temperature, given by (20-8).

Example 20-4

□ **Entropy Change on Heating** Calculate the entropy change accompanying the conversion of 0.50 mol ice at 0°C and 1 atm into 0.50 mol liquid H_2O at 25°C and 1 atm. The molar enthalpy of fusion of ice at 0°C is 6.02 kJ mol^{-1} and $\widetilde{C}_P$ for water is 75 J mol^{-1} K^{-1}.

Solution In order to calculate the entropy change we must devise a reversible path connecting the initial and final states, regardless of what the path may be in the process actually occurring. Here we consider two successive reversible steps:

1. Melting of 0.50 mol ice at 0°C and 1 atm
2. Reversible heating of the resulting 0.50 mol water from 0 to 25°C at a constant pressure of 1 atm

Step 1 is reversible, because ice and water are in equilibrium at 0°C and 1 atm. Thus

$$\Delta S_1 = \frac{q_{rev}}{T} = \frac{n\Delta\widetilde{H}_{fus}}{T} = \frac{(0.50 \text{ mol})(6020 \text{ J mol}^{-1})}{273 \text{ K}}$$
$$= 11.0 \text{ J K}^{-1}$$

For *step 2*, we use (20-8), with $C = n\widetilde{C}_{P,H_2O}$. Thus

$$\Delta S_2 = C \ln \frac{298}{273} = (0.50 \text{ mol})(75 \text{ J mol}^{-1})(0.088)$$
$$= 3.3 \text{ J K}^{-1}$$

Thus, for the entire change,

$$\Delta S = \Delta S_1 + \Delta S_2 = \underline{14.3 \text{ J K}^{-1}} \ \blacksquare$$

Exercise 20-4

□ Calculate the entropy change when 2.0 mol of an ideal gas is cooled at constant volume from 200 to 100°C. Assume that $\widetilde{C}_V$ for the gas is 28 J K^{-1} mol^{-1}. ■

Additional Comments on Entropy By (20-2) and the definition of an adiabatic process ($q = 0$), $\Delta S = 0$ for a reversible isothermal adiabatic process. This is true also for a reversible process in an isolated system, since necessarily $q = 0$ for any change in an isolated system. If the isothermal change is not reversible, the entropy of an adiabatic or isolated system necessarily increases. These same generalizations apply also to processes in which the *temperature changes*. Thus *the entropy of an adiabatic system or an isolated system can never decrease.* It remains constant for reversible processes and increases for irreversible ones, whether or not the temperature is constant.

20-3   The Statistical Interpretation of Entropy

Thermodynamics permits (although it is independent of) interpretations on a molecular basis. These microscopic interpretations are often valuable in providing an intuitive feeling for entropy and its properties.

If all the details that *might* be known about the molecules in a given system *were* known, the macroscopic thermodynamic properties of the system could be calculated by taking appropriate averages of the molecular properties. There is usually an enormous number of microscopic states that have the same average properties. Such states are therefore indistinguishable macroscopically, that is, they each represent the same macroscopic state, with some particular energy, temperature, pressure, density, and other physical characteristics. We shall call the number of microscopic states that correspond to a specific macroscopic state the number of *realizations* of that state, and symbolize this number by W.

Suppose that an isolated system is in a certain macroscopic state 1 that has W_1 realizations and that it is possible for the system to go to another macroscopic state 2 with W_2 realizations. If W_2 is larger than W_1 the system will tend to change from state 1 to state 2 as a result of the randomizing effects of molecular collisions ("temperature motion"). The state of an isolated system always tends to move in the direction of increasing numbers of realizations.

As an illustration, consider a box containing many small red and blue spheres of equal size and mass. At the beginning, all the red spheres are placed in layers at the bottom of the box and all the blue spheres are put on top. The box, which is not full, is then shaken for some time. At the end, the distribution of the colored spheres is random. The many different realizations of the *initial* state are obtained by interchanging all the red spheres among themselves and, separately, all the blue spheres among themselves. The realizations that correspond to the *final* state are obtained by interchanging *all* spheres with each other, regardless of their color. The final state has many more realizations than the initial state—that is, there are many more ways in which the individual spheres can be arranged if one disregards their color than if one must keep all the red spheres together and all the blue spheres together.

There are parallels between this example and the dissolving of a substance. Initially, one has pure solvent and pure solute, but temperature motion eventually causes the solute molecules to be evenly distributed throughout the solvent, just as continued shaking of the box results eventually in a random distribution of the colored spheres. It is extremely unlikely that the solution will ever become unmixed into *pure* solvent and pure solute, just as it is unlikely that shaking the box will ever substantially decrease the randomness of the distribution of the colored spheres.

Both entropy and W increase during spontaneous processes in isolated systems and assume their maximum values when equilibrium has been reached. In fact, Boltzmann showed that for an isolated system the entropy must be proportional to the logarithm of W:

$$S = k \ln W \tag{20-9}$$

The proportionality factor k is the same Boltzmann constant that was introduced in Chap. 4, the gas constant R divided by Avogadro's number N_A:

$$k = R/N_A = 1.381 \times 10^{-23} \text{ J K}^{-1} = 1.381 \times 10^{-16} \text{ erg K}^{-1}$$

Note that (20-9) refers not to an entropy difference but to the entropy itself. The *change* in entropy on going from a state with W_1 realizations to a state with W_2 realizations is given by

$$\Delta S = k \ln \frac{W_2}{W_1} \tag{20-10}$$

This equation implies that if the change leads to an increase in the number of realizations, as in the example of the change from an ordered to a random arrangement of colored spheres, then W_2/W_1 is larger than 1 and ΔS is positive. If $W_2/W_1 < 1$, there has been a decrease in the number of realizations and ΔS is negative. No entropy change is associated with a transition between states with the same number of realizations, i.e., when $W_1 = W_2$.

An interesting application of (20-10) is to crystalline solids at 0 K. For a system that consists of a perfectly ordered crystal at 0 K, there is just one realization, because each molecule in the crystal has a specified location and orientation. Thus for such a system, $S = k \ln 1 = 0$.

Under the influence of intermolecular interactions, most crystalline substances do in fact become highly ordered at low temperatures. In some substances, however, random molecular orientations can persist. An example is N_2O. In an ordered crystal of N_2O the orientations of the molecules would be regular, as indicated in this highly schematic fashion:

NNO	NNO	NNO	NNO	NNO	NNO	NNO	NNO
NNO	NNO	NNO	NNO	NNO	NNO	NNO	NNO
NNO	NNO	NNO	NNO	NNO	NNO	NNO	NNO
NNO	NNO	NNO	NNO	NNO	NNO	NNO	NNO

The ends of an N_2O molecule are very much alike, however, and during crystallization the molecules may be trapped end-for-end randomly:

NNO	ONN	ONN	NNO	NNO	NNO	ONN	NNO
ONN	NNO	NNO	ONN	NNO	ONN	NNO	ONN
ONN	ONN	NNO	ONN	NNO	ONN	ONN	NNO
NNO	ONN	NNO	ONN	ONN	NNO	ONN	NNO

Two orientations are possible for each molecule. Thus for a crystal that contains N_A molecules the number of realizations for a completely random arrangement is

$$W_r = \underbrace{2 \cdot 2 \cdot 2 \cdot 2 \cdot 2 \ldots 2 \cdot 2}_{N_A \text{ times}} = 2^{N_A}$$

The entropy change associated with the transformation of a mole of crystalline N_2O from a random to a completely ordered state is therefore

$$\Delta S = k \ln \frac{1}{W_r} = k \ln \frac{1}{2^{N_A}} = -N_A k \ln 2$$

$$= -R \ln 2 = -5.8 \text{ J K}^{-1} \text{ mol}^{-1}$$

Although this ΔS cannot be measured directly it can be obtained from a combination of heat capacity and spectroscopic data. The entropy change derived from measurements agrees within experimental error with the calculated value.

Table 20-2
Molar Entropies of Fusion
and Vaporization*

	$\Delta \widetilde{S}_{\text{fus}}$/J mol^{-1} K^{-1}	$\Delta \widetilde{S}_{\text{vap}}$/J mol^{-1} K^{-1}
N_2	11.4	72.0
O_2	8.2	75.7
Cl_2	37.2	85.3
CS_2	27.3	83.8
CH_4	10.3	74.5
CH_3Cl	43.9	88.8
CCl_4	9.7	85.8
C_6H_6	35.3	87.1
NH_3	28.9	97.5
Hg	10.0	92.0

*At mp and bp, respectively.

Disorder and Phase Transitions When a solid is melted or sublimed and when a liquid is vaporized, there is an increase in entropy because the state of disorder increases.

Table 20-2 contains a number of entropies of fusion and of vaporization. It will be noted that the entropy of fusion is generally smaller than the entropy of vaporization. This indicates that the increase in randomness is smaller when going from the solid to the liquid than when going from the liquid to the vapor. Note also the orders of magnitudes of these molar entropy changes, from about 8 to some 100 J mol^{-1} K^{-1}. It is *joules* per kelvin that are involved, whereas the corresponding molar enthalpy changes are best expressed in *kilo*joules.

Inspection of Table 20-2 shows that the entropies of vaporization for many liquids are approximately the same, although there is no similar regularity shown by the entropies of fusion. The approximate constancy of $\Delta \widetilde{S}_{\text{vap}}$ is called *Trouton's rule,* according to which

$$\frac{\Delta \widetilde{H}_{\text{vap}}}{T_b} = \Delta \widetilde{S}_{\text{vap}} \approx 88 \text{ J mol}^{-1} \text{ K}^{-1} \tag{20-11}$$

where T_b is the boiling temperature at 1 atm and the approximate value for $\Delta \widetilde{S}_{\text{vap}}$ is an average over many more substances than are listed in the table. The rule holds fairly well for many liquids, but there are exceptions.

20-4   Equilibrium and Spontaneous Processes

The Free Energy We began this chapter with a qualitative discussion of the two factors that determine the direction of spontaneous change—energy and entropy. The American physical chemist J. W. Gibbs formalized these concepts in terms of a function of the system called the free energy, symbolized as G and defined as

$$G = H - TS \tag{20-12}$$

It is a function of state, because H, T, and S are, and it has the dimensions of an energy, as do H and TS. The usefulness of G can be established by considering its behavior for a finite change in state at constant pressure and temperature. For such a change

$$\Delta G = \Delta H - T\Delta S \tag{20-13}$$

because at constant temperature

$$\Delta(TS) = T_2S_2 - T_1S_1 = T(S_2 - S_1) = T\Delta S$$

Conditions for Equilibrium and Spontaneous Processes Three cases can be distinguished for chemical processes that occur *at constant temperature and pressure with only* P,V *work:*

1. A chemical change, or any other process, is at equilibrium when

$$\Delta G = 0 \qquad (20\text{-}14a)$$

as the process is made to proceed in either direction.

2. An actual process must occur in the direction for which

$$\Delta G < 0 \qquad (20\text{-}14b)$$

3. A process for which

$$\Delta G > 0 \qquad (20\text{-}14c)$$

is thermodynamically forbidden.

For chemical processes at constant T and P and with only P,V work, conditions (20-14a) and (20-14b) imply that equilibrium is characterized by a minimum of G. The situation is analogous to that of an object at rest in a gravitational field. The object is in (stable) equilibrium when its potential energy is a minimum.

The Importance of ΔH and ΔS The condition (20-14b) for the spontaneous direction of a reaction at constant temperature and pressure with only P,V work can be written in the form

$$\Delta G = \Delta H - T\,\Delta S < 0 \qquad (20\text{-}15)$$

A spontaneous process is therefore favored not only by a decrease of H (negative ΔH) but also by an increase of S (positive ΔS). The influence of the entropy change becomes more important as the absolute temperature increases; the temperature is a weighting factor for the entropy change relative to the enthalpy change.

Four basically different situations can be envisioned, differing in the signs of ΔH and ΔS for the process under consideration:

1. As long as $\Delta H < 0$ and $\Delta S > 0$ at all T, then ΔG is necessarily negative *at any temperature*. (If the sign of ΔH or ΔS changes as T changes, then ΔG will not be negative at all temperatures and the situation falls under 2 or 3 below.)

2. When $\Delta H < 0$ and $\Delta S < 0$, the process may occur *only* when $|T\,\Delta S| < |\Delta H|$. The process must thus be sufficiently exothermic to overcome the handicap of the entropy decrease. Processes that fall in this category include condensation of a gas or freezing of a liquid. At sufficiently large T, the term $T\,\Delta S$ will overcome the ΔH term and the process can no longer occur. If ΔH and ΔS are known at some temperature, then the temperature T' above which the process becomes prohibited by thermodynamics is, by (20-15), given approximately by $T' = \Delta H/\Delta S$. This relationship is only approximate, however, because both ΔH and ΔS depend to some extent on the temperature.

3. When $\Delta H > 0$ and $\Delta S > 0$, the process tends to occur *only if* $T\,\Delta S > \Delta H$. Here the advantage of the entropy increase must overcome the handicap of the endothermic nature of the process, and for this T must be sufficiently large. Examples include melting and evaporation. The heat required by the process comes, of course, from the surroundings—that is, from the con-

Table 20-3
Influence of Enthalpy and
Entropy Changes on
Chemical Reactions at
Constant P and T with
only P,V Work

Case	ΔH	ΔS	Spontaneous reaction
1	<0	>0	At any T
2	<0	<0	When T sufficiently low
3	>0	>0	When T sufficiently high
4	>0	<0	Not at all

stant-temperature bath needed to maintain the temperature at a constant level. When T is too small the reaction is prohibited.

4. When $\Delta H > 0$ and $\Delta S < 0$, the process is *prohibited thermodynamically at any temperature,* unless a sign change of ΔH or ΔS with temperature occurs in such a way that the sign of ΔG changes also.

Table 20-3 summarizes these four situations. For a reaction taking place at constant T and P with P,V work only, the *condition for equilibrium* is

$$\Delta G = \Delta H - T\,\Delta S = 0$$

or
$$\Delta H = T\,\Delta S \qquad (20\text{-}16)$$

The two terms ΔH and $T\,\Delta S$ are thus exactly equal *at equilibrium.* If they are both positive, the tendency toward greater randomness, measured by ΔS and weighted by multiplication by T, is just sufficient to compensate for the endothermic nature of the reaction. If they are both negative, the exothermic character of the reaction just compensates for the decrease in randomness, again weighted by T. Equations (20-3) and (20-4) are just special cases of (20-16), applying to phase transformations occurring under equilibrium conditions.

At equilibrium there is no change in free energy G when one phase is transformed into another or when reactants are transformed into products, provided that equilibrium concentrations and pressures of all reaction participants are maintained.

Example 20-5

□ **Free-Energy Change during a Phase Transformation** Consider the phase transformation

$$H_2O(l) \longrightarrow H_2O(g)$$

at 25°C and 1 atm. Calculate ΔG from the values of ΔH and ΔS ($\Delta H = 44.0$ kJ, $\Delta S = 118.7$ J K^{-1}) and discuss the result. [The pressure of 1 atm is that applied to the $H_2O(l)$ and $H_2O(g)$.]

Solution We use (20-13)

$$\Delta G = \Delta H - T\Delta S$$

$$= 44.0\text{ kJ} - 298\text{ K} \times 118.7\text{ J K}^{-1} \times \frac{1\text{ kJ}}{1000\text{ J}}$$

$$= (44.0 - 35.4)\text{ kJ} = \underline{8.6\text{ kJ}}$$

At 25°C, $T\,\Delta S$ is smaller than ΔH, and ΔG is positive. For the reverse process, condensation, ΔG is therefore negative, so that water vapor at 1 atm and 25°C is unstable and condenses. (Water vapor *is* stable at 25°C at a low enough pressure; the vapor pressure at 25°C is 0.033 atm.) ■

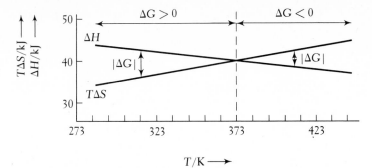

Figure 20-1 **ΔH and $T\,\Delta S$ for the Phase Transformation $H_2O(l) \rightarrow H_2O$ (g, 1 atm)**
At the point where the two curves cross (373 K) there is equilibrium, since $\Delta H = T\,\Delta S$ implies
that $\Delta G = 0$. To the left of this intersection the difference between the two curves is $+\Delta G$ because ΔH is larger than $T\,\Delta S$; on the right the difference is $-\Delta G$ because ΔH is now smaller
than $T\,\Delta S$. Thus ΔG for the reaction $H_2O(l) \rightarrow H_2O(g, 1\text{ atm})$ is positive below 373 K and negative above. Below 373 K, liquid water is more stable than water vapor at 1 atm; above 373 K,
vapor at 1 atm is the more stable; and at 373 K there is equilibrium.

□ Consider the phase transformation

Exercise 20-5

$$I_2(s) \longrightarrow I_2(g)$$

at 25°C and 1 atm. Calculate ΔG from the values of ΔH (62.3 kJ) and ΔS
(114 J K^{-1}) and discuss the result. ■

A more general question than that discussed in the example is this: At what
temperature would equilibrium between vapor at 1 atm and liquid water be
achieved? At equilibrium, ΔH and $T\,\Delta S$ must be equal, and since ΔH is larger
than $T\,\Delta S$ at 298 K, the equilibrium temperature must be higher. The equilibrium temperature T_{eq} may be estimated to be

$$T_{eq} \approx \frac{\Delta H(298)}{\Delta S(298)} = 371 \text{ K} \tag{20-17}$$

on the assumption that ΔH and ΔS do not vary with temperature. The actual
equilibrium temperature is, of course, 100°C or 373 K. Figure 20-1 shows the
actual behavior of ΔH and $T\,\Delta S$ for the vaporization of water at 1 atm, as a
function of temperature. It is seen that ΔH slowly decreases with temperature.
For the approximate relationship (20-17), a constant ΔH was assumed. The
reasonable agreement between the estimated and actual equilibrium temperatures (371 and 373 K) is due in part to a cancellation of errors: ΔS slowly
decreases with temperature also, and the curve $T\,\Delta S$ rises less steeply than it
would if ΔS were constant.

Combination of Free Energies Because the free energy is a function of state,
values of ΔG for different chemical reactions may be combined in the same way
that the chemical equations for the reactions considered may be combined into
the equations for new reactions. All this is analogous to the way in which ΔH
values can be combined (Sec. 19-3).

□ **Combination of ΔG Values** The overall equations of two chemical
reactions, (1) and (2), are given below. Under each is the value of ΔG at

Example 20-6

298 K for transforming as many moles of reactants into products as are indicated by the coefficients in the equations, with ions at concentrations of 1 mol liter^{-1} and gases at partial pressures of 1 atm. What is ΔG for reaction (3), which may be obtained by suitably combining reactions (1) and (2)?

$$16H^+ + 2MnO_4^- + 10Cl^- \longrightarrow 5Cl_2(g) + 2Mn^{2+} + 8H_2O(l) \qquad (1)$$
$$\Delta G_1 = -142.0 \text{ kJ}$$

$$Cl_2(g) + 2Fe^{2+} \longrightarrow 2Fe^{3+} + 2Cl^- \qquad (2)$$
$$\Delta G_2 = -113.6 \text{ kJ}$$

$$8H^+ + MnO_4^- + 5Fe^{2+} \longrightarrow 5Fe^{3+} + Mn^{2+} + 4H_2O(l) \qquad (3)$$
$$\Delta G_3 = ?$$

Solution The third equation can be obtained by adding one-half of the first equation and five-halves of the second equation. Therefore,

$$\Delta G_3 = \tfrac{1}{2}\Delta G_1 + \tfrac{5}{2}\Delta G_2 = [\tfrac{1}{2}(-142.0) + \tfrac{5}{2}(-113.6)] \text{ kJ}$$
$$= \underline{-355.0 \text{ kJ}} \quad \blacksquare$$

Exercise 20-6

☐ Consider the following two reactions and the corresponding values of ΔG:

$$Cr_2O_7{}^{2-} + 6Fe^{2+} + 14H^+ \longrightarrow 2Cr^{3+} + 6Fe^{3+} + 7H_2O(l)$$
$$\Delta G = -325.4 \text{ kJ}$$

$$4Fe^{2+} + O_2(g) + 4H^+ \longrightarrow 4Fe^{3+} + 2H_2O(l)$$
$$\Delta G = -176.8 \text{ kJ}$$

What is ΔG for the reaction

$$Cr_2O_7{}^{2-} + 8H^+ \longrightarrow 2Cr^{3+} + 4H_2O(l) + \tfrac{3}{2}O_2(g) \quad \blacksquare$$

Standard Free Energies Since only differences in free energy are defined by thermodynamics, it is possible, just as in the case of enthalpies, to agree on certain reference states that are assigned zero free energy of formation *by convention*. These are exactly the same reference states that were assigned zero standard enthalpies of formation—the elements at standard conditions in their most stable forms (see Sec. 19-3). The *standard molar free energies of formation* $\widetilde{G}_f^\circ$ of compounds, and of elements in other than their stable modifications, are then their free energies of formation at standard conditions from the elements in their most stable forms. For substances in aqueous solution, standard conditions imply unit concentration (or more precisely, activity) and the standard free energies of formation of ionic species are defined by the convention that the standard free energy of formation of H^+ is zero. Table D-4 contains a representative list of such standard free energies of formation or, more simply, *standard free energies,* as they will usually be called. In many texts these quantities are given the symbols $\Delta\widetilde{G}_f^\circ$, $\Delta\widetilde{G}_f^\circ$, or $\Delta\widetilde{G}f^\circ$. Unless otherwise indicated, standard free energies refer to 25°C or 298 K. Occasionally another temperature T is more useful, and molar standard free energies at T are indicated by the symbol $\widetilde{G}_f^\circ(T)$. Thus $\widetilde{G}_f^\circ(500)$ is the free energy of formation of the substance in question from the elements at standard conditions and 500 K. Again the convention is to take as reference states the elements in their most stable forms at 1 atm and the temperature T, to which are thus assigned zero standard free energies of formation.

The standard free-energy change for any reaction can be found from the standard free energies of all substances involved by the relationship

$$\Delta G^{\ominus} = \Sigma G^{\ominus}_{f,\,\text{prod}} - \Sigma G^{\ominus}_{f,\,\text{react}} \qquad (20\text{-}18)$$

This quantity is meaningful only if the explicit overall balanced equation of the reaction it applies to is given. It refers to the complete conversion into products of as many moles of reactants as are indicated by the coefficients in the chemical equation, all at standard conditions. A temperature of 298 K is implied here also unless another temperature is specifically mentioned.

□ **$\Delta G^{\ominus}$ for Burning NH_3 to NO** What is the standard free-energy change for burning ammonia in oxygen to NO?   **Example 20-7**

$$4NH_3(g) + 5\,O_2(g) \longrightarrow 4NO(g) + 6H_2O(l)$$
$$-16.7 0 86.7 -237.2$$

This is an important industrial reaction, a step in the manufacture of nitric acid from synthetic ammonia. The standard molar free energies of formation of the substances involved (Table D-4) are shown under the chemical symbols.

Solution

$$\Delta G^{\ominus} = [4 \times 86.7 + 6(-237.2) - 4(-16.7)]\,\text{kJ} = \underline{-1010\,\text{kJ}}$$

This value refers, of course, to 298 K or 25°C and to the numbers of moles specified in the reaction equation given. ■

□ What is the standard free-energy change for the reaction   **Exercise 20-7**

$$H_2O(g) + C(\text{graphite}) \longrightarrow CO(g) + H_2(g) \; ■$$
$$-228.6 0 -137.3 0$$

20-5 Chemical Equilibria

The Free Energy for Mixtures of Substances Many systems of chemical interest involve mixtures. The free energy of a mixture is not simply the sum of the free energies of the pure components. This can be demonstrated for the simple case of a mixture of ideal gases. Consider 1 mol of an ideal gas, initially in its standard state: the pressure $P^{\ominus} = 1$ atm at the temperature T. The gas is now mixed at constant temperature with other ideal gases and its partial pressure in the mixture is P_1. Since $\Delta H = 0$ for any isothermal mixing of ideal gases, any change in the free energy of the gas must arise from ΔS, which is given by (20-5):

$$\Delta \widetilde{S}_1 = R \ln \frac{V_1}{V^{\ominus}} = R \ln \frac{P^{\ominus}}{P_1}$$

and thus

$$\Delta \widetilde{G}_1 = \widetilde{G}_1 - \widetilde{G}_1^{\ominus} = -T\,\Delta \widetilde{S}_1 = TR \ln \frac{P_1}{P^{\ominus}} \qquad (20\text{-}19)$$

In an ideal-gas mixture $\Delta \widetilde{S}_1$ is independent of the presence of other components.

The free energy of n_1 mol of the gas in the mixture is

$$G_1 = n_1 \widetilde{G}_1 = n_1 \left[\widetilde{G}_1^\ominus + RT \ln \frac{P_1}{P^\ominus} \right] \qquad (20\text{-}20)$$

Analogous expressions can be written for each of the other components in the mixture. If the pressure in the standard state is chosen as 1 atm, it is customary to write (20-20) in the form (for component gas i)

$$G_i = n_i [\widetilde{G}_i^\ominus + RT \ln P_i] \qquad (20\text{-}21)$$

where it is understood that the partial pressure P_i must be expressed in units of atmospheres.

The General Thermodynamic Equilibrium Expression An important application of the free energy of mixtures is to chemical equilibrium. As an illustration, consider the reaction between hydrogen and nitrogen to produce ammonia:

$$3H_2(g) + N_2(g) \rightleftharpoons 2NH_3(g) \qquad (20\text{-}22)$$

The free-energy change accompanying the transformation from reactants to products in (20-22) is given by

$$\Delta G = \Sigma G_{\text{prod}} - \Sigma G_{\text{react}} \qquad (20\text{-}23)$$

When appropriate expressions of the type (20-21) for the free energies of the three gases are inserted into (20-23) we get

$$\Delta G = 2\widetilde{G}_{\text{NH}_3}^\ominus - 3\widetilde{G}_{\text{H}_2}^\ominus - \widetilde{G}_{\text{N}_2}^\ominus + RT\,(2 \ln P_{\text{NH}_3} - 3 \ln P_{\text{H}_2} - \ln P_{\text{N}_2}) \qquad (20\text{-}24)$$

The combination of $\widetilde{G}^\ominus$ terms in (20-24) is just the standard free-energy change of the reaction, $\Delta G^\ominus$:

$$\Delta G^\ominus = 2\widetilde{G}_{\text{NH}_3}^\ominus - 3\widetilde{G}_{\text{H}_2}^\ominus - \widetilde{G}_{\text{N}_2}^\ominus \qquad (20\text{-}25)$$

The term in (20-24) that includes the factor RT can be rearranged as follows:

$$RT\,(2 \ln P_{\text{NH}_3} - 3 \ln P_{\text{H}_2} - \ln P_{\text{N}_2}) = RT\,(\ln P_{\text{NH}_3}^2 + \ln P_{\text{H}_2}^{-3} + \ln P_{\text{N}_2}^{-1})$$

$$= RT \ln \frac{P_{\text{NH}_3}^2}{P_{\text{H}_2}^3 P_{\text{N}_2}} = RT \ln Q \qquad (20\text{-}26)$$

where Q is the reaction quotient for the chemical equation (20-22). Substituting from (20-25) and (20-26) into (20-24), we obtain

$$\Delta G = \Delta G^\ominus + RT \ln Q \qquad (20\text{-}27)$$

for the free-energy change when (20-22) is carried out at the particular partial pressures specified in the quotient Q. This free-energy change applies for the number of moles of reactants being changed into products that is indicated by the overall chemical equation. The pressures of both reactants and products are assumed to remain constant during the reaction.

   If the ΔG indicated by (20-27) is negative, there is a tendency for the reactants of the corresponding equation to change into products under the conditions specified; if ΔG is positive, the reverse is true. In other words, the reaction tends to proceed in such a direction that the overall free energy decreases.

When the partial pressures in (20-27) are equilibrium pressures, $\Delta G = 0$ so that

$$RT \ln Q_{eq} = -\Delta G^{\ominus} \tag{20-28}$$

Since $\Delta G^{\ominus}$ is a constant at constant temperature, the left side in (20-28) must also be constant under these circumstances. Hence at equilibrium

$$Q_{eq} = K \tag{20-29}$$

and
$$-RT \ln K = \Delta G^{\ominus} \tag{20-30}$$

Equation (20-29) is a shorthand statement of the equilibrium law discussed in Chap. 11. Equation (20-30) gives the relationship between the equilibrium constant K at a given temperature and the *standard* free-energy change for the reaction at that temperature. Although these equations were derived in terms of a specific example, they are of completely general validity for reactions among ideal gases. If the values of $\widetilde{G}_f^{\ominus}$ in Appendix D are used to evaluate $\Delta G^{\ominus}$, the result applies to 298 K. Since the standard states used for molar free energies of formation are at 1 atm, all partial pressures in Q must be expressed in atmospheres.

□ **Equilibrium Constant for the Formation of Ammonia** What are the values for $\Delta G^{\ominus}$ and K at 298 K for the formation of ammonia by reaction (20-22)?

Example 20-8

Solution From Table D-4

$$\Delta G^{\ominus} = 2\widetilde{G}_{f,\mathrm{NH_3}}^{\ominus} - 3\widetilde{G}_{f,\mathrm{H_2}}^{\ominus} - \widetilde{G}_{f,\mathrm{N_2}}^{\ominus}$$
$$= [2(-16.7) - 3 \times 0 - 0]\,\mathrm{kJ} = \underline{-33.4\,\mathrm{kJ}}$$

From (20-30),

$$\ln K = -\frac{\Delta G^{\ominus}}{RT} = \frac{33.4\,\mathrm{kJ} \times \dfrac{1000\,\mathrm{J}}{\mathrm{kJ}}}{8.31\,\mathrm{J\,K^{-1}} \times 298\,\mathrm{K}} = 13.49$$
$$K = e^{13.49} = \underline{7.2 \times 10^5\,\mathrm{atm^{-2}}}$$

The units for K are chosen to agree with its definition

$$K = \frac{P_{\mathrm{NH_3}}^2}{P_{\mathrm{H_2}}^3 P_{\mathrm{N_2}}} \quad ■$$

□ What are the values of $\Delta G^{\ominus}$ and K at 298 K for the reaction

Exercise 20-8

$$2\mathrm{H_2}(g) + \mathrm{O_2}(g) \longrightarrow 2\mathrm{H_2O}(l) \quad ■$$

It is important to recognize the essential distinction between $\Delta G^{\ominus}$ and ΔG. The former quantity refers to a change in state in which the reactants in their standard states are transformed into products in their standard states. In general this will not be a situation in which these participants in the reaction are in equilibrium with one another, except in the unlikely event that the equilibrium constant happens to be equal to 1. It may seem paradoxical that this standard free-energy change, which usually refers to a nonequilibrium situation, is what determines the magnitude of the equilibrium constant, by (20-30). However, as our derivation shows, this is true just because in an equilibrium situation ΔG

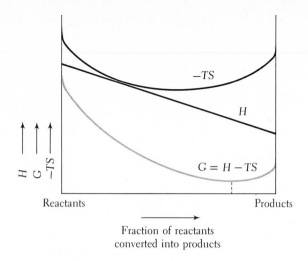

Figure 20-2 The Free Energy of a Typical Reaction Mixture
The graph shows H, $-TS$, and G for situations in which different
fractions of reactants have been converted into products, from reactants only (left edge) to products only (right edge). As the reaction
proceeds from reactants to products, the concentrations or partial
pressures of all participants change in such a way that the free energy of the system is lowered. Equilibrium exists when G is at its
minimum value. If the reaction mixture initially contained only
products, the reaction would proceed toward the left until the same
minimum value of G had been reached. Note the importance of the
term $-TS$ for the existence of a minimum in G.

itself, which applies to the change in state at the concentrations or pressures
actually existing, is necessarily zero.

The ΔG in (20-27) is the change in free energy for a reaction under the
restrictive condition that the partial pressures of products and reactants must be
kept constant. It is instructive to consider the total free energy of the reaction
mixture for changing compositions and partial pressures that correspond to
different degrees of conversion of reactants into products, from no reaction (no
products formed) to complete conversion (no reactants remaining). A typical case
is shown in Fig. 20-2. As the pressures or concentrations of the reactants decrease,
so does the free energy associated with them, while the free energy associated
with the products begins to rise as their pressures or concentrations increase. The
total free energy decreases at first, reaches a minimum value, and then increases
again. The minimum corresponds to the equilibrium composition of the reaction
mixture.

Equilibrium for General Solution Reactions For mixtures of substances other
than ideal gases, the dependence of the free energy on concentrations and partial
pressures cannot be derived from thermodynamics alone. It requires experimental data or sometimes results from the statistical theory of the behavior of large
numbers of molecules. Although no generally applicable relation can be derived,
the composition dependence of the free energy can usually be expressed by
equations analogous in form to (20-21).

For example, the free energy of the solute species in dilute solutions is, to a
good approximation,

$$\widetilde{G}_{i,s} = \widetilde{G}_i^{\ominus} + RT \ln c_i \qquad (20\text{-}31)$$

where c_i is the concentration in moles per liter of the species i—also denoted $[i]$
in reaction quotients—and $\widetilde{G}_i^{\ominus}$ is the standard molar free energy of the species i
in solution. The subscript s is used to differentiate the free energy of the species i
in solution from that in a gas mixture as shown in (20-21); $\widetilde{G}_{i,s}$ depends on the
solvent and on T, as well as on the total pressure P.

Example 20-9

☐ **Equilibrium Constant for a Reaction in Solution** Ferrous ion can be
oxidized by chlorine in aqueous solution:

$$2Fe^{2+} + Cl_2(g) \longrightarrow 2Fe^{3+} + 2Cl^-$$

$$-84.9 \qquad\quad 0 \qquad\qquad -10.5 \quad -131.2$$

What are $\Delta G^{\ominus}$ and K at 298 K for this reaction? The standard free energies $\widetilde{G}_f^{\ominus}$ of the species involved as found in Appendix D are listed beneath the chemical symbols.

Solution

$$\begin{aligned}
\Delta G^{\ominus} &= 2\widetilde{G}_{f,\mathrm{Fe}^{3+}}^{\ominus} + 2\widetilde{G}_{f,\mathrm{Cl}^-}^{\ominus} - (2\widetilde{G}_{f,\mathrm{Fe}^{2+}}^{\ominus} + \widetilde{G}_{f,\mathrm{Cl}_2}^{\ominus}) \\
&= [2(-10.5) - 2(131.2) + 2(84.9) + 0]\,\mathrm{kJ} \\
&= \underline{-113.6\,\mathrm{kJ}}
\end{aligned}$$

$$\ln K = \frac{113.6\,\mathrm{kJ} \times \dfrac{1000\,\mathrm{J}}{\mathrm{kJ}}}{8.31\,\mathrm{J\,K^{-1}} \times 298\,\mathrm{K}} = 45.9$$

$$K = e^{45.9} = \underline{8 \times 10^{19}\,\mathrm{M^2\,atm^{-1}}}$$

The units of K are chosen to be consistent with its definition

$$K = \frac{[\mathrm{Fe}^{3+}]^2[\mathrm{Cl}^-]^2}{[\mathrm{Fe}^{2+}]^2 P_{\mathrm{Cl}_2}} \quad \blacksquare$$

☐ What are $\Delta G^{\ominus}$ and K at 298 K for the reaction

Exercise 20-9

$$4I^- + O_2(g) + 4H^+ \longrightarrow 2I_2(g) + 2H_2O(l) \quad \blacksquare$$

$$-51.7 \qquad 0 \qquad\quad 0 \qquad\qquad 19.4 \qquad -237.2$$

Temperature Dependence of the Equilibrium Constant The fundamental relationship between an equilibrium constant K and the standard free-energy change $\Delta G^{\ominus}$ associated with changing reactants into products under standard conditions is

$$-RT \ln K = \Delta G^{\ominus} = \Delta H^{\ominus} - T\,\Delta S^{\ominus} \qquad (20\text{-}32)$$

Division by $-RT$ yields

$$\ln K = \frac{\Delta S^{\ominus}}{R} - \frac{\Delta H^{\ominus}}{RT} \qquad (20\text{-}33)$$

Thus the temperature dependence of $\ln K$ has the form

$$\ln K = 2.303 \log K = \frac{-\Delta H^{\ominus}}{RT} + \mathrm{const} \qquad (20\text{-}34)$$

and a graph of $\ln K$ against $1/T$ is expected to yield a straight line with the slope $-\Delta H^{\ominus}/R$ (Fig. 20-3) provided that $\Delta H^{\ominus}$ and $\Delta S^{\ominus}$ are constant in the temperature interval considered. It is therefore possible to determine enthalpies of reaction by measuring the equilibrium constant at several temperatures and plotting the measurements as in Fig. 20-3.

Equation (20-34) implies exponential dependence of K on $\Delta H^{\ominus}/RT$,

$$K = \mathrm{const} \times \exp\frac{-\Delta H^{\ominus}}{RT}$$

a temperature-dependence characteristic of many phenomena.

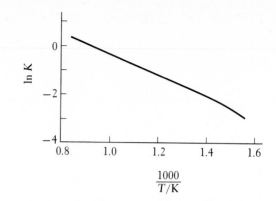

**Figure 20-3 Relation between ln K and $1/T$ for the Reaction
$CO_2(g) + H_2(g) \rightleftharpoons CO(g) + H_2O(g)$**
The slope of the curve is equal to $-\Delta H^\ominus/R$. The value of $\Delta H^\ominus$ (35 kJ at
1000 K) is seen to change slightly with temperature.

Equation (20-33) can also be used to derive a convenient relation between K
and $\Delta H^\ominus$. Suppose we consider two temperatures T_1 and T_2 that are sufficiently
close to each other that $\Delta S^\ominus$ and $\Delta H^\ominus$ can be considered to remain constant as T_1
is changed to T_2. If K_1 and K_2 are the equilibrium constants at these two
temperatures, we have

$$\ln K_1 = \frac{\Delta S^\ominus}{R} - \frac{\Delta H^\ominus}{RT_1} \tag{20-35}$$

and

$$\ln K_2 = \frac{\Delta S^\ominus}{R} - \frac{\Delta H^\ominus}{RT_2} \tag{20-36}$$

Subtraction of the first equation from the second yields

$$\ln K_2 - \ln K_1 = \frac{\Delta H^\ominus}{R}\left(\frac{1}{T_1} - \frac{1}{T_2}\right) = \frac{\Delta H^\ominus(T_2 - T_1)}{RT_1T_2} \tag{20-37}$$

Example 20-10

□ **Temperature Dependence of an Equilibrium Constant** At 25°C the
value of the equilibrium constant for the reaction

$$\tfrac{1}{2}I_2(g) + \tfrac{1}{2}Br_2(g) \rightleftharpoons IBr(g)$$

is 20.5 and $\Delta H^\ominus$ is $-5.29\ \text{kJ mol}^{-1}$. What is the value of K at 100°C?

Solution From (20-37) we have

$$\ln\frac{K_{100}}{K_{25}} = \frac{(-5.29 \times 10^3\ \text{J mol}^{-1})\,(75\ \text{K})}{(8.31\ \text{J mol}^{-1}\,\text{K}^{-1})\,(298\ \text{K})\,(373\ \text{K})}$$

$$= -0.430$$

Hence

$$\frac{K_{100}}{K_{25}} = e^{-0.430} = 0.651$$

$$K_{100} = 0.651 \times 20.5 = \underline{13.3}\ \blacksquare$$

Exercise 20-10

□ At 25°C the value of the equilibrium constant for the reaction

$$Ag^+ + Fe^{2+} \rightleftharpoons Ag(s) + Fe^{3+}$$

is $3.0\ M^{-1}$ and $\Delta H^\ominus$ is $-65.7\ \text{kJ}$. What is the value of K at 100°C? ■

Summary

The direction of spontaneous change in a chemical system is governed by a competition between the tendency of the potential energy to be minimized and a second factor that is the product of the absolute temperature of the system and the change in a state function of the system called the entropy. For any isothermal process the change in the entropy is given by $\Delta S = q_{\mathrm{rev}}/T$, where q_{rev} is the heat transferred to the system along a reversible path leading from the specified initial state to the specified final state and T is the absolute temperature of the system. Since S is a state function, ΔS is the same for any process, reversible or irreversible, that leads from the given initial state to the desired final state.

A general expression for the entropy change as the result of a phase transformation is $\Delta S_{\mathrm{tr}} = \Delta H_{\mathrm{tr}}/T_{\mathrm{tr}}$ in which ΔH_{tr} is the enthalpy of transformation (e.g., enthalpy of fusion) and T_{tr} is the temperature of the transformation under conditions of equilibrium between the phases. For many substances the entropy of vaporization at the normal boiling temperature is about 88 J mol^{-1} K^{-1} (Trouton's rule). When n mol of an ideal gas expands or contracts isothermally from V_1 to V_2, the entropy change is $\Delta S = nR \ln (V_2/V_1)$. For a process in which the temperature of a system with heat capacity C is changed from T_1 to T_2 with no reactions or phase changes, $\Delta S = C \ln (T_2/T_1)$.

The sum of the entropy change in a system and the accompanying entropy change in its surroundings satisfies the relation $\Delta S_{\mathrm{tot}} = (\Delta S_{\mathrm{sys}} + \Delta S_{\mathrm{surr}}) \geq 0$. The condition $\Delta S_{\mathrm{tot}} = 0$ corresponds to reversible processes, and the inequality $\Delta S_{\mathrm{tot}} > 0$ corresponds to irreversible (spontaneous) changes. It follows that the entropy of an adiabatic system or an isolated system can never decrease.

The number of microscopic states of a system that have the same average properties as a given macroscopic state is known as the number of realizations W of the macroscopic state. A statistical interpretation of entropy associates S with W: $S = k \ln W$, where the proportionality factor k is the Boltzmann constant, the gas constant divided by Avogadro's number, R/N_A. In this respect the entropy of a state is a measure of its disorder. A process in which the entropy increases is one in which the disorder has increased; conversely, decreases in entropy can be associated with increased order.

A useful function for the description of equilibria is the free energy, $G = H - TS$. For a change in state at constant T and P, $\Delta G = \Delta H - T\Delta S$. When P and T are constant and there is only P,V work, conditions for equilibrium and for the natural ("spontaneous") direction of change can be stated in terms of ΔG: (1) $\Delta G = 0$ at equilibrium; (2) For a system not at equilibrium, a change must occur in the direction for which $\Delta G < 0$; (3) A process for which $\Delta G > 0$ cannot be spontaneous. Spontaneous processes are favored by a decrease of H (negative ΔH) and by an increase of S (positive ΔS). The effect of temperature on the direction of spontaneous change can be determined by considering the signs and magnitudes of ΔH and ΔS.

The standard free-energy change for any reaction can be found from the standard free energies of formation of the substances involved: $\Delta G^{\ominus} = \Sigma G^{\ominus}_{f,\mathrm{prod}} - \Sigma G^{\ominus}_{f,\mathrm{react}}$. If the substances are not in their standard states, the free-energy change is given by $\Delta G = \Delta G^{\ominus} + RT \ln Q$, where Q is the reaction quotient. Application of the condition of equilibrium leads to $\Delta G^{\ominus} = -RT \ln K$, an expression that can be used to calculate equilibrium constants from tables of thermodynamic data. Changes of K with temperature are related to the standard enthalpy change by the equation $\ln (K_2/K_1) = (\Delta H^{\ominus}/R)(T_2 - T_1)/T_1 T_2$.

Terms and Concepts

Problems and Questions

20-1 Entropy of Vaporization The normal boiling point of methanol is 65.0°C and the enthalpy of vaporization is 35.3 kJ mol^{-1}. What is its entropy of vaporization?

20-2 Entropy Is it possible to decrease the entropy of a fluid (a) by an adiabatic expansion? (b) By an adiabatic compression? Explain.

20-3 Spontaneous Processes If any of the following is impossible, explain why. If not, give one example of each: (a) a spontaneous process in which the entropy of the system decreases; (b) a spontaneous process at constant T and P in which the free energy increases when only P,V work is done; (c) a spontaneous process that is endothermic.

20-4 Entropy Changes Indicate for each of the following pairs of processes which process is accompanied by the larger entropy increase. Explain your reasoning.

    (a) (i) Solid water at 0°C changes to water vapor at 0°C; (ii) liquid water at 0°C changes to water vapor at 0°C.

    (b) (i) A liquid is vaporized to gas at a pressure lower than the vapor pressure at a given temperature; (ii) a liquid is vaporized to gas at a pressure equal to the vapor pressure at a given temperature.

    (c) (i) A gas at 30°C is heated reversibly to 70°C; (ii) the same gas at 30°C is heated irreversibly to 80°C.

20-5 Entropy, a State Function (a) One mole of an ideal gas at 300 K and 1 atm is compressed reversibly and isothermally to half its volume. (b) Another mole of the gas at the same initial conditions is heated reversibly to 600 K at constant volume and then cooled reversibly to 300 K at constant pressure. Assure yourself that the final state is the same in each case and calculate ΔS for the two paths, assuming a constant $\widetilde{C}_P$ of 20.9 J mol^{-1} K^{-1}.

20-6 Heating a Gas Consider an ideal gas for which, in the temperature range of interest, $\widetilde{C}_P$ is constant and equal to 29.3 J mol^{-1} K^{-1}. (a) The temperature of 1.00 mol of the gas, contained in a cylinder with a piston exerting a constant pressure of 1.00 atm, is raised gradually and reversibly from 250 to 500 K. Calculate q, ΔV, w, ΔU, ΔH, and ΔS for this sample of gas. (b) Suppose that the same change in state described in (a) had occurred irreversibly (e.g., by plunging the cylinder into a large "thermostat" at 500 K). Explain how ΔV, ΔU, ΔH, and ΔS for the gas would differ (if at all) from the values obtained in (a).

20-7 Melting Ice Find $\Delta \widetilde{S}$ for the change in state $H_2O(s,\ 0°C) \longrightarrow H_2O(l,\ 25°C)$. The molar enthalpy of fusion of ice at 0°C is 6.02 kJ mol^{-1}; $\widetilde{C}_P$ for $H_2O(l)$ is 75.3 J mol^{-1} K^{-1} and is to be assumed constant.

20-8 Enthalpy of Vaporization Benzene boils at 80°C. Estimate $\Delta \widetilde{H}_{vap}$ from Trouton's rule.

20-9 Crystallization of Supercooled Water The molar enthalpy of fusion of ice at 0°C is 6.02 kJ mol^{-1}; the molar heat capacity of supercooled water is 75.3 J mol^{-1} K^{-1}. (a) One mole of supercooled water at −10°C is induced to crystallize in a heat-insulated vessel. The result is a mixture of ice and water at 0°C. What fraction of this mixture is ice? (b) What is ΔS for the system?

20-10 Chemistry of Molecules The strongest chemical bond known is that in carbon monoxide,

CO, with a bond enthalpy of 1.05×10^3 kJ mol^{-1}. Furthermore, the entropy increase in a gaseous dissociation of the kind AB $\rightleftharpoons$ A + B is about 110 J mol^{-1} K^{-1}. These factors establish a temperature beyond which there is essentially no chemistry of molecules. Show why this is so and find the temperature.

20-11 Two Solid Phases A certain substance consists of two modifications A and B; $\Delta G^{\ominus}$ for the transition from A to B is positive. The two modifications produce the same vapor. Which has the higher vapor pressure? Which is the more soluble in a solvent common to both?

20-12 Vaporization of Ethanol The enthalpy of vaporization of ethanol is 38.7 kJ mol^{-1} at its boiling point, 78°C. Calculate q, w, ΔU, ΔS, and ΔG for vaporizing 1 mol ethanol reversibly at 78°C and 1.00 atm. Neglect the volume of liquid ethanol and assume that the vapor is an ideal gas.

20-13 Iron Oxides From the values in Table D-4 calculate $\Delta H^{\ominus}$ and $\Delta G^{\ominus}$ for the reaction $3Fe_2O_3(s) \rightarrow 2Fe_3O_4(s) + \frac{1}{2}O_2(g)$ at 25°C. Which of the two oxides is more stable at 25°C and $P_{O_2} = 1$ atm?

20-14 Entropy of Vaporization Calculate the entropy of vaporization of ethanol at 1 atm and 50°C from the molar enthalpy of vaporization at the normal boiling point, 78°C, which is 38.7 kJ, and from the constant-pressure heat capacities of liquid and gaseous ethanol, 111 and 65 J K^{-1} mol^{-1}, respectively. Is this change in state spontaneous? Explain how your answer to this question is consistent with the sign of ΔS and the known boiling point.

20-15 Equilibrium Constants Find $\Delta G^{\ominus}$ and the equilibrium constant K for the following reactions by referring to the values of the standard free energies of formation listed in Table D-4:
(a) $4I^-(aq) + O_2(g) + 4H^+ \rightleftharpoons 2H_2O(l) + 2I_2(g)$
(b) $CO(g) + 2H_2(g) \rightleftharpoons CH_3OH(l)$
(c) $3H_2(g) + SO_2(g) \rightleftharpoons H_2S(g) + 2H_2O(l)$
(d) $Ca(s) + CO_2(g) \rightleftharpoons CaO(s) + CO(g)$

20-16 Standard Free-Energy Change and Equilibrium Constant Calculate $\Delta H^{\ominus}$ and $\Delta G^{\ominus}$ for the reaction of 2 mol gaseous SO_2 with O_2 to form gaseous SO_3 at 25°C, and calculate the equilibrium constant for this reaction. Use data in Table D-4.

20-17 Thermodynamic Quantities for an Inaccessible Equilibrium From the data in Table D-4 calculate $\Delta H^{\ominus}$ and $\Delta G^{\ominus}$ for the following reaction at 298 K:

$$6CH_4(g) + \tfrac{9}{2}O_2(g) \longrightarrow C_6H_6(l, \text{benzene}) + 9H_2O(l)$$

20-18 Isomerization Equilibrium The gaseous compounds allene and propyne are isomers with formula C_3H_4. Calculate the equilibrium constant and the standard enthalpy change at 25°C for the isomerization reaction

$$\text{allene}(g) \longrightarrow \text{propyne}(g)$$

from the following data, all of which apply to 298 K:

	$\widetilde{H}_f^{\ominus}$/kJ mol^{-1}	$\widetilde{G}_f^{\ominus}$/kJ mol^{-1}
Allene	192	202
Propyne	185	194

20-19 Thermodynamic Criteria State the conditions for which: (a) $\Delta G = 0$ indicates a reversible process; (b) $\Delta S = 0$ indicates a reversible process; (c) $\Delta G = \Delta H - T\Delta S$; (d) $\Delta H = \Delta U + V\Delta P$; (e) $\Delta H = q + P\Delta V$. (f) What is the sign of ΔG for a spontaneous process under the conditions for which (a) is true? (g) What is the sign of ΔS for a spontaneous process under the conditions for which (b) is true?

20-20 Free-Energy Change (a) Calculate the standard free-energy change and the equilibrium constant for the dimerization of NO_2 to N_2O_4 at 25°C (see Table D-4). (b) Calculate ΔG for this reaction at 25°C when the pressures of NO_2 and N_2O_4 are each held at 0.01 atm. Which way will the reaction tend to proceed?

20-21 Equilibration of Isomers There are two isomeric hydrocarbons with formula C_4H_{10}, butane and isobutane, which we denote here B and I. The standard enthalpies of formation for the gaseous species are -124.7 kJ mol^{-1} for B, -131.3 kJ mol^{-1} for I; the standard free energies of formation are -15.9 kJ mol^{-1} for B, -18.0 kJ mol^{-1} for I. (a) Which is the more stable under standard conditions, and which has the higher entropy? (b) The reaction B $\rightleftharpoons$ I can occur in the presence of a catalyst. Calculate the equilibrium constant at

298 K for the conversion of B to I, and calculate the percentage of B in the equilibrium mixture.

20-22 Solubility Products Find K_{sp} for the following salts by using the values of the standard free energies of formation given below and those listed in Table D-4: (a) AgCl(s), $\widetilde{G}_f^\ominus = -110\,\text{kJ}$ mol^{-1}; (b) AgBr(s), $\widetilde{G}_f^\ominus = -96\,\text{kJ mol}^{-1}$; (c) AgI(s), $\widetilde{G}_f^\ominus = -66\,\text{kJ mol}^{-1}$

20-23 Free Energy of Atoms What is the partial pressure of atomic chlorine that would be at equilibrium with $Cl_2(g)$ at 25°C and 1 atm?

20-24 Equilibrium Pressure of CO_2 From the values of $\widetilde{G}_f^\ominus$ for $CO_2(g)$, $H_2O(l)$, and $H_2CO_3(aq)$ in Table D-4, calculate the partial pressure of CO_2 in equilibrium with a 1 M solution of H_2CO_3.

20-25 Gas Equilibrium At 3500 K the equilibrium constant for the reaction $CO_2(g) + H_2(g) \rightleftharpoons CO(g) + H_2O(g)$ is 8.28. What is $\Delta G^\ominus(3500)$ for this reaction? What is ΔG at 3500 K for transforming 1 mol CO_2 and 1 mol H_2, both held at 0.1 atm, to 1 mol CO and 1 mol H_2O, both held at 2 atm? In which direction would this last reaction run spontaneously?

20-26 Free Energy from Equilibrium Measurements At 35°C and a total pressure of 1 atm, $N_2O_4(g)$ is 27.2 percent dissociated into $NO_2(g)$. What is $\Delta G^\ominus$ at 35°C for the reaction $N_2O_4(g) \rightleftharpoons 2NO_2(g)$?

20-27 Vapor Pressures of Bromine and Iodine Use the data in Table D-4 to calculate the vapor pressure of liquid bromine and solid iodine at 25°C. Express the answers in atmospheres and torr.

20-28 Change of Equilibrium Constant with Temperature At a certain temperature, T', the equilibrium constant for the reaction of Prob. 20-16 is $1.0 \times 10^6\,\text{atm}^{-1}$. Assume that $\Delta H^\ominus$ and $\Delta S^\ominus$ are independent of temperature, and calculate T'.

20-29 Enthalpy of Neutralization and K_w At 24°C, $K_w = 1.0 \times 10^{-14}$. Find the enthalpy of the reaction

$$H_2O(l) \rightleftharpoons H^+(aq) + OH^-(aq)$$

from the values in Table D-4. Assume that this enthalpy of neutralization is temperature-independent and use it to estimate K_w at normal body temperature (about 37°C) and at 100°C.

20-30 Entropy and Enthalpy of Formation The free energy of formation $\widetilde{G}_f^\ominus$ for CO is $-182.3\,\text{kJ}$ mol^{-1} at 800 K and $-200.5\,\text{kJ mol}^{-1}$ at 1000 K. What are the entropy and the enthalpy of formation, $\widetilde{S}_f^\ominus$ and $\widetilde{H}_f^\ominus$, in this temperature range, assuming both to be constant?

20-31 Vapor Pressure of Benzene (a) From the values in Table D-4 find the enthalpy and the free energy of vaporization of benzene, C_6H_6, at 25°C. (b) What is the vapor pressure of benzene at 25°C? (c) Assume that the entropy and enthalpy of vaporization are constant and estimate the normal boiling point of benzene.

20-32 $\Delta G^\ominus$ from Equilibrium Data At 1200 K and in the presence of solid carbon, an equilibrium mixture of CO and CO_2 ("producer gas") contains 98.3 mol percent CO and 1.69 mol percent CO_2 with the total pressure at 1 atm. What are P_{CO} and P_{CO_2}, what is the equilibrium constant, and what is $\Delta G^\ominus$ associated with the reaction

$$CO_2(g) + C(\text{graphite}) \rightleftharpoons 2CO(g)$$

at 1200 K?

20-33 Vaporization of Ammonia The normal boiling point of liquid ammonia is 240 K; the enthalpy of vaporization at that temperature is 23.4 kJ mol^{-1}. The heat capacity of gaseous ammonia at constant pressure is 38 J $\text{K}^{-1}\,\text{mol}^{-1}$. (a) Calculate q, w, ΔH, and ΔU for the following change in state:

2.00 mol $NH_3(l$, 1 atm, 240 K) $\rightarrow$
2.00 mol NH_3 (g, 1 atm, 298 K)

Assume that the gas behaves ideally and that the volume occupied by the liquid is negligible. (b) Calculate the entropy of vaporization of NH_3 at 240 K and then, assuming that $\Delta H^\ominus$ and $\Delta S^\ominus$ are independent of temperature, calculate the vapor pressure of liquid ammonia at 0°C.

Electrochemistry

"In the third series of these Researches, after proving the identity of electricities derived from different sources, and showing, by actual measurement, the extraordinary quantity of electricity evolved by a very feeble voltaic arrangement, I announced a law, derived from experiment, which seemed to me of the utmost importance to the science of electricity in general, and that branch of it denominated electrochemistry in particular. The law was expressed thus: The chemical power of a current of electricity is in direct proportion to the absolute quantity of electricity which passes."
MICHAEL FARADAY, 1834

21

21-1 Introduction

The field of electrochemistry deals with redox reactions, reactions in which electrons are exchanged between two species. It includes a wide range of practical processes and their interpretation in terms of thermodynamics, kinetics, and other general chemical and physical principles. Some spontaneous redox reactions are valuable as sources of electrical energy—for example, the reactions in a lead storage battery (the kind of battery in most automobiles) or a flashlight battery or an efficient fuel cell like those used in spacecraft. Redox reactions are also responsible for the corrosion and consequent malfunction of all manner of man-made objects, from ships and bridges to "tin" cans and microscopic circuits in electronic equipment. The chemistry of corrosion is closely related to the chemistry of batteries.

Many substances that are not stable in the normal atmosphere on the earth's surface can be produced by using electrochemical methods to force appropriate redox reactions to go in a direction opposite to their natural or spontaneous direction. These techniques are used to produce the most reactive elements, including those with broad applications in modern consumer and industrial technology, such as aluminum and magnesium, chlorine and fluorine.

We begin the discussion of electrochemistry with a summary of its historical background and some of its basic concepts, particularly Faraday's law of electrolysis.

Historical Background The first recognition of a relation between electrical and chemical phenomena dates to the end of the eighteenth century. Luigi Galvani, a lecturer in anatomy at the University of Bologna, observed in 1786 that a frog's leg muscle contracted if it was in contact with any of various metals, provided that the metal was in turn in contact with a second metal that touched a nerve of the frog. Galvani attributed this effect to "animal electricity", but his countryman, the physicist Alessandro Volta, showed that it was due to the contact of the two dissimilar metals. This led Volta in 1799 to the invention of the electric battery. His famous "pile" was a sequence of disks of silver and zinc in

441

contact with one another, each pair separated from the next by a sheet of heavy paper moistened with salt water. When the disk of zinc at one end of the pile was connected by a wire to the disk of silver at the other end, a current was generated.

Electrochemical methods were first applied to the isolation of strongly electropositive elements by Sir Humphry Davy, who discovered potassium and sodium in 1807 by electrolyzing melts of their carbonates. Later he produced four of the alkaline-earth metals in a similar way. The success of these experiments confirmed in Davy's mind the idea that chemical and electrical attraction were essentially the same phenomenon, a view developed into a system of electrochemical dualism by Berzelius (Sec. 17-1).

The key quantitative discovery in electrochemistry was made by the self-educated English physicist and chemist Michael Faraday. His studies of electrolysis led him in 1833 to the discovery of a relation between the quantity of electricity passed through a solution and the weight of the substance deposited. He thought, however, that the ions that carried the electricity in the solution were created by the current. It was more than half a century before Arrhenius recognized that ions exist in conducting solutions whether or not there is actual passage of current.

Current and Charge An electric current is a flow of charge. The SI unit of electric current is the ampere (A) (Appendix A). For a constant current I, the amount of charge Q moving past a given point in the time t is given by the expression

$$Q = It \qquad (21\text{-}1)$$

The common unit of charge is the coulomb (C), corresponding to a current of one ampere flowing for one second; that is, 1 coulomb = 1 ampere second.

Example 21-1

□ **Current and Time** A current of 0.70 A flows through a battery-operated radio. How many coulombs pass through the radio during 1 min of operation?

Solution Since $1\,A = 1\,C\ s^{-1}$, we can write, by (21-1), $Q = 0.70\,A \times 60\,s = 0.70\,C\ s^{-1} \times 60\,s = \underline{42\,C}$ ■

Exercise 21-1

□ During exactly 1 hour 1.80×10^3 C flow through a flashlight bulb. What is the average current? ■

The charges that move in a metal when an electric current flows are electrons, but electrons do not exist by themselves in conducting solutions or in molten salts. In these substances, generally called electrolytes,[1] the moving charges are ions. Consequently when an electric current passes from an *electrode* (typically a conducting rod or plate) into or out of an electrolyte, an electron-transfer reaction must occur. Electrons must leave the electrode and combine with some species in the electrolyte or some species in the electrolyte must give up electrons that enter the metal.

At an electrode where electrons are delivered to the electrolyte there is reduc-

---

[1] It is customary in electrochemistry to define electrolytes as the solutions or melts that conduct electric current rather than as just the dissolved substances that cause solutions to conduct (Sec. 9-1).

tion, for example, cupric ion being reduced to metallic copper,

$$Cu^{2+} + 2e^- \longrightarrow Cu(s)$$

More generally, if the oxidized and reduced forms of the species involved are called ox 1 and red 1, then such a reaction can be represented by

$$\text{ox } 1 + \mathbf{n}_1 e^- \longrightarrow \text{red } 1 \qquad (21\text{-}2)$$

where $\mathbf{n}_1$ is the number of electrons transferred. At an electrode where electrons are received from the electrolyte there is oxidation of a reduced species (red 2) to some oxidized form (ox 2),

$$\text{red } 2 \longrightarrow \mathbf{n}_2 e^- + \text{ox } 2 \qquad (21\text{-}3)$$

The flow of electric current through an electrolyte is thus tied to the progress of the electrode reactions. Both reduction and oxidation must occur, even though they occur in separate places.

The electrolyte and the external circuit can remain electrically neutral only if the number of electrons delivered by an electrode to the electrolyte in a given time is matched by the number removed at another electrode. The electrical potentials produced by even a slight departure from electrical neutrality are so enormous[2] that for all practical purposes exact neutrality is maintained at all times on a macroscopic basis. Thus the half-reactions at the two electrodes must be adjusted so that the number of electrons is the same in each. The corresponding equations can be suitably adjusted if (21-2) is multiplied by $\mathbf{n}_2$ and (21-3) by $\mathbf{n}_1$, so that when $\mathbf{n}_1 \mathbf{n}_2$ electrons are delivered to the electrolyte, the same number is also withdrawn from it. If the equations are then added, the electrons cancel and only an equation relating two redox couples remains:

$$\mathbf{n}_2 \text{ ox } 1 + \mathbf{n}_1 \text{ red } 2 = \mathbf{n}_2 \text{ red } 1 + \mathbf{n}_1 \text{ ox } 2 \qquad (21\text{-}4)$$

This approach is indeed one way of balancing redox equations (Chap. 10).

The combination of electrodes and mobile ions in a solution or a molten electrolyte is the essence of what is called an *electrochemical cell*. When the reaction of such a cell occurs in the spontaneous direction, one speaks of a *galvanic* or *voltaic* cell, or sometimes a *battery*, although originally this term was reserved for a sequence of galvanic cells joined together. In some cells, called *electrolytic* cells, the reaction is driven by an external potential applied to the electrodes. The reaction then goes in a direction opposite to the spontaneous one and is termed *electrolysis* (Figs. 1-4 and 21-1).

Chemists find it useful to label electrical terminals by the type of electrode reaction associated with them:

If the reaction is a reduction, the terminal is called a *cathode*.
If the reaction is an oxidation, the terminal is called an *anode*.

The electrons therefore always enter an electrochemical cell at the cathode and leave at the anode. Signs are sometimes associated with the electrodes of a cell, but they are a frequent source of confusion, and we shall not consider them here, except to point out that on commercial batteries, such as lead storage batteries and dry cells, the terminal labeled (+) is the cathode when the battery is acting as a source of electricity (see footnote 5, page 451). More details of electrochemical cells and batteries are considered in Secs. 21-4, 21-5, and 21-6.

---

[2]A free charge of about 10^{-17} mol electrons causes a potential of 1 V at a distance of 1 cm.

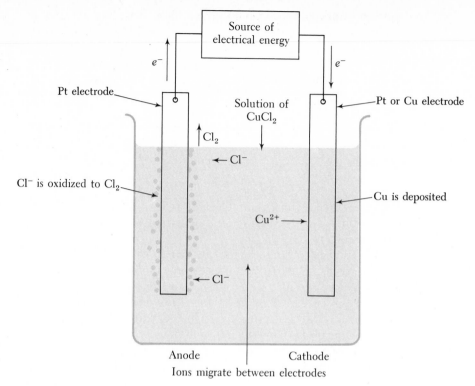

Figure 21-1 Electrolytic Cell
At the anode Cl^- is oxidized to gaseous Cl_2 and at the cathode Cu^{2+} is reduced to solid Cu. The source of direct current and the wires going to the electrodes make up the external circuit.

21-2 Faraday's Law, Electrochemical Cells, and Electrodes

Faraday's Law When a source of direct current is connected to the electrolytic cell shown in Fig. 21-1 in such a way that electrons flow into the right electrode, it is observed that metallic copper plates out on the right electrode and bubbles of chlorine gas appear on the left. The right electrode is the cathode; it is here that the reduction

$$Cu^{2+} + 2e^- \rightarrow Cu(s) \qquad (21\text{-}5)$$

takes place. The oxidation occurring at the anode is

$$2Cl^- \rightarrow Cl_2 + 2e^- \qquad (21\text{-}6)$$

Since the number of electrons passing from the external circuit into the cathode must be the same as the number passing from the anode into the external circuit, one chlorine molecule is formed for every atom of copper deposited. The solution stays electrically neutral at all times. Current is carried through the electrolyte by the movement of Cu^{2+} and Cl^- ions.

The electrons in (21-5) and (21-6) play the roles of reactants and products, and the principles of stoichiometry apply to them. For example, the reduction of one Cu^{2+} ion requires two electrons. Two moles of electrons are needed to produce 1 mol of copper. While it is often convenient to determine the numbers of moles of atoms or molecules involved in reactions by weighing, numbers of electrons are most easily determined by measuring quantities of charge. It is useful to define the faraday ($\mathcal{F}$), which is the charge on one mole of electrons

$$\mathcal{F} = 96{,}485 \text{ C mol}^{-1}$$

This value is usually rounded to $96{,}500 \text{ C mol}^{-1}$, correct to 0.02 percent. If $\mathbf{n}$ is the number of *electrons* participating in the reaction of *one molecule* or ion, then the reaction of *one mole* of substance involves $\mathbf{n}$ *faradays*. Thus, for each faraday of electricity passed, the weight of any substance formed or used up in electrode reactions is equal to its weight per mole divided by the appropriate value of $\mathbf{n}$. More generally, whatever the quantity of electricity that passes, the weights of different substances formed or used up are proportional to their formula weights divided by the corresponding values of $\mathbf{n}$. This is Faraday's law of electrolysis.[3]

☐ **Electrolytic Cell** An electric current passes through an electrolytic cell containing a solution of $AgNO_3$. After the current has flowed for 1 hour it is found that 506 mg silver has deposited on the cathode. What is the average current in amperes?

Example 21-2

Solution The cathode reaction is

$$Ag^+ + e^- \rightarrow Ag(s)$$

and thus $\mathbf{n} = 1$. For every faraday, or $96{,}500$ C, passing through the cell, 1 mol Ag atoms or 107.9 g Ag is plated out. The 506 mg deposited corresponds to

$$\frac{0.506 \text{ g}}{107.9 \text{ g (mol Ag)}^{-1}} \times \frac{1 \text{ mol } e^-}{\text{mol Ag}} \times \frac{96{,}500 \text{ C}}{\text{mol } e^-} = 453 \text{ C}$$

Spread over 3600 s, this corresponds to an average current of

$$\mathbf{I} = \frac{\mathbf{Q}}{t} = \frac{453 \text{ C}}{3600 \text{ s}} = 0.126 \text{ C s}^{-1} = \underline{0.126 \text{ A}} \quad \blacksquare$$

☐ An electric current of 0.445 A passes through an electrolytic cell containing a $CuSO_4$ solution and deposits Cu during exactly 2 hour. What is the weight of the Cu deposited? What weight of Ag would be deposited from a solution of $AgNO_3$ by this same quantity of electricity? $\blacksquare$

Exercise 21-2

Electric Potential and Electric Work A useful analogy to an electric current flowing through a system of wires and electric devices is that of a liquid flowing through pipes and hydraulic equipment (Fig. 21-2). Electric current is measured in amperes, equal to coulombs per second; the hydraulic analog to the current is the rate of liquid flow, which can be expressed in liters per second. The liquid moves as the result of a pressure difference, measured, for example, in atmospheres. Similarly, an electric charge moves as the result of a potential difference, usually measured in volts.

The work performed by a liquid of volume V that is driven through the turbine in Fig. 21-2 by the pressure difference $\Delta P = P_A - P_B$ is $w = V(P_A - P_B) = V\Delta P$.

---

[3]The quantities FW/$\mathbf{n}$ are sometimes called *redox equivalent weights*.

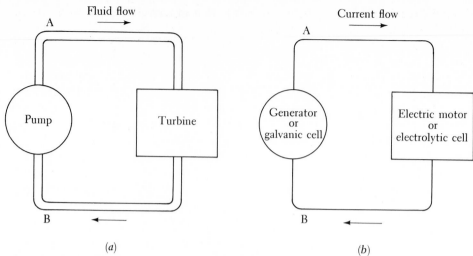

(a)

(b)

Figure 21-2 An Analogy between Electric and Hydraulic Systems
(a) In one type of fluid-drive system a pump forces a liquid through a turbine that is coupled to a drive shaft. The pressure generated by the pump at point A must be larger than the pressure at point B. (b) The electric analog of pressure is electric potential, Φ. The electric potential at A must be larger than that at B for current to flow through the motor.

In complete analogy the electric work performed by a charge $\mathbf{Q}$ that is moved through the potential difference $\Phi_A - \Phi_B = \Delta\Phi$ is

$$w = \mathbf{Q}(\Phi_A - \Phi_B) = \mathbf{Q}\Delta\Phi \qquad (21\text{-}7)$$

If $\mathbf{Q}$ is in coulombs and Φ is in volts the work is in joules. Since an electric current $\mathbf{I}$ that flows for a time t represents a charge $\mathbf{Q} = \mathbf{I}t$ (Eq. 21-1), the work done by a current that is driven by a potential difference $\Delta\Phi$ is

$$w = \mathbf{I}t\Delta\Phi \qquad (21\text{-}8)$$

the customary units being volt ampere seconds = watt seconds (symbolized W s) = joules.

Example 21-3

□ **Charging a Car Battery** To bring a completely discharged 100-A–hour car battery to full charge requires that at a minimum an amount of charge equivalent to 10 A flowing for 10 hour pass through the battery. If electricity costs 10¢ per kilowatthour, what is the minimum cost for charging a 12-V battery?

Solution By (21-8), the work performed is

$$\begin{aligned} w &= 12\text{ V} \times 10\text{ A} \times 10\text{ hour} \\ &= 1.2 \times 10^3 \text{ VA hour} = 1.2 \times 10^3 \text{ W hour} \\ &= 1.2 \text{ kW hour} \end{aligned}$$

Thus at 10¢ per kilowatthour, the minimum cost is $1.2 \times 10 = \underline{12¢}$. ∎

Exercise 21-3

□ One popular electronic calculator can run for 4 hour on a 3.6-V battery pack, an average current of 140 mA being required. How much energy does the battery pack provide? Note that 1 J = 1 W s. ∎

Galvanic Cells How can redox reactions be harnessed to perform electrical work? This must be done by keeping the reacting species apart and allowing them to exchange electrons through an external circuit. For example, consider the reaction of metallic zinc, a bright silvery metal, with a solution of a salt of Cu^{2+}. If a piece of zinc is dropped into a solution of cupric sulfate, there is an immediate reaction. A reddish brown deposit of metallic copper forms on the surface of the zinc, the solution becomes warm, and the intensity of the blue color of the solution (due to hydrated Cu^{2+}) is gradually reduced. Zinc atoms in the metal transfer electrons to cupric ions. The zinc is oxidized to Zn^{2+} and the Cu^{2+} is reduced to Cu:

$$Zn(s) + Cu^{2+} \longrightarrow Zn^{2+} + Cu(s) \qquad (21\text{-}9)$$

When reaction (21-9) proceeds in the manner described, it is useless as a source of electrical energy. However, if the reductant (Zn) and the oxidant (Cu^{2+}) are separated and a path for electron flow between them is provided, the reaction can serve as a source of electrical energy, a chemical battery.

Consider the cell illustrated in Fig. 21-3a. The zinc electrode dips into a zinc sulfate solution, and the copper electrode is surrounded by a copper sulfate solution. The two solutions must be in electrical contact to allow the passage of current between them, yet the solutions must not be allowed to mix, for this would lead to the *direct* reaction of Cu^{2+} with the Zn electrode, and electrons would not flow through the external circuit. The solutions can be separated, as shown, by a thin layer of unglazed porcelain or another porous membrane that allows ions to pass but slows the mixing process.

Another technique for preventing mixing of solutions in a cell while still maintaining electrical contact is the use of a *salt bridge,* shown in Fig. 21-3b. A salt bridge is an inverted U-tube containing a KCl solution given a jellylike consistency by addition of a substance such as agar. The K^+ and Cl^- ions carry the current through the bridge.

When the cells illustrated in Fig. 21-3 produce an electric current, reaction (21-9) occurs, zinc being oxidized to Zn^{2+} and Cu^{2+} being reduced to metallic copper. Thus, the half-reaction

$$Zn(s) \rightarrow Zn^{2+} + 2e^-$$

occurs at the anode, which causes a loss of Zn from the electrode and an increase in the concentration of Zn^{2+} ions (in the left-hand compartment of the cell in Fig. 21-3b). Correspondingly, the weight of the copper electrode increases, and the concentration of Cu^{2+} in the right-hand compartment in Fig. 21-3b decreases as a result of the cathode reaction

$$Cu^{2+} + 2e^- \rightarrow Cu(s)$$

At the same time, in order to maintain electrical neutrality, Cl^- ions from the salt bridge move into the anode compartment to balance the increase in positive charge caused by the increase in $[Zn^{2+}]$, and K^+ ions move from the salt bridge into the cathode compartment to balance the decrease in $[Cu^{2+}]$.[4] The flow of ions between the two compartments exactly parallels the external flow of electrons through the wire. Were there no provision for these ion migrations, the

---

[4]Small quantities of Cu^{2+}, Zn^{2+}, and $SO_4{}^{2-}$ ions also diffuse through the salt bridge, but these quantities are normally negligible.

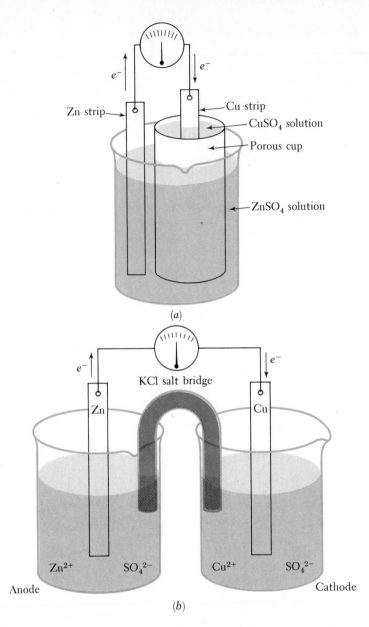

Figure 21-3 Typical Galvanic or Voltaic Cells
This cell is composed of an anode of zinc in a solution containing Zn^{2+} ions and a cathode of copper in
a solution containing Cu^{2+} ions. In (*a*) the two solutions are separated by a porous membrane, which in-
hibits mixing but permits passage of ions. In (*b*) a
KCl salt bridge is used for this purpose.

reaction would stop very quickly as unneutralized charges accumulated around
the electrodes and built up a potential that offset the electrode potential difference.

As the cell reaction proceeds, the concentrations of the ions involved in the
electrode reactions change significantly and the voltage that can be supplied by
this battery invariably decreases, approaching zero as a limit. The battery "runs
down". This is a perfectly general phenomenon, characteristic of all galvanic
cells; some can be recharged by forcing a current to pass through them in the
opposite direction, but this is not always practicable.

Electrodes of Different Kinds   In the examples that have been given, the
electrodes are simply pieces of metal. Many useful electrodes are more compli-

cated. The hydrogen electrode is a typical gas electrode. Hydrogen gas is bubbled over a platinum electrode that has been coated with finely divided platinum (called platinum black) and that is surrounded by a solution containing hydrogen ions. The half-reaction at this electrode is

$$2H^+ + 2e^- \longrightarrow H_2(g) \tag{21-10}$$

The platinum acts as catalyst for the reaction between the H^+ ions and H_2 molecules and acquires a potential characteristic of this reaction. Other gas electrodes may be constructed similarly. Examples include the oxygen electrode, with the half-reaction

$$4H^+ + O_2(g) + 4e^- \longrightarrow 2H_2O$$

and the chlorine electrode, with the half-reaction

$$Cl_2(g) + 2e^- \longrightarrow 2Cl^-$$

Electrodes may be characterized by the redox couple they represent, such as Cu^{2+}/Cu, Zn^{2+}/Zn, H^+/H_2, O_2/H_2O, and Cl_2/Cl^- for the electrodes mentioned so far. We always list the oxidized form first, followed by the reduced form.

Some electrodes represent redox couples in which both oxidized and reduced forms are ionic species in the solution and the electron transfer occurs at an inert electrode. An example is a solution containing Fe^{2+} and Fe^{3+} ions and a gold or platinum electrode. The potential corresponds to the Fe^{3+}/Fe^{2+} couple, with the half-reaction

$$Fe^{3+} + e^- \longrightarrow Fe^{2+}$$

In another type of electrode, the concentration of the cation associated with the electrode metal is kept fixed through the solubility-product principle by the presence of a salt of low solubility. Such electrodes are used primarily as reference electrodes for measurements made with a negligibly small flow of current. A silver surface coated with solid silver chloride, written $AgCl/Ag$, is representative. Since $AgCl(s)$ is present, the solution is saturated with silver chloride, so that at all times $[Ag^+][Cl^-] = K_{sp}$, where K_{sp} is the solubility-product constant. If any Ag^+ ions are reduced to Ag by the gain of electrons, they are replenished by the dissolving of $AgCl(s)$. The dissolving of $AgCl(s)$ also produces Cl^- ions. These reactions may be described by the equations

$$Ag^+ + e^- \longrightarrow Ag(s)$$
$$AgCl(s) \longrightarrow Ag^+ + Cl^-$$

Conversely, if any $Ag(s)$ is oxidized to Ag^+ by loss of electrons, $AgCl(s)$ is precipitated to keep the product $[Ag^+][Cl^-]$ equal to K_{sp}. The half-reaction of this electrode is most simply described as the sum of the foregoing reactions:

$$AgCl(s) + e^- \longrightarrow Ag(s) + Cl^- \tag{21-11}$$

21-3 Free Energy and Cell EMF

As we have noted earlier, the voltage of a galvanic cell decreases as the spontaneous reaction that drives the cell proceeds. The drop in cell voltage parallels the decrease in ΔG that occurs as the spontaneous reaction moves toward equilibrium (Sec. 20-4). It is possible to derive a precise relation between the thermodynamic

449

and electric changes in terms of a quantity called the *cell electromotive force*, or *cell emf*, which we now define.

The movement of electrons through an external circuit is the result of the difference in electric potential between the electrodes. For example, when an electrochemical cell is galvanic, the potential of the cathode, $\Phi_{cathode}$, must be more positive than the potential of the anode, Φ_{anode}, in order to drive the electrons in the external circuit from the anode to the cathode. If $\mathbf{n}$ mol of electrons is involved, the total charge moving is $\mathbf{n}\mathcal{F}$ and the electric work $-w$ performed by the cell on the surroundings is, by (21-7),

$$-w = \mathbf{Q}(\Phi_{cathode} - \Phi_{anode}) = \mathbf{n}\mathcal{F}\mathbf{E} \qquad (21\text{-}12)$$

where $\mathbf{E}$ is called the *cell emf* and is defined to be the potential of the cathode minus the potential of the anode,

$$\mathbf{E} = \Phi_{cathode} - \Phi_{anode} \qquad (21\text{-}13)$$

On the other hand, it can be shown by thermodynamics that the electrical work $-w$ performed by an electrochemical cell on its surroundings under conditions of reversibility and constant temperature and pressure is just equal to the negative of the change in free energy of the cell; that is, $-w = -\Delta G$, or

$$w = \Delta G \qquad (21\text{-}14)$$

Here ΔG is the difference between the free energies of the products and the reactants in the cell reaction, which involves a transfer of $\mathbf{n}$ mol electrons. Combination of (21-12) and (21-14) yields the important relation

$$\Delta G = -\mathbf{n}\mathcal{F}\mathbf{E} \qquad (21\text{-}15)$$

Equation (21-15) can be solved for $\mathbf{E}$:

$$\mathbf{E} = \frac{-\Delta G}{\mathbf{n}\mathcal{F}} \qquad (21\text{-}16)$$

Since ΔG is negative for spontaneous processes, (21-16) implies that if $\mathbf{E}$ is positive the cell is galvanic when the cell reaction proceeds in the direction written. If $\mathbf{E}$ is negative the cell is electrolytic and an external potential is needed to drive the cell reaction in the direction it is written; an external potential of magnitude $|\mathbf{E}|$ would just prevent the reaction from proceeding spontaneously in the opposite direction.

A special case of (21-16) is that of a cell with all the reactants and products at standard conditions, in which case the emf is called the *standard emf* $\mathbf{E}^{\ominus}$ and is related to $\Delta G^{\ominus}$ of the cell reaction by

$$\mathbf{E}^{\ominus} = \frac{-\Delta G^{\ominus}}{\mathbf{n}\mathcal{F}} \qquad (21\text{-}17)$$

The Nernst Equation To derive an expression for the emf $\mathbf{E}$ of an electrochemical cell, we recall the expression for ΔG in terms of $\Delta G^{\ominus}$ and the reaction quotient Q developed in Sec. 20-5, especially Equation (20-27):

$$\Delta G = \Delta G^{\ominus} + RT \ln Q \qquad (21\text{-}18)$$

Insertion of (21-18) into (21-16) gives

$$\mathbf{E} = -\frac{\Delta G^{\ominus} + RT \ln Q}{\mathbf{n}\mathcal{F}}$$

or, by use of (21-17),

$$E = E^{\ominus} - \frac{RT}{n\mathfrak{F}} \ln Q \qquad (21\text{-}19)$$

This is the *Nernst equation*. It expresses the emf of an electrochemical cell in terms of the standard emf, the temperature T, the number n of electrons involved in the overall cell reaction, and the concentrations and partial pressures of the chemical species involved. The conventions that apply to Q in connection with the equilibrium law apply here also.

It is convenient to replace the natural logarithm in the Nernst equation by the base-10 logarithm and to combine the conversion factor between these two logarithms with the value of $RT/\mathfrak{F}$ at 25°C, the temperature commonly assumed in calculations with cell potentials. To do this, we use the following conversions: $1\ \mathfrak{F} = 96,485$ J V^{-1} mol^{-1}; $RT = 8.314 \times 298.15 = 2479$ J mol^{-1}; and $\ln x = 2.303 \log x$. Thus at 298 K,

$$\frac{RT}{n\mathfrak{F}} \ln Q = \left(\frac{2479 \times 2.303}{n \times 96,485} \log Q \right) \text{volt} = \left(\frac{0.0592}{n} \log Q \right) \text{volt}$$

so that, at 25°C,

$$E = E^{\ominus} - \left(\frac{0.0592}{n} \log Q \right) \text{volt} \qquad (21\text{-}20a)$$

$$E \approx E^{\ominus} - \left(\frac{0.06}{n} \log Q \right) \text{volt} \qquad (21\text{-}20b)$$

In applications a temperature of 298 K will always be assumed unless another temperature is explicitly indicated.

□ **The EMF of a Cell**   Write the Nernst equation for a cell like that shown in Fig. 21-3, with the overall reaction $Zn(s) + Cu^{2+} \rightarrow Zn^{2+} + Cu(s)$.

Example 21-4

Solution   Since $n = 2$,

$$E = E^{\ominus} - \frac{0.0592}{2} \log \frac{[Zn^{2+}]}{[Cu^{2+}]} \qquad (21\text{-}21a)$$

$$E \approx E^{\ominus} - 0.03 \log \frac{[Zn^{2+}]}{[Cu^{2+}]} \qquad (21\text{-}21b)$$

where $E^{\ominus}$ is the cell emf at standard conditions. Note that when $[Zn^{2+}]$ and $[Cu^{2+}]$ are both $1\ M$, the logarithmic term in (21-21) vanishes, so that $E = E^{\ominus}$, as it should, since these concentrations correspond to standard conditions. In fact, the logarithmic term vanishes whenever $[Zn^{2+}] = [Cu^{2+}]$. Furthermore, when the cell is *written* in the opposite direction, $E^{\ominus}$ then changes sign and the logarithmic term becomes $\log ([Cu^{2+}]/[Zn^{2+}]) = -\log ([Zn^{2+}]/[Cu^{2+}])$. Since both terms change sign, E retains its magnitude but changes sign, as anticipated.[5] Experimental measurements show that $E^{\ominus}$ for this cell is about 1.1 V. ■

---

[5]Note, however, that for any given cell, that is, a specific pair of electrodes and specified concentrations of electrolytes, the sign (as well as the magnitude) of the voltage difference between the terminals, which could be measured with a voltmeter, is fixed and is completely independent of the direction in which the cell reaction is *written*.

Exercise 21-4

☐ Write the Nernst equation for a cell with the overall reaction $Al(s) + 3Ag^+ \rightleftharpoons Al^{3+} + 3\,Ag(s)$. ∎

Example 21-5

☐ **Number of Electrons Exchanged** What value of **n** must be used in the Nernst equation for each of the following reactions:

(a) $2Fe^{3+} + Pb \longrightarrow 2Fe^{2+} + Pb^{2+}$

(b) $Ag(s) + Cu^{2+} + 2Cl^- \longrightarrow AgCl(s) + CuCl(s)$

(c) $Pt(s) + 6Cl^- + 4H^+ \longrightarrow 2H_2 + PtCl_6^{2-}$

(d) $6H^+ + IO_3^- + 5I^- \longrightarrow 3I_2(s) + 3H_2O$

Solution Since the number of electrons gained must be the same as that lost, the value of **n** can be found by considering either the substance oxidized or that reduced. As a check, we shall consider both.

(a) Each of two ferric ions gains one electron, so **n = 2**; this is consistent with the single Pb atom's losing two electrons to form Pb^{2+}.

(b) $Ag(s)$ combines with one Cl^- to form uncharged $AgCl(s)$, so <u>one electron</u> is lost in this half-reaction. The same value of **n**, 1, is derived from the fact that Cu^{2+} and one Cl^- form uncharged $CuCl(s)$, a process that must be accompanied by the gain of one electron to maintain charge balance.

(c) Each of four protons gains an electron as $4H^+$ forms $2H_2$, so **n = 4**. The single Pt atom combines with $6Cl^-$ to form $PtCl_6^{2-}$ with a loss of four units of negative charge, again consistent with **n = 4**.

(d) Each of five I^- loses an electron in going to I_2, which implies **n = 5.** Simultaneously, the I in IO_3^-, which is in the $+5$ oxidation state, is also changed into iodine in the zero oxidation state, corresponding to a gain of five electrons by each IO_3^-, in agreement with the value **n = 5**. ∎

Exercise 21-5

☐ What value of **n** is associated with each of the following reactions?

(a) $Sn^{2+} + Cl_2(g) \rightleftharpoons Sn^{4+} + 2Cl^-$

(b) $Hg_2Cl_2 + Zn(s) \rightleftharpoons 2Hg(l) + Zn^{2+} + 2Cl^-$

(c) $Co(s) + 6HF \rightleftharpoons CoF_6^{3-} + 3H^+ + \tfrac{3}{2}H_2(g)$

(d) $ClO_3^- + 3H_2SO_3 \rightleftharpoons 3SO_4^{2-} + Cl^- + 6H^+$ ∎

21-4 Applications of the Nernst Equation

Standard Electrode Potentials In order to use the Nernst equation to calculate the voltage of a cell it is necessary to know the cell reaction (and thereby the number of electrons transferred), the concentrations (or pressures) of all reactants and products, and the standard electrode potential. It is clearly inconvenient to determine $E^{\ominus}$ by direct measurement for every cell that must be considered and, in fact, such a multitude of experiments is unnecessary. Each *electrode* of an electrochemical cell can be assigned a standard electrode potential such that the $E^{\ominus}$ for the cell is the difference between the values for the two electrodes. The zero point of the scale of standard electrode potentials is quite arbitrary, since only the differences between the values can be measured experimentally. This arbitrary reference point is conventionally[6] chosen to be the standard hydrogen electrode, a hydrogen electrode in which the H_2 gas is at 1 atm pressure and the hydrogen-ion concentration is $1\,M$.

Consider, for example, a cell for which the reaction is

$$H_2(g) + Zn^{2+} \longrightarrow 2H^+ + Zn(s)$$

By the Nernst equation, the emf is

$$E = E^\ominus - \frac{0.0592}{2} \log \frac{[H^+]^2}{P_{H_2}[Zn^{2+}]} \tag{21-22}$$

We now write $E^\ominus$ as the difference between the standard electrode potential for the cathode and that for the anode

$$E = E^\ominus_{Zn^{2+}/Zn} - E^\ominus_{H^+/H_2} - 0.03 \log \frac{[H^+]^2}{P_{H_2}[Zn^{2+}]} \tag{21-23}$$

where the standard electrode potentials are identified by a subscript.

When products and reactants are in their standard states (21-23) reduces to

$$E = E^\ominus_{Zn^{2+}/Zn} - E^\ominus_{H^+/H_2} \tag{21-24}$$

The choice of the standard hydrogen electrode as a reference corresponds to choosing the value $E^\ominus_{H^+/H_2} = 0.000$, so that the measured cell emf is the value of the Zn^{2+}/Zn standard electrode potential

$$E = E^\ominus_{Zn^{2+}/Zn}$$

Values of Standard Electrode Potentials Values of standard electrode potentials can be obtained from measurement of voltages of appropriate cells and also from thermodynamic data. A representative set of standard electrode potentials is given in Tables 21-1 and D-5. The choice of the standard hydrogen electrode as

---

[6]In parallel with the conventions that $G^\ominus_{f,H^+} = 0$ and $G^\ominus_{f,H_2} = 0$.

Table 21-1
Standard Electrode Potentials

Reaction	$E^\ominus$/volt	Reaction	$E^\ominus$/volt
$F_2(g) + 2e^- = 2F^-$	2.87	$Hg_2Cl_2(s) + 2e^- = 2Hg(l) + 2Cl^-$	0.268
$H_2O_2 + 2H^+ + 2e^- = 2H_2O$	1.77	$AgCl(s) + e^- = Ag(s) + Cl^-$	0.222
$MnO_4^- + 8H^+ + 5e^- = Mn^{2+} + 4H_2O$	1.51	$Cu^{2+} + e^- = Cu^+$	0.153
$PbO_2(s) + 4H^+ + 2e^- = Pb^{2+} + 2H_2O$	1.47	$S(s) + 2H^+ + 2e^- = H_2S$	0.14
$Cl_2(g) + 2e^- = 2Cl^-$	1.359	$HSO_4^- + 3H^+ + 2e^- = SO_2(g) + 2H_2O$	0.14
$Cr_2O_7^{2-} + 14H^+ + 6e^- = 2Cr^{3+} + 7H_2O$	1.33	$S_4O_6^{2-} + 2e^- = 2S_2O_3^{2-}$	0.09
$O_2(g) + 4H^+ + 4e^- = 2H_2O$	1.229	$2H^+ + 2e^- = H_2(g)$	0.000
$2IO_3^- + 12H^+ + 10e^- = I_2(s) + 6H_2O$	1.19	$Pb^{2+} + 2e^- = Pb(s)$	-0.126
$Br_2(l) + 2e^- = 2Br^-$	1.065	$Ni^{2+} + 2e^- = Ni(s)$	-0.23
$Cu^{2+} + I^- + e^- = CuI(s)$	0.85	$Fe^{2+} + 2e^- = Fe(s)$	-0.440
$Ag^+ + e^- = Ag(s)$	0.799	$Zn^{2+} + 2e^- = Zn(s)$	-0.763
$Fe^{3+} + e^- = Fe^{2+}$	0.771	$Al^{3+} + 3e^- = Al(s)$	-1.66
$O_2(g) + 2H^+ + 2e^- = H_2O_2$	0.69	$Mg^{2+} + 2e^- = Mg(s)$	-2.37
$I_3^- + 2e^- = 3I^-$	0.535	$Na^+ + e^- = Na(s)$	-2.698
$I_2(s) + 2e^- = 2I^-$	0.534	$K^+ + e^- = K(s)$	-2.925
$Cu^{2+} + 2e^- = Cu(s)$	0.337	$Li^+ + e^- = Li(s)$	-3.03

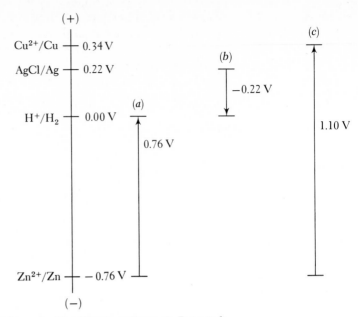

Figure 21-4 Relative Values of a Few Standard Electrode Potentials
The vertical line at the left represents a linear voltage scale; the short horizontal lines on it are spaced in proportion to the relative values of the standard emfs of the indicated electrodes. The differences in standard electrode potentials for three electrode pairs are given beside the arrows labeled (a), (b), and (c). Only *differences* of electrode potentials can be measured experimentally. The emf values given are for cells under standard conditions. Each arrow in the figure points from the electrode *assumed* to be the anode to that *assumed* to be the cathode. Line (a), for example, represents the potential of a cell for which the standard zinc electrode is the anode and the standard hydrogen electrode the cathode. The (+) and (−) at the ends of the scale are the signs conventionally associated with electrode potentials in these regions of the scale. When the standard hydrogen electrode is assigned the reference value $E^\circ = 0.000$ V, the standard value for the Zn^{2+}/Zn electrode is then -0.76 V, that for AgCl/Ag is $+0.22$ V, and that for Cu^{2+}/Cu is $+0.34$ V.

the reference for measurement of electrode potentials is analogous to the choice of mean sea level as the zero point for measurement of heights on the earth's surface; it is merely a convenience. Any other electrode—AgCl/Ag, for example—might have been used as a reference instead. The difference between two heights on the earth need not be measured directly if each has been established relative to mean sea level; their difference may be obtained directly by subtraction, taking into account the sign of each if one is above and the other below sea level. The same is true for electrode potentials (Fig. 21-4).

It is only by convention that we assign a negative sign to the altitudes of points that are on the same side of mean sea level as the center of the earth (points such as the low point of the floor of Death Valley) and a positive sign to the altitudes of mountain peaks and other points "outside" mean sea level. The opposite convention could just as well be used. Similarly, the convention for the signs of electrode potentials is arbitrary. A positive sign is given to the electrode emfs of redox couples for which the oxidized form is a better oxidizing agent—that is, gains electrons more easily, is more easily reduced—than is H^+ in the standard hydrogen electrode. Thus couples containing strong oxidizing agents, such as F_2/F^-, Cl_2/Cl^-, and MnO_4^-/Mn^{2+}, have highly positive standard emfs. Couples containing the best reducing agents, including strongly electropositive metals like aluminum, the alkaline-earth metals, and the alkali metals, have quite negative standard emfs.

□ **Combining Standard Electrode Potentials** Calculate $E^{\ominus}$ for a cell based on the overall reaction

Example 21-6

$$2Ag^+ + Pb(s) \longrightarrow 2Ag(s) + Pb^{2+} \tag{21-25}$$

Solution The two half-reactions that can be combined to give (21-25) are

$$Ag^+ + e^- \longrightarrow Ag(s) \tag{21-26a}$$

$$Pb^{2+} + 2e^- \longrightarrow Pb(s) \tag{21-26b}$$

From Table 21-1 we find $E^{\ominus}_{Ag^+/Ag} = 0.799$ V for (21-26a) and $E^{\ominus}_{Pb^{2+}/Pb} = -0.126$ V for (21-26b). To combine Equations (21-26a) and (21-26b) into (21-25), Equation (21-26a) has to be multiplied by 2. This does not affect the electrode potential associated with it.[7] Equation (21-26b) must be subtracted, which means that the electrode potential associated with it has to be subtracted also. In other words, the rule for combining standard electrode potentials is

$$E^{\ominus} = E^{\ominus}_{cathode} - E^{\ominus}_{anode} \tag{21-27}$$

Insertion of the values provided yields

$$E^{\ominus} = E^{\ominus}_{Ag^+/Ag} - E^{\ominus}_{Pb^{2+}/Pb}$$
$$= 0.799 - (-0.126) = \underline{0.925 \text{ V}} \quad \blacksquare$$

□ Calculate $E^{\ominus}$ for a cell based on the overall reaction

$$2Al^{3+} + 3Cu(s) \rightleftharpoons 2Al(s) + 3Cu^{2+} \quad \blacksquare$$

The value and sign of the cell voltage depend not only on $E^{\ominus}$ but also on the concentrations of the species participating in the cell reaction. Consider a cell with the reaction

$$Cl_2(g) + 2Fe^{2+} \longrightarrow 2Cl^- + 2Fe^{3+} \tag{21-28}$$

The Nernst equation for this cell is $E = E^{\ominus} - (0.06/2) \log Q$

$$E = E^{\ominus} - \frac{0.06}{2} \log \frac{[Cl^-]^2[Fe^{3+}]^2}{P_{Cl_2}[Fe^{2+}]^2} \tag{21-29}$$

where $E^{\ominus} = E^{\ominus}_{Cl_2/Cl^-} - E^{\ominus}_{Fe^{3+}/Fe^{2+}} = 1.359 - 0.771 = 0.588$ V.

What changes occur in E and in the concentrations of the ions in the cell as the cell reaction proceeds? Assume first that the pressure of Cl_2 is held at 1 atm and that the concentrations of Fe^{2+}, Cl^-, and Fe^{3+} are such that E, as calculated from (21-29), is positive. The cell is galvanic and, if current is permitted to flow, the reaction as written in (21-28) proceeds from left to right. As a consequence, the concentration of Fe^{2+} decreases and those of Cl^- and Fe^{3+} increase, so Q increases and E *decreases* as the reaction proceeds. If E is negative initially, then for the concentrations present reaction (21-28) will tend to go spontaneously from right to left; however, the reaction may be driven from left to right by an external voltage. The resulting concentration changes would make Q larger, and E would

---

[7]The voltage of a battery does not depend on its size; a flashlight battery and a standard penlight battery have the same voltage, but for similar use the former lasts much longer.

become still more negative. Under these circumstances the cell is electrolytic, and the external voltage required increases as the electrolysis proceeds. In contrast, the voltage available from a galvanic cell, that is, a cell in which the reaction is proceeding spontaneously, decreases toward zero as the reaction goes on.

Example 21-7

☐ **Cell EMF**   What is the emf at 25°C of a cell for which the reaction is $Zn(s) + Cu^{2+} \rightarrow Zn^{2+} + Cu(s)$ when the concentrations of the ions are $[Zn^{2+}] = 1\,M$, $[Cu^{2+}] = 0.1\,M$?

Solution   By Table 21-1 we note that $E^{\ominus}_{Zn^{2+}/Zn} = -0.76$ V and $E^{\ominus}_{Cu^{2+}/Cu} = 0.34$ V. Furthermore, the Zn electrode is the anode because Zn is oxidized to Zn^{2+} as the cell reaction proceeds. Thus

$$E^{\ominus} = 0.34 - (-0.76) = 1.10 \text{ V}$$

Inserting $[Zn^{2+}]/[Cu^{2+}] = 10$ into the Nernst equation (21-21b) yields

$$E = \left(1.10 - \frac{0.06}{2} \log 10\right) \text{volt}$$

$$= \underline{1.07 \text{ V}}$$

Since **E** is positive, the cell is galvanic. ∎

Exercise 21-7

☐ What is the emf at 25°C of a cell for which the reaction is $2Ag^+ + Cu(s) \rightleftharpoons 2Ag(s) + Cu^{2+}$ when the concentrations of the ions are $[Cu^{2+}] = 0.1\,M$, $[Ag^+] = 0.01\,M$? Standard electrode potentials are given in Table 21-1. ∎

Example 21-8

☐ **Cell Reaction and EMF**   Calculate the cell emf for the reaction

$$5Cl^- + MnO_4^- + 8H^+ \rightarrow \tfrac{5}{2}Cl_2(g) + Mn^{2+} + 4H_2O$$

at 25°C, when $P_{Cl_2} = 1$ atm, $[H^+] = 1.0\,M$, and the concentrations of all other ionic species are $0.1\,M$. The standard electrode potentials are $E^{\ominus}_{Cl_2/Cl^-} = 1.36$ V and $E^{\ominus}_{MnO_4^-/Mn^{2+}} = 1.51$ V.

Solution   For the cell reaction $n = 5$, so that

$$E = \left(1.51 - 1.36 - \tfrac{0.06}{5} \log \frac{P_{Cl_2}^{5/2}[Mn^{2+}]}{[Cl^-]^5[MnO_4^-][H^+]^8}\right) \text{volt}$$

the logarithmic term is equal to $-\tfrac{0.06}{5} \log (0.1)^{-5} = -0.06$, and the final result is $E = \underline{0.09 \text{ V.}}$ Since **E** is positive, the reaction is spontaneous and MnO_4^- is capable of oxidizing Cl^- under the conditions stated. However, the smallness of the value of **E** indicates that the concentrations are not far from equilibrium concentrations, as we shall see in the next section. ∎

Exercise 21-8

☐ Calculate the cell emf for the reaction

$$4Fe^{2+} + O_2(g) + 4H^+ \rightleftharpoons 4Fe^{3+} + 2H_2O$$

at 25°C when $P_{O_2} = 1$ atm, $[H^+] = 0.010\,M$, and the concentrations of all other ionic species are $0.1\,M$. Standard electrode potentials are given in Table 21-1. ∎

As a spontaneous cell reaction proceeds, the cell emf drops toward zero—the battery runs down. This is consistent with the Nernst equation, as illustrated in the discussion of the cell corresponding to reaction (21-28) (Sec. 21-4). When $\mathbf{E}$ finally becomes zero, there is equilibrium between the two redox couples present, at their final concentrations. Of course it may happen that equilibrium exists at the initial concentrations chosen when the cell is prepared; if so, the initial value of $\mathbf{E}$ is zero and the cell is of no use as a battery. The fact that $\mathbf{E} = 0$ at equilibrium is reasonable in light of the relation $\Delta G = -\mathbf{n}\mathfrak{F}\mathbf{E}$ at constant temperature and pressure, for we know that under these conditions the criterion of equilibrium is $\Delta G = 0$.

The equilibrium eventually reached when a spontaneous cell reaction has been allowed to proceed is the *same* as that which would have been reached had the reactants been mixed together directly. When the reaction takes place in a galvanic cell, some of the energy released in the reaction is available as electrical energy as a consequence of the flow of electrons through the external circuit. On the other hand, when the reactants are mixed directly, the energy released is not utilized—it is dissipated, chiefly as thermal energy.

Equilibrium constants for redox reactions can be calculated from the standard electrode potentials of the half-cells that are coupled together in a reaction. It does not matter whether a cell could actually be constructed with the two redox couples considered. As discussed in Chap. 20 an equilibrium constant can be calculated from standard free energies regardless of whether the corresponding reaction has ever been carried out or could be carried out in practice. This must also be true of standard electrode potentials because they are proportional to standard free energies per mole of electrons exchanged.

If we write the Nernst equation (21-19) for a cell for which $\mathbf{E}$ is zero, we may replace Q by K because the reactants are at equilibrium with each other under these conditions:

$$\mathbf{E} = 0 = \mathbf{E}^{\ominus}_{\text{cathode}} - \mathbf{E}^{\ominus}_{\text{anode}} - \frac{RT}{\mathbf{n}\mathfrak{F}} \ln K$$

or

$$RT \ln K = \mathbf{n}\mathfrak{F}(\mathbf{E}^{\ominus}_{\text{cathode}} - \mathbf{E}^{\ominus}_{\text{anode}}) = \mathbf{n}\mathfrak{F}\mathbf{E}^{\ominus} \qquad (21\text{-}30)$$

and at 25°C, with $\mathbf{E}^{\ominus}$ in volts,

$$\log K = \frac{\mathbf{n}\mathbf{E}^{\ominus}}{0.0592} \qquad (21\text{-}31)$$

or

$$K = 10^{\mathbf{n}\mathbf{E}^{\ominus}/0.0592} \qquad (21\text{-}32)$$

Table (21-2) shows the relationships between the characteristic features of $\Delta G^{\ominus}$, $\mathbf{E}^{\ominus}$, K, and the direction in which a cell reaction *under standard conditions* occurs spontaneously. The second case ($\Delta G^{\ominus} = 0$), in which the reactants and products are at standard conditions and at equilibrium at the same time, is a rather unlikely situation.

The relationship between cell emf and ΔG shows the feasibility of obtaining free-energy changes by electrical measurements. Values of ΔG may also be deduced by measurement of appropriate thermal properties and of equilibrium constants, as discussed in Chap. 20.

Table 21-2
Correlation among $\Delta G^\ominus$, $E^\ominus$, K, and the Direction of the Cell Reaction

$\Delta G^\ominus$	$E^\ominus$	K	Cell reaction under *standard conditions* is:
<0	>0	>1	Spontaneous toward right
0	0	1	At equilibrium
>0	<0	<1	Spontaneous toward left

Example 21-9

□ **Equilibrium Constants and Standard Electrode Potentials** What is the equilibrium constant at 25°C for the cell reaction

$$Zn(s) + Cu^{2+} \longrightarrow Zn^{2+} + Cu(s)$$

The standard electrode potentials are 0.337 V for Cu^{2+}/Cu and -0.763 V for Zn^{2+}/Zn.

Solution Since Zn is oxidized in the cell as written, the emf of the Zn^{2+}/Zn couple is to be subtracted from that of the other couple and $E^\ominus = 0.337 - (-0.763) = 1.100$ V. The number of electrons interchanged is two. Thus, by (21-31), $\log K = 2 \times 1.100/0.0592 = 37.2$ and $K = \underline{10^{37.2}} = 1.6 \times 10^{37}$.

Actually, the answer might be written as 2×10^{37} because the last digit in the figure 37.2 is uncertain, and this digit represents the entire mantissa of the logarithm. Hence the antilogarithm of that mantissa is only approximate. (Since the antilog of 0.1 is 1.3, an uncertainty of 0.1 in a logarithm to the base 10 corresponds to an uncertainty of a factor of 1.3, or about 30 percent, in the antilogarithm.) ∎

Exercise 21-9

□ What is the equilibrium constant at 25°C for the cell reaction

$$Pb(s) + PbO_2(s) + 4H^+ \rightleftharpoons 2Pb^{2+} + 2H_2O$$

Standard electrode potentials are given in Table 21-1. ∎

Example 21-10

□ **Equilibrium Constant and Standard Electrode Potentials: The Iodate-Iodide Reaction** What is the equilibrium constant at 25°C for the reaction

$$6H^+ + IO_3^- + 5I^- \longrightarrow 3I_2(s) + 3H_2O$$

In this reaction, iodide is oxidized to iodine by iodate, and the iodate is itself reduced to iodine. The standard electrode potentials are 1.19 V for the IO_3^-/I_2 couple and 0.534 V for the I_2/I^- couple.

Solution In this reaction the direction in which the I_2/I^- couple proceeds is that of an oxidation, so its emf is to be subtracted from that of the IO_3^-/I_2 couple. Thus,

$$E^\ominus = (1.19 - 0.534) \text{ V} = 0.66 \text{ V}$$

The number of electrons interchanged is five (Example 21-5d). Thus, by (21-31),

$$\log K = 5 \times \tfrac{0.66}{0.0592} = 56 \quad \text{and} \quad K = \underline{10^{56}}$$

Note that in this example even the power of 10 is somewhat uncertain because of the limited precision of the $E^\ominus$ value for the IO_3^-/I_2 couple. ∎

What is the equilibrium constant at 25°C for the reaction

$$14H^+ + Cr_2O_7{}^{2-} + 6Fe^{2+} \rightleftharpoons 7H_2O + 2Cr^{3+} + 6Fe^{3+}$$

Standard electrode potentials are given in Table 21-1. ■

□ **Solubility Product and Standard Electrode Potentials** Consider the **Example 21-11**
following two half-reactions and standard electrode potentials at 25°C:

$$Ag^+ + e^- \longrightarrow Ag(s) \qquad E^\ominus = 0.799 \text{ V} \qquad (21\text{-}33)$$
$$AgCl(s) + e^- \longrightarrow Ag(s) + Cl^- \qquad E^\ominus = 0.222 \text{ V} \qquad (21\text{-}34)$$

The reaction in each system may be considered to be the reduction of Ag(I) to Ag(0). Why then is there a difference in the standard emfs? Calculate the solubility product of $AgCl(s)$ from the data given.

Solution Recalling what is meant by standard conditions, we see that in (21-33), $[Ag^+] = 1\ M$, whereas in (21-34), $[Cl^-] = 1\ M$ *in the presence of solid AgCl*. The solubility of silver chloride is, however, so small that when $[Cl^-]$ in a solution is $1\ M$, $[Ag^+]$ in that same solution must be very small. Thus we understand why the standard emfs are so different for (21-33) and (21-34): the concentration of Ag^+ in equilibrium with the silver electrode is very different in the two solutions, $1\ M$ in (21-33) and very small in (21-34).

Next, consider the cell for which (21-33) is the anode reaction and (21-34) the cathode reaction, so that the overall reaction is

$$AgCl(s) \longrightarrow Ag^+ + Cl^- \qquad (21\text{-}35)$$

The equilibrium constant for (21-35) is $K = [Ag^+][Cl^-]$, which we recognize as K_{sp} for $AgCl(s)$. Thus, if we determine K for a cell based on (21-33) and (21-34), we will have determined K_{sp}. For this cell the standard emf is $0.222 - 0.799 = -0.577$ V and $\mathbf{n} = 1$. Applying (21-31), we have

$$\log K = -\tfrac{0.577}{0.0592} = -9.75$$
$$K = 10^{-9.75} = \underline{1.8 \times 10^{-10}} \quad ■$$

□ Calculate the solubility product at 25°C of $PbSO_4(s)$ from the standard
electrode potentials for the related couples $PbSO_4/Pb$ and Pb^{2+}/Pb given in Table D-5. ■

21-6 Technological Applications

Electrolysis Electrolytic methods are used in the manufacture of many of the more reactive elements, as discussed in Secs. 24-1 and 25-1. They are also used in the refining of some impure metals (Sec. 25-1), in the synthesis of various organic and inorganic materials, and in electroplating, which has extensive applications in precise analysis and in the creation of durable and handsome finishes on all kinds of objects.

Some Common Batteries In principle any redox reaction can be used to produce an electric current, but in commercial practice a number of requirements must be satisfied. These requirements include low cost of materials and manufacture (except in very specialized applications such as space exploration or heart-pacer batteries, for which cost is secondary), high energy per unit volume and unit weight, ruggedness, portability, long shelf life, and relatively constant potential during discharge even at high discharge rates. The total energy a battery is capable of delivering is measured in watthours and is sometimes called the capacity of the cell.

In the common *flashlight battery* or *dry cell* (Fig. 21-5a) the electrolyte is a saturated solution of ammonium chloride, also containing zinc chloride. The electrolyte is made into a paste with an excess of solid ammonium chloride and diatomaceous earth or another inert filler. This cell is not really dry, for water is an essential component of the electrolyte.

The cathode of the dry cell is a graphite rod in the center, surrounded by a powdered mixture of manganese dioxide and graphite. The MnO_2 serves as the oxidizing agent, accepting electrons that enter the cell at the cathode; if the MnO_2 were not present, the electrons would be accepted instead by the water of the electrolyte and gaseous H_2 would be formed. This would make it impossible to seal the battery and would also produce a film of gas around the cathode, impeding further reaction and therefore making it impossible to draw more current from the battery. The cathode reaction is complex, but is often represented as

$$MnO_2 + NH_4^+ + e^- \longrightarrow MnO(OH) + NH_3$$

The anode of a dry cell is a zinc cylinder that acts as a container for the electrolyte and is essentially the casing of the battery, except for a heavy paper, plastic, or steel shield wrapped around it. The electrolyte is quite acidic.

In the *alkaline dry cell* ammonium chloride is replaced by potassium hydroxide. It is more expensive than the usual dry cell but its useful life is increased by about 50 percent.

Another important though more expensive modern battery is the *mercury cell* (Fig. 21-5b). The battery is contained in a stainless steel cylinder, with the Zn anode at the center. The strongly alkaline electrolyte is saturated with zinc oxide in the form of $Zn(OH)_4^{2-}$ ions. Powdered carbon provides for adequate conduc-

Figure 21-5   Typical Cells
(a) Dry cell; (b) mercury cell.

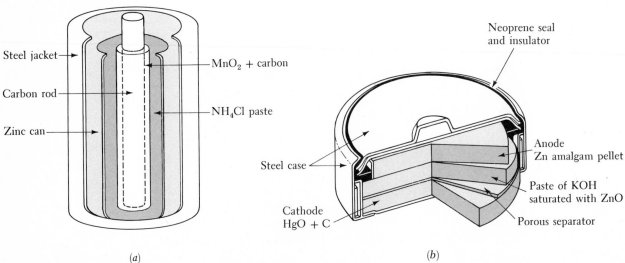

(a)

(b)

tance in the region around the cathode. The electrode reactions are

Anode: $Zn(amalgam) + 2OH^- \longrightarrow ZnO(s) + H_2O + 2e^-$
Cathode: $HgO(s) + H_2O + 2e^- \longrightarrow Hg(l) + 2OH^-$
Overall: $Zn(amalgam) + HgO(s) \longrightarrow ZnO(s) + Hg(l)$

This battery has an excellent voltage characteristic, maintaining a potential of about 1.34 V during its entire life. The reason the voltage remains constant is that the overall cell reaction involves only solid substances and the concentrations of the species associated with solids remain constant (Sec. 11-3); the electrolyte acts solely as an intermediary.

Of great importance are *storage batteries,* which usually consist of three or six identical, rechargeable cells joined together in sequence. One of these is the lead storage battery, which has a potential of about 2.0 V. The electrodes are grids of lead alloy, the interstices being filled with spongy lead at the anode and lead dioxide at the cathode. The reactions that occur as the cell produces current are

Anode [terminal stamped $(-)$]:
$$Pb(s) + SO_4^{2-} \longrightarrow PbSO_4(s) + 2e^- \qquad (21\text{-}36)$$
Cathode [terminal stamped $(+)$]:
$$PbO_2(s) + SO_4^{2-} + 4H^+ + 2e^- \longrightarrow PbSO_4(s) + 2H_2O \qquad (21\text{-}37)$$
Overall:
$$Pb(s) + PbO_2(s) + 2H_2SO_4 \longrightarrow 2PbSO_4(s) + 2H_2O \qquad (21\text{-}38)$$

These reactions produce $PbSO_4$ at each electrode; this salt is only slightly soluble and remains on the plates. Sulfate ions react not only at the cathode (with PbO_2) but also at the anode (with Pb), in spite of their negative charge. Some of the sulfuric acid is used up and water is produced as the spontaneous reaction (21-38) goes on, so that the sulfuric acid electrolyte is gradually diluted as the cell reaction proceeds to the right. It is therefore possible to determine the state of the battery by measuring the density of the liquid in the battery.

During the recharging of a lead storage battery, reaction (21-38) is reversed by the application of an external potential. This can be done with no difficulty when the $PbSO_4$ is freshly precipitated, but aging slowly changes the lead sulfate on the plates to a relatively unreactive form. A battery that remains in a discharged state for weeks can be recharged only partially, if at all.

Another important storage cell is the *nickel-cadmium cell,* with the overall discharge reaction

$$Cd(s) + 2Ni(OH)_3(s) \longrightarrow CdO(s) + 2Ni(OH)_2(s) + H_2O(l)$$

This cell is more expensive than the lead storage cell but it is more rugged, lighter, has a longer life, and may remain uncharged after use with no damage. It is used in cordless electric toothbrushes, electric razors, pocket calculators, and other small portable appliances. The potential of this cell is about 1.35 V.

Fuel Cells The batteries discussed so far are closed systems with fixed amounts of oxidant and reductant, although in storage cells the oxidant and reductant can be regenerated by recharging from an external energy source. Batteries can also be made in which reactants are continuously fed to the electrodes, and such batteries, known as *fuel cells,* are of increasing importance. They are analogous in some ways to ordinary furnaces fed with coal, oil, or gas and to internal-combustion engines that burn gasoline or oil, each of which produces energy on demand from a continuously fed fuel supply.

A number of fuel cells are currently in use. Those best developed are based on the hydrogen-oxygen reaction. The automated source of auxiliary power for the command and service modules of the Apollo space vehicles was a hydrogen-oxygen fuel cell of a type originally designed by F. T. Bacon more than three decades ago. The electrodes in this cell are of porous nickel, and the cell runs at about $200°C$ and at gas pressures below 1 atm. The electrolyte is 80 percent KOH; hydrogen is fed to the anode, oxygen to the cathode.

21-7 Corrosion

Corrosion is the deterioration of a substance as a result of reactions involving the environment. The term is applied almost exclusively to the deterioration of metals, on which the corrosive attack is invariably an oxidation. Corrosion is manifested by such familiar and seemingly inevitable phenomena as the rusting of iron, the tarnishing of silver, or the development of a green patina on copper, brass, and bronze. The damage it causes amounts to billions of dollars each year—and would be far greater if strenuous preventive efforts were not made.

Many factors accelerate corrosion. The presence of both oxygen and water appears to be necessary, and corrosion is markedly accelerated near the sea or in other places where salts can contaminate metal surfaces (e.g., on automobiles driven over roads on which salts have been used to melt ice or prevent dust from blowing). Acids also enhance rates of corrosion, and they are more common in the atmosphere than is usually realized. Carbon dioxide, which dissolves to form the weak acid H_2CO_3, is always present; other volatile acidic oxides, such as NO_2 and SO_2, are also present, especially near industrial areas. Elevated temperatures speed up corrosion, just as they do most other reactions. It is no surprise that corrosion is extremely rapid in tropical jungles, where moisture, elevated temperatures, and acidic products from decaying vegetation act in concert.

Many of the basic features of the process of rust formation apply to other corrosion processes as well. Rust is a hydrated form of iron(III) oxide, $Fe_2O_3 \cdot xH_2O$, with x approximately equal to 3. Iron does not rust in a dry atmosphere or in oxygen-free water. The reactions involved in rust formation are complex and not completely understood, but in the main the steps are thought to be the following. Metallic iron is first oxidized to the dipositive state,

$$Fe(s) \longrightarrow Fe^{2+} + 2e^- \tag{21-39}$$

The resulting electrons usually reduce atmospheric oxygen in a reaction that is facilitated by the slightly acidifying presence of dissolved CO_2,

$$\tfrac{1}{2}O_2(g) + 2H^+ + 2e^- \longrightarrow H_2O \tag{21-40}$$

At the same time, atmospheric oxygen may also combine with the Fe^{2+} ions and H_2O to form rust,

$$\tfrac{1}{2}O_2(g) + 2Fe^{2+} + (2 + x)H_2O \longrightarrow Fe_2O_3 \cdot xH_2O + 4H^+ \tag{21-41}$$

This reaction is itself capable of yielding sufficient H^+ ions to keep reaction (21-40) going.

Reactions (21-39) and (21-40) need not occur at the same position. A galvanic

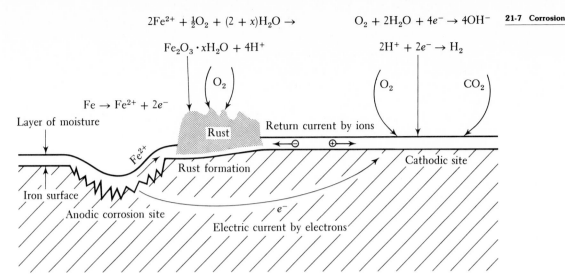

$$2Fe^{2+} + \tfrac{1}{2}O_2 + (2 + x)H_2O \rightarrow$$
$$Fe_2O_3 \cdot xH_2O + 4H^+$$

$$O_2 + 2H_2O + 4e^- \rightarrow 4OH^-$$
$$2H^+ + 2e^- \rightarrow H_2$$

O_2 O_2 CO_2

$$Fe \rightarrow Fe^{2+} + 2e^-$$

Layer of moisture

Rust

Return current by ions

Fe^{2+}

Rust formation

Cathodic site

Iron surface

Anodic corrosion site

e^-

Electric current by electrons

Figure 21-6 Electrochemistry of Rusting

In a common situation ferrous ions are formed at an anodic corrosion site that is protected from oxygen and then diffuse to a location accessible to the atmosphere, where rust is formed. The cathodic site may be at yet another location. Thus the shanks of bolts that fasten iron plates, even though protected from air, may be thinned by corrosion while rust forms on the neighboring plate surface and in other locations. Although the corrosion of many reactive metals, such as Al and Zn, is minimized by a protective oxide layer that adheres tightly to the underlying metal, the formation of rust provides only minor and temporary protection to the underlying iron, because rust flakes off.

cell, in which (21-39) provides the anode reaction and (21-40) that at the cathode, may be set up on the metal surface. The cells formed in this way may be on a microscopic scale, but anodic and cathodic sites may also be separated by macroscopic distances. The electric circuit must be completed by the migration of ions, which explains why the presence of salt water or other good electrolytes accelerates corrosion. Portions of the metal that are under mechanical stress often act as anodic corrosion sites, leading to stress-corrosion cracks that can very markedly weaken the metal. Reaction (21-41) may occur at still another site, because the ions involved may migrate by diffusion. Pitting and rust formation may thus occur in different locations (Fig. 21-6).

Inhibition of Corrosion Corrosion can be curbed in a number of ways. The most obvious method is to cover the metal surface with a coating that keeps out air and moisture. Because corrosion may begin in even microscopic cracks, this coating must itself be completely intact. It might be a layer of paint, plastic, or some similar material, a coating of a strongly adherent oxide, or a layer of some metal that is itself resistant to corrosion.

An interesting phenomenon is *passivation*, in which a metal surface is made inactive by covering it with a very thin oxide layer. Iron can, for example, be passivated by treatment with concentrated nitric acid or dichromate. Once in this state it does not react with acid or reduce cupric ions. Some metals that are inherently reactive in the presence of water and oxygen, including aluminum, magnesium, titanium, and various stainless steels, become passive spontaneously. They acquire a tough and adherent oxide film in air, and this film protects them effectively from serious corrosion damage, except over long periods of time or in unusual atmospheres. Paints that contain oxidizing substances such as "red lead" (Pb_3O_4) or potassium dichromate ($K_2Cr_2O_7$) are especially effective on iron because they make the surface passive. They are regularly used on structural steel exposed to severe weathering—for example, on ships and bridges.

Protective films of metals such as Zn or Sn are used on iron because these metals form self-protective oxide coats. Zinc plating (galvanizing) has the advantage that zinc is more easily oxidized than iron, $E^{\ominus}_{Zn^{2+}/Zn}$ being -0.76 V while $E^{\ominus}_{Fe^{2+}/Fe}$ is -0.44 V. If a scratch exposes the iron, it is still the zinc coating that is attacked. Tin plating, although less expensive than galvanizing, is not nearly as reliable. Tin is less easily oxidized than iron ($E^{\ominus}_{Sn^{2+}/Sn} = -0.14$ V), so that iron corrodes more easily than tin. Thus if a tin coating on iron is damaged, it is the iron that corrodes first. Indeed, the iron sometimes then corrodes more rapidly than it would have without the tin coating, in part because a galvanic cell is set up, the iron being oxidized and the oxide film on the tin being reduced to the metal.

Cathodic protection is an important method of corrosion prevention in which the typical anode reaction (21-39) is inhibited by making the iron the *cathode* of a galvanic cell. This can be done by placing pieces of Zn or Mg as sacrificial electrodes near the object to be protected (pipeline, ship hull) and connecting them to it electrically by direct contact or by heavy wires. The Mg and Zn act as anodes and are consumed while the iron remains uncorroded. The principal cathode reaction at the Fe end of the cell is the reduction of water to hydrogen, but there would also be immediate reduction to Fe of any Fe^{2+} that might have formed. Zinc plating is, in effect, a form of cathodic protection. Another way of achieving cathodic protection is to make the iron the cathode of an *electrolysis* cell. Carbon is used as the anode, and an external voltage is applied to force the iron to be negative relative to the carbon.

If the sacrificial Zn or Mg electrodes were replaced by electrodes of Cu or Sn, the iron would become the sacrificial electrode; instead of being protected, the iron would be made more vulnerable. This is what happens, for example, when iron pipes are connected to copper pipes or brass fittings, and it explains the resulting corrosion problem.

21-8 Ion-Selective Electrodes

Ion-Selective Electrodes Many electrodes for which the emf is indicative of the concentrations of specific ions have been developed in recent years. They are of particular value in analytical applications—for example, in analysis of solutions containing small amounts of many different substances, such as clinical specimens or seawater samples. Such electrodes are also especially adaptable to systems for automatic control of chemical processes on a large scale.

Probably the oldest of such electrodes is the glass electrode (Fig. 21-7a), which is the sensing probe of a pH meter. This device is based on the experimental finding that an emf develops across a very thin glass membrane separating two solutions of different pH and that this emf depends on the difference of the pH values. For some purposes the emf developed across the glass membrane could be determined sufficiently precisely by inserting two wires, one into the bulb and the other into the solution outside, and measuring the difference in emf between

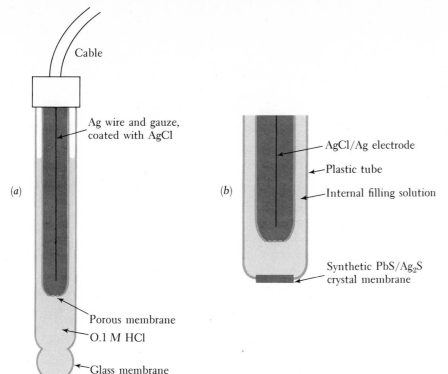

(a)

(b)

Cable

Ag wire and gauze, coated with AgCl

AgCl/Ag electrode

Plastic tube

Internal filling solution

Porous membrane

O.1 M HCl

Glass membrane

Synthetic PbS/Ag_2S crystal membrane

Figure 21-7 Ion-Selective Electrodes
(a) Glass electrode; (b) lead-selective electrode.

them. However, the stability and reproducibility of the emf developed across the interface between the two solutions is far greater if the system contains instead an AgCl/Ag electrode within the bulb, the connection to it being by a liquid junction through a porous wall. For maximum precision, the second electrode in the pH-measuring system is usually a standard calomel reference electrode in which liquid mercury (connected with the outside by a Pt wire) is in contact with solid calomel (Hg_2Cl_2) and a solution containing chloride ions. The half-reaction is

$$Hg_2Cl_2(s) + 2e^- \longrightarrow 2Hg(l) + 2Cl^- \qquad (21\text{-}42)$$

Electrical contact between the calomel electrode and the solution whose pH is to be measured is established through porous membranes and a KCl solution. Because the AgCl/Ag and calomel electrodes develop constant and highly re-producible emf values, the total emf between the leads to the meter depends only on the pH of the solution.

Electrodes with selective sensitivity for more than 20 other ions have also been developed and more are in prospect. The ions include those derived from the alkali and alkaline-earth elements, as well as Ag^+, Pb^{2+}, NO_3^-, ClO_4^-, and S^{2-}, among others. The lead-selective electrode (Fig. 21-7b) is representative. It is sensitive at the parts-per-billion level. Potential applications include the detection of traces of Pb^{2+} in urine and other bodily fluids in the diagnosis of lead poisoning, as well as the analysis of atmospheric samples (after they have been dissolved and concentrated) in studies of air pollution.

Summary

An electric current is a flow of charge. For a constant current $\mathbf{I}$, the amount of charge that flows past a given point in the time t is given by $\mathbf{Q} = \mathbf{I}t$. If the current is driven by a potential difference $\Delta\Phi$, the work done is $w = \mathbf{I}t\,\Delta\Phi$.

Some electrodes are simply pieces of metal; others are more complicated, such as a strip of silver coated with AgCl and immersed in a saturated AgCl solution. The chemical reactions that occur at electrodes can be represented by half-reactions. A reduction takes place at the cathode, the electrode at which electrons are delivered to an electrolyte

$$\text{ox 1} + \mathbf{n}_1 e^- \longrightarrow \text{red 1}$$

There is oxidation at the electrode at which electrons are removed, the anode

$$\text{red 2} \longrightarrow \mathbf{n}_2 e^- + \text{ox 2}$$

An electrochemical cell is a combination of electrodes and electrolyte(s) with appropriate arrangements to allow movement of charge between different sections of the cell. When the cell reaction occurs in the spontaneous direction, driving electrons through an external circuit, the cell is said to be galvanic, and when the cell reaction is driven in the opposite direction by an externally applied potential, the cell is termed electrolytic.

The weights of different substances formed or used up in electrode reactions are proportional to their formula weights divided by $\mathbf{n}$, the number of electrons participating in the reaction. If the reaction of one molecule or ion involves $\mathbf{n}$ electrons, then the reaction of 1 mol involves $\mathbf{n}$ faradays ($1\ \mathfrak{F} \approx 96{,}500$ coulomb). By definition, the cell emf (electromotive force) $\mathbf{E}$ is the potential of the cathode minus the potential of the anode. The free-energy change for the cell reaction is $\Delta G = -\mathbf{n}\mathfrak{F}\mathbf{E}$. Both ΔG and $\mathbf{E}$ change sign when the cell reaction is written in the reverse direction. The Nernst equation relates the cell emf to the reaction quotient Q and the cell emf at standard conditions, $\mathbf{E}^{\oplus}$. At $25\,°C$ it can be written

$$\mathbf{E} = \mathbf{E}^{\oplus} - [(0.0592/\mathbf{n}) \log Q]\ \text{volt}$$

A standard electrode potential can be assigned to every electrode half-reaction by arbitrarily selecting the value of zero volts for the standard hydrogen electrode. The cell emf under standard conditions is then the difference between the standard potentials of the two electrodes:

$$\mathbf{E}^{\oplus} = \mathbf{E}^{\oplus}_{\text{cathode}} - \mathbf{E}^{\oplus}_{\text{anode}}$$

At $25\,°C$ the equilibrium constant for the reaction of a cell is related to the standard emf by the equation $\log K = \mathbf{n}\mathbf{E}^{\oplus}/0.0592$.

The term "battery" is commonly used to describe a practical chemical source of electrical energy. The chemical reactions in storage batteries can be reversed by application of an external potential, i.e., such batteries can be recharged. In fuel cells the reactants are continuously supplied to the electrodes.

Corrosion is largely an electrochemical phenomenon that is favored by the presence of water and oxygen. Water containing electrolytes may act to combine possible cathodic and anodic sites of microscopic or larger size into completed galvanic cells. Corrosion can be inhibited (1) by covering potential corrosion sites with a coating impervious to air and moisture; (2) by passivation, which provides the metal surface with the protection of a very thin oxide layer formed by suitable chemical treatment; (3) by cathodic protection, in which either a sacrificial electrode or an external voltage source is connected to the metal requiring protection.

Ion-selective electrodes, such as the glass electrode used for pH measurements, permit the quantitative determination of the concentrations of specific ions by the measurement of electrical potentials.

Terms and Concepts

Problems and Questions

21-1 Amperes and Coulombs A current of 1.76 A flows in a wire for 75 s. (*a*) How many Coulombs pass through a given cross section of the wire in this time? (*b*) How many faradays? (*c*) How many electrons?

21-2 Deposition of Zinc A current of 0.52 A flowing for 88 min will deposit what weight of zinc?

21-3 Electroplating of Copper During the electroplating of copper from a $CuSO_4$ solution, an ammeter shows that a steady current of 0.50 A is passing through the cell. (*a*) How many electrons pass into or out of the cell per minute? (*b*) What is the rate of deposition of copper in grams per hour?

21-4 Deposition of Iron In the electrolysis of a solution containing Fe^{2+} ions, 0.349 g Fe are plated out using a current of 0.371 A. How much time does this take?

21-5 Production of Oxygen In an electrolytic cell in which silver is plated out at the cathode, oxygen is produced at the anode. What volume of oxygen at STP is liberated when 295 mg Ag are deposited? How long does this take if a current of 0.250 A is used?

21-6 Fuel Cell A current of 7.50 A is produced in a fuel cell when CO is oxidized to CO_2 at the anode while oxygen is reduced at the cathode. How many grams of CO and of O_2 does the cell use up per hour?

21-7 Electrolysis Brass is a copper-zinc alloy that usually contains small amounts of lead. A solution prepared by dissolving brass in nitric acid is electrolyzed, copper being plated out on the cathode, lead being deposited as PbO_2 on the anode, and zinc remaining in solution. Oxygen is also evolved at one of the electrodes. (*a*) Write the anode and cathode reactions. (*b*) After the electrolysis is complete, the weight of the cathode has increased by 0.227 g and that of the anode by 0.045 g. The electrolysis took 67.5 min. What current was used?

21-8 Electrolysis of Magnesium Iodide Hydrogen and iodine are formed at the electrodes when aqueous magnesium iodide is electrolyzed. In a particular experiment starting with 500 ml of 0.200 M magnesium iodide, the volume of hydrogen collected was exactly 1.00 liter when measured over water at 25°C and a pressure of 746 torr. The vapor pressure of water is 24 torr at 25°C. (*a*) How many faradays of electricity had been used? (*b*) If the current flowing was constant at 5.0 A, how long did the electrolysis take? (*c*) What were the final concentrations of I^-, Mg^{2+}, and OH^- in the solution? (Assume that the volume does not change and that no reactions occur among these ions.)

21-9 Ionic Current A current is carried through a nonconducting pipe by the passage of Na^+ and Cl^- ions. The flow of Na^+ ions through the pipe cross section is 0.0501×10^{-3} mol min^{-1}, and the flow of Cl^- ions in the *opposite* direction is 0.0763×10^{-3} mol min^{-1}. What electric current passes through the pipe?

21-10 Electrolysis of KCl Solution In an electrolysis of a KCl solution, hydrogen is evolved at the cathode and a mixture of chlorine and oxygen at the anode, the oxygen accounting for 6.0 percent of the total current. (*a*) What is the molar ratio of O_2 to Cl_2 at the anode? (*b*) What is the ratio of the volumes of gas produced at the anode and cathode?

21-11 Galvanic Cell A galvanic cell is made up of a 3.6-g aluminum rod in an essentially infinite volume of 1.00 M aluminum nitrate and a 6.4-g copper bar in an equally large volume of $1.00\,M$ cupric nitrate. The electrodes are connected externally by a wire, and the solutions are joined through a porous connector to complete the circuit. A steady current of 2.0 A flows through the external circuit. Because the volume of the solutions is so large, concentration changes may be ignored. (a) State what changes occur at each electrode and why; then write the overall chemical reaction that takes place. (b) What maximum voltage will this cell supply? (c) What is the direction of flow of electrons in the external circuit? (d) What is the maximum time for which this cell can supply current? (e) Would the voltage of the cell be increased or decreased by replacing the $1.00\,M$ cupric nitrate solution by $0.0100\,M$ cupric nitrate? Show your reasoning.

21-12 Electrochemical Cells Consider cells made up from the following half-cells, with appropriate electrodes as needed. For each, regard the couple given first as that constituting the anode of the cell. State for each pair the cell reaction and the number of electrons involved.

(a) $Cl^-/Cl_2(g)$ and Fe^{3+}/Fe^{2+}
(b) $HCl(aq)$, $Hg_2Cl_2(s)/Hg$ and $HCl(aq)/H_2(g)$
(c) $H_2SO_4(aq)$, $PbSO_4(s)/Pb$ and $PbO_2(s)/H_2SO_4(aq)$, $PbSO_4(s)$

21-13 Galvanic Cell A battery is made with a cadmium electrode in a solution of $0.100\,M$ $CdSO_4$ and a Zn electrode in a solution of $1.00\,M$ $ZnSO_4$. The two solutions are kept apart by a porous partition. (a) What is the emf of this cell? (b) Which electrode is the anode? (See Table D-5.)

21-14 Cell EMF What is **E** for the reaction $Pb(s) + 2H^+ \longrightarrow Pb^{2+} + H_2(g)$ when the concentrations are $[H^+] = 1.00 \times 10^{-2}\,M$, $[Pb^{2+}] = 0.100\,M$, and $P_{H_2} = 1$ atm?

21-15 Lead Storage Cell A lead storage cell is discharged, which causes the sulfuric acid electrolyte to change from a concentration of 34.6 wt percent (density $= 1.261$ g ml^{-1} at $60°F$) to one of 27.0 wt percent. The original volume of electrolyte is 1.000 liter. How many faradays have left the anode of the battery? Note that water is produced by the cell reaction as H_2SO_4 is used up.

21-16 Peroxydisulfuric Acid This acid, $H_2S_2O_8$, can be prepared by electrolytic oxidation of

H_2SO_4: $2H_2SO_4 \longrightarrow H_2S_2O_8 + 2H^+ + 2e^-$. Hydrogen and oxygen are by-products. In such an electrolysis 9.72 liter H_2 and 2.35 liter O_2 were generated at STP. How many moles of $H_2S_2O_8$ were produced? How long did a current of 1.00 A have to flow to achieve this result? (*Hint*: Consider the reactions at the anode and the cathode separately, noting that the number of faradays involved must be the same at each. There are two separate reactions at the anode and one at the cathode.)

21-17 Equilibrium Position of Redox Reactions Predict whether the following reactions would tend to occur to a significant extent in aqueous solution. Assume that the initial concentrations are about $1\,M$. Explain the basis of your predictions.
(a) $Cl_2 + H_2O_2 \longrightarrow O_2 + 2H^+ + 2Cl^-$
(b) $Sn + Cd^{2+} \longrightarrow Cd + Sn^{2+}$
(c) $2I^- + Sn^{4+} \longrightarrow I_2(s) + Sn^{2+}$
(d) $2Fe^{2+} + Br_2(l) \longrightarrow 2Fe^{3+} + 2Br^-$

21-18 Equilibrium Constants from Electrode Potentials Calculate the equilibrium constants for the reactions given in Prob. 21-17. [Assume that (c) takes place in $1\,M$ HCl.]

21-19 Cell EMF and Equilibrium Constant Consider a cell for which the anode reaction is

$$Pb(s) \longrightarrow Pb^{2+}(10^{-2}\,M) + 2e^-$$

and the cathode reaction is

$$VO^{2+}\,(10^{-1}\,M) + 2H^+\,(10^{-1}\,M) + e^- \longrightarrow V^{3+}\,(10^{-5}\,M) + H_2O$$

The measured emf of the cell is $+0.67$ V. (a) Calculate $\mathbf{E}^{\ominus}$ for the VO^{2+}/V^{3+} couple and (b) calculate the equilibrium constant for the reaction $Pb(s) + 2VO^{2+} + 4H^+ \rightleftharpoons Pb^{2+} + 2V^{3+} + 2H_2O$.

21-20 Predicting Reactions from Redox Couples An imaginative manufacturer, who had seen what happened when a zinc rod was placed in a copper solution, decided to sell $1\,M$ solutions of chromic nitrate, $Cr(NO_3)_3$, for use in "instant chromium plating" of iron objects. His idea was that the iron need only be dipped for a short time in the solution. He shipped his first batch of solutions in aluminum containers; the process did not work. He blamed it on the containers and switched to glass. (a) Was he right that the Al containers might have had some effect? Explain. (b) What are the prospects for his success if his solution is shipped in glass containers?

21-21 Predicting Reactions from Redox Couples (a) An excess of powdered nickel is added to a 2 M solution of silver nitrate. Write an equation for any reaction that will occur, and calculate the equilibrium concentrations of the metal ions present. (b) Give the standard free-energy change for the reaction in (a).

21-22 Complex Formation Constant Use the following data

$$Al^{3+} + 3e^- \rightleftharpoons Al(s) \qquad E^\Theta = -1.66 \text{ V}$$
$$Al(OH)_4^- + 3e^- \rightleftharpoons Al(s) + 4OH^-$$
$$E^\Theta = -2.35 \text{ V}$$

to calculate the equilibrium constant for the reaction $Al^{3+} + 4OH^- \rightleftharpoons Al(OH)_4^-$. (*Hint:* Refer to Example 21-11.)

21-23 Solubility-Product Constant The solubility product of silver iodide is frequently given as 1×10^{-16}. Calculate an approximate value for this quantity from the following data, and comment on any difference from the value given.

$$Ag^+ + e^- \rightleftharpoons Ag \qquad E^\Theta = +0.80 \text{ V}$$
$$AgI(s) + e^- \rightleftharpoons Ag + I^- \quad E^\Theta = -0.15 \text{ V}$$

21-24 Free Energy from Cell Voltage Consider a cell in which the anode is a $AgCl(s)/Ag(s)$ electrode in 0.1 M HCl, and the cathode is gaseous Cl_2 at 1 atm around an inert Pt electrode in the same solution. The cell is galvanic and has the voltage 1.137 V. (a) What is the cell reaction? (b) What are the values of ΔG and ΔG^Θ for this reaction?

21-25 Cell Reaction, Cell EMF, and Free-Energy Change This problem illustrates that the cell emf depends on the number of electrons interchanged as well as on the overall cell reaction, while ΔG depends on the latter alone. Consider three cells made up from the following half-cells, in which all ionic species indicated are at standard concentrations and for which the electrode given first is that at which oxidation is assumed to occur:
(a) Tl^+/Tl and Tl^{3+}/Tl^+
(b) Tl^{3+}/Tl and Tl^{3+}/Tl^+
(c) Tl^+/Tl and Tl^{3+}/Tl
For each cell, find the cell reaction, the number of electrons involved, the cell emf, and ΔG^Θ. The

standard electrode potentials for the three pertinent half-cells are as follows (in volts): Tl^+/Tl, -0.34; Tl^{3+}/Tl, 0.74; Tl^{3+}/Tl^+, 1.28.

21-26 Free Energy of Formation of Ions From the standard electrode potentials given below, calculate the standard free energy of formation of the ions Cu^{2+}, Fe^{2+}, and Fe^{3+}, all at 25°C:

Cu^{2+}/Cu, 0.337 V

Fe^{2+}/Fe, -0.440 V

Fe^{3+}/Fe^{2+}, 0.771 V

21-27 Standard Electrode Potential from Nonelectrical Measurements (a) Find the standard free energy of formation of $Mg(OH)_2$ at 25°C from that of $H_2O(l)$ and the following free energies of reaction at 25°C and under standard conditions: $Mg(s) + \frac{1}{2}O_2(g) \rightarrow MgO(s)$, -569.6 kJ and $MgO(s) + H_2O(l) \rightarrow Mg(OH)_2(s)$, -27.0 kJ. (b) Calculate the standard free energy of $Mg^{2+}(aq)$ from that of $OH^-(aq)$ and from K_{sp} for $Mg(OH)_2(s)$, 1.8×10^{-11}. (c) Calculate E^Θ for the half-cell reaction $Mg^{2+} + 2e^- \rightleftharpoons Mg(s)$.

21-28 Behavior of Electrochemical Cell An electrochemical cell consists of a silver electrode dipping into 0.10 M $AgNO_3$ and a nickel electrode dipping into 0.10 M $Ni(NO_3)_2$:

$$Ni^{2+} + 2e^- \longrightarrow Ni(s) \qquad E^\Theta = -0.23 \text{ V}$$
$$Ag^+ + e^- \longrightarrow Ag(s) \qquad E^\Theta = +0.80 \text{ V}$$

(a) Calculate the initial voltage of this cell. (b) Write the overall cell reaction that occurs as this cell generates electric current, indicate the reaction that occurs at the cathode and that at the anode, and state in which direction electrons flow in the external circuit (from Ag to Ni or from Ni to Ag). Explain. (c) If a steady current of 0.50 A is supplied by this cell for a period of 3.0 hour, what weight of nickel will be *deposited* or *dissolved*? (d) Calculate the equilibrium constant for the overall reaction given in (b). (e) If the volumes of Ni^{2+} and Ag^+ solutions are equal and the reaction is allowed to proceed until equilibrium is reached, what will be the approximate final concentrations of Ni^{2+} and Ag^+?

21-29 Cleaning of Silverware In a procedure for removing Ag_2S tarnish from silverware the silver is placed in a warm Na_2CO_3 solution in an aluminum pan. Explain this procedure with data given in Appendix D and the following: $E^{\ominus}_{Ag_2S/Ag} = -0.71$ V; $E^{\ominus}_{Ag^+/Ag} = 0.80$ V.

21-30 Effects of Local Concentration Changes on Cell Voltage

The running of a galvanic cell for which the anode is a piece of zinc in $1\,M\,ZnSO_4$ and the cathode is a piece of copper in $1\,M\,CuSO_4$ causes the depletion of Cu^{2+} ions in the surround-ings of the cathode to a concentration of $1.0 \times 10^{-4}\,M$. It also causes an increase of $[Zn^{2+}]$ to $1.50\,M$ near the anode (and the appropriate changes in the SO_4^{2-} concentration to maintain an electrically neutral solution). (a) What is the original voltage of the cell? (b) What is the voltage after these local concentration changes (termed "polarization") have taken place?

Chemical Kinetics: The Rates and Mechanisms of Chemical Reactions

"The rate of chemical reactions is a very complicated subject. This statement is to be interpreted as a challenge to enthusiastic and vigorous chemists; it is not to be interpreted as a sign of defeat."

H. S. JOHNSTON,[1] 1966

22

22-1 Introduction

This chapter deals with *reaction kinetics*, that is, with measures of the rates at which chemical reactions take place, with the factors that affect these rates, and with the interpretation of reactions in terms of specific events and structures at the molecular level. It provides the background needed for thinking about such questions as the following. Why does temperature have such a profound effect on the rates of most reactions? Why, for example, does paper "burn" at elevated temperatures, reacting with oxygen to form carbon dioxide and water and giving off much heat and light? At ordinary temperatures it reacts so slowly that no change is noticeable except over a period of many years. Why do changes in concentrations of reactants usually alter reaction rates? Why is it that a solution of hydrogen peroxide (H_2O_2) can be kept unchanged for days or months, but if a tiny amount of the enzyme catalase is added to it the H_2O_2 decomposes to O_2 and H_2O almost instantaneously?

Most of our knowledge of kinetics is empirical, that is, derivable only from experimentation rather than from some fundamental and well-tested theory, for the theoretical foundations of kinetics are still relatively crude. The next two sections of this chapter constitute an introduction to some of the terminology and fundamental concepts of kinetics and to the two general measures of rates of reaction, the instantaneous rate expression (the "rate law") and the corresponding integrated rate expression. These sections are followed by a discussion of homogeneous reactions in the gas phase, which illustrates some of the basic concepts of kinetics, and an introduction to studies of reactions in liquid solutions, with emphasis primarily on the role of the solvent and thus on the differences from reactions in gaseous systems. The chapter concludes with brief discussions of catalysis, reaction mechanisms, and the factors affecting the proportions of different products that may be formed from a given set of reactants.

[1]H. S. Johnston, "Gas Phase Reaction Rate Theory", The Ronald Press Co., New York, 1966.

Mechanisms and Elementary Processes   The goal of chemical kinetics is to obtain an exact description of the sequence of processes that occurs on the molecular level as reactants are converted into products. This microscopic reaction sequence is called the *mechanism* of the reaction. Each of the individual steps in the mechanism is termed an *elementary process,* a phrase implying that the chemical equation representing this particular step includes the exact atomic or molecular species reacting with one another at this stage. The sum of the equations that represent the successive elementary processes making up the entire reaction is necessarily the chemical equation for the overall reaction. However, the sequence of elementary processes involved in any particular reaction can *never* be deduced from the overall stoichiometric equation. In other words, there is *no necessary relation between the stoichiometric equation for a reaction and the actual mechanism of that reaction,* other than that the sum of the equations representing the individual steps in the mechanism must be the overall equation.

To illustrate, the gas-phase reaction of molecular hydrogen with molecular bromine to form hydrogen bromide is represented by the stoichiometric equation

$$H_2(g) + Br_2(g) \longrightarrow 2HBr(g) \tag{22-1}$$

If (22-1) represented an elementary process the equation would imply that this reaction occurred by the direct interaction of one hydrogen molecule with one bromine molecule to form, in a single step, two molecules of hydrogen bromide. Experimental studies have shown, however, that this direct reaction does not occur. Rather, as we discuss later, H_2 and Br_2 react by a series of elementary processes involving first the dissociation of a bromine molecule by collision with some other molecule, subsequent reaction of a resulting bromine atom with a hydrogen molecule to produce hydrogen bromide and a hydrogen atom, and a number of further steps. The important point is that the complex mechanism by which this reaction proceeds could not be deduced from the stoichiometric equation. Indeed, it cannot even be deduced from the most advanced theories currently available, but only from the results of carefully planned experiments.

Nature of the Barrier to Reaction   The range of speeds observed for chemical reactions is enormous. Some reactions are immeasurably slow; others are essentially complete in times as short as 10^{-10} s. Perhaps the most fundamental question that chemical kineticists attempt to answer is, "What is the limiting factor in the rate of reaction?", or put in another way, "What is the barrier to reaction?" If, for example, every collision between reactants leads to reaction, the rate is limited by the rate at which collisions between reactants occur. With such reactions, which are not very common, the barrier is simply the rate at which the reactants diffuse together. In the gaseous or liquid state such a reaction is normally too fast to be observed without sophisticated instrumentation.

More commonly, the rate of reaction is limited by the fact that only a very small fraction of the reacting molecules have enough energy, even when they do collide with one another, to bring about the necessary electronic and atomic rearrangement that characterizes a chemical change. In other words, a certain minimum energy, or "threshold energy", may be needed for a colliding pair to react. The barrier is then an energetic one.

With some reactions, especially those of large polyatomic molecules, it frequently happens that the potential reactants are not properly oriented relative to one another when they do come together, even though their mutual energy may exceed the threshold. Because of this inappropriate mutual orientation, atoms that would be linked together if a product molecule were formed never come sufficiently close for the reaction to take place. In this case the barrier to reaction is said to be configurational or orientational.

22-3 The Rates of Chemical Reactions

The speed of a chemical reaction may be defined as the rate at which one of the reactants disappears or, alternatively, the rate at which one of the products is formed. Normally the rate is expressed in terms of the change with time in the concentration of a reactant or product. Consider the hypothetical reaction

$$A + 3B \longrightarrow 2D \tag{22-2}$$

Its rate could be determined by studying the concentration of one of the species involved as a function of the time. Figure 22-1 represents a possible variation of the concentration of A with time. If the concentration of A at time t is $[A]_t$ and it has become $[A]_t + \Delta[A]$ at $t + \Delta t$, the rate is given by (see Fig. 22-1)

$$\text{Rate} = -\frac{[A]_t + \Delta[A] - [A]_t}{t + \Delta t - t} = -\frac{\Delta[A]}{\Delta t} \tag{22-3}$$

The rate can be expressed in terms of the change in concentration of any reactant

Figure 22-1 The Change in Concentration of a Reactant A with Time t
The concentration of the species A participating in a chemical reaction decreases with time. For this reaction the concentration of A does not vary linearly with time; the rate of the reaction changes as time passes. Consequently the rate at any time t is defined as the limit approached by the ratio $-\Delta[A]/\Delta t$ as Δt becomes smaller and smaller. (The minus sign is used because it is customary to define rates as positive quantities and $\Delta[A]$ is negative if A is a reactant.) In other words, the rate of reaction at any time t is the negative of the slope of the curve at that instant.

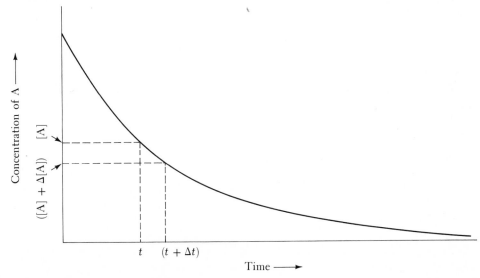

or product. To avoid ambiguity when the numbers of molecules of the various reactants and products differ, it is customary to take the stoichiometric coefficients into account. Thus for reaction (22-2), the rate $-\Delta[A]/\Delta t$ can be expressed equally well as

$$-\frac{1}{3}\frac{\Delta[B]}{\Delta t} \quad \text{or} \quad +\frac{1}{2}\frac{\Delta[D]}{\Delta t}$$

These alternative expressions take into account the fact that by (22-2), three molecules of B disappear for every molecule of A reacting and two molecules of D are formed for each A that is removed. Hence,

$$\text{Rate of reaction} = -\frac{\Delta[A]}{\Delta t} = -\frac{1}{3}\frac{\Delta[B]}{\Delta t} = \frac{1}{2}\frac{\Delta[D]}{\Delta t} \qquad (22\text{-}4)$$

The units of the rate are usually $M\,s^{-1} = \text{mol liter}^{-1}\,s^{-1}$, since concentrations are usually expressed as molarities and time as seconds. Occasionally other units are used. The quantity of reactant or product may be expressed in any units that are proportional to concentration and can be conveniently measured, e.g., the pressure of a gas, units of electrical conductivity, or some optical property.

The rate expressions (22-4) imply that the concentrations of A, B, and D change only because of the reaction considered, that is, reaction (22-2) occurring in the indicated direction. This is not always true, since other reactions involving these species may occur simultaneously. Frequently there is a back reaction in which the products (here D molecules) react to form the reactants again, the reverse of (22-2). In such a situation, we define the net rate as the difference between the rates of the forward and reverse reactions,

$$\text{Net rate} = \text{forward rate} - \text{reverse rate} \qquad (22\text{-}5)$$

At the very beginning of a reaction, when no products are yet present in any appreciable concentration, the reverse rate is negligible so that the net rate is equal to the forward rate. As the forward reaction proceeds, the products accumulate, and, unless the position of equilibrium lies very far toward the product side, the rate of the reverse reaction may need to be taken into account. At equilibrium, the rates of the forward and reverse reactions are equal and the net rate is zero.

From the shape of the curve in Fig. 22-1 it is clear that the rate of the reaction changes with time. There is less change in [A] per unit time as the time increases—the rate decreases continually. Because the reaction rate usually changes with time, the definition given in (22-3) is not adequate; the value of the rate at a given time depends on the size of the time increment Δt. A more precise definition of the rate is obtained by taking the limit of (22-3) as Δt approaches zero[2]

$$\text{Rate} = \lim_{\Delta t \to 0} \frac{-\Delta[A]}{\Delta t} = -\frac{d[A]}{dt} \qquad (22\text{-}6)$$

which is the negative of the slope at a given point of a graph of [A] plotted against t.

---

[2]Students with a knowledge of calculus will recognize that the rate is the negative of the derivative of [A] with respect to time, $-d[A]/dt$, as indicated on the right side of (22-6). See Appendix C.

The Form of the Rate Law The rate at any given time during the course of a reaction, the instantaneous rate, most commonly depends on a number of variables, including the concentration of at least some of the reactants, the temperature, and the nature of the solvent (if any). The first problem in any kinetic study is to find, by appropriate experiments, the form of the functional dependence of the rate on the concentrations of the reactants, which is called the *rate law*. Consider a reaction with stoichiometry such that n_A molecules of A react with n_B molecules of B (and perhaps with some other reactants as well) to form certain products:

$$n_A A + n_B B + \cdots \longrightarrow \text{products} \qquad (22\text{-}7)$$

If there are no interfering reactions that consume or produce A or B and if all the variables that might affect the rate of the reaction except the concentrations of the reactants are held constant, then it is often found (by experiment) that the rate law may be expressed as

$$\text{Rate} = -\frac{1}{n_A}\frac{d[A]}{dt} = k[A]^a[B]^b \cdots \qquad (22\text{-}8)$$

Here the exponents a and b are usually (but not always) integers, with *no necessary relation to the stoichiometric coefficients n_A and n_B in the equation*[3] for the reaction (22-7). The proportionality constant k is called the *rate constant;* it depends not only on the nature of the reaction, but normally also on the temperature, the nature of the solvent, and other possible variables.

Since the *rate* of any reaction has the dimensions concentration time^{-1}, the dimensions of the rate constant k for that reaction depend on the exponents of the concentration terms in the rate law. In particular, if we let p be the sum of the exponents of the concentration terms in the experimental rate law (22-8),

$$p = a + b + \cdots \qquad (22\text{-}9)$$

then the dimensions of k are concentration^{1-p} time^{-1}.

Order of a Reaction The experimental quantity p defined by (22-9) is called the *order of the reaction;* the individual exponents, $a, b, \ldots$ in (22-8) are said to be the orders of the reaction with respect to the individual reactants A, B, ... For example, if in a reaction with two reactants A and B it is found experimentally that $a = 1$ and $b = 2$, the reaction is said to be third order overall and to be first order in A and second order in B.

There are many reactions for which the rate law cannot be expressed in the form (22-8), that is, as a product of concentration terms each raised to some power. For example, the rate may be expressible only as the sum of several terms or even as a ratio of sums. The term *order of reaction* has no meaning for a reaction with such a rate law.

☐ **Concentration Dependence of a Reaction Rate** The rate of oxidation of bromide ions by bromate in an acidic aqueous solution,

$$6H^+ + BrO_3^- + 5Br^- \longrightarrow 3Br_2 + 3H_2O$$

Example 22-1

---

[3] For some reactions the rate law contains concentration terms of species that do not appear in the stoichiometric overall reaction equation.

has been found experimentally to be first order in bromide, first order in bromate, and second order in hydrogen ion; that is,

$$\text{Rate} = -\frac{d[\text{BrO}_3^-]}{dt} = \frac{1}{3}\frac{d[\text{Br}_2]}{dt} = k[\text{Br}^-][\text{BrO}_3^-][\text{H}^+]^2$$

What happens to the rate if, in separate experiments, (a) $[\text{BrO}_3^-]$ is doubled; (b) the pH is increased by 1.0 units; (c) the solution is diluted to twice its volume, with the pH kept constant by use of a buffer?

Solution (a) Since the rate is proportional to $[\text{BrO}_3^-]$, doubling this concentration doubles the rate. (b) Increase of the pH by 1.0 means that $[\text{H}^+]$ is decreased by a factor of 10. The rate is therefore decreased by a factor of 100 since the rate is proportional to $[\text{H}^+]^2$. (c) Both $[\text{Br}^-]$ and $[\text{BrO}_3^-]$ decrease by a factor of 2, so that the rate decreases by a factor of 4 at constant $[\text{H}^+]$. ■

Exercise 22-1

☐ The rate of the reaction

$$3\text{I}^- + \text{H}_3\text{AsO}_4 + 2\text{H}^+ \longrightarrow \text{I}_3^- + \text{H}_3\text{AsO}_3 + \text{H}_2\text{O}$$

is given by the expression

$$\text{Rate} = -\frac{d[\text{H}_3\text{AsO}_4]}{dt} = k'[\text{I}^-][\text{H}_3\text{AsO}_4][\text{H}^+]$$

What happens to the rate if, in separate experiments, (a) $[\text{I}^-]$ is halved; (b) the pH is decreased by 1.0 units; (c) the solution is diluted to three times its volume, at a constant pH? ■

Example 22-2

☐ **Determining a Rate Law** Hypophosphite ion, H_2PO_2^-, decomposes in alkaline solution into hydrogen phosphite ion, HPO_3^{2-}, and hydrogen according to the stoichiometric equation

$$\text{H}_2\text{PO}_2^- + \text{OH}^- \longrightarrow \text{HPO}_3^{2-} + \text{H}_2 \qquad (22\text{-}10)$$

The rate of disappearance of hypophosphite ion at $100°\text{C}$, rate $= -d[\text{H}_2\text{PO}_2^-]/dt$, has been found to depend on the concentrations of the reactants as follows:

$[\text{H}_2\text{PO}_2^-]/M$	$[\text{OH}^-]/M$	Rate/M min^{-1}
0.10	1.0	3.2×10^{-5}
0.50	1.0	1.6×10^{-4}
0.50	4.0	2.5×10^{-3}

(a) Deduce the order of the reaction with respect to each reactant, and write the rate expression.
(b) Calculate the rate constant for the reaction at this temperature.
(c) What is the rate of formation of HPO_3^{2-} in mol min^{-1} in a solution of volume 0.50 liter at the moment at which $[\text{OH}^-] = [\text{H}_2\text{PO}_2^-] = 1.0\,M$?

Solution (a) The first two lines of the tabulated data indicate that when $[\text{OH}^-]$ is kept constant, a fivefold increase in the concentration of H_2PO_2^-

leads to a fivefold increase in the rate. Thus, the rate is proportional to $[H_2PO_2^-]$, that is, the reaction is first order in $H_2PO_2^-$. Comparison of the last two lines of the tabulated data indicates that at constant $[H_2PO_2^-]$ a 4-fold increase in the OH^- concentration leads to an approximately 16-fold increase in the rate. Consequently the reaction is second order in OH^- . Thus, the rate expression is

$$\text{Rate of disappearance of } H_2PO_2^- = k[H_2PO_2^-][OH^-]^2 \quad (22\text{-}11)$$

(b) Substitution of any of the tabulated data—perhaps those of the first line are most convenient—into (22-11) leads to $k = 3.2 \times 10^{-4} \, M^{-2} \, \text{min}^{-1}$. (c) Substitution of the 1.0 M concentrations and the value of k from (b) into (22-11) gives the rate of disappearance of $H_2PO_2^-$ under these conditions as 3.2×10^{-4} (mol liter^{-1} min^{-1}). Since the volume is 0.50 liter and since the stoichiometric equation (22-10) indicates that one HPO_3^{2-} ion is formed for each $H_2PO_2^-$ that disappears, there must be 1.6×10^{-4} mol HPO_3^{2-} ions being formed per minute. ■

□ The reaction

$$H_2O_2 + 3I^- + 2H^+ \longrightarrow I_3^- + 2H_2O$$

Exercise 22-2

is independent of $[H^+]$ at a pH of 3 and higher. At 25°C the observations summarized by the following table can be made.

$[H_2O_2]/M$	$[I^-]/M$	Rate/M min^{-1}
0.10	0.10	6.9×10^{-4}
0.05	0.10	3.4×10^{-4}
0.10	0.20	1.4×10^{-3}

(a) What is the rate expression? (b) Calculate the rate constant for the reaction at this temperature. (c) What is the rate of formation of I_3^- in mol min^{-1} in a solution of volume 0.10 liter at the moment at which $[H_2O_2] = [I^-] = 0.05 \, M$? ■

Integrated Rate Expressions It is frequently inconvenient or even impossible to measure the instantaneous rate of a reaction directly with much precision. This is especially true if the reaction is so fast that the concentrations of the reactants, and hence the rate, change appreciably in the time it takes to make measurements. A common practice in experimental studies of kinetics is to measure the concentration of some reactant (or product) at intervals of time so chosen that a smooth curve can be drawn representing the variation of concentration with time during the course of most of the reaction—for example, a curve like that in Fig. 22-1. If the rate at any instant is desired, it can be found by measuring the slope of the curve at that particular time, as indicated in the legend of Fig. 22-1.

Integrated rate expressions are mathematical representations of curves like that in Fig. 22-1, the name implying that they can be derived from instantaneous rate laws, such as (22-8), by the calculus technique of integration. Instantaneous rate laws describe how the rate changes with concentration; integrated rate expressions describe how the concentration changes with time. Suppose for example, that the concentration of a certain substance X is decreased by a first-order reaction. This implies that

$$\text{Instantaneous rate of decrease of } [X] = -\frac{d[X]}{dt} = k[X] \qquad (22\text{-}12)$$

Let the concentration of X at the time of the initial measurement be $[X]_0$ and let the corresponding time be taken as $t = 0$. Then the integrated rate expression (Appendix C) for the concentration of X at any time t is (with ln x meaning the natural logarithm of x)

$$\ln \frac{[X]_0}{[X]} = -\ln \frac{[X]}{[X]_0} = kt \qquad (22\text{-}13)$$

or
$$\frac{[X]}{[X]_0} = e^{-kt}$$

Thus, the concentration of X at any time t is related to its concentration at $t = 0$, $[X]_0$, by

$$[X] = [X]_0 e^{-kt} \qquad (22\text{-}14)$$

This negative exponential dependence on time is uniquely characteristic of first-order reactions. An enormous variety of chemical and physical processes follow first-order kinetics; one of the most familiar of these is radioactive decay.

If we let $\tau_{1/2}$ represent the *half-life* of a reaction, defined as the time required for the concentration of reactant to fall to half its initial value, then Equation (22-13) implies that the *half-life of a first-order reaction* is independent of the concentration. We have, by (22-13),

$$-\ln \frac{0.5[X]_0}{[X]_0} = -\ln 0.5 = k\tau_{1/2}$$

Since $\ln 0.5 = -\ln 2 = -0.693$,

$$-(-0.693) = k\tau_{1/2}$$

or
$$\tau_{1/2} = \frac{0.693}{k} \qquad (22\text{-}15)$$

Thus it is easy to calculate the half-life of a substance that decomposes by a first-order reaction if the rate constant is known or to calculate the rate constant if the half-life is known. The length of time required for reaction of any other particular fraction, e.g., nine-tenths, of the original sample is also independent of the initial concentration, as implied by (22-13).

Example 22-3

☐ **First-Order Kinetics** Paraldehyde, P, is formed from three molecules of acetaldehyde, A.

At temperatures around 500 K, paraldehyde vapor decomposes by a first-order reaction to acetaldehyde, P → 3A. At 519 K the rate constant for this reaction is $3.05 \times 10^{-4}\,s^{-1}$. What is the half-life of a sample of paraldehyde under these conditions? What is its "tenth-life" (the time required for nine-tenths of the sample to decompose, leaving just one-tenth of the original material)?

Solution The half-life, by (22-15), is

$$\tau_{1/2} = \frac{0.693}{3.05 \times 10^{-4}}\,s = \underline{2.27 \times 10^3\,s}$$

or about 38 min. From (22-13) we know that the time needed for the concentration to fall to one-tenth of its initial value is given by

$$-\ln 0.10 = \ln 10 = k\tau_{0.1}$$

$$\tau_{0.1} = \frac{\ln 10}{k} = \frac{2.303}{3.05 \times 10^{-4}}\,s = \underline{7.55 \times 10^3\,s}$$

(The ratio of the "tenth-life" to the half-life must be the same for all first-order reactions, $2.303/0.693 = 3.32$.) ∎

□ The decomposition of dinitrogen pentoxide, N_2O_5, dissolved in carbon tetrachloride, CCl_4, is first order, with a rate constant of $6.2 \times 10^{-4}\,min^{-1}$ at a certain temperature:

Exercise 22-3

$$2N_2O_5 \longrightarrow 4NO_2 + O_2$$

What is the half-life of N_2O_5 at this temperature? After how many hours is only 20 percent of the original N_2O_5 left? ∎

Integrated rate laws have been derived for rate expressions appreciably more complicated than that for a first-order reaction, (22-12). We shall give only one other, that for a reaction that is second order in a single reactant, for which the rate law is:

$$\text{Instantaneous rate of decrease of } [X] = k[X]^2 \qquad (22\text{-}16)$$

The integrated relation between [X] and the time (Appendix C) is

$$\frac{1}{[X]} - \frac{1}{[X]_0} = kt \qquad (22\text{-}17)$$

where the time is measured from a zero point at which the concentration was $[X]_0$. The time dependence of the concentration of a single reactant[4] that is consumed by a second-order process, Equation (22-17), is quite different from that for a reactant consumed by a first-order process, Equation (22-13) or (22-14). This difference is easily observed experimentally if the reaction is followed for a period significantly longer than the half-life. For a first-order process, the half-life is independent of the initial concentration, as already stressed, but for a sec-

---

[4] Equation (22-17) also applies when two different substances react by a second-order process if, and only if, their initial concentrations are in the same proportion as the stoichiometric coefficients in the equation for the reaction.

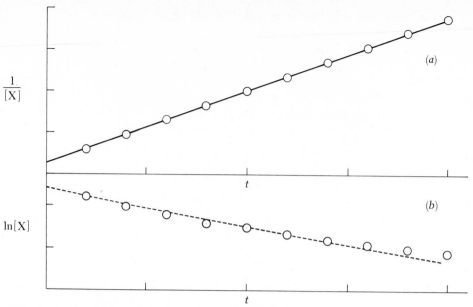

Figure 22-2 Comparison of First- and Second-Order Plots
The points in the graphs represent data for a second-order reaction. As shown in (a), a plot of
$1/[X]$ versus time is linear. A plot of $\ln [X]$ versus time, (b), is nonlinear, an indication that the
reaction is not first order. In any real situation, experimental errors in the data might make the
comparison more difficult to interpret, particularly if measurements were made only for times
short with respect to the half-lives.

ond-order reaction of the kind discussed here, the half-life varies inversely with
the initial concentration. This can be shown by substituting $[X] = [X]_0/2$ (and
$t = \tau_{1/2}$) into (22-17):

$$\frac{1}{[X]_0/2} - \frac{1}{[X]_0} = k\tau_{1/2}$$

$$\frac{2}{[X]_0} - \frac{1}{[X]_0} = k\tau_{1/2}$$

or

$$\tau_{1/2} = \frac{1}{k[X]_0} \qquad (22\text{-}18)$$

Thus, in marked contrast to a first-order reaction, the half-life increases steadily
as the initial concentration decreases.

The distinction between the rate laws expressed in (22-17) and (22-14), or
(22-13), is often most apparent if the experimental data are graphed (Fig. 22-2).
For the first-order reaction, a plot of $\ln [X]$ against the time is a straight line; for
the second-order reaction leading to (22-17), a plot of $1/[X]$ against time is a
straight line. If the fraction of the reactant consumed is small, the data for a
given reaction may appear to give a straight line when plotted either way, but if
the elapsed time is a half-life or more (and the data are precise), only one of the
lines can be straight. Of course, if the reaction is neither first order nor second
order in X, neither line will be straight.

The kinetic[5] theory of gases provides a sound foundation for interpreting the results of experiments involving gaseous molecules. There is, on the other hand, no manageable quantitative theory of liquids or solutions to serve as a guide in attempts to understand experiments in these media. Consequently, in order to show how some of the simplest experimental facts of chemical kinetics can be interpreted theoretically, we begin with a discussion of reactions in the gaseous phase.

Concentration Dependence of the Rate of Molecular Collisions   Collisions between molecules[6] play an essential role in most chemical reactions. For example, at some stage in a reaction in which a new bond is formed, there must be close contact of the species containing the atoms that are eventually joined by the new bond. Clearly the rate of such a reaction cannot exceed the rate at which the reacting molecules come together (although it might be much less than this rate of collision if only a small fraction of the collisions lead to reaction).

The rate at which molecules collide with each other, that is, the number of collisions per second, is directly proportional to the concentrations of the colliding molecules. To see that this is plausible, imagine a box of fixed volume containing molecules of three gases, E, F, and G, moving at speeds characteristic of a temperature T. Consider first the collisions that involve only one molecule of E and one of F. Suppose that we keep the number of molecules of E constant and triple the number of molecules of F, thus tripling the concentration of F. The rate of collisions per liter between E and F will go up by a factor of 3, because there is now three times as great a chance that an E molecule will meet an F molecule, since there are three times as many F molecules in the volume as there were before. Thus the rate of collisions between E and F is proportional to the concentration of F. A similar argument shows that it is also proportional to the concentration of E. Hence the number of collisions between E and F per liter per second, which we shall call "rate of collisions (E,F)", is given by

$$\text{Rate of collisions (E,F)} = c_1[\text{E}][\text{F}] \qquad (22\text{-}19)$$

The proportionality constant c_1, which depends on the temperature and on the sizes and masses of the molecules, can be calculated from kinetic gas theory. As the temperature increases, c_1 increases because the increased molecular speeds result in a higher collision rate. An expression similar to (22-19), with a different proportionality constant, holds when more than two molecules come together. For example, the rate of collisions per liter of one molecule of E, two molecules of F, and one molecule of G is given by

$$\text{Rate of collisions (E, F, F, G)} = c_2[\text{E}][\text{F}]^2[\text{G}] \qquad (22\text{-}20)$$

---

[5] The adjective *kinetic* (Greek: *kinetikos*, moving, pertaining to motion) is used in the phrase *kinetic theory* to imply that the theory is one in which the motions of molecules, and the energies associated with these motions, play a dominant role. It is only indirectly related to the term *kinetics* as applied to the study and interpretation of the rates of chemical reactions.

[6] We are here using the term *molecules* in a general sense to include not only molecules but other possible reacting species, for example, atoms and ions.

The proportionality constant c_2 in (22-20) is much smaller than c_1 in (22-19).

Thus, if the rate of an elementary process is proportional to the number of collisions per unit time between the reactants, then for a general *elementary* process

$$aA + bB + dD + \cdots \longrightarrow \text{products}$$

$$\text{Rate} = k' \left[\text{rate of collisions (A, B, D)}\right] = k'c \, [A]^a[B]^b[D]^d \cdots$$
$$= k[A]^a[B]^b[D]^d \cdots \quad (22\text{-}21)$$

Equation (22-21) has the same form as the empirical expression for the overall rate of reaction, (22-8). For an *elementary* process, however, the exponents a, b, d, ... can be identified as the numbers of molecules involved in the collision, and the rate constant k is the product of the proportionality constant c for the rate of collision and a factor k' that represents the fraction of collisions leading to reaction.

Collisions that involve two molecules, whether of the same or different kinds, are called *bimolecular*; those among three molecules are said to be *termolecular*. The probability of collisions involving more than two molecules simultaneously, which is reflected in the magnitude of the proportionality constants in equations such as (22-20), is far smaller than that of bimolecular collisions because the likelihood that three or more molecules will be in the same place at the same time is very low, especially for gases and for solutes in dilute solutions. Nonetheless, termolecular collisions are important in some reactions, because even though they are rarer, such collisions are for these reactions far more likely to lead to the desired products than are bimolecular collisions. This brings us to a consideration of the fraction of collisions that lead to reaction. For simplicity, we examine this question for bimolecular processes.

Collision Rates for Bimolecular Reactions The rate at which bimolecular collisions occur in a gas can be calculated with fair precision from the kinetic theory of gases. In essence this amounts to evaluating the constant c_1 in (22-19). An "order-of-magnitude" calculation leads to the result that at room temperature c_1 is of the order of 10^{11} liter mol^{-1}s^{-1}. Suppose that two gases A and B are mixed, each at about 0.5 atm pressure, which corresponds to concentrations of about 0.02 mol liter^{-1}. If $c_1 \approx 10^{11}$ liter mol^{-1} s^{-1}, there would be on the order of $10^{11}(0.02)^2 = 4 \times 10^7$ mol of collisions each second between the molecules in a 1-liter sample. Since in this volume there is only 0.02 mol of each reactant to be used up, any reaction in which every collision of A with B yielded products would be complete in a time of the order of 0.02 mol/4×10^7 mol s^{-1} or in less than 10^{-9} s.

Thus such gas reactions would be over almost instantaneously if products were formed in every collision. In fact, very few reactions occur this rapidly. Most known bimolecular rate constants are smaller by many powers of 10 than the value $10^{11} \, M^{-1} \, s^{-1}$. In other words, most of the collisions do not result in reaction; the colliding molecules just bounce off each other and go their separate ways, unchanged except in their direction of motion and, usually, their energy. How can we explain the fact that so small a fraction of the collisions are effective in producing chemical change? We shall see shortly that the explanation is in part tied to the Maxwell–Boltzmann distribution of kinetic energies (Sec. 4-5).

Temperature Dependence of Reaction Rates: Activation Energy The rates of most chemical reactions increase markedly as the temperature is increased. There are exceptions, but they are uncommon. For a typical reaction, an increase of 10 K at temperatures not far from room temperature will increase the reaction rate by a factor of 2 to 3; that is, the rate will increase by 100 to 200 percent. What is the reason for this marked temperature effect? If the volume is constant, the only temperature-dependent term in the collision rate is the molecular velocity. This velocity is proportional to $T^{1/2}$ and enters to the first power in the collision rate. For a 10 K rise in temperature near 300 K (about a 3 percent increase in T or a 1.5 percent increase in $T^{1/2}$), the increase in the collision rate will therefore be only about 1.5 percent, an effect completely negligible in comparison with the usual increase in the reaction rate of 100 percent or more. Thus the temperature dependence of the reaction rate cannot be attributed to the increased collision rate.

Arrhenius observed that for many reactions in gases and in other media the rate constant was related to the absolute temperature exponentially:

$$k = Ae^{-E_a/RT} \tag{22-22}$$

This relation applies to rate constants for elementary processes as well as to overall rate constants. The constant A is called the frequency factor; E_a, which has the dimensions of energy, is called the *activation energy*. For most reactions E_a is a positive quantity; that is, k increases with increasing T (decreasing $1/T$).

Equation (22-22) can be arranged to

$$\ln k = \ln A - \frac{E_a}{R}\frac{1}{T} \tag{22-23}$$

If A and E_a are independent of temperature, a graph of $\ln k$ against $1/T$ should be a straight line with slope $-E_a/R$ and intercept $\ln A$. Such graphs do give remarkably straight lines for many reactions over a significant range of temperature; one is shown in Fig. 22-3. Values of activation energies can thus be derived readily from experimental data. If (22-23) is assumed to hold, then only two experimental points, that is, values of k at two precisely measured temperatures, are needed to find E_a and A. If we let the temperatures be T_2 and T_1 and let the corresponding rate constants be k_2 and k_1, we have, from (22-23),

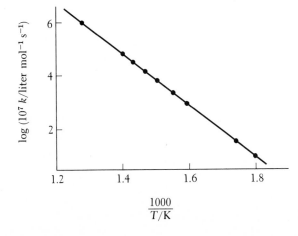

Figure 22-3 Log k as a Function of $1/T$ for the Thermal Decomposition of HI

The reaction whose rate constant was measured at different temperatures is

$$2HI \longrightarrow H_2 + I_2$$

The experimental data have been plotted in a form to test their fit to (22-23). The slope of the curve is proportional to $-E_a/R$. The value of E_a found from the slope is 184 kJ mol^{-1}. Extrapolation of the line to $1/T = 0$ gives an experimental value of A [Equation (22-23)] between 10^{10} and 10^{11} M^{-1} s^{-1}.

$$\ln k_2 - \ln k_1 = -\frac{E_a}{R}\left(\frac{1}{T_2} - \frac{1}{T_1}\right)$$

or

$$\ln \frac{k_2}{k_1} = 2.303 \log \frac{k_2}{k_1} = \frac{E_a}{R}\frac{T_2 - T_1}{T_1 T_2} \tag{22-24}$$

This equation confirms our earlier remark that E_a is positive if k increases with temperature; if T_2 is greater than T_1, then E_a must have the same sign as $\ln (k_2/k_1)$ and so must be positive if k_2 is greater than k_1.

Example 22-4

☐ **Activation Energy** The rate constant of a certain reaction increases by a factor of 18 when the temperature is increased from 20 to 40°C. What is the activation energy for this reaction?

Solution We know that $k_2/k_1 = 18$ when $T_2 = 313$ K and $T_1 = 293$ K. Equation (22-24) then gives

$$\ln 18 = \frac{E_a}{R}\frac{20}{313 \times 293} \text{ K}^{-1}$$

$$E_a = [(2.89)(4.6 \times 10^3)R] \text{ K} = \underline{1.1 \times 10^2 \text{ kJ mol}^{-1}}$$

Note that the precision with which E_a can be determined experimentally with the help of (22-24) depends directly on the precision with which the difference in temperature, $T_2 - T_1$, can be measured, as well as on the precision of the ratio of the rate constants. The activation energy of 110 kJ mol^{-1} found in this example is quite typical, although it is somewhat higher than that corresponding to the rule cited earlier, "a factor of about 2 to 3 in k for a 10 K rise in temperature near room temperature". In this case there is an increase of a factor of more than 4 for each 10 K. ∎

Exercise 22-4

☐ At 30.0°C the rate of decomposition of N_2O_5 (Exercise 22-3) is 1.97 times the rate at 25.0°C. What is the activation energy for this reaction? ∎

Collision Theory and the Arrhenius Equation We now consider the observed temperature variation of the rate constant in terms of the Maxwell-Boltzmann energy distribution. Arrhenius interpreted (22-22) to imply that molecules could not react unless they had sufficient energy to reach an "activated state" with energy E_a above the initial state. This idea is illustrated in Fig. 22-4.

The variable represented as the *reaction coordinate* in this schematic diagram is some measure of the interatomic distance or another parameter that changes smoothly in the progression from reactants to products. At the left the system is in the form of the reactants; at the right the system has been transformed into products. Arrhenius suggested that in order for the reaction to proceed from left to right the system has to move through some intermediate state, the activated state. Thus, for the forward reaction, the activation energy is $E_{a,f}$, the difference in energy between the reactants and the activated state. Unless the reactants have energy equal to or greater than $E_{a,f}$, they cannot progress through the activated state to products.

For the reverse reaction along the same path in the reverse direction, the activation energy is $E_{a,r}$, the difference in energy between the activated state and

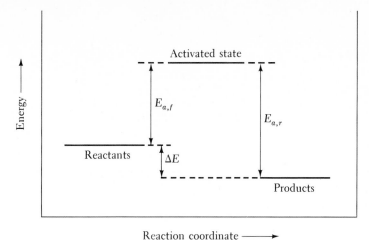

Figure 22-4 Activation Energy
A reaction proceeds in a forward direction along some
path (not shown) from a well-defined initial state (reactants), through an intermediate activated state, to a
well-defined final state (products). For the reverse reaction, the roles of reactants and products are interchanged. The activation energy for either reaction,
which appears in an expression like (22-22) for the rate
constant k, is the energy needed to go from the initial
state for the reaction to the intermediate activated
state.

the products. The overall energy change during the forward reaction is given by

$$\Delta E = E_{a,f} - E_{a,r}$$

To relate the Arrhenius concept of activation to kinetic theory, it is necessary to consider the distribution of kinetic energy among the molecules in a gas or liquid. There is a continual exchange of kinetic energy among the constantly colliding molecules. The distribution of kinetic energies at thermal equilibrium is shown in Fig. 22-5. As shown by the dashed line in the figure, the fraction of molecules with energies far in excess of the average increases very rapidly with increasing temperature, even though the average energy increases only in proportion to T. If the activation energy lies well above the average kinetic energy, only highly energetic molecules can reach the activated state. The rapid increase with temperature of the fraction of highly energetic molecules therefore corresponds to a large increase in the number of molecules with sufficient energy to react, which explains the large increase with temperature of most reaction-rate constants.

Orientational or Steric[7] Effects in Reaction Rates   The activation energy for a particular reaction must be found experimentally from a plot of ln k against $1/T$, as described earlier. Once this has been done, it is possible to test the validity of the collision theory expression for the rate constant. An approximate value of the collision frequency can be calculated for molecules of known diameter at any temperature, and thus a value of k at that temperature can be calculated for any bimolecular reaction of known activation energy. When these calculated k's are compared with the experimental k's, it is usually found that the experimental values are too small, sometimes only slightly, sometimes by many powers of 10. These discrepancies can be explained, at least in part, by the assumption that there is an orientational or steric requirement for reaction as well as an energetic

---

[7] The adjective *steric* (Greek: *stereos,* solid) is used in chemistry to refer to an effect arising as a result of the three-dimensional arrangement of the atoms in a molecule.

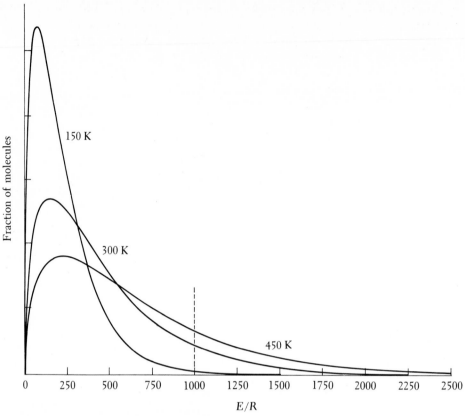

Figure 22-5   Distribution of Kinetic Energies
The fraction of molecules with energy E/R is shown as a function of E/R at 150, 300, and 450 K. The area under each curve to the right of the vertical line represents the fraction of molecules with energies greater than $1000\,R$. It is 0.004 at 150 K, 0.08 at 300 K, and 0.22 at 450 K. (The average molar kinetic energy is $1.5\,RT$—thus it is $225\,R$ at 150 K, $450\,R$ at 300 K, and $675\,R$ at 450 K.)

one. Even though the potentially reacting molecules collide with sufficient energy, if the atoms to be joined by a bond in the new species that might be formed are too far apart at the time of collision, that particular collision cannot result in reaction. Other collisions with similar energy in which the mutual orientation of the reactants is different and the reacting atoms are appropriately positioned can yield product molecules.

Reactions of Hydrogen with Halogens   As examples of gas-phase reactions for which detailed mechanisms have been deduced, we consider the formation of hydrogen halides. The reaction of H_2 with gaseous I_2 to form HI,

$$H_2 + I_2 \longrightarrow 2HI \tag{22-25}$$

was one of the earliest reactions studied kinetically. It was long believed that its second-order rate law

$$\text{Rate} = -\frac{d[I_2]}{dt} = k[H_2][I_2] \tag{22-26}$$

implied the bimolecular mechanism $H_2 + I_2 \rightarrow 2HI$. However, it has been shown by J. H. Sullivan that very little if any HI is produced by direct bimolecular collisions of H_2 with I_2. Rather, at temperatures below 600 K, the mechanism involves an equilibrium between iodine molecules and atoms

$$I_2 \rightleftharpoons 2I \qquad (22\text{-}27)$$

governed by the expression

$$\frac{[I]^2}{[I_2]} = K \qquad (22\text{-}28)$$

Even though the concentration of iodine atoms at equilibrium with I_2 is very small, enough of them are available for an attack on H_2 molecules in which two HI molecules are formed in the termolecular elementary process

$$H_2 + 2I \longrightarrow 2HI \qquad (22\text{-}29)$$

The sum of the two elementary steps (22-27) and (22-29) is just the overall reaction (22-25), as required of any mechanism. Both the forward and the reverse rates of the elementary process (22-27) are rapid, and the corresponding equilibrium is established quickly enough to replenish the I atoms lost through (22-29) and thus to maintain [I] at a level in accord with the equilibrium law (22-28). The rate of the overall reaction is therefore just the rate of the elementary process (22-29). Moreover, $[I]^2$ in the rate law for (22-29) can be replaced by $K[I_2]$ from (22-28). As a result, the rate law for the overall reaction agrees with that observed experimentally,

$$\text{Rate} = k'[I]^2[H_2] = k'K[I_2][H_2] = k[I_2][H_2]$$

with $k = k'K$.

The *stoichiometry* of the reaction of hydrogen with gaseous bromine

$$H_2 + Br_2 \longrightarrow 2HBr \qquad (22\text{-}30)$$

is identical to that for the H_2–I_2 reaction. The rate law, however, is much more complex. It has been determined experimentally from analysis of measurements of the rate for a broad range of partial pressures of the reactants:

$$\text{Rate} = \frac{k[H_2][Br_2]^{1/2}}{1 + k'[HBr]/[Br_2]} \qquad (22\text{-}31)$$

where k and k' are constants. This rate law implies that in the initial stages of the reaction, when [HBr] is so small that the denominator in the right side of (22-31) reduces to approximately 1, the reaction is of three-halves order, first order in H_2 and one-half order in Br_2. Furthermore, as the product of the reaction, HBr, accumulates, the ratio [HBr]/[Br$_2$] increases and so the denominator of the right side of (22-31) increases and the rate falls. Thus HBr is said to be an inhibitor of the reaction.

It took more than a decade before anyone could devise a mechanism to account for the peculiar rate law (22-31), and then three people did it independently and almost simultaneously. They proposed a chain reaction involving bromine atoms and hydrogen atoms as *intermediates,* transient species that exist for times of the order of 10^{-6} s—short-lived by ordinary chemical standards. The term *chain reaction* implies a sequence of steps in some of which there is regeneration of reactants as well as formation of products. A single step to initiate the chain may eventually result in the formation of a large number of product molecules. The proposed steps were

Chain initiation:	$Br_2 \longrightarrow 2Br$		(22-32a)
Chain propagation:	$Br + H_2 \longrightarrow HBr + H$		(22-32b)
	$H + Br_2 \longrightarrow HBr + Br$		(22-32c)
Chain inhibition:	$H + HBr \longrightarrow H_2 + Br$		(22-32d)
Chain termination:	$2Br \longrightarrow Br_2$		(22-32e)

The chain is initiated by bromine atoms formed by the thermal or photochemical dissociation of molecular bromine, reaction (22-32a). In the chain-propagating steps, (22-32b) and (22-32c), two molecules of HBr are formed and a bromine atom is regenerated, so that these steps can be repeated as long as bromine atoms (and the supply of Br_2 and H_2) remain. The inhibiting effect of HBr is accounted for by reaction (22-32d): H atoms produced in (22-32b) are consumed in (22-32d), as well as in (22-32c), forming H_2 and using up HBr. Why was reaction (22-32d) proposed rather than the alternative possible inhibiting reaction, $Br + HBr \to Br_2 + H$? Because, according to the rate law (22-31), the inhibition depends on the ratio $[HBr]/[Br_2]$, implying that HBr and Br_2 compete with each other for some atom. Although by (22-32c) bromine molecules would be expected to compete effectively with HBr for H atoms, they would not compete for Br atoms since there is no driving force for a reaction of Br_2 with Br.

We shall not attempt to derive (22-31) from a combination of the rate laws for the elementary processes represented by Equations (22-32). However, a rate law of exactly the form (22-31) can be derived, in which k and k' represent combinations of rate constants for the elementary processes.

Does the agreement between the derived and experimental rate laws prove that the mechanism represented by (22-32) is correct? It does not. As we have stressed before, it is impossible in science to prove that some model is *the* correct one. All that can be said is that it is consistent with the evidence that has been considered. Other models, not tested, might also be consistent with that evidence. Furthermore, new evidence might be discovered that is inconsistent with the proposed mechanism, and then the mechanism must be revised, as for the $H_2 + I_2$ reaction just discussed.

The reaction of hydrogen with chlorine is similar to that with bromine, although it is more difficult to study experimentally. The contrast between the kinetics of these stoichiometrically identical reactions of hydrogen with the three common halogens is instructive—and should be sobering to anyone contemplating easy generalizations about kinetics.

Rate-Determining Step   When a reaction occurs in a series of successive steps, with some products of one step participating as reactants in a subsequent step, it often happens that the overall rate of the reaction is equal to the forward rate of one of the steps. Such a step is called a *rate-determining step*. An example is the elementary reaction (22-29) between H_2 and 2I in the overall reaction (22-25) between H_2 and I_2 to form HI. In terms of mechanism, a rate-determining step is the first step in a sequence of elementary reactions for which the reverse rate of the step is negligible and does not affect the forward rate of that step. All steps that precede the rate-determining step involve equilibria that are established rapidly. Later steps have no effect on the overall rate.[8]

Since the rates of different steps usually depend on the concentrations of some of the reactants, these rates change as the concentrations change during the course of a reaction. Furthermore the relative rates of different steps can change with other changes in reaction conditions, e.g., in the temperature, the pressure, or the solvent. Thus a step that is rate-determining under one set of conditions may not be under another.

The Transition from Reactants to Products So far we have focused attention on three states in kinetics: the reactants, the products, and the activated state between them. However, the transition from reactants to products involves a succession of many states, from which these three are singled out because they represent, respectively, a position of relative stability at the start, a stable position at the end, and a maximum in potential energy during the transition between the other two states.

Consider what happens as molecules collide and react, for example, in a simple bimolecular reaction

$$XY + Z \longrightarrow X + YZ \qquad (22\text{-}33)$$

When a collision of Z and XY occurs with sufficient kinetic energy and appropriate orientation for reaction, some of the mutual kinetic energy is transferred into vibrational energy of the X-Y bond. The amplitude of these vibrations is thereby greatly increased. As a result the average X-Y bond distance is lengthened and the overlap of the bonding orbitals of X and Y is greatly diminished. Simultaneously, if atom Z is in an appropriate position, it can bond weakly with Y. As the X-Y interaction weakens, the Y-Z interaction can grow stronger. These two effects will eventually be comparable in strength, and we can represent the state at that stage as

$$X\!\!-\!\!-\!\!-Y\!\!-\!\!-\!\!-Z \qquad (22\text{-}34)$$

with the long lines implying greatly stretched and therefore weak bonds. The species (22-34) may pass smoothly over into the products, X plus YZ, or it may dissociate back into the reactants, XY and Z. In any event, it is unstable and will have a very short lifetime. It presumably corresponds to the maximum, or very close to the maximum, in the potential energy during the course of the reaction. Thus (22-34) represents a possible structure for the activated state, or what is often called the transition state, for this reaction. Such specific arrangements of atoms in the activated state are called *activated complexes*. They are in many ways like ordinary molecules with specific geometry and other properties, but differ in that they are extremely reactive, decomposing almost instantly.

An ideal theory of kinetics would permit one to calculate the most likely path by which reactants are converted to products, and it would also give the rate at which reacting species would travel over this path. The various possible paths for

---

[8] An illustrative but very limited analogy to a reaction that occurs in a series of successive steps is an automobile assembly line. Each step in production is necessary before the next one can be performed, and thus each step is essential to the completion of the car. Nothing can come off the line at a rate greater than that of the slowest step in the line. Suppose there is a bottleneck at the stage at which the body of the car is painted. The rates of all subsequent steps (and thus the rate at which cars come off the line) are limited by the rate at which the cars can be painted.

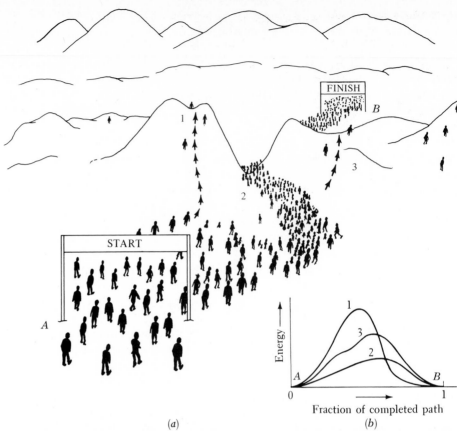

Figure 22-6 (*a*) **Analog to a Potential Energy Surface;** (*b*) **Symbolic Representation of Different Reaction Paths**

In the symbolic picture in (*a*) the varying altitudes of mountains and valleys are a measure of varying gravitational potential energy. This energy varies over the earth's surface and thus is a function of two variables, such as latitude and longitude. The surface might also represent the potential energy of a reaction for which the potential energy depends on just two variables. Reactants are symbolized by the "start" line in one valley, products by the "finish" line in another valley; the wanderings of people symbolize the progress of individual reactions as reactant molecules change into product molecules.

In (*b*) the numbered curves represent the paths with the same numbers in (*a*). The three actual paths are, of course, quite different; placing them on the same two-dimensional graph is only schematic.

any reaction can be thought of as laid out on a multidimensional potential-energy surface. In two dimensions this is analogous to the variation of altitude over the earth's surface. Altitude can be thought of as proportional to gravitational potential energy, and position on the earth's surface can be specified by two variables, e.g., longitude and latitude. In this analogy, valleys represent positions of relative stability (reactants and products) and the mountains between the valleys represent the barriers that must be surmounted in passing from one stable state to another. If reactants in valley *A* (Fig. 22-6) are to form products, they must cross the mountains to valley *B*. There are many paths that can be followed between the two valleys. The path requiring the minimum expenditure of energy is that through the lowest pass, which represents the activated complex. This is the most probable path for the reaction. Some molecules may get to valley B by higher energy paths and some will wander off to other valleys; i.e., other products will be formed by side reactions.

The calculation of multidimensional potential-energy surfaces adequate for prediction of the likely course and activation energy for a reaction is far too complex for any but simple reactions; even then it can be done only approximately. Since detailed calculations are impractical, one must make plausible guesses about possible structures for the transition state in a reaction. The observed rate law provides evidence about the mechanism and thus about the nature of the reactants that come together in the transition state. The principles of structural chemistry then serve as a guide in the formulation of an activated complex that can reasonably be formed from the reactants and can decompose readily into the products.

22-5   Reactions in Liquid Solutions

Most of the qualitative and quantitative concepts developed through analysis of the kinetics of gaseous reactions are also applicable to reactions in solutions. The concentration dependence of reaction rates can be expressed by laws similar to those for gaseous reactions, and frequently these laws are of the particularly simple form of (22-8), a product of concentration terms each raised to some power. The effects of temperature on reaction rates are very frequently expressible in terms of the simple Arrhenius relation, Equation (22-22), and the concepts of activation energy and the activated complex have been extended to reactions in solution. To be sure, the interpretation of these concepts in terms of a molecular mechanism is usually more subtle and uncertain for reactions in liquid media than for those in the gas phase because of the complications caused by the presence of the solvent. It can play many roles, from that of a comparatively passive bystander to that of an active participant in the reaction, and it can change the reaction from what it would have been with no solvent present.

Encounters between Reactants in Solution   If two molecules collide in the gas phase and do not react, they normally move apart again at once and there is little likelihood that this same pair will soon collide again. In contrast, when two potentially reacting solute molecules A and B diffuse together in a solution, they cannot normally move apart again quickly because they are surrounded closely by other molecules, chiefly of solvent. The reactants are, so to speak, in a cage of solvent. It is not a rigid or permanent cage. Sooner or later, after much jostling and agitation, an opening appears in some direction and, if they have not already reacted with each other, either A or B may escape from the cage and move away from its potential reacting partner.

Thus, the *cage effect* causes reactants to remain together far longer in solution than in the gaseous state, long enough for them to undergo tens or hundreds of collisions with each other before they drift apart. The term *encounter* is often used to distinguish this situation, in which potential reactants may remain together long enough to undergo many successive collisions, from the fleeting and nonrecurring collisions characteristic of gases.

For reactants with a high probability of reacting on each collision, the cage effect virtually ensures reaction during each encounter. Hence the rate of such a reaction is limited only by the rate at which the reactants can diffuse together, and the reaction is said to be *diffusion-controlled*. For aqueous solutions, this rate is of the order of $10^{10}\,M\,\mathrm{s}^{-1}$ for unit concentrations, corresponding to a bimolecular rate constant of about $10^{10}\,M^{-1}\,\mathrm{s}^{-1}$—a little smaller for large or massive reactants (which move slowly) and a little larger for reactions between small ions of opposite charge (which are mutually attracted).

This rate constant represents an upper limit for bimolecular reactions in aqueous solutions, for no reaction can proceed faster than the reactants can come together. Various ionic reactions, including many proton-transfer processes, have rate constants of this order of magnitude, as do the recombination reactions of molecular fragments that have unpaired electrons. These reactions typically have activation energies in the range 12 to 20 kJ mol^{-1}, which corresponds to the activation energy for diffusion in most liquids. Thus, the evidence is strong that these reactions are indeed diffusion-controlled.

The cage effect can also play a role in solution reactions that are not diffusion-controlled. For example, for those reactions in which the mutual orientation of the reactants is critical, the relatively long duration of encounters in solution may increase the probability of reaction. Because the reacting partners may collide in many different orientations as they are buffeted by their neighbors, they are more likely to achieve the proper orientation for reaction before they drift apart than if they underwent only a single collision.

Specific Solvent Effects Because of the overriding importance of water as a solvent, we focus attention on the ways it can influence the kinetics of reactions. However, the effects mentioned can be illustrated with many other solvents as well. Only the details would differ.

The most general effects of water on the rate and even the course of a reaction are a consequence of its highly polar nature. When ions are separated by water, their energy of interaction and the force between them are greatly diminished. Furthermore, most ions, especially cations, are hydrated in aqueous solution, as are some neutral polar molecules. These are, of course, just the factors that make water so good a solvent for many ionic and polar compounds; they can markedly affect reaction rates in various ways.

For example, when a hydrated ion or molecule reacts to form an activated complex, it may lose some of its hydration shell. Since hydration of an ion is often a highly exothermic process, dehydration requires a substantial expenditure of energy, and the energy of activation for this path is increased over what it would have been if the hydration shell had been undisturbed. Conversely, molecules that are originally not highly hydrated may react by way of an ionic transition state that is stabilized by hydration. The activation energy for this reaction will then be much lower in water than in some less polar medium, and the rate will be correspondingly much greater.

22-6   Reaction Rates and Equilibria

For any reaction that occurs by a single elementary step, the rate of the forward reaction must, at equilibrium, be just equal to the rate of the reverse reaction. Thus, if in the elementary step A combines with B to form C, with k_f and k_r the rate constants for the forward and reverse reactions, respectively,

$$A + B \xrightarrow{k_f} C \qquad C \xrightarrow{k_r} A + B$$

then at equilibrium

$$k_f[A][B] = k_r[C] \qquad (22\text{-}35)$$

We know that at equilibrium $[C]/([A][B])$ is a constant, which we represent as K; thus, by (22-35),

$$K = \frac{[C]}{[A][B]} = \frac{k_f}{k_r} \qquad (22\text{-}36)$$

The result in (22-36) can readily be generalized for reactions that proceed by a series of steps. The overall stoichiometric equation must be the sum of the equations for the individual steps, and we know from Chap. 11 that the equilibrium constant for a reaction that can be represented as the sum of other reactions is just equal to the product of the equilibrium constants for those other reactions. For a multistep reaction, each elementary step must proceed at the same rate as its reverse when the system is at equilibrium. Thus, if the overall reaction is the sum of three individual steps, denoted by the subscripts 1, 2, and 3, we apply (22-36) to each step and get for the overall equilibrium constant K

$$K = K_1 K_2 K_3 = \frac{k_{f,1}}{k_{r,1}} \frac{k_{f,2}}{k_{r,2}} \frac{k_{f,3}}{k_{r,3}} \qquad (22\text{-}37)$$

22-7 Catalysis

Many reactions are speeded up by substances that are not themselves permanently changed during the reaction. These substances are called *catalysts,* a term introduced by Berzelius early in the nineteenth century. The effects are often dramatic. Reactions that proceed immeasurably slowly in the absence of catalysts may go too rapidly to be followed when a catalyst is present. Many catalysts are effective even when present in almost imperceptibly small amounts. Under certain conditions, for example, one molecule of the enzyme catalase can cause the decomposition of 10^5 molecules of hydrogen peroxide every second. In the absence of a catalyst, hydrogen peroxide and its solutions can be stored for years.

Since catalysts undergo no permanent change during the reaction whose rate they are accelerating, they do not appear in the stoichiometric equation for that reaction and can have no effect on the position of equilibrium. This implies that they must accelerate the reverse reaction by the same factor by which they accelerate the forward reaction, so that the equilibrium constant, the ratio of these rates at equilibrium, remains the same. For example, the hydrolysis of esters, a reaction exemplified by

$$CH_3C\underset{OC_2H_5}{\overset{O}{\diagup}} + H_2O \longrightarrow C_2H_5OH + CH_3C\underset{OH}{\overset{O}{\diagup}} \qquad (22\text{-}38)$$

| Ethyl acetate (an ester) | Ethanol (an alcohol) | Acetic acid (a carboxylic acid) |

is catalyzed by acids and by certain enzymes. Exactly the same catalysts have an equal effect on the rate of *formation* of esters and are used in the laboratory and in nature to speed up each of these reactions. Bases also speed up the hydrolysis of esters, but at the same time they react with the carboxylic acids formed to give salts such as sodium acetate. Thus their role is more than that of a catalyst.

Many catalysts are highly specific, being effective only for a certain reaction or a group of closely related reactions. This specificity is particularly characteristic of enzymes, a generic name for the thousands of different proteins that catalyze reactions in living systems.

The Mode of Action of Catalysts   Catalysts are often classified according to whether they act homogeneously (in a single phase) or heterogeneously (at an interface between phases). Heterogeneous reactions are of great practical importance but are complex and not well understood; we discuss here only homogeneous reactions. Although by definition a catalyst is necessarily unaltered at the end of a reaction, every catalyst participates in the reaction it catalyzes. It interacts in some way with at least one of the reactants, and then in a subsequent step or steps, or perhaps as a continuation of the original interaction, it is regenerated in its original form while more or less simultaneously the final products are formed. Because catalysts are continually regenerated, they can act effectively when present in a very small molar proportion relative to the reactants—for example, 1:1000 or even smaller.

Some catalysts react chemically with one of the initial reactants, changing it to a substance that reacts more rapidly; the catalyst is then returned to its original state by a subsequent chemical change. Other catalysts form a rather loose complex with a reactant, perhaps involving only van der Waals forces and hydrogen bonds rather than any stronger chemical interactions. This complex reacts more rapidly than the original reactant and, as it is consumed, it leaves behind the unaltered catalyst, ready to act again.

In general, catalysts lower the activation energy of a reaction, thereby increasing the rate. Some lower the height of the barrier between reactants and products along a path close to that which the uncatalyzed reaction follows. Others open quite different pathways, which also have lower barriers than the original path but which are for some reason not accessible in the absence of the catalyst.

There are substances that slow down certain reactions, rather than accelerating them. These are called inhibitors or, sometimes, negative catalysts. The latter name is misleading, however, for they are not true catalysts since they are themselves permanently changed at the same time that they are inhibiting the reaction. True negative catalysts do not exist.[9]

We turn now to brief discussions of catalysis in redox reactions, in acid-base reactions, and in biological systems.

Oxidation-Reduction Reactions   Many redox reactions, including oxidations by molecular oxygen, can be greatly accelerated by the addition of small amounts of ions of certain transition metals with two or more oxidation states that are

---

[9]There are substances that can interfere effectively and reversibly with the action of catalysts, especially with enzymes. If they inhibit the action of the catalyst essentially completely, the net effect is to force the reaction to follow its normal uncatalyzed pathway. They do not, however, diminish the rate below that of the uncatalyzed reaction.

accessible under the reaction conditions. Typical catalysts include Ag^+, Cu^+, Cu^{2+}, and Mn^{2+}. These catalysts, like many others, operate by what have some-times been called compensating reactions, being alternately oxidized and reduced so that they are continually regenerated. For example, the oxidation of V^{3+} by Fe^{3+},

$$V^{3+} + Fe^{3+} \longrightarrow V^{4+} + Fe^{2+} \tag{22-39}$$

is catalyzed by either Cu^+ or Cu^{2+}, the rate being proportional to the concentra-tions of V^{3+} and of either copper ion but independent of the concentration of Fe^{3+}. This result is explained by the two-step sequence of elementary processes

$$V^{3+} + Cu^{2+} \longrightarrow V^{4+} + Cu^+ \qquad \text{(rate-determining)} \tag{22-39a}$$
$$Fe^{3+} + Cu^+ \longrightarrow Fe^{2+} + Cu^{2+} \tag{22-39b}$$

whose sum is just (22-39). If Cu^+ is used as the catalyst, it is quickly changed to Cu^{2+} by reaction (22-39b) and the rate-determining step is still (22-39a). Al-though much has been learned during the last two decades about the mecha-nisms of many oxidation-reduction reactions, it is not yet clear why electron transfer from V^{3+} to Cu^{2+} and then from Cu^+ to Fe^{3+} should be so much faster than direct transfer from V^{3+} to Fe^{3+}.

Acid-Base Catalysis A great variety of reactions can be catalyzed by acids or bases, or by both acids and bases, some only by H^+ or OH^-, others by any acid or base present. We have already cited the formation and hydrolysis of esters; many other reactions of organic compounds are similarly catalyzed. In general, the mechanism of acid catalysis involves transfer of a proton to a reactant molecule, and catalysis by bases proceeds via proton transfer from the reactant molecule to the base, resulting in a species of enhanced reactivity in each case.

Enzymes The molecular weights of the proteins known as enzymes that cata-lyze reactions in living systems are usually in the range 10^4 to 10^6. A complex organism functions properly only because of a delicate balance of chemical reactions catalyzed by thousands of different enzymes, each with a distinct function and highly specific structure, essentially identical in all individuals of the same biological species and remarkably similar in different species. We shall defer consideration of any mechanistic details of enzyme action until the chemi-cal nature and structure of proteins in general have been discussed (Chap. 30). However, a few of the facts and concepts relating to enzymes and their action that do not depend on knowledge of the structural details of any particular enzyme are worth discussing at this point.

The substance acted upon by the enzyme is termed the *substrate*. It is observed that the rates of many enzyme-catalyzed reactions have the following charac-teristics: (1) they are first order in the substrate at low substrate concentrations; and (2) at a constant concentration of the enzyme, they become essentially independent of the substrate concentration (zero order) as it becomes very high (Fig. 22-7). The rate is usually first order in the enzyme as well, but this can be studied only at very low concentrations because the solubilities of proteins are low and their molecular weights are high.[10] Michaelis and Menten proposed in

---

[10] For a protein with molecular weight of 10^5, a 1 percent solution, which contains about 10 g protein per liter, has a concentration of $10^{-4}\ M$.

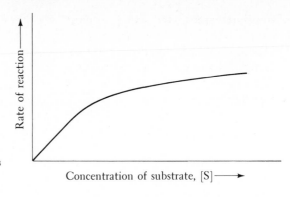

Figure 22-7 Rate of a Typical Enzyme-Catalyzed Reaction at Constant Concentration of Enzyme
The rate of reaction is proportional to the concentration of substrate at low values of [S], indicating a first-order reaction. The rate becomes independent of substrate concentration when [S] is high.

1913 a mechanism that accounts for these facts—a mechanism that is still accepted today and is almost always associated with their names.

The catalytic activity of enzymes is attributed to particular regions of each molecule, called active sites. To simplify this discussion, we assume there is only one active site per enzyme molecule. A substrate molecule S can combine reversibly with the enzyme E, occupying the active site and forming an enzyme-substrate complex ES. The complex ES can then either dissociate to its original components, E and S, or it can decompose to give the products of the reaction that is being catalyzed as well as uncomplexed enzyme, ready to pick up more substrate. These reactions can be summarized as

$$E + S \rightleftharpoons ES \qquad (22\text{-}40a)$$
$$ES \longrightarrow E + products \qquad (22\text{-}40b)$$

This combination of reactions is far faster than the uncatalyzed reaction by which S is changed into products. The equilibrium (22-40a) is established rapidly; the rate-determining step is (22-40b), which is first order in the complex ES. At low concentrations of S, most active sites on the enzyme molecules are unoccupied; [ES] increases linearly with [S], and thus the rate of (22-40b) is proportional to [S]. As [S] increases at constant enzyme concentration, however, more and more of the active sites become occupied. When there are few empty sites left, [ES] is almost unaffected by further increases in [S], and the rate levels off (Fig. 22-7).

22-8 Reaction Mechanisms

We summarize here some ways of learning about the elusive details of what happens as reactants are changed into products.

Postulating a Mechanism The first step in any study of kinetics is the determination of the rate law and its temperature dependence. If the rate law involves fractional orders or a sum of terms, the mechanism is certain to consist of several steps, no one of which is rate-determining. For simple rate laws there are some guidelines to aid in postulating a mechanism from the rate law and the known stoichiometry of the reaction. The key generalization is that the observed rate law must be explainable in terms of the reaction occurring in the rate-determining step. No experimental evidence has ever been obtained for a molecularity greater than 3. Thus, when the overall order is greater than 3, one or more equilibria and intermediates *prior* to the rate-determining step are very likely to be involved in the mechanism.

If the rate law contains an inverse dependence on the concentration of some species, that species must be a product of a rapid equilibrium prior to the rate-determining step and some other product of that same equilibrium must be a participant in the rate-determining step.

For example, iodide reacts with hypochlorite in alkaline solution to produce hypoiodite and chloride:

$$I^- + OCl^- \longrightarrow OI^- + Cl^- \tag{22-41}$$

The rate law is found to be

$$\text{Rate of appearance of } OI^- = k \frac{[I^-][OCl^-]}{[OH^-]} \tag{22-42}$$

The inverse dependence on $[OH^-]$ suggests that hydroxide ion is the product of a rapid equilibrium prior to the rate-determining step. Since OCl^- appears in the numerator of the rate expression, its participation as a reactant in such an equilibrium is consistent with the experimental observations. Thus, a plausible rapid equilibrium is the hydrolysis of OCl^-:

$$OCl^- + H_2O \rightleftharpoons HOCl + OH^- \quad \text{(rapid equilibrium)} \tag{22-43}$$

The observed law is consistent with two additional steps in the mechanism, the sum of all the steps being the stoichiometric equation:

$$I^- + HOCl \longrightarrow HOI + Cl^- \quad \text{(rate-determining)} \tag{22-44}$$
$$OH^- + HOI \rightleftharpoons OI^- + H_2O \quad \text{(rapid equilibrium)} \tag{22-45}$$

The rate law is that of the slow step,

$$\text{Rate of appearance of } HOI = k'[I^-][HOCl] = k'K \frac{[I^-][OCl^-]}{[OH^-]} \tag{22-46}$$

with K the equilibrium constant for (22-43), the hydrolysis constant for OCl^-. Since each molecule of HOI is converted rapidly to one hypoiodite ion, OI^-, the rate of appearance of OI^- is given by the same expression, in agreement with the observed rate law (22-42), the experimental rate constant k being equal to $k'K$.

Testing a Mechanism   A mechanism can never be proved to be correct, but it can be proved wrong by one unambiguous piece of relevant experimental evidence. Sometimes several different mechanisms are consistent with the available evidence. It is then particularly desirable to design experiments that can distinguish among them. Among the useful kinds of experiments are:

1. Careful analysis of reaction mixtures to establish the possible presence of small amounts of products other than the main products.
2. Efforts to detect short-lived intermediates in reaction mixtures; spectroscopic methods are often valuable.
3. The use of isotopes as *tracers* to permit following a reaction in which there is no net chemical change or to give information about the breaking of bonds during a reaction. Radioactive isotopes are often used, but stable isotopes, such as ^2H (deuterium) and ^{18}O, are also of great value. They can readily be detected and estimated precisely with a mass spectrometer (Fig. 1-8).

4. The use of isotopes to provide mechanistic evidence by their influence on the rate of the reaction itself. This may be possible if an appropriate isotope can be substituted for one of the atoms that participates directly in bond breaking or bond formation during the rate-determining step of a reaction. The effect is discernible only for isotopes of relatively low atomic weight, no more than about 30, because it depends on the ratio of the masses of the normal isotope and that which is used in its place.

5. Variation of the nature of the medium in which the reaction is carried out, which can provide important clues about the mechanism, e.g., comparing solvents of quite different polarity.

6. Study of changes in the stereochemistry of the molecule that has been altered, that is, in the three-dimensional arrangement of its bonds. This method is particularly useful for molecules that are not superimposable upon their mirror images (Chaps. 27, 29, and 30).

22-9 Equilibrium Control or Kinetic Control of Product Distribution

In some reactions equilibrium is established so quickly that the relative amounts of reactants and products can be predicted solely on thermodynamic grounds, from equilibrium constants. The products of such a reaction are said to be *equilibrium-controlled*. Sometimes several different products are formed from a given reaction mixture by alternative pathways. If such a reaction is equilibrium-controlled, the proportions of the different products are entirely independent of the mechanisms and the relative rates of the reactions by which the various products are formed.

Most acid-base reactions and some others, including certain oxidation-reduction processes and acid-catalyzed reactions, are equilibrium-controlled. However, the great majority of the common reactions of organic, inorganic, and biological chemistry are usually not run under conditions in which equilibration of the possible products can occur. The relative amounts of the different products formed from a particular reaction mixture are therefore quite independent of their relative stabilities and depend only on their relative rates of formation. The distribution of products is said to be *kinetically controlled*.

Many of the reactions of carbon-halogen compounds (Chap. 29) are kinetically controlled, for example, the hydrolysis of *t*-butyl chloride to *t*-butyl alcohol,

$$(CH_3)_3CCl + H_2O \longrightarrow (CH_3)_3COH + H^+ + Cl^- \qquad (22\text{-}47)$$

In an alkaline solution, there are several possible reactions in addition to the formation of the alcohol. One of them results in the formation of a hydrocarbon containing a carbon-carbon double bond (an alkene):

$$CH_3-\underset{\underset{Cl}{|}}{\overset{\overset{CH_3}{|}}{C}}-CH_3 + OH^- \longrightarrow CH_3-\overset{\overset{CH_3}{|}}{C}=CH_2 + Cl^- + H_2O \qquad (22\text{-}48)$$

The analogous reactions for *t*-butyl bromide and *t*-butyl iodide also occur, but at quite different rates. Significantly, the proportion of the alkene is the same regardless of which halide is used. Thus, at room temperature in aqueous ethanol, about one-sixth of each halide is converted into the alkene. This is not the equilibrium proportion of the alkene, as can be shown from the known standard free energies of the alkene, the alcohol, and the other possible products.

The fact that the different halides react at different rates to give the alkene in the same proportions suggests strongly that the reactions proceed through an intermediate common to all three halides. The formation of this intermediate from the starting material is the rate-determining step and thus the rate depends on which halide is used. However, the proportion of products depends only on the ratio of the rates of the different possible reactions of this common intermediate and hence is independent of the starting material. The intermediate is believed to be the t-butyl cation, $(CH_3)_3C^+$. When it adds water and then loses a proton,

$$(CH_3)_3C^+ + OH_2 \longrightarrow \left[(CH_3)_3CO \begin{matrix} H \\ \diagdown \\ H \end{matrix} \right]^+ \longrightarrow (CH_3)_3COH + H^+ \qquad (22\text{-}49)$$

the product is t-butyl alcohol. When it loses a proton directly, the alkene is formed. Since all the protons in $(CH_3)_3C^+$ are equivalent, only one alkene can be formed from it by loss of a proton. With some cations, however, loss of a proton can occur in several distinct ways to give isomeric alkenes. Under some conditions the proportions in which these are formed are kinetically controlled, depending only on the relative rates of loss of the different protons. Under other conditions the proportions may be equilibrium-controlled, the final ratio of products then being determined solely by the relative stabilities of the different alkenes that can be formed.

22-10   Concluding Remarks

Most of our present understanding of kinetics is derived from studies of the average behavior of large numbers of molecules with a range of energy states and modes of motion, and often in very complex environments. Because of the averaging, any resulting view is necessarily imprecise when it gets down to molecular details. However, there is every reason to believe that a revolution in the understanding of the dynamics of chemical change is now under way, a revolution that will eventually lead to extraordinarily detailed information about and control over the specific molecular interactions that occur as different kinds of reactions take place.

No real understanding of chemical dynamics could be contemplated until structural chemistry was on a firm foundation, both experimentally and theoretically, for chemical dynamics is essentially structural chemistry with the dimension of time added. Such a foundation has been provided in the last four or five decades. The rapid advances in electronics and instrumentation in the last two decades have made it possible to design and carry out detailed experimental studies of elementary gaseous reactions by methods that employ intersecting beams of molecules, reactions in special mass spectrometers, ultrashort pulses of radiation, and related techniques. The effects of varying the mutual energies of the reactants, and of distributing this energy in varying proportions among translational, rotational, vibrational, and electronic modes, can be observed and analyzed, as can the ways in which energy and momentum are distributed among the products. These experiments are complex and expensive, but they should lead to valuable generalizations that can extend our understanding and control over the reactions of more complex molecules.

Summary

The microscopic reaction sequence by which a chemical reaction occurs is called the mechanism of the reaction. Each individual step in the mechanism is termed an elementary process and can be described by a chemical equation involving the exact atomic or molecular species reacting with one another at this stage. No relationship need exist between the overall stoichiometric equation for a reaction and the mechanism other than that the sum of the equations representing the individual steps must be the overall equation.

The rate of a reaction usually changes during the course of the reaction. Often the instantaneous rate can be represented by an expression of the form

$$\text{Rate} = k[A]^a[B]^b[C]^c \ldots$$

where A, B, C, . . . are reactants; a, b, c, . . . are typically integers; and the proportionality constant k is called the rate constant for the specific reaction. The exponent associated with the concentration of a particular reactant is known as the order with respect to the reactant, and the sum of the exponents is the overall order. Integrated rate expressions are mathematical descriptions of the relation between the concentration of a reactant or product and time. Such expressions can be derived from the equation for the reaction rate.

The order of a reaction can sometimes be identified by characteristic features. For example, the half-life, the time required for the concentration of a reactant to fall to half its initial value, is independent of the concentration if the reaction is first order, but is inversely proportional to the initial concentration for a second-order reaction. For a first-order reaction a plot of $\ln c$ versus t is linear; for a reaction that is second order in a single reactant, a plot of $1/c$ versus t is linear.

Arrhenius observed that the rate constants for many reactions depend exponentially on the absolute temperature: $k = Ae^{-E_a/RT}$, where A is called the frequency factor and E_a is known as the activation energy. Plots of $\ln k$ versus $1/T$ are often nearly straight lines. The slope of such a graph is $-E_a/R$. The rates of most reactions increase with increasing temperature, a situation implying a positive activation energy.

The kinetic theory of gases can be used to interpret the experimental facts of the chemical kinetics of gas-phase reactions if it is assumed that rates are determined by the number of "productive" collisions between reacting molecules per second. In a productive collision the molecules have sufficient energy and are properly oriented. In this model the activation energy is the minimum energy required to bring the reactants to an intermediate state between products and reactants, the activated state.

The mechanism for the formation of HBr from H_2 and Br_2 is an example of a chain reaction. In such a reaction there is a sequence of steps in some of which there is regeneration of reactants as well as formation of products. Agreement between an experimental rate law and one derived from a mechanism does not prove that the mechanism is correct. A different mechanism may lead to the same rate expression.

When reactions take place in solutions, the solvent molecules tend to keep the reacting species together much longer than in gas reactions (cage effect), substantially increasing the probability of reaction. The presence of solvent molecules may also have more specific consequences and may influence reaction paths and products by stabilizing one possible transition state over others.

At equilibrium, the rate of every elementary process involved in a reaction is proceeding at the same rate as its reverse. The equilibrium constant for an elementary process is equal to the ratio of the rate constant for the forward reaction to that for the reverse reaction, k_f/k_r. Similarly, K for the overall reaction is the product of the equilibrium constant expressions for all the steps, $(k_{f,1}/k_{r,1})$ $(k_{f,2}/k_{r,2})$ . . . . Sometimes when a reaction occurs in a series of successive steps, the overall rate is equal to the forward rate of one of the steps, the rate-determining step. When there is a rate-determining step all steps that precede it reach equilibrium rapidly; the reverse rate of the step must be negligible.

Reactions can often be accelerated by catalysts, substances that are not themselves permanently changed by the reaction. In general, a catalyst acts by lowering the activation energy of the reaction, which may be achieved by lowering the energy of an activated complex or by the opening of new pathways.

A given reaction mixture may often produce several products along different pathways. When equilibria are established quickly, as in most acid-base reactions and in some oxidation-reduction processes, the relative amounts of different products can be predicted on the basis of thermodynamic stabilities, and the reaction is termed equilibrium-controlled. In the majority of common reactions, no equilibration of possible products occurs; the relative amounts of products depend on the relative rates of formation, and the reactions are said to be kinetically controlled.

Terms and Concepts

Problems and Questions

22-1 Determination of Order of Reaction Nitric oxide, NO, reacts with chlorine to form nitrosyl chloride, NOCl. At 22°C, with all reactants gaseous, the following data were obtained:

[NO]/M	[Cl$_2$]/M	$-\dfrac{d[\text{Cl}_2]}{dt}/M\,\text{s}^{-1}$
0.100	0.100	8.0×10^{-3}
0.50	0.100	2.0×10^{-1}
0.100	0.50	4.0×10^{-2}

(a) What is the order of the reaction with respect to each reactant? (b) What is the rate constant for the reaction? (c) Suppose 0.30 mol NO and 0.60 mol Cl$_2$ are mixed in a 3.0-liter flask. How much chlorine disappears during the first 0.5 s?

22-2 Reaction Rate At constant hydrogen-ion concentration the rate expression for the reaction of bromide and bromate to give bromine,

$$5\text{Br}^- + \text{BrO}_3^- + 6\text{H}^+ \longrightarrow 3\text{Br}_2 + 3\text{H}_2\text{O}$$

501

can be written as

$$\frac{d[\text{Br}_2]}{dt} = k\,[\text{Br}^-][\text{BrO}_3{}^-]$$

The rate constant k is 2.1×10^{-4} liter $\text{mol}^{-1}\,\text{s}^{-1}$. A solution is prepared by dissolving 1.00×10^{-2} mol KBr in 100 ml water, which is mixed with 100 ml of a solution that is $1.00 \times 10^{-2}\,M$ in KBrO$_3$ and 1 M in H$_2$SO$_4$. (a) At what rate is bromine appearing in the solution after exactly half the BrO$_3{}^-$ has reacted? (b) What is the molar concentration of Br$_2$ in the solution at this point? (c) State qualitatively the effect on the rate, at this point, of doubling the concentration of Br$_2$ by adding liquid bromine to the solution.

22-3 Second-Order Reaction A certain compound, A, decomposes in solution according to a second-order rate law. When the initial concentration is $0.100\,M$ at $25°$C, the concentration falls to $0.075\,M$ in 2.5 hour. If the initial concentration is $0.500\,M$, what will the concentration be after 5.0 hour?

22-4 Order and Rate of a Reaction The following data have been obtained for the instantaneous rate of this reaction occurring in solution:

$$3\text{A} + 2\text{B} \longrightarrow \text{A}_3\text{B}_2$$

[A]/M	[B]/M	$(-d[\text{A}]/dt)/(M\,\text{s}^{-1})$
0.10	0.10	6.0×10^{-4}
0.30	0.30	5.4×10^{-3}
0.50	0.10	1.5×10^{-2}
0.50	0.50	1.5×10^{-2}

(a) Calculate the order of the reaction with respect to A and with respect to B, and calculate the value of k (including its units).
(b) Suppose that you have 2.0 liter of a solution that is $0.50\,M$ in A and $0.50\,M$ in B. How many moles of A$_3$B$_2$ are formed in the first second of reaction?
(c) How long will it take for the concentration of A to fall to $0.05\,M$ from the initial conditions described in (b)?

22-5 Mechanism and Rate Law A reaction proceeds by the following mechanism:

$$m\text{A} + n\text{B} + p\text{C} \longrightarrow \text{products}$$

with m, n, and p all positive integers. Doubling the concentration of A, B, and C increases the overall rate of reaction by a factor of 16. Tripling the concentration of C has the same effect as tripling the concentration of A. An increase in the concentration of B has a larger effect than an increase in the concentration of A. What are m, n, and p? What is unrealistic about this mechanism?

22-6 First-Order Kinetics A substance X decomposes by a first-order reaction with half-life t_1. About how much time would be needed for the concentration of X to fall to 0.1 percent of its initial value?

22-7 Determination of Order of Reaction A certain compound A decomposes in solution to form two other substances, B and C, under conditions such that the reverse reaction and competing reactions are negligible. The following kinetic data were measured:

Time/min	[A]/M	Time/min	[A]/M
0	0.100	58	0.037
10	0.084	92	0.020
20	0.071	140	0.009
36	0.054		

(a) What is the order of the reaction? (b) What is the rate constant?

22-8 Determination of Order of Reaction A certain hydrocarbon A is unstable in the gas phase at temperatures above $300°$C. Measurements of its concentration spectroscopically showed that, at $350°$C and an initial pressure of 1.0 atm, the concentration of A fell to half its original value in 1.5 min; if the initial pressure was 0.10 atm at the same temperature, the half-life was 15 min. What can you deduce about (a) the order of the reaction by which A disappears, (b) the specific rate constant for this reaction, and (c) the activation energy of this reaction?

22-9 Quenching of a Reaction Some reactions in solution may be brought essentially to a halt by diluting the reaction mixture with an inert solvent at the same temperature. Does this possibility of quenching a reaction depend on the rate law? If so, how?

22-10 Half-Life of a Fast Reaction The rate constant for the transfer of a proton from H$_3$O$^+$ to

OH⁻ is $1.4 \times 10^{11}\,M^{-1}\,s^{-1}$. If one assumes perfect mixing, how long would it take for half of a $10^{-3}\,M$ HCl solution to be neutralized by $10^{-3}\,M$ NaOH?

22-11 First-Order Kinetics The gas-phase decomposition of sulfuryl chloride, SO_2Cl_2, into sulfur dioxide and chlorine is first order, with $k = 2.2 \times 10^{-5}\,s^{-1}$ at 320°C. If a sample of sulfuryl chloride is heated for 10.0 hour at 320°C, what fraction of it will still be present as SO_2Cl_2 at the end of this time?

22-12 Activation Energy for First-Order Reaction Gaseous sulfuryl chloride, SO_2Cl_2, decomposes into SO_2 and Cl_2 by a first-order reaction with $k = 2.2 \times 10^{-5}\,s^{-1}$ at 320°C. If the activation energy for this reaction is 125 kJ mol^{-1}, how long will it take at 340°C for the pressure of SO_2Cl_2 to fall from 1.00 atm to 0.25 atm because of this decomposition?

22-13 First-Order Reaction Peroxydisulfate ion decomposes by a first-order reaction when heated in aqueous solution:

$$S_2O_8{}^{2-} + H_2O \longrightarrow 2HSO_4{}^- + \tfrac{1}{2}O_2$$

The half-life at 70°C is 7.2 hour; at 90°C it is 0.72 hour. (*a*) What is the rate constant for this reaction at 70°C, and what is it at 90°C? (*b*) What is the activation energy for this reaction? (*c*) If a $0.50\,M$ solution of $K_2S_2O_8$ (a strong electrolyte) is heated at 70°C for 25 hour, what will be the concentration of peroxydisulfate ion in the solution at the end of this time, and what will be the concentration of $HSO_4{}^-$?

22-14 Activation Energy The rate constant for the disappearance of chlorine in the third-order reaction of NO with Cl_2 to form NOCl is $4.5\,M^{-2}\,s^{-1}$ at 0°C and $8.0\,M^{-2}\,s^{-1}$ at 22°C. What is the activation energy for this reaction?

22-15 Activation Energy At high temperatures nitrogen dioxide decomposes into NO and O_2 by the second-order rate law

$$\text{Rate} = -\frac{d[NO_2]}{dt} = k[NO_2]^2$$

At 592 K the rate constant is 4.98×10^{-1} liter mol^{-1}s^{-1}, and at 656 K it is 4.74 liter mol^{-1}s^{-1}. Calculate an activation energy that accounts for these values.

22-16 Activation Energy Two second-order reactions have identical values of A in the Arrhenius expression, Equation (22-22). The activation energy of reaction 2 is 20 kJ mol^{-1} greater than that of reaction 1. Calculate the ratio of their rate constants (*a*) at 27°C and (*b*) at 327°C.

22-17 Implications of a Mechanism The following observations have been made about a certain reacting system: (i) When A, B, and C are mixed at about equal concentrations in neutral solution, two different products are formed, D and E, with the amount of D about 10 times as great as the amount of E. (ii) If everything is done as in (i) except that a trace of acid is added to the reaction mixture, the same products are formed, except that now the amount of D produced is much smaller than (about 1 percent of) the amount of E. The acid is not consumed in the reaction. The following mechanism has been proposed to account for some of these observations and others about the order of the reactions:

(1) $A + B \underset{k_{-1}}{\overset{k_1}{\rightleftharpoons}} F$ (rapid equilibrium)

(2) $C + F \overset{k_2}{\longrightarrow} D$ (negligible reverse rate)

(3) $C + F \overset{k_3}{\longrightarrow} E$ (negligible reverse rate)

(*a*) Explain what this proposed scheme of reactions implies about the dependence (if any) of the rate of formation of D on the concentrations of A, of B, and of C. What about the dependence (if any) of the rate of formation of E on these same concentrations? (*b*) What can you say about the relative magnitudes of k_2 and k_3? (*c*) What explanation can you give for observation (ii) in view of your answer to (*b*)?

22-18 Reaction Mechanism Ferrous ion is oxidized by chlorine in aqueous solution, the overall equation being

$$2Fe^{2+} + Cl_2 \longrightarrow 2Fe^{3+} + 2Cl^-$$

It is found experimentally that the rate of the overall reaction is decreased when either the ferric-ion or the chloride-ion concentration is increased. Which of the following possible mechanisms is consistent with the experimental observations?

(a) (1) $Fe^{2+} + Cl_2 \underset{k_{-1}}{\overset{k_1}{\rightleftharpoons}} Fe^{3+} + Cl^- + Cl$
(rapid equilibrium)

(2) $Fe^{2+} + Cl \xrightarrow{k_2} Fe^{3+} + Cl^-$
(negligible reverse rate)

(b) (3) $Fe^{2+} + Cl_2 \underset{k_{-3}}{\overset{k_3}{\rightleftharpoons}} Fe(IV) + 2Cl^-$
(rapid equilibrium)

(4) $Fe(IV) + Fe^{2+} \xrightarrow{k_4} 2Fe^{3+}$
(negligible reverse rate)

where Fe(IV) is Fe in the (+IV) oxidation state.

22-19 Rate of a Proton-Transfer Reaction
Eigen and his coworkers found that the specific rate of proton transfer from a water molecule to an ammonia molecule in a dilute aqueous solution is $k_1 = 2 \times 10^5 \, s^{-1}$. The equilibrium constant for the dissociation of "ammonium hydroxide" is $1.8 \times 10^{-5} \, M$. What, if anything, can be deduced from this information about the rate of transfer of a proton from NH_4^+ to a hydroxide ion? Write equations for any reactions you mention, making it clear to which reaction(s) any quoted constant(s) apply.

22-20 Rate and Equilibrium Consider the reaction

$$A + B \rightleftharpoons C + D$$

with all reactants and products gaseous (for simplicity) and an equilibrium constant K. (a) Assume that the elementary steps in the reaction are those indicated by the stoichiometric equation (in each direction), with specific rate constants for the forward reaction and the reverse reaction, respectively, k_f and k_r. Derive the relation between k_f, k_r,

and K. Comment on the general validity of the assumptions made about the relation of elementary steps and the stoichiometric equation and also on the general validity of K. (b) Assume that the reaction as written is exothermic. Explain what this implies about the change of K with temperature. Explain also what it implies about the relation of the activation energies of the forward and reverse reactions and how this relation is consistent with your statement about the variation of K with temperature.

22-21 Decomposition by Different Paths A certain molecule, X, decomposes simultaneously in two different ways, forming Y with first-order rate constant k_1 and forming Z with first-order rate constant k_2:

(1) $X \xrightarrow{k_1} Y$

(2) $X \xrightarrow{k_2} Z$

At 25°C, $k_1/k_2 = 100$; at 100°C, $k_1/k_2 = 0.10$. The standard molar free energies of formation of X, Y, and Z, which are all gases, are:

$\widetilde{G}_f^\circ/(kJ \, mol^{-1})$ X, -50 Y, -250 Z, -275

(a) Which decomposition process has the higher activation energy? (b) What is the difference in activation energy for the two processes? (c) Calculate the ratio of the pressure of Y to the pressure of Z after net reaction ceases at 25°C if the decomposition of A is equilibrium-controlled, and compare this ratio with the value that would be found if the decomposition at 25°C were kinetically controlled.

Hydrogen, Oxygen, Nitrogen, and the Noble Gases

23

"It strikes a sympathetic chord, I think, to learn that 700 years ago . . . the then Queen of England moved out of the city to Nottingham where she was residing because of the insufferable smoke; and that some 300 years later the brewers of Westminster offered to use wood instead of coal because of Queen Elizabeth's allergy to coal smoke. But it was only about the end of her reign that feeling began to lead to action; and then there was a prohibition—probably ineffective—of the use of coal in London while Parliament was sitting!"

HUGH E. C. BREWER,[1] 1955

An enormous body of descriptive chemistry underlies the theoretical aspects of the science that have been discussed in the earlier chapters. Most often it is only after regularities in nature have been discovered by observation and experiment that a body of theory is developed to explain the experimental facts. Systematic studies that turn up unexpected deviations from "normal" behavior provide the impetus for modifications of existing theories and sometimes disprove theories entirely.

This chapter is the first of eight concerned with the descriptive chemistry of some of the elements, an essential part of the science of chemistry. Much current research is concerned with developing new materials and finding new pathways for the synthesis and modification of known substances. To accomplish this the chemist must have an intuition for the way substances react and how their physical properties depend on their structures. This intuition is developed by studying in a systematic way the body of chemical information that has been collected through the years.

We have not provided an encyclopedic coverage of descriptive chemistry. The aim of these chapters is rather to give the reader a feeling for the variety of chemical compounds and reactions and the importance of some of them in our lives and our society. This chapter deals with three important elements, hydrogen, oxygen, and nitrogen, as well as the noble gases. It includes a discussion of the chemistry of water, water pollution, and air pollution. While it is true of any active experimental science that some of today's "facts" will be found in the future to be in error or incomplete, this admonition applies especially to the sections of this chapter dealing with the environment, which should be considered to be in the nature of a status report of our current understanding of some rather complex chemistry.

---

[1]Hugh E. C. Brewer, in F. S. Malette (ed.), "Problems and Control of Air Pollution," Van Nostrand Reinhold Co., New York, 1955.

Occurrence and Physical Properties Hydrogen appears to be the most abundant element in the universe. In those parts of the earth that are most accessible—the crust, the oceans, and the atmosphere—hydrogen ranks ninth in abundance on a weight basis (Table 23-1), although there are more atoms of hydrogen than of any other element except oxygen and silicon. Virtually all the hydrogen is combined with other elements; any free hydrogen in the atmosphere is eventually lost because the earth's gravity is not sufficiently strong to retain this very light molecule.

The normal form of hydrogen is H_2 because of the large enthalpy change associated with the dissociation reaction

$$H_2 \longrightarrow 2H \qquad \Delta H^\oplus = 436 \text{ kJ} \tag{23-1}$$

Atomic hydrogen can be formed at very high temperatures, for example in electric arcs, but even at 3000 K the degree of dissociation is only about 8 percent.

The van der Waals interactions between H_2 molecules are very weak. Hence the gas condenses to a liquid at extremely low temperature (20.4 K at 1 atm) and a very soft molecular solid forms at 14.1 K. Liquid hydrogen is the lightest of all liquids—its density is 0.070 g cm^{-3}, only 7 percent the density of water—and solid hydrogen has the lowest density of all solids, 0.088 g cm^{-3}.

Deuterium, the hydrogen isotope of mass number 2, constitutes 1 part in 5000 of naturally occurring hydrogen; tritium, ^3H or T, occurs to a much smaller extent in nature. The physical properties of isotopic molecules differ significantly only when there is a large percentage difference in the masses. Since the masses of H_2, HD, D_2, DT, and T_2 are in the ratio $2:3:4:5:6$, the variation in the physical properties through the series is unusually large. For example, the boiling points of HD and D_2 are, respectively, 2.1 and 2.2 K higher than that of H_2, and at 20 K the molar volumes of the liquids are 9 and 17 percent smaller.

Isotopic differences are also observable in the spectra of molecules containing hydrogen. The frequency of vibration of a bonded pair of atoms X—D is lower than the X—H frequency. Thus, when D is substituted for H in a compound, there is a characteristic shift in the frequency of the spectral lines associated with the X—H bond vibrations and the lines can be readily identified. For example, a comparison between the spectra of CH_3OH and CH_3OD (Table 23-2) shows that the frequency shift is smallest for lines attributed to CH_3, which indeed is little affected by the substitution. The shift is only slightly larger for CO vibrations, but there is a large difference for OH and OD vibrations. Such changes in spectra

**Table 23-1
Abundance of the Elements
on the Earth's Surface***

Oxygen	O	49.5		Chlorine	Cl	0.19	
Silicon	Si	25.7		Phosphorus	P	0.12	
Aluminum	Al	7.5		Manganese	Mn	0.09	
Iron	Fe	4.7		Carbon	C	0.08	
Calcium	Ca	3.4	99.2	Sulfur	S	0.06	0.7
Sodium	Na	2.6		Barium	Ba	0.04	
Potassium	K	2.4		Chromium	Cr	0.033	
Magnesium	Mg	1.9		Nitrogen	N	0.030	
Hydrogen	H	0.87		Fluorine	F	0.027	
Titanium	Ti	0.58		Zirconium	Zr	0.023	

All others < 0.1

*Percentages by weight in the earth's crust, the oceans, and the atmosphere.

Table 23-2
Some Infra-Red Lines of
CH_3OH and CH_3OD

Group to which observed line is attributed	WAVE NUMBERS OF LINES OBSERVED, $\tilde{\nu}/cm^{-1}$	
	CH_3OH	CH_3OD
CO	1034	1040
OH or OD	1340	823
CH_3	1430	1427
CH_3	1455	1459
CH_3	1477	1480
CO	2053	2065
CH_3	2844	2849
CH_3	2977	2964
OH or OD	3682	2720

can be very useful keys to the determination of structural features of molecules by the analysis of complex spectra.

Nuclear Magnetic Resonance Spectroscopy Another important tool for the determination of molecular structure is also based on a physical property of the hydrogen atom, its nuclear spin. Like the electron, the proton has a spin and behaves like a tiny magnet. In a magnetic field the proton can be in either of two energy states distinguished by the orientation of its magnetic moment with respect to the field direction (Fig. 23-1). The difference in energy between the states is directly proportional to H, the strength of the field:

$$\Delta\epsilon = \beta H \qquad (23\text{-}2)$$

A proton in the lower state can absorb radiation of frequency $\nu = \Delta\epsilon/h = \beta H/h$ and be excited to the upper state; protons in the upper state can drop to the lower state by emitting radiation of the same frequency. It is customary to describe these processes as *spin flips* between the two orientations with respect to the field.

Measurements of the spectral lines associated with nuclear spin flips can provide information about molecular structure because the magnetic field that determines the absorption or emission frequency for a given proton is the sum of the applied field, produced by the laboratory magnet, and the local field, produced by the presence of other atoms in the immediate surroundings of the proton. In effect, every proton in a molecule can act as a magnetic sensor of its atomic environment. Protons that have identical environments absorb or emit

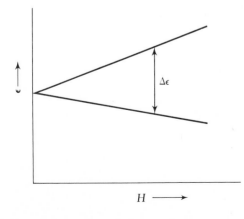

Figure 23-1 Schematic Energy Levels for a Proton in a Magnetic Field
In the absence of a magnetic field the two states of the proton have equal energy. The energy difference between the states, $\Delta\epsilon$, is directly proportional to the field strength H.

507

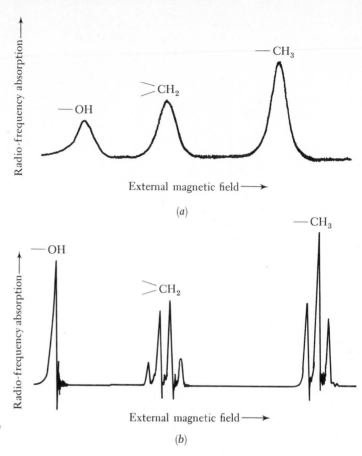

Figure 23-2 Nmr Spectrum of Ethanol

Graph (a) shows the nmr absorption spectrum of ethanol, CH_3CH_2OH, at low resolution; (b) shows the "fine structure" at higher resolution.

radiation at the same frequency; those in other environments are characterized by slightly different frequencies. This is the basis of nuclear magnetic resonance (nmr) spectroscopy.

A representative nmr spectrum is shown in Fig. 23-2. Even for very strong magnetic fields, the energy change associated with the flip of a proton spin is small, and the corresponding frequency is in the radio range. Nuclear magnetic resonance spectroscopy is therefore also known as radio-frequency spectroscopy. The sample spectrum is for ethanol, CH_3CH_2OH, which has protons in three different environments. Three protons are part of the —CH_3 group. They are magnetically equivalent; that is, they experience the same local field in their environment. The two protons in the $>CH_2$ group constitute another distinct group, and the third environment is that of the proton in —OH. Under high resolution (Fig. 23-2b) each of the three absorption lines shows fine structure that is characteristic of the group to which the protons belong. The area under each peak (or group of peaks in the high-resolution spectrum) is proportional to the number of structurally identical atoms. In principle, nmr spectra can be observed for all nuclei with spins. These include ^{19}F, ^{17}O, ^{2}H, and ^{14}N. Proton nmr is most commonly studied because hydrogen is present in so many compounds and because of the high signal strengths characteristic of proton spectra.

Preparation and Industrial Uses of Hydrogen Hydrogen of the highest purity is produced by the electrolysis of water, but this is a relatively expensive process and only about 1 percent of the hydrogen used in the United States is made in this way. There are various commercial methods of preparation, the most common of which is the reaction of methane from natural gas with steam at temperatures in the neighborhood of 800°C:

$$CH_4 + H_2O \rightleftharpoons 3H_2 + CO \qquad (23\text{-}3)$$

The hydrogen is separated from the CO by mixing the products with steam and passing them over a catalyst at elevated temperature to convert the CO to CO_2:

$$CO + H_2O \rightleftharpoons H_2 + CO_2 \qquad (23\text{-}4)$$

Carbon dioxide is very soluble in water; hydrogen is only slightly soluble. Hence the CO_2 can be removed by passing the product mixture under pressure into water. The loss of even the small quantity of hydrogen that dissolves is uneconomical, however, and many other schemes have been developed for the separation. One process uses hot carbonate solutions to absorb the CO_2:

$$CO_2 + CO_3{}^{2-} + H_2O \rightleftharpoons 2HCO_3^-$$

At the low pressure and high temperature used a negligible amount of H_2 dissolves.

The use of hydrogen as an energy storage and transmission medium has been proposed, in what has been termed the *hydrogen economy*. Water might be decomposed by reactions like (23-3) and (23-4), or it could be electrolyzed at times when electrical power demand for other purposes is low. The hydrogen produced could be stored, perhaps in liquid form, and transported to areas in which the need for power is high. Energy would be recovered by the direct combustion of hydrogen or by oxidation in a fuel cell (Sec. 21-6).

Only 7 percent of the hydrogen produced in the United States is marketed as H_2, the rest being used immediately in other processes. The major industrial application is in hydrogenation reactions, the direct combination of hydrogen with other molecules. Examples are the synthesis of methanol

$$2H_2 + CO \longrightarrow CH_3OH \qquad (23\text{-}5)$$

and the production of NH_3, Equation (23-20). Almost 50 percent of the hydrogen produced is consumed in these two processes, the preparation of ammonia representing the single largest use. Carbon-carbon double bonds can be converted to single bonds by hydrogenation, Equation (29-1). This is the process by which "unsaturated" vegetable oils are changed into "saturated" solid fats. Some hydrogen is used as a fuel in oxyhydrogen torches, which produce temperatures higher than 2000°C, and hydrogen-oxygen fuel cells have provided electrical power on space missions.

Bonding of Hydrogen Although the oxidation state of hydrogen in many compounds is +1, corresponding at least formally to the loss of an electron, hydrogen can also gain an electron to form the hydride ion, H^-. This process occurs in the reaction of hydrogen with highly electropositive metals. An example is the reaction of molecular hydrogen with sodium at elevated temperatures:

$$H_2(g) + 2Na \longrightarrow 2NaH \qquad (23\text{-}6)$$

The compounds formed are typical colorless crystalline salts known as *saltlike hydrides*. The hydride ion reacts vigorously with water:

$$H^- + H_2O \longrightarrow H_2 + OH^- \qquad (23\text{-}7)$$

Hydrogen also forms many binary covalent molecules with elements of groups IV through VII, known as *covalent hydrides*. Boron, aluminum, and gallium, members of group III, form *complex hydrides,* ions with the general formula XH_4^-. A typical complex hydride is lithium aluminum hydride, which can be made by the reaction of lithium hydride and aluminum chloride in dry ether:

$$4LiH + AlCl_3 \longrightarrow LiAlH_4 + 3LiCl \qquad (23\text{-}8)$$

Complex hydrides find wide use as reducing agents.

Bridge Hydrogens Although a typical covalent bond involves the sharing of two electrons between a pair of atoms, there is a class of compounds, termed *electron deficient,* in which the number of electron pairs available for bonding is less than the number of conventional covalent bonds. Many of the known electron-deficient molecules contain boron and hydrogen; the simplest is diborane, B_2H_6. It contains 12 electrons, 3 from each boron atom and 1 from each hydrogen atom. A conventional valence bond structure would require seven bonds, six between B and H atoms and one connecting the boron atoms.

Structural investigations have shown that there are two distinct types of H atoms in diborane (Fig. 23-3a). Four of them are terminal H atoms, each associated with a single B atom; the others are bridging atoms that are equidistant from the two B atoms. The bonding in B_2H_6 is depicted in Figs. 23-3b and c. The 2s and 2p orbitals of boron are hybridized to sp^3 orbitals, and the bonds between the boron and the terminal hydrogen atoms can be thought of as arising from the overlap of the hybrid orbitals with the 1s orbitals of each hydrogen atom. These four bonds require four electron pairs. Each of the bridging hydrogen atoms is associated with *one* electron pair that bonds it to the two B atoms by what is called a *three-center bond.* This bond may be considered to result from the combination of one sp^3 orbital from each boron and the s orbital of a hydrogen atom.

Figure 23-3 Structure and Bonding in Diborane, B_2H_6
(*a*) Structural formula. The wedge-shaped bonds imply that the H atoms are above the plane of the page; the dashed bonds imply that the H atoms are below the plane of the page. (*b*) Diagram depicting the 2s and 2p orbitals of the boron atoms, hybridized to sp^3 and overlapping with the 1s orbitals of the hydrogen atoms. (*c*) Diagram showing these overlapping orbitals schematically combined into bond orbitals, each occupied by an electron pair.

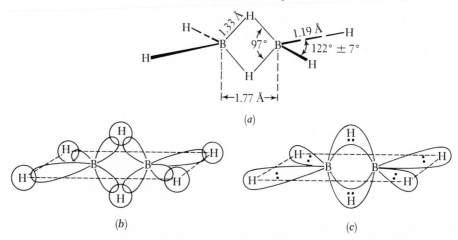

510

Occurrence and Physical Properties The most abundant of the chemical elements on earth is oxygen. Elemental oxygen constitutes 23 percent by weight (21 vol percent) of the atmosphere, and oxygen in a combined form comprises almost 50 percent by weight of the crust of the earth and 89 percent of the oceans. It is a constituent of all living matter.

The low boiling and melting points of O_2 (90 K and 54 K, respectively) are evidence of the weakness of the van der Waals interactions between the molecules. The gas is odorless, tasteless, and colorless, but liquid oxygen is blue. Oxygen is slightly soluble in water; a saturated solution at 20°C with $P_{O_2} = 1$ atm has an oxygen concentration of $1.4 \times 10^{-3}\ M$. Since the solubility of oxygen is about twice that of nitrogen, the ratio of dissolved O_2 to N_2 in water is $1:2$, rather than $1:4$ as it is in air. The electronic structure and paramagnetism of O_2 were discussed in Chap. 18; the paramagnetism is observed in the gaseous, liquid, and solid states and vanishes only at very low temperatures.

Production and Industrial Uses Oxygen is obtained commercially from the liquefaction and subsequent distillation of air. Small quantities are also produced by electrolysis of water. Three isotopic species of oxygen occur in nature: ^{16}O (99.76 percent), ^{17}O (0.04 percent), and ^{18}O (0.20 percent). Both ^{17}O and ^{18}O can be concentrated by fractional distillation of water. Highly enriched $H_2^{17}O$ and $H_2^{18}O$ are commercially available; the latter is commonly used as a tracer, particularly in biochemical applications.

In 1976 the industrial production of oxygen in the United States was over 10^{10} kg. Only three chemicals were produced in larger quantity: H_2SO_4, CaO, and NH_3 (in decreasing order). The steel industry is the largest consumer of oxygen. A preliminary step in steel production (Sec. 26-4) is the oxidation of impurities in molten iron with oxygen. Oxygen is also used in oxyacetylene and oxyhydrogen torches for welding, and to oxidize kerosene in rocket propulsion.

Ozone Ozone, O_3, is a naturally occurring but less familiar molecular form of oxygen. At room temperature it is a gas with a characteristic pungent odor (Greek: *ozein,* smell) that forms a blue liquid at 162 K and solidifies to a dark violet solid at 80 K. Urban atmospheres contain ozone at concentrations as high as[2] 0.7 ppm on very smoggy days. It is an irritant and also plays an integral part in the complex series of reactions that produce photochemical smog (Sec. 23-6). Nevertheless ozone is essential to the presence of all life on earth; an ozone layer in the upper atmosphere, formed by photochemical reaction of oxygen at an altitude of about 25 km, filters the biologically lethal ultraviolet radiation from the sun. The concentration of ozone in this protective layer can be diminished significantly by reaction of O_3 with other reactive species, such as NO or halogen atoms. Below 20 km the principal ozone-destroying reactions involving NO are

$$NO + O_3 \longrightarrow NO_2 + O_2$$
$$NO_2 + O_3 \longrightarrow NO_3 + O_2$$
$$NO_3 + h\nu \longrightarrow NO + O_2$$

---

[2]ppm stands for *parts per million.*

which sum to

$$2 O_3 \longrightarrow 3 O_2$$

There are similar reactions involving Cl. A given NO molecule or Cl atom causes the destruction of a large number of O_3 molecules, because at the end of the reaction chain with O_3 the NO or the Cl is regenerated. There is evidence that damagingly high concentrations of Cl atoms are produced photochemically in the upper atmosphere from the normally inert chlorofluorocarbon[3] propellant gases used in some aerosol sprays. The use of such propellants in the United States has recently been banned.

The ozone molecule is similar to SO_2 in geometry and electronic structure. It is bent, with a bond angle of 117°. Its electronic structure can be represented by

the superposition of two resonance formulas, and, correspondingly, the bond distance lies between that of a single and a double O—O bond.

Ozone is generally prepared by passing oxygen through an electric discharge:

$$3 O_2 \longrightarrow 2 O_3 \qquad \Delta H^{\ominus} = 285 \text{ kJ} \tag{23-9}$$

As one might expect from (23-9), energy-rich O_3 is a much more powerful oxidizing agent than O_2.

Bonding of Oxygen Binary compounds of oxygen with all the elements save He, Ne, Ar, and Kr are known. Bonding in these oxides ranges from ionic to covalent.

There are three types of ionic oxides. The most important are those that contain the oxide ion, O^{2-}. Although considerable energy must be expended to form O^{2-} from molecular oxygen, the strong electrostatic interactions of the small O^{2-} ion with cations stabilize oxide crystal lattices. Oxide ion reacts vigorously in solution to form hydroxide ion:

$$O^{2-} + H_2O \longrightarrow 2OH^- \tag{23-10}$$

Compounds containing the O_2^{2-} ion are called *peroxides*. Examples are sodium peroxide, Na_2O_2, and barium peroxide, BaO_2, which react with water or dilute acid to form hydrogen peroxide, H_2O_2 (Sec. 23-3). A small class of ionic oxides contain O_2^-, the superoxide ion; an example is potassium superoxide, KO_2. Peroxides and superoxides find use as strong oxidizing agents.

The oxides of the nonmetals are all covalent. We discuss some of them in subsequent sections. The acidic, basic, and amphoteric properties of oxides have been treated in Sec. 9-4.

23-3 Compounds of Hydrogen and Oxygen

Hydrogen Peroxide This reactive compound, H_2O_2, is a colorless liquid that freezes at $-0.4°C$ and boils at 150°C. The molecular structure of H—O—O—H is

[3]Compounds containing just C, Cl, and F. Some are easily condensed gases that are used in refrigerating systems.

depicted in Fig. 17-1(b), the torsion angle being approximately $90°$ and the bond angle O—O—H approximately $100°$. Hydrogen peroxide has acid-base properties similar to those of H_2O, its acid constant being somewhat larger than that of H_2O.

Disproportionation of H_2O_2 and the Use of Latimer Diagrams One important property of H_2O_2 is that it is a strong oxidizing agent and also a moderate reducing agent. It contains O in the $(-I)$ oxidation state[4] and is capable of spontaneous disproportionation (Sec. 10-4) to molecular oxygen [oxidation state (0)] and water, containing O in the $(-II)$ oxidation state.

A general understanding of the relationship between different oxidation states and the possible disproportionation reactions involving them is facilitated by a kind of diagram popularized by the American chemist W. M. Latimer, which shows the electrode potentials relating the different levels.

Consider, for example, the half-reactions and $\mathbf{E}^\ominus$ values relating the three oxidation states of oxygen mentioned above:

$$O_2 + 2H^+ + 2e^- \longrightarrow H_2O_2 \qquad \mathbf{E}^\ominus = 0.69 \text{ V} \qquad (23\text{-}11)$$

$$H_2O_2 + 2H^+ + 2e^- \longrightarrow 2H_2O \qquad \mathbf{E}^\ominus = 1.77 \text{ V} \qquad (23\text{-}12)$$

The Latimer diagram is a device for summarizing these data:

$$O_2 \xrightarrow{\ 0.69\ } H_2O_2 \xrightarrow{\ 1.77\ } H_2O \qquad (23\text{-}13)$$

By convention, the species of highest oxidation state is shown on the far left.

How can the electrode potentials be used to show that H_2O_2 does indeed have a tendency to disproportionate into oxygen and water? To do this, we combine the half-cells corresponding to (23-11) and (23-12) into a cell (Sec. 21-4), with reaction

$$2H_2O_2 \longrightarrow O_2 + 2H_2O \qquad (23\text{-}14)$$

In combining the half-reactions to form (23-14), reaction (23-12) is the reduction (cathode) reaction and reaction (23-11) is the oxidation (anode) reaction; therefore the overall standard emf corresponding to (23-14) is

$$\mathbf{E}^\ominus = \mathbf{E}^\ominus_{\text{cathode}} - \mathbf{E}^\ominus_{\text{anode}} = 1.77 - 0.69 = 1.08 \text{ V}$$

The fact that $\mathbf{E}^\ominus$ is positive demonstrates that (23-14) is spontaneous in the indicated direction. The equilibrium constant is, in fact, very large, about $10^{2 \times 1.08/0.06} = 10^{36}$.

In general terms, we may denote the successive oxidation states as A, B, and C, A being the highest and C the lowest, so that the Latimer diagram is

$$A \xrightarrow{\ \mathbf{E}_{AB}\ } B \xrightarrow{\ \mathbf{E}_{BC}\ } C \qquad (23\text{-}15)$$

Disproportionation of B into A and C occurs if $\mathbf{E} = \mathbf{E}_{BC} - \mathbf{E}_{AB}$ is positive, that is, if $\mathbf{E}_{BC} > \mathbf{E}_{AB}$. Conversely, if $\mathbf{E}_{BC} < \mathbf{E}_{AB}$, no significant disproportionation can take place.

This result can be visualized and remembered by studying the $\mathbf{E}$ values shown in the Latimer diagram (23-15) and recalling that $\mathbf{E}$ measures the tendency of a

---

[4]Here, as in Appendix B, we use Roman numerals [and zero (0)] to designate oxidation states, in accord with systematic inorganic nomenclature.

reaction to go to the right. We see that when E_{BC} is larger than E_{AB} the reaction leading from B to C is more strongly favored than that leading from A to B, so that the disproportionation would tend to occur. Conversely, if $E_{BC} < E_{AB}$ the reaction from A to B is more strongly favored than that from B to C and there can be no significant disproportionation.

The tendency of a substance to disproportionate will, of course, depend on the concentrations of the various species involved in the half-reactions. It is often useful, however, to get a general feeling of the likelihood for disproportionation, and $E^{\ominus}$ values can be used for this purpose.

Example 23-1

☐ **Disproportionation of Cu⁺ and of Fe²⁺** Consider the Latimer diagrams of the oxidation states of copper and iron:

$$Cu^{2+} \xrightarrow{0.153} Cu^+ \xrightarrow{0.521} Cu$$

$$Fe^{3+} \xrightarrow{0.771} Fe^{2+} \xrightarrow{-0.440} Fe$$

Predict whether Cu^+ disproportionates into Cu^{2+} and Cu and whether Fe^{2+} disproportionates into Fe^{3+} and Fe.

Solution For the copper system we have

$$E = 0.521 - 0.153 = 0.368 \text{ V}$$

and for the iron system

$$E = -0.440 - 0.771 = -1.211 \text{ V}$$

Reasoning along the lines just stated therefore leads to the conclusion that Cu^+ ion is expected to (and indeed does) disproportionate. Ferrous ion is stable, however, and cannot disproportionate significantly. ■

Exercise 23-1

☐ Consider the Latimer diagrams of the oxidation states of cobalt and plutonium:

$$Co^{3+} \xrightarrow{1.82} Co^{2+} \xrightarrow{-0.28} Co$$

$$Pu^{4+} \xrightarrow{0.97} Pu^{3+} \xrightarrow{-2.03} Pu$$

Predict whether Co^{2+} and Pu^{3+} disproportionate into the appropriate species given in the diagrams. ■

Water Water is doubtless the most thoroughly studied compound. Virtually every aspect of its chemical and physical behavior has been studied experimentally and treated by theory. One might expect, then, that there would be little more to be learned about this simple triatomic molecule, but this is not so. Interpretation of the special properties of water, particularly those relating to hydrogen bonding, is exceedingly complex and still presents significant challenges. We shall not review the chemistry of water here; it has been dealt with earlier in the discussions of equilibria, hydrogen bonding, hydration of ions, ionization, and acids and bases. The present focus is on physical properties.

Most liquids become more dense as their temperature is lowered. Water shows this behavior down to 4°C, but below that temperature its density *decreases* as the temperature is lowered. The density decreases still further when water freezes, another unusual effect shown by comparatively few substances. These

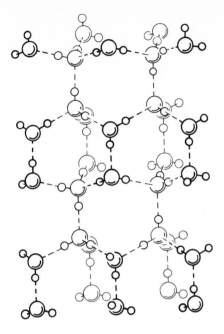

Figure 23-4 The Arrangement of Water Molecules in Ice
Each H_2O molecule is surrounded tetrahedrally by four other H_2O molecules, with which it forms strong hydrogen bonds. Other crystalline forms of ice exist at high pressures; they are more dense than liquid water.

unusual properties are of profound significance to life in bodies of water in cold climates. Consider what happens when a lake, originally above 4°C, cools in the winter. At first the coldest water sinks to the bottom because it is more dense than the warmer liquid. But once the temperature of the water reaches 4°C, the coldest water begins to accumulate at the surface and eventually freezes. The ice, which is less dense than the liquid, floats on the surface and slows the rate of cooling of the liquid underneath. As a result, large lakes do not freeze solidly even in the coldest winters and aquatic life survives.

The unusual behavior of the density of water is caused by strong hydrogen bonding between the molecules. As can be seen in Fig. 23-4, the structure of ice is very open. Each water molecule is surrounded tetrahedrally by four nearest neighbors with which it forms hydrogen bonds. When ice melts, the tetrahedral structure is partially destroyed and the water molecules crowd somewhat more closely together, which accounts for the increase in density on melting. Some residual order due to hydrogen bonding persists in the liquid. As the temperature is raised a little, hydrogen bonds are broken and a more disordered, denser liquid is obtained. A rise in temperature also increases the thermal motion of the molecules, the origin of the normal thermal expansion of liquids. At 4°C the effect of thermal motions overcomes the effect of loss of structure and above 4°C the temperature dependence of the density becomes normal.

The breaking of hydrogen bonds as the temperature is raised is also responsible for the abnormally high heat capacity of water. Energy must be spent to disrupt these bonds as well as to increase the average kinetic energy of the molecules. An important consequence of the high heat capacity of water is the stabilizing influence of large bodies of water on the climate; temperatures are always more moderate near the ocean than they are in the nearby interiors of continents. The oceans and lakes are efficient reservoirs of energy. They absorb energy from their surroundings during warm weather and release it to the surroundings in cold weather.

515

Heavy Water Although there is very little D_2O (heavy water) present in natural water, high concentrations of D_2O can be obtained by electrolysis. Under suitable conditions ordinary hydrogen is liberated roughly six times more readily than D_2; the unelectrolyzed water is therefore enriched in D_2O.

The molecular weight of water is sufficiently low that isotopic substitution can have a marked effect on its physical properties. For example, D_2O freezes at $3.82°C$ and boils at $101.4°C$. The dielectric constant[5] of heavy water is lower than that of H_2O and its viscosity is higher. The chemical properties of the isotopic species are not identical. At $25°C$ the ion product of D_2O is 0.3×10^{-14}, and its heat of formation is about 3 percent larger than that of ordinary water. The properties of HDO are intermediate between those of ordinary water and D_2O.

Water Quality Water is of paramount importance to life, not only life in the narrow biological sense but life as applied to the functioning of our society. When we speak of water in this sense, we are not restricting ourselves to the chemically pure substance H_2O. Instead, we refer to a complex chemical system of varying composition that we find in the oceans and rivers, in municipal water systems, in wastes, in the atmosphere—virtually everywhere in our environment.

Some of the constituents of natural waters are listed in Table 23-3. Even water

---

[5]The interaction energy of two ions is given by $\epsilon = \mathbf{Q}_1\mathbf{Q}_2/\alpha Dr$, where D is the *dielectric constant* of the medium. For a vacuum, $D = 1$ [see Eq. (6-1)]. The larger D, the smaller is the interaction between ions. For water, $D = 80$, a large value.

**Table 23-3
Constituents of
Natural Waters**

Source	PARTICLE SIZE CLASSIFICATION				
			DISSOLVED		
	Suspended	Colloidal	Molecules	Positive ions	Negative ions
Atmosphere	←Dusts→		CO_2 SO_2 O_2 N_2	H^+	HCO_3^- SO_4^{2-}
Mineral soil and rock	←Sand→ ←Clays→ ←Mineral soil particles→		CO_2	Na^+ K^+ Ca^{2+} Mg^{2+} Fe^{2+} Mn^{2+}	Cl^- F^- SO_4^{2-} CO_3^{2-} HCO_3^- NO_3^- Various phosphates
Living organisms and their decomposition products	Algae Diatoms Bacteria ←Organic soil (topsoil)→ Fish and other organisms	Viruses Organic coloring matter	CO_2 O_2 N_2 H_2S CH_4 Various organic wastes, some of which produce odor and color	H^+ Na^+ NH_4^+	Cl^- HCO_3^- NO_3^-

that we might call "pure", such as the water in a mountain stream, contains many of these substances, at least in small amounts. The distinction between brackish water and the water in a mountain stream depends largely on the relative proportions of the contaminants. If Table 23-3 were not limited to natural waters but included industrial effluents and municipal sewage systems as well, the list of impurities and sources would have to be greatly expanded. In the sections that follow we consider the effect of some of the constituents of water on its properties and describe ways in which water can be treated to improve its quality.

Hard Water Water that contains appreciable quantities of any of the ions Ca^{2+}, Mg^{2+}, and Fe^{2+} is said to be "hard". These ions, and others, are picked up by water percolating through even slightly soluble minerals or flowing over them in rivers and streams. For example, slightly acidic groundwater percolating through limestone ($CaCO_3$) or dolomite ($CaCO_3 \cdot MgCO_3$) reacts to form the soluble HCO_3^- salts $Ca(HCO_3)_2$ and $Mg(HCO_3)_2$. Some soils provide Fe^{2+} and Mg^{2+} in the form of the soluble salts $FeSO_4$ and $MgSO_4$.

When soaps are used in hard water, a grayish curdy scum is formed. It is a precipitate produced by the reaction of a metal ion with the active ingredient in soap, the sodium salt of a long-chain organic acid such as stearic acid, $C_{17}H_{35}COOH$:

$$M^{2+} + 2Na^+(C_{17}H_{35}COO^-) \longrightarrow 2Na^+ + M^{2+}(C_{17}H_{35}COO^-)_2 \quad (23\text{-}16)$$

The scum settles on the wash and is aesthetically unpleasing. More importantly, it reduces the cleaning power of the soap by removing the emulsifying agents, the organic acid ions, from solution.

Hard Water Scale in a Pipe. (*Courtesy of The Permutit Company.*)

Boiler scale, a hard deposit that coats the interior of boilers, hot water pipes, heat exchangers, and tea kettles, is formed when hard water containing appreciable quantities of HCO_3^- ion is heated:

$$M^{2+} + 2HCO_3^- \longrightarrow MCO_3 + CO_2 + H_2O \quad (23\text{-}17)$$

The scale consists of the carbonates of Ca, Mg, and Fe; $CaSO_4$ may be present as well if the water contains SO_4^{2-}. The deposits reduce the efficiency of heat transfer and constrict pipes so that the flow of water and steam is impeded.

The effects of hard water present a serious economic problem, and a number of methods have been devised to soften water or to inhibit the action of the metal ions. Precipitation of soap during laundering can be eliminated by the use of synthetic detergents. The active ingredient in these substances is an emulsifying organic ion that does not form an insoluble compound with divalent metal ions. Another component of detergents is a chelating agent such as sodium trimetaphosphate, $Na_3P_3O_9$, which forms soluble complexes with metal ions. Unfortunately, as we shall see later, the phosphates in many detergents have serious environmental effects in certain areas of the country. Washing soda, $Na_2CO_3 \cdot 10H_2O$, can be added to water prior to washing to precipitate the divalent metal ions in an acceptable granular form; it is a component of some low-phosphate or phosphate-free washing products. Care must be exercised in using washing soda, however, because it is highly alkaline.

An elegant way to soften water is to replace the objectionable divalent metal ions by Na^+. This may be accomplished with ion exchangers such as zeolites, which are complex aluminum silicates. These solids are built up of a rigid, negatively charged framework of AlO_4 and SiO_4 tetrahedra (Sec. 24-3) and contain channels into which positive ions can fit. If the zeolite is placed in a concentrated NaCl solution, the channels fill with Na^+ ions. Divalent metal ions bind more strongly to the zeolite than do the sodium ions. Thus, if hard water is passed through a vessel containing a sodium-filled zeolite, the sodium ions are displaced. If a singly charged site on the zeolite anion is represented by Z, the exchange reaction may be written

$$2NaZ + M^{2+} \rightleftharpoons 2Na^+ + MZ_2 \qquad (23\text{-}18)$$

When all the sodium ions have been exchanged, the zeolite can be regenerated (even though the equilibrium lies far to the right) by placing the zeolite in contact with a concentrated salt solution. The reaction is thereby driven backward in accordance with Le Châtelier's principle.

Zeolites are no longer used in most commercial water softeners; they have been replaced by more efficient organic ion-exchange resins, networks of polymer molecules containing groups that react with ions. Resins can be synthesized that exchange either positive or negative ions, and they can be made highly selective. Ion-exchange techniques are widely used for analysis as well as for purification. For example, the highly similar lanthanide ions can be separated from each other by ion-exchange chromatography.

23-4 Water Pollution

Although the divalent metal ions that cause hardness are undesirable constituents of water, they are not normally classified as pollutants. The word *pollutant* is usually reserved for contaminants present in high enough concentration to affect human beings adversely. This definition is obviously not precise, and even when agreement can be reached on definitions of the terms, it is usually very difficult to determine precisely how high the concentration of a substance must be in order to produce adverse effects. For example, do we mean adverse effects that occur immediately, or in weeks, months, years, or generations? Adverse effects may also be indirect. Our ecosystem (Greek: *oiko*, habitation) consists of many interrelated subsystems. An adverse effect on some small subsystem may be amplified as it moves to other subsystems, eventually emerging as a serious pollution problem.

Standards for Drinking Water The 1974 Safe Drinking Water Act required the establishment of new federal standards for the allowable levels of impurities in drinking water. After a review of more than 3000 items of testimony, the Environmental Protection Agency set standards in December 1975 that took effect in June 1977. We will review some of these standards to illustrate the complexities in the definition of pollution and the related economic, governmental, and scientific problems.

Maximum permissible levels of inorganic contaminants in drinking water are given in Table 23-4. Consider the standard for nitrate ion. The level of 45 mg liter^{-1} is generally thought to be well within safe limits for adults. Infants, however, have bacteria in their digestive tracts that convert NO_3^- to NO_2^- (nitrite ion), and nitrite ions react with hemoglobin and thereby reduce the oxygen-carrying capacity of the blood. Nitrite poisoning is the cause of a poten-

Substance	Concentration/mg liter^{-1}
Arsenic	0.05
Barium	1.0
Cadmium	0.01
Chromium	0.05
Cyanide	0.2
Lead	0.05
Mercury	0.002
Nitrate	45
Selenium	0.01
Silver	0.05

**Table 23-4
Federal Standards for Inorganic Chemical Substances in Drinking Water**

tially fatal disease, infantile methemoglobinemia, the blue-baby syndrome. In Minnesota between 1947 and 1950 there were 139 cases of this disease and 14 deaths, all associated with well water containing nitrates and in some cases with NO_3^- levels below the federal standard.

Why then is the nitrate standard not lower? First, some studies appear to demonstrate that 45 mg liter^{-1} is a safe standard even for infants. Such investigations involve the analysis of health records (one obviously cannot feed nitrates to infants and measure the survival rate!) and the interpretation is often difficult. Cost also plays a significant role in the setting of the standard. The most practical method for achieving nitrate concentrations below 45 mg liter^{-1} is ion exchange, a relatively expensive process. Even meeting the current standard is costly. The establishment of every standard involves a decision in which benefits are balanced against costs.

Establishing standards for organic pollutants has proved very difficult. There are several thousand potentially toxic organic compounds that might find their way into water supplies. Little is known about the health effects of these substances at low concentrations, and practical analytical methods are also lacking. Standards have been set, however, for specific pesticides such as DDT.

Other sections of the federal standards deal with the level of bacterial contamination and physical characteristics such as taste, smell, color, and turbidity. Omitted from these standards are limits on the concentrations of viruses. There is evidence that disease-causing viruses do get into water supplies and may cause outbreaks of viral disease, but no routine tests for waterborne viruses have yet been developed.

Both O_3 and Cl_2 (and more recently BrCl) are used for disinfecting water. Ozone has the advantage that taste and odor problems are minimized, but it is expensive and, since it cannot conveniently be stored and transported, it must be produced where it is to be used. Chlorine decomposes in solution much less rapidly than O_3 and so remains available to combat bacterial contamination that may be introduced after the water is treated. It may, however, form harmful products by reaction with organic substances.

Sources of Water Pollution Water is a recyclable resource. It is drawn from rivers, lakes, and wells, utilized in homes, on farms, and by industry, and then returned to the environment. Each time that water is used, its quality—its suitability for use—is altered. Sometimes the quality of water is significantly improved between uses, either by natural purification processes (such as filtration through soil layers) or by treatment in sewage plants, but often the intensity of water usage is so great that little purification takes place.

	Waste water* (10^6 m^3)
INDUSTRY	
Food and related products	2,800
Textile mill products	560
Paper and related products	7,600
Chemical and related products	15,000
Petroleum and coal	5,200
Rubber and plastics	640
Primary metals	17,000
Machinery	600
Electrical machinery	360
Transportation equipment	960
All other manufacturing	1,800
All manufacturing	52,520
DOMESTIC	
Served by sewers (120 million people)	20,000†

*For comparison the average annual flow of the Colorado River is $22,000 \times 10^6 \text{ m}^3$.

†Number of persons $\times$ 0.5 m³ per person per day $\times$ 365 days per year.

The volume of waste water from industry and sewage plants is staggering. Estimated yearly quantities for the United States are shown in Table 23-5; not included are the volumes of water used for agricultural purposes, mining, runoff from rainfall, and the sewage from the 37 percent of the population who were not served by sewers at the time the estimates were made. We consider here only a few of the major contaminants that arise from this intensive use of water.

Degradable Substances Given sufficient time, natural water systems are able to cleanse themselves of organic impurities and certain nitrogen compounds by the action of bacteria and other microorganisms. These organisms consume dissolved oxygen as they function. A measure of the concentration of degradable substances in water is the weight of dissolved oxygen utilized in bacterial action during a given period, the *biological oxygen demand* (BOD). If the BOD is so high that the oxygen is used at a rate greater than the rate at which it can be replenished by aeration or furnished by plants growing under water, the character of the water changes. Aquatic life dies out, organic debris collects, and microorganisms that can survive without oxygen begin to multiply. They feed on organic matter and release carbon dioxide, methane, hydrogen sulfide, and foul-smelling organic sulfur compounds. The hydrogen sulfide reacts with metal ions to form black precipitates that float as a scum on the surface.

To maintain the BOD at acceptable levels it is necessary to control the discharge of organic materials into natural waters. This can be accomplished by proper treatment of sewage, not only in urban areas but also on farms and feedlots (waste production by farm animals in the United States has been estimated to be about 20 times that of the human population). This action is not sufficient, however. A high BOD can be produced by the decay of plants and algae whose growth has been encouraged by the presence of nitrogen- and phosphorus-containing plant nutrients present in waste water. Nitrates and phosphates in fertilizers are carried into groundwater during irrigation. Soluble phosphates come from human and animal wastes, and as much as three-fourths of the soluble phosphates in urban sewage can come from detergents. Effective control of plant nutrients in the water system requires a reformulation of detergents and removal of soluble phosphates in the treatment of sewage.

Nondegradable Organic Substances To an increasing extent new chemical substances have been developed to perform certain functions. Some of these persist for long times in the ecosystem because they are chemically very stable and are also not microbially degradable. Prime examples are the chlorinated hydrocarbon insecticides, such as DDT, dieldrin, and aldrin,

DDT   Dieldrin   Aldrin

and the polychlorinated biphenyls (PCBs)

(X = sites for possible chlorine substitution)

which have been used in paints, in hydraulic systems, as heat-exchange fluids, and as additives in plastics. Chlorinated hydrocarbons degrade very slowly in the soil and are washed into the water system. Although the levels of these compounds in the water are not very high, there is a natural amplification system that concentrates them. Fish scales are coated by a fatty layer. Since chlorinated hydrocarbons dissolve more readily in fats than in water, they are in effect extracted from the water by the fatty layer. When small aquatic animals are eaten, the insecticide is further concentrated in the body fat of larger animals. It works its way along the food chain and eventually reaches man.

Although DDT has been linked to metabolic disorders in birds, there has been no conclusive evidence that it constitutes a health hazard to human beings. As a precautionary measure, however, its use in the United States was banned in 1972. Since then, some scientists have argued that the ban on DDT is potentially more harmful than its continued use. They feel that careful, controlled usage is necessary to control insect-borne diseases such as malaria. PCBs are chemically similar to DDT. They have spread into the environment because of accidental spills and improper disposal of wastes. Their use has now been restricted to sealed systems; the disposal of PCBs is also controlled. The sale and manufacture of dieldrin and aldrin in the United States were halted in 1974 because these insecticides were shown to produce liver tumors in rats. There is as yet no evidence that they can produce cancer in human beings. The use of two other chlorinated hydrocarbon pesticides, heptachlor and chlordane, was restricted in 1975.

Acid Mine Drainage and Coal Wastes Domestic coal has a sulfur content of 6 to 8 percent, and one-fifth to three-fifths of this may come from the mineral pyrite (iron disulfide, FeS_2). The iron disulfide remaining in the mines and left in heaps of coal wastes after coal has been processed oxidizes gradually in moist air to produce sulfuric acid, sulfates, iron oxides, and other compounds:

$$FeS_2 + O_2 + H_2O \longrightarrow H_2SO_4 + Fe_2O_3 + \text{other compounds} \quad (23\text{-}19)$$

Water that enters the mine or washes through the wastes dissolves the oxidation products and carries the resulting acid solution to surface waters.

In 1971 there were over 800 waste piles covering 12,000 acres in the anthracite region of Pennsylvania alone, and it is estimated that acid mine drainage pollutes 10,000 miles of streams in the United States. The best current solution to this problem is a program of prevention—sealing abandoned mines to keep out air and reducing water seepage. Attempts have also been made to control acid drainage by neutralization with CaO.

23-5 Nitrogen

About 78 vol percent of the earth's atmosphere consists of N_2, a colorless, odorless, and tasteless gas. Nitrogen in combined form is also present to a small extent in the earth's crust. Soluble nitrogen salts are important components of fertile soils, and nitrogen makes up about 16 percent by weight of all proteins. Liquid nitrogen boils at 77 K and freezes at 63 K. The element is obtained commercially by the distillation of liquid air; unless it is further purified, it usually is contaminated with small amounts of oxygen.

Because of the strength of the N-N triple bond, N_2 is very unreactive at room temperature. The processes by which N_2 can be converted to more reactive and directly useful compounds are therefore important. The removal of nitrogen from the atmosphere and its conversion to soluble compounds is called *nitrogen fixation*. Some nitrogen is fixed when it reacts with oxygen in lightning discharges, and certain soil bacteria associated with the roots of legumes (alfalfa, clover, beans, peas) are also able to fix nitrogen. Commercially the most common first step for the utilization of N_2 is the preparation of ammonia.

Bonding of Nitrogen Nitrogen is covalently bonded in most of its compounds, although it does form the nitride ion, N^{3-}, when it reacts at high temperatures with the most electropositive elements, such as lithium. Li_3N is a saltlike material, but most nitrides are covalent—for example, boron nitride (Fig. 7-4). Nitrogen normally forms three single covalent bonds, as in the pyramidal molecules NH_3 and NF_3, or multiple bonds, as in N_2 and HCN. The nitrogen octet can also be completed by the gain of an electron and the formation of only two covalent bonds. The amide ion, NH_2^-, which is isoelectronic with water, is an example of this mode of bonding. Amides can be formed by highly electropositive metals; the most common is sodium amide or sodamide, $NaNH_2$, used as a reagent in organic synthesis.

Loss of an electron leaves nitrogen isoelectronic with carbon, and thus N^+ forms four covalent bonds, as in ammonium ion (NH_4^+) and substituted ammonium ions (such as $CH_3NH_3^+$, methylammonium ion).

Ammonia The most important compound of nitrogen and hydrogen is ammonia. It can be synthesized by the direct reaction of gaseous N_2 and H_2:

$$N_2 + 3H_2 \rightleftharpoons 2NH_3 \qquad \Delta H^\circ = -92 \text{ kJ} \qquad (23\text{-}20)$$

At 25°C the equilibrium constant for reaction (23-20) is quite large, $K = 7.1 \times 10^5 \text{ atm}^{-2}$; the rate of reaction, however, is negligible. Raising the temperature increases the rate, but as indicated by the standard enthalpy of formation, the reaction is exothermic and an increase in temperature decreases K. Thus, at 450°C, the reaction rate is only slightly more favorable and K has fallen to 6.5×10^{-3}—the equilibrium no longer favors the product.

In spite of these apparently mutually exclusive requirements for satisfactory yield and rate, reaction (23-20) is the basis of the most important industrial method for the production of ammonia. The German chemist Haber made the process viable through the use of a catalyst to promote the rate and by careful choice of the conditions. As shown in Fig. 23-5, even at temperatures as high as 400 to 600°C acceptable yields can be obtained if the reaction is run at elevated pressures. This result can be predicted by application of Le Châtelier's principle, since the number of moles in the reaction mixture is decreased when product is formed. A further enhancement in the yield is obtained by removing the ammonia as it is formed, thereby driving the reaction toward completion.

There are many similarities between ammonia and water. Like water, liquid ammonia is self-associated by hydrogen bonding. Its boiling point, -33.4°C, is abnormally high when compared with the other hydrides in its group, such as PH_3 and AsH_3, and it has a high dielectric constant. Analogies also exist between the chemistry of water and ammonia. Like water, ammonia can undergo a self-ionization equilibrium:

$$2NH_3 \rightleftharpoons NH_4^+ + NH_2^- \qquad K = [NH_4^+][NH_2^-] \approx 10^{-30} \text{ at } -50°C \quad (23\text{-}21)$$

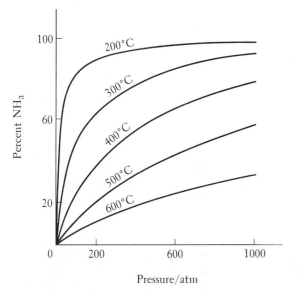

Figure 23-5 **Percentage Conversion of a 1:3 Mixture of N_2 and H_2 into NH_3 at Equilibrium for Different Temperatures and Pressures.**

523

Thus the ammonium ion can be thought of as the analog of the hydronium ion, while the amide ion can be regarded as the analog of the hydroxide ion.

Ammonia is one of a group of compounds that are the chief materials of the chemical industry and are referred to collectively as *heavy chemicals*. The term *heavy* refers to the amounts of the substances produced and not to their densities. A sizable portion of our industrial capacity is connected with the production of heavy chemicals, but they draw little attention because they do not appear in finished products. Huge quantities of ammonia are produced—over 16 million tons in the United States alone during 1977. Much of this ammonia is used as a fertilizer, either directly or in the form of ammonium compounds.

Nitric Acid The production of an important heavy chemical, nitric acid, requires large quantities of ammonia. Gaseous NH_3 is catalytically oxidized to nitric oxide at temperatures between 500 and 1000°C in the Ostwald process:

$$4NH_3 + 5O_2 \xrightarrow{\text{platinum catalyst}} 4NO + 6H_2O \qquad (23\text{-}22)$$

Without the presence of the platinum catalyst, the oxidation stops at elemental nitrogen:

$$4NH_3 + 3\,O_2 \longrightarrow 2N_2 + 6H_2O \qquad (23\text{-}23)$$

Nitric oxide readily combines with atmospheric oxygen at room temperature to give nitrogen dioxide:

$$2NO + O_2 \longrightarrow 2NO_2 \qquad (23\text{-}24)$$

The gaseous NO_2 is then dissolved in water to form nitric acid,

$$3NO_2 + H_2O \longrightarrow 2HNO_3 + NO \qquad (23\text{-}25)$$

and the NO produced in this step is collected and reoxidized to NO_2.

Nitric acid is a strong oxidizing agent; it will oxidize all metals except the noble metals gold, platinum, rhodium, and iridium. As indicated by two of the possible half-reactions that involve reduction of NO_3^-, the oxidizing power of nitric acid depends on the hydrogen-ion concentration.

$$NO_3^- + 2H^+ + e^- \rightleftharpoons NO_2 + H_2O \qquad E^\ominus = 0.80 \text{ V} \qquad (23\text{-}26)$$
$$NO_3^- + 4H^+ + 3e^- \rightleftharpoons NO + 2H_2O \qquad E^\ominus = 0.96 \text{ V} \qquad (23\text{-}27)$$

Whether NO or NO_2 is produced depends largely on the concentration of the acid. Dilute HNO_3 primarily yields NO, as in the reaction

$$3Cu + 8H^+ + 2NO_3^- \longrightarrow 3Cu^{2+} + 2NO + 4H_2O \qquad (23\text{-}28)$$

while concentrated acid gives mainly NO_2:

$$Cu + 4H^+ + 2NO_3^- \longrightarrow Cu^{2+} + 2NO_2 + 2H_2O \qquad (23\text{-}29)$$

Of course the products of reduction also depend on the strength of the reducing agent. For example, the powerful reducing agent zinc ($E^\ominus_{Zn^{2+}/Zn} = -0.76 \text{ V}$) reduces nitric acid all the way to ammonium ion:

$$4Zn + 10H^+ + NO_3^- \longrightarrow 4Zn^{2+} + NH_4^+ + 3H_2O \qquad (23\text{-}30)$$

Nitric acid reacts with concentrated sulfuric acid to form nitronium ion, NO_2^+:

$$HNO_3 + H_2SO_4 \longrightarrow NO_2^+ + HSO_4^- + H_2O \qquad (23\text{-}31)$$

This reaction is of great importance in organic chemistry because of the ease with which NO_2^+ reacts with aromatic molecules, e.g.,

$$\langle \bigcirc \rangle + NO_2^+ \longrightarrow \langle \bigcirc \rangle - NO_2 + H^+ \qquad (23\text{-}32)$$

Hydrazine and Hydroxylamine Ammonia and the nitrides are compounds in which nitrogen is in a ($-$III) oxidation state. Nitrogen can also be found in two other negative oxidation states, ($-$I) and ($-$II), as in hydroxylamine, NH_2OH, and hydrazine, N_2H_4, respectively.

Hydroxylamine can be thought of as derived from NH_3 by the substitution of an OH group for one of the H atoms. The structure of hydrazine is related to that of H_2O_2. The two NH_2 groups are not in an eclipsed position (Fig. 17-2) but are rotated with respect to each other with a torsion angle φ reported to be 90 to 95° (Fig. 17-1). When free of water, hydrazine is a colorless liquid that boils at 144°C and freezes at 2°C. It is a weak base capable of accepting two protons:

$$H_2NNH_2 + H_2O \rightleftharpoons H_2NNH_3^+ + OH^- \qquad K_1 = 1.0 \times 10^{-6}$$
$$H_2NNH_3^+ + H_2O \rightleftharpoons H_3NNH_3^{2+} + OH^- \qquad K_2 = 9 \times 10^{-16}$$

Oxides of Nitrogen Two of the oxides of nitrogen have already been introduced, nitric oxide (NO) and nitrogen dioxide (NO_2). Four other oxides are known; their names, formulas, and properties are given in Table 23-6.

Nitrous oxide, N_2O, is the only nontoxic oxide of nitrogen. It causes hysteria when it is breathed for a short time and for this reason is sometimes called laughing gas. Prolonged inhalation produces unconsciousness and the gas is used as an anesthetic. Canisters of instant whipped cream contain cream and N_2O under pressure. The gas readily dissolves in the fat and when the pressure is released it forms many tiny bubbles that whip the cream.

NO is unusual in that it contains an odd number of electrons. There is no unique Lewis structure for the molecule, but a resonance description based mainly on two formulas may be written

$$\left\{ :\ddot{N}{=}\dot{O}:, \ :\dot{N}{=}\ddot{O}: \right\}$$

Formula	Name	Color	Remarks
N_2O	Nitrous oxide	Colorless	Rather unreactive
NO	Nitric oxide	Gas, colorless; liquid and solid, blue	Moderately reactive
N_2O_3	Dinitrogen trioxide	Blue solid	Extensively dissociated as gas
NO_2	Nitrogen dioxide	Brown	Rather reactive
N_2O_4	Dinitrogen tetroxide	Colorless	Extensively dissociated to NO_2 as gas and partly dissociated as liquid
N_2O_5	Dinitrogen pentoxide	Colorless	Unstable as gas; ionic solid
NO_3; N_2O_6			Not well characterized and quite unstable

Table 23-6
Oxides of Nitrogen

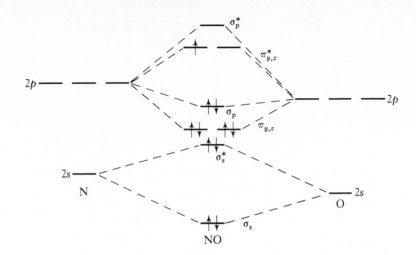

Figure 23-6 Molecular Orbitals of NO
The energies of the atomic orbitals of O are lower than those of N because O is more electronegative (see also Fig. 18-10).

The presence of an unpaired electron makes nitric oxide paramagnetic.

The molecular orbitals for NO are shown in Fig. 23-6. There are 4 pairs of electrons in bonding orbitals and $1\frac{1}{2}$ electron pairs in antibonding orbitals, resulting in $2\frac{1}{2}$ bonds. The molecular orbital diagram is in agreement with the fact that the bond distance in NO is 1.15 Å, while that in the NO^+ ion, which has one less antibonding electron, is 1.06 Å. It explains the ease with which NO loses an electron to form NO^+ and the related fact that the bond energy in NO^+ is 1.6 times greater than that in NO.

Nitrogen dioxide, a red-brown paramagnetic gas, exists in equilibrium with the colorless, diamagnetic N_2O_4. Like NO, NO_2 is an odd molecule; the N_2O_4 dimer necessarily has an even number of electrons. As predicted from the Latimer diagram

$$NO_3^- \xrightarrow{\ 0.80\ } NO_2 \xrightarrow{\ 1.07\ } HNO_2$$

NO_2 and N_2O_4 disproportionate in water to give a mixture of nitrous acid (HNO_2) and nitric acid:

$$2NO_2 + H_2O \longrightarrow HNO_2 + H^+ + NO_3^- \tag{23-33}$$

23-6 Air Pollution

The oxides of nitrogen have recently gained notoriety because of their involvement in air pollution. Standard equipment for automobiles in the United States now includes devices to limit the emissions of so-called NO_x, mixtures of nitrogen oxides. The oxides are formed in internal-combustion engines when the air drawn into the cylinders is subjected to high temperatures and pressures during the combustion process.

Originally the term *smog*, which is a combination of the words smoke and fog, was used to describe the air pollution caused primarily by the combustion of smoky and sulfur-containing fuels. In cities such as London and Pittsburgh, dense, dirty fogs containing fine particles of soot and ash were once very common. When fuels with high sulfur content are burned, the sulfur is oxidized to sulfur dioxide, a poisonous gas with a characteristic irritating odor:

$$S + O_2 \longrightarrow SO_2$$

The SO_2 can be further oxidized to sulfur trioxide:

$$2SO_2 + O_2 \longrightarrow 2SO_3$$

This gas is also very poisonous and damaging to the lungs; it reacts with water to form sulfuric acid. A decrease in the pH of rain in many areas of the world has been associated with air pollution by sulfur oxides but nitrogen oxides probably also play a role.

Smog laden with SO_2 and SO_3 can be literally deadly. In December 1952 heavy smog covered London for 4 days and was responsible for around 4000 deaths. The smog that hovered over Donora, Pennsylvania, in 1948 caused 20 deaths among the 14,000 inhabitants and made almost half the population ill.

Improved smoke-abatement measures and the use of low-sulfur fuels have helped to rid many cities of sulfur-laden and sooty smog. Since 1964 there has been a steady decline in atmospheric SO_2 concentrations in urban areas of the United States. Unfortunately, some regions are plagued by another kind of smog, more difficult to eliminate. It is called photochemical smog, a form of air pollution linked heavily but not exclusively to the automobile. Before discussing the chemistry of photochemical smog, we consider briefly the manner in which light can be involved in chemical reactions.

Photochemical Reactions Figure 23-7a shows schematically the changes in potential energy that may occur when a diatomic molecule AB in its ground electronic state absorbs a photon of frequency ν_1 (energy $h\nu_1$) and undergoes a transition to an excited electronic state AB*. The process can be described by the equation

$$AB + h\nu_1 \longrightarrow AB^* \tag{23-34}$$

Figure 23-7 Schematic Energy Changes Accompanying the Absorption of a Photon
In (a) a molecule absorbs a photon of energy $\Delta\epsilon_1 = h\nu_1$ and undergoes a transition from the ground state AB to an excited electronic state AB*. In (b) the molecule absorbs a photon of different energy $\Delta\epsilon_2$ and is brought to an electronic state AB** that is unstable with respect to dissociation into A + B.

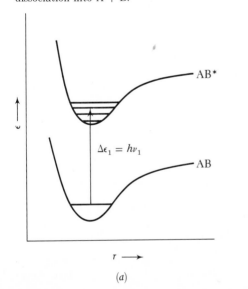

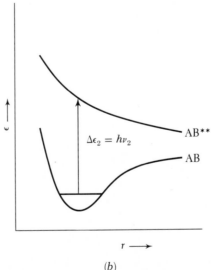

(a)   (b)

The lower curve in Fig. 23-7a represents the way in which the potential energy of AB varies with internuclear separation in the ground state. Even in the ground state of the molecule there is vibrational motion, and the lowest vibrational state of the molecule is indicated by the horizontal line. The upper curve, which represents the potential energy curve for the excited molecule AB*, is similar to the lower curve but not identical because the equilibrium separation and the bond strength depend on the electronic structure. As long as the molecule remains in this excited state, its potential energy will be that represented by the upper curve. The horizontal lines show some possible vibrational states.

Now consider the situation shown in Fig. 23-7b. Here again the molecule starts out in the ground state, but now it absorbs a photon of energy $h\nu_2$, which takes it to a *different* excited electronic state, AB**. Unlike the excited state AB*, this state is not stable to dissociation: there is no minimum in the graph of potential energy against internuclear separation. The potential energy of an A atom and a B atom in this state is lowest when the atoms are very far apart, and there is no bonding. Thus, the equation describing the light absorption process is

$$AB + h\nu_2 \longrightarrow [AB^{**}] \longrightarrow A + B \qquad (23\text{-}35)$$

The absorption of the photon causes the molecule to dissociate.

In this simple example, the photochemical reaction is produced by a photon of a very specific frequency, ν_2. Often in real systems molecules may be excited to a number of very closely spaced levels so that a photochemical reaction can be produced by light in a rather broad range of wavelengths. The extent of the reaction depends on the concentration of ground-state molecules and the intensity of the light, i.e., on the number of photons that strike the sample per second.

Reactions Producing Photochemical Smog There are three necessary ingredients for the production of photochemical smog: light, oxides of nitrogen, and hydrocarbons. The mechanism of smog formation is very complex and not completely understood. It involves many compounds present in the atmosphere at concentrations of only a few parts per million. In one attempt to describe the kinetics of smog formation, 60 elementary reactions were included in the analysis and 120 reactions were considered as possible contributors. Some reactions, however, are unquestionably essential to any mechanistic scheme.

Photochemical smog begins when nitrogen dioxide absorbs light of wavelength shorter than 430 nm and dissociates photochemically:

$$NO_2 + h\nu \longrightarrow NO + O \qquad (23\text{-}36a)$$

The atomic oxygen formed is very reactive. It can combine rapidly with molecular oxygen in the presence of a third molecule to form ozone:

$$O + O_2 \xrightarrow{\text{M}} O_3 \qquad (23\text{-}36b)$$

The third molecule, M, absorbs some of the energy in the $O\text{-}O_2$ collision. Without it the O_3 molecule formed would be highly excited and would immediately revert to the reactants. Atomic oxygen can also react with hydrocarbon vapors to produce radicals, reactive species that contain unpaired electrons:

$$O + \text{hydrocarbons} \longrightarrow R\cdot + RCO\cdot$$

In these formulas the symbol R represents a hydrocarbon part of the molecule, such as CH_3CH_2- (Chap. 29), and the dot signifies an unpaired electron. The identity of R depends on the particular hydrocarbon in the atmosphere encountered by the O atom.

Ozone can also oxidize hydrocarbons:

$$O_3 + \text{hydrocarbons} \longrightarrow RCHO + RCO_2\cdot$$

Ozone and the aldehydes, RCHO, formed in this reaction are components of smog. Ozone irritates mucous membranes and oxidizes double bonds in rubber and plastics, causing the materials to crack and weaken. Aldehydes sting the eyes and can attack the central nervous system.

The radicals produced by the reactions of O and O_3 can react further, as typified by

$$RCO_2\cdot + NO \longrightarrow NO_2 + RCO\cdot \qquad (23\text{-}37)$$
$$RCO\cdot + O_2 \longrightarrow RCO_3\cdot \qquad (23\text{-}38)$$
$$RCO_3\cdot + NO_2 \longrightarrow RCO_3NO_2 \qquad (23\text{-}39)$$

Reaction (23-37) increases the concentration of NO_2 in the atmosphere and thereby increases the amount of O_3 produced by (23-36a) and (23-36b). Peroxyacyl nitrates (PAN), which are especially strong eye irritants, are formed by (23-39). Other reactions that contribute to smog involve other organic species, as well as OH radicals and CO.

The approach to the control of photochemical smog is based on limiting the emissions of the reactants to the atmosphere. Primary sources of oxides of nitrogen are internal-combustion engines and power plants. Hydrocarbons enter the atmosphere from incomplete combustion and from the evaporation of solvents and fuels. The automobile is a major source of both types of reactants. To meet federal standards, manufacturers have changed the conditions of combustion to minimize the formation of nitrogen oxides and to reduce the concentration of unburned hydrocarbons in the exhaust, and catalytic converters have been added to the exhaust systems to oxidize hydrocarbons and carbon monoxide that remain after combustion. In some autos a second converter has been introduced to decompose oxides of nitrogen to N_2 and O_2. New standards are aimed at reducing the release of hydrocarbons from the evaporation of gasoline, for example the vapors displaced by liquid as fuel tanks are filled.

23-7   Group 0

The elements in group 0 have been known as the noble gases, the rare gases, and the inert gases. The latter two terms are in part misnomers. Argon is far from rare; it constitutes almost 1 percent of the atmosphere. Helium is found in concentrations of up to 7 percent in certain natural gases. Since 1962 it has been known that not all the elements in this group are completely inert. The story behind this discovery is an interesting episode in the history of science.

In 1894 Lord Rayleigh and Sir William Ramsay announced that a new element more inert than nitrogen had been isolated from air, and in 1895 Ramsay read a paper at the Royal Society in which he described in detail his unsuccessful efforts to form compounds of the new substance. The name *argon* (Greek: *argos,* inert, not working) seemed appropriate. The five other members of the group were all discovered within the next five years, and during the decades that followed no evidence of compound formation for any of the congeners was uncovered. During the twenties and early thirties it was suggested that the heavier noble gases might form compounds, and Pauling, on theoretical grounds, predicted that a compound of xenon and fluorine should be stable. An attempt to synthesize xenon fluoride and xenon chloride was reported in 1933; its lack of success stifled further experimentation.

It was not until 1962 that Neil Bartlett, after preparing an ionic compound of O_2^+ and the PtF_6^- ion, realized that the ionization energy of Xe is nearly identical to that of O_2. He mixed Xe with PtF_6 vapor and immediately obtained a solid product that was eventually identified as having the composition $XePt_2F_{10}$.

Once the first noble-gas compound was reported, a substantial number of others were made. These include XeF_4, XeF_2, KrF_2, XeO_3, XeO_4, XeO_6^{4-}, $XeCl_2$, and $FXeN(SO_2F)_2$, the first compound containing a Xe—N bond. No compounds of He, Ne, or Ar have been synthesized.

With the exception of helium and radon, the noble gases are obtained from the distillation of liquid air. The world's supply of helium comes mainly from natural gas supplies in the United States and the Soviet Union. Argon is used commercially as an inert shielding gas to prevent oxidation during welding, in the manufacture of high-purity stainless steel, and to provide an inert atmosphere in incandescent light bulbs. It decreases the rate at which the tungsten filament evaporates and therefore allows the filament to be run at a higher temperature to produce a whiter light. Neon finds use in electric signs and other discharge tubes. Because of its very low boiling point (4.2 K), helium is important as a refrigerant for cryogenic studies. Plans to use superconducting metals in technical applications such as electrical transmission lines would require that large quantities of liquid helium be produced.

Summary

Hydrogen appears to be the most abundant element in the universe; it is the third most abundant element on earth in terms of numbers of atoms and the ninth on a weight basis. High-purity hydrogen is obtained by the electrolysis of water but much hydrogen is made commercially by the reaction of methane (CH_4) and steam at high temperature. Major uses of hydrogen include the hydrogenation of vegetable oils, the synthesis of methanol, and the production of ammonia. The binary covalent molecules of hydrogen with the elements of groups IV through VII are known as covalent hydrides. Saltlike hydrides are formed from the reaction of H_2 with highly electropositive elements. Three-center bonds involving hydrogen are found in electron-deficient compounds such as B_2H_6. Because of its nuclear spin, the energy levels of the proton are altered by magnetic fields. In nuclear magnetic resonance spectroscopy (nmr) this dependence is utilized to determine molecular structure.

The most abundant of the chemical elements on earth is oxygen. Elemental oxygen, O_2, is prepared by the distillation of liquid air. Its chief industrial use is for the purification of iron in steel making. A layer of the highly reactive substance ozone, O_3, is present in the upper atmosphere, where it filters ultraviolet radiation. Ionic oxides contain the O^{2-} ion; salts containing O_2^{2-} (peroxides) and O_2^- (superoxides) are also known. Nonmetallic oxides and peroxides are covalent.

Hydrogen peroxide, H_2O_2, is both a strong oxidizing agent and a moderate reducing agent. It can disproportionate spontaneously into O_2 and H_2O. The tendency of a substance to disproportionate can be deduced by inspection of a Latimer diagram.

Liquid water has a maximum density at 4°C; thus the density *decreases* if water is cooled to lower temperatures. There is a further decrease in density when water freezes. This unusual behavior is a consequence of the strong hydrogen bonding between water molecules, which tends to produce a tetrahedral arrangement of neighbors around each water molecule. Water that contains appreciable quantities of divalent metal ions is said to be "hard". Deposits are produced when it is heated; it forms a curd with soap. Water can be softened by the formation of metal complexes and by ion exchange. Water pollutants include: degradable substances, such as nitrates and phosphates; nondegradable substances, of which chlorinated hydrocarbon insecticides are prime examples; and acids that arise from the oxidation of sulfides.

Elemental nitrogen, N_2, constitutes about four-fifths of the atmosphere and is usually obtained by distilling liquid air. Nitrogen is unreactive at ordinary temperatures because of the strength of the triple bond. Most nitrogen compounds are covalent, although a few saltlike nitrides containing the N^{3-} ion are known. The most important compound of N and H is ammonia, prepared in huge quantities by the Haber process. Large amounts of ammonia are converted into HNO_3, a strong acid and a strong oxidizing agent. The process involves oxidation to nitric oxide, NO, followed by further oxidation to nitrogen dioxide, NO_2. Other simple nitrogen compounds include hydrazine, N_2H_4, and hydroxylamine, NH_2OH.

Both NO and NO_2 contain an odd number of electrons. These compounds play an essential role in the formation of photochemical smog. Other participants in the complex reactions that take place in polluted atmospheres are ozone, hydrocarbons, and organic compounds containing oxygen and nitrogen. In areas in which sulfur-containing fuels are burned, SO_2 and SO_3 are major pollutants.

The group 0 elements (the noble gases) are neither as rare nor as unreactive as they once were thought to be. Compounds of xenon include fluorides and oxides.

Terms and Concepts

Problems and Questions

23-1 NMR and Structure Give the number of peaks and the relative peak areas that should be observed in low-resolution proton magnetic resonance spectra of the following molecules: $CH_3CH_2CH_2CH_3$; CH_3OCH_3; CH_3NHCH_3.

23-2 Molecular Weights of Boron Hydrides The empirical formulas and densities of two gaseous boron hydrides are as follows: BH_3 has a density of 0.629 g liter^{-1} at 17°C and 406 torr; B_2H_5 has a density of 0.1553 g liter^{-1} at 19°C and 52 torr. What are the respective molecular weights of these two compounds? What are their correct molecular formulas?

23-3 Bond Energies of O-O Bonds Use data in Table D-4 to find the enthalpy change for the reaction $3O \rightarrow O_3$, and calculate the "bond enthalpy" of the O-O bond in O_3. Repeat these calculations for the O-O bond in H_2O_2, using an $\widetilde{H}_f^\circ$ value of -136 kJ mol^{-1} for $H_2O_2(g)$ as well as values for $H_2O(g)$, etc., from Table D-4. Compare these values with that given for the O-O single bond in Table 17-2 (which is actually an average "bond enthalpy" rather than a bond energy).

23-4 Bonding in Boron Hydrides The structure of the B_4H_{10} molecule (Fig. 23-8) can be described in terms of two single-bonded BH groups and two BH_2 groups with single B-H bonds. The four B atoms are arranged in two triangles that share a side. The shared side consists of the BH groups, linked by a B-B single bond. Along the remaining four sides of the triangles are BHB three-center bonds. Give the total number of valence electrons in B_4H_{10} and show their distribution on a schematic diagram.

Figure 23-8

23-5 Association of Water Molecules In the vapor phase water is believed to be dimerized to a small extent. Calculate the enthalpy change associated with the dimerization reaction $2H_2O(g) \rightleftharpoons H_4O_2(g)$ from the following data: at 40°C, mole fraction $H_4O_2 = 1.6 \times 10^{-3}$, $P_{H_2O} = 55$ torr; at 100°C, mole fraction $H_4O_2 = 5.0 \times 10^{-3}$, $P_{H_2O} = 760$ torr.

23-6 Hydrogen Bonds Define the term *hydrogen bond* and explain the bearing of the geometry of hydrogen bonding on the relative densities of water and ice.

23-7 Solid Solutions of NH_4F in H_2O Ammonium fluoride tends to form solid solutions with ice and is the only known substance that does this. Explain this behavior in terms of the ice structure.

23-8 Properties of a Phosphate Substitute A number of phosphate-free detergents use sodium metasilicate, Na_2SiO_3, as a replacement for phosphate compounds. Sodium metasilicate is the salt of metasilicic acid, H_2SiO_3, for which $K_1 = 2 \times 10^{-10}$ and $K_2 = 1 \times 10^{-12}$. When the recommended quantity of detergent is used, about 1.5 g Na_2SiO_3 is added to each liter of water. What is the pH of the resulting solution? Should any special precautions be taken when storing such products in the home?

23-9 Water Hardness The hardness of water is usually expressed in terms of milligrams per liter of equivalent $CaCO_3$, that is, the concentration of $CaCO_3$ that would by itself produce the hardness observed for that water. During 1974, the Los Angeles Water Department supplied 8.5×10^{10} gal water to residences, and the average hardness was 97 mg liter^{-1} of $CaCO_3$. (*a*) Assume that 10 percent of this water was used for washing and that soap can be represented as sodium stearate, $C_{17}H_{35}COONa$. What weight of soap would have been wasted by precipitation as calcium stearate if all washing had been done with soap? (*b*) If $Na_3P_3O_9$ was used to soften this water by complexing the Ca^{2+} as $CaP_3O_9^-$, what weight of softener would have been required? (1 gal = 3.79 liter.)

23-10 Ammonia Synthesis The theoretically possible equilibrium yield of NH_3 by the Haber synthesis from N_2 and H_2 is higher at 1000 atm than at 1 atm total pressure. Why? It is lower at 500°C than at 25°C. How is this fact related to the sign of ΔH for the reaction? Is the reaction endothermic or exothermic?

23-11 Liquid Ammonia For solutions in liquid ammonia the role of the pH of aqueous solutions is played by $pNH_4 = -\log [NH_4^+]$. Estimate pNH_4 of pure NH_3 at -50°C from Equation (23-21). Explain how it is possible to titrate a solution of NH_4NO_3 in liquid ammonia with a solution of known concentration of KNH_2 (potassium amide) in liquid ammonia. How are acid and base constants of a conjugate acid/base pair related in liquid ammonia?

23-12 Oxidation of Hydrazine In an acid solution, hydrazine (H_2NNH_2) is oxidized quantitatively to N_2 by iodate, Cl_2, Br_2, and I_2. (*a*) Write a balanced equation for the reaction with IO_3^-, which goes to I^-. Note that hydrazine is a weak base and that the predominant species in an acid solution is hydrazinium ion, $N_2H_5^+$. (*b*) Suppose that 10.0 ml hydrazine solution yields 30.7 ml N_2 at

STP. What is the concentration of hydrazine in moles per liter? What is the minimum weight of KIO_3 needed in this oxidation?

23-13 Oxidation and Reduction of Nitrous Acid Describe the following processes by balanced equations. (a) In an acid solution nitrous acid is oxidized to nitrate ion by Br_2, by MnO_4^-, or by $Cr_2O_7^{2-}$, the latter two species yielding Mn^{2+} and Cr^{3+}. (b) Nitrite ion is reduced to NO by I^- going to I_2 and is reduced to NH_3 by HSO_3^- going to SO_4^{2-}. (c) HNO_2 reacts with $NH_3(aq)$ to yield N_2. (d) Use data from Table D-5 to calculate the equilibrium constant for the oxidation of HNO_2 by $Br_2(l)$.

23-14 Lewis Formulas Write possible Lewis formulas for (a) N_2O, NO^+, NO_2^+, NO_2^-, NO_3^-; (b) NH_3, N_2H_4, NH_2OH, CN^-. (c) Indicate the geometric configuration of each molecule or ion.

23-15 Solution of Hydrazine (a) From the data in Table D-3 calculate the pH of a 0.10 M hydrazine solution in water. (b) What is the concentration of $N_2H_5^+$ in a solution prepared by dissolving 0.10 mol hydrazine in water and diluting to 1.0 liter while adjusting the pH to 8.0?

23-16 Photochemistry In the photochemical decomposition of HI,

$$2HI(g) \longrightarrow H_2(g) + I_2(g)$$

the absorption of 500 J of energy in the form of light of a wavelength of 254 nm is found to decompose 2.12×10^{-3} mol HI. How many photons per molecule of HI does this correspond to?

23-17 Molecules of Ar_2? A. D. Buckingham has defined a molecule as "a grouping of atoms that can survive interactions with its environment". The potential energy curve associated with two Ar atoms has a well of depth $\epsilon/k = 100$ K (Fig. 6-7). Under what conditions might one be able to speak of Ar_2?

23-18 The Pauling Oxygen Meter The Pauling oxygen meter measures the O_2 content of a gas by responding to the paramagnetism of the gas. Would the presence of NO or NO_2 falsify the readings obtained? Would O_3 be included as "oxygen" in the readings?

Some Families of Nonmetals

24

"I should be very glad to know what doctrine you teach now with regard to oxymuriatic acid [Cl_2]. Are you yet a convert to chlorine? I am impatient to see Lussac's paper on iodine, in particular to learn how far the facts respecting that substance go to confirm the new views of chlorine. Lussac appears to be a convert to Davy's sentiments, and certainly the acquisition of one who so strenuously opposed them must be accounted a *very* flattering occurrence."

THOMAS CHARLES HOPE, IN A LETTER TO WILLIAM ALLEN, 1814

The preceding chapter was devoted primarily to the chemistry of three of the most common and important individual elements. We could continue to break down the fabric of chemistry into discussions of each of the elements in turn and, indeed, there are many monographs treating each element in depth. Fortunately, however, it is often possible and instructive to consider and contrast the properties of all the elements in one column at one time, rather than following a piecemeal approach. This is the method used in this chapter and Chap. 26.

We begin with a discussion of the halogens, a family of nonmetals that is especially regular in properties. Carbon and silicon, considered in Secs. 24-2 and 24-3, also show many similarities. The oxygen-family elements, sulfur, selenium, and tellurium, are discussed next, and the chapter concludes with an introduction to the chemistry of phosphorus and arsenic, congeners of nitrogen. The metallic elements in columns IVA and VA, tin, lead, antimony, and bismuth, are considered in Chap. 26.

24-1 The Halogens

F
Cl
Br
I
At

Occurrence and Production No member of this highly reactive family is found as the element. Chlorine, the most common halogen, constitutes about 0.05 percent of the earth's solid crust as chloride ion. Concentrated deposits of NaCl (rock salt) are mined, and salt is also obtained from brines pumped out of wells and from the evaporation of seawater, in which it occurs to an extent of about 3 percent by weight. Most Cl_2 is prepared by the electrolysis of concentrated salt solutions, a procedure discussed in Sec. 26-1.

Electrolysis is also the common method of production of F_2. The element, occurring as fluoride ion, is about half as abundant in the earth's crust as chlorine; chief among its ores is fluorite, CaF_2. Fluoride-containing minerals are treated with H_2SO_4 to release HF, the gas is dissolved in molten KF, and F_2 is obtained by electrolysis of the mixture. Although KF itself melts at 846°C, the electrolysis can be performed at temperatures as low as 100°C because low-melting compounds are formed with formulas $KF \cdot HF$ (mp 250°C) and $KF \cdot 2HF$

(mp 72°C), in which the anions are the hydrogen-bonded species FHF^- and $FHFHF^-$. To combat the highly corrosive character of the melt, electrolysis cells are constructed only from metals such as nickel, copper, or monel (a Ni–Cu–Fe alloy), which become covered by protective fluoride layers, and nonreactive electrodes of graphite or nickel are used.

The concentration of bromide ion in seawater is about 70 ppm, sufficiently high to make the extraction of bromine economically feasible. In the industrial process, bromide ion in acidified seawater is oxidized by the relatively inexpensive Cl_2:

$$2Br^- + Cl_2 \longrightarrow Br_2 + 2Cl^- \tag{24-1}$$

Air is bubbled through the solution to sweep out the bromine. The bromine is then concentrated by passing the air through a solution of sodium carbonate in which the Br_2 is converted to bromide and bromate, BrO_3^-:

$$3Br_2 + 6CO_3^{2-} + 3H_2O \longrightarrow BrO_3^- + 5Br^- + 6HCO_3^- \tag{24-2}$$

The bromine is finally obtained by acidifying the solution, which in essence reverses reaction (24-2):

$$5Br^- + BrO_3^- + 6H^+ \longrightarrow 3Br_2 + 3H_2O \tag{24-3}$$

Iodides obtained from kelp (a seaweed) or oil-well brines are also oxidized with Cl_2 to produce I_2, but the major source of iodine is sodium iodate, $NaIO_3$, which is found as an impurity in deposits of sodium nitrate in Chile. The iodate is usually reduced to iodine with hydrogen sulfite ion:

$$2IO_3^- + 5HSO_3^- \longrightarrow I_2 + 5SO_4^{2-} + 3H^+ + H_2O \tag{24-4}$$

Physical Properties of the Halogens All the halogens exist as diatomic molecules that interact with each other by weak van der Waals forces. The variations in their physical properties (Table 24-1) show the regularity that parallels the increase in intermolecular forces accompanying increasing size and polarizability. For example, the melting points, boiling points, and critical temperatures all show a steady increase from fluorine to iodine. At room temperature F_2 and Cl_2 are gases, Br_2 is a liquid, and I_2 is a solid. As expected, the ionization energy and the electron affinity decrease as one goes down the group and the covalent and ionic radii increase. Closer inspection of the trends shows that the changes in

**Table 24-1
Some Properties of the Halogens**

Property	F	Cl	Br	I
Atomic number	9	17	35	53
Stable isotopes, mass numbers	19	35, 37	79, 81	127
Covalent radius/Å	0.72	0.99	1.14	1.33
Ionic radius/Å	1.36	1.81	1.95	2.16
Melting point/°C	−223	−102	−7	113
Boiling point/°C	−188	−35	58	183
Color	Pale yellow	Yellow-green	Red-brown	Gray-black
$E^\ominus$/V for $X_2 + 2e^- \longrightarrow 2X^-$	2.87	1.36	1.06	0.54
$\Delta H^\ominus$/kJ for $X_2 \longrightarrow 2X$	159	239	190	149

properties between F_2 and Cl_2 are significantly larger than those between any other two members of the family. An important irregularity is the abnormally low heat of dissociation of F_2. It has been suggested that this low value is the result of electrostatic repulsions that occur at short range within the small F_2 molecule.

Bonding and Reactions of the Halogens The principal modes of bond formation in the halogens involve the addition of one electron to the s^2p^5 configuration to complete the octet. The ease with which the halide ions, X^-, may be formed is indicated by the high values of the electron affinities of halogen atoms (Table 16-6) and the standard potentials for reduction of the halogen molecules (Table 24-1):

$$X_2 + 2e^- \longrightarrow 2X^- \tag{24-5}$$

The stability of the diatomic halogen molecules is also evidence of the tendency to achieve a noble-gas configuration.

The formation of positive ions is precluded by the very high ionization energies, but with the exception of fluorine, the halogens exhibit positive oxidation states in covalent compounds. Examples are the oxygen acids $HOCl$, $HClO_2$, $HClO_3$, and $HClO_4$, in which chlorine has oxidation states $(+I)$, $(+III)$, $(+V)$, and $(+VII)$, respectively. The formula for the first acid has been written $HOCl$ to show that in this compound the halogen forms only one covalent bond. The other acids are conventionally written as shown, but they also contain the $HO-$ group and have, respectively, two, three, and four $Cl-O$ bonds. Binary covalent compounds of the halogens with each other are known as interhalogen compounds. These include compounds with the general formula XX' (such as $BrCl$, bromine monochloride, an effective disinfectant), as well as those of formula XX'_3 (such as BrF_3, bromine trifluoride) and the compounds BrF_5, IF_5, and IF_7, in which the bonding involves d electrons.

The similarity in properties of the halogens makes it possible to describe much of their chemistry by general group reactions such as those given in Table 24-2. Differences between the congeners become evident when one considers the extent of reaction or the reaction rate. A case in point is the disproportionation reaction

$$X_2 + H_2O \Longleftrightarrow H^+ + X^- + HOX \tag{24-6}$$

The equilibrium constant for (24-6) falls off drastically from chlorine to iodine. Fluorine also reacts according to (24-6), but the HOF is difficult to isolate because it reacts rapidly and vigorously with water. The overall reaction usually observed gives oxygen,

$$2F_2 + 2H_2O \longrightarrow 4HF + O_2 \tag{24-7}$$

Table 24-2
Group Reactions of the Halogens

$nX_2 + 2M \longrightarrow 2MX_n$	
$3X_2 + 2P \longrightarrow 2PX_3$	
$X_2 + PX_3 \longrightarrow PX_5$	not with I_2
$X_2 + H_2 \longrightarrow 2HX$	
$X_2 + H_2O \Longleftrightarrow H^+ + X^- + HOX$	not with F_2 (see text above)
$X_2 + 2X'^- \longrightarrow X'_2 + 2X^-$	X_2 a better oxidizing agent than X'_2, $F_2 > Cl_2 > Br_2 > I_2$
$X_2 + H_2S \longrightarrow 2HX + S$	
$3X_2 + 8NH_3 \longrightarrow 6NH_4X + N_2$	not with I_2

as well as ozone and other oxidation products. When the equilibrium constant for reaction (24-7) is calculated from the standard electrode potentials for the half-reactions

$$F_2 + 2e^- \longrightarrow 2F^- \qquad E^\ominus = 2.87 \text{ V}$$
$$O_2 + 4H^+ + 4e^- \longrightarrow 2H_2O \qquad E^\ominus = 1.23 \text{ V}$$

and K_a for HF, a value of about 10^{124} is obtained. Similar calculations for Cl_2 lead to a smaller, but still large, value of K, 10^9, which is large enough to make HOCl highly unstable. Thus, the observed difference in behavior between F_2 and Cl_2 is a kinetic rather than a thermodynamic effect. Further evidence of this fact is the observation that solutions of Cl_2 release oxygen on standing.

Another reaction that illustrates differences between the halogens is the oxidation

$$X_2 + 2X'^- \rightleftharpoons 2X^- + X'_2 \qquad (24-8)$$

that is used to liberate Br_2 from seawater [Equation (24-1)]. It is apparent from the standard emfs in Table 24-1 that a halogen will displace from solution any congener that lies below it in the periodic table. Thus, F_2 can oxidize Cl^-, Br^-, and I^-, and Cl_2 can in turn oxidize Br^- and I^-.

Halides Many of the most familiar and important chemical compounds are halides. With the exception of He, Ne, and Ar, all the elements in the periodic system form halides. Most metallic halides can be formed by direct interaction of the elements:

$$nX_2 + 2M \longrightarrow 2MX_n \qquad (24-9)$$

Because these reactions are highly exothermic, they can be reversed at sufficiently high temperatures (Fig. 24-1). If the metal is highly electropositive and does not have a very small ionic radius, e.g., an alkali metal or the alkaline earths except Be, the halide is ionic. Metals with relatively high ionization energies and small atomic radii form covalent halides. These trends can be noted in the series of compounds KCl, $CaCl_2$, $ScCl_3$, and $TiCl_4$, in which there is a progression from a high-melting ionic solid to a molecular liquid. If a metal forms more than one halide with a particular halogen, the compound in which the metal has the lower oxidation state has a greater degree of ionic character. For example, $SnCl_4$ is a molecular liquid that freezes at $-30°C$ whereas $SnCl_2$ is a crystalline solid that conducts electricity when molten.

The special character of fluorine is also observable in the halides. Melting points of metallic fluorides are markedly higher than those of other halides. Calcium fluoride melts above $1300°C$; the melting points of the chloride, bromide, and iodide lie $600°C$ lower.

Hydrogen Halides Although the hydrogen halides can be formed by a direct reaction between the elements, they are usually prepared by the reaction of a halide with a nonoxidizing acid, e.g.,

$$CaF_2 + H_2SO_4 \longrightarrow 2HF + CaSO_4 \qquad (24-10)$$

The term *nonoxidizing acid* is relative. Sulfuric acid, an inexpensive reagent, will serve for the production of HF and HCl, but it will oxidize bromides and iodides to Br_2 and I_2. To prepare HBr and HI, therefore, a poorer oxidizer, such as

537

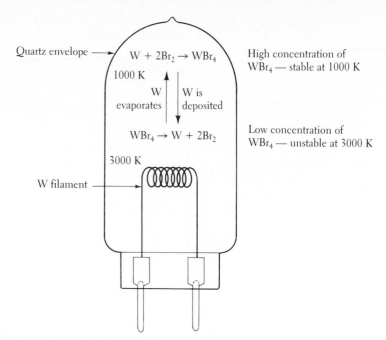

Quartz envelope →

$W + 2Br_2 \rightarrow WBr_4$

1000 K

W evaporates ↑ | ↓ W is deposited

$WBr_4 \rightarrow W + 2Br_2$

3000 K

W filament →

High concentration of WBr_4 — stable at 1000 K

Low concentration of WBr_4 — unstable at 3000 K

Figure 24-1 Quartz-Halogen Projector Lamp
Tungsten that evaporates from the filament at 3000 K diffuses to the cooler wall region where it reacts with Br_2 or Br to give volatile WBr_4. The tungsten bromide diffuses back to the center where it decomposes and deposits on the filament. The bulb must be made of quartz to withstand the high temperature. Were it not for the presence of the halogen, the filament would soon be damaged by the loss of tungsten by sublimation, and a dark film of tungsten would appear on the walls.

phosphoric acid, must be used:

$$NaBr + H_3PO_4 \longrightarrow HBr + NaH_2PO_4 \qquad (24\text{-}11)$$

All the hydrogen halides are gases at room temperature. The physical properties of HCl, HBr, and HI show a smooth progression (Table 24-3) but HF stands alone, as noted earlier in Fig. 6-9. Its melting point, boiling point, enthalpy of fusion, and enthalpy of vaporization are abnormally high as a result of strong hydrogen bonding in the solid and liquid. Hydrogen fluoride is even associated in the gaseous state, where aggregates as large as $(HF)_5$ have been observed.

The acid strengths of HCl, HBr, and HI can be distinguished in nonaqueous solvents: HI is the strongest acid, HCl the weakest. All three acids are almost completely dissociated in water, however. Once again, HF is unique. It is a surprisingly weak acid ($K = 7 \times 10^{-4}$), the low acid strength being primarily a consequence of the strength of the H-F bond. In concentrated solutions (5 to 15 M), HF becomes a strong acid because of the formation of complexes, especially the stable hydrogen difluoride ion HF_2^-:

$$F^- + HF \longrightarrow HF_2^- \qquad K = 5.1 \qquad (24\text{-}12)$$

Reaction (24-12) uses up F^- and thereby enhances the ionization of HF.

The well-known ability of hydrofluoric acid to etch glass is a result of a unique reaction between HF and silicates, forming the gas SiF_4:

$$CaSiO_3 + 6HF \longrightarrow CaF_2 + SiF_4 + 3H_2O \qquad (24\text{-}13)$$

Table 24-3
Properties of the Hydrogen Halides

Property	HF	HCl	HBr	HI
Melting point/°C	−83	−115	−87	−51
Boiling point/°C	20	−85	−67	−35
$\Delta \widetilde{H}_{vap}$/kJ mol^{-1}	30.3	16.1	17.6	19.7
$\Delta \widetilde{H}_{fus}$/kJ mol^{-1}	4.58	2.11	2.40	2.87

Oxyacids and Their Salts Table 24-4 lists some of the oxyacids of the halogens and gives their names and the names of their salts. Hypofluorous acid, not given in the table, has also recently been isolated but is quite unstable. The solution chemistry of these substances is rather complex; the course of reaction depends sensitively on temperature and pH, and mixtures of products are often produced. As examples we consider some reactions involving the hypohalous acids and hypohalites.

In principle, hypohalous acids can be prepared by reaction (24-6), but the equilibrium yield is low and halide ion is also a product. A better procedure is to pass the halogen into an aqueous suspension of mercuric oxide, HgO:

$$2X_2 + 2HgO + H_2O \longrightarrow \{HgO\}HgX_2 + 2HOX \qquad (24\text{-}14)$$

The formula $\{HgO\}HgX_2$ is used here to indicate that the HgX_2 is removed from solution by adsorption on the solid HgO. This loss of HgX_2 drives the reaction to the right. Salts of hypohalous acids can be obtained by the disproportionation of a halogen in basic solution, e.g.,

$$Cl_2 + 2OH^- \longrightarrow Cl^- + ClO^- + H_2O \qquad (24\text{-}15)$$

The hypohalite ion, however, can itself disproportionate further,

$$3ClO^- \longrightarrow 2Cl^- + ClO_3^- \qquad (24\text{-}16)$$

as can be seen from a Latimer diagram (Sec. 23-3) relating the various species in basic solution (at pOH = 0):

$$ClO_3^- \xrightarrow{0.50} ClO^- \xrightarrow{0.40} Cl_2 \xrightarrow{1.36} Cl^- \qquad (24\text{-}17)$$
$$\underset{0.88}{\rule{4cm}{0.4pt}}$$

Corresponding Latimer diagrams for bromine and iodine in basic solution are given in Example 24-1, and diagrams that pertain to acidic solution are given in Exercise 24-1. To see why potentials in acidic and basic solution may differ, consider the half-reaction relating the species ClO_4^- and ClO_3^-. In acidic solution it is written

$$ClO_4^- + 2H^+ + 2e^- \rightleftharpoons ClO_3^- + H_2O \qquad (24\text{-}18a)$$

Table 24-4
Some Oxyacids of the Halogens

Name of acid	Name of salt	Chlorine	Bromine	Iodine
Hypohalous	Hypohalite	HOCl	HOBr	HOI
Halous	Halite	HOClO		
Halic	Halate	HOClO$_2$	HOBrO$_2$	HOIO$_2$
Perhalic	Perhalate	HOClO$_3$	HOBrO$_3$	HOIO$_3$

Since H^+ cannot exist to an appreciable extent in basic solution, at pH = 14 the half-reaction is written

$$ClO_4^- + H_2O + 2e^- \rightleftharpoons ClO_3^- + 2OH^- \tag{24-18b}$$

Equation (24-18b) can be derived by adding the equation

$$2H_2O \rightleftharpoons 2H^+ + 2OH^- \tag{24-19}$$

to (24-18a); hence the values of $\Delta G^\ominus$ for the two reactions differ by $\Delta G^\ominus$ for (24-19), $\Delta G_{19}^\ominus = -2RT \ln K_w$, and the $E^\ominus$'s must also differ. If a half-reaction involves a weak acid or base, the change in $\Delta G^\ominus$ from acid to basic solution will involve the dissociation constant as well (Prob. 24-18).

Example 24-1

□ **Disproportionation of Halogens in Basic Solution** Predict on thermodynamic grounds the products of the disproportionation of Br_2 and I_2 in basic solution. The Latimer diagrams showing the standard reduction potentials for the various species are

$$BrO_3^- \xrightarrow{\;0.54\;} BrO^- \xrightarrow{\;0.45\;} Br_2 \xrightarrow{\;1.06\;} Br^- \tag{24-20}$$

with 0.61 spanning BrO_3^- to Br_2, and 0.76 spanning BrO^- to Br^-.

$$IO_3^- \xrightarrow{\;0.14\;} IO^- \xrightarrow{\;0.45\;} I_2 \xrightarrow{\;0.53\;} I^- \tag{24-21}$$

with 0.26 spanning IO_3^- to I_2, and 0.49 spanning IO^- to I^-.

Solution Since the potential for the Br_2/Br^- couple is more positive than the potential for the BrO^-/Br_2 couple, Br_2 will spontaneously disproportionate to BrO^- and Br^-. Examination of (24-20) shows that BrO^- has no tendency to disproportionate to Br_2 and BrO_3^-, but the potential of the BrO^-/Br^- couple is more positive than that of BrO_3^-/BrO^-; hence BrO^- will tend to disproportionate further to BrO_3^- and Br^-. No further disproportionation is possible. Similar reasoning shows that in basic solution I_2 is unstable with respect to I^- and IO^- and that IO^- will disproportionate further, forming IO_3^- and, eventually, I^-. ■

Exercise 24-1

□ Predict on thermodynamic grounds the stability of Cl_2, Br_2, and I_2 in acidic solution. The pertinent Latimer diagrams are given below; they apply at pH = 0.

$$HClO_3 \xrightarrow{\;1.43\;} HClO \xrightarrow{\;1.63\;} Cl_2 \xrightarrow{\;1.36\;} Cl^-$$

with 1.47 spanning $HClO$ to Cl^-.

$$HBrO_3 \xrightarrow{\;1.49\;} HBrO \xrightarrow{\;1.59\;} Br_2 \xrightarrow{\;1.06\;} Br^- \tag{24-22}$$

with 1.51 spanning $HBrO$ to Br^-.

$$HIO_3 \xrightarrow{\;1.14\;} HIO \xrightarrow{\;1.45\;} I_2 \xrightarrow{\;0.53\;} I^- \quad ■$$

with 1.20 spanning HIO to I_2.

Even though ClO^- is thermodynamically unstable with respect to Cl^- and ClO_3^-, reaction (24-15) is a useful method for the production of hypochlorites because (24-16) is slow at or below room temperature. The analog of (24-16) for BrO^- is rapid at room temperature, so solutions containing hypobromite ion can only be prepared and kept at around 0°C. Hypoiodites are not found because the disproportionation of IO^- is very fast at all practicable temperatures.

The magnitudes of the standard potentials for the reduction of the oxygen-containing halogen species in basic solution indicate that all these substances are strong oxidizers; this generalization is also true for the couples in acid solution. Solutions of sodium hypochlorite are used as a household bleach because of their oxidizing power and low cost. On a commercial scale the solutions are produced by electrolysis of concentrated NaCl solutions. The process is identical to that used to produce Cl_2 and NaOH (Fig. 26-1), but the products obtained at the anode and cathode are allowed to mix and ClO^- is formed by (24-15). Bleach containers carry a warning not to mix the contents with household ammonia (to prepare a strong cleaning solution); the NH_3 reduces the OCl^- and poisonous Cl_2 gas is produced.

Potassium chlorate, the most common halate, is prepared by passing Cl_2 into hot KOH solution [Equations (24-15) and (24-16)]. When heated, chlorates decompose to oxygen and a chloride:

$$2KClO_3 \xrightarrow{\text{MnO}_2} 2KCl + 3O_2 \qquad (24\text{-}23)$$

Manganese dioxide catalyzes the reaction. When chlorates are placed in contact with easily oxidizable materials such as carbon, sugar, or phosphorus, they may react explosively.

Perchloric acid is one of the strongest acids known. Unlike the other oxyacids of chlorine, $HClO_4$ can be isolated in pure form. The Latimer diagram (24-17) can be extended to show the relation of ClO_4^- to other oxygen-containing chlorine species:

$$ClO_4^- \xrightarrow{\ 0.36\ } ClO_3^- \xrightarrow{\ 0.50\ } ClO^- \qquad (24\text{-}24)$$

Chlorate ion is clearly unstable with respect to disproportionation to ClO^- and ClO_4^-, but the reaction occurs so slowly that it is not useful for the preparation of perchlorate salts. Instead, perchlorates are commonly prepared by the electrolytic oxidation of chlorates. Almost all perchlorate salts are very soluble in water, and the perchlorate ion forms almost no complexes; thus perchlorates are commonly used for experiments in which the properties of cations in solution are to be studied. Aqueous solutions containing perchlorates and organic substances can explode when concentrated by heating.

Pure crystals of periodic acid with the formula H_5IO_6 can be obtained. In solution, there are pH-dependent equilibria between several species: H_5IO_6 (paraperiodic acid), $H_4IO_6^-$, $H_3IO_6^{2-}$, and IO_4^- (metaperiodate ion). Perbromic acid and perbromates were predicted to be stable on thermodynamic grounds, but attempts to prepare them in pure form were unsuccessful until 1968. Solutions containing BrO_4^- can be obtained by oxidation of BrO_3^-, electrolytically and with F_2. Solutions of $HBrO_4$ can be prepared from solutions of $NaBrO_4$ by ion exchange.

Rules for Estimating the Strengths of Inorganic Oxygen Acids The acid constants of the oxyacids of the halogens increase with the oxidation state of the halogen, and their values can be estimated by two simple rules. These rules can be applied to any inorganic acid containing oxygen if the structure of the acid is known. In the form given here, as proposed by Pauling, these rules apply to acids with formula $(HO)_m XO_n H_p$ in the uncharged form, where X is the central atom and where the hydrogens of each OH group are acidic and the other hydrogens, which are attached directly to the central atom X, are not acidic. The strength of the acid is found to depend almost entirely on the number (n) of oxygen atoms to which no hydrogen atom is attached.

Rule 1   The first acid constant has approximately the value $K_1 \approx 10^{5n-7}$. For example, K_1 for sulfurous acid, H_2SO_3, which has the structure $(HO)_2SO$, with $n = 1$, is predicted to be about $10^{5 \times 1 - 7} = 10^{-2}$. It is actually 4×10^{-2}.

Rule 2   For polyprotic acids, the ratios of successive acid constants are approximately equal to 10^{-5}. Thus if there are three acidic hydrogens attached to different oxygen atoms, the three acid constants are in the approximate relationship $K_1 : K_2 : K_3 = 1 : 10^{-5} : 10^{-10}$.

As an example of application of the first rule, consider chlorous and chloric acids, $HClO_2$ and $HClO_3$. The structures are $(HO)ClO$ and $(HO)ClO_2$ so that $n = 1$ and 2. The estimated constants are 10^{-2} and 10^3. The measured value of the first of these constants is 1.0×10^{-2}; the second constant is difficult to measure but is known to be very large. An example using both rules is provided by phosphoric acid, H_3PO_4 [or $(HO)_3PO$], for which $n = 1$. The constants are predicted to be $K_1 \approx 10^{-2}$, $K_2 \approx 10^{-7}$, and $K_3 \approx 10^{-12}$; the measured values are 7.1×10^{-3}, 6.2×10^{-8}, and 4.4×10^{-13}.

Since acid constants can range over many powers of 10, any estimate that comes within one or two powers of 10 can be at least qualitatively very useful in indicating whether a given acid is likely to be strong, moderately weak, or very weak. The rules are good to about this degree.

No simple rules are available for estimation of the acid constants of inorganic acids other than oxygen acids with the structure discussed here.

Example 24-2

□ **Acid Constant of Phosphorous Acid**   Phosphorous acid (H_3PO_3) has one hydrogen atom directly bonded to phosphorus,

$$
\begin{array}{c}
\quad\quad O \\
\quad\quad \| \\
H-P-O \\
\quad\quad | \quad\quad \backslash \\
\quad\quad O \quad\quad H \\
\quad\quad | \\
\quad\quad H
\end{array}
$$

and, since such hydrogen atoms are not acidic, H_3PO_3 is diprotic rather than triprotic. Estimate its acid constants.

Solution   A glance at the structural formula just given indicates that $n = 1$. Rule 1 therefore yields $K_1 = 10^{-2}$ and rule 2 yields $K_2 = 10^{-7}$. The measured values are 1.0×10^{-2} and 2.6×10^{-7}. ■

Exercise 24-2

□ Hypophosphorous acid, H_3PO_2, has the structure

$$
\begin{array}{c}
\quad\quad O \\
\quad\quad \| \\
H-P-O \\
\quad\quad | \quad\quad \backslash \\
\quad\quad H \quad\quad H
\end{array}
$$

and is therefore monoprotic rather than triprotic. Estimate its acid constant. ■

24-2 Carbon

Elemental carbon occurs in nature as diamond and graphite. Soot, lampblack, and charcoal consist primarily of microcrystals of graphite. Coal is a mineral rich in carbon; little of the carbon in soft (bituminous) coal is uncombined, but hard (anthracite) coal contains more than 90 percent elemental carbon. Compounds of carbon are the basic structural materials of living matter.

The structures of graphite and diamond have already been discussed (Sec. 7-2). Diamond, the denser form, is less stable than graphite at low pressure, but its conversion to the more stable form does not occur under ordinary conditions. At very high pressures ($\sim 10^5$ atm), diamond is more stable than graphite. Thus, if the conversion is made rapid by raising the temperature and by the use of metal catalysts, diamonds can be produced by compressing graphite. The stones produced are small and not of gem quality, but synthetic diamonds are as hard as natural diamonds and are used industrially as abrasives.

Naturally occurring carbon contains 1.1 percent ^{13}C. Unlike ^{12}C, this isotope has a nuclear spin, and nmr spectroscopy (Sec. 23-1) with ^{13}C as a probe has become an important tool in structure determinations of complex molecules. Another isotope, ^{14}C, which is radioactive, is present naturally in trace amounts and is useful in "radiocarbon dating" (Sec. 28-4).

Most of the chemistry of carbon is classified as organic chemistry, the subject of Chap. 29. A number of compounds of carbon, however, are considered inorganic. The classification is not clear-cut or necessarily logical—some "inorganic" compounds of carbon play an essential role in life processes—but the distinctions are commonly made. We discuss here the chemistry of some of the inorganic compounds of carbon.

Carbon Oxides When carbon is heated in a limited supply of oxygen, carbon monoxide is formed:

$$2C + O_2 \longrightarrow 2CO \tag{24-25}$$

Carbon monoxide is isoelectronic with and similar in physical properties to N_2. It is a colorless, odorless, and tasteless gas that is only very slightly soluble in water. Carbon monoxide is very toxic because it forms a complex with hemoglobin, the oxygen-carrying component of blood. The CO-hemoglobin complex is much more stable than the complex with O_2. When CO is inhaled, hemoglobin binds it in preference to oxygen and the transport of oxygen to body cells is diminished. Fortunately, the CO-hemoglobin reaction is reversible. If the concentration of CO in the lungs is decreased, CO-hemoglobin complexes will dissociate, and the oxygen transport can return to normal if no major damage to the organism has yet occurred because of the lack of O_2.

Carbon monoxide can combine directly with halogens to form compounds with the formula COX_2. An example is the reaction with chlorine to produce the poisonous gas phosgene, $COCl_2$:

$$CO + Cl_2 \longrightarrow COCl_2 \tag{24-26}$$

Transition metals also react with CO to form metal carbonyls (Sec. 25-1). Most metal carbonyls are very hazardous to work with, in part because they dissociate easily to form CO.

Although CO is quite inert at room temperature, it burns readily when heated, forming carbon dioxide:

$$2CO + O_2 \longrightarrow 2CO_2 \tag{24-27}$$

Because carbon dioxide is very stable and denser than air, it is commonly used as a fire extinguisher. Cylinders containing liquid CO_2 under pressure can be used to produce a blanket of inert gas to smother fires. In other extinguishers CO_2 generated by the reaction of bicarbonate and acid,

$$HCO_3^- + H^+ \longrightarrow CO_2 + H_2O \tag{24-28}$$

is used to spray liquid out of a reservoir. The reaction is initiated by inverting the extinguisher, which mixes an acid with a bicarbonate solution.

The acidic properties of solutions of CO_2, which is the cheapest and most abundant weak acid, were discussed in Sec. 12-6. This acid forms many salts, both carbonates and bicarbonates. Carbon dioxide constitutes about 0.03 percent of the atmosphere and about 0.014 percent of the oceans, where it is present chiefly as HCO_3^-. Because the mass of the oceans is much greater than that of the atmosphere, the amount of CO_2 in the oceans is considerably greater than that in the atmosphere, and large quantities are also present in combined form in various carbonate minerals, principally as $CaCO_3$ (Sec. 26-2). During the process of photosynthesis green plants remove CO_2 and H_2O from the atmosphere, form carbohydrates from them, and simultaneously return oxygen to the atmosphere.

When CO_2 reacts with ammonia under high pressure, urea is produced (see also Fig. 7-1):

$$CO_2 + 2NH_3 \longrightarrow OC(NH_2)_2 + H_2O \tag{24-29}$$
$$\text{Urea}$$

Urea is used as fertilizer. It is the principal end product of protein metabolism in the human body and is the chief nitrogen-containing component of urine. This association with life processes made the synthesis of urea from strictly inorganic substances by Wöhler in 1828 a milestone that removed any apparent boundary between inorganic and organic chemistry. It disproved the existence of a mysterious "vital force", a *vis viva,* once believed to be essential in the production of organic substances.

Other Carbon Compounds The direct reaction of carbon and sulfur at high temperature gives carbon disulfide:

$$C + 2S \longrightarrow CS_2 \tag{24-30}$$

This pale yellow volatile liquid has an unpleasant odor and is toxic; it is used as a solvent. Carbon tetrachloride, another common solvent, is prepared by the reaction of CS_2 and Cl_2:

$$CS_2 + 3Cl_2 \longrightarrow CCl_4 + S_2Cl_2 \tag{24-31}$$

Carbon tetrachloride is toxic and can cause serious liver damage. It was once used in some fire extinguishers because it has a dense vapor that can smother flames. It is not recommended for this purpose, however, because at high temperatures CCl_4 reacts with oxygen or water to produce phosgene.

Carbon forms binary compounds, called carbides, with many metals. The carbides of the most electropositive metals are saltlike and contain ions such as C^{4-}, C_2^{2-}, and C_3^{4-}. An important saltlike carbide is calcium carbide, CaC_2, which is made by the reduction of CaO at very high temperature:

$$CaO + 3C \longrightarrow CaC_2 + CO \qquad (24\text{-}32)$$

Carbides containing the C_2^{2-} ion are also called acetylides because they react with water to produce acetylene, C_2H_2, a compound important in organic synthesis:

$$CaC_2 + 2H_2O \longrightarrow Ca(OH)_2 + C_2H_2 \qquad (24\text{-}33)$$

The close-packed arrays of atoms in a metal contain octahedral holes (see Fig. 17-5 and Prob. 17-4). If the atoms have a radius of at least 1.3 Å (for example, Ti, Mo, W), the holes are sufficiently large to contain carbon atoms without distortion of the crystal structure. Carbides in which carbon occupies such sites are called *interstitial carbides*. The presence of the carbon does not fundamentally alter the characteristic metallic properties such as electrical conduction; however, melting points and hardness are enhanced because the C atoms are significantly involved in the bonding (page 572).

An important compound that contains C and N is hydrogen cyanide, HCN, a highly poisonous gas used sometimes as a fumigant. In aqueous solution it is a very weak acid. Cyanide ion forms a large number of very stable complexes, such as $Ag(CN)_2^-$, $Fe(CN)_6^{3-}$, and $Fe(CN)_6^{4-}$. The poisonous action of cyanides is the result of complex formation: certain oxidative enzymes and other biological molecules are inactivated when the transition metals they contain are removed as cyanide complexes.

Two elements, silicon and boron, form completely covalent carbides. Silicon carbide, SiC, is known as carborundum. It exists in several crystalline forms, one of which has the diamond structure (Fig. 7-6). Carborundum is unusually hard and has a very high melting point. Boron carbide, B_4C, is one of the hardest materials known. Some metals of small radius form carbides intermediate in properties between the interstitial and covalent carbides. Important among these is iron; its carbide plays an essential role in steel making (Sec. 26-4).

24-3 Silicon

Our discussion of the chemistry of the halogens highlighted the strong similarities between the elements in this family. A comparison of the properties of silicon and carbon also shows many similarities. As we shall see, however, some of these similarities are superficial, and there are many significant differences between these elements.

Silicon is the second most abundant element in the earth's crust. Most of the rocks that constitute the crust are silicates, compounds of silicon with the most abundant element, oxygen. Pure silicon has the diamond structure: each Si is bonded to four other Si atoms in a tetrahedral arrangement. No silicon analog of graphite is known. Silicon is obtained by the reduction of quartz sand (silicon dioxide, SiO_2) with carbon in an electric furnace. Much of the demand for Si is for electronic components that require materials of very high purity. Silicon too impure for such use is refined by converting it to the tetrachloride,

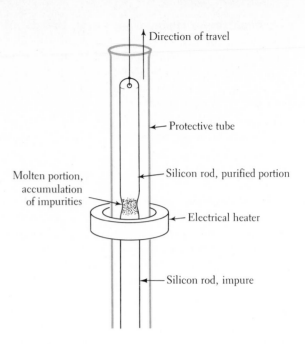

Figure 24-2 Zone Refining of Silicon
The silicon rod is drawn extremely slowly through the heated zone.
Surface tension keeps the molten section of the rod intact.

$$Si + 2Cl_2 \longrightarrow SiCl_4 \qquad (24\text{-}34)$$

and purifying the liquid $SiCl_4$ by fractional distillation. The purified $SiCl_4$ is then reconverted to Si by reaction with hydrogen in a hot tube:

$$SiCl_4 + 2H_2 \longrightarrow Si + 4HCl \qquad (24\text{-}35)$$

The final stage in purification is zone refining (Fig. 24-2). In this process a short segment of a long rod of the element is heated until it melts. The impurities are more soluble in the molten Si than in the solid and therefore concentrate in the liquid. The heat source is slowly moved so that the narrow molten zone traverses the length of the rod, sweeping the impurities into the liquid. When the impure region is brought to the end of the rod, it is allowed to cool and is cut off. The process may be repeated. Zone refining makes it possible to produce silicon with only 1 part in 10^{11} of impurity.

Comparisons between Si and C Like carbon, silicon forms four tetrahedral bonds. The tetrahalides, SiX_4, are analogous in structure to the carbon tetrahalides; however, the silicon compounds are much more reactive. For example, $SiCl_4$ reacts vigorously with water:

$$SiCl_4 + 4H_2O \longrightarrow 4HCl + H_4SiO_4 \qquad (24\text{-}36)$$

The product H_4SiO_4 is called silicic acid; its formula might better be written $Si(OH)_4$. Reaction (24-36) gives both $Si(OH)_4$ and the products of reactions such as

$$2Si(OH)_4 \longrightarrow (HO)_3Si\text{—}O\text{—}Si(OH)_3 + H_2O \qquad (24\text{-}37)$$

Very large molecules connected by Si-O-Si linkages can be formed.

Silicon also forms silanes, hydrides that are analogs of alkanes (Chap. 29): SiH_4, Si_2H_6, ..., Si_6H_{14}. These compounds are more reactive than their carbon counterparts; for example, they burn spontaneously in air:

$$Si_2H_6 + \tfrac{7}{2}O_2 \longrightarrow 2SiO_2 + 3H_2O \qquad (24\text{-}38)$$

Compounds of silicon containing double Si-Si or double Si-C bonds are known only as highly unstable intermediates. Quite generally π bonding between atoms beyond F is rare. This is in part attributed to the repulsion between completely filled inner shells of such atoms (Ne and higher shells). Because of the repulsion, bonded atoms are further apart (their "size" is larger), decreasing the π-π overlap responsible for π bonding. In addition, the increase in atomic size allows an increase in coordination number, permitting electrons that might have participated in π bonding to participate in forming σ bonds.

Silicon dioxide, commonly known as silica, is the analog of CO_2; however, it bears little resemblance to the carbon compound. It is commonly found in nature in three crystalline forms: quartz, tridymite, and cristobalite. Quartz is the form that is stable at low temperature. At atmospheric pressure, quartz and tridymite are in equilibrium at 870°C and tridymite and cristobalite are in equilibrium at 1470°C. However, the transformation from one form to another is usually slow; hence crystals of tridymite and cristobalite, although rarer than quartz, exist at ordinary temperatures, presumably because they were cooled rapidly.

The basic structural unit of silica is an SiO_4 tetrahedron in each of the crystalline forms, as well as in glassy silica. The tetrahedra are linked together differently in the different forms, but a common feature of all is that every oxygen atom is common to two tetrahedra. Thus every silicon is bonded to four oxygen atoms and every oxygen to two silicon atoms, which accounts for the empirical formula SiO_2.

When molten silica is cooled it usually does not crystallize. Instead it becomes increasingly viscous until it hardens and can no longer flow. The solid is non-crystalline and can be described as a supercooled liquid or glass. On the microscopic scale it consists of linked SiO_4 tetrahedra that have local order but lack the long-range order of a crystal. Fused silica or quartz glass finds use in special laboratory equipment where its low thermal expansion and transparency to ultraviolet light may be of importance.

Silicates The silicates are silicon-oxygen compounds of various metals, chiefly the alkali and alkaline-earth elements, aluminum and iron. About 95 percent of the earth's solid crust consists of silicate minerals, and they are major components of meteorites and lunar rocks as well. The preponderance of silicates accounts for the fact that oxygen and silicon are the two most abundant elements on earth, together constituting three-fourths of the mass of the crust (Table 23-1).

More than 1000 different natural silicates have been characterized, and such synthetic materials as bricks, cement, glass, and ceramics are also silicates. All silicates are based on structures derived from SiO_4 tetrahedra. The simplest, such as the mineral zircon, $ZrSiO_4$, contain the SiO_4^{4-} ion. Much more commonly, however, the SiO_4 tetrahedra are linked together through Si-O-Si bonds to form large anions in one, two, or three dimensions. Three common chain and sheet structures built up from SiO_4 tetrahedra are illustrated in Fig. 24-3, and the formulas of some representative silicate minerals are given in Table 24-5.

The principles discussed in Chap. 7 concerning the formation of chains, two-dimensional networks, and three-dimensional frameworks apply to silicate structures as well. When each SiO_4 tetrahedron is linked to just two others, long chains (Fig. 24-3a) or isolated rings (not illustrated) are formed. The pyroxene minerals, of which diopside (Table 24-5) is typical, contain single chains of composition $(SiO_3)_n^{2n-}$. In the amphibole minerals, pairs of such chains are connected by Si-O-Si cross-links, Fig. 24-3b. Tremolite (Table 24-5), one form of asbestos, is a typical amphibole; some forms of it readily yield fibers that can be matted or woven into materials that are valuable for fireproof thermal insulation.

Layered structures are obtained when each SiO_4 tetrahedron is linked to three neighboring tetrahedra to form interconnected planar rings (Fig. 24-3c). Most minerals of this type also contain hydroxyl ions and Mg^{2+} or Al^{3+} that connect the sheets in rather complicated ways. The layered structures are characteristic of clays, talc (Table 24-5), and micas. In talc, for example, pairs of layers, like those in Fig. 24-3c, are held together by Mg^{2+} ions. The resulting double layers are electrically neutral and interact with each other only by relatively weak van der Waals forces and occasional hydrogen bonds. The layers therefore slip by one another readily, giving talc its soft, somewhat greasy, feel. In a typical mica, such as muscovite, one out of every four SiO_4 tetrahedra has been replaced by an AlO_4 tetrahedron. Because Al is in the ($+$III) oxidation state, whereas the Si it

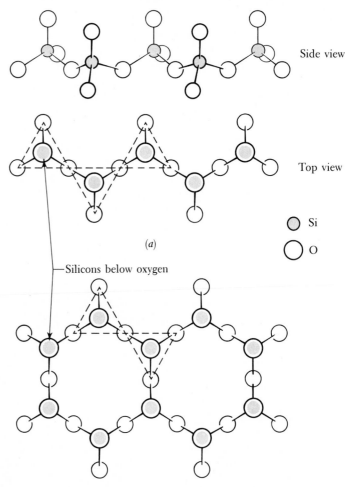

Figure 24-3 Structural Units of Silicates
Each unit is composed of linked SiO_4 tetrahedra. Linear chains (a) are formed when an oxygen is shared by neighboring tetrahedra. In double chains (b), half the silicon atoms share three oxygens with other silicons and half share only two. Layers of planar rings (c) are obtained when each silicon shares three oxygens. The dashed lines outline a few of the SiO_4 tetrahedra.

replaces is in the ($+$IV) state, an extra positive ion, usually K^+ or Na^+, must be incorporated into the structure for each AlO_4 tetrahedron. These extra ions are situated between the double layers, which are negatively charged, and serve to bind the layers together. Micas are consequently harder than talc; however, because of their layer structure, they can be cleaved readily into large sheets.

The extraordinary variety of silicates arises not only because SiO_4 tetrahedra can be linked together in many different arrangements but also because of the ease with which ions of similar size can replace each other in the structures. For example, there is a continuous series of minerals of formula M_2SiO_4 containing $SiO_4{}^{4-}$, in which the metal ion M can be Mg^{2+} or Fe^{2+} or any mixture of the two. Olivine, a common green or yellow mineral in this group, contains about one Fe^{2+} for every ten Mg^{2+}. In some closely related mineral structures, two ions of charge $+1$ in one compound are replaced by one of charge $+2$ in another, or three of charge $+2$ substitute for two of charge $+3$. Because of differences in ion size and placement, substitution of one ion for another usually introduces some strain, so

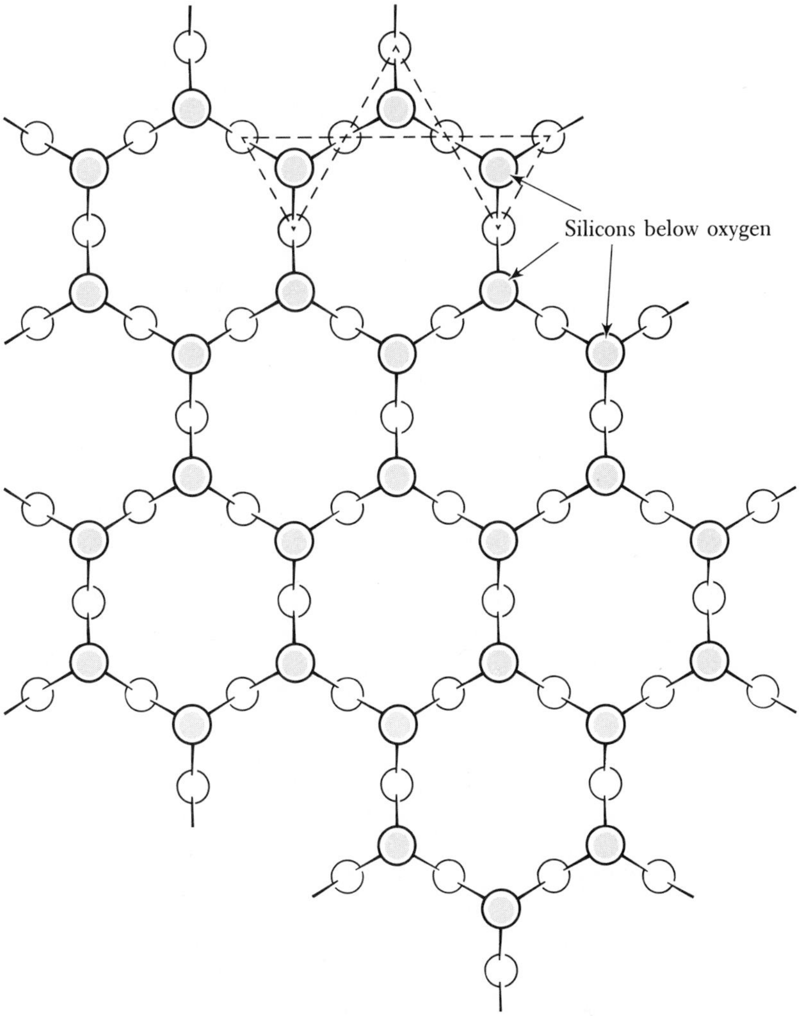

Silicons below oxygen

(c)

Table 24-5
Some Silicate Minerals

Name	Formula	Arrangement of SiO_4 tetrahedra
Zircon	$ZrSiO_4$	Individual $SiO_4{}^{4-}$ groups
Diopside	$CaMg(SiO_3)_2$	Chains of tetrahedra (Fig. 24-3a), $(SiO_3)_n{}^{2n-}$
Tremolite	$Ca_2Mg_5(Si_4O_{11})_2(OH)_2$	Two parallel chains linked together (Fig. 24-3b), $(Si_4O_{11})_n{}^{6n-}$
Talc	$Mg_3Si_4O_{10}(OH)_2$	Double sheets of tetrahedra like those of Fig. 24-3c, $(Si_2O_5)_n{}^{2n-}$, with Mg^{2+} holding pairs of sheets together
Orthoclase	$KAlSi_3O_8$	Three-dimensional framework of linked tetrahedra; one-fourth of SiO_4 tetrahedra replaced by AlO_4 tetrahedra and K^+

a given structural arrangement can tolerate only a limited number of substitutions and remain stable.

The most abundant silicates are the feldspars, a group of substances comprising about 60 percent of the earth's crust. They are aluminosilicates, with the general formula $M(Al,Si)_4O_8$, where M can be Na, K, Ca, or Ba, and the notation (Al,Si) stands for a ratio of Al to Si varying from $1:3$ to $2:2$. Note that the ratio of the number of oxygen atoms to the total of silicon and aluminum atoms is just 2, as it is in SiO_2. The feldspars have three-dimensional framework structures, with SiO_4 and AlO_4 tetrahedra linked by the sharing of all four oxygen atoms on each tetrahedron, just as the SiO_4 tetrahedra are linked in the various forms of SiO_2. Orthoclase (Table 24-5) and anorthite, $CaAl_2Si_2O_8$, are typical feldspars.

Cement and Glass Portland cement is a complex mixture consisting mainly of Ca_2SiO_4, Ca_3SiO_5, $Ca_3Al_2O_6$, and Ca_2AlFeO_5. It is produced by "burning"[1] finely ground limestone and aluminosilicate clay with coal, oil, or gas fuel and crushing the product into a powder. When the cement is mixed with water the $Ca_3Al_2O_6$ hydrolyzes, forming calcium hydroxide and aluminum hydroxide, which react with the silicates to form intermeshed aluminosilicate crystals. Concrete is a mixture of cement, sand, and rock. Although one commonly thinks of cement as drying, the setting process does not in fact involve dehydration. Thus, cement and concrete harden even under water.

The fused silica glass previously described is brittle and difficult to work. Common glass, which has better properties, is a mixture of sodium and calcium silicates. It is produced by melting a mixture of sodium carbonate or sodium sulfate, limestone, and sand. The molten glass is poured into forms or flattened into sheets. Plate glass is ground flat and polished; it is being replaced by float glass, which is produced by floating a layer of molten glass on top of a molten metal, usually tin. A uniform glass surface is thereby achieved and expensive grinding and polishing is unnecessary.

Like the silicate minerals, glasses can be made with an enormous range of compositions. The glass varieties differ in properties such as color, transparency, index of refraction, and thermal expansion. If part of the SiO_2 is replaced by boric oxide, B_2O_3, the resulting borosilicate glass has a very low thermal expansion and is therefore well suited for use in cooking dishes. This glass was originally developed for train signal lamps, which must be capable of withstanding sudden

---

[1] Only the organic impurities in the feed mixture burn. Other processes occurring are the decomposition of the limestone and the reaction of CaO and SiO_2.

temperature changes without breaking. A common trade name for borosilicate glass is Pyrex. Glass containing lead oxide has a high index of refraction that makes it very effective in dispersing white light into its colored components. Such glass is used for so-called crystal[2] glassware and jewelry.

Silicones Although silicon forms no long-chain compounds connected only by Si-Si bonds, polymers can be formed that are based on Si-O-Si linkages. They are called *silicones* and have the general formula

$$-O-\underset{\underset{R}{|}}{\overset{\overset{R}{|}}{Si}}-O-\underset{\underset{R}{|}}{\overset{\overset{R}{|}}{Si}}-O-\underset{\underset{R}{|}}{\overset{\overset{R}{|}}{Si}}-O-$$

where R represents a carbon-containing group such as $-CH_3$ or $-C_6H_5$. Silicone polymers have relatively high thermal and chemical stabilities; the Si-O bond energy (430 kJ mol^{-1}) is more than twice that of the Si-Si bond (190 kJ mol^{-1}) and significantly greater than that of the C-C bond (340 kJ mol^{-1}). Their properties depend on the chain length and the nature of the R groups. Short-chain methylsilicones ($R = -CH_3$) are oils and are used for lubrication and for hydraulic fluids; longer-chain-length methylsilicones find use as greases. Silicone oils have a very low temperature coefficient of viscosity and therefore remain more fluid at low temperatures and more viscous at high temperatures than typical hydrocarbon oils.

Solid, rubberlike materials can be produced when adjacent silicone chains are interconnected by cross-links, as in

$$-O-\underset{\underset{R}{|}}{\overset{\overset{R}{|}}{Si}}-O-\underset{\underset{O}{|}}{\overset{\overset{R}{|}}{Si}}-O-\underset{\underset{R}{|}}{\overset{\overset{R}{|}}{Si}}-O-\underset{\underset{R}{|}}{\overset{\overset{R}{|}}{Si}}-O-\underset{\underset{R}{|}}{\overset{\overset{R}{|}}{Si}}-O-$$

$$-O-\underset{\underset{R}{|}}{\overset{\overset{R}{|}}{Si}}-O-\underset{\underset{R}{|}}{\overset{\overset{R}{|}}{Si}}-O-\underset{\underset{R}{|}}{\overset{\overset{R}{|}}{Si}}-O-\underset{\underset{R}{|}}{\overset{\overset{R}{|}}{Si}}-O-\underset{\underset{R}{|}}{\overset{\overset{R}{|}}{Si}}-O-\underset{\underset{R}{|}}{\overset{\overset{R}{|}}{Si}}-$$

Silicone rubber is used for electrical insulation and as a replacement for ordinary rubber in applications where chemical inertness or stability at high temperatures is required.

24-4 Sulfur, Selenium, and Tellurium

The elements in group VIA exhibit the general trends already noted in group VIIA. Sulfur, selenium, and tellurium show many similarities, and their chemical and physical properties change in a uniform and predictable manner with increasing atomic number. Oxygen, the first-period member of the family, has unique characteristics associated with its high electronegativity and the lack of *d* orbitals in its valence shell. The ($-\text{II}$) oxidation states of oxygen and the other

S
Se
Te

---

[2]The glassware known as crystal is not crystalline; i.e., it has no long-range structural order. Polished cut pieces of the glass sparkle much like crystals of high refractive index, such as diamonds.

members of the group show at least formal similarities; for example, all the analogs of H_2O—H_2S, H_2Se, and H_2Te—are known. Unlike oxygen, however, the other group VIA elements have an extensive chemistry in the (+IV) and (+VI) oxidation states as well. Their physical properties and elemental forms are also distinctly different from those of oxygen.

Occurrence and Physical Properties The name chalcogen (Greek: *khalkos,* copper; *genes,* born) is sometimes applied to the elements of group VIA because they can be obtained from copper ores, such as chalcocite (Cu_2S). Many other metal ores are also sulfides (for example, PbS and ZnS); they often contain traces of metal selenides and tellurides. About 30 percent of the sulfur produced in the United States is derived from these ores when the metal is extracted, and selenium and tellurium are also recovered as by-products. An increasingly important source of sulfur is the hydrogen sulfide component in some natural gas supplies.

Deposits of elemental sulfur also occur. Those in Texas and Louisiana are veins or pockets located in limestone at depths of 150 to 400 m. The sulfur is brought to the surface by the Frasch process, which avoids costly and dangerous mining operations. A hole is bored to the deposit and three concentric pipes are sunk. Water at a temperature of about 160°C (at which its vapor pressure is 6 atm) is pumped down the outer pipe and melts the sulfur, which collects in a pool at the base of the well. Compressed air is forced down the center pipe, mixes with the molten sulfur, and forms a froth that rises to the surface through the middle pipe. The mixture flows into enormous bins where the sulfur solidifies and the water runs off. When full each bin contains up to $\frac{1}{2}$ million tons of 99.5 percent pure sulfur. Molten sulfur is sometimes pumped directly into heated railroad tank cars for shipping to chemical plants.

Allotropy, the existence of an element in more than one form in the same physical state, is much more extensive in sulfur than in oxygen; at least 10 allotropic modifications of solid sulfur have been characterized, and almost 50 have been reported. The most important are the yellow crystalline forms called orthorhombic and monoclinic sulfur, in each of which the structural unit is an S_8 molecule in the form of a puckered ring (Fig. 7-2). They differ in the way the molecules are packed in the crystal.

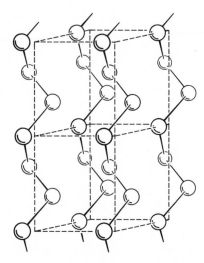

Figure 24-4 The Structure of Metallic Selenium

Each selenium (Se) atom is bonded to two others to form helical chains of indefinite extension. The distance between nearest-neighbor atoms in adjacent chains is significantly greater than that between the bonded atoms in each chain. The dashed lines enclose two "unit cells" of the crystal. The actual crystal is a repetition of such unit cells in all directions. The properties of the crystal in the direction of the chains are markedly different from the properties perpendicular to the chain directions. (See also Fig. 7-2b.)

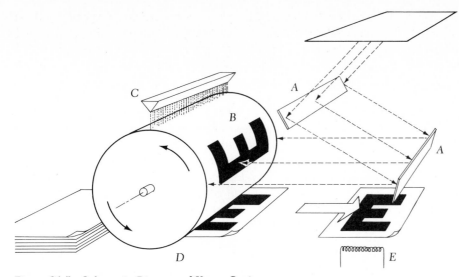

Figure 24-5 Schematic Diagram of Xerox Copier
Optical system (A) focuses image on (B) charged Se-coated drum. Dry ink powder dusted on
drum (C) clings to charged (unilluminated) areas of drum. Ink is transferred to paper (D) and
fused by heating (E).

Orthorhombic sulfur is the form stable at low temperature; above 95.6°C the
monoclinic form, which melts at 119.5°C, is stable. The transitions between the
crystalline forms occur very slowly; hence it is possible to heat orthorhombic
sulfur to its melting point, 112.8°C, before any appreciable conversion to monoclinic sulfur occurs. Among the other solid allotropes of sulfur are a purple form
that consists of S_2 molecules, an unstable form containing S_6 rings (Fig. 7-2), and
a soft rubbery material, known as plastic sulfur, that contains long helical chains
of sulfur atoms.

Liquid sulfur is a complex material. The orthorhombic and monoclinic solids
melt to a mobile yellow liquid that for the most part consists of S_8 rings. When
the liquid is heated above 160°C, it darkens and becomes increasingly viscous. By
about 200°C it has become dark red-brown and is so thick that it cannot be
poured from the container. These transformations are associated with the
breaking of the S_8 rings and the subsequent reaction of the S_8 units to form
long-chain molecules. If the viscous liquid is rapidly cooled in water, the chains
persist and plastic sulfur is obtained. The viscosity of the liquid reaches a
maximum at 200°C; further heating leads to an increase in fluidity because the
chains break into smaller units. Sulfur vapor consists of a mixture of species,
including S_8 and S_2 molecules.

Selenium also can exist in several allotropic modifications. A gray crystalline
form, called metallic selenium because of its luster, contains infinite chains of
selenium atoms (Fig. 24-4). A red form, which is less stable at room temperature,
consists of Se_8 rings. Only one crystalline form of tellurium is known.

Metallic selenium conducts electricity poorly in the dark, but its conductivity
in light is markedly higher and increases in proportion to the light intensity. This
property of *photoconduction* is utilized in selenium photocells, found in some
photographic light meters. It is also the basis of the dry copying process called
xerography (Greek: *xeros,* dry) (Fig. 24-5). In this process a layer of photocon-

ductive material deposited on an electrically conducting backing is electrostatically charged. When an image is focused on the layer, the illuminated areas become conductive and their charge flows away through the backing. Unilluminated areas, which remain charged, are allowed to pick up an ink powder and the powder is then transferred to paper, where it is set by heat.

Sulfides and Hydrogen Sulfide Like oxygen, sulfur forms binary compounds by direct reaction with most of the elements. The alkali-metal and alkaline-earth sulfides (for example, Na_2S and BaS) are ionic and readily soluble in water. No transition-metal sulfide is predominantly ionic and, as one would predict from the relative electronegativities of oxygen and sulfur, the metal sulfides tend to have much less ionic character than the corresponding metal oxides. In some metal sulfides the distances between metal atoms are so short that there is very probably some direct metal-metal bonding. In fact, alloylike behavior—metallic luster, variable composition, and electrical conductivity—is found, among others, in the sulfides of iron, cobalt, and nickel. As a consequence of these differences between oxides and sulfides, the formulas of the sulfides of a metal frequently differ from those of the oxides. Compare, for example, FeO, Fe_2O_3, and Fe_3O_4 with FeS and FeS_2. Even when the formulas are analogous, the oxide and sulfide usually have different structures (for example, SnS_2 and SnO_2).

The insolubility of transition-metal sulfides is utilized to advantage in qualitative and quantitative analysis of metal ions. The hydrolysis of S^{2-},

$$S^{2-} + H_2O \rightleftharpoons HS^- + OH^- \tag{24-39}$$

allows the concentration of sulfide ion in solution to be controlled by adjustment of the pH. Since K_2 for H_2S is about 10^{-15}, the equilibrium constant for (24-39), which is equal to K_w/K_2, has a value of about 10, so S^{2-} is extensively hydrolyzed unless the solution is very alkaline. The sulfide-ion concentration can be varied over a range of at least 10^{22} when the pH is varied between 14 and 0. Thus, in an appropriately buffered solution the concentration of S^{2-} can be set at a level at which a specific insoluble sulfide precipitates, leaving more-soluble sulfides in solution.

Hydrogen sulfide, H_2S, is formed when metal sulfides are treated with dilute nonoxidizing acids such as HCl. Similar reactions of acids with selenides and tellurides produce H_2Se and H_2Te. All three hydrides are colorless and highly poisonous gases with extremely unpleasant odors; their properties are compared with those of H_2O in Table 24-6. Each molecule has an angular structure and, with the exception of water, a bond angle close to 90°. The heats of formation become more positive with increasing size of the chalcogen. The formation of H_2Se and H_2Te from the elements requires *expenditure* of free energy, and these

Table 24-6
Properties of the Hydrides of the Group VIA Elements

Property	H_2O	H_2S	H_2Se	H_2Te
Boiling point/°C	100	-61	-42	-2
$\tilde{H}_f^{\ominus}$/kJ mol^{-1}	-286	-20	77	143
Bond angle/deg	104	92	91	90
Ionization constants, 25°C:				
$\quad K_1$	$1.8 \times 10^{-16*}$	9.1×10^{-8}	1.9×10^{-4}	2.3×10^{-3}
$\quad K_2$	$\sim 10^{-24}$	1.2×10^{-15}	10^{-11}	10^{-11}

*See footnotes concerning H_2O in Table 12-2.

compounds slowly decompose on standing. Acid strength increases in a parallel fashion from H_2O to H_2Te—the weaker the chalcogen-hydrogen bond, the stronger the acid. As noted before (Sec. 6-6), hydrogen bonding accounts for the anomalously high boiling point of H_2O while the steady increase in boiling point from H_2S to H_2Te correlates well with the increase in van der Waals interaction associated with increasing molecular size.

If solutions of sulfides are boiled with excess sulfur, polysulfide ions are formed:

$$S^{2-} + (n-1)S \longrightarrow S_n^{2-} \tag{24-40}$$

Ions containing chains of from two to six sulfur atoms are known. Examples of crystalline polysulfides are BaS_3, BaS_4, Cs_2S_6, and FeS_2 (pyrite, "fool's gold").

Selenides and tellurides closely resemble sulfides but tend to be more covalent. Polyselenide and polytelluride ions have also been characterized.

Oxides and Oxyacids Both the dioxides and the trioxides of sulfur, selenium, and tellurium are known, but only SO_2 and SO_3 are of importance. The dioxides can be prepared by burning the elements in air. Large quantities of SO_2 are produced when sulfides are heated in air or when sulfur-containing fuels are burned.

At room temperature SO_2 is a colorless gas with an unpleasant, irritating odor. SeO_2 is a white solid composed of infinite chains,

and TeO_2 exists in two forms, both of which are white solids with structures similar to those of metallic oxides.

Sulfur dioxide is quite soluble in water and forms acid solutions that contain sulfurous acid, H_2SO_3. Salts containing the SO_3^{2-} ion are called sulfites; those containing HSO_3^- are known as bisulfites or hydrogen sulfites. The alkali-metal sulfites and bisulfites are the most common. Sulfur dioxide is often used as a reducing agent and is also employed commercially as a preservative for prunes and other dried fruits because it destroys fungi and bacteria.

If solutions containing sulfites are boiled with sulfur, thiosulfate ion ($S_2O_3^{2-}$) is formed. The prefix *thio* is applied to species in which sulfur replaces oxygen. The thiosulfate ion can be thought of as derived from the tetrahedral sulfate ion by replacing one of the oxygen atoms by a sulfur atom. Sodium thiosulfate, $Na_2S_2O_3 \cdot 5H_2O$, is commonly known as *hypo* and is used in the photographic process (page 622). After the film has been through the developer, undeveloped grains of silver bromide are removed by forming the soluble silver thiosulfate complex

$$AgBr + 2S_2O_3^{2-} \rightleftharpoons Ag(S_2O_3)_2^{3-} + Br^- \tag{24-41}$$

Thiosulfate is a mild reducing agent and can be oxidized by iodine to tetrathionate ion, $S_4O_6^{2-}$:

$$(24\text{-}42)$$

a reaction analogous to the joining of two SO_4^{2-} tetrahedra to give peroxydisulfate ion, $S_2O_8^{2-}$ (Fig. 3-2). Reaction (24-42) is the basis of a procedure for the quantitative analysis of oxidizing agents. The oxidizer is allowed to react with an excess of iodide ion and the I_2 produced is titrated with a standard solution of thiosulfate, with a starch solution as indicator. Starch gives a characteristic blue color with iodine.

Almost 90 percent of the sulfur produced is converted to SO_2 and used in the production of sulfuric acid, for which it must be oxidized further to SO_3:

$$2SO_2 + O_2 \rightleftharpoons 2SO_3 \qquad (24\text{-}43)$$

This reaction occurs very slowly at low temperature. It is also appreciably exothermic; hence raising the temperature, which increases the reaction rate, decreases the equilibrium yield of product. A viable commercial process is made possible by the use of vanadium pentoxide or platinum catalysts.

The SO_2 is mixed with air and passed over the catalyst at a temperature of 400 to 450°C. The resulting SO_3 is dissolved in sulfuric acid to produce $H_2S_2O_7$, known as oleum or fuming sulfuric acid:

$$SO_3 + H_2SO_4 \longrightarrow H_2S_2O_7 \qquad (24\text{-}44)$$

Oleum is then diluted with water to make sulfuric acid of any desired concentration:

$$H_2S_2O_7 + H_2O \longrightarrow 2H_2SO_4 \qquad (24\text{-}45)$$

This roundabout procedure is used because the direct reaction of SO_3 and water produces a fine fog of acid that is difficult to condense.

It is interesting to note that the quantity (more than 100 million tons) of SO_2 produced each year by the oxidation of sulfur impurities in the coal and petroleum fuels used in power generating stations in the United States is more than sufficient to replace the sulfur required for the production of sulfuric acid. Most of the processes currently used to reduce the atmospheric SO_2 pollution from stack gases, however, do not allow for economic recovery of the sulfur.

At room temperature pure H_2SO_4 is a dense, oily liquid that is highly corrosive and hazardous. It is relatively cheap strong acid and as such is one of the most important industrial chemicals. Sulfuric acid has a great affinity for water, a property that makes it useful as a dehydrating agent and as a solvent for reactions in which water must be released. A striking example of the dehydrating power of H_2SO_4 is its ability to break down sucrose, ordinary sugar, to carbon and water:

$$C_{12}H_{22}O_{11} \xrightarrow{H_2SO_4} 12C + 11H_2O \qquad (24\text{-}46)$$

Dilute sulfuric acid is not a good oxidizing agent, but hot concentrated acid is highly effective. For example, it can dissolve copper and it oxidizes carbon to carbon dioxide:

$$Cu + 2H_2SO_4 \longrightarrow CuSO_4 + 2H_2O + SO_2 \qquad (24\text{-}47)$$
$$C + 2H_2SO_4 \longrightarrow CO_2 + 2H_2O + 2SO_2 \qquad (24\text{-}48)$$

Selenic acid, H_2SeO_4, is a white crystalline solid that is similar in acid strength and molecular structure to H_2SO_4. It is a tetrahedral molecule in which the central selenium atom is surrounded by two oxygen atoms and two hydroxyl groups. The corresponding oxyacid of tellurium, telluric acid, has the formula H_6TeO_6 and is an octahedral molecule, containing six —OH groups. It is a weak diprotic acid with a pK_1 of 7.8.

24-5 Phosphorus and Arsenic

The group VA elements straddle the line between the metals and the nonmetals. Nitrogen and phosphorus are distinctly nonmetallic, arsenic is classed as a metalloid, and antimony and bismuth are primarily metallic. It is not surprising, then, that there are few strong family characteristics. One similarity that can be noted is the structure of the hydrides: all the group VA elements form gaseous hydrides with molecular formulas MH_3. All these hydride molecules are pyramidal; only NH_3 is appreciably basic. Another group characteristic is the existence of oxides with the *empirical* formula M_2O_5, all of which are acidic. We deal here with phosphorus and arsenic.

Occurrence and Physical Properties Phosphorus is derived from the relatively abundant phosphate minerals. A prime source is apatite, a group of minerals with compositions represented by the formula[3] $3Ca_3(PO_4)_2 \cdot CaX_2$, where X can be OH (hydroxyapatite) or can be either F or Cl or both, in any ratio (fluorapatite). The term *phosphate rock* is applied to ores that are rich in calcium phosphate and that may range in composition from pure $Ca_3(PO_4)_2$ to hydroxyapatite. Hydroxyapatite is the main mineral constituent of bones and teeth.

Phosphorus is prepared commercially by heating a mixture of phosphate rock, silica, and coke in an electric furnace. The process is thought to occur in two stages: formation of P_4O_{10} vapor,

$$2Ca_3(PO_4)_2 + 6SiO_2 \longrightarrow P_4O_{10} + 6CaSiO_3 \qquad (24\text{-}49)$$

and its subsequent reduction by carbon,

$$P_4O_{10} + 10C \longrightarrow P_4 + 10CO \qquad (24\text{-}50)$$

The resulting gaseous mixture of P_4 and CO is passed through water, where the phosphorus condenses to a solid.

The soft, waxy, yellowish white crystals obtained in the commercial process contain tetrahedral P_4 molecules (Fig. 7-2) and are known as white phosphorus. This allotropic form of the element does not dissolve in water but is soluble in nonpolar solvents such as benzene and carbon disulfide. It is extremely poisonous and causes very painful and slow-healing burns. At temperatures higher than 40°C, white phosphorus ignites spontaneously in air; it oxidizes slowly at room temperature and gives off a white light, the phenomenon from which phosphorus derives its name (Greek: *phosphoros,* light-bearing).

When white phosphorus is heated it changes to a more stable form, red phosphorus. The rate of conversion is increased by exposure to light or by the addition of small amounts of iodine. Unlike the white form, red phosphorus is

---

[3]Formulas such as this for minerals are used to represent the composition and do not necessarily indicate structural features. Thus, for example, hydroxyapatite contains no distinct $Ca(OH)_2$ units and its formula might better be written as $Ca_5(OH)(PO_4)_3$.

relatively nonpoisonous and unreactive; it will not ignite in air at temperatures below about 240°C and is practically insoluble in most liquids. Another crystalline form, black phosphorus, which is even more stable than the red, can be obtained by subjecting white phosphorus to high pressures. Commercial red phosphorus is amorphous; various crystalline red forms have been reported, but their structures are unknown. Black phosphorus has a layer structure consisting of sheets of puckered hexagons.

Most of the elemental phosphorus produced is oxidized to P_4O_{10} and used to prepare phosphoric acid. White phosphorus was once used for making matches but may no longer legally be used for this purpose because its toxicity led to serious health problems among workers in match factories. The head of a modern safety match is composed of a mixture of antimony sulfide, Sb_2S_3, which serves as a source of sulfur; potassium dichromate, which acts as an oxidizing agent; and glue as a binder. A mixture of red phosphorus, powdered glass, and glue comprises the strip on which the matches must be struck. When the match is rubbed against this strip, heat generated by friction raises the temperature of bits of phosphorus sufficiently that they are oxidized by the dichromate. The heat released by this reaction ignites the antimony sulfide.

Minerals containing arsenic have been known since early times. The bright yellow orpiment, As_2S_3, and the red realgar, AsS, were used by primitive people to decorate faces and bodies and to paint earthenware. Arsenic can be obtained from these ores by roasting to produce As_4O_6 and subsequent reduction by carbon. Most arsenic, however, comes from various iron, nickel, and cobalt arsenic sulfides (for example, $FeAsS$ and $CoAsS$), which contaminate metal ores.

The common form of the element is "metallic" arsenic, a steely-gray lustrous crystal that contains puckered sheets of As atoms and conducts electricity poorly. At 610°C gray arsenic sublimes at a pressure of 1 atm to tetrahedral As_4 molecules. If arsenic vapor is rapidly cooled, yellow arsenic, an unstable crystalline form consisting of As_4 molecules, is obtained. On standing, yellow arsenic soon reverts to the gray form.

Arsenic is a constituent of numerous alloys and its compounds are employed as insecticides and preservatives for hides. Some of the earliest drugs that were effective in combating microbial infections contained organic compounds of arsenic as the active ingredients.

Hydrides and Halides Phosphine, PH_3, is an extremely poisonous gas. It cannot be made by direct combination of the elements but can be produced by the reaction of white phosphorus with warm aqueous solutions of strong bases, which also yields hypophosphite ion, $H_2PO_2^-$:

$$4P + 3OH^- + 3H_2O \longrightarrow 3H_2PO_2^- + PH_3 \qquad (24\text{-}51)$$

Although pure phosphine is not spontaneously flammable in air, the gas produced by reaction (24-51) will often ignite because of the presence of impurities such as P_4 vapor or diphosphine, P_2H_4, which is very reactive.

Phosphine is only slightly soluble in water and does not yield basic solutions—its pK_b is estimated to be 25. When PH_3 is mixed with gaseous HCl, HBr, or HI, phosphonium halides, which contain the PH_4^+ ion, are formed. These salts are much less stable than the corresponding ammonium compounds and they decompose completely in solution, releasing PH_3.

All the trihalides of phosphorus and pentahalides other than PI_5 are known.

With the exception of PF_3 they can be made by direct combination, e.g.,

$$2P + 3Br_2 \longrightarrow 2PBr_3 \tag{24-52}$$

The pentahalides can also be prepared from the trihalides,

$$PX_3 + X_2 \longrightarrow PX_5 \tag{24-53}$$

The trihalides are pyramidal molecules similar in structure to phosphine; in the vapor state the pentahalides have a trigonal bipyramidal structure. Crystals of the pentachloride and pentabromide are ionic: PCl_5 consists of tetrahedral $[PCl_4]^+$ and octahedral $[PCl_6]^-$ ions; PBr_5 consists of tetrahedral $[PBr_4]^+$ and Br^-.

All the halides react with water to give hydrogen halides and oxyacids of phosphorus. Trihalides give phosphorous acid, H_3PO_3,

$$PX_3 + 3H_2O \longrightarrow H_3PO_3 + 3H^+ + 3X^- \tag{24-54}$$

and the pentahalides undergo a two-step reaction giving first the oxyhalide, POX_3, and then phosphoric acid, H_3PO_4:

$$PX_5 + H_2O \longrightarrow POX_3 + 2H^+ + 2X^- \tag{24-55}$$
$$POX_3 + 3H_2O \longrightarrow H_3PO_4 + 3H^+ + 3X^- \tag{24-56}$$

Arsine can be formed by the reaction of zinc in acid solution with any soluble arsenic compound—for example, arsenite ion, AsO_2^-:

$$AsO_2^- + 3Zn + 7H^+ \longrightarrow AsH_3 + 3Zn^{2+} + 2H_2O \tag{24-57}$$

This reaction is the basis of the very sensitive Marsh test for the presence of arsenic. A substance suspected of containing arsenic is treated with zinc and acid; any arsine evolved is decomposed by heat to arsenic and hydrogen. The arsenic is detected by allowing it to condense as a shiny mirror on a cool smooth surface.

Arsenic forms all the trihalides, but only one pentahalide is known, AsF_5. The reactions of the trihalides with water are analogous to (24-54); unlike the phosphorus trihalides, however, the arsenic trihalides do not react completely and the reaction can be easily reversed by the addition of HX.

Oxides and Oxyacids The oxides and oxyacids of phosphorus all have structures involving tetrahedral bonding. When phosphorus is burned in an excess of air, the chief product is phosphorus(V) oxide,[4] P_4O_{10}, shown in Fig. 24-6a. With a limited supply of air, phosphorus(III) oxide, P_4O_6, is obtained (Fig. 24-6b). Both oxides are white crystalline solids. Phosphorus(III) oxide melts at about room temperature, 24°C, whereas the higher oxide sublimes at 1 atm at 258°C as discrete P_4O_{10} molecules.

Reaction of P_4O_6 with water produces phosphorous acid:

$$P_4O_6 + 6H_2O \longrightarrow 4H_3PO_3 \tag{24-58}$$

As noted in the discussion of the strengths of the oxyacids of phosphorus in Sec. 24-1 (Example 24-2 and Exercise 24-2), hydrogen atoms directly bonded to phosphorus are not acidic. Phosphorous acid, which contains two —OH groups,

---

[4]For historical reasons, the empirical formulas P_2O_5 and P_2O_3 are sometimes used for the phosphorus oxides, which are referred to as phosphorus pentoxide and phosphorus trioxide.

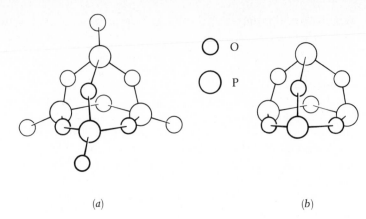

Figure 24-6 Structures of Phosphorus Oxides
(*a*) The phosphorus atoms in P_4O_{10} are situated at the corners of a tetrahedron, and six of the oxygens lie along the edges. The P-O-P bond angles are 124°; the O-P-O angles involving only bridging oxygens are 99°; those also involving the peripheral oxygens are 120°. (*b*) The four peripheral oxygen atoms of P_4O_{10} are missing in P_4O_6, and the P-O-P angle is slightly larger, 128°.

(*a*)

(*b*)

is therefore diprotic.

$$\overset{\displaystyle H}{\underset{\displaystyle O}{HO-P-OH}} \qquad (24\text{-}59)$$

Phosphorus(V) oxide has an extremely high affinity for water and is much used as a drying agent. The reaction

$$P_4O_{10} + 6H_2O \longrightarrow 4H_3PO_4 \qquad (24\text{-}60)$$

gives orthophosphoric acid (frequently called simply phosphoric acid). Phosphoric acid is a poorly oxidizing triprotic acid.

Phosphate, hydrogen phosphate (HPO_4^{2-}), and dihydrogen phosphate ($H_2PO_4^{-}$) salts are known. In general, the dihydrogen phosphate salts are the most soluble. Among the phosphates, only the alkali-metal and ammonium salts are appreciably soluble in water; their solutions are strongly alkaline as a result of hydrolysis and they are widely used as cleaning agents. Fertilizers to supply the phosphate essential for all plant growth are made by treating insoluble phosphate rock with sulfuric acid to produce soluble calcium dihydrogen phosphate:

$$Ca_3(PO_4)_2 + 2H_2SO_4 \longrightarrow 2CaSO_4 + Ca(H_2PO_4)_2 \qquad (24\text{-}61)$$

Sufficient water is added to convert the calcium sulfate to gypsum, $CaSO_4 \cdot 2H_2O$, and the product mixture is marketed as "superphosphate of lime". Pure $Ca(H_2PO_4)_2$ is known as "triple phosphate" fertilizer because it contains three times as much phosphate per calcium as $Ca_3(PO_4)_2$. It is made by the reaction of H_3PO_4 with phosphate rock.

Other oxyacids of phosphorus include pyrophosphoric acid, $H_4P_2O_7$, which is obtained by loss of water from orthophosphoric acid,

$$2H_3PO_4 \longrightarrow H_4P_2O_7 + H_2O \qquad (24\text{-}62)$$

and the metaphosphoric acids, which have the composition $(HPO_3)_n$, with $n = 3$, $4, 5, \ldots$, and are produced when orthophosphoric acid is heated to 325–350°C:

$$nH_3PO_4 \longrightarrow (HPO_3)_n + nH_2O \qquad (24\text{-}63)$$

Organic phosphates play an essential role in the chemistry of life processes

(Chap. 30); consequently, pollution of water by phosphates (e.g., from detergents)
can greatly affect growth of organisms in the water (Sec. 23-4).

The oxides of arsenic are similar in structure to those of phosphorus. When arsenic is burned in air, arsenic(III) oxide, As_4O_6, is obtained. Aqueous solutions of this oxide contain arsenious acid and are weakly acidic ($K = 6 \times 10^{-10}$). This oxide is somewhat amphoteric, with $K_b \approx 10^{-15}$. Arsenic(V) oxide, whose structure is not well established, cannot be prepared by direct reaction of arsenic with oxygen, but it can be made by boiling As_4O_6 with nitric acid. It is very soluble in water and gives solutions of arsenic acid (H_3AsO_4), a moderately strong oxidizing agent, comparable in acid strength to H_3PO_4. Many of the salts of arsenic acid resemble the corresponding phosphates in properties. Arsenates are toxic because arsenate can replace phosphate in some organic phosphates, but the resulting organic arsenates do not behave metabolically like the phosphates.

Summary

The halogens are highly reactive. Fluorine and chlorine are produced electrolytically; bromine and iodine are prepared by oxidation of the corresponding halides with chlorine. There are general trends in properties within the halogens and their compounds; the differences in properties between fluorine and chlorine are larger than those between other adjoining members of the family. Bonding in the halogens usually involves the completion of the octet by the addition of one electron. Halogen molecules are easily reduced to halides, F_2 being the strongest oxidizing agent (most easily reduced) and I_2 the weakest. With the exception of fluorine, the halogens exhibit positive oxidation states in covalent compounds. Examples are the oxyacids, such as HOCl, $HClO_2$, $HClO_3$, and $HClO_4$, in which chlorine has oxidation states I, III, V, and VII, respectively. Many interhalogen compounds exist, such as ICl, BrF_3, and IF_7. The reaction of F_2 with H_2O produces O_2 and HF as well as O_3, whereas the other halogens give HOX and X^-. Halides of alkali metals and alkaline-earth metals are ionic; metals with relatively high ionization energies and small atomic radii form covalent halides. The hydrogen halides HCl, HBr, and HI are strong acids but HF is moderately weak. All oxygen-containing halogen species are strong oxidizers. Solutions of sodium hypochlorite, NaOCl, are used as a household bleach.

Latimer diagrams that pertain to acid solution are based on $E^\ominus$ values at pH = 0 and on reactions written in terms of H^+ rather than OH^-. Diagrams for alkaline solution refer to $E^\ominus$ values at pH = 14 and equations in terms of OH^-. The acid constants of many inorganic oxyacids can be estimated by the use of two simple rules.

Elemental carbon occurs in nature as diamond (very hard) and graphite (soft). Although diamond is the less stable form at low pressure, conversion to graphite does not occur at ordinary conditions. The chief oxides of carbon are the colorless gases CO and CO_2. Carbon monoxide, which is poisonous, reacts directly with Cl_2 to produce phosgene, $COCl_2$. Urea, $(NH_2)_2CO$, a crystalline solid used as a fertilizer, is formed by the direct reaction of CO_2 with NH_3. Other notable inorganic carbon compounds are CS_2 and CCl_4, which are used as solvents, and HCN and its derivatives, the cyanide salts and complexes. Metal carbides are either saltlike or interstitial compounds; silicon and boron form extremely hard covalent carbides.

Elemental silicon is prepared by the reduction of SiO_2 with C. Silicon halides and hydrides are more reactive than their carbon analogs. Silicon is the second most abundant element in the earth's crust. Most rocks in the crust are silicates, compounds built up of SiO_4 tetrahedra that share O atoms. Chains, two-dimensional networks, and three-dimensional frameworks may be present, differing in the proportion of shared oxygens. In aluminosilicates, some of the Si atoms are replaced by Al atoms and an appropriate number of cations to compensate for the difference in oxidation state between Si(IV) and Al(III). Portland cement and glass are materials made from silicates.

Polymers in which carbon-containing groups are

bonded to the silicon atoms in —Si—O—Si— chains are called silicones. The physical properties of silicones depend on the nature of the groups and on the chain length.

The chemistry of the (II) state of sulfur, selenium, and tellurium resembles in some ways that of their congener, oxygen. These elements differ from oxygen, however, by exhibiting the oxidation states (IV) and (VI) as well. Allotropy, the existence of an element in more than one form in the same physical state, is extensive in both S and Se. Most elements form sulfides; the low solubility of transition-metal sulfides is utilized in chemical analysis. With the exception of H_2O, the hydrides of the group VIA elements show regular trends in their chemical and physical properties. Both the dioxides and trioxides of S, Se, and Te are known. Sulfur dioxide forms acid solutions in water that contain sulfurous acid, H_2SO_3. Sulfuric acid, the commonest strong acid, is prepared from SO_3.

Phosphorus and arsenic form similar hydrides and oxides. The commercial production of phosphorus involves the reduction of phosphates with carbon. One allotropic form of phosphorus, white phosphorus, ignites spontaneously in air, but red phosphorus is relatively unreactive. Aqueous solutions of phosphine, PH_3, an extremely poisonous gas, are not basic. Most phosphorus trihalides and pentahalides can be made by direct combination of the elements. The oxides of phosphorus, P_4O_6 and P_4O_{10}, can also be made directly. They react with water to give, respectively, phosphorous acid, H_3PO_3, and phosphoric acid, H_3PO_4, which, like the oxides, have structures involving tetrahedral bonding. Phosphates are constituents of bones and teeth and are essential plant nutrients.

Terms and Concepts

Problems and Questions

24-1 Reactions of Halogens and Halides The following pairs of substances are mixed in aqueous solution: $NaCl + Br_2$; $NaBr + I_2$; $NaI + Cl_2$; $NaI + Br_2$. (*a*) Write balanced equations for any reactions that occur. (*b*) Use $E^{\ominus}$ values given in Table 24-1 to evaluate the equilibrium constant for the first mixture.

24-2 Inorganic Reactions Explain the following observations. (*a*) Chlorine is more soluble in cold NaOH solutions than in pure water at the same temperature. (*b*) Pure HBr(*aq*) can be prepared by distillation from a mixture of KBr and concentrated H_3PO_4 but not from KBr and concentrated H_2SO_4.

24-3 Halide Solutions Dilute aqueous solutions of NaCl are neutral but dilute solutions of NaF are slightly basic. Explain.

24-4 Entropies of Fusion and Vaporization Use the data in Table 24-3 to calculate the entropies of fusion and vaporization of the hydrogen halides. What factors might account for the major trends in the entropies?

24-5 Solutions of Chlorine When gaseous chlorine is bubbled through water at 20°C, 2.26 liter of the gas (measured at STP) dissolve per liter of water. The Cl_2 reacts with H_2O according to (24-6) to give HOCl; the equilibrium constant is $K = 4.7 \times 10^{-4} M^2$. Calculate the concentration of HOCl in a solution saturated with chlorine at 1 atm.

24-6 Acid Constants and Structure Predict the successive dissociation constants of H_3AsO_4 and H_2SO_4. Compare them with the experimental values in Table D-3 and Table 12-2.

24-7 Solubility of Limestone From data in Tables D-2 and D-3 calculate the solubility of limestone in water of a pH that is maintained at 7.00. How much limestone could be carried away in a year by a river with the average annual flow of the Colorado River ($22,000 \times 10^6 \ m^3$), if the pH is 7.00?

24-8 Equilibrium between C, CO, and CO_2 From Table D-4 find the values of $\Delta G^{\ominus}$ and $\Delta H^{\ominus}$ for the reaction $2CO(g) \longrightarrow C(graphite) + CO_2(g)$. (a) Which side would be favored at 298 K and a total pressure of 1 atm if equilibrium were established; which side at very high temperatures? (b) Assume $\Delta H^{\ominus}$ and $\Delta S^{\ominus}$ to be independent of T and estimate the temperature at which the partial pressures of CO and CO_2 are both 0.500 atm.

24-9 Structures of Silicates Using models or sketches, show how SiO_4 tetrahedra can be linked together to form linear, two-dimensional, and three-dimensional structures.

24-10 Titration of a Peroxide Peroxides such as Na_2O_2 and BaO_2 are highly reactive oxidizing agents. For example, they liberate I_2 from solutions of I^-, while dioxides such as PbO_2 and MnO_2 are more inert and do not oxidize I^-. A 0.250-g sample of impure Na_2O_2 was allowed to react with an I^- solution and the resulting I_2 was titrated with a $0.0637 M$ thiosulfate solution [Equation (24-42)]. A volume of 47.8 ml thiosulfate solution was required. What is the purity of the Na_2O_2 sample? (Assume that there are no oxidizing impurities in the Na_2O_2.)

24-11 Thiosulfates Thiosulfates are readily prepared by boiling elemental sulfur with a sulfite solution. The addition of acid to a thiosulfate solution essentially reverses this reaction, causing the formation of S and H_2SO_3. When radioactive elemental sulfur is used to prepare thiosulfate in this manner and the thiosulfate is subsequently decomposed, the H_2SO_3 recovered is not radioactive. Explain.

24-12 Group VA Halides Phosphorus forms both trihalides and pentahalides, but nitrogen forms only trihalides. Discuss a possible explanation for this difference.

24-13 Separation by Sulfide Precipitation A solution is $1.0 \times 10^{-2} M$ in both $ZnSO_4$ and $CdCl_2$. It is desired to precipitate as much of the cadmium as possible, without precipitating the zinc, by bubbling H_2S at 1 atm through the solution at a pH controlled by a buffer. (a) What is the highest pH at which no ZnS would precipitate? (b) What fraction of the Cd would remain unprecipitated at that pH? Data: $[H_2S] = 0.10 M$ in a solution saturated by H_2S at 1 atm. K_{sp} is $1.6 \times 10^{-24} M^2$ for ZnS and $7.9 \times 10^{-27} M^2$ for CdS.

24-14 Phosphorus Acids (a) What is the oxidation state of phosphorus in each of the following acids: H_3PO_4, $H_4P_2O_7$, $(HPO_3)_3$, H_3PO_3 (one H bonded to P), H_3PO_2 (two H bonded to P). (b) Give structural formulas for these acids. (c) Name the acids. Use Appendix B or any other reference source.

24-15 Double and Triple Bonds It is often stated that double or triple bonds involving elements beyond the second row of the periodic table are less stable than those involving second-row elements. Cite at least one specific example that supports or opposes this generalization, including references to the structures of at least two elements or compounds.

24-16 pH of Washing Aids Many washing aids give an alkaline reaction in aqueous solution, the more alkaline the less "mild". Soap is the sodium salt of an organic acid for which $pK = 4.8$. Indicate the order of increasing alkalinity (increasing pH) of solutions of soap, sodium phosphate, sodium carbonate, sodium borate (which may be assumed for this purpose to be a salt of boric acid), and, for purposes of comparison, sodium sulfate and sodium chloride. Relevant acid constants may be found in Tables 12-2 and D-3. Assume the molarity of all solutions to be the same.

24-17 Disproportionation of I_2 When iodine is dissolved in an aqueous solution of a strong base, it is completely converted into I^- and IO_3^-. If the solution is acidified, however, the iodide and iodate are converted completely into iodine again. Explain these facts with the help of any pertinent $E^\ominus$ values from Table D-5.

24-18 Change in Half-Cell Potential with pH From the half-cell potential for the ClO^-/Cl_2 couple at pOH = 0 given in the Latimer diagram, Equation (24-17), 0.40 V, and that for the $HClO/Cl_2$ couple given in Table D-5, 1.63 V, together with K_w, calculate an approximate value for the dissociation constant of hypochlorous acid, HClO. The value given in Table 12-2 is 1.1×10^{-8}.

Metals and the Metallic State

"Although those who burn, roast and calcine the ore, take from it something which is mixed or combined with the metals; and those who crush it with stamps take away much; and those who wash, screen and sort it, take away still more; yet they cannot remove all which conceals the metal from the eye and renders it crude and unformed. Wherefore smelting is necessary, for by this means earths, solidified juices, and stones are separated from the metals so that the metals obtain their proper color and become pure, and may be of great use to mankind in many ways."

GEORGIUS AGRICOLA, "DE RE METALLICA", 1556

25-1 Characteristic Properties and Methods of Production

This chapter and the two that follow are concerned with the chemistry of metals. A general discussion of metals and the metallic state serves as an introduction. We begin with a consideration of the properties of metals and methods by which metals can be produced from their compounds. Typical atomic arrangements in metals and certain kinds of imperfections found in most metallic crystals are discussed in the second section. The final section is concerned with the electronic structure of metals. Chapter 26 deals with the chemistry of some important metallic elements and their compounds, and Chap. 27 treats complexes of transition metals.

Some Physical Properties The characteristic properties of metals (Sec. 7-4) include metallic luster, high thermal conductivity, and high electrical conductivity that decreases with increasing temperature. Pure metals normally have a silvery appearance, except for copper, which is red-brown, and gold, which is yellow; finely divided metals often appear black. It is not easy to distinguish between elemental metals and alloys by appearance or by superficial tests. Alloys that contain small amounts of nonmetals, such as carbon, silicon, or phosphorus, exhibit typical metallic properties.

Only three metals are liquid at or near room temperature: mercury melts at $-39°C$, cesium at $28°C$, and gallium at $30°C$. The highest melting points, boiling points, and densities are found near the middle of the periodic table, chiefly among the transition metals. "Heavy" metals are much denser than other common substances, representative values being, in grams per cubic centimeter, 7.9 (iron), 11.3 (lead), 19.3 (tungsten), and 21.4 (platinum); the most dense metal at 1 atm is osmium (22.5). In contrast, most common rocks have densities in the range 2 to 3.

Most metals can be deformed readily without breaking; they can be drawn into wires and hammered or rolled into sheets. A few, however, such as chromium and bismuth, are brittle or have brittle modifications, but they have a metallic appearance and are good conductors of electricity and heat.

Metals are widely used for structural purposes. Other important applications, particularly of copper, silver, and gold, are based on their exceptional electrical conductivity. Many uses are tied to more specific properties. For example, antimony and some of its alloys are used for casting type because they expand on solidification and therefore permit exact replication of the fine details of the mold. Tungsten, because of its very high melting point ($\sim 3400\,^\circ$C), is particularly suited for use as light-bulb filaments and electrodes in spark plugs. The high-temperature reaction vessels in chemical laboratories are frequently made from the relatively unreactive metal platinum.

Some Chemical Properties Metals are virtually insoluble in inorganic or organic solvents, unless there is a chemical reaction. They are soluble, however, in the melts of other metals. Trace quantities of mercury dissolve in water; alkali metals dissolve in liquid ammonia and, in trace amounts, in water. The dissolved alkali-metal atoms react almost immediately with the water, however, and the metal atoms eventually react with the ammonia (Sec. 26-1). Metals are monatomic in the vapor state and when dissolved in the melts of other metals (as can be shown, e.g., by the freezing-point depression).

The chemical reactivities of metals vary tremendously. Gold and the platinum metals (Ru, Rh, Pd, Os, Ir, and Pt) are chemically quite inert, whereas the alkali metals are extremely reactive. The most general manifestation of this reactivity is the tendency to form positive ions. This tendency increases toward the left and downward in the periodic table. It is strongest for cesium.

Production Methods The natural sources of metals are ores, minerals from which a metallic constituent can be extracted. Some ores contain metals in the elemental state—for example, gold, platinum, silver, and copper. However, most metals occur naturally in positive oxidation states, often as oxides, sulfides, or binary compounds with other group VI elements, or as hydroxides or carbonates. The metal is then obtained from the ore by chemical or electrolytic reduction.

Metals in the zero-oxidation or *native* state can be separated from rocky contaminants by hand or other simple means. If the ore contains only a relatively small amount of the metal (a low-grade ore), chemical extraction methods are often used. Gold is sometimes extracted from ores with mercury. The mercury-gold solution, called a gold *amalgam*, is separated from the ore and the mercury is then distilled away. Alternatively, the ore may be treated with a cyanide solution in the presence of air. A soluble gold-cyanide complex is formed:

$$4\text{Au} + 8\text{CN}^- + \text{O}_2 + 2\text{H}_2\text{O} \longrightarrow 4\text{Au(CN)}_2{}^- + 4\text{OH}^- \qquad (25\text{-}1)$$

Although gold is not normally oxidized by oxygen, the stability of the cyanide complex helps to drive this reaction to the right. Similar methods are used with silver ores, including both those containing native silver and those containing its compounds, such as the mineral argentite, Ag_2S.

The least electropositive metals can sometimes be obtained from sulfide or oxide ores by heating the ore in air; for example:

$$\text{HgS} + \text{O}_2 \longrightarrow \text{Hg} + \text{SO}_2 \qquad (25\text{-}2)$$
$$\text{Cu}_2\text{S} + \text{O}_2 \longrightarrow 2\text{Cu} + \text{SO}_2 \qquad (25\text{-}3)$$
$$2\text{HgO} \longrightarrow 2\text{Hg} + \text{O}_2 \qquad (25\text{-}4)$$

Most other sulfide ores are converted into oxides when heated in air. The result-

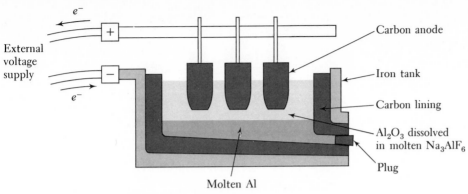

Figure 25-1 Electrolytic Production of Aluminum

In the Hall process, a melt of Al_2O_3 in Na_3AlF_6 is electrolyzed at a temperature near 1000°C. Sufficient heat to keep the electrolyte liquid is provided by the passage of current. Liquid aluminum is produced at the cathodic carbon lining by the reaction $Al^{3+} + 3e^- \rightarrow Al$. It collects at the bottom and may be drawn off. The carbon anodes are consumed by the anode reaction $C + 2O^{2-} \rightarrow CO_2 + 4e^-$.

ing oxides, as well as naturally occurring oxide ores, hydroxide ores, and carbonate ores, are reduced to the metal in various ways. One common method, used in the production of iron, nickel, copper, zinc, cadmium, and tin, is to heat the oxide with carbon. The carbon may reduce the oxide directly, with the formation of metal and either CO or CO_2, or some of the carbon may react first with oxygen to produce CO, which is itself an effective reducing agent. Since CO is gaseous, it can make much better contact with the solid oxide than can solid carbon. The reduction of iron oxide by CO in the blast furnace is illustrative (Fig. 26-4, Sec. 26-4).

Some oxides are reduced with aluminum or magnesium instead of with carbon, in part because of a tendency of certain metals, particularly transition metals, to react with carbon to form carbides.

Highly electropositive metals, such as the alkali and alkaline-earth elements and aluminum, are usually produced by the electrolysis of melts. Sodium is obtained in this way from NaOH or NaCl, calcium from $CaCl_2$ (with some CaF_2), and aluminum from Al_2O_3 dissolved in molten cryolite, Na_3AlF_6 (Fig. 25-1).

Some ores are too low grade to be directly useful, but methods have been developed for concentrating the desired material. The most common of these is *flotation,* used especially with sulfide ores, which are wetted by oil but not by water. For example, 95 percent of the copper may be recovered from ores that contain only about 2 percent of Cu_2S. The finely crushed ore is first ground with a mixture of oil and water; then a foaming agent is added and air is bubbled through the mixture. The sulfide particles are concentrated, with the oil, in the foam that rises to the top, while the remainder of the ore, chiefly silicates, settles to the bottom.

Metals as first obtained from their ores must often be purified further for particular applications. Many transition metals form volatile molecules on reaction with carbon monoxide, and these can be used to obtain very pure samples of the metal. In the Mond process for refining nickel, CO is passed over impure nickel; nickel carbonyl, $Ni(CO)_4$, a volatile and highly toxic liquid, is formed. It is purified by distillation and then decomposed into pure Ni and CO by heating. Iron can be similarly purified through the formation of the volatile

567

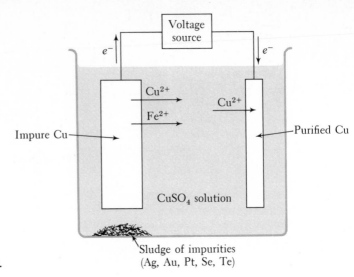

Figure 25-2 Electrolytic Refining of Copper.

Fe(CO)$_5$. Electrolytic methods are also used in the refining of some impure metals, notably copper, as described next.

Electrolytic Refining of Copper The impure copper, formed into slabs directly from the crude metal as it is obtained from its ores, is made the anode in an electrolytic cell and a sheet of pure copper is used as the cathode (Fig. 25-2). The electrolyte is a dilute solution of a salt of Cu^{2+}. As electricity passes through the cell, copper is oxidized to Cu^{2+} at the impure copper anode and Cu^{2+} is reduced to metallic copper at the pure copper cathode, which is usually coated with a thin layer of graphite so that the fresh copper deposit can be easily removed. Metals more noble than copper—that is, metals that are harder to oxidize, such as silver, gold, and platinum—will not be oxidized to their cations in the presence of Cu. If a cation of any such noble metal were formed, or were introduced, it would immediately be reduced again by the copper, with formation of Cu^{2+} and the corresponding metal. Hence such metals fall to the bottom of the cell as the anode dissolves and they can be recovered from the sludge that accumulates there. Metals that are more easily oxidized than copper do indeed dissolve as the anode dissolves, but they are not plated out at the cathode. No metal more easily oxidized than copper can be plated from a solution containing a significant concentration of Cu^{2+}; if it were formed, it would at once reduce Cu^{2+} to Cu and itself be oxidized to its cation. Electrolytically refined copper is typically about 99.95 percent pure.

25-2 Atomic Arrangements in Metallic Crystals

Close-Packed Structures The atoms in crystals of about two-thirds of the metallic elements are packed as closely as possible in regular three-dimensional arrays. To describe this three-dimensional close packing of atoms, which we idealize as spheres, we begin with the close-packed layer illustrated in Fig. 25-3. Layers of this type are stacked together so that each sphere in a given layer is in contact with three spheres of the layer below and three spheres of the layer

Layer

Figure 25-3   Closest Packing of Equal Spheres in a Plane

Each sphere (atom) is in contact with six neighboring spheres in a hexagonal pattern and is surrounded by six interstices (spaces), three marked by crosses (x) and three by circles (o). Additional spheres can be placed on top of this layer by nesting them in the niches just above the interstices. An entire layer of spheres can be placed on top of the interstices marked by crosses, or those marked by circles, because the distances between the interstices in either set are exactly those between the centers of the original spheres. A new layer can also be nested against the original one from below. To describe the nesting of more than two layers (Figs. 25-4 and 25-5), it is convenient to designate the original layer and all layers whose spheres lie directly above or below those of the original layer as type *A*, all layers whose spheres lie above or below interstices marked with circles as type *B*, and those with spheres above or below crosses as type *C*.

above. The spheres fit into niches of the adjoining layers. An ambiguity regarding the stacking of any *three* layers arises because the spheres of the outside layers may be fitted, from above and below, into the *same* or into *different* indentations of the layer in the middle (Fig. 25-4). A periodic repetition of the first mode of stacking is called *hexagonal close packing;* a periodic repetition of the second mode is termed *cubic close packing.*

In terms of the layer types described in Fig. 25-3, hexagonal close packing corresponds to the sequence . . . *ABABAB* . . . and cubic close packing to the sequence . . . *ABCABC* . . . (Fig. 25-5). The repeat in the first sequence is two

Figure 25-4   Hexagonal and Cubic Stacking of Layers

In hexagonal close packing, the spheres of the first and the third layers are seated above and below the *same* interstices of the second layer; in cubic close packing, *different* interstices are utilized. Side views of the two ways of stacking are shown in Fig. 25-5. Cubic close packing corresponds to a face-centered cubic arrangement of atoms, although this is not apparent in these views.

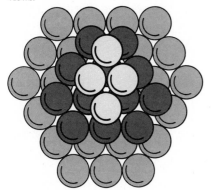

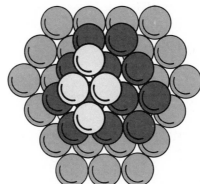

Hexagonal   Cubic

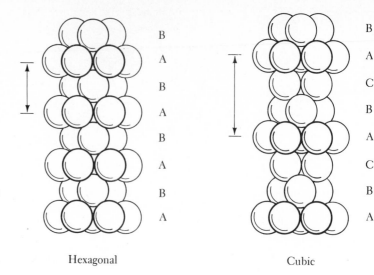

B
A
B
A
B
A
B
A

Hexagonal

B
A
C
B
A
C
B
A

Cubic

Figure 25-5   Side Views of Hexagonal and Cubic Close Packing

The layers of hexagonal and cubic close packing shown are periodic repetitions of two layers A and B or of three layers A, B, and C, respectively. The double-headed arrows indicate the repeat distances.

layers; in the second it is three layers. The descriptions ... $BCBCBC$ ... and ... $ACACAC$ ... (for hexagonal) and ... $ACBACBACB$ ... (for cubic close packing) are equivalent. Stacking with other layer sequences (such as ... $ACABACABACAB$ ... with a four-layer repeat) and random stacking (... $ABCACBCABACB$ ...) are also possible, but the cubic and hexagonal modes are the most important ones.

The Body-Centered Cubic Structure and Other Structures   In another simple structure exhibited by metallic elements, atoms are located at the corners and the center of a cubic unit cell (Fig. 25-6); the structure is called *body-centered cubic*. Atoms are packed less efficiently in this structure than in cubic close packing. Some 70 percent of all metallic elements have cubic or hexagonal close-packed structures,[1] about 25 percent are body-centered cubic, and 5 percent

---

[1] Crystals of the noble-gas elements also contain atoms packed in hexagonal and cubic close-packed arrays.

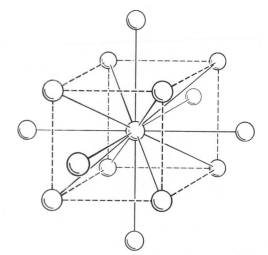

Figure 25-6   Body-Centered Cubic Structure

The atom at the center of the unit cube (shown in dashed outline) has eight nearest neighbors at the corners of the cube. They are at a distance $d_1 = a\sqrt{3}/2$, where a is the cube edge. It has six next-nearest neighbors at the centers of adjoining unit cubes, forming the corners of an octahedron. They are at the distance $d_2 = a = 2d_1/\sqrt{3} = 1.15d_1$, 15 percent larger than d_1. The environment of all atoms is the same.

Table 25-1
The Crystal Structures of
Some Metals and Metalloids

Li bcc; ccp; hcp	Be hcp; bcc	B comp.											C graphite; diam.
Na bcc; hcp	Mg hcp	Al ccp											Si diam.
K bcc	Ca ccp; bcc	Sc hcp; bcc	Ti hcp; bcc	V bcc	Cr bcc	Mn dist. ccp; comp.	Fe bcc;	Co ccp; hcp	Ni ccp	Cu ccp	Zn hcp	Ga comp.	Ge diam.
Rb bcc	Sr ccp; hcp; bcc	Y hcp; bcc	Zr hcp; bcc	Nb bcc	Mo bcc	Tc hcp	Ru hcp	Rh ccp	Pd ccp	Ag ccp	Cd hcp	In dist. ccp	Sn diam.; comp.
Cs bcc	Ba bcc	La ccp; bcc	Hf hcp; bcc	Ta bcc	W bcc	Re bcc	Os hcp	Ir ccp	Pt ccp	Au ccp	Hg dist. ccp	Tl hcp; bcc	Pb ccp

ccp = cubic close packing; hcp = hexagonal close packing; bcc = body-centered cubic; diam. = diamond structure; dist. = distorted; comp. = complex structure. When several forms are listed, the first is the stable form at room temperature and 1 atm.

have structures that are more complex (Table 25-1). Many alloys also have complex structures. Except for Mg and Al, all metallic elements on the left side of the periodic system have body-centered cubic structures near their melting points.

The structures of tin and its congeners illustrate the relationship between structure and metallic character. The common form of tin, called white tin, is metallic. Each atom has six nearest neighbors in a deformed octahedral arrangement. A nonmetallic form of tin, called gray tin, is stable below 13°C. It has the diamond structure (Fig. 7-6), each atom being surrounded by four others in a regular tetrahedral arrangement at an average distance about 10 percent smaller than in the metallic phase. The metallic form is about 27 percent denser than gray tin because of the greater number of near neighbors, even though each neighbor is somewhat further away. The conversion from white to gray tin is slow at 13°C, but it becomes faster at lower temperatures. In very cold climates, objects made of tin suffer from the so-called tin pest or tin disease—they begin to disintegrate into a powder of the gray form. This structural change has many dramatic practical consequences, for example, the gradual disintegration of tin in organ pipes in unheated churches.

Silicon and germanium, the elements just above tin in the periodic table, are metalloids. They have the diamond structure under normal conditions, but each can be transformed into a metallic form with the white tin structure at pressures of a few thousand atmospheres. It may be possible to transform diamond into a metallic form also, but it has been estimated that a pressure of some 2 million atm would be needed.

Alloys As discussed briefly in Sec. 7-4, alloys are metallic substances that are intimate mixtures of two or more elements, at least one of which is a metal. The mixture may be homogeneous, or there may be an intermingling of crystallites with different compositions and atomic arrangements. In other words, a solid alloy may consist of one phase or it may be heterogeneous and contain several solid phases. Many alloys have desirable properties that differ substantially from those of the parent elements. For example, addition of tin to copper, both of which are soft, produces hard and tough bronzes. Small amounts of carbon change soft, pure iron to steel.

For many alloy phases the crystal structure is that of the more abundant parent element, with an appropriate number of original atoms replaced by atoms of the admixed element. Other alloys have different and unique structures. For example, in alpha brass (a Cu alloy with less than about 35 percent Zn) some of the atoms in the copper structure are replaced by Zn atoms, while eta brass (consisting of more than about 95 percent Zn) is based on the structure of zinc, with some of the Zn atoms replaced by Cu atoms. There are three other intermediate structures among the brasses, each associated with a specific composition: $CuZn$ (beta brass), Cu_5Zn_8 (gamma brass), and $CuZn_3$ (epsilon brass). In brasses that contain somewhat more Cu than corresponds to one of these formulas, Zn atoms in the appropriate structure are replaced by Cu atoms. When Zn is in excess the opposite substitution takes place.

Interstitial Compounds Even in the close-packed structures of Fig. 25-4, about 26 percent of the space is unfilled, corresponding to the interstices (spaces) between the spheres. Atoms such as H, B, C, N, and even O are small enough to fit into the interstices between the atoms in crystals of some transition metals without causing considerable distortion. Such substances are called interstitial compounds.[2] The small atom does not simply fill a hole in the structure; there is strong chemical bonding between metal and nonmetal atoms. This bonding is evidenced, for example, by the fact that the compounds are harder and higher melting than the pure metal. Interstitial compounds are often nonstoichiometric, as might be expected from the foregoing description.

The bonding in interstitial compounds is of the electron-deficient kind; i.e., there are more bonds than there are pairs of bonding electrons. It is similar to the bonding in the boranes (Sec. 23-1) and in metals (Sec. 25-3). Interstitial compounds usually have metallic properties, such as good electrical conductivity and silvery sheen, but they are often brittle.

One unusual interstitial compound is palladium hydride. When there is less than one hydrogen atom for every two palladium atoms, which corresponds to less than 0.5 percent hydrogen by weight, the material is a good electrical conductor. With more hydrogen, conductivity is poor. Gaseous H_2 can diffuse through a hot membrane of palladium. The rate of diffusion is proportional to $(P_{H_2})^{1/2}$, so that the diffusing species is atomic hydrogen. Since nothing else diffuses so rapidly through palladium, hydrogen can be purified by allowing it to diffuse through palladium.

Some substances that are referred to as interstitial compounds have a structure different from that of the metal to which they are related, a structure that cannot be explained simply in terms of filling interstices with small atoms. An example is cementite, Fe_3C, grains of which are in part responsible for the toughness of some steels.

Dislocations The regularity of the sequence of atoms in crystals is usually not perfect. Among the different kinds of imperfections are dislocations, one type of which is shown schematically in Fig. 25-7. In the neighborhood of such an edge

---

[2] Not all transition-metal compounds of the nonmetallic elements mentioned are of the interstitial type, particularly not all oxides.

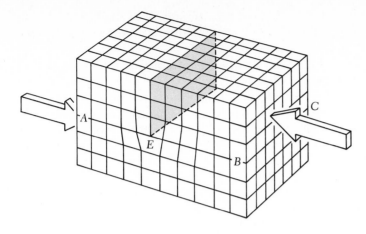

Figure 25-7 An Edge Dislocation
An edge dislocation can be described as the presence of an extra plane of atoms (indicated by the shading) in part of the crystal. The edge of this plane (just above E) is called the dislocation line. There is maximum distortion of the structure near this line—compression above it and expansion below.

dislocation, the bonding between the atoms above and below the plane through A, B, and C is weakened because of the distortion in the structure—compression above the plane and expansion below. As a result, when a stress is applied to the crystal in the manner symbolized by the arrows, the atoms below the ABC plane may slip, as a unit, to the right with respect to the atoms above this plane. In effect the dislocation moves to the left under stress.[3] Dislocations therefore facilitate slippage and allow a metal to be deformed. Another type of dislocation, the screw dislocation (Fig. 25-8), is important for crystal growth as well as for mechanical properties. For a screw dislocation, slippage occurs easily in a direction parallel to the dislocation axis.

There are other more complex sorts of dislocations. Edge and screw dislocations may interact to form complicated patterns. Moving dislocations eventually pile up at regions where impurity atoms are concentrated, or at the boundaries of crystal grains. Such pile-ups often mark the locations and the beginnings of

---

[3]Charles Kittel gives an analogy for the motion of an edge dislocation: "The motion of an edge dislocation through a crystal is analogous to the passage of a wrinkle across a rug: the wrinkle moves more easily than the whole rug, but passage of the wrinkle across the rug does amount to sliding the rug on the floor."

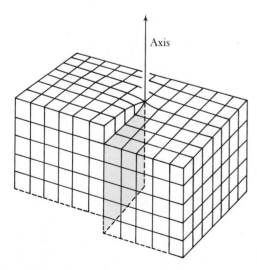

Figure 25-8 A Screw Dislocation
In a screw dislocation a single layer of atoms winds like a spiral staircase around a line, the dislocation axis. There is maximum strain in the immediate environment of the axis.

mechanical failure. This is true for crystals of nonmetals as well as of metals.

Repeated deformation of a metal, e.g., by hammering or bending, is called *cold working* and tends to produce dislocations. While a few dislocations promote slippage of crystal planes, a high concentration of dislocations hinders motion of planes because the dislocations disrupt the structural regularity. The result is a mechanical strengthening and toughening of the metal called *work hardening*. The majority of the dislocations can be removed and the metal can be made soft again by annealing, that is, by heating to about half the melting temperature. For example, an annealed piece of copper tubing is easy to bend, but bending becomes progressively more difficult as it is continued. The original softness can be restored by heating the copper red hot.

Crystals may also be strengthened by the addition of a small concentration of impurity atoms that inhibit the movement of dislocations. This is often why alloys have better mechanical properties than the component metals, which may be soft in the pure state. For example, pure iron is relatively soft, whereas mild carbon steels (containing less than about 1 mol percent carbon) are strong, in part because carbon atoms distort the structure and reduce the mobility of dislocations.

Whiskers Calculations show that metals should be hundreds of times stronger than they are ordinarily observed to be. The reason is the presence of dislocations: the calculations are made for perfect crystals. It has been found that theoretical strength is approached by *whiskers,* extremely thin threads of metals that sprout from metal surfaces under certain conditions. The whiskers are single crystals and are virtually free of dislocations, excepting the screw dislocations responsible for their growth. Iron whiskers with a strength more than 400 times that of ordinary iron have been made.

25-3 Electronic Structure of Metals

There are two distinct ways of describing the electronic structure of metals: (1) the resonance picture, which is most effective in describing the chemical bonding that holds the atoms of a metal together; (2) the molecular orbital picture, which accounts in a natural way for the electrical properties of metals as well as of electrical insulators and of materials called semiconductors. The two theories are complementary and together provide a deeper insight into the metallic state than does either alone.

The Resonance Approach Pauling has described the bonding in metals as a superposition of or a resonance among many formulas that involve the sharing and transfer of electrons. Consider the alkali metals, where there is one valence electron per atom so that on the average only one bond can be formed for every two atoms. The two-dimensional resonance structures shown in Fig. 25-9 are indicative of the possible three-dimensional resonance structures that can be drawn to represent the bonding. The bonding in other metals can be represented by similar formulas, but the average number of electrons involved in bonding per atom, the so-called metallic valence, will be different. An important requirement for formulas in which some of the atoms show extra bonds—such as formulas (*c*) through (*f*) in Fig. 25-9—is the availability of one low-energy unoccupied orbital

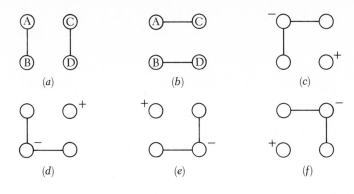

(a)   *(b)*   *(c)*

(d)   *(e)*   *(f)*

Figure 25-9 Resonance Formulas for Four Alkali-Metal Atoms

In formulas (*c*), (*d*), (*e*), and (*f*) of this schematic two-dimensional representation, the positively charged atom has lost its valence electron while the atom with two covalent bonds has gained an extra electron. The bonding between the four atoms can be described as a superposition of the formulas shown.

per atom, enabling the atom to accept the extra bonds. Such extra orbitals are indeed generally present for metal atoms, as Pauling has shown.

The resonance description of metallic bonding is consistent with the fact that most metals can readily be deformed without breaking. New bonds are easily formed during the deformation, when atoms torn from their original neighbors find themselves moved close to new atoms. In a qualitative manner this picture can, for example, explain the relative brittleness of transition metals, lanthanides, and actinides as being caused by the high directionality of the *d* and *f* orbitals involved in the bonding. In contrast, alkali metals and metals on the right side of the periodic system are soft because the bonding involves nondirectional *s* orbitals (or *p* orbitals, which are less highly directed than *d* and *f* orbitals).

Metallic Valence and Properties of Metals   As a starting point for the discussion of metallic valence, we consider the "efficiency" of the packing of atoms in metals by examining densities. If the packing in all metals were equally efficient, the density of a metal would be proportional to its atomic weight. Variations from metal to metal in the ratio of density to atomic weight therefore yield information about packing efficiency. Figure 25-10 shows this ratio, scaled by a constant factor and called the *adjusted density,* for the metals in the first long period. Considering the many different crystal structures represented by these metals

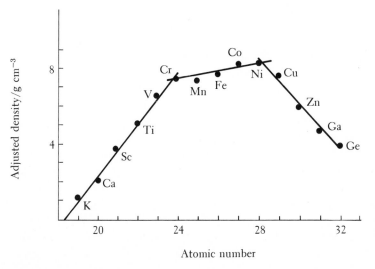

Figure 25-10 Adjusted Densities of Some Metals

The adjusted densities of the metals of the first long period are shown. They have been calculated by dividing the actual density by the atomic weight and multiplying by 54, the average atomic weight of the metals shown. The adjusted densities reflect the efficiency of the packing of the atoms.

(Table 25-1) the regularities apparent in the figure are remarkable. For the first six metals on the left the adjusted density increases with the atomic number, indicating more efficient packing, which implies an increase in the strength of interatomic bonds, regardless of whether the improved packing is achieved by a decrease in interatomic distances, by an increase in the number of nearest neighbors, or by both. The adjusted density stays relatively constant for the elements chromium through copper and decreases again for still higher atomic numbers, paralleled presumably by a similar behavior of the bond strength. Other properties of the metals shown, such as their melting points, hardness, and tensile strengths, vary similarly.

According to Pauling the variation in these properties is an indication of the metallic valence. In potassium each atom has one valence electron available for bonding, while in calcium there are two, in scandium three, and so on, until chromium, with six valence electrons, has been reached. It is this rise in metallic valence that accounts for the increase in the adjusted density, the melting point, and other properties in the sequence of metals discussed. From chromium through nickel there are no substantial changes in these properties, which is interpreted by Pauling to mean that the metallic valence remains at about 6 even though the number of electrons beyond the filled argon shell rises to 10. Beyond nickel the metallic valence decreases again, in steps of one, accompanied by a decrease in the adjusted density, in hardness, and in other properties. The metallic valences in the sequence discussed are thus, at least approximately, the following:

K	Ca	Sc	Ti	V	Cr	Mn	Fe	Co	Ni	Cu	Zn	Ga	Ge
1	2	3	4	5	6	6	6	6	6	5	4	3	2

The congeners of the metals shown exhibit similar metallic valences.

The leveling off and subsequent fall in the metallic valence is related to the fact that it is difficult to form more than six good bonding hybrids with s, p, and d orbitals. This also explains why there are at most six ligands in complexes formed by transition metals in the first long period. As an illustration of the reduced metallic valence for elements beyond nickel, consider zinc. It has 12 valence electrons. There are nine orbitals in its valence shell (one $4s$, five $3d$, and three $4p$). One of the orbitals must be reserved to allow resonance formulas such as (c) to (f) in Fig. 25-9, a feature that is postulated to be essential in metallic bonding. Four of the eight remaining orbitals must then contain pairs of electrons, and there are four unpaired electrons available for bonding. The result is a metallic valence of 4.

The transition metals in the higher long periods and the lanthanides and actinides have f orbitals available in addition to s, p, and d orbitals, and more than six good bonding orbitals can now be formed. Complexes of such metals do indeed sometimes have more than six ligands, and metallic valences greater than 6 are also found.

The Molecular Orbital Approach Another useful model pictures metals as composed of electrically neutral aggregates of positive ions held in relatively fixed positions and highly mobile electrons. The electrons are free to roam within the confines of the crystal and are thus able, for example, to act as carriers of electric

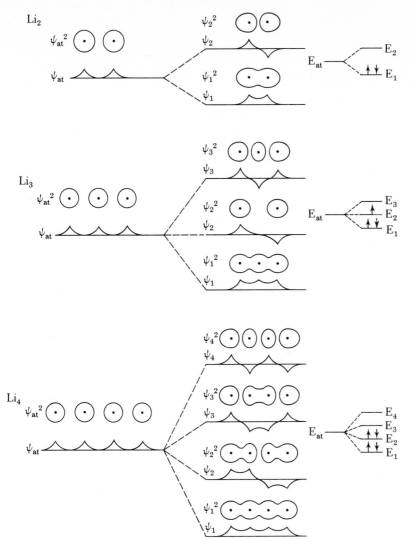

Figure 25-11 Schematic Representation of Orbitals and Energies for Aggregates of Lithium Atoms

On the left are schematic side views of the atomic wave functions for two, three, and four separate lithium atoms (ψ_{at}), as well as top views of the corresponding electron distributions (ψ_{at}^2). No details about the shapes of these functions are implied by the diagrams. To the right of the diagrams of individual atoms are schematic views of the molecular orbitals (ψ_i) obtained from the atomic orbitals; also shown are the resulting electron distributions (ψ_i^2), again viewed from above. On the far right are the atomic and molecular energy states, the lowest states being filled with the valence electrons of the atoms. The number of molecular states is equal to the number of contributing atomic orbitals; in each of the three cases shown, the energy increases with the number of nodes in the corresponding molecular orbital.

current. This is the *electron gas* or *electron-in-a-box* model, which was largely conceived and developed by Sommerfeld in the late twenties. Refinement of this approach by Bloch and others shows that the electrons can assume energy values in certain ranges only, a result that can be understood from molecular orbital theory.

Consider a sequence of molecules, such as Li_2, Li_3, Li_4, ..., Li_N, where N is a very large number, so that Li_N represents a metallic crystal. In each of these molecules the atomic $2s$ orbitals of the individual atoms may be combined into molecular orbitals and the resulting energy levels worked out. Figure 25-11 shows the situation for short chains containing two, three, and four lithium atoms.

Two features of the figure remain correct when the number of atoms is increased to macroscopic proportions and when three-dimensional rather than linear aggregates of atoms are considered. The number of states remains equal to the original number of atomic orbitals involved, and the spacing of the new states becomes closer as the number of atoms increases (Fig. 25-12). For macroscopic

Individual atomic states   Energy bands in crystal

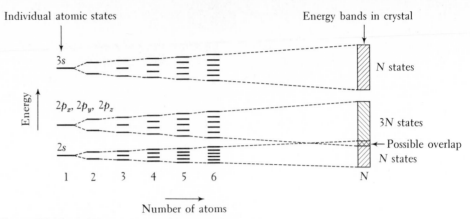

Figure 25-12 Schematic Energy States in Aggregates of Atoms
Each of the interacting states of the N original atoms gives rise to a band of N states with energies that are, for all practical purposes, spaced continuously.

crystals the spacing is so dense that the levels virtually coalesce into continuous *energy bands* of permitted states, each band containing as many states as there are contributing atomic orbitals.

Bands may result from any atomic states that interact to form molecular orbitals (i.e., from overlap of suitable orbitals, such as s, p, d, or even hybridized orbitals of the separated atoms). There may or may not be overlap between different bands, depending on the energy separation of the original atomic states and the degree of interaction, which controls the widths of the bands.

Metals, Semiconductors, and Insulators The details of the band structure of the energy states available to the valence electrons of a solid determine whether the solid has metallic properties. Figure 25-13 illustrates in an idealized way the energy level structure for three different types of electrical behavior. In all three cases the electrons fill the lowest states possible, except for a small fraction of thermally excited electrons that may be in higher states. In the absence of an external electric field the distribution of electron velocities in the three types of solids is such that for every electron moving in a specified direction there is always an electron moving in the opposite direction. Hence, no current flows. Application of an electric field causes preferential flow in one direction, *provided that levels are available to accommodate the increased energy of these electrons.* This is how the three cases differ.

The good electrical conductivity of metals is explained by the availability of easily accessible states in the incompletely filled valence band, the band produced by the interactions of valence orbitals. Each state in an energy band may, by Pauli's principle, hold two electrons with antiparallel spins. Thus, for example, the alkali metals, with one valence electron, have half-filled valence bands (one state for each atom and two electrons per state). For the alkaline earths, the N states of the lowest band would be exactly filled by the $2N$ valence electrons available (two per atom) were it not for the fact that the band produced by the interaction of the p orbitals overlaps the band originating from the s orbitals. The decrease in electrical conductivity when the temperature is raised can be related to the increased thermal motion of the atoms and the increase of the average interatomic distances, which diminish the overlaps between the orbitals and hence reduce the conductivity.

The availability of low-energy unfilled states also favors good thermal conductivity. Electrons that are readily mobile absorb energy from vibrating atoms at sites of vigorous atomic motion, transport it to places where the atoms are moving less vigorously, and there transfer the energy to these atoms, thereby increasing their motion.

In *semiconductors* the electrons in the valence band are unable to act as carriers because the band is completely filled. However, a few electrons are thermally excited to the next higher band, which is otherwise unoccupied. These electrons can act as carriers of current because close-lying energy states are available to them, and the band they are in is therefore called the *conduction band*. Additional conduction is provided by the vacancies left in the valence band by the thermally promoted electrons. These vacancies are called *holes* and behave like mobile positive charges, because an electron in a nearby filled state may move to fill one hole, thereby creating another hole elsewhere. The electrical conductivity of semiconductors is much poorer than that of metals because many fewer carriers participate. It increases with temperature, in contrast to the conductivity of metals, as a result of the promotion of additional electrons to the conduction band.

In *insulators* (also called *dielectrics*) the gap between the valence band and the conduction band is so large that valence electrons cannot be boosted to the higher states by thermal energy. These substances are therefore not conductors.

The difference between semiconductors and insulators is one of degree. At elevated temperatures an insulator may become a semiconductor; at very low temperatures a semiconductor ceases to conduct and becomes an insulator. Thermal conductivity is notably poor in both semiconductors and insulators because of the absence of freely movable electrons.

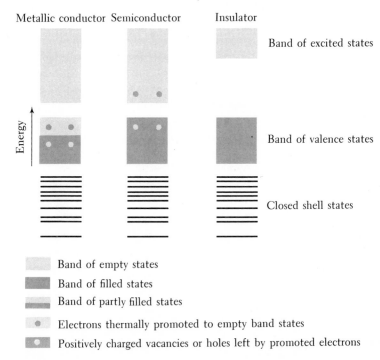

Figure 25-13 Schematic Energy States for Solids of Different Electrical Behavior
The low-lying closed-shell states of the individual atoms are relatively unaffected by the presence of other atoms and thus remain separated, whereas there is orbital overlap and band formation for the higher states. The band containing the valence electrons is only partly filled in metals, but it is completely filled in semiconductors and insulators. In semiconductors there is a gap of the order of a few kT between the valence band and the empty band lying above it; in insulators the gap is large compared to kT. Electrons that have been thermally promoted from one band to another are shown schematically, together with the positively charged vacancies that are created by this promotion.

579

When light strikes some semiconductors, such as selenium (Sec. 24-4) or heat-treated zinc sulfide, electrons may be promoted to the conduction band with a resulting increase in the electrical conductivity. This effect is called *photo-conduction*. The inverse effect also exists, in which electrons are promoted to excited states by an electric field and return to the valence band with the emission of light. This is basically the mechanism of operation of the many *light-emitting diodes* (LED) now used in instrument panels and for numerical display in small calculators.

Summary

Most metals are prepared by the reduction of their oxide ores, or of oxides obtained by the heating of sulfide ores in air. Common reduction methods are heating with carbon, high-temperature reaction with aluminum, and electrolysis of melts. A few of the least reactive metals are found naturally in the metallic or "native" state. Low-grade ores are often concentrated first by flotation.

The atomic arrangement in crystals of about 70 percent of the metallic elements is that of cubic or hexagonal close packing. For most others it is body-centered cubic, while a few metals have more complex structures. Most solid alloy phases have complex structures, and some atoms in these structures can often be replaced more or less randomly by atoms of a different kind. In interstitial compounds, small atoms, including H, B, C, N, and sometimes O, may be situated in interstices between larger metal atoms and may participate in the bonding. Other structural imperfections of crystals include localized distortions, such as edge dislocations and screw dislocations. The mechanical properties of crystals are affected by dislocations. Metallic whiskers owe their great strength to the virtual absence of dislocations.

One approach to understanding bonding in metals is to consider a superposition of many resonance formulas that involve the sharing as well as the transfer of electrons. An important requirement for such structures is the availability of one empty low-energy orbital per atom that permits the formation of new bonds. The considerable deformation that metals can undergo without breaking can be understood by the ease with which new bonds can be formed. Correlations exist between the number of electrons that participate in the bonding in metals and physical properties such as density, melting point, hardness, and tensile strength.

Bonding in metals can also be described in terms of molecular orbitals that extend over the entire crystal. Energy levels associated with the orbitals coalesce into bands of states. The details of the band structure of a crystal and the degree to which the states in these bands are populated by electrons determine whether the crystal has metallic properties. If there are vacant states in the valence band, the band formed by the interaction of the valence orbitals on each atom, the crystal is metallic. The high electrical and thermal conductivities of metals reflect the virtually free mobility of electrons in the unfilled valence band. An insulator has a completely filled valence band, and there is no close-lying vacant band to which electrons can be promoted by absorption of thermal energy. In semiconductors, on the other hand, there is such a band. Electrons excited into this band (the conduction band) can act as carriers of current. Vacancies left in the valence band contribute to the conduction by acting like mobile positive charges. The conductivity of a semiconductor increases with temperature because additional electrons are promoted to the conduction band.

Terms and Concepts

Problems and Questions

25-1 Reduction of Metal Oxides by Carbon
Consider the reduction by carbon of (a) Fe_2O_3,
(b) Fe_3O_4, and (c) MnO_2. Find the values of ΔH,
ΔG, $T\Delta S$, and ΔS for the reactions at 298 K of
these oxides with C (graphite) to produce CO and
to produce CO_2. (d) Comment on the effect on ΔS
that one of the products is a gas. (e) Give rough
estimates of the temperatures above which the
reductions considered might be feasible. (f) In
what way would your considerations be modified
by the fact that at temperatures actually used for
reductions the metals are liquid while the metal
oxides and carbon are mainly solid?

25-2 Discovery of Metals In the biblical book
Numbers (chapter 31, verse 22), the priest Eleazar
classifies the following as substances that can be
purified by fire: "the gold, and the silver, the brass,
the iron, the tin, and the lead." (The word *brass* is
more properly translated as bronze, a copper-tin
alloy.) What characteristics of these six metals
(Au, Ag, Cu, Sn, Fe, Pb) might have led to their
discovery and use in such early times? Lithium,
sodium, and potassium were not recognized as elements until the early years of the nineteenth century; rubidium and cesium were first isolated in
1860 and 1861. To what properties might the relative lateness of the discovery of the alkali metals be
attributed?

25-3 Stabilization of Au(I) (a) Use the Latimer
diagram relating the (0), (I), and (III) oxidation
states of gold

$$Au^{3+} \xrightarrow{1.5} Au^+ \xrightarrow{1.7} Au$$

to explain why Au(I) is not found in aqueous solution. (b) The half-reaction

$$Au(CN)_2^- + e^- \rightleftharpoons Au + 2CN^-$$

has an electrode potential $\mathbf{E}^{\ominus} = -0.60$ V. By combining this reaction with the reduction

$$Au^+ + e^- \rightleftharpoons Au$$

show quantitatively how the formation of the
$Au(CN)_2^-$ complex stabilizes Au(I) in solution.

25-4 Aluminum Production The overall reaction for the electrolytic production of aluminum in
the Hall process can be written

$$Al_2O_3 + \tfrac{3}{2}C \longrightarrow 2Al + \tfrac{3}{2}CO_2$$

for which $\Delta G^{\ominus} = 650$ kJ at 970°C. (a) What is the
minimum voltage required to produce aluminum
electrolytically by the Hall process at 970°C?
(b) In the actual process the voltage applied is 4.5
times the ideal value. Calculate the energy required to produce 500 g of aluminum from Al_2O_3
by the Hall process (recall that the energy expended in the operation of an electrochemical cell
is $Q\,\Delta\Phi$, where Q is the charge that is moved and
$\Delta\Phi$ is the potential difference). For comparison, a
typical automobile requires the input of about
10^4 kJ of energy per mile.

25-5 Refining of Copper In the electrolytic
method of refining copper (Fig. 25-2), the impure
copper is made the anode of an electrolytic cell and
the purified metal deposits on the cathode. This

works because some impurities in the anode never dissolve while others never get plated out again. Which among the following metals belong in the first category and may be recovered from the sludge that accumulates at the bottom of the cell? Which belong in the second category? Possible impurities: Pt, Pb, Fe, Ag, Zn, Au, Ni.

25-6 Impurity Atoms A cubic crystal contains a mole fraction of 1×10^{-5} impurity atoms. Estimate the average distance between impurity atoms in terms of d', the average nearest-neighbor distance characteristic of the crystal. (Note that each cube of 100,000 atoms contains on the average one impurity atom.)

25-7 Common Alloys Look up (e.g., in a good encyclopedia or a chemical handbook) and report the compositions of the following alloys: brass, phosphor bronze, type metal, plumber's solder, pewter, white gold.

25-8 Mechanical Properties Discuss structural interpretations of the characteristic ductility and malleability of many metals. Why are alloys frequently harder than pure metals?

25-9 Metallic Valence Consider the electrons available for bonding, and predict the relative melting points and the relative hardnesses of the metals rubidium, strontium, and yttrium.

25-10 Group I Metals Why are group IA metals, such as potassium and rubidium, much softer than group IB metals, such as copper and silver?

25-11 Purification of Copper The electrolytic purification of copper is described in Fig. 25-2. Assume that among the impurities are $Ag(s)$ and $Fe(s)$ and that $[Cu^{2+}] = 1.00\ M$ in the electrolyte. (a) What is the equilibrium ratio of the concentrations of Fe^{3+} and Fe^{2+} during the electrolysis? (b) What is the maximum Ag^+ concentration that can be achieved by the dissolving of $Ag(s)$ under the conditions given? (c) What is the theoretical value of $[Fe^{2+}]$ required for $Fe(s)$ to be plated out? Use the emf values given in Appendix D.

25-12 Copper Production A strip mine in Utah produces 1.8×10^8 kg of copper each year. The ore, which contains chalcopyrite, $CuFeS_2$, is only about 0.75 percent copper by weight. (a) If the density of the ore is 3 g cm^{-3}, what volume of ore is removed from the mine each year? If this volume were in the shape of a cube, what would be the length of one side? (b) What weight of SO_2 is produced when one year's output is converted to crude copper by roasting? (Assume $CuFeS_2$ is the only source of sulfur.) (c) What is the minimum energy required to refine the copper by electrolysis [see Prob. 25-4(b)] if the voltage used is 0.25 V?

25-13 Metallic Valence The densities in grams per cubic centimeter of the metallic elements in the second long period of the periodic table are (in order of increasing atomic number): 1.5, 2.5, 4.4, 6.5, 8.6, 10.2, 11.5, 12.4, 12.0, 10.5, 8.7, 7.3, 5.8. (a) Prepare a graph similar to Fig. 25-10 in which you plot density divided by atomic weight, scaled by multiplying by 100, against atomic number. (b) Discuss the general variation in metallic valence within this period. Suggest reasons for the trends.

25-14 Electrical Conductivity Give definitions of the terms *insulator, semiconductor,* and *metallic conductor.* Describe how the band structure of these materials influences their electrical conductivities.

25-15 Nonrenewable Resources It has been estimated that at the current rate of use world reserves of the following metals will essentially be depleted within 40 years: gold, mercury, silver, tin, zinc, lead, copper, and tungsten. Consider what direct and indirect effects the lack of availability of these metals would have on your life. What steps might be taken to conserve these resources? What materials might function as substitutes for them?

Metallic Elements and Their Compounds

"Went to the Royal Institution to see Davy. . . . Pepys went with me. He showed us his new experiments on the decomposition of potash [K_2CO_3] and soda [Na_2CO_3]. From the oxygen, or zinc end of a combination of troughs, pure potash was decomposed . . . and a new substance produced in little globules, which has the properties of a metal. . . . The globule explodes and ignites in contact with water. . . . Pepys and I concluded we would have cheerfully walked fifty miles to see the experiment. Here is another grand discovery in chemistry."

FROM THE DIARY OF WILLIAM ALLEN, NOVEMBER 16, 1807

26

26-1 The Alkali Metals

The highly reactive alkali metals (column IA) can be prepared by electrolysis of their molten hydroxides or chlorides. They are comparatively soft and low-melting (Table 26-1) and their densities are low: lithium, sodium, and potassium are less dense than water. The electrical and thermal conductivities of the alkali metals are high; liquid sodium is used as a coolant in nuclear reactors, and the feasibility of using sodium for underground power transmission, sheathed in suitable cables, has been considered.

Li
Na
K
Rb
Cs
Fr

The chemistry of the alkali metals is dominated by their having a single outer electron and a low ionization energy. Lithium is the best reducing agent in an aqueous medium (Table 26-1), even though cesium has a lower ionization energy (Fig. 16-2). The ionization energy refers to a change in state involving gaseous atoms and ions, not a change from the metal to a hydrated ion, and the hydration energy of the small lithium ion is sufficiently more negative than that of Cs^+ that it more than makes up the difference in ionization energy (Prob. 26-1).

The alkali metals are highly electropositive and most of their compounds are ionic, although covalent bonds can be formed, as in methyl sodium, $NaCH_3$. The metals are usually stored in an inert medium, such as kerosene, to protect them from reaction with oxygen, water vapor, and carbon dioxide in air.[1] These reactions would lead, respectively, to the formation of peroxides, hydroxides, and bicarbonates or carbonates:

$$2Na + O_2 \longrightarrow Na_2O_2 \tag{26-1}$$
$$2Na + 2H_2O \longrightarrow 2NaOH + H_2 \tag{26-2}$$
$$NaOH + CO_2 \longrightarrow NaHCO_3 \tag{26-3}$$

Reaction of the alkali metals with liquid water can be extremely hazardous. Even with a tiny piece of metal, the heat generated may be sufficient to ignite the

---

[1] A fresh lithium surface will darken rapidly in nitrogen as a result of the formation of lithium nitride, Li_3N. The other alkali metals do not react directly with N_2.

Table 26-1
Some Properties of Alkali
Metals

	Li	Na	K	Rb	Cs	Fr
Density/g cm^{-3}	0.53	0.97	0.86	1.53	1.87	
Melting point/°C	179	98	64	39	28	27
Boiling point/°C	1317	883	758	700	670	
E°/V for $M^{+}(aq) + e^{-} \longrightarrow M$	-3.03	-2.70	-2.92	-2.99	-2.95	

hydrogen produced; under some circumstances the hydrogen-air mixtures that are formed explode when ignited. Metallic sodium is sometimes used as a reducing agent in preparative chemistry.

The alkali metals dissolve in liquid ammonia (normal boiling point, $-33°C$) to form solutions that are very good conductors of electricity and that appear to contain solvated electrons, that is, free electrons associated loosely with one or more ammonia molecules. The solutions are blue when dilute and copper-colored at higher concentrations of the alkali metal; the more concentrated solutions have high conductivity and metallic luster and resemble liquid metals. The solutions decompose slowly, with the evolution of hydrogen and the formation of (dissolved) amides (for example, $Na^{+}NH_{2}^{-}$).

The alkali metals and their compounds impart characteristic colors to flames and have characteristic line spectra (Fig. 14-10). The sodium-vapor lamps commonly used for highway illumination have a yellow-orange color that is produced by the same electronic transition in sodium atoms responsible for the color in flames.

Almost all salts of the alkali metals are water-soluble. The most important is sodium chloride. Seawater now contains about 3 wt percent of $NaCl$, a figure that has gradually increased over the eons as rivers have carried soluble salts into the ocean basins. The importance of sodium chloride in animal diets is doubtless related to the fact that much of early evolution took place in the oceans.

Two of the most widely used alkaline compounds, in both industry and the laboratory, are $NaOH$ and $Na_{2}CO_{3}$. Sodium hydroxide is prepared by electrolysis of an aqueous solution of $NaCl$ with a mercury cathode (Fig. 26-1). The sodium that is generated dissolves in the cathode, forming an amalgam. Reaction of the dissolved sodium with water yields $NaOH$ and hydrogen; chlorine is liberated at the anode.

Sodium carbonate is used in the manufacture of soaps, laundering agents, and

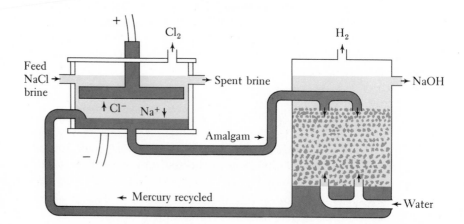

Figure 26-1 An Electrolysis Cell for the Production of Chlorine and Sodium Hydroxide.

glass. Much of it is made by the *Solvay process,* in which a solution of NaCl is saturated with NH_3 and CO_2. The carbon dioxide reacts with the weakly alkaline solution to form the bicarbonate ion, HCO_3^-,

$$NH_3 + H_2O + CO_2 \longrightarrow NH_4^+ + HCO_3^- \qquad (26\text{-}4)$$

and the relatively insoluble salt $NaHCO_3$ (baking soda) precipitates. The sodium carbonate is formed by heating the dried bicarbonate,

$$2NaHCO_3 \longrightarrow Na_2CO_3 + H_2O + CO_2 \qquad (26\text{-}5)$$

and the CO_2 produced in this step is recycled to form additional bicarbonate ion. Sometimes, calcium hydroxide, $Ca(OH)_2$, obtained from the $CaCO_3$ that is the original source of the CO_2 (Sec. 26-2), is added to the remaining solution so that the ammonia can also be recovered and recycled. Calcium chloride is produced as a by-product:

$$Ca(OH)_2 + 2NH_4Cl \longrightarrow CaCl_2 + 2NH_3 + 2H_2O \qquad (26\text{-}6)$$

Sodium peroxide, the principal product formed by the reaction of sodium with (dry) air, is the active ingredient of some bleaching powders. Its effectiveness is a result of a reaction with water in which oxygen is produced:

$$2O_2^{2-} + 2H_2O \longrightarrow 4OH^- + O_2 \qquad (26\text{-}7)$$

When potassium, rubidium, and cesium react with air, the chief product is the corresponding superoxide, containing the ion O_2^-. Both potassium superoxide (KO_2) and Na_2O_2 have been used in space capsules to regenerate oxygen from water vapor by reactions analogous to (26-7) and from reaction with CO_2 as well:

$$4KO_2 + 2CO_2 \longrightarrow 2K_2CO_3 + 3O_2 \qquad (26\text{-}8)$$

The properties of potassium compounds are generally quite similar to those of the corresponding sodium compounds, but they are more expensive and much less widely used. Potassium nitrate is found in some desert regions, where extensive deposits of it were left by evaporation of earlier seas. It is used in fertilizers and as an oxidizing agent in gunpowder.

Potassium and sodium ions have essential roles in some biological processes, including the transmission of signals by nerves. An important part of a nerve cell is a slender extension, the axon, along which the nerve impulses are propagated from the main body of the cell to the synapse, which is the junction of the axon with another nerve cell or a muscle cell. At the synapse the impulse is transmitted by release of a molecule called acetylcholine; however, transmission along the axon involves changes in the concentrations of Na^+ and of K^+ on the inside and the outside of the axon. In the resting state of a nerve, the potassium-ion concentration is higher inside an axon than outside; conversely, the sodium-ion concentration is higher outside than inside. The concentration differences across the axon membrane produce a potential difference, the inside of the axon being 60 to 95 mV more negative than the outside. These concentration differences are maintained by a mechanism referred to as the *sodium pump,* which is believed to involve large chelating molecules that carry Na^+ to one side and K^+ to the other side of the cell membrane enclosing an axon. During transmission of a nerve impulse, temporary local breakdowns of this mechanism occur at a given location, starting at the main body of the nerve cell and moving progressively toward the synapse at a speed of the order of 100 m s^{-1}. As a consequence of ion movement, the potential difference across the membrane is altered, the inside of the axon becoming about 50 mV more *positive* than the outside. It is this electrical phenomenon that corresponds to the nerve impulse.

The principal use of rubididum and cesium is in the manufacture of photoelectric cells that respond efficiently to light in the visible region of the spectrum.

26-2 The Alkaline-Earth Metals

Be
Mg
Ca
Sr
Ba
Ra

The elements of column IIA are prepared chiefly by electrolysis of molten halides. They are less reactive than the alkali metals, having greater ionization energy, and also are denser, harder, higher melting, and higher boiling (Table 26-2). There are two electrons in the valence shell, and even though the energies required to remove these two electrons are large, the hydration energies of the $+2$ ions are so great that these ions are characteristic in aqueous solution (and in crystals). For example, the sum of the first and second ionization energies of Ca is 1735 kJ mol^{-1}, but the hydration energy of Ca^{2+} is around $-1561 \text{ kJ mol}^{-1}$, so that the net energy required to form $Ca^{2+}(aq)$ from a gaseous Ca atom is only 10 percent of that needed to form $Ca^{2+}(g)$. The metals and their compounds give characteristic colors to flames, a property made use of in the manufacture of fireworks.

The alkaline-earth elements react with water to form hydroxides, the reaction at room temperature being slow for beryllium and magnesium and rapid for the others. They all form stable oxides, and calcium, strontium, and barium form ionic peroxides.

The chemistry of beryllium is quite different from that of the other alkaline earths. Beryllium is less electropositive and has a greater tendency than its congeners to form covalent bonds, e.g., in $BeCl_2$. Some of its compounds contain three-center bonds, as in the polymeric solid $(BeH_2)_n$, which can be represented schematically as

$$\text{Be} \diagup_{H}^{H} \text{Be} \diagup_{H}^{H} \text{Be} \diagup_{H}^{H} \text{Be} \diagup_{H}^{H} \text{Be} \diagup_{H}^{H} \qquad (26\text{-}9)$$

Beryllium hydroxide is amphoteric whereas the hydroxides of the other alkaline earths are strong bases, although their solubility is limited.

Magnesium is widely used in low-density structural alloys. It is a valuable reducing agent in the production of other strongly electropositive metals and is used to make the important Grignard reagent in organic chemistry (Chap. 29). Because it burns with a brilliant and very hot flame, the metal is used, in powdered or ribbon form, in incendiary devices and has been used in photoflash

Table 26-2
Some Properties of Alkaline-Earth Metals

	Be	Mg	Ca	Sr	Ba	Ra
Density/g cm^{-3}	1.85	1.74	1.55	2.54	3.5	5.0
Melting point/°C	1280	650	850	770	725	700
Boiling point/°C	2970	1107	1487	1384	1140	
E°/V for $M^{2+}(aq) + 2e^- \longrightarrow M$	-1.85	-2.37	-2.87	-2.89	-2.90	-2.92

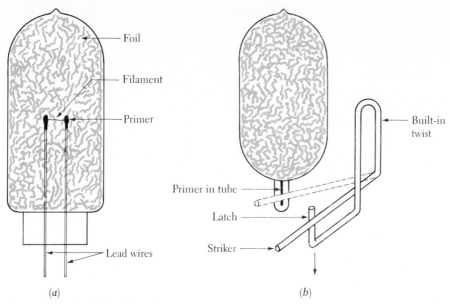

Foil

Filament

Primer

Lead wires

(a)

Built-in twist

Primer in tube

Latch

Striker

(b)

Figure 26-2   Two Photoflash Lamps
Both bulbs contain oxygen and finely shredded Zr foil (formerly Mg and Al foil). Also present is a "primer" that is exploded and scatters burning particles throughout the bulb, thus igniting the foil evenly. The primer is exploded electrically in lamp (a) and mechanically in lamp (b). The mechanical arrangement gets around the possibility of battery failure.

bulbs, although zirconium foil is now much more common (Fig. 26-2). The magnesium ion has a number of vital biochemical roles. It is present in chlorophyll, the green pigment in plants that initiates photosynthesis, the process by which plants convert carbon dioxide and water into carbohydrates and oxygen. Magnesium ion is also involved in the activation of smooth muscle and in the reactions by which energy is provided for many biological processes, reactions involving the breakdown of the molecule adenosine triphosphate (ATP) (Chap. 30).

Calcium is widespread in nature, chiefly as $CaCO_3$, which is essentially the sole component of limestone and marble and is the principal ingredient of chalk, of coral reefs, and of many shells. When calcium carbonate is roasted at high temperature, it decomposes to carbon dioxide and calcium oxide (quicklime) (Sec. 13-4, Figs. 13-2 and 13-3):

$$CaCO_3 \longrightarrow CaO + CO_2 \qquad (26\text{-}10)$$

Calcium hydroxide, *slaked lime,* is the least expensive strong base, so it is widely used in industry. It is produced from the oxide by the addition of water. Slaking of lime is a rather dramatic procedure, since part of the water added is turned to steam by the high temperature generated by the reaction. Common mortar is a 3:1 mixture of sand and slaked lime, which hardens by absorption of CO_2 from air and evaporation of water:

$$Ca(OH)_2 + CO_2 \longrightarrow CaCO_3 + H_2O \qquad (26\text{-}11)$$

587

Calcium ion participates in many biochemical reactions, including those that lead to muscle contraction and to the clotting of blood. Calcium phosphate is the major component of teeth and bones.

The chemical similarity between strontium and calcium makes the radioactive ^{90}Sr from the fallout of nuclear weapons tests in the atmosphere especially dangerous. The ^{90}Sr replaces calcium in bones, where the radiation it produces can damage the red blood corpuscles being formed in the bone marrow. Barium sulfate is highly insoluble and is used for quantitative analysis of barium and of sulfate. Because the high atomic number of barium makes it relatively opaque to X rays, suspensions of barium sulfate are widely used in medical X-ray diagnosis to make features of the intestinal tract more visible. Soluble barium compounds are poisonous, so that the insolubility of the sulfate is an important factor in this technique. Barium sulfate is used as a whitening agent for pigments and leather and also to add weight to drilling mud to help maintain pressure at the bottom of holes drilled for oil.

26-3 Some Other Representative Metals: Groups IIIA, IVA, and VA

B	C	N
Al	Si	P
Ga	Ge	As
In	Sn	Sb
Tl	Pb	Bi

Aluminum and Its Congeners Aluminum is the most abundant metallic element in the earth's crust, occurring widely in aluminosilicate minerals. The other metals of column IIIA, gallium, indium, and thallium, are less common and less important. All four of these metals may be prepared by electrolysis of suitable melts, as described for aluminum in Fig. 25-1, and all but aluminum may also be obtained by electrolysis of aqueous solutions. Aluminum and gallium are comparatively hard; indium and thallium are softer. Gallium and indium are unusual in being liquid over a range of about 2000 K, the widest ranges for any known substances; gallium melts just above room temperature.

Aluminum is the most reactive of these metals and would react rapidly with the oxygen or moisture in air if it were not protected by a thin, transparent, and strongly adherent layer of aluminum oxide. All four elements form ions of charge $+3$; thallium also forms stable compounds containing the thallous ion, Tl^+, many of whose salts resemble the corresponding salts of the alkali metals. The salts of Tl(I) are, however, extremely poisonous.

Aluminum has low density for a metal, 2.7 g cm^{-3}, and high tensile strength. It is easily malleable, can be rolled into thin foils, and is an excellent electrical conductor. Its conductivity is only a little more than half those of copper, silver, and gold for wires of equal diameter, but because the density of aluminum is less than one-third that of any of these other elements, the conductivity of wires of equal mass per unit length is greatest for aluminum. Since it is by far the least expensive of any of these excellent conductors, aluminum is widely used in high-voltage transmission lines. Aluminum wiring is generally not used in homes because the oxide coating may eventually produce a high contact-resistance at connections at some outlets and switches, which may then overheat and cause fires.

Aluminum and its alloys find broad application in aircraft and space vehicle manufacture, in the construction industry, and in household utensils. It is the metallic ingredient of "silver" paint. A mixture of powdered aluminum with iron oxide is used in the *thermite process* for welding iron:

$$2Al + Fe_2O_3 \longrightarrow 2Fe + Al_2O_3 \qquad (26\text{-}12)$$

Molten iron is produced by this highly exothermic reaction, which is often started with a mixture of barium peroxide and powdered aluminum. A temperature of about 3000°C is reached and the white-hot liquid iron formed by the reduction can be directed into the space between the pieces to be welded.

Despite its strong reducing power ($\mathbf{E}^{\ominus} = -1.66$ V), aluminum does not normally react with even hot water because it is protected by the tough oxide film. However, amalgamated aluminum reacts readily:

$$2Al(\text{amalgam}) + 6H_2O \longrightarrow 2Al(OH)_3 + 3H_2 \qquad (26\text{-}13)$$

Since aluminum oxide is amphoteric [Secs. 9-4 and 12-6, Equation (12-44)], it is soluble in strong acids and strong bases, and thus these materials will normally[2] dissolve metallic aluminum readily:

$$2Al + 2OH^- + 6H_2O \longrightarrow 2Al(OH)_4^- + 3H_2 \qquad (26\text{-}14)$$

$$2Al + 6H^+ \longrightarrow 2Al^{3+} + 3H_2 \qquad (26\text{-}15)$$

Aluminum oxide, Al_2O_3, occurs naturally as the very hard mineral corundum (used as an abrasive) and as various gemstones that have the same structure as corundum but contain traces of transition-metal ions. These ions impart characteristic colors; for example, traces of Cr(III) color aluminum oxide a deep red and traces of Ti(IV) or Fe(II) and Fe(III) color it blue, producing the gems ruby and sapphire, respectively. Synthetic corundum and synthetic rubies and sapphires are made by melting precipitated aluminum hydroxide, with appropriate additives, in an electric furnace and then cooling the melt slowly under carefully controlled conditions.

The fact that $Al(H_2O)_6^{3+}$ is acidic ($K \approx 10^{-5}$) and thus readily "hydrolyzes" is the basis for several important commercial uses of the alums, which are salts containing this ion. They have the general formula $MAl(SO_4)_2 \cdot 12H_2O$, with M representing an alkali-metal cation (other than Li^+), NH_4^+, Ag^+, or Tl^+. Ammonium and potassium alums, the most important ones, are widely used both in baking powders, in which they furnish the acid that causes the release of CO_2 from sodium bicarbonate, and as mordants in dyeing. Mordants bind the dye to the material. They are substances that attach themselves to fibers of fabrics and also can easily adsorb dyes. In the dyeing process the fabric is immersed in a solution of the alum, and aluminum hydroxide is precipitated on the fibers by the addition of sodium carbonate or calcium hydroxide. The dye is added later. Alums are also used to clarify water. The gelatinous $Al(OH)_3$ that is formed by hydrolysis carries down suspended material as it slowly precipitates.

Aluminum chloride, $AlCl_3$, forms ionic crystals in which each aluminum atom is octahedrally coordinated. On the other hand, when the chloride is dissolved in nonpolar solvents, or is melted or vaporized, it has a quite different structure, consisting of discrete nonionic molecules of formula Al_2Cl_6. These molecules have a bridged structure, with each aluminum atom surrounded by four halogens in an

---

[2] Nitric acid, however, renders aluminum passive (Sec. 21-7) and thus can be shipped in aluminum containers.

approximately tetrahedral arrangement:[3]

$$\begin{array}{ccc} & \text{Cl} & \\ \text{Cl} & & \text{Cl} \\ & \text{Al} \quad \text{Al} & \\ \text{Cl} & & \text{Cl} \\ & \text{Cl} & \end{array} \qquad (26\text{-}16)$$

Aluminum bromide and iodide have the same structure, even in the crystalline state.

Aluminum, like its congener boron, also forms electron-deficient compounds. An example is $Al_2(CH_3)_6$, the geometric structure of which resembles that of Al_2Cl_6, the carbon atoms of the methyl groups being situated where the chlorine atoms are in (26-16). In each of these molecules the peripheral atoms are covalently bonded to the aluminum atoms. In the methyl compound, however, there is only one electron pair available for bonding each bridging methyl group to the two aluminum atoms, one bonding electron being provided by the methyl group and one by an aluminum atom. There are thus two three-center bonds involved in the bridging, each using one electron pair:

$$\begin{array}{ccc} & \text{H} \quad \text{H} \quad \text{H} & \\ & \text{C} & \\ \text{Al} & & \text{Al} \\ & \text{C} & \\ & \text{H} \quad \text{H} \quad \text{H} & \end{array} \qquad (26\text{-}17)$$

In the halides, on the other hand, each bridging halogen atom has enough unshared electron pairs that a normal single bond can be formed to each aluminum atom. Each bridging atom is thus held by two covalent bonds:

$$\begin{array}{ccc} & \text{Cl} & \\ \text{Al} & & \text{Al} \\ & \text{Cl} & \end{array} \qquad (26\text{-}18)$$

Tin and Lead These two metals of group IVA are congeners of carbon, silicon, and the metalloid germanium, and indeed tin behaves as a metalloid at low temperatures (p. 571). Both tin and lead are relatively low-melting and are soft and malleable. Copper and tin were known thousands of years before metallic iron was discovered; their ores are much more easily reduced than those of the more electropositive iron. Before 3000 B.C. it was discovered in the Middle East that copper could be hardened by alloying it with tin. The resulting bronzes could be forged into utensils, tools, and weapons so superior to others then available that they eventually produced a major revolution in technology, ushering in the Bronze Age.

The chief ores of tin and lead are cassiterite, SnO_2, and galena, PbS. The

---

[3] The wedge-shaped bonds imply that the Cl atoms are above the plane of the page; the dashed bonds imply that the Cl atoms are below the plane of the page.

metals, which are readily made by reduction of these ores, are good reducing agents in alkaline or strongly acid solutions. When pure they do not readily liberate hydrogen from dilute acids. Both metals form compounds of oxidation states (II) and (IV), including covalent as well as ionic substances. Stannous ion, Sn^{2+}, is a fairly good reducing agent and is also quite acidic in water because of extensive hydrolysis,

$$Sn^{2+} + H_2O \longrightarrow SnOH^+ + H^+ \qquad K \approx 10^{-2} \qquad (26\text{-}19)$$

Stannic oxide, SnO_2, is amphoteric, dissolving in hot NaOH to form the octahedral $Sn(OH)_6^{2-}$ ion and in concentrated HCl to form $SnCl_6^{2-}$. The chief uses of tin are in the manufacture of bronzes and other alloys—including type metal (Pb–Sb–Sn), pewter (Sn–Sb–Cu), and solders (low-melting Sn–Pb alloys)—and as the coating on tin cans.

Pb(II) salts are more stable than Pb(IV) compounds; both are important in the lead storage battery. Considerable quantities of lead are consumed in the manufacture of storage batteries and in the production of the gasoline additive tetraethyllead, $Pb(CH_2CH_3)_4$, a covalent compound. When this compound is heated, it liberates ethyl radicals, CH_3CH_2, which help to keep the heated gasoline-air mixture from igniting before the proper time in the engine cycle and thus to minimize "knocking". To prevent lead deposits from forming in the engine, dibromoethane ($C_2H_4Br_2$) is also added to the fuel. It reacts with the lead to form lead bromide, a volatile compound that becomes part of the exhaust gases. Other important uses of lead are in paint pigments, of which "white lead" [essentially $2PbCO_3 \cdot Pb(OH)_2$], "chrome yellow" ($PbCrO_4$), and "red lead" (Pb_3O_4) are especially important.

Lead compounds are poisonous, chiefly because Pb^{2+} reacts with the essential —SH groups of proteins to form insoluble precipitates. Metallic lead was used for pipes and for wine vessels as far back as Roman times, and some historians claim that widespread lead poisoning played a significant role in the decline of the Roman empire. More recently paints containing lead have sometimes been a source of chronic poisoning. They are now banned by law from being used on objects such as toys or baby furniture that are likely to be chewed by infants. Much effort is being made to find suitable substitutes for tetraethyllead as an antiknock fluid, in order to reduce lead pollution in the atmosphere and in water contaminated by atmospheric runoff.

Since lead rapidly destroys the effectiveness of the catalytic converters used to control emission of hydrocarbons in automobile exhausts, lead-free gasoline must be used for cars that have such a converter.

Antimony and Bismuth The congeners of these elements of group VA are nitrogen, phosphorus, and arsenic, all nonmetals. Although arsenic forms steel-gray lustrous crystals, they are brittle and are poor electrical conductors. Antimony is a metalloid, occurring in a silvery brittle metallic form and in several nonmetallic modifications; bismuth exists only as a metal. Antimony and bismuth may both be prepared by reduction of their oxides, obtained by roasting sulfide ores in air. The elements are purified electrolytically.

Each element forms ionic as well as covalent compounds, with oxidation states (III) and (V). Antimony is stable in air at ordinary temperatures, but it burns when heated, forming Sb_2O_3. At high temperatures this compound consists of molecules of formula Sb_4O_6, containing a tetrahedron of antimony atoms sharing oxygen atoms along each edge, like the phosphorus and arsenic analogs (see Fig. 24-6). Bismuth is also stable in air; at high temperature it reacts to form Bi_2O_3. Both antimony and bismuth react readily with chlorine, forming $SbCl_5$ and $BiCl_5$, respectively. Antimony pentachloride is a powerful reagent for introducing chlorine atoms into other molecules; BiF_5 is an equally effective fluorinating agent.

Many of the alloys of antimony and bismuth, as well as the pure metals themselves, expand upon freezing and thus are useful for castings. A number of bismuth alloys melt below the boiling point of water—e.g., Woods metal (mp 71°C), containing Bi, Pb, Sn, and Cd in proportions 4:2:1:1. Such alloys are used in automatic fire sprinklers, in safety plugs in steam boilers to guard against overheating, and in various kinds of fuses.

26-4 Some Transition Metals

The transition metals include many of the most important metals of industry and commerce. Iron is the central metal of modern civilization—we are still living in the Iron Age that began about 3000 years ago. Many other transition metals are used to impart desired qualities to iron alloys (steels). The coinage metals, copper, silver, and gold, are important not only commercially but also because of special properties, such as their high electrical conductivity, comparative inertness to attack by the atmosphere, and the sensitivity of the silver halides to light, the basis of the photographic process. The unique properties of many transition metals lead to other important applications, e.g., tungsten in light-bulb filaments, platinum and palladium as catalysts for hydrogenation and other reactions, and iron, cobalt, nickel, and their alloys in magnets. Transition-metal compounds, sometimes only in trace quantities, are essential for almost all known forms of life (Chap. 30).

In this section we discuss a few selected and representative transition metals and some of their compounds.

Sc
Y
La Ce $\cdots$ Lu
Ac Th $\cdots$ Lr

Group IIIB Elements, the Lanthanides, and the Actinides Lanthanum and actinium, the congeners of scandium and yttrium in column IIIB, give their names to the series of 14 elements referred to as the lanthanides and actinides, which correspond, respectively, to the filling of the 4f and 5f subshells. The chemistry of the lanthanides is very similar to that of Sc and Y, oxidation state (III) predominating. Just as Sc and Y readily lose three electrons to form ions of charge +3 (losing the electrons in the 4s and 3d, or 5s and 4d, orbitals, respectively), so each of the lanthanides forms a tripositive ion with no electrons in the 6s and 5d orbitals. The 4f subshell in these ions is progressively increased in population, from no electrons in La^{3+} to 14 in Lu^{3+}. The only other common oxidation states found for any of these elements can be explained in terms of the special stability of filled or half-filled subshells: Ce^{4+} (xenon configuration), Tb^{4+} and Eu^{2+} (half-filled 4f subshell), and Yb^{2+} (filled 4f subshell).

The lanthanides are difficult to separate from one another because of their nearly identical chemical properties, but ion-exchange chromatography of appropriate complexes provides efficient separations of almost all of them. These elements were formerly called the *rare earths* (the term *earth* implying oxide) but they are not unusually scarce. Some have important applications. Yttrium vanadates, with an added trace of Eu(III), serve as red phosphors for color television. (The blue and green phosphors are usually ZnS with small quantities of added silver and copper, respectively.) Ceric oxide, CeO_2, mixed with a large proportion of ThO_2, is used in the white-glowing mantles of gasoline and propane lanterns.

The radii of the tripositive lanthanide ions decrease steadily with increasing atomic number, from 1.06 Å for La^{3+} to 0.85 Å for Lu^{3+}. This *lanthanide contraction* arises because of poor screening of the nuclear charge by the 4f electrons, so that there is a progressive increase in the effective nuclear charge experienced by all electrons as the atomic number increases. A similar contraction occurs in the actinide series, for the same reason. An important consequence of the lanthanide contraction is that it approximately offsets the increase in ionic radius that might have been expected within a given *column* of the periodic table because of the increase in the number of electrons and the higher principal quantum numbers of the valence shell. Such an increase is found between the first and second transition series, which involve no f electrons. In group IVB, for example, the radii of the +4 ions are 0.68 Å for Ti^{4+} (first transition series), 0.79 Å for Zr^{4+} (second series), and 0.78 Å for Hf^{4+} (third series). The radii of Zr and Hf are nearly identical in other oxidation states as well, and atoms of these elements readily substitute for one another in minerals and other substances. Zirconium and hafnium are very difficult to separate; their chemical properties are more similar than those of any other pair of congeners. This similarity between elements in the second and third rows of the transition elements (and in the same column) becomes less pronounced as one moves to the right in the periodic table.

The actinides are radioactive; we leave the discussion of their chemistry to Chapter 28, which deals with the properties of radioactive substances.

Chromium and Manganese The transition elements in columns IVB to VIIB (Ti, V, Cr, Mn, and their congeners) exhibit many oxidation states, exemplified in Table 26-3. The highest oxidation state of each element is identical with the number characterizing its column and is usually found in compounds or anions containing oxygen or fluorine. We discuss here only chromium and manganese.

Cr	Mn
Mo	Tc
W	Re

The Latimer diagrams for the different oxidation states of chromium and manganese are typical of the relationships among oxyanions and hydrated cations. When the equilibria involve H^+ or OH^- it is necessary to consider acidic and basic solutions separately. The Latimer diagrams for Cr are

Acidic solution:

$$Cr_2O_7{}^{2-} \xrightarrow{+1.33} Cr^{3+} \xrightarrow{-0.41} Cr^{2+} \xrightarrow{-0.91} Cr(s) \qquad \text{(26-20)}$$

with -0.74 spanning Cr^{3+} to $Cr(s)$.

Table 26-3
Oxidation States in Some Binary Compounds

	Ti	V	Cr	Mn
F	III, IV	III, IV, V	II, III, IV, V, VI	II, III, IV
Cl	II, III, IV	II, III, IV	II, III, IV	II
Br	II, III, IV	II, III, IV	II, III, IV	II
I	II, III, IV	II, III	II, III	II
O	II, III, IV	II, III, IV, V	II, III, IV, VI	II, III, IV, VII

Basic solution:

$$CrO_4^{2-} \xrightarrow{-0.13} Cr(OH)_3(s) \xrightarrow{-1.1} Cr(OH)_2(s) \xrightarrow{-1.4} Cr(s) \qquad (26\text{-}21)$$

The corresponding diagrams for Mn are

Acidic solution:

$$MnO_4^- \xrightarrow{+0.56} MnO_4^{2-} \xrightarrow{+2.26} MnO_2(s) \xrightarrow{+0.95} Mn^{3+} \xrightarrow{+1.51} Mn^{2+} \xrightarrow{-1.18} Mn(s) \qquad (26\text{-}22)$$

$$+1.69 \qquad\qquad\qquad +1.23$$

Basic solution:

$$MnO_4^- \xrightarrow{+0.56} MnO_4^{2-} \xrightarrow{+0.60} MnO_2(s) \xrightarrow{-0.2} Mn(OH)_3(s) \xrightarrow{+0.1} Mn(OH)_2(s) \xrightarrow{-1.55} Mn(s) \qquad (26\text{-}23)$$

$$+0.59 \qquad\qquad\qquad -0.05$$

The differences between diagrams in acidic and basic solution are explained in Sec. 24-1. Some of the $E^\ominus$ values given are known only with low precision because some of the measurements are experimentally difficult, e.g., those involving gelatinous, slightly soluble hydroxides such as $Cr(OH)_2$ or $Mn(OH)_3$.

Chromium is a rather hard, high-melting, silvery metal. It is comparable to zinc in reducing power, as implied by its $E^\ominus$ values, and dissolves readily in nonoxidizing strong acids such as HCl and H_2SO_4. It does not, however, dissolve in oxidizing media such as HNO_3 because it is rendered passive (Sec. 21-7). The metal is also extremely resistant to atmospheric attack, because of formation of an adherent oxide film. Thus it is extensively used for corrosion protection, especially of iron, which is first electroplated with copper and then with chromium. Chromium is also an important component of many steels: the most common stainless steel contains about 18 percent Cr and 6 percent Ni, and a typical tool steel contains about 1 percent each of Cr, Mn, W, and C.

The three common oxidation states of chromium, other than that of the element itself, are Cr(II), which is basic, Cr(III), which is amphoteric, and Cr(VI), which is acidic. The chromous ion, Cr^{2+}, is a strong reducing agent, easily oxidized by atmospheric oxygen and even slowly liberating hydrogen from water. Green chromic oxide, Cr_2O_3, which can be obtained by heating the metal to high temperatures in air, is used as a pigment. Cr(III) has a great tendency to form hexacoordinate complexes, and thousands of these are known. They include complexes with water, ammonia, halide ions, and many other anions and neutral molecules.

The aqueous chemistry of Cr(VI) is influenced strongly by equilibria involving the yellow chromate ion, CrO_4^{2-}, and the orange dichromate ion, $Cr_2O_7^{2-}$:

$$HCrO_4^- \rightleftharpoons H^+ + CrO_4^{2-} \qquad K = 3.2 \times 10^{-7} \qquad (26\text{-}24)$$
$$2HCrO_4^- \rightleftharpoons H_2O + Cr_2O_7^{2-} \qquad K = 40 \qquad (26\text{-}25)$$

In alkaline solutions, most Cr(VI) exists as CrO_4^{2-}; in strongly acidic solutions, $Cr_2O_7^{2-}$ predominates. Many Cr(VI) compounds are generally good oxidizing agents; aqueous solutions of sodium dichromate, $Na_2Cr_2O_7$, are used to retard rust formation in automotive cooling systems, and a solution of this same salt in concentrated sulfuric acid is a potent cleaning agent for laboratory glassware.

This solution contains dichromic acid, $H_2Cr_2O_7$, and sometimes deposits red needlelike crystals of chromium trioxide, CrO_3.

Manganese is one of the more abundant transition metals in the earth's crust, exceeded in quantity only by iron and titanium. Although it is stable in air unless heated to high temperatures, it is readily soluble in dilute acids, forming salts of Mn(II), and even dissolves slowly in cold water, producing $Mn(OH)_2$. The metal is important principally as a component of steels, but it is also used in many alloys of aluminum and magnesium to improve corrosion resistance and mechanical properties.

Manganese has five common positive oxidation states: Mn(II) and Mn(III) are basic, Mn(IV) is amphoteric, and Mn(VI) and Mn(VII) are acidic. As implied by the Latimer diagram (26-22), Mn^{3+} is unstable in aqueous solution, disproportionating even at low concentrations to form Mn^{2+} and insoluble MnO_2:

$$2Mn^{3+} + 2H_2O \rightleftharpoons Mn^{2+} + MnO_2(s) + 4H^+ \qquad (26\text{-}26)$$

Manganese dioxide, MnO_2, which is found naturally as the lustrous black mineral pyrolusite, the principal ore of manganese, is extensively used in dry cells (Fig. 21-5a). The green manganate ion, MnO_4^{2-}, is stable in aqueous solution only if the solution is basic; in acidic or neutral solution it disproportionates to MnO_2 and permanganate ion, MnO_4^-. The latter ion is intensely purple and is a widely used oxidizing agent, chiefly in the form of the potassium salt. There are a number of similarities in the chemistry of Mn(VII) and Cl(VII); for example, both form unstable liquid heptoxides (Mn_2O_7 and Cl_2O_7), and the tetrahedral permanganate and perchlorate ions are almost the same size and can substitute for one another in many crystals.

Iron, Cobalt, and Nickel These elements, with their congeners, constitute the three columns of the periodic table jointly labeled VIII. Iron commonly assumes oxidation states (II) and (III), and rarely (VI), as in K_2FeO_4, a powerful oxidizing agent. Oxidation states (II) and (III) are also common for cobalt; for nickel, oxidation state (II) is the most important, with (III) and (IV) occurring only in a few compounds. The congeners of these elements are all noble metals, not readily attacked chemically. They form many complexes, and higher oxidation states than those observed for Fe, Co, and Ni are common. Thus, for ruthenium and osmium, the congeners of iron, states (VI) and (VIII) are well known; OsO_4, a volatile and poisonous compound, is a useful oxidizing agent in organic chemistry. Palladium and platinum, the congeners of nickel, form many stable complexes of oxidation state (IV) as well as (II); some of these are mentioned in Secs. 27-1 and 27-2.

Fe	Co	Ni
Ru	Rh	Pd
Os	Ir	Pt

Iron is the second most abundant metallic element in the earth's crust (after aluminum) and is believed to form a significant fraction of the earth's core. It is also common in meteorites. It is a high-melting silvery metal that easily corrodes in air containing water vapor (Sec. 21-7) and dissolves readily in dilute acids, but it is rendered passive by strong oxidizing agents. Cobalt and nickel are somewhat more inert; polished nickel retains its luster in air, although the powdered metal reacts with the oxygen in air. Both cobalt and nickel dissolve slowly in dilute acids but not in concentrated nitric acid, which makes them passive also.

All three of these elements form stable ionic compounds and a large number of complexes (Chap. 27). The common oxides of iron, FeO, Fe_3O_4, and Fe_2O_3 all show some tendency to be nonstoichiometric, especially FeO (Sec. 7-7). Each of these oxides may be regarded as consisting of a close-packed array of O^{2-}, with either Fe^{2+} or Fe^{3+} or both present in differing proportions and occupying different kinds of sites, with different numbers of surrounding oxide ions.

One important Fe(II) complex is the pale green hexaaquo ion, $Fe(H_2O)_6^{2+}$, which occurs in aqueous solutions and in a number of crystalline salts. In solution it is rather easily oxidized to Fe(III) by the oxygen in air. Another noteworthy complex of Fe(II) is the ferrocyanide ion, $Fe(CN)_6^{4-}$. When ferric ion, Fe^{3+}, is added to an aqueous solution of the sodium salt of $Fe(CN)_6^{4-}$, a deep blue precipitate, "Prussian blue", forms. It has the composition $NaFe_2(CN)_6$, and the same substance is formed if a solution of Fe^{2+} is added to a solution containing Na^+ and ferricyanide ion, $Fe(CN)_6^{3-}$.

The acidic properties of $Fe(H_2O)_6^{3+}$ have been mentioned in Sec. 9-6; it is so strong an acid ($pK\sim3$) that at least 1 percent of it is hydrolyzed at any pH above zero. The resulting solutions have a yellow or brown tinge; the hydrated ferric ion itself is pale violet. Iron(III) forms especially strong complexes with ligands that coordinate through oxygen atoms, for example, PO_4^{3-} and many organic compounds. The strongly colored red complexes with SCN^- (thiocyanate) provide a very sensitive test for Fe^{3+}.

Fe(II) and Fe(III) have a number of important biological roles. Their complexes with the large heme group (Fig. 26-3) are essential to the functioning of the proteins hemoglobin and myoglobin, which are responsible for the transport and storage of oxygen in many animals, including humans. When the protein is not holding an oxygen molecule, the iron atom is octahedrally coordinated by five nitrogen atoms and one water molecule. If the oxygen pressure is sufficiently high (e.g., in the lungs) and the pH is appropriate, the water molecule is replaced by O_2, which can then be transported by hemoglobin through the blood stream to sites where it is needed. Other iron complexes are involved in the enzymes that catalyze many biological oxidation reactions (Fig. 30-6).

Both Co(II) and Co(III) form complexes with a great variety of ions and molecules. Cobalt(III) is present in vitamin B_{12}, where it forms a complex with a group similar to heme. This vitamin is essential for the prevention and treatment of pernicious anemia, a progressive disease affecting red blood cells. Co(III) forms numerous other stable complexes, but $Co(H_2O)_6^{3+}$ is unstable in solution, being a sufficiently powerful oxidizing agent to oxidize water, forming O_2 and Co^{2+}.

The red-pink color of the hydrated cobaltous ion, $Co(H_2O)_6^{2+}$, is used to advantage as an indicator of the state of drying agents. When the pressure of water vapor is sufficiently low, certain Co^{2+} salts are anhydrous and blue; if the

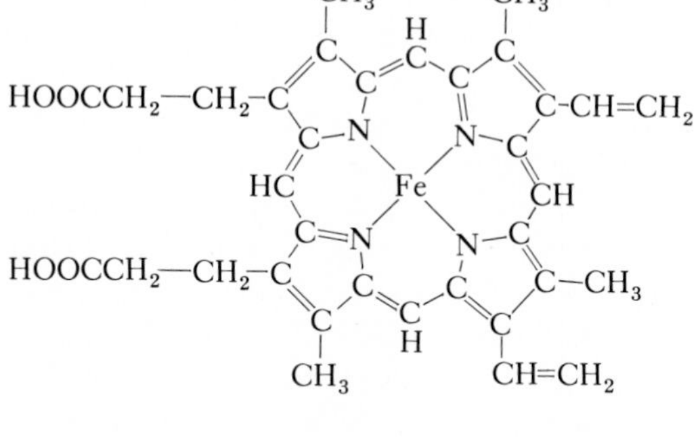

Figure 26-3 The Heme Group
This group is found in hemoglobin, myoglobin, and certain enzymes such as the cytochromes (Sec. 30-2).

Heme group

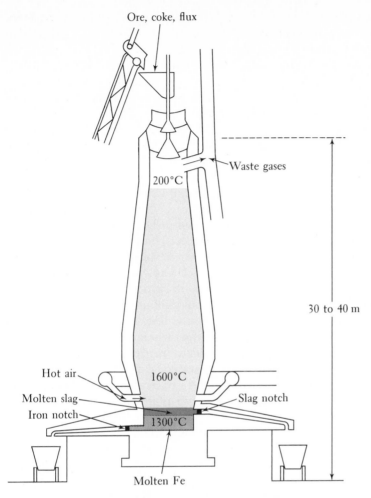

Ore, coke, flux

Waste gases

200°C

30 to 40 m

Hot air

Molten slag

Iron notch

1600°C

1300°C

Slag notch

Molten Fe

Figure 26-4 Blast Furnace for Smelting Iron Ores
The furnace is fed from the top, alternately with coke (chiefly carbon) and with a mixture of iron ore and a flux that produces liquid slag with ore components. With ores rich in silica (SiO_2), the flux is limestone; with basic ores, rich in $CaCO_3$ or $MgCO_3$, the flux is rich in SiO_2. Preheated air is blown in near the bottom. The solid materials are completely converted to gases and two liquids as they descend in the furnace. The reactions, somewhat simplified, are

$$2C + O_2 \longrightarrow 2CO$$
$$3CO + Fe_2O_3 \longrightarrow 2Fe(l) + 3CO_2$$
$$CaCO_3 \longrightarrow CaO + CO_2$$
$$CaO + SiO_2 \longrightarrow CaSiO_3 \quad \text{(molten slag)}$$

pressure of water vapor gets too high (because the drying agent present is no longer effective), the cobaltous salt turns red as the hydrated ion is formed.

The stability of oxidation state II relative to III increases as one progresses through the transition metals from Sc to Ni, as implied in Table 26-3 and by the chemistry of iron and cobalt. Compounds in which the oxidation state of nickel is greater than II are uncommon and usually quite reactive; the hydroxide (or oxide) of Ni(III) is the oxidizing agent in the Ni-Cd storage cell (Sec. 21-6). The common compounds of nickel are those of Ni(II) and include simple salts and many complexes with coordination numbers 4, 6, and even 5. The hydrated hexaaquo Ni^{2+} ion is green, the color characteristic of many Ni(II) salts. One important compound of nickel in the zero oxidation state is $Ni(CO)_4$, used in the purification of the metal (Sec. 25-1).

Steel Pure iron is quite soft, but its mechanical properties are considerably altered by the presence of small amounts of carbon and other elements. The molten iron from a blast furnace (Fig. 26-4) contains 3 to 4 percent carbon from the coke with which it has been in contact, and also smaller amounts of silicon, manganese, phosphorus, and sulfur. When this melt is cooled quickly, as in making castings, the resulting *cast iron* is hard and brittle. It contains large amounts of the compound Fe_3C, cementite. Cast iron, the most inexpensive form of iron, can be converted by heat treatment into a more malleable form, and if refined so that the content of carbon and other impurities is only a few tenths of a percent, the product is called *wrought iron*. This material, which is much stronger and tougher than cast iron, can readily be welded and forged and was widely used until this century. It has now largely been replaced by mild steel.

Steels are purified alloys of iron that contain carefully specified amounts of carbon and other elements. More than 600 million tons of steel are made in the world each year, by a complex and varied technology. Impure liquid iron is first refined to obtain the desired carbon content and remove other impurities; then the desired alloying elements are added and the product is handled under carefully controlled conditions to obtain the proper composition and texture for a particular application.

There are two broad commercial classes of steels: carbon steels and alloy steels. Carbon steels, which constitute about 90 percent of United States steel production, contain from about 0.03 to 1.5 percent carbon, sometimes as much as 1.5 percent manganese, and very small quantities of other elements. They are further classified according to increasing carbon content (which is also the order of increasing hardness and strength) as low-carbon, mild, medium, or high-carbon steels. Carbon steels are used extensively for appliances, containers, automobile bodies, ships, machinery, rails, and the structural framework of buildings.

The most important alloying elements other than carbon are nickel, chromium, and manganese, but molybdenum, tungsten, vanadium, titanium, and others are also used. Low-alloy steels, containing a total of no more than about 5 percent of elements other than iron, are used when special properties such as stress resistance or corrosion resistance are needed but cost is important. Applications include aircraft landing gears and other highly stressed parts of machines, such as axles, shafts, and gear wheels. High-alloy steels contain more than 5 percent of one alloying element and are used for their special properties: toughness and heat resistance for cutting tools, which contain tungsten or molybdenum, or corrosion resistance and appearance in stainless steels, used for tableware and kitchen utensils, chemical equipment, and jet engines.

Iron has very pronounced and almost unique magnetic properties, properties classified as *ferromagnetic*. Compounds with unpaired electrons are paramagnetic, and when samples of such compounds are placed in a nonuniform magnetic field they are pulled into the region of stronger field. In ferromagnetic substances there is very strong interaction between the unpaired electron spins on neighboring atoms, which causes the spins to be oriented parallel to each other throughout large regions of the sample, called magnetic domains. The result is an enormous reinforcement of the magnetism inherent in the individual atoms containing unpaired electrons. Ferromagnetic substances are pulled very strongly into magnetic fields and also remain magnetized even after they are removed from magnetic fields.

Ferromagnetism is restricted to solid substances and requires special conditions on the atomic level, so that few substances are ferromagnetic. The effect is not, however, restricted to metals; for example, the mineral magnetite, Fe_3O_4, one of the chief ores of iron, is ferromagnetic. It was probably the first known magnetic substance, referred to in Greek writings as early as 800 B.C. and known to the English as the lodestone. Metallic cobalt and nickel are also ferromagnetic, and some Fe–Co–Ni alloys are important technically in the manufacture of magnets for industrial and scientific applications—e.g., alnico, which contains about 51 percent Fe, 24 percent Co, 14 percent Ni, 8 percent Al, and 3 percent Cu.

Copper, Silver, and Gold These metals of group IB are all quite soft, malleable, and ductile, and are excellent conductors. Their handsome appearance, resistance to corrosion, and comparative scarcity has made them valued for use in jewelry and other ornamentation and as money. They are often called the coinage metals. Silver coins manufactured in the United States until the mid-1960s were made of an alloy containing 90 percent silver and 10 percent copper, and a copper-nickel alloy containing 75 percent copper has long been used for United States coins. Copper is a component of many important alloys, notably brass and bronze.

Cu
Ag
Au

The valence shells of atoms of copper, silver, and gold all have ten d electrons and one s electron, and thus a predominance of the ($+$I) oxidation state in their compounds might be expected. This is found, however, only for silver. Most copper compounds are those of Cu(II), with many fewer known for Cu(I). Au(III) is more stable than Au(I), although both are found in solution only in complexes. A few compounds of Ag(II), Ag(III), and Cu(III) have been reported, but they are rare and unusual.

When copper is heated in air it first forms Cu_2O and then CuO. Long exposure to moist air at ordinary temperatures gives copper and bronze a green patina, consisting of basic copper carbonates, mixtures of $Cu(OH)_2$ and $CuCO_3$. Silver is attacked chiefly by traces of H_2S in air, which produce a tarnish, a thin layer of black Ag_2S. None of these group IB metals is attacked by dilute nonoxidizing acids, but both copper and silver dissolve in nitric acid, which is reduced to NO and NO_2. Hot concentrated sulfuric acid is a good enough oxidizing agent to dissolve copper, being itself in part reduced to SO_2:

$$Cu + 2H_2SO_4 \longrightarrow CuSO_4 + SO_2 + 2H_2O \qquad (26\text{-}27)$$

Gold can be dissolved by a mixture of concentrated HNO_3 and concentrated HCl (aqua regia), as can platinum and some other noble metals. This action results from a combination of the strongly oxidizing properties of the nitric acid and the strong complexing properties of the chloride ion present. The stable complex $AuCl_4^-$ is formed (and $PtCl_6^{2-}$ for platinum):

$$Au + 4HCl + HNO_3 \longrightarrow AuCl_4^- + H^+ + NO + 2H_2O \qquad (26\text{-}28)$$

Many other strong complexes are formed by each of these elements, such as deep blue-violet $Cu(NH_3)_4^{2+}$, blue $Cu(H_2O)_4^{2+}$, colorless $Ag(NH_3)_2^+$, and the complexes of Ag^+ with thiosulfate ion that are used in photography (Secs. 24-4, 27-6, and 27-7).

The most important salt of copper is the sulfate, which is white when anhydrous but blue as the pentahydrate, $CuSO_4 \cdot 5H_2O$ (Fig. 7-8). Copper sulfate is used in copper plating, in dye and pigment manufacture, as an insecticide in vineyards [mixed with $Ca(OH)_2$ and water], in tanning, in engraving, and as an algicide in water treatment.

Silver nitrate is the most widely used soluble silver compound, important for the manufacture of silver halides for photography, as an antiseptic, in silver plating and the manufacture of mirrors, in hair dyes, and as a laboratory reagent. The silver halides, except for the fluoride, are insoluble in water. They are all photosensitive, darkening by decomposition into the elements when exposed to light of sufficiently short wavelength.

Zn
Cd
Hg

Zinc is the most reactive of the metals of group IIB; it burns in air and dissolves readily in acidic and basic solutions. It is, however, protected from atmospheric attack by an adherent oxide-carbonate film. Cadmium also burns in air, but is dissolved only with difficulty by nonoxidizing acids and not at all by strong bases. Mercury is quite noble and is stable in air, although it reacts with chlorine; it dissolves in nitric acid, forming mercuric nitrate, $Hg(NO_3)_2$.

All three elements form compounds in oxidation state (II). Mercury also forms compounds of Hg(I), containing two covalently bonded mercury atoms, in such molecules as Hg_2Cl_2 (which is linear) and in the mercurous ion, Hg_2^{2+}. Zinc hydroxide is amphoteric, dissolving readily in both acids and bases. Cadmium hydroxide is basic, as is mercuric oxide, HgO; mercury forms no hydroxides.

Both zinc and cadmium are used for plating to protect against corrosion; galvanized iron is iron coated with zinc (Sec. 21-7). All three metals form many alloys. Brass is a hard, yellow, comparatively inert alloy containing copper and zinc in proportions that may vary rather widely. Alloys of mercury with other metals are called *amalgams;* some of these have definite composition, corresponding to the existence of specific compounds. Mercury reacts vigorously with some metals, such as sodium and potassium, forming compounds like Hg_2Na and Hg_2K. However, it has no effect on others, such as iron, and iron containers are used to store and ship mercury.

Most compounds of mercury are highly toxic. The metal itself can be harmful, particularly in the presence of decaying organic matter because of the formation of CH_3Hg^+, which is easily ingested and highly poisonous. Even though the vapor pressure of mercury is extremely low, about 2×10^{-10} atm at $20°C$, the vapor is sufficiently toxic that prolonged exposure to the vapor in rooms in which liquid mercury is handled can cause severe poisoning. Calomel (Hg_2Cl_2) was long used as a cathartic but because of its potential toxicity its use has been largely discontinued. The toxicity of mercury compounds is due to their irreversible combination with proteins. Very dilute solutions of $HgCl_2$, which is little dissociated, are toxic to bacteria and are used as disinfectants.

Summary

The highly reactive and strongly electropositive alkali metals can be prepared by electrolysis of their molten chlorides or hydroxides. Their reactions usually involve the loss of the single outer electron. Almost all alkali-metal compounds are water-soluble. Sodium hydroxide, NaOH, is produced by electrolysis of brine. The Solvay process, used for the production of sodium carbonate, Na_2CO_3, involves $NaHCO_3$ as an intermediate. The bicarbonate is formed by the reaction of CO_2 with NaCl in an ammoniacal solution. Chemical properties of potassium compounds are generally quite similar to those of the corresponding sodium compounds. Potassium and sodium ions are essential in the transmission of signals in nerves.

The alkaline-earth metals, having higher ionization energies, are less reactive than the alkali metals. They are also denser, harder, higher melting, and higher boiling. The metals react with water to give the hydroxides, which, with the exception of the amphoteric $Be(OH)_2$, are strong bases. Beryllium is less electropositive and has a greater tendency than its congeners to form covalent bonds. Some of its compounds, such as the polymeric $(BeH_2)_n$ contain three-center bonds. When $CaCO_3$ (limestone, marble, shells) is roasted at high temperature, calcium oxide (quicklime) is formed. Calcium hydroxide (slaked lime) is the least expensive strong base.

Aluminum and its congeners, gallium, indium,

and thallium, form stable +3 ions; thallium forms +1 ions as well. Aluminum is a strong reducing agent, but its reactivity is diminished because it is covered with a tough oxide film. Like boron, aluminum forms electron-deficient compounds, an example being $Al_2(CH_3)_6$. Since Al_2O_3 is amphoteric, it is soluble in strong acids and strong bases, and thus these materials will normally dissolve aluminum readily. Aluminum oxide occurs naturally as the very hard mineral corundum, which is used as an abrasive, and is the principal constituent of rubies and sapphires. Alums have the general formula $MAl(SO_4)_2 \cdot 12H_2O$, where M represents an alkali-metal cation (except Li^+), NH_4^+, Ag^+, or Tl^+. Ammonium and potassium alums are used in baking powders.

Tin and lead are low-melting and soft. Their compounds of oxidation states II and IV include covalent as well as ionic substances. Stannous ion, Sn^{2+}, is a fairly good reducing agent. Tin is used mainly as a component of alloys and to coat tin cans. Lead is employed in storage batteries, in leaded gasoline, and in pigments. Lead salts are poisonous.

Antimony and bismuth form both covalent and ionic compounds in oxidation states III and V. Antimony is a metalloid, bismuth a metal. Antimony pentachloride is a powerful chlorinating agent, and bismuth pentafluoride is equally effective for fluorination. Many alloys of Sb and Bi are employed for castings; bismuth is a principal component of low-melting alloys.

The lanthanides are the 14 elements whose position in the periodic table corresponds to the filling of the 4f subshell. A similar series, the actinides, corresponds to the filling of the 5f level. It is difficult to separate the lanthanides from one another because of their nearly identical chemical properties. The radii of the tripositive lanthanide ions decrease steadily with increasing atomic number (the "lanthanide contraction"). As a result, the radii of the elements in the third transition series are little different from those of the corresponding second-transition-series elements.

Chromium and manganese exhibit many oxidation states, II, III, and VI being the most common for Cr and II, III, IV, VI, and VII for Mn. The low oxidation states tend to be basic, the high oxidation states acidic, and those in between amphoteric. Chromium is a hard silvery metal, used extensively for plating because it is protected by an oxide film. It is an important component of many steels, as is manganese. In alkaline solutions, Cr(VI) exists as chromate ion, CrO_4^{2-}; in strongly acidic solutions dichromate ion, $Cr_2O_7^{2-}$, predominates. Both species are good oxidizing agents. The intensely purple permanganate ion, MnO_4^-, is also a widely used oxidizing agent, chiefly in the form of the potassium salt.

The common oxidation states of iron and cobalt are II and III; for nickel, state II is the most important. Iron is the second most abundant metal in the earth's crust. It corrodes easily in moist air and dissolves readily in dilute acids. Cobalt and nickel are somewhat more inert. All three elements form stable ionic compounds and a large number of complexes. An important Fe(II) complex is the pale green hexaaquo ion, $Fe(H_2O)_6^{2+}$, which occurs in aqueous solutions. The Fe(III) hexaaquo complex is acidic, with $K \approx 10^{-3}$. Heme complexes of iron play an essential role in the functioning of the proteins hemoglobin and myoglobin, and Co(III) is present in vitamin B_{12}. Iron, cobalt, and nickel are ferromagnetic, that is, they are very strongly attracted into magnetic fields. Steels are purified alloys of iron that contain carefully specified amounts of carbon and other elements, such as Ni, Cr, and Mn.

Copper, silver, and gold are soft, malleable, and ductile metals and are excellent conductors. They are relatively unreactive. All three metals are found in oxidation state I; other common states are II for copper and III for gold. None of these elements is attacked by dilute nonoxidizing acids, but copper and silver dissolve in HNO_3 and gold in aqua regia, a mixture of concentrated hydrochloric and nitric acids. Copper has extensive electrical applications, and is used in alloys (e.g., brass and bronze) and in coins. The chief use of silver is in photography.

Zinc is the most reactive of the group IIB metals, zinc, cadmium, and mercury. All three metals exhibit oxidation state II. Mercury, in addition, forms compounds of Hg(I) in which two Hg atoms are bonded together, as in Hg_2Cl_2 and mercurous ion, Hg_2^{2+}. Zinc hydroxide is amphoteric; $Cd(OH)_2$ and HgO are basic. The three metals form many alloys, brass (Zn–Cu) being the most important. Mercury alloys are called amalgams. Zinc and cadmium are widely used in plating to protect against corrosion. Mercury compounds are toxic.

Terms and Concepts

Problems and Questions

26-1 Reducing Power of Alkali Metals The enthalpy change for the half-reaction

$$M(s) \longrightarrow M^+(aq) + e^-$$

is a measure of the reducing power of an alkali metal M. Since enthalpy is a state function, the enthalpy change for this reaction can be obtained by calculating the enthalpy change for the following equivalent three-step process:

(I) Sublimation of the solid

$$M(s) \longrightarrow M(g)$$

(II) Ionization of the monatomic vapor

$$M(g) \longrightarrow M(g)^+ + e^-$$

(III) Hydration of the gaseous ion

$$M(g)^+ \longrightarrow M^+(aq)$$

(Both the overall reaction and step III involve H_2O implicitly.) Use this thermodynamic process to compare the reducing powers of lithium and cesium. What is the effect of hydration energy on the relative reducing powers? (Enthalpies of sublimation: Li, 160 kJ mol^{-1}; Cs, 78 kJ mol^{-1}. First ionization energies: Li, 520 kJ mol^{-1}; Cs, 376 kJ mol^{-1}. Enthalpies of hydration: Li, -506 kJ mol^{-1}; Cs, -259 kJ mol^{-1}.)

26-2 Balanced Chemical Equations Write balanced equations for the following reactions: (a) a blue solution of sodium in liquid ammonia becomes colorless by the formation of sodium amide ($NaNH_2$) and hydrogen; (b) permanganate ion, MnO_4^-, is reduced to Mn^{2+} in acidic solution by ethanol, CH_3CH_2OH, which is oxidized to acetic acid, CH_3COOH; (c) ethanol is oxidized in acidic solution to acetaldehyde, CH_3CHO, by dichromate ion, $Cr_2O_7^{2-}$, which is reduced to Cr^{3+}; (d) a melt of MnO_2 in KOH is oxidized by air to potassium manganate, K_2MnO_4; (e) manganate ion is oxidized to permanganate ion by chlorine in alkaline solution.

26-3 Acidic Properties of Metal Oxides Arrange the metal oxides in each of the following groups in order of decreasing acidity and explain the reasons for your ranking: (a) Al_2O_3, Na_2O, MgO; (b) Cr_2O_3, CrO_3, CrO.

26-4 Oxidations in Alkaline and Acidic Solution It is often asserted that it is much easier to oxidize an element to an oxyanion of high oxidation number in an alkaline solution than in an acidic solution. Analyze this assertion by calculating the fraction of Mn^{2+} oxidized to MnO_4^- by gaseous oxygen ($P = 1$ atm) at pH = 0 and at pH = 14. Use $E^{\oplus}$ values from Table D-5. If the assertion seems generally true, what is its explanation? (*Hint:* Consider why and how the half-cell potential for oxidation to form a typical oxyanion depends on the pH.)

26-5 Lanthanides Explain the predominance of the ($+$III) oxidation state in lanthanide compounds. Why are lanthanides difficult to separate from each other?

26-6 Latimer Diagram The Latimer diagram below pertains to the oxidation states IV, V, and VI of uranium in acid solution. Is the oxidation state V stable?

$$UO_2^{2+} \xrightarrow{\ 0.05\ } UO_2^+ \xrightarrow{\ 0.62\ } U^{4+}$$

26-7 Stability of Manganate (*a*) Discuss the relative stability of MnO_4^{2-} in acidic and in alkaline solutions by using the Latimer diagrams given in the text. (*b*) What is the equilibrium constant for the equilibrium $3MnO_4^{2-} + 2H_2O \rightleftharpoons MnO_2(s) + 2MnO_4^- + 4OH^-$? What is $[MnO_4^{2-}]$ at equilibrium with $[MnO_4^-] = 0.0100\ M$ and $[OH^-] = 1.00\ M$?

26-8 Precipitation of Chromate In an acid solution $Cr(+VI)$ exists predominantly as $Cr_2O_7^{2-}$. When Ba^{2+} is added to such a solution, practically all the chromium is precipitated as $BaCrO_4$. Explain in terms of equilibria, making whatever reasonable assumptions are needed.

26-9 Manganese in Steel In order to determine the manganese content in a sample of steel, a 0.200-g sample of the steel is dissolved in an acidic solution and the manganese is oxidized to permanganate ion. The permanganate is then reduced to Mn^{2+} by addition of an excess of ferrous ion, 50.0 ml of 0.0625 M ferrous sulfate being used. The excess of ferrous ion is determined by titration with 0.0100 M dichromate, 18.7 ml being required. The dichromate is reduced to Cr^{3+}. What is the percentage of Mn in the steel?

26-10 Behavior of $FeSO_4$ Solutions When air is bubbled through aqueous $FeSO_4$, both the pH and the color of the solution change. In which direction does the pH change? Explain, giving equation(s) for any reaction(s) mentioned.

26-11 Relative Stability of Oxidation States Consider the disproportionation of a 1 M solution of Fe^{2+} into $Fe(s)$ and Fe^{3+}. (*a*) Use the values of $\widetilde{G}_f^{\oplus}$ given in Table D-4 to decide whether the reaction is spontaneous. (*b*) Calculate (at 298 K) the equilibrium constant involved.

26-12 Steel Manufacture In the Bessemer process for making steel, oxygen is bubbled through impure molten iron. What is the effect of this procedure on the equilibrium $Fe_3C \rightleftharpoons 3Fe + C$ involving the cementite (Fe_3C) component of the impure iron?

26-13 Use of Carbonates in Steel Making In steel making, small quantities of nonmetallic impurities such as phosphorus, sulfur, and silicon are oxidized and removed from molten iron. This process is often carried out in furnaces lined with $MgCO_3$ and $CaCO_3$, which decompose at high temperatures to MgO and CaO. Explain with equations how the presence of this lining can help in the removal of the oxides of nonmetallic impurities from the iron.

26-14 Balanced Chemical Equations Write balanced equations for the following reactions: (*a*) copper dissolves in dilute sulfuric acid in the presence of air; (*b*) hot concentrated sulfuric acid is reduced to SO_2 by copper; (*c*) gold dissolves in aqua regia to give $AuCl_4^-$ and NO_2; (*d*) zinc dissolves in a KOH solution to give $Zn(OH)_4^{2-}$.

26-15 Stability of Cu^+ Ion in Aqueous Solution (*a*) Use $E^{\oplus}$ values from Table D-5 as an aid in predicting whether Cu^+ in aqueous solution might disproportionate to Cu^{2+} and Cu. (*b*) Estimate the equilibrium constant for the reaction $2Cu^+ \rightleftharpoons Cu^{2+} + Cu(s)$.

26-16 $E^{\oplus}$ Values in Acidic and Basic Solutions (*a*) Two different $E^{\oplus}$ values are given for the $MnO_4^{2-}/MnO_2(s)$ couple in the Latimer diagrams (26-22) and (26-23) that apply, respectively, to acidic and basic solutions. Relate the two values given to K_w. (*b*) Similarly relate the $E^{\oplus}$ values for the $Cr^{2+}/Cr(s)$ and $Cr(OH)_2(s)/Cr(s)$ couples given in diagrams (26-20) and (26-21) to the solubility product of $Cr(OH)_2(s)$. What value of K_{sp} for $Cr(OH)_2$ is in accord with these data?

27

"If we think of the metal atom as the center of the whole system, then we can most simply place six molecules connected with it at the corners of an octahedron."
A. WERNER, 1893

"With due acknowledgment to G. B. Shaw for the inspiration, it might be said that theories of chemical bonding—neglecting quite a few which are entirely valueless—fall into one of two categories: those which are too good to be true and those which are too true to be good. 'True' in this context is intended to mean 'having physical validity' and 'good' to mean 'providing useful results, especially quantitative ones, with a relatively small amount of computational effort'. The proper, rigorous wave equation for any molecular situation represents a theory of that situation which is too true to be good. Most theories which are too good to be true are those in which the real problem per se is not treated, but rather an artificial analogue to the real problem, contrived so as to make the mathematics tractable, is set up and solved.

"The electrostatic crystal field theory, using the point charge approximation, is such a theory—at its best, or, perhaps, at its worst."
F. ALBERT COTTON[1]

There are qualitative distinctions between the properties of the transition metals and those of the representative metals, such as the alkali and alkaline-earth elements, aluminum, and tin. Representative metals usually have only one, or at most two, stable oxidation states and form colorless diamagnetic compounds. In contrast, transition metals usually have several stable oxidation states and their compounds are frequently colored and paramagnetic. The reactions of representative metal ions in solution occur virtually instantaneously, while transition-metal ions often react so slowly that it takes minutes or even hours for their reactions to go to completion.

Much of the recent research in inorganic chemistry has been concerned with transition-metal complexes. These compounds are of great practical importance in analytical chemistry, industrial processes, agriculture, and biochemistry. They offer challenges to the chemist interested in preparing new and unusual compounds, and the significant progress that has been made in understanding their properties is an example of the successful application of quantum mechanics to complex multiatomic systems.

We begin our consideration of the chemistry of transition-metal complexes with some historical background, focusing on the work of Alfred Werner. Next we examine the stereochemistry and spectra of complex ions, and the crystal-field and molecular orbital theories that provide interpretations of the properties of

---

[1] *Journal of Chemical Education*, vol. 41, p. 466, 1964.

complexes. The chapter concludes with a discussion of equilibria between different complex species, and of some applications that illustrate the technological importance of these complexes.

27-1 Historical Background

Some of the properties and structural features of representative complex ions have been described briefly and illustrated in earlier chapters, especially in Secs. 7-5 and 9-6. The modern picture of the geometry, bonding, and reactions of complexes is a direct outgrowth of the research of an Alsatian-Swiss chemist, Alfred Werner, who worked during the late years of the nineteenth century and the early years of this one. Although many of what are now called coordination compounds (Sec. 9-6), a term originated by Werner, were known before his day, few systematic studies of their properties and reactions had been made and ideas concerning their structure were almost completely fallacious. The reasons for the comparatively late development of a correct structural view of these relatively simple inorganic compounds are of some interest. They provide insight into the ways in which fashions in one area of science inhibit progress in other areas, even those quite closely related, a sad testimony to the limitations of the imaginations of all but a few rare individuals in any period.

Around the middle of the nineteenth century organic chemistry had come to dominate the chemical scene, attracting many of the best minds with a bent for chemistry and, in fact, for science generally. By the mid-1870s, the structural theory of organic chemistry was well established and even some three-dimensional aspects of structure were understood. As complicated a molecule as that of the blue dyestuff indigo (p. 390), which had been used for thousands of years, had been synthesized in the laboratory. During the last quarter of the century the structures of a great many physiologically important organic compounds were determined or were under vigorous investigation, including sugars, amino acids (the building blocks of proteins), fragments of nucleic acids, and chlorophyll.

In contrast, relatively little progress was made toward a real understanding of the structures of any but the simplest inorganic compounds. The very success of the structural principles applicable to organic compounds led to attempts to apply these principles to the inorganic field. However, they were of limited value for compounds of elements other than carbon, because the chief bonding characteristic of carbon atoms, their tendency to join together to form chains and rings, is rare in the chemistry of other elements. Modification of the prevalent structural ideas of the time, as well as the creation of new ones, was a necessary prerequisite to the development of an understanding of inorganic complexes. Werner almost single-handedly provided these new insights, on the basis of years of careful and ingenious experimental work.

When Werner began his work, *incorrect* structural formulas such as

$$
\begin{array}{c}
\text{Cl} \qquad \text{NH}_3\text{—Cl} \\
\diagdown \quad \diagup \\
\text{Pt} \\
\diagup \quad \diagdown \\
\text{Cl—NH}_3 \qquad \text{NH}_3\text{—NH}_3\text{—NH}_3\text{—Cl}
\end{array}
\qquad (27\text{-}1)
$$

were common. This formula, strongly influenced by the concept of chains of

Compound	Fraction of chloride reacting with Ag⁺	Modern formula	Number of conducting species	Relative electrical conductivity of solution*
$PtCl_4 \cdot 6NH_3$	$\frac{4}{4}$	$Pt(NH_3)_6^{4+}, 4Cl^-$	5	523
$PtCl_4 \cdot 5NH_3$	$\frac{3}{4}$	$Pt(NH_3)_5Cl^{3+}, 3Cl^-$	4	404
$PtCl_4 \cdot 4NH_3$	$\frac{2}{4}$	$Pt(NH_3)_4Cl_2^{2+}, 2Cl^-$	3	229
$PtCl_4 \cdot 3NH_3$	$\frac{1}{4}$	$Pt(NH_3)_3Cl_3^+, Cl^-$	2	97
$PtCl_4 \cdot 2NH_3$	$\frac{0}{4}$	$Pt(NH_3)_2Cl_4$	0	~0
$PtCl_4 \cdot NH_3 \cdot KCl$	$\frac{0}{5}$	$K^+, Pt(NH_3)Cl_5^-$	2	109
$PtCl_4 \cdot 2KCl$	$\frac{0}{6}$	$2K^+, PtCl_6^{2-}$	3	256

*Arbitrary scale. Conductivity measured for solutions of same molarity.

bonded atoms, common in organic chemistry, represented the compound with composition $PtCl_4(NH_3)_5$ or $PtCl_4 \cdot 5NH_3$, one of a series of compounds of platinum with chlorine and ammonia (Table 27-1). It was known that the addition of $AgNO_3$ to solutions of compounds such as those in this series does not necessarily cause the precipitation as AgCl of all the chlorine they contain. For example, only three-quarters of the chlorine of the second compound in the table reacts with Ag⁺. The incorrect formula (27-1) explained this result if it was assumed that the Cl atom attached directly to the Pt atom does not react with Ag⁺ to form AgCl, whereas the Cl atoms attached to the NH_3 groups do. As more experimental facts were established—not all the evidence in Table 27-1 and indeed not all the compounds were known before Werner's day—it became clear that formulas such as (27-1), attributing just four covalent bonds to the platinum atom in these compounds, could not be correct.

Werner used many different methods, including syntheses of series of compounds such as that in Table 27-1 and a variety of physical measurements, to establish the modern picture of the structure of complexes. The modern formulas listed in Table 27-1, which are essentially those he proposed, show that in each compound all the NH_3 molecules are attached directly to the Pt, together with enough chlorine atoms to make the total number of ligands six. The remaining chlorine atoms, if any, are present as Cl⁻. The oxidation state of Pt in each compound is +IV. The first four complexes listed are positively charged and are thus cationic; the fifth is electrically neutral, and the last two are anionic. The measurements of electrical conductivity listed in the final column of the table were made by Werner. They provide a rough indication of the number of conducting, and therefore charged, species in each compound and are in good accord with the formulas Werner suggested.

27-2 Stereochemistry of Complexes

Werner realized the importance of three-dimensional structural features of complexes. He was able to deduce the configuration of the ligands in complexes by clever experiments and reasoning. His arguments were based on the existence and properties of different isomeric complexes: the number of different possible structures corresponding to a given formula and the symmetry of these structures, as revealed by their influence on plane-polarized light, a phenomenon referred to as optical activity, discussed briefly in Sec. 29-3.

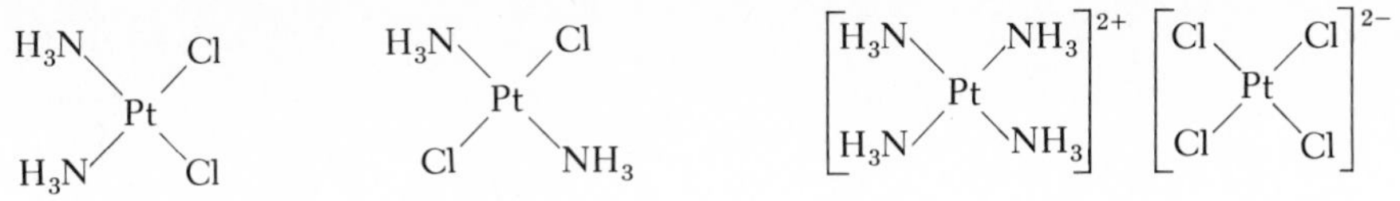

Figure 27-1 Examples of Square Planar Complexes

The two compounds on the left have the same composition as the compound on the right, but half the formula weight. The first two molecules are isomers. Their relationship with the third compound is not strictly one of isomerism, although it has been considered as such historically.

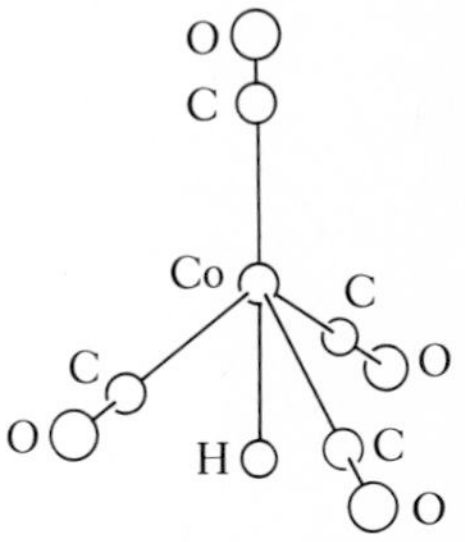

Newer experimental methods, such as infra-red, Raman, and nuclear magnetic resonance spectroscopy and what is perhaps the most fruitful technique, direct structure determination by X-ray diffraction, have confirmed many of Werner's deductions. The determination of the spatial arrangement of the ligands around the central metal atom in a complex—the stereochemistry of the complex—is now a standard procedure in the characterization of new complexes.

We have already discussed some features of the stereochemistry of complexes (Sec. 9-6, Table 9-2) and the coordination geometries in ionic crystals (Sec. 17-3, Fig. 17-5). Some features of a few common structures are described below.

$CoH(CO)_4$

Coordination Numbers 2 and 4 A number of transition metals form complexes with just two ligands. Typical examples are the linear silver (I) complexes $Ag(NH_3)_2^+$, $Ag(CN)_2^-$, and $AgCl_2^-$. Those with coordination number 4 are more common. Some have tetrahedral structures, as in $FeCl_4^-$, $CoCl_4^{2-}$, $Ni(CO)_4$, and HgI_4^{2-}; others, such as $Ni(CN)_4^{2-}$, $PdCl_4^{2-}$, $PtCl_4^{2-}$, and $Pt(NH_3)_4^{2+}$, are square planar. Figure 27-1 shows some examples of isomers for square planar complexes with two different ligands.

Octahedral Complexes Many transition-metal complexes have sixfold coordination and an octahedral geometry. An important feature of this spatial arrangement is the existence of isomers for complexes with certain general formu-

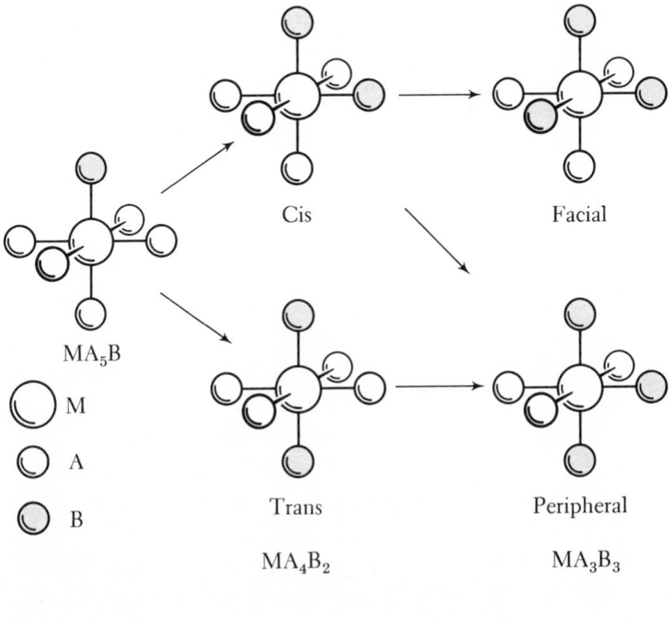

Figure 27-2 Isomers Expected for an Octahedral Ion or Molecule

In MA_6 all six ligands occupy the corners of a regular octahedron, with M at the center. Since the six corners of the octahedron are equivalent, there is only one species MA_5B. For the formula MA_4B_2 there are two isomers, one with the two B ligands at adjoining corners of the octahedron (cis) and the other with the two B's at diagonally opposite corners (trans). The complex of formula MA_3B_3 also has two isomers, in which all three B's either are at the corners of the same face or define a plane that also goes through M. The arrows show from which precursor a given isomer can be made by substituting B for A.

607

Mirror plane

(a)   (b)   (c)

Figure 27-3   Isomers of Co(en)$_2$Cl$_2{}^+$
There are three isomers of this complex, in which en represents ethylenediamine, H$_2$NCH$_2$CH$_2$NH$_2$. Although the mirror image of isomer (a), in which the two Cl atoms are in *trans* position, is superimposable on the original ion, this is not true for isomers (b) and (c), in which the Cl atoms are in *cis* position. These two isomers are mirror images of each other and are not superimposable.

◯ Co

◯ Cl

⬭ ≡ (ethylenediamine structure with H H / H H on carbons C—C, and H—N, H—N with H atoms)

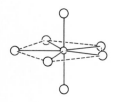

TaF$_7{}^{2-}$, NbF$_7{}^{2-}$

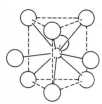

ZrF$_7{}^{3-}$, UF$_7{}^{3-}$

Nd(H$_2$O)$_9{}^{3+}$, ReH$_9{}^{2-}$

Fe$_2$(CO)$_9$

las, such as MA$_4$B$_2$ or MA$_3$B$_3$, where M represents the central metal atom and A and B are two kinds of *unidentate* ligands (i.e., ligands capable of forming one bond to the central atom) (Fig. 27-2).

Structural arguments lead to the conclusion that there must also exist pairs of isomers that are mirror images of each other and cannot be superimposed, as shown in Fig. 27-3. They are related as are a right hand and a left hand. Such isomers can be distinguished by their interaction with polarized light (Sec. 29-3), but they are identical in most other respects and cannot be distinguished on the basis of physical properties such as boiling point and melting point.

Complexes with coordination numbers other than 4 and 6 are relatively rare.

Polynuclear Complexes   Although all the complexes described so far contain only one central metal atom, it is possible to prepare coordination compounds called polynuclear complexes that contain several metal atoms (Sec. 9-6). An example is the condensation of the yellow-colored chromate ion to orange dichromate ion. Chromate ion is stable in aqueous solution at high pH. If acid is added, however, the binuclear dichromate ion is formed (Sec. 26-4):

$$2\begin{bmatrix} O \\ | \\ O-Cr-O \\ | \\ O \end{bmatrix}^{2-} + 2H^+ \longrightarrow \begin{bmatrix} O & O \\ | & | \\ O-Cr-O-Cr-O \\ | & | \\ O & O \end{bmatrix}^{2-} + H_2O \qquad (27\text{-}2)$$

If the solution is sufficiently concentrated and acidic, Cr$_2$O$_7{}^{2-}$ reacts with another CrO$_4{}^{2-}$ to form trichromate ion, Cr$_3$O$_{10}{}^{2-}$. The complexes can grow still larger, a process leading eventually to a form of chromium trioxide, (CrO$_3$)$_n$O^{2-} with n very large, a red polynuclear complex that consists of essentially infinite chains of oxygen-bridged chromium atoms similar in structure to the SO$_3$ chains shown in Fig. 7-2e. The same sequence of reactions also occurs with OH as the ligand, as in Equation (9-32).

Many polynuclear carbon monoxide complexes are known; in some the structural units are connected only by metal-metal bonds and in others the metal atoms are bridged by CO ligands as well. Heavy transition metals such as molybdenum and tantalum form ions containing clusters of many metal atoms. (Cluster compounds of representative metals are also known. For example, bismuth forms an ion with the formula Bi_9^{5+}.)

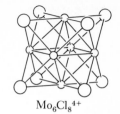

$Mo_6Cl_8^{4+}$

Sandwich Compounds The preparation of the compound $(C_5H_5)_2Fe$, commonly known as ferrocene, was reported in 1951. It was first assumed to have the structure

$$\text{◯—Fe—◯}$$

However, chemical and physical studies showed, and X-ray crystallographic studies confirmed, that this structure is incorrect. The iron atom is sandwiched between the hydrocarbon rings as shown in Fig. 27-4a. The molecule may be considered to be composed of two 5-membered ring anions (cyclopentadienyl ions, $C_5H_5^-$, each of which contains six π electrons) and Fe^{2+}. Many other sandwich compounds have been synthesized, including complexes of other metals such as chromium and ruthenium with cyclopentadienyl anions, as well as complexes involving other hydrocarbon rings such as benzene and cyclo-octatetraene, C_8H_8.

The bonding in sandwich compounds involves the π electrons of the hydrocarbon ring and is best described in terms of molecular orbitals that are combinations of the π orbitals of the two rings and the valence-shell orbitals of the metal atom. Most of these complexes satisfy the rule that the total number of electrons involved in the bonding is 18, so that the valence shell of the metal atom can in a sense be regarded as filled. In dibenzene chromium (Fig. 27-4b), for example, each of the two benzene molecules has 6 π electrons and the chromium atom has 6 valence electrons, a total of 18.

Complexes involving π bonding need not be of the symmetric sandwich form. As shown by the example in Fig. 27-4d, they can contain a single hydrocarbon in conjunction with other ligands. Nonaromatic hydrocarbons with π electrons, such as ethylene, also form complexes with some transition metals (Fig. 27-4e).

Complexes Based on Porphin Several biologically important complexes are based on derivatives of porphin, a planar molecule made up of four 5-membered

Figure 27-4 Sandwich Compounds and Related Molecules
The metal atoms in compounds (a) to (c) are sandwiched between parallel rings. The CO ligands in (e) occupy five of the six corners of an approximate octahedron, the sixth corner of which is at the center of the C-C double bond.

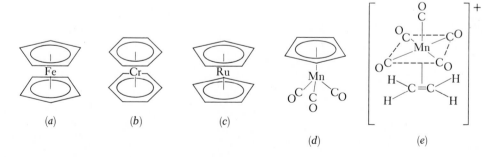

609

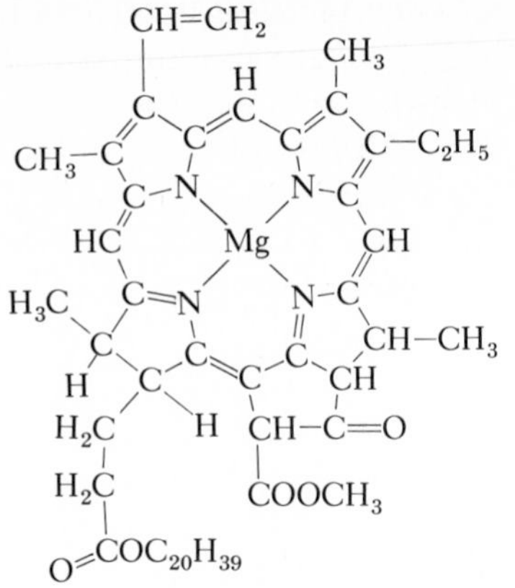

$$CH=CH_2$$

(chemical structure of Chlorophyll a with groups labeled: $CH=CH_2$, CH_3, H, C, CH_3-C, C_2H_5, $C-N$, $N-C$, HC, Mg, CH, H_3C, $C=N$, $N-C$, $CH-CH_3$, H, CH, H_2C, H $CH-C=O$, H_2C, $COOCH_3$, $O=COC_{20}H_{39}$)

Chlorophyll a

Figure 27-5 Chlorophyll a

Chlorophyll b has the same structure as chlorophyll a, except that the —CH_3 group at the upper right is replaced by a —CHO group. Just one of the possible resonance formulas is shown.

rings of carbon and nitrogen atoms. The four nitrogen atoms near the center of the molecule are positioned so that they are capable of chelating a metal atom.

Naturally occurring molecules with the porphin framework are called porphyrins. The green pigments in plants are chlorophylls a (Fig. 27-5) and b, which are porphyrin chelates of magnesium. These pigments are found in the chloroplasts, the primary sites of photosynthesis. Another important porphyrin chelate is the heme group, shown in Fig. 26-3. This group is a structural unit in hemoglobin and myoglobin, the biological molecules that are involved in the transport and storage of oxygen in many mammals, including humans.

27-3 Crystal-Field Theory

Octahedral Surroundings of an Ion The crystal-field theory[2] of bonding in metal complexes focuses on the five d orbitals of the central metal atom or ion. When an atom is undisturbed by an external field these orbitals have the same energy—there is fivefold degeneracy. If the atom is in a spherically symmetric electrostatic field this degeneracy is maintained, but in a less symmetric field the degeneracy may be partly or completely broken.

Consider a specific case in which an electrostatic field of octahedral symmetry is created by six identical negative ions placed at the corners of an octahedron surrounding an atom or ion. For example, we may picture the hexafluorocobaltate(III) ion, CoF_6^{3-}, as a Co^{3+} ion situated in the field produced by six F^- ions (Fig. 27-6a).

The Energy Changes of the d States We need to know how the energies of electrons in the d orbitals are altered by the octahedral electrostatic field. To assess these changes qualitatively, we note that an electron occupying any of the five d orbitals will be repelled by the six F^- ions and that the energy associated with this repulsion—the quantity we wish to know—depends on the orbital being occupied. Consider, for example, the $3d_{xy}$, $3d_{yz}$, and $3d_{zx}$ orbitals (Fig. 27-6c). The

---

[2] So called because it was first applied by H. A. Bethe to ions in crystals, but the theory can be applied to isolated complexes as well.

lobes of the $3d_{xy}$ orbital are directed at a point halfway between the F$^-$ ions in the x, y plane and those of the $3d_{yz}$ point halfway between the ions in the y,z plane; those of the $3d_{zx}$ orbital have a corresponding orientation. This similarity in orientation of the lobes with respect to the surrounding ions requires that electrons in any of these orbitals experience the same energy change (an increase because they approach the F$^-$). Next, consider an electron in the $3d_{x^2-y^2}$ orbital, the lobes of which point toward the F$^-$ ions in the x, y plane (Fig. 27-6b). On the average such an electron will be closer to some of the ions than electrons in any of the three orbitals just discussed, so its energy will be even higher. Turning to the $3d_{z^2}$ orbital, we see from its shape that there is a high probability that an electron occupying it will be near one of the F$^-$ ions in the $+z$ or $-z$ directions and that the probability of finding the electron near one of the ions in the x,y plane is also high. Thus, the increase in energy caused by electrostatic repulsion is also higher for this electron than it is for electrons in the $3d_{xy}$, $3d_{yz}$, or $3d_{zx}$ orbitals. More detailed considerations show that the energy increase for an electron in a $3d_{z^2}$ orbital is exactly the same as that for an electron in a $3d_{x^2-y^2}$ orbital.

The foregoing geometric arguments apply quite generally to all octahedral complexes. If the ligands were uncharged species such as NH$_3$ molecules, the

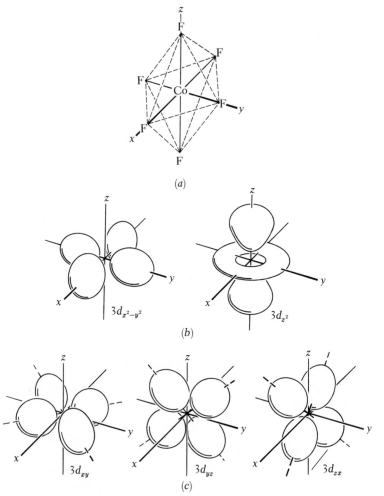

(a)

(b)

$3d_{x^2-y^2}$

$3d_{z^2}$

(c)

$3d_{xy}$

$3d_{yz}$

$3d_{zx}$

Figure 27-6 An Octahedral Species and the Five d Orbitals

(a) The environment of Co(III) in CoF$_6^{3-}$. The Co atom is surrounded by six F$^-$ ions that occupy the corners of an octahedron. (b) and (c) The five d orbitals. Electrons occupying these orbitals are most likely to be found in the directions of the lobes.

611

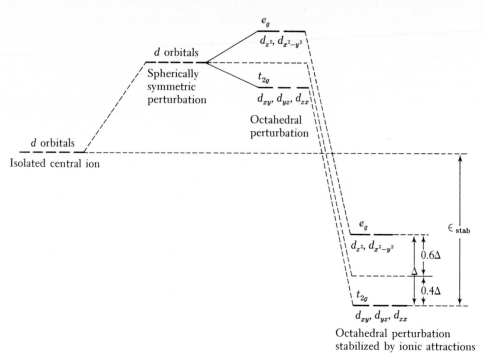

Figure 27-7   Energies of _d_ Orbitals
The change in the energy of an electron in a _d_ orbital by the interactions between six octahedral ligand anions and the positive central metal ion may be considered to be the sum of three effects: (1) the average repulsion of the electron by a negative charge equal to the charge on the ligands but smeared evenly over the surface of a sphere; (2) the change in repulsions when the negative charge is apportioned equally to the corners of an octahedron; (3) the overall stabilization caused by the attraction between ligands and central ion, an attraction that constitutes the bonding between ligands and metal ion.

electrostatic field could be thought of as originating from the partial negative charge on the nitrogen resulting from the polar N—H bonds. The electrostatic field produced by these ligands would differ in strength from the field generated by the fluoride ions, but it would have the same geometry. The _pattern_ of splitting of the _d_ orbitals of the central metal ion would therefore be identical.

In summary, the interaction of an octahedral electrostatic field with the _d_ orbitals of the metal results in an overall increase in the energy of the five _d_ states and in a splitting of the states into two sets: three states of relatively lower energy (d_{xy}, d_{yz}, d_{zx}) and two of relatively higher energy ($d_{x^2-y^2}$ and d_{z^2}). The lower energy states are usually referred to collectively as t_{2g} states and the upper ones as e_g states. Chemical bonding between the ligands and the central metal ion produces a decrease in the energy of the orbitals with respect to the isolated metal atom, an attraction called the _stabilization energy_ of the complex.

A schematic energy-level diagram showing the changes in electron energy caused by the interactions in an octahedral complex is shown in Fig. 27-7. The energy difference between the t_{2g} and the e_g states is called the _crystal-field splitting_ and is often designated by Δ. The level splitting is not symmetric: the three t_{2g} states lie 0.4 Δ below the average energy of the _d_ orbitals and the two e_g states lie 0.6 Δ above the average. The values of Δ and the stabilization energy depend on the nature of the complex. They can be calculated in some special cases but they are usually obtained from experiments, commonly from spectroscopic measurements.

Electronic Structure of Octahedral Complexes The schematic energy-level diagram in Fig. 27-7 can be used to predict the d-electron structure of octahedral complexes by the same principles used to predict the electronic structures of atoms (Sec. 16-2) and molecules (Sec. 18-3). To begin, one must know the total number of d electrons in the complex. Spectroscopic observations show that all the valence electrons in transition-metal *ions* are d electrons. In other words, while the $4s$ state, for example, lies below the $3d$ states in *atoms* of the transition elements, the sequence of states is reversed in the corresponding ions—the d states are filled before the s states. Thus, once the charge on a transition-metal ion is known, it is a simple matter to determine the number of d electrons from the atomic number, as shown in Example 27-1.

☐ **Electron Configurations of Ions** What is the d-electron configuration of the ions Fe^{2+}, Fe^{3+}, and Pt^{4+}?

Example 27-1

Solution The atomic number of iron is 26 and the next lower noble gas is argon, for which $Z = 18$, so iron has eight valence electrons, Fe^{2+} has six, and Fe^{3+} has five. The d electron configurations of the ions are therefore d^6 and d^5, respectively. The easiest way to arrive at the answer for Pt^{4+} is to note that platinum is in the same column of the periodic table as nickel (element 28), which has 10 valence electrons. The loss of four electrons to form the Pt^{4+} ion leaves a d^6 configuration. The same result can be obtained from the atomic number of platinum, 78, which indicates that the atom has 24 more electrons than krypton, the closest noble gas. Of these electrons, 14 are in the $4f$ orbitals, leaving $10d$ electrons in the neutral atom. ∎

☐ What are the d-electron configurations of the ions Cu^+, V^{2+}, and Re^{2+}? ∎

Exercise 27-1

Now consider the sequence in which the d states in an octahedral complex are filled by the available electrons. The first three electrons go into the three t_{2g} orbitals. Each of the electrons goes into a separate orbital because this is the configuration in which the electrostatic repulsion between the electrons is at a minimum (Hund's rule, Sec. 16-2, Example 16-1). There are two alternatives for configurations involving four electrons (Fig. 27-8). The fourth might be paired, with opposite spin, with one of the electrons in the t_{2g} orbitals; or it might go, without change of spin, into one of the e_g orbitals, a configuration that minimizes electron repulsion but also requires the expenditure of energy because of the use of the higher-energy e_g orbital. Both d orbital configurations are found. The first (pairing in lower-energy orbitals before any higher-energy orbitals are occupied) occurs when the crystal-field splitting Δ is large compared to the energy of repulsion between the electrons; this is called the *strong-field* case. The second is found when Δ is small compared to the repulsion energy (the *weak-field* case).

Similar considerations apply to five-electron configurations. If Δ is small, each electron occupies a different orbital; if Δ is large, two of the t_{2g} levels contain paired electrons and one is singly occupied. Diagrams showing the configurations for from 1 to 10 d electrons are sketched in Fig. 27-8. The strong-field and weak-field configurations differ when there are four, five, six, or seven electrons. As we shall see in the following sections, these differences have a significant effect on the properties of complexes.

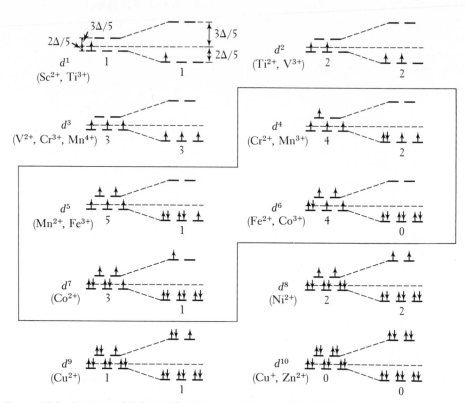

Figure 27-8 Patterns of Filling d Orbitals in Octahedral Complexes

The pattern of filling d orbitals is shown for values of the crystal-field splitting, Δ, that are large and small compared to the electron-repulsion energy. Also shown for each configuration is N, the number of unpaired electrons. The total spin S is $N/2$ because each electron has spin $\frac{1}{2}$. Only for the d^4 through d^7 configurations (inside box) is there a difference between the weak-field and strong-field cases. For these configurations large S is associated with small Δ and small S with large Δ.

Although the discussion of crystal-field splitting has been confined to octahedral complexes, similar considerations apply to other geometries as well. For ions with tetrahedral configuration, for instance, it is found that there are again three t_{2g} levels and two e_g levels. For such ions, however, the e_g levels are more stable than the t_{2g}.

Complexes that contain unpaired electrons are *paramagnetic*. As mentioned in Sec. 18-3, paramagnetic substances are pulled into nonuniform magnetic fields. In contrast, *diamagnetic* substances, which do not contain unpaired electrons, are forced out of such fields. A relatively simple technique for measuring the magnetic properties of a substance is the Gouy method, in which a tube containing the sample is suspended from a balance and placed partway into the field between the poles of an electromagnet. The sample is weighed with the magnet off and with it on. Any difference in weight is due to the magnetic properties of the sample. A paramagnetic substance "weighs" more when the field is on than when it is off; a diamagnetic substance "weighs" less.

It is usually possible to determine the number of unpaired electrons associated with an ion from a measurement of its paramagnetism. For the d^4 through d^7 configurations there is a difference in total spin for the weak-field and strong-field cases (Fig. 27-8); thus, magnetic measurements can be used to determine the character of the splitting of energy states for such ions. Because of this relation between paramagnetism and electron configuration, ions are often described as being *high spin* (small Δ) or *low spin* (large Δ).

□ **Magnetic Criteria for Weak-Field or Strong-Field Behavior** The magnetic properties of the following octahedral ions suggest that they contain the number of unpaired electrons indicated in parentheses: $Fe(CN)_6^{3-}$ (1); FeF_6^{3-} (5); $Co(NO_2)_6^{3-}$ (0); CoF_6^{3-} (4). Determine whether the ions are examples of weak- or strong-field splitting, and indicate the electron configuration for each.

Example 27-2

Solution The iron complexes contain Fe(III), which has five d electrons; the cobalt complexes contain Co(III), which has six d electrons. In the weak-field case no electrons are paired until all five orbitals are singly filled. Thus, in a weak field a d^5 ion will have five unpaired electrons ($t_{2g}^3 e_g^2$); for d^6 there will be four unpaired electrons ($t_{2g}^4 e_g^2$), since one of the t_{2g} orbitals must contain an electron pair. In the strong-field case there is electron pairing in the three t_{2g} orbitals before any of the e_g orbitals becomes populated. A d^5 ion will therefore have one unpaired electron (t_{2g}^5) and a d^6 will have none (t_{2g}^6). Applying this reasoning to the ions listed above, we see that the F$^-$ complexes must be weak-field cases and the others must be strong-field. We might also have noted that for a given metal ion the high-spin case always corresponds to weak-field splitting and the low-spin case to strong-field splitting. ∎

□ The magnetic properties of the following octahedral ions suggest that they contain the number of unpaired electrons indicated in parentheses: (*a*) [Ni(NH$_3$)$_6$]$^{2+}$ (2); (*b*) [Mn(CN$_6$)]$^{4-}$ (1); (*c*) [Mn(NH$_3$)$_6$]$^{2+}$ (5); (*d*) [Cr(CN)$_6$]$^{4-}$ (2). Determine whether the ions are examples of weak- or strong-field splitting, and indicate the electron configuration for each. ∎

Exercise 27-2

27-4 The Spectra of Complexes

The Spectrochemical Series While it is possible to make theoretical estimates of the crystal-field splitting, in practice Δ is adjusted to fit experimental data. This procedure is less satisfying than direct calculations, but it is meaningful because the values of Δ obtained from different experiments are in substantial agreement and form a general pattern for the different transition-metal ions.

Typical values of Δ lie between 80 and 400 kJ mol^{-1}. They depend on the metal and ligand atoms that form the complex and on the oxidation state of the metal atom, and are much larger for high oxidation states than for low ones. This dependence on oxidation state is shown, for instance, by complexes of cobalt and ammonia: the weak-field case applies for Co(II) and the strong-field case for Co(III).

$$(t_{2g})^5(e_g)^2 \qquad (t_{2g})^6$$
$$Co(NH_3)_6^{2+} \qquad Co(NH_3)_6^{3+}$$

This effect may be thought of as related to the decrease in size of the central ion as its charge is increased—ionic radii decrease with increasing oxidation state—so that ligands can approach the metal atom more closely and interact with it more strongly.

Many common ligands can be arranged in a sequence called the *spectrochemical series* that indicates the relative magnitude of the splitting they produce. A partial listing, in order of decreasing power to split d levels, is

$$CN^- \gg NH_3 > O^{2-} \approx OH^- \approx H_2O > F^- > Cl^- > Br^- > I^-$$

For example, CoF_6^{3-} and $Co(H_2O)_6^{3+}$ are weak-field complexes and have four unpaired electrons whereas $Co(NH_3)_6^{3+}$ and $Co(CN)_6^{3-}$ are strong-field complexes and have no unpaired electrons:

$$(t_{2g})^4(e_g)^2 \qquad (t_{2g})^6$$
$$Co(H_2O)_6^{3+}, CoF_6^{3-} \qquad Co(NH_3)_6^{3+}, Co(CN)_6^{3-}$$

The crystal-field splitting is similar for the transition-metal atoms in a given column of the periodic table. However, the repulsion energies between electrons of antiparallel spin are much larger for the elements in the first long period (Sc through Cu) than in the next two periods (Y through Ag and Hf through Au), in part because $4d$ and $5d$ orbitals are spatially much more extended than $3d$ orbitals. Low-spin complexes are therefore quite common for metals in the second and third long periods.

The order of ligands in the spectrochemical series clearly demonstrates one of the weaknesses of crystal-field theory. If the interaction between the metal ion and the ligands were purely electrostatic, then surely a ligand of high charge density such as fluoride ion would produce a larger splitting than neutral ammonia or water. This inconsistency between experiment and theory is only one of a number of observations that lead to the conclusion that in many complexes there is overlap between the d orbitals of the metal and orbitals associated with the ligands.

Ligand-field theory is a modification of crystal-field theory that allows for some degree of covalent bonding between metal and ligand in addition to the purely electrostatic interaction. In many respects it is merely a method for patching up the deficiencies of the electrostatic theory, and it is therefore of limited utility. A molecular orbital theory of metal complexes (Sec. 27-5) can in principle provide a more realistic picture of metal-ligand bonding. Both the molecular orbital and ligand-field treatments lead to level-splitting schemes that agree qualitatively with the predictions of the simpler electrostatic model. Qualitative arguments based on crystal-field theory are thus little changed by the use of more realistic theories.

Absorption Spectra The characteristic colors of many transition-metal complexes result from electronic transitions involving d electrons. The absorption spectrum in the visible region of a pale red-purple solution containing the ion $Ti(H_2O)_6^{3+}$ is shown in Fig. 14-9. It consists of a band with a maximum at about 20,000 cm^{-1}, corresponding to a wavelength of about 5000 Å. The titanium ion in the complex has a d^1 configuration, and the absorption of light is thought to result in the transition of the d electron from a t_{2g} to an e_g state:

The value of Δ necessary to account for the wavelength of this transition is approximately 240 kJ mol^{-1} [$=20 \times 10^3$ cm$^{-1}(1.2 \times 10^{-2}$ kJ mol^{-1}/cm$^{-1})$]. Visible light that passes through the solution with little absorption falls into two wavelength ranges, one around 7000 Å (red light) and the other around 4000 Å (purple light), which accounts for the red-purple color.

When a complex contains several d electrons, the number of possible electronic transitions increases and the corresponding absorption spectrum has several bands. While the analysis of complex spectra is not always unambiguous, a great deal of information about crystal-field splitting has been obtained from spectroscopic studies. It has also been possible to explain many general observations about colors of ions in terms of the behavior and number of d electrons. For example, the absence of color of the alkali-metal and alkaline-earth ions and of ions such as Sc^{3+}, Y^{3+}, Ti^{4+}, and Zr^{4+} is readily associated with the absence of d electrons in the valence shell. Similarly, Zn^{2+}, Cd^{2+}, Hg^{2+}, and Ag^+ are colorless because they have a complete d shell and no readily accessible states of appropriate energy into which d electrons can be promoted.

The intensity of the color associated with an ion depends on the concentration of the ion and also on the intrinsic strength of the absorption. Solutions of $Mn(H_2O)_6{}^{2+}$ are faintly pink, rather than strongly colored, because the electronic transition associated with the absorption of radiation has a low probability of occurring. Manganese(II) has five d electrons and the octahedral environment of the six water molecules is known to produce weak-field splitting, so the electron configuration of the complex is $(t_{2g})^3(e_g)^2$. Transitions to low-lying excited states involve the promotion of electrons from the t_{2g} to the e_g orbitals; for example,

Every conceivable transition involves a change in the total number of unpaired electrons and, for reasons explained by quantum mechanics, transitions that require a change in the total spin occur only rarely.

If cyanide ions are added to a solution containing Mn(II), the deep violet $Mn(CN)_6{}^{4-}$ complex is formed. Cyanide ion is at the top of the spectrochemical series, indicating that it forms strong-field complexes. The five d electrons in this complex are therefore in a $(t_{2g})^5$ configuration and it is possible to have transitions to the e_g state that do not require a change in electron spin; for example,

Not all the transitions that produce color in transition-metal complexes are between d levels, so-called d–d transitions. Another mechanism involves the transfer of an electron to the metal ion from one of the ligands. This type of electronic excitation, called charge transfer, produces very intense color.

617

27-5 The Molecular Orbital Approach

While our discussion of crystal-field theory has centered mainly on octahedral ions, the behavior of ions of other geometries is also well understood. There are many other applications than those discussed; in particular the theory has shed much light on details of the crystal structures of transition-metal compounds. Nevertheless, as we have already pointed out, crystal-field theory is not entirely satisfactory because of its undue emphasis on purely electrostatic interactions.

Like the crystal-field theory, the molecular orbital theory of bonding in transition-metal complexes predicts that the d states of the metal ion are grouped in different sets, such as the t_{2g} and e_g orbitals. In addition it gives a more realistic picture of the bonding. It is sufficiently flexible to deal with a variety of metal-ligand interactions: electrostatic, covalent, and even multiple bonding. The theory also yields at least semiquantitative values for the observed splittings and therefore explains the spectrochemical series. However, it is quite compli-

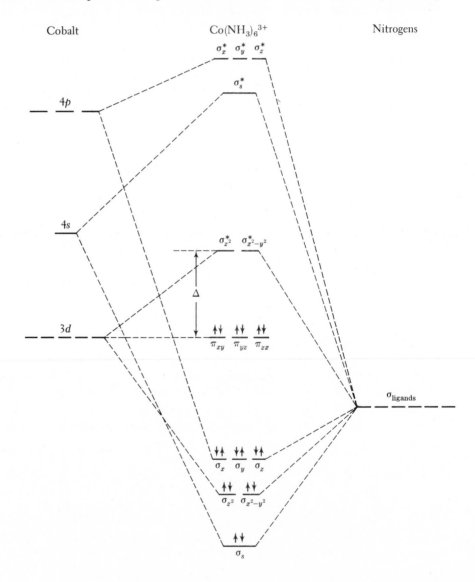

Figure 27-9 Schematic Energy Levels of Cobalt, Nitrogen, and $Co(NH_3)_6{}^{3+}$

cated, and our discussion of it will be concerned only with a qualitative energy-level diagram predicted by the theory for the $Co(NH_3)_6^{3+}$ complex (Fig. 27-9).

We know from crystal-field theory that $Co(NH_3)_6^{3+}$ is a strong-field complex, in which the configuration of the six d electrons is $(t_{2g})^6$. The molecular orbital treatment leads to an equivalent picture in which three molecular states with t_{2g} characteristics are filled before two e_g states when the 18 valence electrons in the complex are accommodated in the lowest-lying orbitals. There are 18 valence electrons to be considered because the Co^{3+} ion contributes six electrons and each of the six ammonia molecules contributes its unshared pair.

The energy level diagram is similar to those shown for molecular orbitals in Sec. 18-3. Approximate energies of the valence states of a cobalt atom are shown at the left, and the $2s$ states of six nitrogen atoms are on the right. The nitrogen states lie below those of cobalt because nitrogen is more electronegative than cobalt. The center of the diagram shows the approximate relative energies of the molecular orbitals obtained by suitable combinations of the cobalt and nitrogen atomic orbitals, as shown by the dashed lines.

The symbols used to designate the molecular orbitals are similar to those used in Chap. 18. The subscripts of the states π_{xy}, π_{yz}, and π_{zx} are meant to imply that these orbitals have the character of the cobalt d_{xy}, d_{yz}, and d_{zx} orbitals; those on the $\sigma_{x^2-y^2}^*$ and $\sigma_{z^2}^*$ orbitals refer to the $d_{x^2-y^2}$ and d_{z^2} orbitals. These π and σ^* orbitals correspond to the t_{2g} and e_g orbitals of the crystal-field theory, and the energy difference between the molecular states is identical to Δ. In this case, however, Δ can be obtained from theory and need not be chosen to fit experiments.

27-6 Equilibria Involving Complexes

We turn now to a consideration of equilibria involving complexes. In recent decades, careful studies of such equilibria, and the evaluation of equilibrium constants involving different complex species, have provided much information about coordination compounds.

The Stepwise Attachment of the Ligands Complexes may be formed by the reaction of ligands with metal ions in solution. Usually the ligands are attached successively, and for each concentration of ligands and metal ions there are corresponding equilibrium concentrations of species that contain different numbers of ligands. For example, the reaction between silver ion and thiosulfate ion, important in the fixing of a photographic image on film, can be described by the following equations and the complex *formation* constants related to them:

$$Ag^+ + S_2O_3^{2-} \rightleftharpoons AgS_2O_3^- \qquad K_1 = 6.6 \times 10^8 \qquad (27\text{-}3)$$

$$AgS_2O_3^- + S_2O_3^{2-} \rightleftharpoons Ag(S_2O_3)_2^{3-} \qquad K_2 = 4.4 \times 10^4 \qquad (27\text{-}4)$$

$$Ag(S_2O_3)_2^{3-} + S_2O_3^{2-} \rightleftharpoons Ag(S_2O_3)_3^{5-} \qquad K_3 = 4.9 \qquad (27\text{-}5)$$

The constant for the overall reaction,

$$Ag^+ + 3S_2O_3^{2-} \rightleftharpoons Ag(S_2O_3)_3^{5-} \qquad (27\text{-}6)$$

is the product of the individual constants, since (27-6) is the sum of the preceding three chemical equations:

$$K_{\text{tot}} = K_1 K_2 K_3 = 1.4 \times 10^{14} \qquad (27\text{-}7)$$

In reactions of this kind the ligands do not react with a bare metal ion but rather with one that is hydrated, as emphasized in Sec. 9-6. In other words, the ligands replace H_2O in an aquo complex. However, because the concentration of water is constant in dilute solutions it can be omitted from the equilibrium expressions (and the chemical equations). This is usually done, partly for simplicity and partly because the number of water molecules is frequently unknown.

When consulting tables of equilibrium constants for complex ions, it is essential to note carefully just how the constants are defined. They may be given for *each step separately,* as in (27-3) to (27-5), or the *overall* constant may be given, as in (27-7). Furthermore, quite frequently, especially in textbooks, *dissociation* constants are quoted rather than *formation* constants. If stepwise dissociation constants are considered for a complex with several ligands, it must be remembered that the dissociation of the first ligand is the reverse of the addition of the last one. For example, silver ion forms two stable complexes with ammonia, $AgNH_3^+$ and $Ag(NH_3)_2^+$. The equations for the formation of these complexes, with the corresponding constants subscripted f to identify them as formation constants, are

$$Ag^+ + NH_3 \rightleftharpoons AgNH_3^+ \qquad K_{1f} = 1.6 \times 10^3 \qquad (27\text{-}8)$$

$$AgNH_3^+ + NH_3 \rightleftharpoons Ag(NH_3)_2^+ \qquad K_{2f} = 6.8 \times 10^3 \qquad (27\text{-}9)$$

Dissociation of one ammonia molecule from $Ag(NH_3)_2^+$ is then the reverse of (27-9).

$$Ag(NH_3)_2^+ \rightleftharpoons AgNH_3^+ + NH_3$$

and thus its equilibrium constant is $K_{1d} = 1/K_{2f} = 1.5 \times 10^{-4}$, the subscript $1d$ implying dissociation of the first ligand.

It is perhaps easier to think in terms of dissociation constants because this is how we have discussed acids and bases (Chap. 12). A small dissociation constant implies a very stable complex, just as a small acid dissociation constant implies a very stable acid with little tendency to transfer its proton to the base water.

Table 27-2
Dissociation Constants for Representative Complexes*

Complex	$K_{tot,d}$	K_{1d}	K_{2d}	K_{3d}	K_{4d}
$Ag(NH_3)_2^+$	9×10^{-8}	1.5×10^{-4}	6×10^{-4}		
$Cu(NH_3)_4^{2+}$	2×10^{-13}	7×10^{-3}	1.3×10^{-3}	3×10^{-4}	7×10^{-5}
$Ni(NH_3)_4^{2+}$	1.1×10^{-8}	7×10^{-2}	1.9×10^{-2}	6×10^{-3}	1.6×10^{-3}
$Zn(NH_3)_4^{2+}$	9×10^{-10}	9×10^{-3}	4×10^{-3}	5×10^{-3}	5×10^{-3}
$Co(NH_3)_6^{2+}$	8×10^{-6}	4	6×10^{-1}	1.8×10^{-1}	9×10^{-2}
$Co(NH_3)_6^{3+}$	7×10^{-36}	4×10^{-5}	9×10^{-6}	2.5×10^{-6}	8×10^{-7}
$Zn(OH)_4^{2-}$†	3×10^{-16}	5×10^{-2}	6×10^{-5}	1.4×10^{-6}	7×10^{-5}
$Fe(SCN)_4^-$	4×10^{-4}	1.3	2.5	1.0×10^{-2}	1.1×10^{-2}
$Ag(S_2O_3)_3^{5-}$	7×10^{-15}	2.0×10^{-1}	2.3×10^{-5}	1.5×10^{-9}	
$Ca(OH)^+$	3×10^{-2}				
$Sr(OH)^+$	1.5×10^{-1}				
$Ba(OH)^+$	2.3×10^{-1}				

*The subscript d denotes dissociation, to avoid confusion with equilibrium constants for the formation reactions. K_{tot} is the product of the individual stepwise K's; for the two cobalt complexes only the first four of these stepwise K's are listed here.

†The constants K_{2d} and K_{3d} for $Zn(OH)_4^{2-}$ refer to equilibria involving the dissolved neutral species $Zn(OH)_2$. Solid zinc hydroxide, $Zn(OH)_2(s)$, is almost insoluble ($K_{sp} = 7 \times 10^{-18}$); the concentration of undissociated $Zn(OH)_2$ in equilibrium with the solid is about $10^{-7}\ M$.

Table 27-2 gives values of both overall ($K_{tot,d}$) and stepwise dissociation constants for a few representative complexes.

Complexometric Titrations The question may be raised as to whether the formation of stable complexes ML_n permits the titration of solutions of M with titrant containing a standard solution of the ligand L. For many ligands the answer is no, because if several different complexes are formed (as they usually are), the individual equilibrium constants for the successive stages frequently differ by so little that the titration curve shows no well-developed step or steps.

An ingenious way to remedy this situation is to link several ligands together, giving a multidentate complexing group. Such a group can form a chelate (Sec. 9-6) if the different potential ligands have the proper geometry so that all can coordinate with the metal ion simultaneously. One of the most widely used molecules of this kind is ethylenediaminetetraacetic acid (abbreviated EDTA), one form of which has the formula

$$\begin{array}{c} \text{HOOC}-\text{H}_2\text{C} \\ \text{HOOC}-\text{H}_2\text{C} \end{array} \!\! N-\text{CH}_2-\text{CH}_2-N \!\! \begin{array}{c} \text{CH}_2-\text{COOH} \\ \text{CH}_2-\text{COOH} \end{array} \qquad (27\text{-}10)$$

EDTA has four acidic hydrogen atoms, and we shall abbreviate its formula as H_4Y. When it forms chelate complexes with metal ions, the complexing agent is actually $EDTA^{4-}$ or Y^{4-}. Both nitrogen atoms of this anion and one oxygen atom of each of its four $-CH_2COO^-$ (acetate) groups can coordinate with a single metal atom. The EDTA anion is thus a sexidentate (Latin: *sex,* six) complexing agent. It forms stable complexes with many metal ions, coordinating the central ion octahedrally. The structure of one such complex, as found by X-ray crystallography, is illustrated in Fig. 27-10.

The formation constants for complexes between Y^{4-} and metal ions are usually very large. The complex ZnY^{2-} is typical:

$$Zn^{2+} + Y^{4-} \rightleftharpoons ZnY^{2-} \qquad K_f = 3.2 \times 10^{16} \qquad (27\text{-}11)$$

Because this is a single-step process with no intermediate forms, and because K_f is so large, it is possible to titrate Zn^{2+} with EDTA. Many other metal ions can equally well be titrated. The metal-ion concentration changes abruptly in the region of the equivalence point, so that a sharp end point can be observed with a suitable indicator.

27-7 Some Applications of Metal Complexes

Complex ions are frequently employed in analytical chemistry, as described in Sec. 27-6. They also have many nonscientific applications, most of which involve control of the solubility of metal ions. Laundry detergents usually contain complexing agents to prevent the precipitation of calcium and magnesium ions (Sec. 23-3). The iron in plant foods is often in the form of its EDTA complex [Formula (27-10) and Fig. 27-10]; large quantities of iron can damage plants and the slow dissociation of this soluble complex provides a controlled supply of trace amounts of the metal.

Control of the concentration of metal ions by complexing is a technique also

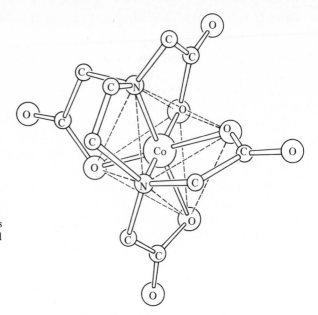

Figure 27-10   Structure of the EDTA Complex of Co(III) in Crystals of Rb[CoEDTA]·2H₂O

The structure of this complex was determined by X-ray crystallography. The molecular geometry of EDTA⁴⁻—bond lengths and angles and the flexibility allowed by rotation about single bonds—is such that each of six of the atoms of the anion, two nitrogens and four oxygens, can be simultaneously at one of the six different corners of an octahedron about the central metal atom and at a distance from that atom appropriate for bonding. The dashed lines are an aid to visualizing the octahedral surroundings of the Co atom.

used in electroplating. When a metal is plated with a more noble metal, e.g., when copper is silver-plated, the plating reaction occurs spontaneously, i.e., with no applied voltage. As a consequence, the rate of formation of the metal layer cannot be controlled by adjusting the voltage and current, and the metal finish produced is of poor quality. If the plating is carried out in a solution containing a suitable ligand such as cyanide ion, however, the concentration of free metal ion can be so drastically reduced that, as predicted from the Nernst equation (Sec. 21-3), the plating process is no longer spontaneous. Plating will occur only if electrical work is performed, and the plating can therefore be controlled.

Complex formation is also an essential part of the photographic process. The light-sensitive emulsion that coats photographic film consists of a thin layer of gelatin in which sulfur and very fine grains of silver bromide are suspended. When the film is exposed, the AgBr grains that have been struck by light undergo chemical reaction, probably the formation of a small amount of Ag_2S on their surfaces. Such grains can be attacked and reduced to silver by treatment in the dark with a developer, typically an alkaline solution of an organic reducing agent. Grains that have not been exposed to light do not react with the developer and are removed from the emulsion with a solution of "hypo", sodium thiosulfate. Chemically, this part of the photographic process, which is called *fixing,* is the formation of soluble, very stable thiosulfate complexes such as $[Ag(S_2O_3)]^-$ and $[Ag(S_2O_3)_2]^{3-}$. The resulting film is a negative of the original image—black in the regions where silver has been deposited because of exposure to light and transparent in unexposed regions, from which AgBr grains have been removed.

Some recent applications of metal complexes involve their reactions with molecular oxygen and nitrogen. Complexes of several transition metals combine reversibly with O_2 and may be useful for storing oxygen in spacecraft and submarines. It may be possible to develop complexes to which nitrogen can be bound, reduced to ammonia, and then released. Such compounds could function like the nitrogen-fixing soil bacteria that are associated with nodules on the roots of legumes.

Summary

The geometrical arrangement of the ligands around the central atom of a complex ion or molecule is tetrahedral or square planar for four ligands, and usually octahedral for six ligands. Isomers characteristic of these arrangements may exist if the ligands are of different kinds or are polydentate. Complexes with coordination numbers other than four and six are less common. Polynuclear complexes contain several metal atoms that are linked by metal-metal bonds, bridging ligands, or both. Aromatic rings, such as the $C_5H_5^-$ anion or benzene, can form complexes in which the bonding to the metal atom involves the π electrons of the ring. Biologically important complexes include chlorophyll and the heme group, which are porphyrins.

The crystal-field theory of transition-metal complexes focuses on the effect of the electric field produced by the ligands on the energy levels of the d orbitals of the central atom. In complexes with octahedral symmetry, the field raises the energy of the $d_{x^2-y^2}$ and d_{z^2} orbitals (collectively called e_g) more than it does the energy of the d_{xy}, d_{yz}, and d_{zx} orbitals (collectively called t_{2g}). The energy difference between the e_g and t_{2g} orbitals is the crystal-field splitting, Δ. For metal atoms with d^4 through d^7 electron configurations, the pattern of filling of the d orbitals depends on the value of Δ. If Δ is large (strong-field case) the e_g states are filled only after the t_{2g} levels have been completely filled. Since this pattern maximizes the number of paired electrons, it leads to a "low-spin" state. If Δ is small (weak-field case) each orbital is filled with a single electron before any electrons are paired. This pattern produces a "high-spin" state. High- and low-spin states can be distinguished experimentally by their magnetic properties.

Depending on the metal and ligand atoms and on the oxidation state of the metal, typical values of Δ lie between 80 and 400 kJ mol^{-1}. The relative magnitudes of the splittings produced by two ligands can be deduced from the positions of the ligands in the sequence called the spectrochemical series. The characteristic colors of many transition-metal complexes result from electronic transitions involving d electrons; hence the visible absorption spectra can often be related to the crystal-field splitting.

Bonding in transition-metal complexes can also be described in terms of molecular orbital theory. The energy level diagrams are comparable to those obtained with the crystal-field approach. In crystal-field theory, the splitting must be treated as an adjustable parameter; however, with molecular orbital theory, Δ can be predicted.

Complexes may be formed by the reaction of ligands with metal ions in solution. The ionic equilibria involve a number of related species whose concentrations depend on the ligand concentration in much the same way that the concentrations of related polyprotic acid species depend on the pH. Titrations of metal ions can be performed with multidentate ligands such as EDTA. Complexing agents are used in many products to control the solubility of metal ions. "Hypo", sodium thiosulfate, is employed in the photographic process to remove unexposed grains of silver bromide from the film emulsion.

Terms and Concepts

Problems and Questions

27-1 Isomers of Cobalt Complexes Two compounds with the formula $Co(NH_3)_5ClBr$ can be prepared. Use structural formulas to show how they differ.

27-2 Isomers of Cobalt Complexes Draw structural formulas for the isomers of the octahedral complex CoA_4BD^+, with $A = NH_3$, $B = OH^-$, and $D = Cl^-$.

27-3 Isomers (a) Two isomers with the molecular formula $Pt(NH_3)_2Cl_2$ are known. Is this number of isomers consistent with an arrangement of the ligands (1) at the corners of a planar square? (2) At the corners of a tetrahedron? (b) Three isomers with the molecular formula PtABCD are known, where A, B, C, and D symbolize four different atoms. Discuss the possibilities for isomerism with a square planar configuration and with a tetrahedral configuration about the Pt atom.

27-4 Isomers Draw all possible isomers of a compound of the formula MA_2XYZ, where M is located in the center of a symmetric square pyramid and the atoms A, A, X, Y, Z are at the five corners.

27-5 Isomers Draw structures for the six possible isomers with formula $CrCl_3(H_2O)_6$ (not all of them are known). [*Hint:* Cr(III) is octahedrally coordinated in all the complexes.]

27-6 Color Change A solution of $Cr(NO_3)_3$ is violet. Addition of HCl changes the color to green. Give a possible explanation in terms of complex ions.

27-7 Inorganic Reaction Explain the following observation: When cupric hydroxide is shaken with pure water, the solvent remains practically colorless; however, if the water contains dissolved ammonium chloride, a blue solution results.

27-8 Redox Couples Explain the following experimental observations. (a) The reducing power of the Fe^{3+}/Fe^{2+} couple is considerably increased by the addition of phosphate ions, known to form strong complexes with Fe^{3+}. (b) Ba^{2+} combines with MnO_4^{2-} to form sparingly soluble $BaMnO_4$; in the presence of Ba^{2+} the alkaline couple MnO_4^-/MnO_4^{2-} becomes much more strongly oxidizing.

27-9 Stable Carbonyls Count the number of electrons in the valence shell of the central atom, including those contributed by the ligands, in each of the following known species, and compare this number with the number associated with a filled valence shell of the central atom: $V(CO)_6^-$, $Cr(CO)_6$, $Mo(CO)_6$, $W(CO)_6$, $Ru(CO)_5$, $Os(CO)_5$, $MnH(CO)_5$, $TcH(CO)_5$, $ReH(CO)_5$, $Fe(CO)_5$, $CoH(CO)_4$.

27-10 Pi Complexes Establish the number of electrons involved in the bonding to the central atom of the following species: $V(C_5H_5)(CO)_4$, $Mn(C_5H_5)(CO)_3$, $Co(C_5H_5)(CO)_2$.

27-11 Energies of d Orbitals Draw two diagrams showing how the five $3d$ energy states of a transition-metal ion are split in an electric field of octahedral symmetry. In the first diagram, show where the $3d$ electrons of a Fe^{2+} ion would be found if the electric field induced by the ligands were very weak. In the second, show the disposition of these same $3d$ electrons if the ligand field were very strong. If the electronic structures you show in the two diagrams are different, what sort of experiment will distinguish between them?

27-12 Occupancies of d Levels Consider, separately, octahedral complexes of V(III), Fe(II), and Fe(III). Show the splitting of the $3d$ levels of the central metal atom and the occupancies of these levels in the weak-field and the strong-field situations. If the two situations differ, indicate which is the low-spin and which the high-spin complex.

27-13 Color of Ions of Transition Elements Which of the following hydrated ions would you expect to be colorless: Ti^{3+}, V^{4+}, Mn^{2+}, Sc^{3+}, Cu^+, Hf^{4+}? Explain your answer.

27-14 Paramagnetism Explain why the species CoF_6^{3-} is paramagnetic while $Co(NH_3)_6^{3+}$ is not. Similarly, why does the paramagnetism of $Fe(H_2O)_6^{3+}$ correspond to five unpaired electrons and that of $Fe(CN)_6^{3-}$ to only one?

27-15 High-Spin and Low-Spin Cases Predict the numbers of unpaired electrons in the strong-field and weak-field situations for octahedral species containing Mn(III), V(III), Cr(II), or Rh(III).

27-16 Crystal-Field Theory (*a*) Why is the crystal-field theory not concerned with *s* states? (*b*) Predict the effect of the ligands on the p_x, p_y, and p_z states of the Co atom of Fig. 27-6*a*. Draw a diagram similar to that of Fig. 27-7. Is the degeneracy of *p* states broken by an octahedral field? (*c*) Consider the effect of a tetrahedral field on the *p* states.

27-17 Crystal-Field Splitting for an Elongated Octahedron Consider an atom surrounded octahedrally by ligands, two of which, at opposite corners of the (distorted) octahedron, are more distant than the other four. Simulate the effect of the ligands on the central atom by negative charges, two of which are located along the $+z$ and $-z$ axes at a distance a and the other four along the $+x$, $-x$, $+y$, and $-y$ axes at a distance b, with a larger than b (Fig. 27-11). Consider the effect of this charge distribution on the energy of each of the five *d* states of the atom, and draw a diagram that shows qualitatively the relative values of the energies of these states.

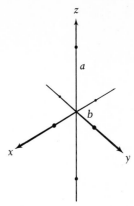

Figure 27-11

27-18 Solubility of AgBr in a Thiosulfate Solution Calculate the solubility of AgBr ($K_{sp} = 5.2 \times 10^{-13}\, M^2$) in a solution in which sufficient sodium thiosulfate has been dissolved that, at equilibrium, the concentration of $S_2O_3^{2-}$ is 0.2 M. Use constants for the silver-thiosulfate system from Table 27-2. {*Hint:* This problem is analogous to the polyprotic acid problems of Sec. 12-6. Note that the equilibrium value of $[S_2O_3^{2-}]$ is just equal to K_{1d} for $Ag(S_2O_3)_3^{5-}$, so that the concentrations of the trithiosulfato and dithiosulfato complexes are equal and that of the monothiosulfato complex is negligible.} Use the resulting solubility of AgBr in this solution and the known distribution of the complex species, to calculate what the *initial* concentration of $Na_2S_2O_3$ must have been before any AgBr dissolved.

Nuclear Chemistry

28

" . . . we have obtained a substance whose activity is about 400 times that of uranium.

"We believe . . . that the substance that we have extracted from the pitchblende contains a metal not yet observed, closely related to bismuth in its analytical properties. If the existence of this new metal is confirmed, we propose to call it *polonium*, after the name of the native country of one of us."

P. CURIE AND M. S. CURIE, 1898

"The problem of transmutation and the liberation of atomic energy to carry on the labour of the world is no longer surrounded with mystery and ignorance, but is daily being reduced to a form capable of exact quantitative reasoning. It may be that it will remain forever unsolved. But . . . it [is] probable that one day we will see its achievement.

"Should that day ever arrive, let no one be blind to the magnitude of the issue at stake, or suppose that such an acquisition to the physical resources of humanity can safely be entrusted to those who in the past have converted the blessings conferred by science into a curse."

FREDERICK SODDY, 1920

28-1 Introduction

Although nuclear chemistry became an established chemical field only during the second world war, it is in many ways a direct extension of the work of the earliest chemists. Transmutation of the elements, the goal of the alchemists, occurs in nuclear reactions, and the work of Mendeleev and Meyer continues in the production and characterization of new elements. Much of the research and many of the applications in nuclear chemistry are involved with methods of analysis, but at levels of sensitivity unimagined by the early chemists.

We begin the chapter with the story of the first investigations of radioactivity. The transformations involved in radioactive decay and radioactive decay rates are then discussed. Nuclear stability and the energy relations in nuclear reactions are considered next, with special emphasis on fission and fusion reactions. Section 28-3, which deals with the synthesis of elements, is, in effect, a continuation of the descriptive chemistry discussed in the preceding five chapters. The chapter concludes with a consideration of some applications.

Early History The discovery of the radioactivity of uranium by Becquerel in 1896 was briefly described in Chap. 1. Becquerel also observed that, like X rays, the emanations from uranium could ionize air. As a part of her thesis research, Marie Sklowdowska Curie used this property to measure accurately the intensity of the radiation and to characterize it. Her husband, Pierre Curie, had developed an electrometer with which the ionization could be measured with precision.

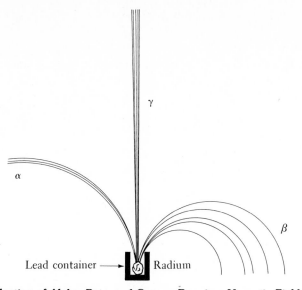

Figure 28-1 Deflection of Alpha, Beta, and Gamma Rays in a Magnetic Field
A magnetic field is perpendicular to the plane of the drawing. Alpha and beta rays are deflected
in directions that indicate they are, respectively, charged positively and negatively. Gamma rays
are not deflected, indicating the absence of electric charge.

Madame Curie tested most of the known elements for radioactivity and found
that only uranium and thorium produced significant ionization. She showed that
the radiation was a characteristic of the element, independent of its chemical or
physical state. An important discovery was that the radiation from the uranium
ore pitchblende was more intense than that from pure uranium. This she interpreted as evidence that the pitchblende contained a new, highly radioactive
substance.

The Curies performed countless chemical separations to isolate trace constituents of the ore. In July 1898 they reported the discovery of a new radioactive
element, which was associated with the bismuth fraction of the pitchblende.
They named it polonium, in honor of Mme. Curie's native Poland. Five months
later the Curies announced that another radioactive constituent had been found.
After two more years of work, one-tenth gram of this new element, radium, was
isolated by fractional crystallization from the barium fraction of the ore. This
was the yield from two tons of pitchblende.

Studies of the radiation from radioactive substances by a number of investiga-
tors showed that there are three distinct kinds of rays, which can be characterized
by their depth of penetration into metal foils and their response to electric and
magnetic fields (Fig. 28-1). Alpha rays are helium nuclei; beta rays are high-speed
electrons. Diffraction experiments demonstrated that gamma rays are short-
wavelength electromagnetic radiation. Rutherford and Soddy understood by 1903
that radioactivity was associated with the spontaneous transformation of one
chemical element into another. This was almost a decade before Rutherford
proposed his nuclear model of the atom, which made it clear that radioactivity is
associated with the nucleus.

Radioactive Decay   Radioactivity is a manifestation of nuclear instability. In beta emission, that is, when a nucleus emits an electron, the atomic number Z increases by 1 but the mass number A does not change. In effect, one neutron in the nucleus has been converted into a proton and an electron. In representing such a nuclear reaction by an equation, our concern is only with the nuclei involved, not with any accompanying changes involving extranuclear electrons:

$$_Z^A\text{X} \longrightarrow {}_{Z+1}^{A}\text{Y} + {}_{-1}^{0}e \tag{28-1}$$

When a nucleus decays by emission of an alpha particle, A must decrease by 4 and Z by 2:

$$_Z^A\text{X} \longrightarrow {}_{Z-2}^{A-4}\text{Q} + {}_2^4\text{He} \tag{28-2}$$

All nuclear reactions conform to the same general principles: there is conservation of mass-energy, conservation of charge, and conservation of nucleons (in the absence of antiprotons and antineutrons, the usual situation). Other quantities are conserved as well, but they are not of concern here. The only difference from the conservation laws for ordinary chemical reactions is that atoms are conserved in chemical reactions but not in nuclear reactions. In ordinary chemical reactions it is possible to speak of conservation of mass by itself, because the energy changes involved are so small that the mass differences corresponding to them (through Einstein's relation $E = mc^2$) are of the order of 1 part in 10^{10} or even smaller, and are thus normally undetectable. In nuclear reactions the energy changes are often so large (Sec. 28-3) that mass changes are readily detectable, but the sum of Δm and $\Delta E/c^2$ is always zero, i.e., mass-energy is conserved.

Example 28-1

☐ **Radioactive Decay**   The plutonium isotope with mass number 241, which has a half-life of 13 year, decays by beta emission. The product of this decay is itself radioactive and decays by emission of an alpha particle. What are the atomic and mass numbers of the products of the beta and alpha decays?

Solution   We find from either the table of atomic weights (inside the front cover) or a periodic table that the atomic number of plutonium is 94. The first decay can be written

$$_{94}^{241}\text{Pu} \longrightarrow {}_{95}^{241}\underline{\text{X}} + {}_{-1}^{0}e$$

where the mass number and atomic number of the product follow from the requirement that mass and charge are conserved (that is, $241 = 241 + 0$ and $94 = 95 - 1$). Applying the same conservation principles to the second decay gives

$$_{95}^{241}\text{X} \longrightarrow {}_{93}^{237}\underline{\text{Y}} + {}_2^4\text{He}$$

The element of atomic number 95 is americium and that of atomic number 93 is neptunium. ∎

Exercise 28-1

☐ The uranium isotope $_{92}^{233}\text{U}$ is a product of beta decay. It is itself radioactive and decays by alpha emission. From what nuclide does $_{92}^{233}\text{U}$ arise and what is the product of its decay? ∎

Some nuclei decay only by emission of a photon (a gamma ray); neither A nor Z changes. The original nucleus is in an excited state, corresponding to some unstable configuration of its nucleons. As a more stable arrangement is achieved, a photon is emitted corresponding to the difference in energy of the two nuclear states, just as a photon is emitted by an atom in an excited electronic state that decays to some more stable arrangement of the electrons. Nuclear energy levels are typically separated by energies around 10^6 eV (1 Mev), about 1 million times the separation of the outer electronic energy levels in atoms. Gamma rays have wavelengths of the order of 10^{-2} Å.

Radioactive Decay Rates Because nuclear energy levels are so widely spaced relative to ordinary thermal and chemical-bond energies, the rates of radioactive disintegrations and other nuclear reactions generally do not depend on temperature, physical state, or chemical form. The activity, X, of a sample, which is defined as the number of disintegrations per unit time, is directly proportional to N, the number of atoms of the radioactive substance present

$$\text{Rate} = -\frac{d[N]}{dt} = k[N]$$

Radioactive decay is thus a classic example of a first-order rate process (Sec. 22-3). If the activity of the sample is X_0 at time $t = 0$ and X at time t (for example, the present), then from (22-13) and (22-15), with $\tau_{1/2}$ the half-life of the radioactive atom,

$$\ln \frac{X}{X_0} = \ln \frac{N}{N_0} = -0.693 \frac{t}{\tau_{1/2}} \tag{28-3}$$

☐ **Americium Decay** The americium isotope $^{241}_{95}$Am is an alpha emitter with a half-life of 458 year. Capsules containing americium are used to ionize air in some smoke detectors. How many years does it take for the americium activity to fall by 5 percent?

Example 28-2

Solution The relation between activity, half-life, and time is given by (28-3):

$$\ln \frac{X}{X_0} = \ln(0.95) = -0.693 \left(\frac{t}{458 \text{ year}} \right)$$

$$t = \frac{(458 \text{ year})(-0.051)}{-0.693} = \underline{34 \text{ year}} \blacksquare$$

☐ Assume that when the earth was formed, about 5×10^9 year ago, there existed 10^{20} mol of a radioactive isotope of an element. What half-life must the isotope have if only 1000 atoms of it remain today? ∎

Exercise 28-2

Nuclear Stability Consideration of the properties of stable nuclei shows that the presence of neutrons in the nucleus is associated with nuclear stability. Every stable nucleus containing more than two protons has at least as many neutrons as protons. Furthermore, the ratio of the number of neutrons, N, to the number of protons, Z, in stable nuclei increases as Z increases. Figure 28-2 shows graphically the relation between Z and $N (= A - Z)$ for stable nuclei. Up to $Z = 20$ there are numerous stable nuclei with $N = Z$, but beyond that N gradually becomes greater than Z. For the heaviest known stable nucleus, ^{209}Bi, $N = 1.52Z$; there are no stable nuclei with Z greater than 83.

Nuclear radii vary from about 1×10^{-5} to 8×10^{-5} Å. At these distances there are very strong coulomb repulsive forces between pairs of protons as a consequence of their positive charge; at a distance of 1×10^{-5} Å the proton-proton coulomb repulsive energy amounts to more than 10^8 kJ mol^{-1}. Very strong short-range attractive forces between nucleons (neutron-neutron, neutron-proton, and proton-proton), arising through the exchange of subatomic particles called π mesons, can provide enough binding energy to offset these repulsions for nuclei of moderate size. However, these attractive forces are of much shorter range than the coulomb repulsions, and as the number of protons continues to increase, the repulsions tend to dominate and the nucleus becomes unstable. Addition of more neutrons cannot compensate for this because Pauli's principle that no two particles of the same kind may occupy the same state applies to nucleons (protons and neutrons) as well as to electrons. The added neutrons would therefore have to occupy high energy levels and thus could not confer added stability.

When an unstable nucleus is created that is outside the band of stability indicated in Fig. 28-2, it will change to a more stable nucleus in such a way as to return to the band of stability. If the N/Z ratio is too high, for example, emission of an electron from the nucleus will decrease N and increase Z, thus lowering the N/Z ratio. Conversely, if that ratio is too low, the nucleus may capture one of the atom's extranuclear electrons or emit a *positron* (a positive electron); in either case N increases by 1 and Z decreases by 1, raising the N/Z ratio. Alternatively, a nucleus of high atomic number for which N/Z is too low may emit an alpha particle, thereby changing into a more stable nucleus with a higher N/Z ratio.

Table 28-1 illustrates some of these nuclear transformations. It shows the series of disintegration products arising from the long-lived isotope ^{238}U, whose half-life, 4.5 billion year, is comparable to the age of the earth. Successive radioactive disintegrations occur by either electron or alpha-particle emission until the stable nucleus ^{206}Pb is formed. All mass numbers in this series differ from that of the parent nucleus, ^{238}U, by an integral multiple of 4. There are three other similar

Figure 28-2   Relation between the Number of Protons and the Number of Neutrons in Stable Nuclei

This broad curve, which tapers at each end, represents the general trend of the relation between N and Z for the stable nuclei. There are, however, many unstable nuclei within this "band of stability", in particular most of those with both N and Z odd, and some with either N or Z odd. The majority of the approximately 280 known stable nuclei have both N and Z even. Only five stable nuclei have odd Z and odd N (^{2}H, ^{6}Li, ^{10}B, ^{14}N, and ^{180}Ta; in addition ^{50}V is almost stable, with a half-life of 4×10^{14} year, so that it has scarcely disintegrated during the lifetime of the earth). Some elements of even atomic number have many stable isotopes—for example, Sn (10) and Xe (9). Many odd-Z elements have no more than one stable isotope, although many unstable isotopes are known for some (for example, 19 for iodine).

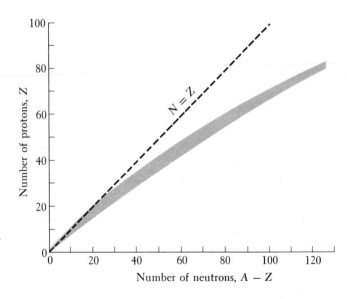

Table 28-1
The Uranium-Radium
Radioactive Decay Series*

Z / A	82	83	84	85	86	87	88	89	90	91	92
238											U 4.5×10^9 y
234									Th 24.5 d	Pa 1.1 m	U 2×10^5 y
230									Th 8×10^4 y		
226							Ra 1622 y				
222					Rn 3.8 d						
218			Po 3 m								
214	Pb 27 m	Bi 20 m	Po 1.5×10^{-4} s								
210	Pb 22 y	Bi 5 d	Po 140 d								
206	Pb Stable										

*The arrows diagonally to the left correspond to the emission of an alpha particle by a nucleus; the shorter arrows pointing to the right correspond to nuclear emission of an electron. The figures below the symbols for the elements represent the half-lives of the corresponding isotopes; abbreviations are as follows: y, years; d, days; h, hours; m, minutes; s, seconds. A few of the nuclei shown have alternative modes of decay, but they are of minor importance, occurring only about once in a thousand disintegrations.

series, two of which originate with naturally occurring long-lived radioactive isotopes, ^{235}U and ^{232}Th; the third was discovered during World War II and is named for its longest-lived member, ^{237}Np, with a half-life of 2.2 million year.

Binding Energy The stability of different nuclei can be compared by calculation of the *binding energy* for each. This quantity is defined for a given nucleus as the energy equivalent (by the relation $E = mc^2$) to the difference between the sum of the masses of the individual nucleons present and the mass of the nucleus itself. Further comparison is made in terms of the binding energy per nucleon, that is, the total binding energy divided by the total number of nucleons (which is the mass number, A).

☐ **Binding Energy** The nucleus of an ordinary helium atom contains two neutrons and two protons. Its mass, and those of a neutron and a proton, are, respectively, 6.64462×10^{-27} kg, 1.67495×10^{-27} kg, and 1.67265×10^{-27} kg. Calculate the binding energy for ^{4}He.

Example 28-3

Solution The difference between the sum of the masses of the nucleons constituting the helium nucleus and the mass of the nucleus itself is

$$\Delta m = (2 \times 1.67495 + 2 \times 1.67265 - 6.64462) \times 10^{-27} \text{ kg}$$
$$= 5.058 \times 10^{-29} \text{ kg}$$

Hence by the Einstein relation the difference in mass corresponds to an energy

$$E = \Delta m \, c^2 = 5.058 \times 10^{-29} \text{ kg} \times (2.9979 \times 10^8 \text{ m s}^{-1})^2$$

$$= 4.546 \times 10^{-12} \text{ kg m}^2 \text{ s}^{-2} \times \frac{1 \text{ J}}{\text{kg m}^2 \text{ s}^{-2}} = \underline{4.546 \times 10^{-12} \text{ J}}$$

This is a huge energy for the formation of just one nucleus. For 1 g of ^{4}He nuclei, with atomic weight 4.00, the binding energy is

$$\frac{4.546 \times 10^{-12} \text{ J}}{\text{nucleus}} \times \frac{6.022 \times 10^{23} \text{ nuclei}}{1 \text{ mol}} \times \left(\frac{1 \text{ mol}}{4.00 \text{ g}}\right) 1 \text{ g} = \underline{6.84 \times 10^8 \text{ kJ}} \quad \blacksquare$$

Exercise 28-3

☐ The mass of the nucleus of $^7_3$Li is 11.65063×10^{-27} kg. Calculate the binding energy for 1 mol (7.00 g) of ^{7}Li nuclei. ∎

Fission and Fusion The binding energy per nucleon rises rapidly with increasing A up to a maximum for A values in the range 55 to 65, and then it falls slowly (Fig. 28-3). Thus energy is released in a nuclear reaction in which low-A nuclides combine to form a nucleus of somewhat higher A value (a process called nuclear *fusion*) or in a reaction in which a nucleus with a high value of A splits to form nuclides of intermediate A values (a *fission* reaction).

When certain very heavy nuclei, notably ^{235}U and ^{239}Pu, absorb slow neutrons, fission occurs. The products are nuclei of intermediate mass numbers, ranging from about 75 to 155; on the average, two to three neutrons are produced as well

Figure 28-3 Binding Energy per Nucleon for Stable Nuclei
The maximum binding energy per nucleon, and thus maximum stability, occurs for mass numbers near 60. The slow decrease in binding energy per nucleon as A increases beyond about 60 is due to increasing mutual coulomb repulsion of the protons in the nucleus as the atomic number, Z, increases.

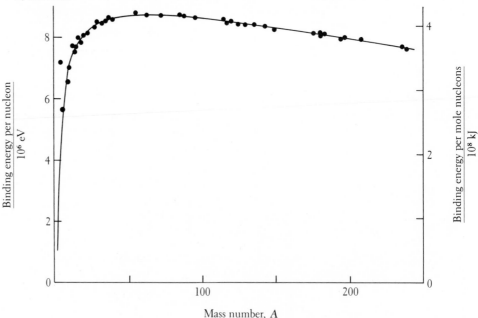

for each nucleus undergoing fission. A representative fission reaction for ^{235}U is

$$^{235}_{92}\text{U} + ^{1}_{0}n \longrightarrow [^{236}_{92}\text{U}] \longrightarrow 3^{1}_{0}n + ^{94}_{36}\text{Kr} + ^{139}_{56}\text{Ba} \qquad (28\text{-}4)$$

Because the N/Z ratios of the product nuclei are higher than those for stable nuclei in this range of A values, fission products are usually intensely radioactive, emitting highly energetic electrons and gamma rays (and occasionally further neutrons) before they decay into stable nuclides. The chief problem in utilizing the energy released in fission as an important energy source is the safe disposal of these intensely radioactive products.

The energy released in a fission reaction of ^{235}U is about 2×10^{10} kJ mol^{-1}, about 10^{7} times the energy released in a typical chemical reaction of this same quantity of uranium (its oxidation or fluorination, for example). Because neutrons are emitted during each fission, they can (if slowed down) be absorbed by other ^{235}U nuclei, causing them in turn to undergo fission. Since each fission produces more than one neutron, the splitting of one nucleus can trigger the fission of two or three nuclei. A chain reaction can therefore be generated, and the entire sample can react in a very small fraction of a second. If the sample is of even modest size and if all of the energy is released in a small volume within a very short time, a powerful explosion will result. The first nuclear weapons were fission bombs.

Chain reactions do not occur in ordinary uranium samples[1] because ^{238}U, which constitutes 99.3 percent of the naturally occurring element, does not undergo fission. Neutrons produced by fission of the 0.7 percent ^{235}U have little chance of colliding with fissionable nuclei so a chain reaction cannot take place. Since the number of neutrons absorbed by a sample is proportional to its volume, and the number that leave the sample increases with increasing surface area, there is a critical sample volume-to-surface ratio (which is shape-dependent) below which chain reactions will not occur. The ratio of volume to surface increases with increasing sample size; hence there is a minimum quantity of a substance (the "critical mass") required to sustain a chain reaction. Such a critical system can be operated in a controlled manner if the energy produced is properly dissipated, as in a nuclear reactor.

Fusion reactions provide the energy of the sun and other stars. The principal reaction in the sun is the fusion of four protons to form a single ^{4}He nucleus. This reaction, which predominates at temperatures below about 20 million degrees, has been pictured as occurring by the following scheme (in which, $^{0}_{1}e$ denotes a positron):

$$^{1}_{1}\text{H} + ^{1}_{1}\text{H} \longrightarrow ^{2}_{1}\text{H} + ^{0}_{1}e$$
$$^{2}_{1}\text{H} + ^{1}_{1}\text{H} \longrightarrow ^{3}_{2}\text{He} + h\nu \qquad (28\text{-}5)$$
$$^{3}_{2}\text{He} + ^{3}_{2}\text{He} \longrightarrow ^{4}_{2}\text{He} + 2\,^{1}_{1}\text{H}$$

At considerably higher temperatures further fusion reactions occur, producing carbon and heavier elements.

Fusion reactions are effected in an explosive way in thermonuclear weapons

---

[1] Up until about 400 million years ago a controlled chain reaction (Sec. 28-4) went on in deposits of uranium ore in West Africa. The chain reaction could be sustained because the abundance of ^{235}U at that time was much higher than it is today. The fraction of U that is ^{235}U decreases with time because the half-life of ^{235}U is about six times shorter than that of ^{238}U.

(hydrogen bombs). In these awesomely frightening devices, deuterium and tritium are fused, with ^{6}Li present also, to form ^{4}He and a neutron. The reaction is initiated by a fission bomb, which raises the temperature to around 10^7 K in a matter of microseconds. Highly exothermic fusion reactions then begin, further increasing the temperature and, thereby, the rate of fusion. The energy released is of the order of 10^3 or more times that of a fission bomb.

28-3 Elemental Synthesis

"Artificial" Radioactivity The discovery of the element rhenium in 1925 left four gaps in the periodic table. Elements 43, 61, 85, and 87 were still missing despite extensive efforts to isolate them from elements thought to have similar properties. In 1934, Frédéric and Irène Joliot-Curie[2] reported that boron, aluminum, and magnesium could be made radioactive by bombardment with α particles, the radioactivity continuing after the source of α particles was removed. This discovery showed that radioactive species not found in nature could be produced. The Joliot-Curies had discovered nuclear reactions for the synthesis of new, unstable isotopes of nitrogen, phosphorus, and silicon:

$$^{10}_{5}B + ^{4}_{2}He \longrightarrow ^{13}_{7}N + ^{1}_{0}n$$
$$^{27}_{13}Al + ^{4}_{2}He \longrightarrow ^{30}_{15}P + ^{1}_{0}n$$
$$^{24}_{12}Mg + ^{4}_{2}He \longrightarrow ^{27}_{14}Si + ^{1}_{0}n$$

A new approach to the discovery of elements had been uncovered—elements might be synthesized.

Technetium (Greek: *tekhnetos*, artificial), the first previously unknown element produced artificially, was isolated in 1936 from a molybdenum plate containing a mixture of Mo isotopes that had been bombarded by deuterons (deuterium nuclei) in a particle accelerator (known as a cyclotron) for several months:

$$^{A}_{42}Mo + ^{2}_{1}H \longrightarrow ^{A+1}_{43}Tc + ^{1}_{0}n$$

Later large quantities of a long-lived isotope $^{99}_{43}$Tc ($\tau_{1/2} = 2.1 \times 10^5$ year) were produced by neutron irradiation of ^{98}Mo ("neutron capture") in a nuclear reactor:

$$^{98}_{42}Mo + ^{1}_{0}n \longrightarrow [^{99}_{42}Mo] \longrightarrow ^{99}_{43}Tc + ^{0}_{-1}e \qquad (28\text{-}6)$$

Astatine (Greek: *astatos*, unstable), element 85, was also first prepared in a cyclotron. Bismuth 209 was bombarded with α particles

$$^{209}_{83}Bi + ^{4}_{2}He \longrightarrow ^{211}_{85}At + 2\,^{1}_{0}n$$

A preliminary report of the discovery was made in 1940 but the second world war halted further publication and definitive measurements were not published until 1947.

Attempts to produce element 61 (promethium) by cyclotron bombardment of its rare-earth neighbors, neodymium and samarium, produced new radioactivity that was attributed to isotopes of the new element. Promethium's existence was

---

[2] Irène Curie, daughter of Marie and Pierre Curie, married Frédéric Joliot, who worked as her mother's assistant.

proved conclusively only when several of its isotopes were found in the products of fission of uranium 235. An isotope of francium, $^{223}_{87}$Fr, was first detected in very small amounts as a product of a very rare α decay of $^{227}_{89}$Ac. Later, the isotope $^{221}_{87}$Fr was found to be a part of the decay series that starts with the transuranium element neptunium.

No stable isotopes of elements 43, 61, 85, or 87 have been found or are likely to be; all the known isotopes of these elements are relatively short-lived. Any of these elements that may have been present at the time of the formation of the earth disappeared long ago (see Exercise 28-2).

Transuranium Elements The techniques of elemental synthesis that were successfully applied in filling the gaps in the periodic table have also been used to extend the known elements beyond uranium. In 1934, Enrico Fermi, an Italian physicist, attempted to prepare element 93 by irradiating uranium with neutrons, a procedure analogous to (28-6), the production of technetium by neutron capture. Much new radioactivity appeared in the sample after the neutron irradiation, and Fermi and his coworkers attributed it to the synthesis of elements beyond uranium. Five years later, however, Hahn and Strassman reported that they had found that most of the new radioactivity was associated with isotopes of elements (for example, barium) of much lower atomic weight than uranium. A new phenomenon had been discovered—the uranium had undergone fission.

Under proper conditions neutron irradiation *can* be used to prepare transuranium elements. Neptunium ($Z = 93$), which was discovered in 1940, was made by neutron capture as was plutonium ($Z = 94$), first prepared in 1941. Other transuranium elements have been produced by bombardment of high-Z nuclei with more massive particles, such as helium, boron, and carbon ions. Table 28-2 is a list of the known transuranium elements, the year of their discovery, and their first method of synthesis.

Table 28-2
Transuranium Elements

Z	Symbol	Year synthesized	First synthetic method	Half-life of first isotope produced*
93	Np	1940	^{238}U + n	2.35 d
94	Pu	1941	^{238}U + n	86.4 y
95	Am	1945	^{239}Pu + n	458 y
96	Cm	1944	^{239}Pu + n	162.5 d
97	Bk	1949	^{241}Am + ^{4}He	4.5 h
98	Cf	1950	^{242}Cm + ^{4}He	44 m
99	Es	1952	^{238}U + n (thermonuclear explosion)	20 d
100	Fm	1953	^{238}U + n (thermonuclear explosion)	22 h
101	Md	1955	^{253}Es + ^{4}He	1.5 h
102	No	1958	^{246}Cm + C	3 s
103	Lr	1961	^{252}Cf + B	8 s
104	[Rf]†	1969	^{249}Cf + C	4.5 s
105	[Ha]‡	1970	^{243}Am + ^{22}Ne	1.4 s
106	—	1974	^{249}Cf + ^{18}O	$<10^{-2}$ s

*See footnote to Table 28-1 for abbreviations.

†Rutherfordium; an alternative name is kurchatovium.

‡Hahnium; an alternative name is bohrium.

Over 100 isotopes of transuranium elements have been synthesized. Neptunium and plutonium have been prepared by the ton; other elements have been produced only in submicrogram quantities. Many arguments have arisen concerning priority of discovery of transuranium elements. For example, Russian researchers claim to have been the first to produce elements 104 and 105 and, as discoverers, propose that the elements be called kurchatovium and bohrium. A counterclaim is made by American scientists who have used different preparative techniques and question the validity of the Russian results. They have suggested the names rutherfordium and hahnium. Both groups claim the discovery of element 106, which has not yet been named.

The Actinides Very elegant and sensitive techniques have been developed to determine the chemical and physical properties of minute amounts of the transuranium elements. In some instances, only a few atoms of the elements have been produced, and determination of properties is essentially impossible. It is now clear that the 14 elements beyond actinium form a series analogous to the lanthanides that corresponds to the filling of the $5f$ subshell.

The actinides do not resemble each other as much as the lanthanides do, because the $5f$ electrons are less effectively shielded from neighboring atoms by the $6s$ and $6p$ electrons than are the $4f$ by the $5s$ and $5p$ electrons. Although the actinides resemble the lanthanides in some ways, their chemistry is more complex. Many actinides have several stable oxidation states, ranging from (III) to (VI). The most stable oxidation state of the most important actinide, uranium, is (VI)—exemplified in the uranyl ion, $UO_2{}^{2+}$, which forms a number of salts, and in UF_6, a highly reactive molecule that forms volatile yellow crystals with a vapor pressure of 1 atm at 57°C. The volatility of UF_6 made possible the gaseous-diffusion separation of uranium isotopes at Oak Ridge, Tennessee, during World War II (Sec. 4-4).

All isotopes of the actinides are radioactive. Actinium, thorium, protactinium, and uranium occur naturally; the other actinides occur on earth only because they are made by nuclear reactions.

Further Extensions of the Periodic Table Can we expect that the periodic table will continue to expand as still more elements are synthesized? A study of the half-lives of the transuranium elements in Table 28-2 leads to the conclusion that it will be difficult to produce new elements. The increasingly shorter half-lives are indications of the decrease in stability with higher atomic number.

The stability of nuclei can be predicted by theories of nuclear structure. It is known, for example, that nuclei containing certain "magic numbers" of protons or neutrons (2, 8, 20, 28, 50, 82, 126) are particularly stable. Elements for which Z is a magic number generally have a large number of stable isotopes when compared with their immediate neighbors in the periodic table. They tend also to be the most abundant nuclei in their mass range. "Doubly magic" nuclei, those that have magic numbers of both protons and neutrons, are especially stable. A prime example is $^{208}_{82}Pb$, which has 82 protons and 126 neutrons. These magic numbers are reminiscent of the numbers associated with complete shells of electrons and, indeed, a very successful shell theory of the nucleus, with complete shells of nucleons at the "magic numbers", has been developed.

It has been predicted that the next nuclear shell closure will be reached near element 114. A "superheavy" element with $Z \approx 114$ might therefore be relatively stable and have a half-life sufficiently long to allow it to be prepared. Attempts to produce superheavy elements by the collision of heavy nuclei in accelerators have so far failed. Searches for traces of superheavy elements in nature have also been unsuccessful.

28-4   Some Applications

^{14}C Dating   A radioactive isotope of carbon, ^{14}C, has a natural abundance of about 10^{-10} percent. Carbon 14 has a half-life (5730 year) that is short on a geological time scale. It is found in nature despite its continuous radioactive decay because it is continually formed in the upper atmosphere by the reaction of neutrons, produced by cosmic rays, with the very abundant ^{14}N. The pertinent equations for formation and decay are

$$^{14}_{7}N + ^{1}_{0}n \longrightarrow ^{14}_{6}C + ^{1}_{1}H \qquad (28\text{-}7)$$

$$^{14}_{6}C \longrightarrow ^{14}_{7}N + ^{0}_{-1}e \qquad (28\text{-}8)$$

The concentration of ^{14}C in natural carbon is a steady-state concentration—the number of ^{14}C nuclei disintegrating by (28-8) just matches the number formed from ^{14}N by (28-7).

The ^{14}C formed in the atmosphere by (28-7) is oxidized to CO_2 and mixes with ordinary CO_2, so that atmospheric CO_2 reaches a steady-state level of radioactivity: one of every 10^{12} molecules of CO_2 contains a radioactive ^{14}C atom. The atmospheric CO_2 may be transformed into cellulose and other carbon compounds by plants, and the plants may in turn serve as food for animals or human beings. The carbon compounds in living tissue thus reach the same steady-state level of one ^{14}C atom for every 10^{12} carbon atoms. Upon death of the plant or animal, however, the ^{14}C that decays by (28-8) is no longer replaced. The concentration of ^{14}C in dead wood, in cloth, and in other carbonaceous material derived from living matter diminishes slowly, falling to half the original steady-state level in 5730 year, to a quarter of this level in twice 5730 year, and so on. By measuring the ^{14}C content of ancient dead plant or animal matter, it is thus possible, if the measurements are sufficiently sensitive, to determine how long ago death occurred.

□ **Dating of Fossils**   Some fossil mammalian bones found in a certain stratified layer in the Olduvai Gorge in Africa have a ^{14}C activity about 0.285 times that existing in comparable living matter today. If it is assumed that the atmospheric level of ^{14}C was the same at the time of death of the animal as today, what is the approximate age of the sample?

Example 28-4

Solution   The half-life of ^{14}C is 5730 year. Thus, given $X_0/X = \frac{1}{0.285} = 3.51$, we get, from (28-3)

$$\ln 3.51 = 0.693 \frac{t}{5730 \text{ year}} = 1.255$$

or

$$t = \frac{1.255(5730)}{0.693} \text{ year} = \underline{10.4 \times 10^3 \text{ year}} \quad \blacksquare$$

Exercise 28-4   ☐ The ^{14}C activity in a piece of wood from a chair found in King Tut's tomb is about 0.671 times that existing in comparable living matter today. What is the approximate age of the wood? ∎

A basic assumption in the early applications of carbon dating was that the steady-state level of the ratio of ^{14}C to ordinary ^{12}C in the atmosphere is the same now as it was at the time of death of the plant or animal from which the material was taken. Comparisons of radiocarbon dates of historical and prehistorical objects with dates established by historical accounts and geological research suggested that there may have been significant variations in the level of atmospheric ^{14}C, and recent dating of wood samples by tree-ring counting has supported these results. It now appears that radiocarbon dates earlier than 500 B.C. are systematically in error. The deviation between the radiocarbon date and true date becomes progressively larger as the sample gets older, a radiocarbon date of 4700 B.C. corresponding to a true date of 5400 B.C. Smaller systematic deviations have also been observed for dates later than 500 B.C., but these errors are not greater than about 100 year.

Controlled Nuclear Reactions   If a chain reaction in fissionable material is controlled so that each fission produces exactly one more fission, there can be no sudden growth in the number of nuclei splitting per unit time, and the energy produced can be harnessed. Nuclear fission provides an enormously valuable source of energy. It produces power at a steady rate under easy and precise control for a considerable length of time before more fuel is needed. A nuclear reactor (Fig. 28-4) is a device for carrying out fission under such controlled conditions. Disposal of the extremely hazardous radioactive fission products in a permanently safe way is a problem that is becoming increasingly critical as reactors become more widely used. The future of nuclear power as a source of electrical generating capacity for industry and consumers is also clouded by public concern for the safety of reactors.

A serious limitation on the large-scale development of nuclear power is the limited supply of fissionable uranium. "Breeder" reactors, however, can produce more nuclear fuel than they consume, thereby postponing the time when the world's supply of fissionable material will have been exhausted. In one form of breeder reactor, a "blanket" of ^{238}U surrounds the core. Neutrons from the fission of the ^{235}U fuel are absorbed by the ^{238}U and initiate a decay scheme that leads to fissionable plutonium 239:

$$^{238}_{92}U + {}^{1}_{0}n \longrightarrow {}^{239}_{92}U \longrightarrow {}^{239}_{93}Np + {}^{0}_{-1}e$$
$$^{239}_{93}Np \longrightarrow {}^{239}_{94}Pu + {}^{0}_{-1}e$$

The doubling time for a breeder reactor, that is, the time required to produce as much fissionable material as was originally present in the reactor, is typically of the order of 7 to 10 year. Throughout this time the reactor would also generate power. The practicability of breeder reactors has been proven. Their future development is uncertain, however, because of fears that fuels such as plutonium that are produced might be used for the manufacture of nuclear weapons.

Nuclear reactors are used not only as energy sources but also for the manufacture of isotopes of almost every element, a few directly from the fission products and most indirectly by neutron irradiation of stable isotopes. In addi-

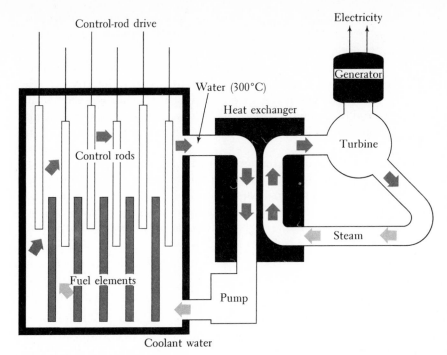

Control-rod drive

Water (300°C)

Heat exchanger

Electricity

Generator

Turbine

Control rods

Steam

Fuel elements

Pump

Coolant water

Figure 28-4 Schematic Diagram of a Nuclear Reactor
The reactor is designed to permit carrying out a self-sustaining fission reaction as a source of power. The fuel elements contain some nuclide, such as ^{235}U, that undergoes fission after absorption of a slow neutron, with the emission of additional neutrons. These neutrons may cause other nuclei to undergo fission or may be absorbed by the control rods, which are made of cadmium, boron, or some other material that strongly absorbs neutrons. By moving the control rods into or out of the matrix of fuel elements, the number of nuclei undergoing fission per unit time, and thus the power output of the reactor, can be controlled.

tion, intense beams of neutrons are available from reactors and are of great value in many kinds of experiments, including determination of molecular structure and molecular motion in solids (and even some liquids) by diffraction and scattering of neutrons, and precise qualitative and quantitative analysis by neutron activation (described below).

The possibility of using nuclear *fusion* rather than fission as a practicable power source is the subject of intense research in many countries. Controlled fusion would have the advantage over present nuclear power sources that the primary products are not radioactive and that the needed fuel, chiefly isotopes of hydrogen, is available in virtually unlimited supply. Most of the fusion reactions available, do however, produce neutrons so that the production of highly radioactive species as a result of the neutron bombardment will present problems. Moreover, the technical difficulties in achieving the temperatures necessary for fusion, confining the reactants adequately, and carrying out the process in a safe and controlled manner are enormous.

Other Applications The presence of radioactivity in a substance can be detected with instruments such as Geiger and scintillation counters. Radioactive counting techniques are very sensitive—under certain conditions it is possible to detect as little as 10^{-14} g of an element.

Radioactive isotopes (radioisotopes) of every element can be produced in nuclear reactors, either as fission products or by neutron irradiation of materials inserted into the reactor. Many radioisotopes of light elements have half-lives sufficiently long that samples are commercially available and are routinely shipped throughout the world. They are used in research laboratories for tracer studies in which molecules are synthesized containing specific atoms that are "labeled" because they are radioactive. The path of the labeled atom in chemical

639

reactions can be traced by determining the distribution of radioactivity in the products.

Radioisotopes also have important uses in medicine. Radiation treatment of the thyroid, for example, is easily accomplished by injections of a solution containing radioactive iodide. Since iodine concentrates in the thyroid, the radiation is highly localized and has little effect on other tissue. The radioactive isotope ^{60}Co, produced by irradiation of stable ^{59}Co with neutrons, has a half-life of 5.3 year and is widely used in cancer therapy as a source of highly penetrating gamma rays.

Each radioisotope has its own characteristic decay signature, which can be distinguished by half-life, nature and energy of the radiation emitted, and decay products. Thus measurements of radioactivity can be used for qualitative and quantitative analysis. Trace analysis of nonradioactive species can be performed by neutron activation. The sample to be studied is irradiated with neutrons, causing identifiable radioactive isotopes to be formed.

Summary

In any nuclear decay there is conservation of mass-energy, of charge, and of nucleons. When a nucleus decays by beta emission, it emits an electron, causing Z to increase by 1 while A remains unchanged. The emission of an alpha particle (an He nucleus) causes Z to decrease by 2 and A by 4. In gamma decay Z and A are unchanged because gamma rays are electromagnetic radiation. Energy changes in nuclear reactions are frequently so large that mass changes (through the relation $E = mc^2$) are readily detectable.

Radioactive decay is a first-order process. The rates of radioactive disintegrations and other nuclear reactions generally do not depend on temperature, physical state, or chemical form. With the exception of $^1_1$H and $^3_2$He, every stable nucleus has at least as many neutrons as protons; beyond $Z = 20$ stability is found only when the number of neutrons is greater than the number of protons. Nuclei with an odd number of either neutrons or protons (or both) tend to be unstable. Certain numbers (magic numbers) of neutrons and protons confer special stability. The binding energy is a measure of relative stability of a nucleus. It is the energy equivalent to the difference between the sum of the masses of the individual nucleons present in a nucleus and the mass of the nucleus.

Fission is a process in which a nucleus with a high value of A splits to form nuclides of intermediate A values. In fusion, two low-A nuclides combine to form a nucleus of higher A value. When certain very heavy nuclei absorb slow neutrons, fission occurs. If more neutrons are emitted during each fission than are absorbed, a chain reaction may occur. Bombardment of stable nuclei by alpha particles or other atomic species may cause nuclear reactions that produce unstable nuclei (artificial radioactivity). Numerous isotopes of existing elements have been made in this way, and the technique has also been used to produce elements not found in nature.

The 14 elements beyond actinium form a series analogous to the lanthanides that corresponds to the filling of the $5f$ subshell. All the actinides are radioactive; those beyond uranium (the transuranium elements) are not found in nature. There is reason to believe that elements with Z near 114 might someday be prepared.

Archaeological objects that contain wood, cloth, or other carbonaceous material derived from plants or animals can be dated by measuring their ^{14}C content relative to the steady-state level in living matter. Other applications of nuclear chemistry include neutron activation analysis and the use of radioactive tracers. Nuclear reactors are devices for carrying out chain reactions in fissionable materials under controlled conditions. In a breeder reactor, more fissionable material is produced than is consumed.

Terms and Concepts

Problems and Questions

28-1 Nuclear Reactions Give the chemical symbols, atomic numbers, and mass numbers of X, Y, and Z in the following nuclear reactions: (*a*) a neutron reacts with ^{6}Li to form an α particle plus a particle X; (*b*) an α particle combines with ^{14}N to yield a proton and a particle Y; (*c*) a proton combines with ^{12}C to form a particle Z and a photon.

28-2 Strontium 90 The nuclide ^{90}Sr is produced in significant amounts by fission. It decays by emitting electrons with an energy of about 0.54 million electronvolt (MeV) and has a half-life of 28 year. (*a*) If the ^{90}Sr is produced by fission of ^{235}U, together with three neutrons, what other nuclide is also formed? (*b*) What is the nuclide formed when ^{90}Sr decays by beta emission? (*c*) What fraction of the ^{90}Sr formed in a nuclear explosion will still be present after 100 year? (*d*) About how many chemical bonds of average energy 400 kJ mol^{-1} could be ruptured by one electron of energy 0.54 MeV from ^{90}Sr if all its energy were available for bond breaking?

28-3 Survival of a Nuclide Suppose the elements of the earth were created 5×10^9 year ago (a very conservative estimate). What fraction of a nuclide with a half-life of 1×10^8 year would still remain today?

28-4 Disintegration of ^{50}V Ordinary vanadium contains about 0.25 percent ^{50}V, which has a half-life of 4×10^{14} year. Suppose that you have a 1.00-g sample of vanadium that contains no other radioactive material. What will be the average number of nuclei disintegrating per minute?

28-5 Mass Change in a Chemical Reaction The formation of 2 mol of liquid water from gaseous hydrogen and gaseous oxygen at 25°C and 1 atm is accompanied by the release of 572 kJ. What is the loss in weight (in grams) during this reaction? What is the percentage loss in weight?

28-6 Fission Products Explain why the products of nuclear fission are highly radioactive and why they are likely to emit electrons, neutrons, and gamma rays. (*Hint:* Consider Fig. 28-2.) Explain also why the products of a fusion reaction are less likely to be radioactive.

28-7 ^{14}C Dating A piece of charred wood found in a cave believed to have been inhabited by prehistoric people was found to have a ^{14}C activity 0.123 times that in wood growing now. What is the approximate age of the charred wood?

28-8 Carbon Dating and Tree Rings In 1951, wood from two sequoia trees was dated by the ^{14}C method. In one tree, clean borings located between the growth rings associated with the years A.D. 1057 and 1087 (i.e., wood known to have grown 880 ± 15 year prior to the date of measurement) had a ^{14}C activity about 0.892 of that of wood growing in 1951. A sample from a second tree had an activity about 0.838 of that of new wood, and its age was established as 1377 ± 4 year by tree-ring counting. (*a*) What ages does carbon dating associate with the wood samples? (*b*) What values of $\tau_{1/2}$ can be deduced if the tree-ring dates given are used as the starting point? (*c*) Discuss assumptions underlying the calculations in (*a*) and (*b*), and indicate in what direction failures of these assumptions might affect the calculations.

28-9 The Curie The rate of disintegration of a radioactive sample is often given in terms of the curie unit: 1 curie = 3.700×10^{10} disintegrations per second. This unit corresponds roughly to the disintegration rate of the radon in steady-state "equilibrium" with 1 g radium. A sample containing ^{32}P, a beta emitter of half-life 14.2 day, has an activity of 1 microcurie (1×10^{-6} curie). What weight of ^{32}P is in the sample?

28-10 Fusion Reactions Several reactions have been proposed as likely candidates for controlled fusion. Among them are

$$_{1}^{2}H + {}_{1}^{3}H \longrightarrow {}_{2}^{4}He + {}_{0}^{1}n \qquad (1)$$

$$_{1}^{2}H + {}_{1}^{2}H \longrightarrow {}_{2}^{3}He + {}_{0}^{1}n \qquad (2)$$

$$_{1}^{2}H + {}_{1}^{2}H \longrightarrow {}_{1}^{3}H + {}_{1}^{1}H \qquad (3)$$

Which of these reactions releases the largest amount of energy? Nuclear masses/10^{-27} kg: ^{1}n, 1.6750; ^{1}H, 1.6727; ^{2}H, 3.3435; ^{3}H, 5.0075; ^{3}He, 5.0065; ^{4}He, 6.6446.

28-11 Ancient Isotopic Abundance of Uranium The natural abundance of ^{238}U is 142 times that of ^{235}U in ores that are mined today. The half-life of ^{238}U is 4.5×10^9 year and that of ^{235}U is 7.1×10^8 year. What was the ratio of the abundances 400 million year ago?

28-12 Change in Relative Abundance with Time An element has two radioactive isotopes A and B. The half-life of A is 5 times that of B. A sample is prepared containing both isotopes. In a period of time t, the concentration of A falls to half its initial value. What fraction of the initial concentration of B remains?

28-13 Measurement of Long Half-Lives The half-life of ^{40}K as a result of beta decay is 1.39×10^9 year. (a) What change would be observed in the number of beta particles emitted per second from a sample of ^{40}K over a period of 1 month? ($e^{-x} = 1 - x$ for small x.) (b) For long-lived nuclides, the half-life can be determined from the instantaneous rate expression (22-12) if the number of atoms of the radioactive nuclide in a sample is known. A 25.0-mg sample of ^{40}KCl emits 1.98×10^5 beta particles per minute. Calculate the half-life of ^{40}K from this information.

28-14 Methods of Power Generation A typical electrical generating plant has a capacity of 500 megawatt (MW; 1 MW = 10^6 J s^{-1}) and an overall efficiency of about 25 percent. (a) The combustion of 1 kg of bituminous coal releases about 3.2×10^4 kJ and leaves an ash residue of 100 g. What weight of coal must be used to operate a 500-MW generating plant for 1 year, and what weight of ash must be disposed of? (b) Enriched fuel for nuclear reactors contains about 4 percent ^{235}U, fission of which gives 1.9×10^{10} kJ per mole of ^{235}U. What weight of ^{235}U is needed to operate a 500-MW power plant, assumed to have 25 percent efficiency, for 1 year, and what weight of fuel must be reprocessed to remove radioactive wastes? (c) The radiation from the sun striking the earth's surface on a sunny day corresponds to a power of 1.5 kW m^{-2}. How large must the collection surface be for a 500-MW solar-power generating plant? (Assume that there are 6 hour of bright sun each day and there are storage facilities in order to continue producing power at other times. The efficiency for solar-power generation would be about 15 percent.)

Organic Chemistry

"Organic chemistry just now is enough to drive one mad. It gives me an impression of a primeval tropical forest, full of the most remarkable things, a monstrous and boundless thicket, with no way to escape, into which one may well dread to enter."
F. WÖHLER, IN LETTER TO J. J. BERZELIUS, 1835

"The world has been in an 'Organic Chemical Age' for almost thirty years, and the era is just reaching maturity. Multimillion-dollar industries based on organic compounds have taken their places alongside the other commercial giants of our civilization. Farm products, wood, coal, petroleum, and natural gas have been chemically tailored to provide organic products that vitally affect our environment. Fabrics, dyes, paints, coatings, and structural materials literally surround us. Rubber takes the violence out of movement, gasoline fuels our travel, gas both warms and cools our houses, plastics fashion our implements, and drugs guard our health and prolong our lives. Unprecedented opportunities to exercise the imagination are available in organic research, and much still remains to be done."
D. J. CRAM AND G. S. HAMMOND,[1] 1958

Organic chemistry is the chemistry of the compounds of carbon.[2] More than a million organic compounds are now known, almost all of them containing hydrogen as well as carbon and the great majority containing other elements as well, most commonly oxygen and nitrogen. No other element forms as many stable compounds as carbon. Its uniqueness arises from the fact that it forms stable covalent bonds with itself as well as with many other elements and that each carbon atom forms four bonds, which makes possible extended three-dimensional structures (Chap. 7) and almost innumerable possibilities for isomerism.

Many organic compounds are found, as the name implies, in living organisms, but a significant fraction of the organic substances now known have been created in the laboratory. The quotations at the start of this chapter suggest the enormous progress and changes that occurred in organic chemistry in a little more than a century. In 1835 the concepts of chemical bonding and structure were unknown and only a handful of organic substances had been characterized. Now the chemist can determine the detailed three-dimensional molecular architecture of compounds containing thousands of atoms and devise laboratory syntheses for complex molecules, including many that are designed and created to serve a specific purpose. In this chapter we discuss some of the simplest classes of organic compounds and reactions and suggest the kinds of considerations that govern the relation between structure and properties. We begin by examining some of the important types of hydrocarbons and then introduce the concept of functional groups, which greatly simplifies the chemistry of compounds that contain elements in addition to carbon and hydrogen. The importance of molecular struc-

---

[1] "Organic Chemistry," p. 3, McGraw-Hill Book Co., New York, 1959.
[2] For historical reasons, CO and CO_2 and substances derived from CO_2, such as carbonates and CS_2, are not considered organic, nor are HCN and other cyanides and related compounds.

ture in determining the properties of compounds is emphasized in the next two sections, which deal with optical isomerism and polymers. Some of the characteristic properties and reactions of typical organic compounds are discussed in Sec. 29-5. The following chapter is concerned with a few of the essential compounds and reactions that are found in living organisms, the realm of biochemistry.

29-1   Hydrocarbons

The simplest organic substances are the hydrocarbons, compounds containing only C and H. Representative hydrocarbons include ethane (CH_3CH_3), ethylene ($CH_2{=}CH_2$), and benzene (C_6H_6), the structures of which were discussed in Chaps. 17 and 18. These three molecules differ in the nature of their carbon-carbon bonds; ethane is the simplest molecule containing a C-C single bond, ethylene is the simplest containing a double bond, and benzene is the prototype aromatic hydrocarbon, with a completely delocalized π-bond system (Sec. 18-3). The hydrocarbons provide a logical starting point for a discussion of organic chemistry because they contain only two elements, their chemistry is comparatively simple, and they serve as the basis for the systematic naming of all classes of organic compounds. A systematic scheme of nomenclature is essential because of the structural complexities possible with organic compounds; some of the simpler rules for naming organic substances are given in Appendix B and in Secs. 29-1 and 29-5 of this chapter. We will utilize the systematic names primarily but widely used common names of some compounds will be given in parentheses.

   Some of the molecular formulas of hydrocarbons and other types of organic compounds on the following pages may appear bewildering at first. However, a little practice and attention to a few fundamental principles established in earlier chapters will quickly reveal the simple and regular patterns followed by the formulas and structures. The most important points to keep in mind are these:

1.   The number of covalent bonds normally formed by an atom of each element and the number of unshared pairs remaining on that bonded atom (Table 29-1). A common error of beginners is to draw organic structural formulas with three or five bonds to a carbon atom, rather than four, or with two bonds to a hydrogen atom rather than one.

2.   The geometric arrangement of the bonded neighbors around an atom that has no unshared pairs and has four, three, or two bonded neighbors (Table 18-1), as does carbon in different common bonding situations. These consid-

Table 29-1
Number of Covalent Bonds and Unshared Electron Pairs for Representative Bonded Atoms[*][†]

	C	H	N	N$^+$	O	O$^+$	F, Cl, Br, I
Number of covalent bonds	4	1	3	4	2	3	1
Number of unshared pairs on bonded atom	0	0	1	0	2	1	3
Examples	CH_4 $H_2C{=}CH_2$	CH_4	NH_3	$NH_4{}^+$	H_2O $(CH_3)_2C{=}O$	H_3O^+	CH_3F CBr_4

[*] Sulfur and phosphorus in some organic compounds conform to the pattern indicated here for their congeners O and N, but they may also form more bonds, e.g., in the molecules SF_6 and PCl_5 and in the groups $-SO_2OH$, $-OSO_2OH$, and $-OPO(OH)_2$.

[†] Note that N$^+$ is isoelectronic with C and O$^+$ is isoelectronic with N.

erations are important in the determination of the number and the nature of possible isomers. Don't be confused by the fact that flat drawings are used to represent three-dimensional structures.

3. The fact that at ordinary temperatures there is normally almost free rotation about single bonds and no rotation about double bonds. These facts are important in the determination of the possibilities of isomerism.

Alkanes Methane, ethane, and propane are the three simplest representatives of this class of compounds:

$$
\begin{array}{ccc}
\underset{\text{Methane}}{\text{H}-\overset{\displaystyle\text{H}}{\underset{\displaystyle\text{H}}{\text{C}}}-\text{H} \quad \text{or} \quad \text{CH}_4}
&
\underset{\text{Ethane}}{\text{H}-\overset{\displaystyle\text{H}}{\underset{\displaystyle\text{H}}{\text{C}}}-\overset{\displaystyle\text{H}}{\underset{\displaystyle\text{H}}{\text{C}}}-\text{H} \quad \text{or} \quad \text{CH}_3\text{CH}_3}
&
\underset{\text{Propane}}{\text{H}-\overset{\displaystyle\text{H}}{\underset{\displaystyle\text{H}}{\text{C}}}-\overset{\displaystyle\text{H}}{\underset{\displaystyle\text{H}}{\text{C}}}-\overset{\displaystyle\text{H}}{\underset{\displaystyle\text{H}}{\text{C}}}-\text{H} \quad \text{or} \quad \text{CH}_3\text{CH}_2\text{CH}_3}
\end{array}
$$

The two ways of writing molecular formulas shown here were discussed briefly in Sec. 3-3. Although those shown on the right do not display each bond explicitly, they are much more convenient typographically and hence more common. Their implications with regard to the number of C-C and C-H bonds for each carbon atom must be carefully noted. In these *condensed* formulas, the hydrogen atoms are usually written immediately to the right of the carbon atom to which they are bonded; C-C bonds are shown explicitly only if they are in a vertical direction on the page (as in later examples).

Each of these three hydrocarbons has only single bonds, and their molecular formulas differ successively by CH_2 (CH_4, C_2H_6, and C_3H_8). These molecules are the first three members of a series with general formula C_mH_{2m+2}, where m is a positive integer. Hydrocarbons with this general formula are called *alkanes* or saturated hydrocarbons, the general term *saturated* implying that they contain no multiple bonds.

Ethane may be regarded as derived from methane by (mentally) replacing one of the four equivalent hydrogen atoms of methane by a CH_3 (methyl) group. Similarly, propane can be considered to be derived from ethane by the replacement of one of the six equivalent[3] hydrogen atoms of ethane by CH_3. The replacement of one H of propane by CH_3 can lead to either of two possible molecules because there are two nonequivalent kinds of bonding environments for hydrogen atoms in propane. The two hydrogen atoms attached to the central carbon atom are equivalent to one another but are quite distinct from the six attached to the end carbon atoms. The latter six are in turn equivalent to one another because there is essentially free rotation about the C-C bonds. Thus there are two isomeric molecules of formula C_4H_{10} (butanes):

$$
\underset{\text{Butane (\textit{n}-butane)}}{\text{H}-\overset{\displaystyle\text{H}}{\underset{\displaystyle\text{H}}{\text{C}}}-\overset{\displaystyle\text{H}}{\underset{\displaystyle\text{H}}{\text{C}}}-\overset{\displaystyle\text{H}}{\underset{\displaystyle\text{H}}{\text{C}}}-\overset{\displaystyle\text{H}}{\underset{\displaystyle\text{H}}{\text{C}}}-\text{H} \quad \text{or} \quad \text{CH}_3\text{CH}_2\text{CH}_2\text{CH}_3}
$$

2-Methylpropane (isobutane): CH_3CHCH_3 with CH_3 group, central carbon bonded to CH_3 below.

---

[3] Inspection of Fig. 17-2 or a model of ethane will show that the six hydrogen atoms are equivalent to one another, although they may not at first seem to be in the two-dimensional formulas used above to represent the structure of this molecule.

Table 29-2
Properties of Some Alkanes

Name	Molecular formula	Molecular weight	mp/°C	bp/°C	Number of isomers
Methane	CH_4	16	−183	−162	1
Ethane	C_2H_6	30	−172	−89	1
Propane	C_3H_8	44	−187	−42	1
Butane	C_4H_{10}	58	−135	0	2
Pentane	C_5H_{12}	72	−130	36	3
Hexane	C_6H_{14}	86	−94	69	5
Heptane	C_7H_{16}	100	−91	98	9
Octane	C_8H_{18}	114	−57	126	18
Nonane	C_9H_{20}	128	−54	151	35
Decane	$C_{10}H_{22}$	142	−30	174	75
Eicosane	$C_{20}H_{42}$	282	36	*	3.66×10^5

*Eicosane boils at about 205°C at a pressure of 15 torr.

The letter n in the common name signifies *normal* and implies an unbranched chain of carbon atoms. Isobutane is the simplest branched alkane, the term *branched* implying that at least one carbon atom is bonded directly to three or four other carbons. The systematic name of isobutane is 2-methylpropane, a name derived by choosing the longest chain of carbon atoms in the molecule, numbering the atoms consecutively in this chain starting at one end, and indicating the point of attachment of the substituent as shown (see Appendix B).

The physical and chemical properties of isobutane are similar to but nonetheless distinct from those of butane. For example, its melting temperature is 10 K lower, it boils 11 K lower, and it has a slightly different density in condensed states.

The alkanes with more than four carbon atoms are named by combining the ending *-ane* with a Greek root that signifies the number of carbon atoms in the longest chain in the molecule (*pent* for 5, *hex* for 6, and so on), as indicated in Table 29-2. There are three isomeric pentanes, with significantly different physical properties:

$$CH_3CH_2CH_2CH_2CH_3 \qquad CH_3CHCH_2CH_3 \qquad CH_3CCH_3$$

Pentane   2-Methylbutane   2,2-Dimethylpropane
(*n*-pentane)   (isopentane)   (neopentane)
(bp 36°C)   (bp 28°C)   (bp 10°C)

The number of possible isomers rises rapidly for the higher alkanes, as indicated in Table 29-2. The lower alkanes are colorless, odorless gases under ordinary conditions, and those from C_5 through C_{18} are colorless liquids. The higher alkanes are waxy solids. All alkanes are nonpolar and water-insoluble. The liquids and solids have densities between about 0.6 and 0.8 g ml^{-1} and thus float on water.

Example 29-1

□ **Implications of Formulas** Which of the following formulas represent the same molecules and which are incorrect in their implications about bonding?

(a) $(CH_3)_2CHCH_2CH_3$

(d) $CH_3CH_2\overset{\displaystyle CH_3}{\underset{\displaystyle |}{C}}HCH_3$

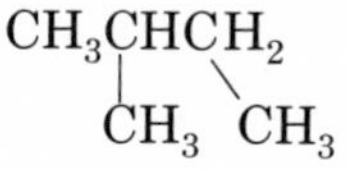

(b) $CH_3\underset{\displaystyle CH_3}{\underset{\displaystyle |}{C}}HCH_2\underset{\displaystyle CH_3}{\overset{\displaystyle \diagdown}{}}$

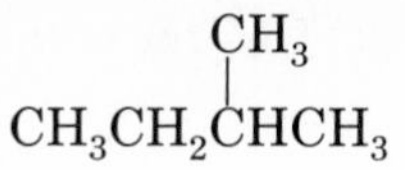

(e) $(CH_3)_2CH(CH_3)_2$

(c) $CH_3CH_2CH(CH_3)_3$

Solution Formulas (a), (b), and (d) all represent the same molecule, 2-methylbutane, written in different ways. This is apparent if one locates the longest chain of carbon atoms and finds the position of the methyl substituent on this chain. A formula such as (b), with a bend in the chain, is unconventional; since there is freedom of rotation about a C-C single bond, the methyl group written at the lower right might just as well have been written on the same line as the upper three carbon atoms. Formulas (c) and (e) are incorrect; in (c) the third carbon from the left has five attached groups (CH_2, H, and three CH_3) and in (e) the central carbon atom has five attached groups. ■

☐ Which of the following formulas represent the same molecules? Which are incorrect in their implications about bonding?

Exercise 29-1

(a) $(CH_3)_3CCH_3$
(b) $CH_3C{\equiv}CH$
(c) $(CH_3)_2C(CH_3)_2$

(d) $H_3CC(CH_3)_2$
(e) $H_2CCH{=}CH_2$ ■

☐ **Isomers of Hexane** Write the formulas for the isomeric hexanes, C_6H_{14}, and name the molecules.

Example 29-2

Solution A systematic approach is to start with the three isomers of C_5H_{12}, substitute CH_3 for H at each unique position of each molecule, and then eliminate the duplicate molecules. For simplicity we show here merely the carbon skeletons, with the newly introduced CH_3 in boldface type.

From pentane, we get three C_6-isomers

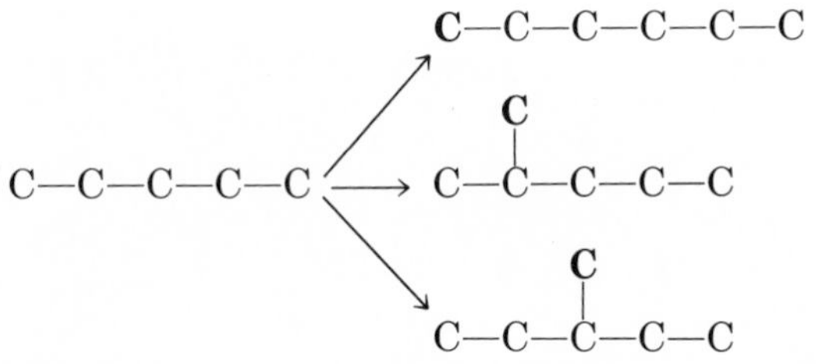

Substitution on the fourth carbon of pentane is equivalent to substitution on the second; substitution on the fifth is equivalent to that on the first.

647

Substitution of CH_3 for H on each of the carbon atoms of 2-methylbutane gives

$$
\begin{array}{c}
\quad\quad\quad\quad\quad\quad\quad\;\; \text{C} \\
\quad\quad\quad\quad\quad\quad\quad\;\; | \\
\text{C}-\text{C}-\text{C}-\text{C}-\text{C} \quad\quad \text{3-methylpentane (again)}
\end{array}
$$

$$
\begin{array}{c}
\quad\quad\;\; \textbf{C} \\
\quad\quad\;\; | \\
\quad\quad\;\; \text{C} \\
\quad\quad\;\; | \\
\text{C}-\text{C}-\text{C}-\text{C} \quad\quad \text{3-methylpentane (again)}
\end{array}
$$

$$
\begin{array}{c}
\quad\;\; \text{C} \quad\quad\quad\quad\quad\quad \text{C} \\
\quad\;\; | \quad\quad\quad\quad\quad\quad | \\
\text{C}-\text{C}-\text{C}-\text{C} \;\longrightarrow\; \text{C}-\text{C}-\text{C}-\text{C} \quad\quad \text{2,2-dimethylbutane} \\
\quad\quad\quad\quad\quad\quad\quad\quad\quad\;\; | \\
\quad\quad\quad\quad\quad\quad\quad\quad\quad\;\; \textbf{C}
\end{array}
$$

$$
\begin{array}{c}
\quad\quad\; \text{C}\;\; \textbf{C} \\
\quad\quad\; | \;\; | \\
\text{C}-\text{C}-\text{C}-\text{C} \quad\quad \text{2,3-dimethylbutane}
\end{array}
$$

$$
\begin{array}{c}
\quad\quad\; \text{C} \\
\quad\quad\; | \\
\text{C}-\text{C}-\text{C}-\text{C}-\textbf{C} \quad\quad \text{2-methylpentane (again)}
\end{array}
$$

The fact that the first two of these are the same reflects the equivalence of the two methyl groups on the second carbon atom of 2-methylbutane; do not be confused by the writing of the five-carbon chain "around a corner" in the second formula.

Since all 12 hydrogen atoms of 2,2-dimethylpropane are equivalent to each other, there is only one possible substitution product:

$$
\begin{array}{c}
\quad\; \text{C} \quad\quad\quad\quad\quad\quad \text{C} \\
\quad\; | \quad\quad\quad\quad\quad\quad | \\
\text{C}-\text{C}-\text{C} \;\longrightarrow\; \text{C}-\text{C}-\text{C}-\textbf{C} \quad\quad \text{2,2-dimethylbutane (again)} \\
\quad\; | \quad\quad\quad\quad\quad\quad | \\
\quad\; \text{C} \quad\quad\quad\quad\quad\quad \text{C}
\end{array}
$$

Inspection of these formulas and names shows that there are five unique isomeric hexanes, in agreement with Table 29-2. ∎

Exercise 29-2

☐ Write the formulas for the isomeric heptanes, C_7H_{16}, and name the molecules. ∎

Natural gas and petroleum are composed principally of alkanes and constitute their chief natural sources. A typical natural gas contains about 80 percent methane, with about 10 percent ethane and smaller amounts of propane, butanes, and pentanes. Methane is also produced by the bacterial degradation of cellulose in vegetable matter in the absence of oxygen and thus is a primary constituent of the gas formed under water in marshes and below the surface of garbage dumps. It is found as well in significant quantities in some coal mines, where it can form dangerously explosive mixtures with air.

Some alkanes are also found in small amounts in plant materials. Heptane occurs in the turpentine from certain pines, and the waxy coatings on pears, apples, and cabbage leaves contain the C_{29} n-alkane. Beeswax contains alkanes in the C_{27} to C_{31} range.

Alkanes are normally quite inert chemically, being unaffected by hot acids, bases, metals, and most oxidizing and reducing agents. They do, however, react with oxygen when heated, forming CO_2 and H_2O with an excess of air or oxygen and forming CO and H_2O when the quantity of O_2 is limited. They also react with chlorine and bromine under the influence of light and with fluorine even in the dark. For example, methane reacts with chlorine to form HCl and a mixture of products with from one to four chlorine atoms substituted for the hydrogen atoms of methane: CH_3Cl, CH_2Cl_2, $CHCl_3$, and CCl_4.

Cycloalkanes These hydrocarbons contain at least one ring of carbon atoms and no double or triple bonds. Those with only one ring have the general formula C_mH_{2m}; cyclopropane (C_3H_6) is the simplest representative of this class. Cyclopentane and cyclohexane are found in some samples of petroleum, as are several other alkanes containing five- and six-membered rings:

$$
\begin{array}{cc}
\begin{array}{c}
CH_2 \\
CH_2 \quad CH_2 \\
CH_2 - C(CH_3)_2
\end{array}
&
\begin{array}{c}
CHCH_3 \\
CH_2 \quad CHCH_3 \\
CH_2 \quad CH_2 \\
CHCH_3
\end{array}
\end{array}
$$

1,1-Dimethylcyclopentane   1,2,4-Trimethylcyclohexane

The chemical and physical properties of the cycloalkanes are very similar to those of the alkanes.

Alkenes Alkenes contain one carbon-carbon double bond and have the general formula C_mH_{2m}, with $m = 2$ or more. Those with m greater than 2 are isomeric with cycloalkanes containing only one ring; e.g., propene, $CH_3CH=CH_2$, is isomeric with cyclopropane. The simplest alkene is ethylene, $CH_2=CH_2$. There is only one propene but there are four different butenes:

$$CH_3CH_2CH=CH_2 \qquad
\begin{array}{c}
H \qquad\quad H \\
C=C \\
CH_3 \qquad CH_3
\end{array}
\qquad
\begin{array}{c}
H \qquad\quad CH_3 \\
C=C \\
CH_3 \qquad H
\end{array}
\qquad
\begin{array}{c}
CH_3 \\
H_2C=C \\
CH_3
\end{array}$$

1-Butene   *cis*-2-Butene   *trans*-2-Butene   2-Methylpropene (isobutene)

The two forms of 2-butene illustrate the phenomenon of *geometric isomerism*. It arises because of the impossibility of rotation about the double bond at ordinary temperatures. The term *cis* is Latin for "on this side" while *trans* is Latin for "on the other side". In *trans*-2-butene the two methyl groups lie diagonally across from each other, while in the *cis* compound the methyl groups are on the same side of the double bond. The naming of alkenes is discussed briefly in Appendix B.

The most important reactions of alkenes are those in which the double bond is converted to a single bond. For example, hydrogen reacts with an alkene (an unsaturated hydrocarbon) in the presence of an appropriate catalyst to form the corresponding alkane (a saturated hydrocarbon):

$$CH_3CH=CHCH_3 + H_2 \xrightarrow{\text{Pt or Ni}} CH_3CH_2CH_2CH_3 \qquad (29\text{-}1)$$

Bromine reacts with alkenes to form the dibromoalkane with one bromine atom attached to each carbon atom originally involved in the double bond, a reaction often used to test for the presence of a C-C double bond:

$$(CH_3)_2C{=}CH_2 + Br_2 \longrightarrow (CH_3)_2CBrCH_2Br \qquad (29\text{-}2)$$

2-Methylpropene   1,2-Dibromo-2-methylpropane

Each of these reactions may be considered to involve an attack on and opening of the π bond in the alkene (Sec. 18-3). The electrons originally paired in this bond are thereby made available for the formation of new covalent bonds.

Alkynes These hydrocarbons contain a carbon-carbon triple bond and have the general formula C_mH_{2m-2}. Ethyne (acetylene), $HC{\equiv}CH$, is the simplest alkyne. Alkynes undergo most of the reactions characteristic of alkenes. Those with the triple bond at the end of a chain, that is, with the grouping $-C{\equiv}CH$, are very weakly acidic; the terminal hydrogen atom reacts with metallic sodium to form H_2 and with Ag^+ to produce an insoluble silver salt of the alkyne. For example,

$$CH_3C{\equiv}CH + Na \longrightarrow CH_3C{\equiv}C^-Na^+ + \tfrac{1}{2}H_2 \qquad (29\text{-}3)$$

$$CH_3C{\equiv}CH + Ag^+ \longrightarrow CH_3C{\equiv}CAg + H^+ \qquad (29\text{-}4)$$

Benzene
C_6H_6

Naphthalene
$C_{10}H_8$

Aromatic Hydrocarbons This is an extensive and important group of compounds, of which benzene, C_6H_6, is the simplest example, the π-electron system extending over all six carbon atoms. In the formulas used conventionally for aromatic rings, the corners of the rings denote carbon atoms. A hydrogen atom is assumed to be attached to each such corner if no other atom or group is shown, and a circle inside the hexagon indicates the presence of a delocalized π-electron system. In naphthalene, it extends over all 10 carbon atoms. There are many other aromatic systems, with even more rings "fused" together as in naphthalene, and also numerous hydrocarbons derived from benzene, naphthalene, and the other ring systems. An example is toluene (methylbenzene), $C_6H_5CH_3$, whose trinitro derivative is the common high explosive TNT:

$$CH_3$$

O_2N NO_2

$$NO_2$$

2,4,6-Trinitrotoluene
(TNT)

Aromatic compounds react much less readily than alkenes and alkynes, despite their unsaturation, but nonetheless do undergo many reactions.

Petroleum Petroleum is a naturally occurring mixture of hydrocarbons. Some crude oils consist chiefly of alkanes; others may have as much as 40 percent of cycloalkanes and aromatic hydrocarbons. Some sources of petroleum contain up to about 1 percent of sulfur compounds, which are ecologically objectionable because, when burned, they produce sulfur dioxide, an irritating and noxious gas. Petroleum is refined principally by a series of fractional distillations. Gasoline is the fraction distilling in the temperature range from about 35 to about 210°C; it contains hydrocarbons with from 5 to about 12 carbon atoms. The fraction distilling from about 200 to 275°C is kerosene; less volatile fractions find use as lubricating oils, petroleum jelly, and paraffin. These higher fractions can be *cracked* by various methods to break them down to hydrocarbons of lower molecular weight, suitable for inclusion in gasoline. These methods include both pyrolysis (decomposition by heating) at temperatures up to 700°C, often at high pressure, and treatment with special catalysts at somewhat lower but still elevated temperatures. The development of successful cracking methods has more than doubled the yield of gasoline from petroleum.

29-2 Alkyl Groups and Functional Groups

Most organic compounds contain not only carbon and hydrogen but other elements as well, in various common groupings such as —OH, —NH$_2$, —Br, and —COOH. The unconnected line emanating from each group is *not* a negative charge; it merely denotes the potential for forming a covalent bond. Simple examples of compounds with such groups are

CH_3CH_2OH	CH_3COOH	$CH_3CH_2CH_2CH_2NH_2$	CH_3CHFCH_3
Ethanol	Acetic acid	1-Aminobutane	2-Fluoropropane

Each of these molecules may be regarded as derived from an alkane by substitution of an atom or group of atoms for one of the hydrogen atoms of the alkane. Because of the chemical inertness of the alkanes, the reactions of the derivative molecules depend to a great extent upon the nature of the substituted groups, which are therefore referred to as *functional groups*. The hydrocarbon fragment formed from an alkane by substituting for one of the hydrogen atoms is called an *alkyl group*—for example, CH_3— (methyl) or $CH_3CH_2CH_2$— (propyl).

Alkyl groups are necessarily uncharged (since they correspond to the removal of a neutral atom from a neutral molecule), and they contain an unpaired electron (since they contain an odd number of electrons). Species that contain unpaired electrons are called *radicals*.[4] If we symbolize an alkyl radical as R— and a functional group as Y—, then we can represent compounds like those above as RY. The simplest alkyl groups, together with two other common hydrocarbon radicals, are depicted in Table 29-3; the common functional groups are indicated,

---

[4] Although most radicals are extremely reactive and have very short lifetimes under normal conditions, some are more stable. Among the common molecules that contain unpaired electrons, and thus are radicals, are NO and NO$_2$ and, in its most stable state, O$_2$.

Table 29-3 Some Common Hydrocarbon Radicals

CH_3—	CH_3CH_2—	$CH_3CH_2CH_2$—	$CH_3CH_2CH_2CH_2$—
Methyl	Ethyl	Propyl	Butyl

$\begin{array}{c}H_3C\\[-2pt]\\H_3C\end{array}$ CH—	$\begin{array}{c}H_3C\\[-2pt]\\H_3C\end{array}$ CH—CH$_2$—	CH_3—CH—CH$_2$—CH$_3$	$\begin{array}{c}H_3C\\[-2pt]H_3C\\[-2pt]H_3C\end{array}$ C—
Isopropyl	Isobutyl	sec*-Butyl	tert*-Butyl

CH_2=CH—	C_6H_5— or ⬡		
Vinyl	Phenyl		

*The abbreviations *sec* and *tert* (or *s* and *t*) stand for secondary and tertiary. They relate to the fact that a carbon atom with four bonded neighbors is referred to as primary, secondary, or tertiary, depending on whether it is directly linked to one, two, or three other carbon atoms.

Table 29-4
Some Common Classes of Organic Compounds*

Compound class	General formula†		Comment	Specific example
Haloalkanes (alkyl halides)	R—**X**	(R**X**)	X may be F, Cl, Br, or I	CH_3I Iodomethane (methyl iodide)
Alcohols	R—**O**—**H**	(R**OH**)		$CH_3CH_2CHCH_3$ OH 2-Butanol
Ethers	R—**O**—R′	(R**O**R′)		$CH_3CH_2OCH(CH_3)_2$ Ethyl isopropyl ether
Amines	R—**N**(**H**)(**H**)	(R**NH**$_2$)	One or both of the H's on the N may be replaced by an alkyl radical	$CH_3CH_2CH_2CH_2NH_2$ 1-Aminobutane $CH_3CH_2N(CH_3)_2$ *N,N*-Dimethylaminoethane
Aldehydes	R—**C**(**H**)(=**O**)	(R**CHO**)		CH_3CHO Ethanal (acetaldehyde)
Ketones	R—**C**(R′)(=**O**)	(R**CO**R′)	Note that there is no O—R′ bond	CH_3COCH_3 2-Propanone (acetone) $CH_3CH_2COCH_3$ 2-Butanone
Carboxylic acids	R—**C**(**O**—**H**)(=**O**)	(R**COOH**)		$CH_3CH_2CH_2CH_2COOH$ Pentanoic acid
Esters	R—**C**(**O**—R′)(=**O**)	(R**COO**R′)		$CH_3COOCH(CH_3)_2$ Isopropyl acetate
Amides	R—**C**(**N**(**H**)(**H**))(=**O**)	(R**CONH**$_2$)	One or both of the H's on the N may be replaced be an alkyl radical	$CH_3CH_2CONH_2$ Propanamide $CH_3CONHCH_2CH_2CH_2CH_3$ *N*-Butyl acetamide

* Not including hydrocarbons. The functional group characteristic of each class is shown in **boldface** in the general formula.

† The formula given in parentheses is that normally used for typographical convenience. R and R′ signify alkyl radicals (Table 29-3) and may be the same or different.

in representative compounds, in Table 29-4. The names of the hydrocarbon radicals in Table 29-3 are used not only in the common names of many compounds with comparatively few carbon atoms but also in the systematic names of branched-chain higher hydrocarbons and other more complex molecules.

The fact that the properties of organic compounds are, to a first approximation, a weighted sum of the properties attributable separately to their hydrocarbon portions and their functional groups provides a great simplifying feature in learning and understanding organic chemistry. It is not necessary to treat each new compound as a new entity. Rather, it is possible to predict the kinds of reactions a molecule will undergo by considering its component parts. The reactivity is governed largely by the nature of the functional groups present.

Physical Properties of Organic Compounds Most organic compounds of relatively low molecular weight are gaseous or liquid for the reasons discussed in Chaps. 6 and 7: the intermolecular attractions are not strong. Substances in which there is considerable opportunity for hydrogen bonding are both less volatile and more viscous; the trihydroxy compound glycerol, for example, is a syrupy high-boiling liquid, although its molecular weight is about the same as that of hexane, which boils at 69°C. Amino acids of even low molecular weight are relatively high-melting solids, because there are strong forces between the dipolar molecules (page 672).

The solubility of organic substances in different solvents varies widely. A general rule is that compounds tend to dissolve in substances of similar structure and composition and not to dissolve in substances of very different structure and composition. Thus, compounds in which the proportion of oxygen atoms (and especially of —OH groups) or nitrogen atoms is high tend to be soluble in water and insoluble in hydrocarbons. Substances that consist primarily of a hydrocarbon chain or other nonpolar portion tend to dissolve in hydrocarbons and not in water.

CH_2OH
$CHOH$
CH_2OH

1,2,3-Propanetriol
(glycerol; glycerin)

29-3   Optical Isomerism

A number of kinds of isomerism have been discussed in Chap. 3 and in earlier sections of this chapter. One additional widespread type of isomerism, of particular significance in biochemistry, is that of two molecules that are mirror images of one another but are not superimposable, even in thought, like a right hand and a left hand. Such isomers are called *optical isomers* or *enantiomers*; the property of being not superimposable upon (congruent with) one's own mirror image is called *chirality* (Greek: *cheir,* hand).

Molecules that are not congruent with their own mirror images are said to be *optically active* because their solutions will rotate the plane of plane-polarized light (Fig. 29-1). Two of the isomers of $Co(en)_2Cl_2^+$ illustrated in Fig. 27-3 have this property. Any molecule containing a tetrahedrally substituted carbon atom joined to four different atoms or groups of atoms, such as bromochlorofluoromethane (Fig. 29-2b), is chiral and therefore optically active. In fact, it was the optical activity of certain naturally occurring compounds containing a carbon atom bonded to four different groups[5] that led van't Hoff and Le Bel (quite independently) to postulate in 1874 that the four single bonds at a carbon atom were arranged tetrahedrally. Their brilliant insight was not confirmed by direct structural evidence until the arrangement of the carbon atoms in diamond crystals was determined more than 40 years later.

The physical properties of optical isomers are identical when they are measured by methods that have either no directional implications (e.g., density, melting point, enthalpy) or no chirality associated with a directional property (such as dipole moment). Their chemical properties are also identical, except when they involve interaction with other chiral species. Thus the rates of reac-

---

[5] Such a carbon atom is referred to as an *asymmetric carbon atom.*

tion of two optically isomeric acids with a nonchiral alcohol, such as ethanol, would necessarily be identical, but their reaction rates with a chiral alcohol, such as one of the optical isomers of 2-butanol, might be quite different. Because all organisms contain numerous chiral molecules, the interactions of optical isomers with organisms almost invariably differ. For example, a particular chiral molecule may be essential for the proper functioning of an organism whereas its mirror image molecule cannot be used at all by that same organism.

A molecule whose internal symmetry is such that one-half of it is the mirror image of the other half is congruent with its own mirror image and hence cannot be optically active. Such a molecule may have a plane of symmetry. An example is bromochloromethane (Fig. 29-2c), whose symmetry plane passes through the centers of the Br, C, and Cl atoms and bisects the H–C–H angle. Molecules without internal mirror planes and without certain other symmetry elements are necessarily chiral. Since a helix is an inherently chiral object (Fig. 29-3), all helical molecules are chiral. When they are synthesized in an environment that is itself chiral, as are such biologically important helical molecules as DNA and proteins containing helical regions (Chap. 30), then there is normally a preference for one sense of the helix, either "right-handed" or "left-handed". Otherwise, both kinds of helices are created in equal numbers.

Figure 29-1   Rotation of the Plane of Plane-Polarized Light by an Optically Active Sample
Light is electromagnetic radiation and as such is associated with transverse oscillations of electric and magnetic fields (i.e., oscillations perpendicular to the direction of propagation). In this figure unpolarized light enters from the left; this is light in which transverse oscillations occur in all directions. A vertically oriented polarizer permits the passage of the vertical component of the electrical oscillation only, producing plane-polarized light. The optically active sample rotates the plane of polarization clockwise from the orientation BB and $B'B'$ to the orientation CC by an angle α that is proportional to the length of the tube and to the concentration of the active substance in it. The surface described by the oscillation of the electric field of the light in the sample looks like a twisted ribbon. The orientation of the analyzer shown permits maximum passage of the light with the new orientation of its plane of polarization.

If all the molecules of the optically active sample were replaced by their mirror images, the figure would also have to be replaced by its mirror image. In this new situation, the polarization plane of the light passing through the sample would be rotated by $-\alpha$. Molecules that are superimposable on their mirror images are not optically active (Fig. 29-2).

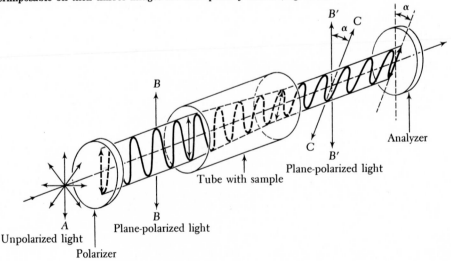

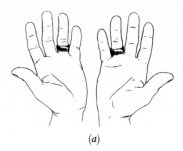

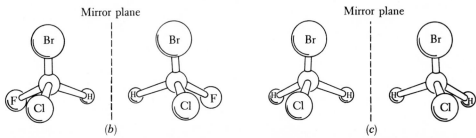

Figure 29-2 (*a*) **A Left Hand and its Mirror Image, a Right Hand;** (*b*) **an Optically Active Molecule and its Mirror Image; (C) an Optically Inactive Molecule and its Mirror Image**
A mirror plane, viewed here edge on, has the effect that a mirror appears to have—(*a*) changing a right hand held a certain distance from the mirror into a left hand that is equally distant on the opposite side of the mirror. The two molecules in (*b*) are both bromochlorofluoromethane; they are nonsuperimposable mirror images of one another, or optical isomers. Study of the drawings will show that if the Br, C, and Cl atoms are superimposed, the F atom of one molecule will fall on top of the H atom of the other, and conversely. These molecules are chiral; that is, they represent "right-handed" and "left-handed" forms. A solution of the right-handed form will rotate the plane of polarized light by the same amount, but in the *opposite direction,* as a solution of the left-handed form of the same concentration in the same solvent (Fig. 29-1).

The two molecules of bromochloromethane in (*c*) are also mirror images of each other, but they are congruent, that is, superimposable and thus indistinguishable. This substance is not optically active.

29-4   Organic Polymers

The term *polymer* (Greek: *poly,* many; *meros,* part) refers to any very large molecule, with molecular weight of the order of 10^4 or more, that is made by combining many identical or very similar small molecules (monomers). Naturally occurring polymers are not only the principal ingredients of most biological structures but also play central roles in metabolism and in genetics. Synthetic polymers have been tailor-made by chemists during the last half century to fill specific needs in industry, commerce, and the home, often as substitutes for less durable or more expensive materials.

The preparation and properties of polymers involve the same types of reactions and the same general principles of structure and reactivity that are applicable to smaller molecules (Sec. 29-5). Nevertheless, there was little real progress in understanding the detailed chemical nature or structures of most polymeric materials, especially those that are naturally occurring, until the middle third of this century because of the great difficulty in characterizing most of them.

Vinyl Polymers   Many alkenes can be polymerized, and since most of these compounds contain the vinyl group, $CH_2=CH-$, the products are known gen-

Mirror plane

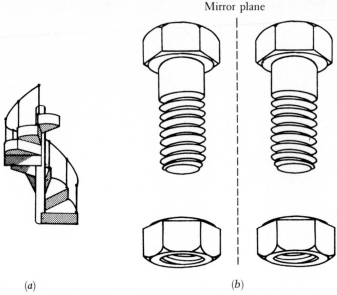

(a)   (b)

Figure 29-3 Some Helices

(a) A spiral staircase; (b) right-handed and left-handed bolts and nuts. A helix is the curve described by a point rotating about a straight line and simultaneously advancing parallel to the line through a distance proportional to the angle of rotation. Helices have a chirality associated with the sense of the rotation; the two shown in (b) are mirror images of one another. It is impossible to thread the right-handed nut onto the left-handed bolt, and vice versa. Turning either of the nuts (or bolts) end over end does not change its chirality; each nut still fits the same bolt that it did before. Some important helical structures in large molecules are shown in Figs. 29-4, 30-2b, and 30-7a.

erally as vinyl polymers. For example, ethylene polymerizes to form polyethylene, and the related molecule tetrafluoroethylene undergoes a similar reaction to form the polymer polytetrafluoroethylene (Teflon):

$$\text{Many } CH_2{=}CH_2 \longrightarrow \text{\textasciitilde\textasciitilde} CH_2CH_2CH_2CH_2CH_2CH_2CH_2CH_2 \text{\textasciitilde\textasciitilde}$$
$$\text{Many } CF_2{=}CF_2 \longrightarrow \text{\textasciitilde\textasciitilde} CF_2CF_2CF_2CF_2CF_2CF_2CF_2CF_2 \text{\textasciitilde\textasciitilde}$$

Although a few monomers are so reactive that in the absence of refrigeration and inhibitors they may polymerize spontaneously, polymerization is usually initiated by catalysts [such as a mixture of $TiCl_4$ with $Al(C_2H_5)_3$] or by addition of substances that form radicals. Radical-initiated polymerization is normally done with a small quantity of an organic peroxide,[6] which readily splits to form radicals:

$$ROOR \longrightarrow 2RO\cdot \tag{29-5}$$

One of the radicals so formed then attacks the π bond of a monomer molecule, $CH_2{=}CHY$ [such as those in (29-1) and (29-2)], producing a new radical:

$$RO\cdot + CH_2{=}CHY \longrightarrow ROCH_2\dot{C}HY \tag{29-6}$$

This radical can then in turn attack another monomer molecule:

---

[6]The organic peroxide is often termed a catalyst, even though its mode of action is not strictly that of a catalyst. The substances mentioned above may also not be catalysts by strict definition.

$$ROCH_2\dot{C}HY + CH_2=CHY \longrightarrow ROCH_2CHYCH_2\dot{C}HY \qquad (29\text{-}7)$$

This process continues in a chain reaction and the polymer grows rapidly; polymer growth is terminated only when two radicals combine, which happens on the average after 10^3 or 10^4 steps. Thus most of the resulting polymer molecules are extremely large and the effects of the RO— end groups are negligible. The resulting polymer molecules usually have a spread of molecular weights in the range 10^4 to 10^5. Hydrocarbon polymers are chemically inert, and polytetrafluoroethylene is especially unreactive because of the great stability of the C-F bond.

Polymerization of a vinyl monomer, $CH_2=CHY$, can give rise to structural variations in the product not possible with polyethylene, because in the polymer the groups Y may be arranged either randomly or in various ordered ways along the newly formed chain. For example,

Isotactic (Greek: *iso,* same; *taxis,* arrangement)

$$-CH_2\overset{\displaystyle Y}{\underset{|}{C}}HCH_2\overset{\displaystyle Y}{\underset{|}{C}}HCH_2\overset{\displaystyle Y}{\underset{|}{C}}HCH_2\overset{\displaystyle Y}{\underset{|}{C}}HCH_2\overset{\displaystyle Y}{\underset{|}{C}}HCH_2\overset{\displaystyle Y}{\underset{|}{C}}HCH_2\overset{\displaystyle Y}{\underset{|}{C}}HCH_2\overset{\displaystyle Y}{\underset{|}{C}}H- \qquad (29\text{-}8)$$

Syndiotactic (alternating)

$$-CH_2\overset{\displaystyle Y}{\underset{|}{C}}HCH_2\underset{\displaystyle Y}{\overset{|}{C}}HCH_2\overset{\displaystyle Y}{\underset{|}{C}}HCH_2\underset{\displaystyle Y}{\overset{|}{C}}HCH_2\overset{\displaystyle Y}{\underset{|}{C}}HCH_2\underset{\displaystyle Y}{\overset{|}{C}}HCH_2\overset{\displaystyle Y}{\underset{|}{C}}HCH_2\underset{\displaystyle Y}{\overset{|}{C}}H- \qquad (29\text{-}9)$$

Atactic (random, disordered)

$$-CH_2\overset{\displaystyle Y}{\underset{|}{C}}HCH_2\overset{\displaystyle Y}{\underset{|}{C}}HCH_2\overset{\displaystyle Y}{\underset{|}{C}}HCH_2\underset{\displaystyle Y}{C}HCH_2\overset{\displaystyle Y}{\underset{|}{C}}HCH_2\underset{\displaystyle Y}{C}HCH_2\underset{\displaystyle Y}{C}HCH_2\overset{\displaystyle Y}{\underset{|}{C}}H- \qquad (29\text{-}10)$$

Radical-initiated polymerization invariably leads to random or atactic polymers. Catalytic polymerization can lead to stereoregular polymers—e.g., the isotactic form exemplified by (29-8) or the alternating arrangement of (29-9).

The difference in physical properties between polymers of random structure and those of ordered structure can be enormous. Comparison of models of atactic and isotactic polypropylene indicates the reason. The regularity of isotactic polymers permits efficient packing of the chains and thus gives rise to "crystalline" (highly ordered) regions in the solid polymer and hence to stronger interchain forces and a consequent greater rigidity and higher melting or softening temperature. We consider this topic further below.

In addition to ethylene, propene, and tetrafluoroethylene, many other unsaturated molecules serve as monomers from which common vinyl polymers are made. Some of these are listed in Table 29-5. The polymers can be represented by formulas similar to (29-8) to (29-10). Sometimes two different monomers are mixed before polymerization. The resulting product is called a copolymer and usually has properties significantly different from those of the polymer formed from either monomer alone. Saran and similar kitchen wraps are copolymers of vinyl chloride or acrylonitrile with vinylidene chloride, $CH_2=CCl_2$.

Table 29-5
Some Other Vinyl Polymers

MONOMER		
Common name	Formula	Trade names of polymer
Styrene	$CH_2{=}CHC_6H_5$	Polystyrene, Styron, Lustron
Vinyl chloride*	$CH_2{=}CHCl$	PVC, Koroseal, Geon
Acrylonitrile	$CH_2{=}CHCN$	Orlon, Acrilan
Methyl methacrylate	$CH_2{=}\overset{\displaystyle CH_3}{\overset{\displaystyle |}{C}}COOCH_3$	Lucite, Plexiglas, Perspex

* Long exposure to vinyl chloride *monomer* has been correlated with above-normal incidence of certain kinds of cancer.

Natural rubber and most synthetic rubbers are closely related to vinyl polymers. Natural rubber is a stereoregular polymer with the repeating unit

$$-CH_2\overset{\displaystyle CH_3}{\overset{\displaystyle |}{C}}{=}CHCH_2-.$$ It can be made from the monomer 2-methyl-1,3-butadiene

(isoprene), $CH_2{=}\overset{\displaystyle CH_3}{\overset{\displaystyle |}{C}}CH{=}CH_2$. When this diene polymerizes, the two double bonds are converted to single bonds, with introduction of a new double bond in the center of the molecule and, in effect, an unpaired electron at each end that permits the isoprene units to be joined together by single bonds. The reaction of the first two molecules can be represented as

$$2CH_2{=}\overset{\displaystyle CH_3}{\overset{\displaystyle |}{C}}CH{=}CH_2 \longrightarrow 2\ -CH_2\overset{\displaystyle CH_3}{\overset{\displaystyle |}{C}}{=}CHCH_2- \longrightarrow$$

$$-CH_2\overset{\displaystyle CH_3}{\overset{\displaystyle |}{C}}{=}CHCH_2CH_2\overset{\displaystyle CH_3}{\overset{\displaystyle |}{C}}{=}CHCH_2- \quad (29\text{-}11)$$

$$\overset{\displaystyle Cl}{\overset{\displaystyle |}{CH_2{=}C}}CH{=}CH_2$$
2-Chloro-1,3-butadiene

Neoprene, a common synthetic rubber, is a polymer of 2-chloro-1,3-butadiene and is more polar than natural rubber. Natural rubber has a tendency to swell in and be dissolved by hydrocarbons; neoprene does not. Consequently, neoprene is substituted for natural rubber whenever the material may be in contact with hydrocarbons, e.g., in hoses for gasoline and oils.

Condensation Polymers As the preceding examples show, the polymerization of unsaturated molecules proceeds by the addition of one molecule to another, all atoms of the monomers being retained in the polymer. In condensation polymerization there is simultaneously a formation of covalent bonds linking monomers together and an elimination of a low-molecular-weight compound, such as water, an alcohol, or a halide. The monomers must contain two functional groups (a "bifunctional" monomer), and in the process of polymerization functional groups on different monomers react with one another. For example, the common synthetic fiber known as Dacron or Terylene is made by heating a bifunctional alcohol with a bifunctional acid:

$$HOCH_2CH_2OH + HOC\!\!-\!\!\langle\bigcirc\rangle\!\!-\!\!COH \longrightarrow$$
$$\quad\quad\quad\quad\quad\quad\quad O\quad\quad\quad\quad O$$

1,2-Ethanediol   1,4-Benzenedicarboxylic acid

$$H\left(-OCH_2CH_2OC\!\!-\!\!\langle\bigcirc\rangle\!\!-\!\!C\!\!-\right)_{\!n}\!\!OH + n\,H_2O \quad (29\text{-}12)$$

Dacron

The formula of the product is intended to imply that the portion in parentheses is repeated n times, where n is some large integer. Such polymers are called polyes-

$$\quad\quad\quad\quad\quad\quad\quad\quad\quad O$$
ters because they contain the $-\overset{\|}{C}-O-C-$ linkage typical of esters (Table 29-4).

Nylon, one of the first and most successful of the synthetic fibers, is made by the reaction of a six-carbon diamine with a six-carbon dicarboxylic acid

$$\overset{O}{\overset{\|}{HOCCH_2CH_2CH_2CH_2}\overset{O}{\overset{\|}{C}OH}} + H_2NCH_2CH_2CH_2CH_2CH_2CH_2NH_2 \longrightarrow$$

$$HOC(CH_2)_4CN(CH_2)_6NC(CH_2)_4CN(CH_2)_6NC \ldots \quad\quad or$$

$$HO\left[-C(CH_2)_4CN(CH_2)_6N-\right]_n\!\!H \quad (29\text{-}13)$$

$$\quad\quad\quad\quad\quad\quad\quad\quad\quad O$$
Nylon is a polyamide because an amide group $-\overset{\|}{C}-\underset{H}{N}-$ is formed at each

junction between monomers. Proteins are naturally occurring polyamides. However, each of the "monomers" from which they are made has one amino group and one carboxyl group rather than two similar groups. They are amino acids (Sec. 29-5), whose general structure may be represented as $RCHCOOH$, NH_2

so that a segment of a protein chain may be represented as

$$\quad\quad R\quad\quad R'\quad\quad R''\quad\quad R'''$$
$$H_2NCHCONHCHCONHCHCO\cdots NHCHCOOH \quad (29\text{-}14)$$

Each of the thousands of different proteins is characterized by a highly specific sequence of amino acids, corresponding to the different R groups in (29-14), as discussed in the following chapter.

Physical Properties of Polymers The physical properties of common polymeric substances vary enormously. Some are soft but somewhat elastic; others are hard and brittle. Some of those that are hard at ordinary temperatures soften or melt on heating; others do not. Some are soluble in solvents such as benzene, acetone, or ethyl acetate, and some in water; others are unaffected by solvents. These differences in properties can be explained by considering the chemical nature of the materials and the general principles of structure and interatomic interactions discussed in earlier chapters, especially Chaps. 6, 7, and 17. There are no new kinds of forces in polymers. Different polymer molecules and different regions of the same molecule interact with each other by van der Waals forces and, when appropriate groups are present, by hydrogen bonding. Different molecules may even be cross-linked by covalent bonds, or by ionic interactions if oppositely charged groups are present.

As the temperature of a sample of polymer is raised, the component atoms and groups of atoms begin to vibrate increasingly. When the forces between the chains are weak, as with hydrocarbon chains and other comparatively nonpolar portions of molecules, not much thermal energy is needed to overcome intermolecular attractions. However, the energy required does depend significantly on how closely the chains fit together, since van der Waals forces are extremely sensitive to interatomic distance and are thus very dependent on the fit of one molecule to its neighbors. Atactic polymers, which have no ordered arrangement of side groups, pack together much less effectively than do the highly regular isotactic polymers. Consequently, appreciably more thermal energy is needed to overcome the interchain attractions in isotactic materials and they have higher softening and melting points.

The flexibility of polymer chains, that is, their ability to adopt different conformations by rotation of adjoining segments about the bonds between them, also plays an important role. A chain in which the barrier to rotation about the bonds is comparatively low will be flexible at ordinary temperatures. On the other hand, double bonds will stiffen a chain, and if the substituted groups are relatively bulky, rotation even about single bonds may be highly hindered except at elevated temperatures. This effect is manifested in the much greater flexibility of polyethylene objects, e.g., tubing, than of those made from Teflon. Teflon is very stiff. Its polymer chains are comparatively inflexible because the rotation of one CF_2 group past that adjacent to it requires appreciable energy, much more than does the rotation of one CH_2 relative to the next in a polyethylene chain. The van der Waals radius (packing radius) of a fluorine atom is only 15 to 20 percent greater than that of a hydrogen atom, but this difference has an appreciable effect (Fig. 29-4).

The packing of chains in polyamides such as nylon is stabilized by the formation of strong hydrogen bonds between the $>NH$ groups of one chain and the $>C=O$ groups of an adjacent one, $>NH \text{---} O=C<$. When nylon is first made, the individual chains are randomly oriented. The polymer is melted and extruded into filaments, which are then stretched to four or five times their original length.

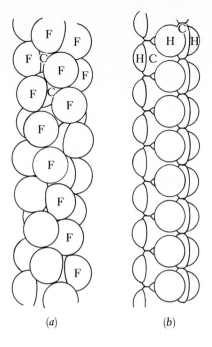

(a)   (b)

Figure 29-4   Packing Models of Chains in Teflon and Extended Polyethylene
The increased bulkiness of the fluorine atoms in the Teflon chain (a), relative to that of the hydrogen atoms in polyethylene (b), stiffens the Teflon because of the barrier to rotation of one CF_2 group relative to the next. It also imparts a twist to the Teflon chain, making it necessarily helical. Polyethylene is much more flexible. It can adopt the extended zigzag configuration illustrated here, with 180° torsion angles around the C-C bonds, or can coil up, with the most favorable torsion angle then being 60° (Sec. 17-2).

This orients the molecules more or less parallel to the filament axis. The tensile strength of these fibers is greater than that of the original unoriented polymer, because of the hydrogen bonding and the more efficient packing of the oriented chains. Hydrogen bonds between $>$NH and $>$C$=$O groups play a crucial role in the structure of the fibers of silk and wool as well; these are both natural protein fibers.

We pointed out in Chap. 7 that elements whose atoms can bond to only two neighbors form long-chain molecules (or rings) (Fig. 7-2) but not two- or three-dimensional networks. This same principle applies to monomers that can form two bonds. They produce long-chain polymers, either by opening a double bond or by condensation reactions involving two functional groups. In elements whose atoms can bond to three or more neighbors, two-dimensional or three-dimensional structures can be built by linking chains together. Similarly, when a monomer can form more than two bonds, it can bind chains together to give extended networks or three-dimensional structures. If these bonds are strong, the resulting polymer will consist of only one or a few giant molecules, rather than many separate chains, and will be hard and insoluble. Only a few such cross-links can produce an appreciable change in the properties of a polymer. For example, addition of less than 0.1 percent of divinylbenzene to styrene before polymerization will produce a polymer that is insoluble in benzene, whereas ordinary polystyrene readily dissolves. The divinylbenzene provides cross-links between chains; one of the vinyl groups forms a segment of one chain, the second a component of a second chain, and the benzene ring forms the bridge linking the chains together:

661

$$
\begin{array}{cc}
\overset{|}{C}H_2 & \overset{|}{C}H_2 \\
C_6H_5-CH & CH-C_6H_5 \\
CH_2 & CH_2 \\
CH-C_6H_4-CH & \\
CH_2 & CH_2 \\
C_6H_5-CH & CH-C_6H_5 \\
\overset{|}{C}H_2 & \overset{|}{C}H_2
\end{array}
$$

The vulcanization of rubber by heating it with sulfur, which converts the rubber from a soft gummy material into a product of varying hardness depending on the amount of sulfur used, involves the creation of cross-links that consist of —S—S— groups.

$$
\begin{array}{c}
\overset{\displaystyle CH_3}{|} \\
-CH_2-\overset{|}{\underset{|}{C}}-CH-CH_2- \\
 H \;\; S \\
 S \quad H \\
-CH_2-CH-\overset{|}{\underset{|}{C}}-CH_2- \\
 CH_3
\end{array}
$$

This same disulfide cross-linkage binds together different chains, or different regions of one chain, of some proteins and thus helps to maintain the highly specific chain conformations that are essential to the proper biological functioning of the protein: The disulfide links in proteins are brought about naturally by oxidation of —SH groups in nearby segments of the protein chains.

Synthetic polymers with extensive cross-linking are classified as *thermosetting* polymers, the term implying that heat is needed to bring about the polymerization but that once the polymer is finally formed it remains hard and rigid. *Thermoplastic* polymers, more often called just plastics (Greek: *plastikos,* that can be molded), soften upon heating and can then be shaped into any desired form. They will retain that form upon being cooled sufficiently and will soften again when reheated. Thus scraps or broken objects can be reused by reheating and molding them into any desired shape.

At a sufficiently low temperature, any plastic material is, like glass, hard and brittle. When heated it becomes first somewhat flexible and pliable and then may reach a state in which it can be deformed but will return to its original shape if the deformation is not maintained for too long a period—that is, it shows some elasticity. Finally, on continued heating it softens and then melts to a viscous liquid.

Different plastics exhibit these various properties at quite different temperatures because of variations in the nature and arrangements of their chains and the side groups on the chains. In the glassy state there are often highly ordered microcrystalline regions of the polymer where some chain segments are packed together in a regular array, although the length and the flexibility of the chains prevents the formation of large regions in which three-dimensional order persists. As the temperature is raised, the increasing mobility of chain segments leads to a more flexible material. Elasticity arises when the chains, normally twisted and coiled, are sufficiently flexible that they can be straightened to some degree when tension is applied but are not free to slip past one another. When the tension is relaxed, the chains coil up again and the material contracts once more to its original shape—provided that the tension has not been applied for so long a time that the chains have been able to slip by one another, a phenomenon that becomes increasingly likely as the temperature is raised. At still higher temperatures, the chains can move more or less independently of one another and the polymer softens and melts, although because of the length of the chains, which ensures their remaining to some degree tangled, the resulting liquid is always viscous.

At ordinary temperatures, thin sheets of polyethylene are sufficiently flexible to be used for raincoats, packaging materials, and the like whereas a polymer such as Lucite is in the glassy state. Polyethylene softens below the temperature of boiling water; hence polyethylene kitchenware put in a dishwasher often sags and loses its shape. Polypropylene softens at around $120°C$, Lucite near $160°C$ or higher, and polystyrene at around $200°C$.

Variations in the chemical behavior of polymers, such as their stability in the presence of acids and bases or on heating, can also be understood in terms of the chemical nature of the polymer. Since both amides and esters are readily hydrolyzed by acids and bases, it is not surprising that a drop of acid will make a hole in a silk or nylon blouse or that sodium hydroxide will disintegrate a sweater made of Dacron. Polyethylene containers are understandably inert to acids and bases, but they will burn readily, as other hydrocarbons do. Teflon, on the other hand, is stable at relatively high temperatures, because of the great stability of the C-F bond.

By varying the nature of the polymeric mix—the structures of the monomer and the catalyst, the possible presence of cross-linking agents or of a second monomer that can form a copolymer—chemists have succeeded in tailor-making polymers with an enormous range of characteristics. More than half of all chemists employed in industry work on various facets of polymer chemistry.

29-5 Properties and Reactions of Common Types of Organic Compounds

In this section some of the characteristic properties and reactions of typical compounds containing common functional groups are considered in the order in which they appear in Table 29-4, which should be used as a guide in reading the section.

Most organic compounds are not ionic, in contrast to many of the common inorganic materials discussed in earlier chapters. Thus, most of the organic reactions that we shall consider involve changes in bonds that are predominantly covalent. This does not imply, however, that organic substances are nonpolar. Because fluorine, oxygen, nitrogen, and chlorine are appreciably more electronegative than carbon, bonds between carbon and these elements have significant ionic character, with carbon the more positive atom. This bond polarity plays an important role in many organic reactions.

Organic reactions are usually much slower than most familiar ionic reactions in aqueous solutions. Furthermore, most organic reactions are under kinetic control rather than equilibrium control (Sec. 22-9). Careful specification of such reaction conditions as solvent and temperature is frequently necessary in order to maximize the yield of desired product, which is often far less than 100 percent because a number of different reactions compete with each other.

R—X
(X = F, Cl, Br, I)

Haloalkanes The simpler haloalkanes (alkyl halides) are usually named in terms of the alkyl radical present, but more complicated ones are named systematically in a fashion parallel to that used for hydrocarbons (Appendix B):

$$(CH_3)_2CHBr \qquad CH_3CHICHCH_2CH_3$$
$$\overset{\displaystyle CH_3}{|}$$

Isopropyl bromide   2-Iodo-3-methylpentane

They may be prepared by direct substitution of a halogen atom for a hydrogen atom in an alkane, as mentioned earlier, or by addition of HX to an appropriate alkene:

$$(CH_3)_2C{=}CH_2 + HCl \longrightarrow (CH_3)_3CCl \qquad (29\text{-}15)$$

2-Methylpropene   2-Chloro-2-methylpropane

Although 1-chloro-2-methylpropane (isobutyl chloride), $(CH_3)_2CHCH_2Cl$, is another possible product of (29-15), it is formed in much smaller quantity. The general rule is that in addition of HX to a double bond, the hydrogen atom normally becomes attached to the doubly bonded carbon atom already richer in hydrogen. Haloalkanes may also be formed by reaction of the corresponding alcohol with the appropriate hydrohalic acid; for some alcohols, a catalyst such as the corresponding zinc halide, ZnX_2, must also be present:

$$ROH + HX \xrightarrow{ZnX_2} RX + H_2O \qquad (29\text{-}16)$$

The chief importance of the haloalkanes, especially the bromides, lies in their use in the preparation of the extremely useful *Grignard reagents* by reaction with metallic magnesium in dry diethyl ether:

$$RBr + Mg \longrightarrow RMgBr \qquad (29\text{-}17)$$

These reagents are among the most valuable in organic chemistry; a few applications are mentioned below.

Another important reaction of haloalkanes is removal of the elements of HX (dehydrohalogenation) from adjacent carbon atoms under the influence of KOH dissolved in alcohol, with the resultant formation of an alkene, as exemplified by

$$CH_3CHICH_3 \xrightarrow[\text{KOH}]{\text{alcoholic}} CH_3CH{=}CH_2 \qquad (29\text{-}18)$$

Isopropyl iodide   Propene

R—O—H

Alcohols Alcohols contain the hydroxyl group, —OH, joined to a (nonaromatic) carbon atom to which no other oxygen atoms are bonded. They are named systematically as derivatives of the longest carbon-atom chain to which the

—OH group is attached, with that group given the lower possible number if there is a choice. The suffix -ol denotes the —OH group:

$$
\begin{array}{cc}
\text{Cyclopentanol} & \text{4-Methyl-3-hexanol} \\
& \text{(not 3-methyl-4-hexanol)}
\end{array}
$$

$$
CH_3CH_2CHCHCH_2CH_3
$$

Alcohols with three or fewer carbon atoms are completely miscible with water because they can readily form hydrogen bonds with the water, both as donors and acceptors, without great disruption of the water structure. As the size of the hydrocarbon portion increases further, however, the solubility in water decreases.

Methanol and ethanol, the simplest and most important alcohols, are volatile liquids that boil at 65 and 78°C, respectively. Methanol (methyl alcohol, wood alcohol) is extremely poisonous, causing damage to brain tissue and being oxidized readily in the body to formic acid, which attacks the optic nerves and causes blindness. Methanol was first made by heating dried wood at high temperatures, but it is now produced industrially by the direct combination of CO and H_2 at 350 to 400°C and about 300 atm in the presence of a chromic oxide–zinc oxide catalyst. It is one of the crucial starting materials for the synthetic organic chemical industry.

CH_3OH
Methanol

Ethanol (ethyl alcohol, grain alcohol, "alcohol") has been made since antiquity by fermentation of various grains, molasses, and other materials containing carbohydrates (Sec. 30-1). In moderate quantities ethanol is intoxicating, but slightly larger doses can be fatal through depression of the respiratory center in the brain. About 10^9 kg ethanol is now produced annually in the United States, the bulk of it for industrial purposes, by the hydration of ethylene with an acid catalyst at elevated temperatures and pressures.

CH_3CH_2OH
Ethanol

It is important to note that alcohols do not ionize significantly in water; they are nonelectrolytes. They may be considered as alkyl derivatives of water, with one of the hydrogen atoms of water replaced by an alkyl group. They are at once extremely weak bases and extremely weak acids, much weaker than water. They accept protons only from the strongest acids to form ROH_2^+ (analogous to H_3O^+) and they react with the most reactive metals (e.g., sodium) to produce hydrogen and a salt:

$$CH_3OH + Na \longrightarrow CH_3O^- + Na^+ + \tfrac{1}{2}H_2 \qquad (29\text{-}19)$$

The resulting alkoxide ions, RO^-, are stronger bases than OH^-. They therefore cannot exist in appreciable concentrations in aqueous solutions, because they react with water to form the alcohol ROH and OH^-.

The reactions of alcohols are those of the —OH group or of its hydrogen atom [as in (29-19)]. Dehydration of an alcohol results in an alkene by removal of the —OH from one carbon atom and of a hydrogen atom from an adjoining carbon atom:

$$(CH_3)_3COH \xrightarrow{\text{catalyst}} (CH_3)_2C{=}CH_2 + H_2O \qquad (29\text{-}20)$$

$$
\begin{array}{cc}
\text{2-Methyl-2-propanol} & \text{2-Methylpropene} \\
(t\text{-butyl alcohol}) & \text{(isobutene)}
\end{array}
$$

The catalyst may be sulfuric acid or aluminum oxide. The hydroxyl group may

665

be replaced by a halogen atom [Equation (29-16)], and the alcohol may react with a carboxylic acid to form an ester:

$$CH_3CH_2CH_2C\overset{OH}{\underset{O}{\diagdown}} + CH_3CH_2OH \xrightarrow[catalyst]{acid}$$

Butanoic acid
(butyric acid)   Ethanol

$$CH_3CH_2CH_2C\overset{OCH_2CH_3}{\underset{O}{\diagdown}} + H_2O \quad (29\text{-}21)$$

Ethyl butanoate
(ethyl butyrate)

Another example of an esterification reaction is (29-12), the synthesis of the polyester fiber Dacron. Alcohols can also react with inorganic acids containing —OH groups to form esters:

$$CH_3OH + (HO)_3PO \longrightarrow CH_3OPO(OH)_2 + H_2O \quad (29\text{-}22)$$

Phosphoric acid   Methyl dihydrogen phosphate
(or just methyl phosphate)

$$\begin{array}{l} CH_2ONO_2 \\ | \\ CHONO_2 \\ | \\ CH_2ONO_2 \end{array}$$

1,2,3-Propanetriol trinitrate
(nitroglycerin)

Phosphate esters are essential components of many biochemical systems (Chap. 30). The sodium salts of sulfate esters of long-chain alcohols are important synthetic detergents, and nitrate esters such as 1,2,3-propanetriol trinitrate (nitroglycerin) are ingredients of some explosives.

Ethers These compounds contain an oxygen atom bonded to two carbon atoms. They are comparatively inert chemically and are used chiefly as solvents. Ordinary "ether", the first substance used as a general anesthetic for operations (in the 1840s), is diethyl ether. It has a high vapor pressure, boiling at about 35°C, and, since it is flammable and its mixtures with air are explosive, it has generally been superseded as an anesthetic. It is, however, still a common, and somewhat hazardous, laboratory solvent. Ethers are somewhat less soluble in water than are alcohols with the same number of carbon atoms (which are isomeric with them); because ethers have no —OH group, they can only act as acceptors, not donors, in hydrogen-bond formation.

R—O—R′

Amines Amines are derivatives of ammonia, with one, two, or all three of the hydrogen atoms of ammonia replaced by hydrocarbon radicals. The group —NH$_2$ is called the *amino* group. Simple amines are commonly named by attaching the names of the radicals to the word *amine,* as in isopropylamine, $(CH_3)_2CHNH_2$. In systematic nomenclature the name is derived from the parent hydrocarbon and the prefix *amino.* When two or more different radicals are bonded to the nitrogen atom, the compound is named as a derivative of the largest radical, with the prefix *N* used to designate other radicals substituted on the amino group, as in *N,N*-dimethylaminoethane (Table 29-4).

$$R-N\overset{H}{\underset{H}{\diagdown}}$$

$$R-N\overset{R'}{\underset{H}{\diagdown}}$$

$$R-N\overset{R'}{\underset{R''}{\diagdown}}$$

The most characteristic property of amines is their basicity, a consequence of the availability of the unshared pair of electrons on the nitrogen atom. Alkyl amines are comparable in base strength to ammonia, reacting slightly with water. For example,

$$(CH_3CH_2)_2NH + H_2O \rightleftharpoons (CH_3CH_2)_2NH_2^+ + OH^- \quad (29\text{-}23)$$

Diethylamine   Diethylammonium
ion

Similarly, amines form salts with acids stronger than water; for example, trimethylamine, $(CH_3)_3N$, reacts with acetic acid to give trimethylammonium ion $[(CH_3)_3NH^+]$ and acetate ion.

Amines can be prepared by treating an appropriate alkyl halide with ammonia, which produces an alkylammonium salt, and then liberating the amine by treating the salt with a strong base such as NaOH:

$$RX + NH_3 \longrightarrow RNH_3^+X^- \xrightarrow{OH^-} RNH_2 \qquad (29\text{-}24)$$

Ammonia reacts with alkyl halides because the C-X bond polarity makes the carbon atom slightly positive, and ammonia is said to be a good *nucleophilic* reagent, that is, it tends to seek positive reaction centers (because of its unshared electron pair). Detailed kinetic and mechanistic studies indicate that in reactions of this kind the ammonia molecule attacks the halo-substituted carbon atom from the side opposite the C-X bond, displacing the halogen atom with its bonding pair as a halide ion. For example,[7]

$$H_3N\!:\; + \begin{array}{c} CH_3 \\ | \\ C\!-\!Cl \\ H\diagup \;\diagdown \\ CH_3CH_2 \end{array} \longrightarrow \left[\begin{array}{c} CH_3 \\ | \\ H_3N\!-\!C \\ \diagdown H \\ CH_2CH_3 \end{array} \right]^+ + Cl^- \qquad (29\text{-}25)$$

The reactions are typical nucleophilic displacements (or substitutions). **Amines with more than one alkyl substituent on the nitrogen can be prepared in similar fashion by treating an alkyl halide with an appropriate amine.** For example,

$$CH_3NH_2 + CH_3CH_2CH_2Br \longrightarrow [CH_3NH_2CH_2CH_2CH_3]^+Br^- \xrightarrow{OH^-}$$

$$CH_3NHCH_2CH_2CH_3 \qquad (29\text{-}26)$$

Amines with few carbon atoms are water-soluble and volatile, but as the hydrocarbon portion of the molecule becomes larger, the volatility and water solubility decrease. Methylamine, dimethylamine, and trimethylamine are all found in brine in which herring have been soaked, and amines generally tend to have a somewhat fishy odor.

Amines and related compounds play a major role in many biochemical systems, and we consider them again in Chap. 30.

Aldehydes and Ketones Aldehydes and ketones have very similar reactions; the chemistry of these compounds is that of the carbonyl group, $>C=O$. In aldehydes, the carbonyl group is joined to at least one hydrogen atom (and in formaldehyde, $H_2C=O$, to two hydrogen atoms); in ketones, it is joined to two other carbon atoms. Aldehydes are named systematically with the aid of the suffix *-al*, but the simplest ones are usually given common names based on the name of the corresponding acid (especially formaldehyde and acetaldehyde). Ketones are named systematically with the help of the suffix *-one* and numbers designating the position of the carbonyl group in the longest carbon chain. Alternatively, the names of the radicals attached to the carbonyl group and the word *ketone* may be used.

---

[7] The wedge-shaped and dashed bonds imply that the CH_3CH_2 group is above and the H atom below the plane of the page.

$$\underset{\text{Propanal}}{CH_3CH_2CHO} \qquad \underset{\substack{\text{3-Pentanone}\\ \text{(diethyl ketone)}}}{CH_3CH_2COCH_2CH_3} \qquad \underset{\text{3-Chloro-2-butanone}}{\overset{\overset{\displaystyle Cl}{|}}{CH_3CHCOCH_3}}$$

Note that a too casual inspection of these formulas for ketones might suggest that they are ethers; that this is not so is evident when one counts the bonds to the carbon atom written to the left of the oxygen atom.

The double bond from carbon to oxygen is quite polar, with the oxygen atom the more negative, and the reactions of the carbonyl group reflect this polarity. One of the most useful reactions is that with a Grignard reagent, RMgX [Equation (29-17)], for this provides a means of lengthening a carbon chain or adding side chains. Because of the highly electropositive character of magnesium, the C-Mg bond in RMgX is polar:

$$-\overset{\displaystyle |}{\underset{\displaystyle |}{C}}\overset{\delta-}{}\!-\!\overset{\delta+}{Mg}-$$

Consequently the carbon atom attached to the magnesium atom is nucleophilic and becomes bonded to the more positive atom of the carbonyl group, the carbon atom, while the —MgX group becomes attached to the oxygen atom. The product is a magnesium salt of an alcohol, and treatment with water readily frees the alcohol. For example, with acetaldehyde,

$$RMgBr + \underset{H}{\overset{CH_3}{\diagup}}C=O \longrightarrow H-\underset{R}{\overset{CH_3}{\overset{|}{C}}}-OMgBr \xrightarrow{H_2O} H-\underset{R}{\overset{CH_3}{\overset{|}{C}}}OH \qquad (29\text{-}27)$$

The product is a secondary alcohol, that is, an alcohol with two other carbon atoms joined to the carbon bearing the —OH. One of these atoms is in the methyl group from acetaldehyde; the other is in the R— group from the Grignard reagent. With formaldehyde instead of acetaldehyde, the product would be a primary alcohol, RCH_2OH. With a ketone, the product is a tertiary alcohol:

$$RMgBr + \underset{R''}{\overset{R'}{\diagup}}C=O \longrightarrow \xrightarrow{H_2O} R-\underset{R''}{\overset{R'}{\overset{\diagup}{\underset{\diagdown}{C}}}}OH \qquad (29\text{-}28)$$

Aldehydes and ketones can easily be reduced to give alcohols with the same number of carbon atoms. This can be done with hydrogen and a suitable catalyst, such as Pt or Ni, or with a reducing agent like lithium aluminum hydride ($LiAlH_4$) or sodium borohydride ($NaBH_4$). An aldehyde gives a primary alcohol, and a ketone gives a secondary alcohol:

$$\underset{\substack{\text{2-Methylpropanal}\\ \text{(isobutyraldehyde)}}}{(CH_3)_2CHCHO} \xrightarrow{LiAlH_4} \underset{\substack{\text{2-Methylpropanol}\\ \text{(isobutyl alcohol)}}}{(CH_3)_2CHCH_2OH} \qquad (29\text{-}29)$$

$$\underset{\substack{\text{2-Propanone}\\ \text{(acetone)}}}{(CH_3)_2C=O} \xrightarrow{LiAlH_4} \underset{\substack{\text{2-Propanol}\\ \text{(isopropyl alcohol)}}}{(CH_3)_2CHOH} \qquad (29\text{-}30)$$

Conversely, primary and secondary alcohols can be oxidized under appropriately mild conditions to produce the corresponding aldehydes and ketones. A suitable reagent is aqueous potassium dichromate, $K_2Cr_2O_7$. Further oxidation of the aldehyde will give the corresponding carboxylic acid,

$$RCHO \xrightarrow{ox} RCOOH \tag{29-31}$$

and indeed it is sometimes difficult to stop the oxidation of a primary alcohol at the stage of the aldehyde. Most ketones are resistant to mild oxidation; like most organic compounds, they can be oxidized to CO_2 and H_2O under more severe conditions.

□ **Isomers of C_3H_8O** (a) Write the structural formulas for the isomeric molecules with molecular formula C_3H_8O. (b) Mild oxidation of one of these isomers gives a carboxylic acid. What is the isomer?

Example 29-3

Solution (a) The formula indicates that the molecules must be alcohols or ethers without a ring or a double bond. The only possibilities with three carbon atoms are 1-propanol, 2-propanol, and methyl ethyl ether:

$$CH_3CH_2CH_2OH \qquad CH_3\underset{\underset{OH}{|}}{C}HCH_3 \qquad CH_3OCH_2CH_3$$

(b) Of these three molecules, only 1-propanol would form a carboxylic acid (propanoic acid) on mild oxidation. Acetone would be formed from 2-propanol, and the ether is inert under mild oxidizing conditions. ■

□ There are two molecules with formula C_2H_4O that do not contain any —OH group. (a) Write their structural formulas. (b) In one of these isomers all protons are equivalent (as shown by its nmr spectrum). What is the isomer? ■

Exercise 29-3

□ **Synthesis of 3-Hexanol from Propanol** Devise a scheme for synthesizing 3-hexanol from propanol and any needed inorganic substances.

Example 29-4

Solution In devising a synthetic scheme, it is usually simplest to start at the end and work backward. The desired product, 3-hexanol, $CH_3CH_2\underset{\underset{OH}{|}}{C}HCH_2CH_2CH_3$, is a secondary alcohol that could be made by the addition of a Grignard reagent to an aldehyde, reaction (29-27). The corresponding aldehyde and alkyl halide could be propanal and 1-bromopropane:

$$CH_3CH_2CH_2MgBr + CH_3CH_2CHO \longrightarrow \xrightarrow{H_2O} CH_3CH_2\underset{\underset{OH}{|}}{C}HCH_2CH_2CH_3$$

The available starting material, propanol, could be converted into propanal by mild oxidation and into 1-bromopropane by appropriate treatment with HBr [Equation (29-16)]:

$$CH_3CH_2CH_2OH \xrightarrow{ox} CH_3CH_2CHO$$

$$CH_3CH_2CH_2OH \xrightarrow[ZnBr_2]{HBr} CH_3CH_2CH_2Br \xrightarrow[\substack{dry \\ ether}]{Mg} CH_3CH_2CH_2MgBr \quad ■$$

□ Devise a scheme for synthesizing 2-butanol from ethanal and any needed inorganic substances. ■

Exercise 29-4

The carbonyl group is common in various natural products, including sugars, which are polyhydroxy aldehydes or polyhydroxy ketones (Sec. 30-1). The pungent odor of formaldehyde is familiar to zoology students since its aqueous solution (formalin) is used in preserving specimens. The simplest ketone, acetone, is a common solvent in the laboratory, in industry, and in some common household glues and similar products.

$$R-C\underset{OH}{\overset{O}{<}}$$

Carboxylic Acids[8] The simpler carboxylic acids have common names that reflect their origin: formic acid is the active ingredient in the sting of red ants (Latin: *formica,* ant), acetic acid is the acidic component of vinegar (Latin: *acetum,* vinegar), and butyric (or butanoic) acid is present in rancid butter (Latin: *butyrum,* butter). The systematic names, now usually used for acids with three or more carbon atoms, involve the root hydrocarbon name with the final *e* changed to *oic* and the word *acid* added. The carbon atom of the —COOH group is included in counting the number of carbon atoms for the purpose of naming the longest chain.

$$HCOOH \qquad BrCH_2CH_2\underset{\underset{CH_3}{|}}{C}HCOOH$$

Formic acid   2-Methyl-4-bromobutanoic acid

Most carboxylic acids are weakly acidic, like the familiar acetic acid. Those with four or more carbon atoms have limited solubility in water, but they are soluble even in as weakly basic a solution as one containing sodium bicarbonate because they are converted to salts:

$$RCOOH + HCO_3^- \longrightarrow \underset{\text{A carboxylate ion}}{RCO_2^-} + H_2O + CO_2 \qquad (29\text{-}32)$$

Carboxylic acids can be prepared by controlled oxidation of primary alcohols, RCH_2OH, or aldehydes, $RCHO$, with the same number of carbon atoms as the acid, as just discussed. The carboxyl group may also be introduced into a molecule by adding CO_2 to a Grignard reagent. For example,

$$\underset{\substack{\text{Ethylmagnesium}\\\text{bromide}}}{CH_3CH_2MgBr} + CO_2 \longrightarrow CH_3CH_2COOMgBr \overset{H_2O}{\longrightarrow} \underset{\substack{\text{Propanoic}\\\text{acid}}}{CH_3CH_2COOH} \quad (29\text{-}33)$$

Finally, acids may be produced by the reaction of water with (hydrolysis of) an ester or an amide:

$$RCOOR' + H_2O \longrightarrow RCOOH + R'OH \qquad (29\text{-}34)$$
$$RCONH_2 + H_2O \longrightarrow RCOOH + NH_3 \qquad (29\text{-}35)$$

A strong base is generally used to catalyze the hydrolysis of esters. Since most esters are insoluble in water, the base also increases the extent of hydrolysis by converting the product acid to RCO_2^-, which is thereby removed from the reaction mixture because it dissolves in the aqueous phase. Amides are normally hydrolyzed in an acidic medium, which catalyzes the reaction and converts the ammonia or amine that is produced into the corresponding ammonium salt.

An important type of compound related to an acid is the corresponding acid chloride, $RC\underset{O}{\overset{Cl}{<}}$. Acid chlorides are valuable intermediates for the ready preparation of esters, amides, and other acid derivatives.

---

[8] The typographic convenience of the representation RCOOH should not obscure the fact that there is no —O—O—H (hydroperoxy) group in these molecules.

Esters The formation of carboxylate esters from alcohols and carboxylic acids was considered earlier in the discussion of alcohols [Equation (29-21)]. Equation (29-34) represents the reverse of this reaction, the hydrolysis of an ester. Esters are named by combining the name of the radical corresponding to the alcohol with the name of the acid, altered by replacing the ending *-ic* by *-ate*:

$$R-C\overset{O}{\underset{O-R'}{}}$$

$$HCOOCH_2\underset{\underset{CH_3}{|}}{C}HCH_2CH_3 \qquad (CH_3)_2CHCOOCH_2CH_2CH_3$$

2-Methylbutyl formate   Propyl 2-methylpropanoate
                               (propyl isobutyrate)

Esters are important as laboratory and commercial solvents. The most common is ethyl acetate; about 200 million pounds of it are used in the United States annually, 90 percent of this as a solvent, chiefly for coatings, from synthetic rubbers to nail polishes. The ester linkage is common in many naturally occurring and biologically important compounds. For example, fats are triesters of the trihydroxy alcohol glycerol (1,2,3-trihydroxypropane) with primarily long-chain carboxylic acids, C_{16} and C_{18} acids predominating. It is interesting that when these acids are unsaturated, the fats are normally liquid at ordinary temperatures (rather than solid, like beef tallow). The presence of double bonds leads to bent molecules that do not pack together readily. As a result, interactions between adjacent side chains are weak and the solid has a low melting point. These unsaturated fats predominate in the fatty tissue of aquatic mammals that live in polar regions, such as arctic seals and polar bears, permitting this tissue to remain mobile at low temperatures. Unsaturated fats are also the major constituents of plant oils, such as olive, corn, peanut, and safflower oils.

$$CH_2OH$$
$$CHOH$$
$$CH_2OH$$
Glycerol

Natural waxes, such as beeswax and the protective waxy coating on leaves, feathers, and wool, consist in part of esters of long-chain acids with long-chain alcohols. Many fruits owe their odor and flavor to simple esters. For example, bananas contain 3-methylbutyl-3-methylbutanoate and raspberry flavoring consists of a mixture of up to seven different esters, with from three to nine carbon atoms.

Amides The naming of amides is illustrated in Table 29-4. Although these molecules contain an amino group, they have no significant basicity, because of delocalization of the unshared pair formally on the nitrogen atom by resonance with the adjacent carbonyl group:

$$R-C\overset{H}{\underset{\overset{N}{O}}{}}{}^H$$

$$\left\{ R-C\overset{NH_2}{\underset{O}{}} \, , \, R-C\overset{NH_2^+}{\underset{O^-}{}} \right\} \qquad (29\text{-}36)$$

$$R-C\overset{H}{\underset{\overset{N}{O}}{}}{}^{R'}$$

Amides are most readily prepared in the laboratory by reaction of an acid chloride, RCOCl, with ammonia or with an appropriate amine that has at least one hydrogen atom bonded to the nitrogen:

$$R-C\overset{R'}{\underset{\overset{N}{O}}{}}{}^{R''}$$

$$\underset{\substack{\text{Acetyl}\\\text{chloride}}}{CH_3COCl} + NH_3 \longrightarrow \underset{\text{Acetamide}}{CH_3CONH_2} + HCl \qquad (29\text{-}37)$$

$$\underset{\substack{\text{Propanoyl}\\\text{chloride}}}{CH_3CH_2COCl} + \underset{\substack{\text{Dimethyl-}\\\text{amine}}}{HN(CH_3)_2} \longrightarrow \underset{\substack{N,N\text{-Dimethyl-}\\\text{propanamide}}}{CH_3CH_2CON(CH_3)_2} + HCl \qquad (29\text{-}38)$$

The amide linkage, —CONH—, is an essential structural feature of all protein molecules (Sec. 30-2) and also of synthetic polymers like nylon, discussed in Sec. 29-4.

Other Kinds of Organic Substances Although alkyl, or occasionally cycloalkyl, radicals have been used exclusively in the examples in this section so far, aromatic or alkenyl radicals such as phenyl or vinyl (Table 29-3) might also have been used. In most instances the properties of the resulting compounds would be very similar to those described. The greatest difference in properties occurs for compounds in which a hydroxyl group is substituted directly on an aromatic ring. The simplest of these substances is C_6H_5OH, which has the common name *phenol*; the general class of compounds with hydroxyl groups attached to aromatic rings is called *phenols*. Phenols are weak acids, which will lose their proton to a base such as CO_3^{2-}, although not usually to HCO_3^-. Thus phenols will dissolve in a solution of sodium carbonate:

OH
Phenol
C_6H_5OH

$$C_6H_5OH + CO_3^{2-} \longrightarrow C_6H_5O^- + HCO_3^- \qquad (29\text{-}39)$$
$$\text{Phenol} \qquad\qquad\qquad \text{Phenolate}$$
$$\text{ion}$$

Aromatic amines are much weaker bases than ammonia or alkylamines. For example, K_b for aniline, $C_6H_5NH_2$, is smaller by a factor of 10^5 than K_b for ammonia because of partial delocalization of the unshared electron pair from the nitrogen atom to the aromatic ring.

NH$_2$
Aniline
$C_6H_5NH_2$

We have mentioned so far only ring systems containing nothing but carbon atoms. However, there are many important *heterocyclic* compounds, that is, compounds with other atoms in the ring in addition to carbon, most notably nitrogen and oxygen (and occasionally sulfur). Some of these will be encountered in Chap. 30.

Only a few molecules containing more than one functional group have been mentioned; they include CCl_4, TNT, glycerol and its esters, and sugars. One very important class of molecules with more than one functional group is the amino acids, which contain an amino group and a carboxyl group. Just 20 of these naturally occurring molecules are used, in highly patterned ways, to construct most protein molecules in every known organism. These 20 amino acids (Table 30-3) all have the carboxyl group and the amino group joined to the same carbon atom; they can be represented generally as

$$\overset{\displaystyle NH_2}{\underset{\displaystyle}{\mid}} \\ RCHCOOH \qquad (29\text{-}40)$$

where R now may stand for H or for a radical derived from various possible substituted alkanes, aromatic molecules, or heterocyclic molecules. Actually, because an ammonium ion is a much weaker acid than a carboxylic acid (that is, ammonia is a stronger base than a carboxylate ion), these molecules are more correctly represented as

$$\overset{\displaystyle NH_3^+}{\underset{\displaystyle}{\mid}} \\ RCHCOO^- \qquad (29\text{-}41)$$

Thus they are "internally ionized" and exist in a dipolar form. Amino acids other than these 20 occur in various natural sources and have a variety of biochemical roles, although they are generally not incorporated into proteins.

A Classification of Organic Reactions There are hundreds of known reactions and the mechanisms of many have been studied carefully. In many of these, a bond is broken in such a way that one of the fragments formed retains both the bonding electrons. Such reactions may be classified in one of the following four categories: displacement (or substitution), elimination, addition, and rearrangement.

In a displacement reaction, a species with an unshared pair of electrons (a nucleophile) becomes attached to an alkyl group by means of the electron pair and another atom or group is displaced from the alkyl group, taking with it the electrons that formerly bound it to the carbon atom. Examples include reactions (29-16), (29-25) and (29-26), and (29-37) and (29-38). In an elimination reaction, a small fragment is ejected from a molecule and multiple bonding is introduced. Examples include reactions (29-18) and (29-20). An addition reaction involves, in essence, the reverse of an elimination, the degree of multiple bonding being reduced. Examples include reactions (29-1), (29-2), and (29-15). Reactions (29-27) and (29-28) involve a sequence of processes, addition and then displacement. In rearrangement reactions, not considered here, the atoms of a molecule are rearranged into a new bonding pattern without the substitution, elimination, or addition of any atoms.

Some reactions proceed by means of a bond cleavage in which one electron is left with each of the two fragments. This leads to radicals, and hence such reactions are called radical reactions. Examples include the halogenation of alkanes and, under some circumstances, the polymerization of unsaturated molecules, as discussed in Sec. 29-4.

29-6 Conclusion

We have given only a glimpse of the kinds of compounds and reactions with which organic chemists are concerned, and hardly any hint at all about what they do. Refined synthetic schemes, taking into account the preferred conformations of the starting materials and of the desired products, as well as what is known about the mechanisms of organic reactions under different circumstances, permit the creation of hitherto unknown molecules. Some of these might be new polymers of potential value. Others may be synthesized that differ in minor ways from materials of known biological activity, in the hope that the new compounds may be even more desirable than the natural materials—perhaps less toxic, less expensive, more selective, more stable, or more soluble. The product might be a new insecticide, a dye, an antibiotic, a contraceptive, or a flavoring agent. It might be a molecule with no potential application but of quite novel structure, dreamed up first by an imaginative chemist and then synthesized to permit a test of some point of theory.

Proof of the structure of a molecule isolated from some natural source or from a reaction mixture is essential. The separation methods discussed in Chap. 8, most especially the various forms of chromatography, distillation, and crystallization, are used for isolation and purification. Then a combination of chemical tests for various functional groups, molecular-weight determination, and spectroscopic methods, including nmr, infra-red, and ultraviolet, as well as high-resolution mass spectroscopy, will usually provide clues to all or most of the structure unless the molecule is very complex. If suitable crystals (a fraction of a millimeter on an edge) can be obtained, diffraction of X rays by these crystals can be used to give detailed information about the molecular geometry, even of protein molecules with thousands of atoms.

Summary

Organic chemistry is the chemistry of the compounds of carbon. The simplest organic compounds, hydrocarbons, contain only carbon and hydrogen. Hydrocarbons with general formula C_mH_{2m+2}, where m is a positive integer, are called alkanes and are saturated hydrocarbons, i.e., they contain no multiple bonds. Cycloalkanes are saturated hydrocarbons that contain at least one ring of carbon atoms. The alkanes are normally quite inert chemically but will react with oxygen when heated, forming H_2O and CO_2 (or CO).

Isomeric species exist for all alkanes beyond propane, C_3H_8. In the "normal" isomers, each carbon atom is bonded to no more than two other carbons, whereas "branched" isomers contain at least one carbon atom that is bonded to three or four other carbons. The number of possible isomers rises rapidly as the number of carbon atoms in the molecule increases.

Alkenes contain one carbon-carbon double bond and have the general formula C_mH_{2m}, with $m = 2$ or more. The most important reactions of the alkenes are those in which the double bond is converted to a single bond. Carbon-carbon triple bonds are found in the alkynes, compounds of general formula C_mH_{2m-2}. Acetylene, C_2H_2, the first compound in the series, is a typical alkyne. Aromatic hydrocarbons, typified by benzene (C_6H_6) and naphthalene ($C_{10}H_8$), contain π-electron systems that extend over many atoms.

Petroleum is a naturally occurring mixture of hydrocarbons, including alkanes, cycloalkanes and aromatic compounds. In the refining processes it is separated into various fractions, primarily on the basis of volatility.

Most organic compounds contain functional groups that largely determine their behavior. Examples of organic molecules with functional groups are halides (RX, with X being F, Cl, Br, or I), alcohols (ROH), ethers (ROR'), amines (RNH_2), aldehydes (RCHO), ketones (RCOR'), carboxylic acids (RCOOH), esters (RCOOR'), and amides ($RCONH_2$).

At ordinary temperatures there is almost free rotation about single bonds. The absence of rotation about *double* bonds makes possible the phenomenon of geometric isomerism. Another form of isomerism is found in substances that can have right- and left-handed forms. Molecules that are mirror images of one another but are not superimposable (a property called chirality) are identical in most of their physical properties but can be distinguished by certain measurements that are sensitive to chirality, such as rotation of the plane of polarized light. The chemical properties of such "optical isomers" may also differ in reactions with other chiral species.

Polymers are very large molecules, with molecular weights of the order of 10^4 or more. They are made by combining many identical or very similar small molecules known as monomers. The products of polymerizing ethylene and tetrafluoroethylene are polyethylene and Teflon, respectively. Polymerization is typically initiated by radicals formed by the splitting of organic peroxides. Other common man-made polymers are polystyrene, PVC, Orlon, and Plexiglas. Rubber is a natural polymer of isoprene. Copolymers, such as Saran, are the result of using a mixture of two kinds of monomers. In condensation polymerization, the monomers contain two functional groups and are linked to each other by the elimination of small molecules such as water. Dacron and nylon are condensation polymers.

The physical properties of different polymers may be widely different. They reflect the properties of the individual chains, such as their flexibility or stiffness, the ease with which the chains can pack together, and the interactions between chains, which include covalent bonding, hydrogen bonding, ionic interactions between charged groups, and van der Waals forces. The packing of chains may be affected by the arrangement of side chains, which may be regular or irregular. Efficient packing leads to relatively rigid structures with high softening points, as does cross-linking, while inefficient packing results in rubberlike behavior. Cross-linking also decreases the solubility of a polymer in organic solvents. The vulcanization of rubber consists in cross-linking by disulfide bonds. Synthetic polymers with extensive cross-links are often thermosetting and therefore remain hard and rigid after polymerization. Thermoplastic polymers can be shaped after being softened by heating.

The functional-group concept makes it possible to develop an understanding of many organic reactions from a knowledge of the chemical and physical properties of the common classes of organic compounds listed in Table 29-4. Halides can be prepared by the direct replacement of H atoms by halogens, by adding HX to a carbon-carbon double bond, $>C=C<$, and by reaction of HX with an alcohol. Reaction of RBr with Mg in dry ether yields the Grignard reagent RMgBr. In alcoholic solutions of KOH, haloalkanes may lose a molecule of HX, a double bond being formed.

Important alcohols are methanol or wood alcohol (toxic) and ethanol, the common grain alcohol. Dehydration of ROH may introduce a double bond. Reaction with carboxylic or inorganic acids yields esters, examples of which are ethyl acetate (an important solvent), nitroglycerin, and the biologically important phosphate esters. Fats are triesters of glycerol with long-chain carboxylic acids.

Amines are basic and form salts. They can be prepared by the reaction of NH_3 with a haloalkane. The carbon-oxygen double bond in aldehydes and ketones can react with a Grignard reagent, producing a primary, secondary, or tertiary alcohol, depending on the starting materials. Aldehydes and ketones can be reduced to primary and secondary alcohols, respectively, and these two kinds of alcohols can in turn be oxidized to aldehydes or ketones. For aldehydes the oxidation can be carried further to yield carboxylic acids. The most common ketone, acetone, is a widely used solvent.

Carboxylic acids, the simplest of which are formic and acetic acids, can be made by oxidation of primary alcohols or of aldehydes, by adding CO_2 to a Grignard reagent, and by hydrolysis of an ester or an amide. Replacement of the OH group in a carboxylic acid by a chlorine atom yields an acid chloride, RCOCl. Carboxylic acids that contain an amino group are called amino acids. They exist in an internally ionized, dipolar form. Proteins are made of amino acids that have the carboxyl and amino groups attached to the same carbon atom. Characteristic of proteins is the amide linkage, —CONH—, which occurs also in some man-made polymers, such as nylon. Amides can be made by the reaction of an acid chloride with NH_3 and are not significantly basic. Aromatic amines are much weaker bases than alkylamines, and phenols, compounds with a hydroxyl group attached to an aromatic ring, are weakly acidic, in contrast to the nonacidic alkyl alcohols. Heterocyclic rings contain at least one atom other than carbon, chiefly N, O, and S.

Terms and Concepts

Problems and Questions

29-1 Structural Formulas Write formulas for the following compounds: heptane; 2-amino-ethanol; 1,1,2-trichloroethane; 2,3-hexanediol; 4-methyl-2-pentanone; 3-ethyl-2,3,5-trimethylhexane; 3-methylbutanal; isopropyl *t*-butyl ether; ethyl butanoate; 2-pentyne; 3-bromocyclobutanol.

29-2 Names and Formulas of Isomers Write formulas and give systematic names for the eight unique compounds of formula $C_5H_{11}Cl$, not considering optical isomers.

29-3 Isomeric Dibromopropanes Write structural formulas and names for all possible dibromopropanes, $C_3H_6Br_2$. Do not repeat identical structures.

29-4 Names and Formulas of Isomers Write structural formulas and give names for all compounds having the general formula (*a*) $C_4H_{10}O$; (*b*) $C_2H_2Cl_2$; (*c*) C_4H_8.

29-5 Intermolecular Interactions Explain why the boiling point of ethanol (78°C) is much higher than that of its isomer dimethyl ether (−25°C) and why the boiling point of CH_2F_2 (−52°C) is far above that of CF_4 (−128°C).

29-6 Formula of an Amine A compound $C_4H_{11}N$ is known from its reactivity and spectroscopic properties to have no hydrogen atoms attached directly to the nitrogen atom. Write all structural formulas consistent with this information.

29-7 Formula of an Unknown Compound A compound $C_4H_{10}O$ does not react with sodium. When the compound reacts with chlorine in light, three, and only three, monochloro derivatives, C_4H_9OCl, are formed. What structural formulas for the original compound are consistent with this information?

29-8 Formula of an Unknown Compound A compound C_3H_6O has a hydroxyl group but no double bonds. Write a structural formula consistent with this information.

29-9 Formulas of Benzene Derivatives Indicate with structural formulas the number of different isomers of (*a*) chlorobenzene, C_6H_5Cl; (*b*) dichlorobenzene, $C_6H_4Cl_2$; (*c*) bromochlorobenzene, C_6H_4BrCl; (*d*) trichlorobenzene, $C_6H_3Cl_3$; (*e*) bromodichlorobenzene, $C_6H_3BrCl_2$.

29-10 Synthesis of a Ketone Devise a reaction sequence for the synthesis of 2-butanone from ethanal (acetaldehyde) and ethyl bromide. Assume the availability of any inorganic reagents and of organic solvents.

29-11 Esterification Acetyl chloride, CH_3COCl, reacts with the hydroxyl groups of alcohols to form ester groups with the elimination of HCl. When an unknown compound X with formula $C_4H_8O_3$ reacted with acetyl chloride, a new compound Y with formula $C_8H_{12}O_5$ was formed. (*a*) How many hydroxyl groups were there in X? (*b*) Assume that X is an aldehyde and write a possible structure for X

676

and a possible structure for Y consistent with your structure for X.

29-12 Geometric Structure Which of the following molecules are linear, which are planar, and which are neither linear nor planar: CH_2O, ClCN, BrCCH, N_2H_4, C_6H_6, H_2CCCH_2?

29-13 Cis-trans Isomerism For which of the following compounds are isomeric cis and trans forms possible: (*a*) propene; (*b*) 2-pentene; (*c*) 1-butene; (*d*) 2-methyl-2-butene; (*e*) 3-methyl-2-pentene; (*f*) 1,2-dichloropropene; (*g*) 2-butyne?

29-14 Reaction Products Write structures for the products of the following reactions: (*a*) bromine reacts with cyclopentene in the dark; (*b*) HBr reacts with 1-butene; (*c*) isopropyl alcohol is dehydrated; (*d*) 2-methyl-2-butanol reacts with acetic acid; (*e*) trimethylamine reacts with HI; (*f*) 2-pentanone is reduced with $LiAlH_4$; (*g*) 2-methylpropanal is oxidized mildly; (*h*) an amide is formed from diethylamine and the acid chloride of formic acid; (*i*) vinyl chloride is polymerized.

29-15 Syntheses Devise a sequence of reactions for synthesis of each of the following molecules, starting with ethyl alcohol and any necessary in-organic materials and organic solvents: 2-butanol; 2-methylbutanoic acid; 3-methyl-3-bromopentane.

29-16 Optical Isomerism Which of the following molecules are chiral?

(*a*) $CH_3CH_2CHCH_3$
$\qquad\qquad\quad |$
$\qquad\qquad\ \ OH$

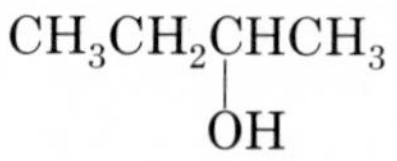

(*b*) $CH_3CH_2CHCH_3$
$\qquad\qquad\quad\ |$
$\qquad\qquad\ \ CH_3$

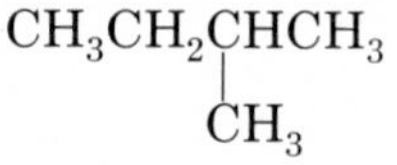

(*c*) 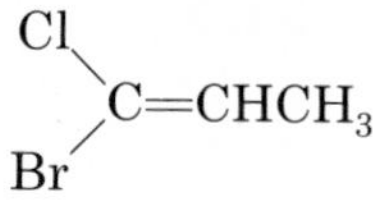 Cl
$\qquad\ \ \diagdown$
$\qquad\qquad C{=}CHCH_3$
$\qquad\ \diagup$
$\quad\ \ Br$

(*d*) H_2C
$\qquad\quad CHCl$
$\qquad\ \diagup|$
$\qquad\ \diagdown CHBr$

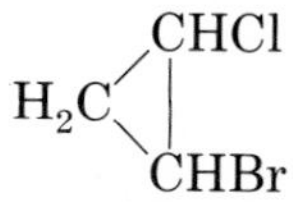

29-17 Polymers (*a*) Define the terms vinyl monomer, isotactic polymer, atactic polymer, co-polymer, condensation polymer, and thermoplastic polymer. (*b*) Which of the following materials is likely to be damaged by spilled acid or base? Teflon, Lucite, polyvinyl chloride, polystyrene, nylon, polyester, Orlon. Explain.

Biochemistry

"It has become clear that the basic metabolic processes of all living cells are very similar. A number of identical compounds, mechanisms, structures, and reaction pathways are found in all living things so far observed, including such diverse cells as those of bacteria, begonias, bees, birds, and biochemists.

"The biopolymers, the synthetic pathways, and the small molecules that are common to all living things have been preserved almost unchanged for more than three billion years. The proto-organism would have had a degree of complexity sufficient to include all these common components shared by present living things. . . . The inferred biochemical structure of this primitive cell provides evidence that it was itself the product of many evolutionary steps, very similar to ones which have occurred in its descendants."

M. O. DAYHOFF AND R. V. ECK,[1] 1972

Biochemistry has flowered remarkably in the last three decades. New tools and new methods have provided detailed information at the molecular and atomic level about the structures of the various components of and the reactions that occur in living organisms. As with any science, the earlier stages were chiefly descriptive. They involved the accumulation of information about the chemical nature and the distribution of the thousands of molecules and ions present in living systems, a task that is still incomplete because of the complexity of some of the molecules and molecular interactions in cells and because of their sensitivity to changes in conditions. However, as even partial information became available, it began to be possible to understand in some detail the chemical changes that accompany and indeed are responsible for many of the processes and activities that, taken together, characterize life—digestion, energy utilization, reproduction, growth, muscle contraction, nerve-impulse transmission, immunity, response to various drugs and other stimuli, and so on. Much is now known about some of these topics, little about others. Only a few of them are considered in this chapter. Our aim is to indicate the ways in which chemistry is able to provide some explanations for and insight into biological phenomena.

30-1 Biochemical Uniformity

There is a quite remarkable uniformity, indeed a near universality, in the general chemical makeup of all living matter on the earth and a striking similarity in many of the sequences of metabolic reactions in the most diverse organisms. Even the specific details of some structures and reactions are nearly universal, al-

[1]M. O. Dayhoff (ed.), "Atlas of Protein Sequence and Structure", vol. 5, National Biomedical Research Foundation, Bethesda, Md., 1972.

Table 30-1
Some Minor Elements in Living Systems*

Element	Predominant form and role
Fluorine	F^-. In teeth and some similar hard bony structures
Sodium	Na^+. Principal extracellular cation
Magnesium	Mg^{2+}. Essential for the activity of many enzymes; in chlorophyll
Phosphorus	Phosphate. Component of nucleic acids and some lipids; essential for all biosynthetic and energy-transfer processes
Sulfur	Usually $S(-II)$ or $S(-I)$, sometimes as sulfate. In many proteins and other molecules
Chlorine	Cl^-. Principal mobile anion
Potassium	K^+. Principal intracellular cation
Calcium	Ca^{2+}. Principal cation in bone and in shells; role in muscle contraction
Manganese	$Mn(II)$. In some enzymes
Iron	$Fe(II)$, $Fe(III)$. Oxygen transport and storage in hemoglobin and myoglobin; in key enzymes involved in oxidation-reduction reactions
Cobalt	$Co(III)$. In vitamin B_{12}
Copper	$Cu(II)$. In certain oxidative enzymes; in oxygen-transport protein in some marine organisms
Zinc	$Zn(II)$. In several enzymes and in insulin
Molybdenum	Mo–Fe–S cluster. In nitrogenase, a nitrogen-fixing enzyme
Iodine	$I(-I)$. Essential for thyroid activity

* Not including C, H, N, and O. The elements are listed in order of increasing atomic number. The cations of the transition elements are often present only in trace amounts (in coordination complexes attached to large molecules).

though many of them vary in at least minor ways from one living system to another. In this section, we examine some of the universal biochemical themes, devoting only minor attention to the variations on them.

Living matter is composed primarily of four elements: carbon, hydrogen, oxygen, and nitrogen. Another 15 or 20 also play essential roles, although together they constitute only 1 or 2 percent of the atoms present. Some of these minor but essential elements are indicated in Table 30-1. A few others are of occasional importance; none of these has atomic number above 30. It is noteworthy that silicon, a congener of carbon, is of very minor significance—it is present only in the skeletons of certain algae called *diatoms* and possibly in some other plants—despite the fact that silicon atoms are 100 times more abundant than carbon atoms on the earth's surface. The lack of silicon compounds in living systems is related to the weakness of the Si-Si bond as compared with the C-C bond.

Common Molecules of Biological Systems There are four general classes of relatively small molecules that constitute the building blocks of biological macromolecules, as well as being important in their own right. One class is the amino acids, described briefly in the previous chapter; they are the "monomers" from which proteins are built. Amino acids and proteins are discussed in Sec. 30-2. The other three groups are sugars and the large molecules made from them, which together are called carbohydrates; nucleotides and their polymers, the nucleic acids; and lipids.

Carbohydrates This class of compounds includes various simple sugars (also called monosaccharides), which are polyhydroxyaldehydes or ketones containing from three to seven carbon atoms, and macromolecules that yield sugars as products of hydrolysis. Starch, cellulose, and glycogen are typical macromolecules of this sort (polysaccharides). The names of most carbohydrates end in *-ose*; the sugars with five and six carbon atoms, to which we give most attention, are called pentoses and hexoses.

Sugar molecules contain unbranched chains of carbon atoms, and although their structures are often represented in a straight-chain or "open-chain" form, most sugars occur predominantly with a more stable cyclic structure. The conversion of the open chain to the cyclic form results from the reaction of the sugar carbonyl group with an alcoholic hydroxyl group in the same molecule to form a product called a *hemiacetal*:

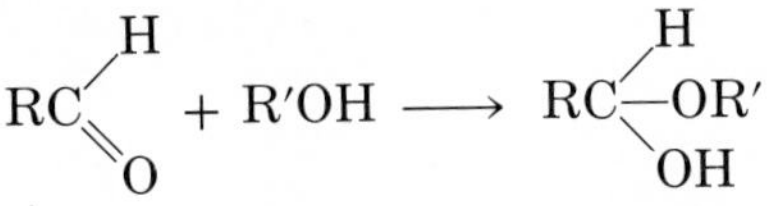

$$
RC\overset{H}{\underset{O}{\diagdown}} + R'OH \longrightarrow RC\overset{H}{\underset{OH}{-}}OR' \tag{30-1}
$$

A hemiacetal

For example, the open-chain form of the most common hexose, glucose, has an aldehyde group at one end of a six-carbon chain, with a hydroxyl group on each of the other carbon atoms. The hydroxyl group on the fifth carbon atom reacts with the aldehyde group, forming a hemiacetal with a six-membered ring that includes one oxygen atom (the circled numbers identify corresponding carbon atoms in the two forms).

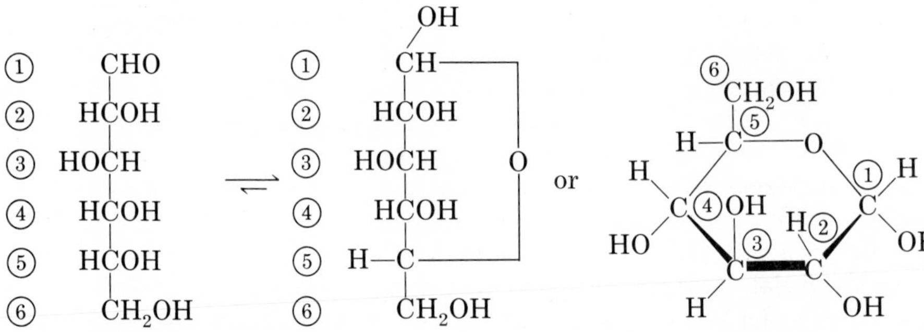

(30-2)

D-Glucose
(open-chain form)

α-D-Glucose
(one cyclic form;
a hemiacetal)

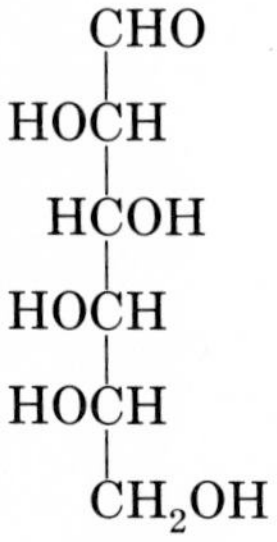

L-Glucose

The six-membered ring is so much more stable than the noncyclic form that under most circumstances the equilibrium between them strongly favors the cyclic form. This is true as well of most other hexoses and pentoses, and we usually represent them in the cyclic form.

There are four asymmetric carbon atoms in the open-chain form of glucose—the four (C-2, 3, 4, and 5) between the —CHO and the —CH$_2$OH groups. D-Glucose has a specific configuration about each of these asymmetric centers;

its mirror image, L-glucose,[2] has the opposite configuration about *all four* asymmetric carbon atoms. A molecule with a different configuration about only some, but not all, of these atoms is a molecule of a different sugar, with different properties. For example, if the positions of the H and the OH are interchanged on the fourth carbon atom, the structure is that of a molecule of the sugar galactose, which also exists in both D and L forms, both cyclic and noncyclic. These hexoses contain four asymmetric carbon atoms, each of which can have two possible configurations. Thus, there are $2^4 = 16$ possible aldehyde hexoses, or eight enantiomeric pairs. The two optical isomers of glucose and the two of galactose are two of these eight pairs.

Cyclization of one of these hexoses, as in (30-2), introduces a fifth asymmetric center at the carbon atom that was formerly in the aldehyde group, C-1 in (30-2). Hence two possible forms of each cyclic molecule exist; they are distinguished by the prefixes α- and β-, rather than by separate names. Only the α form of cyclic D-glucose is illustrated in (30-2); in the β form, the positions of the H and the OH on C-1 would be interchanged.

The cyclic forms of sugars are found in many larger molecules—for example, combined with other sugar molecules in polysaccharides. These compounds are formed by reactions parallel to (30-1); the hemiacetal produced in that reaction might combine, for example, with another molecule of an alcohol (or a phenol) to form an *acetal*:

$$\underset{\text{A hemiacetal}}{RC\overset{H}{\underset{OH}{-}}OR'} + R''OH \longrightarrow \underset{\text{An acetal}}{RC\overset{H}{\underset{OR''}{-}}OR'} + H_2O \qquad (30\text{-}3)$$

An acetal made from the cyclic hemiacetal form of sugar is called a *glycoside*; the specific name for a particular glycoside is derived from the names of the alcohol and the sugar involved. For example,

$$CH_3CH_2OH + \alpha\text{-D-glucose} \longrightarrow \quad + H_2O \qquad (30\text{-}4)$$

Ethyl α-D-glucoside
(an acetal)

Amines with at least one hydrogen atom on the nitrogen atom can react with sugars to form similar products; the nucleotides, discussed below, have this linkage. Many naturally occurring alcohols, phenols, and amines are found as glycosides, in combination with various sugars.

A disaccharide is a glycoside that is formed by combining two monosaccharides, the hemiacetal group of one combining with a hydroxyl group of the second as in

---

[2] The prefix D- (Latin: *dexter,* right) is often used to designate one of the two enantiomeric forms of an optically active molecule. The prefix L- (Latin: *laevus,* left) denotes its mirror image.

CHO
|
HOCH
|
HCOH
|
HCOH
|
HOCH
|
CH₂OH

L-Galactose
(open-chain form)

Equation (30-3). Sucrose and lactose are representative disaccharides:

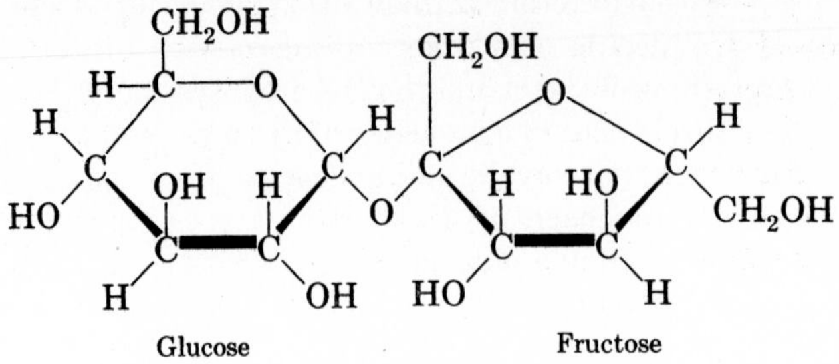

CH₂OH ... Glucose ... CH₂OH ... Fructose

Glucose   Fructose

Sucrose

(30-5)

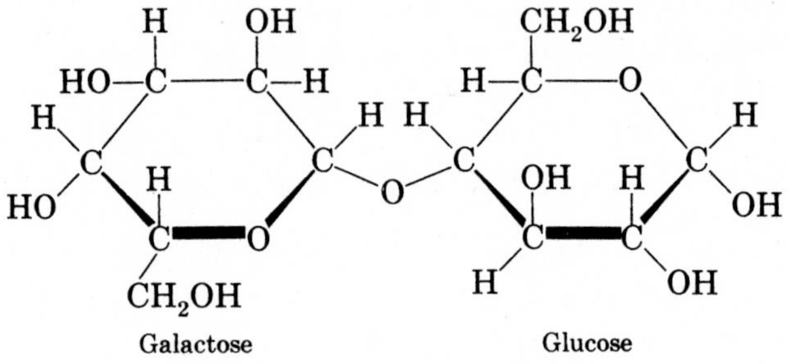

Galactose   Glucose

Lactose

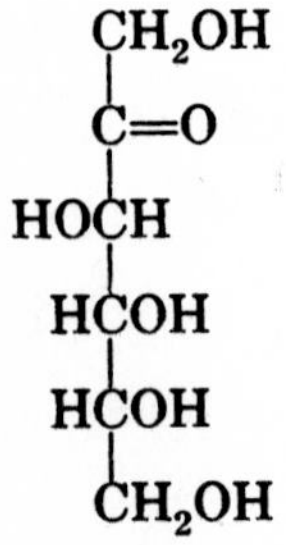

CH₂OH
C=O
HOCH
HCOH
HCOH
CH₂OH

D-Fructose
(open-chain
form)

Sucrose is the common sugar of the home and of commerce; it is widely distributed in all photosynthetic plants, although it is produced commercially only from sugarcane and sugar beets. It contains one glucose unit joined by an α-acetal linkage from its first carbon atom to the second carbon of fructose, a hexose that has a ketone group at C-2 in its open-chain form. Lactose is a compound of galactose and glucose joined by a β-acetal linkage from C-1 of galactose to C-4 of glucose—a so-called β-1,4 linkage. This disaccharide, sometimes called "milk sugar", constitutes as much as 5 percent of the milk of most mammals.

In polysaccharides, hundreds or even many thousands of monosaccharide units, which may be the same or different, are joined together to form linear or branched polymers. Polysaccharides serve both as structural components of living matter and as forms of storage of sugars. Glucose is the most common monosaccharide unit in these macromolecules, but galactose and others are found as well.

Cellulose is the most abundant polysaccharide and indeed is the most abundant organic compound on the earth. It is the major structural element of plants, whose cell walls must provide structural strength since plants lack a bony skeleton. Wood consists of about half cellulose and half a noncarbohydrate heterogeneous polymer called lignin. Such plant fibers as cotton and flax are more than 90 percent cellulose. Cellulose is a linear polymer of glucose units, joined by β-1,4 linkages; the molecular weight is of the order of 10^6.

It might seem that a substance that consists of nothing but linked glucose units should be an excellent foodstuff, since glucose itself is a prime fuel for organisms. However, cellulose is insoluble and must be broken down into small units before it can be utilized as food. The enzymes present in the saliva and intestines of many animals, including human beings, cannot catalyze the hydrolysis of the β-1,4 linkage in cellulose, so cellulose is not a practical foodstuff[3] for these species. Grazing animals and insects like termites do have in their intestines bacteria that have the proper enzymes, so they can digest cellulose readily.

About half the total energy requirement of organisms is supplied by carbohydrates. The energy-storage reservoirs in plants and animals contain starch and glycogen, respectively. These polysaccharides are also composed entirely of glucose units, but differ from cellulose chiefly in that they contain primarily α-1,4 linkages. Since the enzymes present in the digestive tracts of animals can readily hydrolyze these α linkages, the plant polysaccharide starch is a valuable animal food. There are several forms of starch, some unbranched and some with occasional linkages involving other carbon atoms that provide branches and cross-links. Glycogen, the animal counterpart of starch, is even more highly branched and is often bound to proteins. Deposits of it occur in various tissues and organs, especially liver and muscle.

There are many polysaccharides besides cellulose that play structural roles, including chitin, the material that makes up the shells of crustaceans (such as lobsters and shrimp) and the scales of insects. Chitin is rather similar to cellulose, except that it has an acetylamino group, $-NHCOCH_3$, on C-2 of the glucose units instead of an $-OH$ group. Complex polysaccharides containing this N-acetylglucosamine group are also present in bone, skin, and various connective tissues, as well as in the cell walls of bacteria (where they are cross-linked by short chains of amino acids) and in the coatings of many animal cells. Many other polysaccharides are found in both plants and animals.

Nucleotides and Nucleic Acids Mononucleotides are molecules that consist of three portions: (1) a weakly basic nitrogen-containing ring compound, joined by a glycosidic N-C bond to (2) a five-carbon sugar (ribose or deoxyribose), one hydroxyl group of which is esterified by (3) phosphoric acid. Base-sugar-phosphate molecules such as those in Fig. 30-1a and b polymerize in highly patterned sequences to form the various kinds of nucleic acids. Simple mono- and dinucleotides, sometimes with additional phosphate groups attached, participate in all energy-transfer processes in organisms, act as catalysts for metabolic oxidation-reduction reactions (Fig. 30-1c), and play other vital biochemical roles.

Adenosine 5′-phosphate (Fig. 30-1a) is one of the four nucleotides that are combined together in various sequences and proportions to make up molecules of the different ribonucleic acids (RNA). It consists of the base adenine, the sugar ribose, and a phosphate group attached to the fifth carbon atom (numbered 5′) of ribose; the combination of adenine and ribose is called adenosine. Adenosine triphosphate, or ATP, has two additional phosphate groups joined to that already present:

$$\text{Adenine-ribose}\!-\!\text{O}\!-\!\overset{\displaystyle O}{\underset{\displaystyle O^-}{\overset{\displaystyle \|}{\underset{|}{P}}}}\!-\!\text{O}\!-\!\overset{\displaystyle O}{\underset{\displaystyle O^-}{\overset{\displaystyle \|}{\underset{|}{P}}}}\!-\!\text{O}\!-\!\overset{\displaystyle O}{\underset{\displaystyle O^-}{\overset{\displaystyle \|}{\underset{|}{P}}}}\!-\!\text{O}^- \qquad (30\text{-}6)$$

$$
\begin{array}{c}
\text{CHO} \\
|\ \\
\text{HCOH} \\
|\ \\
\text{HCOH} \\
|\ \\
\text{HCOH} \\
|\ \\
\text{CH}_2\text{OH}
\end{array}
$$

D-Ribose

---

[3] A few animals, such as dogs, have so high a concentration of HCl in their stomachs that they can hydrolyze cellulose at a significant rate even though they lack the proper enzymes.

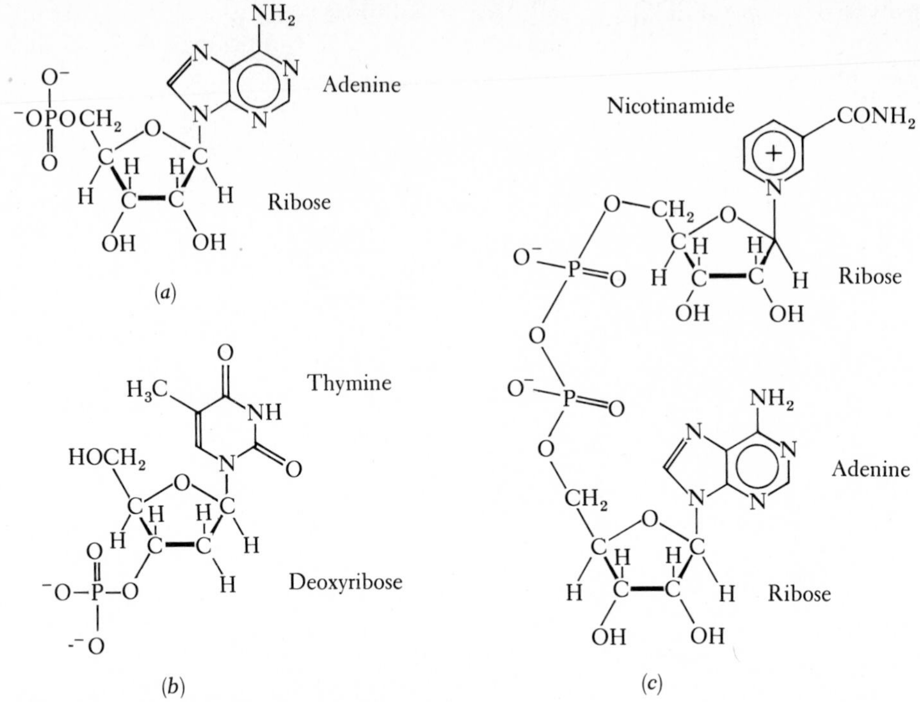

Figure 30-1 Formulas of Three Nucleotides

(a) Adenosine 5'-phosphate; (b) thymidine 3'-phosphate; (c) nicotinamide adenine dinucleotide. Adenosine 5'-phosphate, also known as adenosine monophosphate (AMP), is a ribonucleotide containing the sugar ribose. Thymidine 3'-phosphate is a deoxyribonucleotide, its sugar being 2-deoxyribose, i.e., ribose without the hydroxyl group on the second carbon atom. Nicotinamide adenine dinucleotide (NAD) contains two base-ribose-phosphate combinations, joined through their phosphate groups. This dinucleotide is an important oxidizing agent in many biochemical reactions; its reduced form, NADH, is readily oxidized again. The nicotinamide portion of the molecule is the part that is reversibly reduced and oxidized.

ATP is remarkable in its universality as the common currency for the exchange or transformation of energy in all living cells. No biosynthesis occurs, no nerve impulse travels, no heart beats, no eye opens, no firefly glows without the participation of ATP, by reactions considered further below.

Thymidine 3'-phosphate (Fig. 30-1b) is one of the four nucleotide components of deoxyribonucleic acid (DNA). It consists of the base thymine; the sugar

Table 30-2
Major Bases Present in Nucleic Acids*

Name	Common abbreviation	Present in
Adenine	A	DNA, RNA
Guanine	G	DNA, RNA
Cytosine	C	DNA, RNA
Thymine	T	DNA only
Uracil†	U	RNA only

*The structures of A, G, C, and T are illustrated in Fig. 30-3.
†Uracil has the same structure as thymine (Fig. 30-3a) except that it has a hydrogen atom in place of the methyl group of thymine. Its hydrogen-bonding potentiality is almost identical to that of thymine, so T and U are interchangeable in base pairs, such as that in Fig. 30-3a.

deoxyribose, which differs from ribose only in that the hydroxyl group on the second carbon atom (numbered 2′) is missing; and a phosphate group, joined to the third carbon atom, 3′. The four bases present in most DNA are adenine, thymine, guanine, and cytosine, abbreviated A, T, G, and C, respectively (Table 30-2). Three of these four bases, A, G, and C, are also major components of RNA; the fourth base in RNA is usually uracil, abbreviated U (Table 30-2).

The polymerization of mononucleotides to form the nucleic acids occurs by linking of the sugar molecules through phosphate groups, joining the C-5′ of one sugar to the C-3′ of the next. This linkage scheme is the same in DNA, in which the only pentose present is deoxyribose, and in RNA, in which the only sugar molecules are ribose. It is illustrated schematically in Fig. 30-2a.

As Fig. 30-2b shows, molecules of DNA consist of two individual polynucleotide strands wound around a common axis in a spiral, with an outer diameter of about 20 Å. This is the famous double helix discovered by Watson and Crick in 1953. At the perimeter are the two sugar-phosphate chains; the bases lie inside these chains, with their planes perpendicular to the axis of the helix. Each base on one chain is paired by hydrogen bonding with a base on the other chain (Fig. 30-3).

Although there are four bases in DNA—A, T, G, and C—only two kinds of hydrogen-bonded base pairs are found: A-T and G-C. These are called *complementary base pairs* and are the only pairs bound by strong hydrogen bonds that fit properly into the DNA helix; they are of the right length, and the glycosidic N-C bonds (the bonds between the bases and the sugars) project from them at angles appropriate for the sugar residues in the sugar-phosphate chains. Other possible base-pair combinations (for example, A-G, C-T, A-C, and so on) are too bulky, or not large enough, or otherwise inappropriate for the steric requirements of this remarkably precisely engineered helix.

The fact that only two of the many possible base pairs are actually present has profound implications: it means that every time the base A occurs in one strand, the base T must occur at the same level in the second strand, and every time G occurs in one, C must be present in the second. Thus the sequence in one strand is determined completely by that in the other. The two strands are complementary; if the sequence is . . . A-G-C-A . . . in one, it must be . . .T-C-G-T. . . at the corresponding place in the other.

In *replication* of DNA during cell reproduction, the paired strands first separate. Then free deoxyribonucleotides, present in the cell as their triphosphates, are polymerized by special enzymes present and two new strands are created, each one complementary to and paired with one of the original strands. Thus there are now two DNA molecules for every one present initially, each identical to the original one.

Protein synthesis is effected by *transcription* of DNA. Molecules of DNA may contain from about 10^4 to more than 10^6 base pairs, depending on the nature of the organism. The particular sequence of bases in each strand constitutes the unique genetic information repository, the specific coded instructions that determine protein structure and synthesis and, thereby, all the characteristics of each unique organism. The marvelous mechanism by which this code is interpreted and protein synthesis effected begins with the creation of many special RNA molecules, called *messenger RNA* (mRNA), each complementary to an appropriate section of the DNA. In RNA, the base complementary to A is U rather than T. Thus the base pairs formed as RNA is synthesized on DNA are A-U, T-A, G-C, and C-G, where the bases associated with DNA are given first and

685

those associated with RNA given second in each pair. These messenger RNA molecules, patterned uniquely by the particular DNA present during their formation, then initiate protein synthesis (considered briefly at the end of Sec. 30-2).

Figure 30-2 *(a)* **Schematic Representation of a Portion of a Single Chain of a Nucleic Acid Molecule;** *(b)* **the Double Helix of DNA, Represented in Different Ways**

At the top of *(b)*, the base pairs are represented by the crossbars and the phosphate-sugar chains by the spiraling ribbons. In the center is a somewhat more detailed representation. The P and S represent the phosphate and sugar groups; the bases are denoted by A, G, T, and C. A space-filling model is depicted at the bottom.

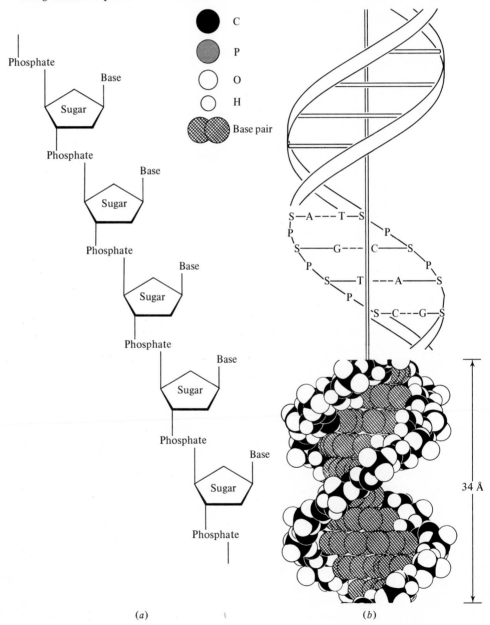

(a)   *(b)*

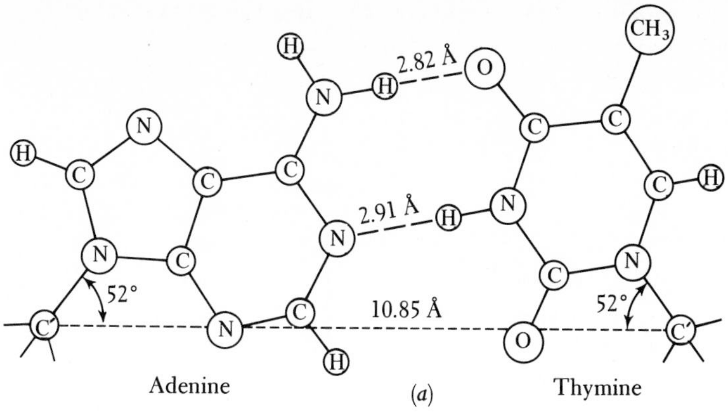

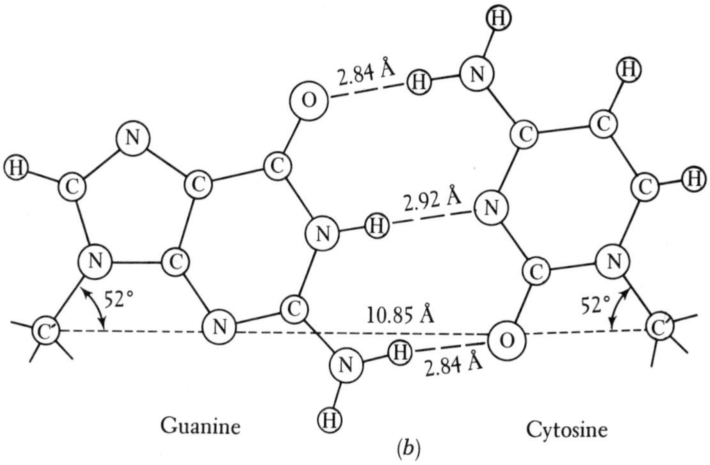

2.82 Å

2.91 Å

52°   10.85 Å   52°

CH$_3$

Adenine   (a)   Thymine

2.84 Å

2.92 Å

52°   10.85 Å   52°

2.84 Å

Guanine   (b)   Cytosine

Figure 30-3   Base Pairs in DNA
(a) Adenine-thymine (A-T); (b) guanine-cytosine
(G-C). In each of these pairs, the hydrogen bonds
are nearly linear and of the appropriate length for
strong interaction. Although other pairings of the
four bases in DNA with reasonable hydrogen bonding and mutual disposition of the glycosidic N-C′
bonds are conceivable, they are less favorable than
the two shown, which are, as far as is known, the
only pairs that normally occur in DNA.

Lipids   The lipids constitute a heterogeneous group of compounds, the only
common feature of which is that they are all soluble in benzene and ether and are
relatively insoluble in water. The most abundant subclass of lipids includes esters
(and some amides) of long-chain carboxylic acids with glycerol and with other
alcohols and amino alcohols. Some of these compounds contain phosphate groups
as well. Long-chain acids are often called *fatty acids* because their triesters with
glycerol (triglycerides) constitute common animal and plant fats and oils (Chap.
29). Among the other materials classed as lipids are long-chain hydrocarbon
waxes, certain kinds of animal and plant pigments, and the steroids, a group of
compounds containing four rings that includes a number of hormones and related
substances.

We restrict this brief discussion to lipids containing fatty acids. Most of these
acids have an even number of carbon atoms; the great majority have unbranched
chains, with C_{16} and C_{18} acids predominating. Some of these acids have one or
more double bonds, as mentioned in the previous chapter; they are found especially in seeds and in cell membranes, chiefly as esters.

Lipids have two principal natural functions, serving for energy storage and as
components of macroscopic structures. Fats (triglycerides) are the main source of
stored energy in animals and in the seeds of some plants. At least 10 percent of
the body weight of most mammals is normally in the form of triglycerides, which
can be utilized rapidly to meet demands for energy. No other form of food can be

stored in comparable quantities; for example, the capacity of an animal to store carbohydrate in the form of glycogen is usually less than a tenth as great. Furthermore, oxidation of the hydrocarbon chains of lipids produces a higher energy yield than is available from other foods.

Lipids also have an important structural role—particularly in cell membranes and at other interfaces, where they are often bound to proteins. Phospholipids are especially abundant in membranes and in nerve and brain tissue. The vital role of lipids in various membranes is attributable to the fact that they contain both nonpolar and polar regions (Fig. 30-4). The long hydrocarbon chains are *hydrophobic* (Greek: *hydor,* water; *phobos,* fear, flight)—i.e., they tend to avoid water, showing a preference for hydrocarbons or ether. In contrast, the polar portions of lipid molecules, the ester and phosphate groups, show an affinity for an aqueous medium; they are *hydrophilic.*

Thus the lipid molecules are easily arrayed in layers or double layers (Fig. 30-5), which may be given additional stability if the molecules are somehow bound together, as by interaction with the nonpolar portions of proteins. A schematic model of a membrane is depicted in Fig. 30-5. Membranes are not passive, static structures. They are flexible and somewhat elastic and perhaps even at times somewhat fluid, and there is usually a continual and highly selective transport of ions and molecules through them as the metabolism of the cell proceeds. Lipid structures play a key role in these processes. Much still remains unknown about the organization, structure, and properties of membranes, which are the subject of a great deal of current research.

Lipids serve in part as thermal insulators—for example, in fat deposits beneath the skin in aquatic mammals—and they also function as buffers against mechanical shock by forming deposits around sensitive organs.

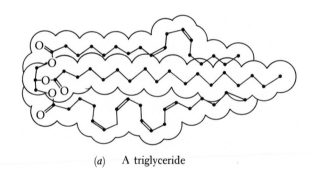

(a)   A triglyceride

Figure 30-4   Drawings of Models of Two Lipids
The black dots represent carbon atoms, with the appropriate number of attached hydrogen atoms. The lines between the dots denote single or double bonds, and the curved contours outline the overall shape of the molecule. In both structures, the saturated fatty acid chains are more regular in shape than the unsaturated ones, which have all double bonds in the cis configuration. Lipid (*a*) is a triglyceride, with three fatty acid chains. Lipid (*b*) is also a triester of glycerol, but one of the hydroxyl groups of glycerol has been esterified by phosphoric acid, which is esterified as well by the amino acid serine, $HOCH_2CH(NH_3^+)COO^-$.

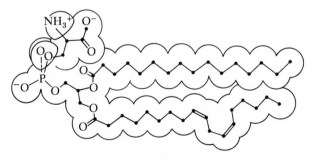

(b)   Phosphatidyl serine

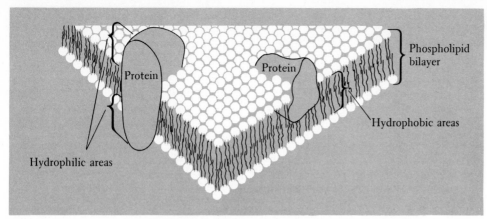

Figure 30-5   A Highly Schematic Representation of a Portion of a Membrane
According to this model, some membranes consist of two layers of phospholipid molecules, with
their polar layers outward and their hydrophobic tails in contact. The different molecules in this
double-layer sandwich are held together in part by proteins, presumably through interaction at
their hydrophilic ends as well as by nonpolar interactions.

30-2   Amino Acids and Proteins

As mentioned in the previous chapter, the naturally occurring amino acids from
which proteins are formed can be written generally as

$$R-\underset{\underset{\text{H}}{|}}{\overset{\overset{\text{NH}_3{}^+}{|}}{\text{C}}}-\text{COO}^- \qquad (30\text{-}7)$$

The amide linkages joining amino acids together in proteins, illustrated in
formula (29-14), are called *peptide bonds* because they are broken during digestion (Greek: *peptein,* to digest). The general name *polypeptides* is given to
molecules containing many amino acids so linked, while terms like dipeptide and
tripeptide are used for molecules with just two or three amino acids. The word
protein is reserved for polypeptides containing more than about 50 amino acids.

It is a remarkable fact that every molecule of a pure, biologically active protein
or other polypeptide from a given source is identical—the same amino acids,
joined together in precisely the same sequence to form a chain of the same length,
or sometimes several distinct chains, assembled together to make one protein
molecule. When some variation occurs—e.g., the substitution of one amino acid
for another—it occurs because there has been a change in the code that dictates
the specific amino acids to be used, in sequence, to build that polypeptide or
because there has been an error in reading the code.

In this section we first describe the formulas of representative amino acids and
a few simple peptides and examine some typical three-dimensional structural
features of proteins. We then consider some of the many different roles played by
proteins and discuss their synthesis briefly. Their catalytic function as enzymes is
the topic of Sec. 30-3.

Table 30-3
Representative Amino
Acids of Proteins*

General formula: $\overset{\overset{\displaystyle NH_3{}^+}{|}}{\underset{\underset{\displaystyle H}{|}}{R\,C\,COO^-}}$

R—	Name of amino acid	Common abbreviation	Other amino acids with similar chemical nature
H—	Glycine	Gly	
CH_3—	Alanine	Ala	
$(CH_3)_2CHCH_2$—	Leucine	Leu	Valine (Val)
			Isoleucine (Ile, Ilu)
†	Proline	Pro	
$C_6H_5CH_2$—	Phenylalanine	Phe	Tryptophan (Try)
$CH_3SCH_2CH_2$—	Methionine	Met	
$HSCH_2$—	Cysteine	Cys	
$HOCH_2$—	Serine	Ser	Threonine (Thr)
HO—⬡—CH_2—	Tyrosine	Tyr	
$H_2NCOCH_2CH_2$—	Glutamine	Gln	Asparagine (Asn)
$HOOCCH_2CH_2$—	Glutamic acid	Glu	Aspartic acid (Asp)
$H_2NCH_2CH_2CH_2CH_2$—	Lysine	Lys	Arginine (Arg)
$HC{=}CCH_2$— (imidazole ring)	Histidine	His	

*Arranged with the less polar side chains near the top of the list and the more polar ones near the bottom.

† Proline contains a ring that incorporates the amino-nitrogen atom and so cannot be represented by the general

formula above. It is (proline structure)

$CH_2{-}CHCOO^-$
$HOCH \quad NH_2{}^+$
$\quad CH_2$

4-Hydroxyproline

Amino Acids and Simple Peptides The 20 amino acids that are commonly found in proteins and for which genetic code words are known to exist vary in the nature of the side chain R in (30-7). Table 30-3 gives the common names and accepted abbreviations for 13 representative amino acids and the formulas of their side chains. It also lists the names of the other protein amino acids that are chemically similar to those whose formulas are given; their structures are not of concern here.

As Table 30-3 shows, some side chains are nonpolar, containing only carbon and hydrogen. Others contain —OH, —SH, amide, or acidic or basic groups, all of which have hydrogen-bonding potentialities. Furthermore, at physiological pH, somewhat above 7, glutamic and aspartic acids are negatively charged because of ionization of the second carboxyl group; lysine and arginine bear a net positive charge because functional groups in their side chains are sufficiently strong bases to pick up a proton even at higher pH values. When these acidic and basic amino acids are incorporated into proteins, the side chains retain their charges under normal physiological conditions and hence attract oppositely charged ions or segments of proteins.

The 20 amino acids in Table 30-3 are found in a wide variety of proteins. A few other amino acids occur in particular proteins; for example, 4-hydroxyproline constitutes about 12 percent of collagen, an important protein in connective tissue in animals. All the amino acids found in proteins except glycine are optically active and all have the same configuration at the asymmetric carbon atom bearing the amino and carboxyl groups, arbitrarily designated the L configuration. No D-amino acids, those with opposite chirality, are found in any protein, although some do occur in a few smaller peptides. For example, D-phenylalanine is a component of gramicidin-S, a cyclic decapeptide that is an effective antibiotic. Other naturally occurring peptides include the tripeptide glutathione (a widely distributed biological reducing agent that has the sequence Glu-Cys-Gly) and a number of hormones—ranging from the relatively simple nonapeptide oxytocin, which stimulates lactation and uterine contraction, to the adrenocorticotropic hormone (ACTH), containing 39 amino acid residues, which stimulates the surface layers or cortex of the adrenal gland.

The metabolic systems of plants are able to manufacture all the amino acids needed to build the essential plant proteins. However, human beings can synthesize in adequate quantity only 12 of the amino acids they require and must therefore get the other 8 (Val, Leu, Ile, Phe, Try, Met, Thr, Lys) in the diet, either in proteins or in other sources from which the amino acids can readily be obtained. If these eight indispensable amino acids are not present in adequate quantity in the diet, the body is unable to manufacture essential enzymes and other proteins containing these amino acids. This malnutrition leads first to general debility and eventually to death. Other animals must also have certain amino acids in their diets.

Proteins Proteins are such complex molecules, with molecular weights usually in the tens or hundreds of thousands, that until the 1950s no one was sure it would ever be possible to learn the details of the chemical structure and three-dimensional architecture of any of them. There was serious debate about whether all molecules of a particular protein from a given source, e.g., hemoglobin isolated from the blood of a particular individual, were identical in chemical composition and structure or whether there might be some variations. It is now known that they are indeed all the same if they are programmed by identical genes during their biosynthesis.

With the improvement in methods of purification of proteins and the development of chromatographic methods for the separation and quantitative estimation of amino acids and simple peptides in the 1940s and 1950s, precise analyses became possible and the exact amino acid composition could be determined even for a protein of high molecular weight. Techniques were then worked out for discovering the exact sequence of the amino acids in each protein chain. This sequencing is now done chiefly by breaking specific bonds by enzyme action and identifying the fragments present. The amino acids at the ends of the original chains, or the ends of the chains in each fragment, can be identified because their free amino or free carboxyl groups react with special reagents. The process of deducing from these data the original sequences of amino acids in the intact protein is in some ways like putting together a giant jigsaw puzzle—with a unique solution possible if and when all the pieces have been properly found and identified. In the last two decades, the precise sequence of amino acids has been determined for each of around a thousand pure proteins. This sequence, including the location of any disulfide bridges that may link cysteine residues within one chain or between nearby chains, is called the *primary* structure of the protein.

Proteins play two quite distinct roles: some serve only as structural materials; others are biologically active as catalysts or regulators of metabolic reactions, as antibodies to protect the organism from the effects of foreign materials, or in

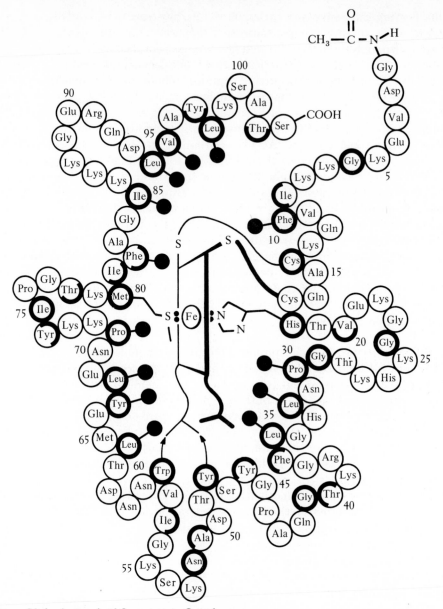

Figure 30-6 Amino Acid Sequence in Cytochrome c

This schematic diagram shows the sequence of the 103 amino acid residues in tuna reduced cytochrome c, an oxidation-reduction enzyme. A circle ringed in black represents an amino acid in
the interior of the molecule; those without such rings indicate amino acids on the outside. Note
that the nonpolar amino acids (Table 30-3) tend to be in the interior whereas the polar ones
congregate on the exterior. The square group in the middle of the figure represents a heme group
(Fig. 26-3), a group of four heterocyclic rings joined together in such a way that their four nitrogen atoms can coordinate to a central iron atom. Octahedral coordination of the iron atom is
completed in this molecule, as indicated, by one of the nitrogen atoms of a His residue and a
sulfur atom in a Met residue. The heme group is bound to the protein by covalent bonds and
hydrogen bonds. The iron atom in this complex is oxidized and reduced when cytochrome c participates in the electron-transport chain by which ATP is synthesized. (*Reproduced by courtesy
of R. G. Dickerson.*)

other ways. Some structural proteins have protective functions (such as the proteins in skin), some are parts of connective tissue, and still others are part of the motive machinery that constitutes muscle. These structural proteins are almost always fibrous in nature, composed of long-chain molecules joined together by hydrogen bonding or other cross-linking into extended filaments, which may in turn be coiled together to make insoluble fibers and other macroscopic structures.

Biologically active proteins are usually ellipsoidal in shape. Their polypeptide chains are folded and twisted to form a globular mass rather than an extended one. The biological activity of most globular proteins is carried on in aqueous biological fluids. The folding of the polypeptide chains is such that the amino acids with polar side chains tend to be on the outside of the molecule, where these polar side chains make the molecule water-soluble, while the nonpolar amino acids, especially those with large side chains, are usually in the interior of the molecule (Fig. 30-6).

Certain specific hydrogen-bonded arrangements of the polypeptide chains are characteristic of fibrous proteins—for example, the α helix and several sheetlike structures (Fig. 30-7). The α helix is the predominant structural feature of the protein α-keratin, which is present in wool, feathers, claws, and similar materials. This stable helical arrangement is also found to varying extents in many globular proteins. About 80 percent of each polypeptide chain in the oxygen-transport blood protein hemoglobin and the related muscle protein myoglobin (Fig. 30-8) is in α-helical form, in eight separate segments of varying length; on the other hand, less than 10 percent of the electron-transport agent cytochrome *c* is helical. Silk fibroin has the sheet structure illustrated in Fig. 30-7*b*, which is also found to a limited extent in a wide variety of globular proteins.

These helix and sheet structures, and other similar specific hydrogen-bonded arrangements involving the polypeptide backbone, are referred to as *secondary*

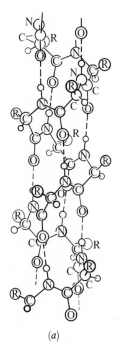

(*a*)

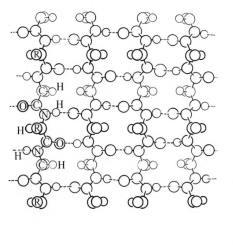

(*b*)

Figure 30-7 (*a*) **The Right-Handed α Helix, Found in Many Proteins;** (*b*) **a Portion of a Sheet Structure, Found in Silk and Other Proteins**

The side chains, denoted by R, extend out from the helix and above and below the sheet. Hydrogen bonds between NH and CO groups in successive turns of the helix help to stabilize this structure. Similar N—H----O hydrogen bonds between different chains hold the sheet structure together.

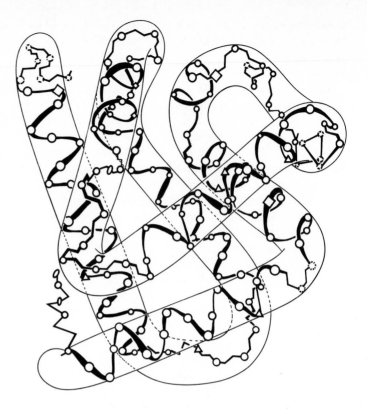

Figure 30-8   The Tertiary Structure of Myoglobin and Many Other Globins

The different circles, diamonds, and triangles represent the —CHR— groups of different amino acids in the myoglobin chain. The links joining them represent the peptide bonds, —NHCO—. The extension of side chains outside the helix is not indicated. Note the distinct helical regions, joined by randomly coiled segments of the polypeptide chains. Myoglobin, an oxygen-storage protein in muscle, is closely related to the blood protein hemoglobin and other globins in a great variety of organisms. Although their amino acid sequences are in some cases quite distinct, certain key positions in the sequences are invariant in all these molecules, and the other substitutions do not affect the tertiary structure in more than minor ways. Doubtless all these molecules have a common ancestor. The preservation of this intricate molecular shape indicates that the structural features found in these molecules are those essential to their biological functioning. [*From M. O. Dayhoff (ed.), "Atlas of Protein Sequence and Structure", p. 18, National Biomedical Research Foundation, Bethesda, Md., 1967–1968.*]

structures in proteins. The intricate folding and convoluting of the polypeptide chain of each globular protein that is brought about by interactions involving the side chains and that gives the protein its characteristic overall shape (for example, that shown in Fig. 30-8) are called the *tertiary* structure. The secondary and tertiary structural features of different proteins seem to be determined entirely by the primary structure. In other words, the formation of helical regions and hydrogen-bonded sheets and the folding and twisting of the chains in globular proteins are dictated by the sequence of amino acids, which is in turn determined by the sequence of nucleotide base pairs in the DNA of the organism. When a genetic mutation occurs, it produces a change in one of the base pairs and hence in general a substitution of one amino acid by another. If this change merely replaces one amino acid by another of similar chemical structure, the secondary and tertiary structure of the protein may be unaltered and its biological functioning remain unchanged. If, on the other hand, the mutation replaces a nonpolar amino acid by a polar one, the secondary and tertiary structure of the protein may be drastically altered and hence its biological activity may be changed in a major way, since this activity depends on the three-dimensional architecture of the molecule.

The effect of a change in a single amino acid in a protein chain on the properties of a globular protein is dramatically illustrated by the contrast in properties between hemoglobin molecules (hemoglobin A) in the blood of normal adult humans and those (hemoglobin S) in the blood of people who suffer from sickle-cell anemia. Each hemoglobin molecule has four polypeptide chains. The only difference between hemoglobins A and S is that in two of the four chains a glutamic acid residue in A has been replaced by a valine in S. There are 574 amino acid residues in the hemoglobin molecule. This alteration of just two of them, from acidic to nonpolar, changes the properties of the molecule sufficiently in the deoxygenated state that the solubility of S is decreased by a factor of 25 relative to that of A. The abnormal molecule polymerizes into microfilaments that combine to form a rodlike insoluble mass that distorts the red blood cell into a sickle shape. The sickled cells cannot easily pass through the capillaries, so vital organs are deprived of their normal blood supply and gradually deteriorate. Individuals who have only hemoglobin S seldom live beyond the age of 40, and about half of them die in infancy. On the other hand, those who have half normal hemoglobin A and half hemoglobin S (because they received from their parents one A-forming gene and one S-forming gene, rather than two A genes or two S genes) usually do not have the serious problems of those with sickle-cell anemia, although sickling of their red cells may occur at very low oxygen pressure (for example, at high altitudes).

The survival of any protein molecules as potentially lethal as hemoglobin S in a population suggests that in some environment they must confer an evolutionary advantage. This turns out indeed to be true. When the parasite that causes malaria is developing in the red blood cells, the oxygen pressure is reduced sufficiently to cause sickling of many cells containing hemoglobin S. Such cells are then attacked by other cells called phagocytes, whose function is to engulf and destroy alien particles. The malarial parasite is destroyed at the same time. Individuals who have only normal hemoglobin A do not have this protective mechanism against malaria, so that the population in malarial regions tends to be enriched in individuals with one A gene and one S gene. Mating of two of these individuals can produce in the next generation individuals with gene combinations SS and AA as well as AS. This explains why the incidence of sickle-cell anemia in the United States is highest in the black population descended from individuals who came from malarial regions of Africa.

As mentioned in Sec. 30-1, protein synthesis is initiated by messenger RNA molecules that are complementary to the DNA in the appropriate genes. An ingenious series of experiments has shown that each *codon,* the unit that specifies which amino acid is to be added to the growing chain at a particular point, is a sequence of three bases in the messenger RNA molecule. For example, glutamic acid is coded either by the triplet G-A-A or by the triplet G-A-G; if the central base in either of these triplets is changed to U, giving G-U-A or G-U-G, the amino acid specified by the codon is valine. Thus the genetic change responsible for the production of hemoglobin S in place of hemoglobin A must have been a change at one point in the DNA such as to change the central base in this triplet at one point in the complementary RNA from A to U, with the result that a particular glutamic acid in each of two chains of each hemoglobin molecule is replaced by valine.

Since there are four different bases in RNA, there are $4^3 = 64$ different possible triplets of bases to be used to code for the 20 different amino acids that occur in most proteins. Most of these amino acids can be specified by several different codons, as in the examples just cited. The complete code has been deciphered. Sixty-one of the sixty-four possible triplets are used for encoding amino acids; the other three are used for terminating or interrupting protein synthesis.

Much progress has been made in understanding the complex process by which information in DNA is translated into a synthesized protein, but knowledge of some details is still incomplete. The process involves the orderly interactions of more than 100 different macromolecules (including various kinds of RNA and proteins) and many small molecules and ions (including nucleoside triphosphates and Mg^{2+}).

30-3   Enzymes

Without the catalytic effects of enzymes, life would be impossible because almost all the reactions that occur in organisms would proceed at negligible rates under the mild conditions of temperature, pressure, and pH characteristic of normal cells. Indeed, most common poisons act by interfering with just one, or a few, of the thousands of enzymes in the body, diminishing the catalytic activity sufficiently that the corresponding reaction is effectively stopped. Some aspects of the kinetics of enzyme-catalyzed reactions were discussed in Sec. 22-7, with emphasis on the essential role of the complex formed by the reacting (substrate) molecule with the enzyme at the so-called active site. In this section we consider briefly the nature of enzymes, their often remarkable specificities, and ways in which their activity can be inhibited.

The Nature of Enzymes   More than 1000 enzymes have been characterized and more than 100 of them have been studied in considerable detail. All are proteins; many require for effective catalytic activity the presence of one or more additional components, called *cofactors*. Some cofactors are firmly bound to the protein—for example, a heme group in some enzymes. Other cofactors are small organic molecules, easily separable from the protein and referred to as *coenzymes*. Many coenzymes are closely related to vitamins; if they cannot be synthesized by the body, they must be in the food supply, although the amounts need only be small since they act catalytically in concert with the enzyme. Still other cofactors are metallic ions, such as K^+, Ca^{2+}, or Mn^{2+}, which may be bound to the protein through carboxyl, hydroxyl, amino, or phosphate groups.

Enzymes may be divided into three general categories on the basis of their size and complexity. The simplest, which constitute a rather small group, consist of only a single polypeptide chain, with molecular weight typically in the range 10^4 to 4×10^4. All have hydrolytic functions; some break down carbohydrates, others proteins, and still others nucleic acids. A second, larger group consists of *multisubunit enzymes*—that is, enzymes containing several independent polypeptide chains, some of which may be identical. The enzyme activity is associated only with the assembled entity of subunits; the separated chains are catalytically inactive. Finally, there are in many cells *multienzyme complexes*—highly structured groups of enzymes arranged so as to facilitate easy transfer of substrates from one enzyme to another during a sequence of reactions within a cell.

The catalytic activity of enzymes is quite remarkable in a number of ways. As indicated, they act under mild conditions, often thousands of times more effectively than the best nonenzyme catalysts for the same reactions. Most enzymes will cause the reaction of around 100 to 1000 substrate molecules every second for each molecule of enzyme present, but some can process (one at a time) as many as 10^6 or even 10^7 molecules per second. Among the fastest are catalase, which catalyzes the decomposition of hydrogen peroxide to water and oxygen and prevents the accumulation of H_2O_2 in cells, and carbonic anhydrase, which promotes the equilibration of H_2CO_3, CO_2, and water, a process important in respiration.

Like all catalysts, enzymes are needed only in small amounts because they are continually regenerated as the reaction they catalyze proceeds. An enzyme or any other protein that loses its biological activity without breaking of the polypeptide chain is said to be *denatured*. Denaturation is likely to occur upon change of pH or temperature because such changes easily disrupt the tertiary (and sometimes the secondary) structure of the protein (e.g., by breaking intra-molecular hydrogen bonds) and the activity is sensitively dependent on the details of the tertiary structure. Most mammalian enzymes are optimally active at physiological pH and at temperatures near $37°C$. Enzymes in organisms adapted to environmental extremes sometimes have optimal temperatures for their catalytic activities admirably suited to their environment. For example, algae in hot springs contain enzymes that function well at temperatures that would destroy most mammalian enzymes whereas certain enzymes in bacteria that grow in the arctic are most active near $0°C$.

Enzyme Specificity Unlike many catalysts, enzymes are often highly specific. Some have what is called absolute specificity: they will catalyze the reaction of only one particular molecule, and not others, no matter how closely those others resemble the usual substrate in structure. Urease, for example, will decompose only urea, and glucokinase catalyzes only the transfer of a phosphate group from ATP to D-glucose to form glucose 6-phosphate. Other enzymes will catalyze the reactions of a group of compounds more or less closely related in structure. Hexokinase promotes the transfer of phosphate from ATP to glucose and to certain other hexoses as well, but not to any other sugars less closely similar to glucose. One of the digestive enzymes will break down almost any protein, although it will not degrade other amides; other digestive enzymes are more selective, breaking peptide bonds only if they occur adjacent to particular amino acid residues.

The specificity of an enzyme is a consequence of its highly intricate tertiary structure. The folding of the polypeptide chain brings together amino acid residues from quite different regions of the chain, thus providing at once an environment that can activate the substrate for reaction and a uniquely shaped template into which only a molecule of the proper shape and proper polar character can fit snugly. Because enzymes are themselves chiral objects, composed of L-amino acids, their specificity extends to what has been termed *chiral recognition*. They distinguish clearly between optical isomers, favoring those found in the natural environment with which they have been evolved to deal. Glucokinase will not catalyze the transfer of phosphate to L-glucose, and no digestive enzyme will attack a polypeptide made of D-amino acids.

Enzyme Inhibition Substances that interfere with the catalytic effects of enzymes are termed inhibitors. Some are molecules that do not react like the normal substrate but resemble it so closely in shape that they can attach themselves to the enzyme molecules at the active sites and thus compete with the normal substrate for these sites. As the concentration of such an inhibitor is increased, it competes more and more effectively for the active sites on the enzyme and thus reduces more and more the rate of reaction of the substrate. With some inhibitors, increase of substrate concentration will counteract this trend by Le Châtelier's principle. Sulfanilamide, one of the first of the sulfa drugs that were so important in treating bacterial infections in the 1930s and 1940s, competes with a molecule of similar shape and hydrogen-bonding properties—4-aminobenzoic acid—that bacteria normally need for biosynthesis of essential amino acids and nucleic acid bases.

Many inhibitors act by altering enzymes in such a way that their biological activity is destroyed. These include such toxic materials as CN^-, which combines with transition metals in various enzymes, and heavy-metal ions such as Ag^+ or Pb^{2+}, which inactivate the —SH groups of cysteine residues in proteins. Nerve gases, which are alkylfluorophosphates, $(RO)_2POF$, react with the hydroxyl groups of serine residues in the enzyme acetylcholine esterase (among others), thereby preventing the regeneration of the resting state of nerve fibers after a nerve has transmitted an impulse. The consequence is that all muscular control is lost, including that over such vital functions as breathing and heart action.

Competitive inhibition of enzyme activity plays a key role in the regulation of metabolism in all organisms. One of the most important regulatory mechanisms is feedback inhibition: a product of a long sequence of reactions is an inhibitor for an enzyme that catalyzes one of the initial reactions in the sequence. Thus, as this product accumulates in the organism, its rate of synthesis is diminished; as it is depleted, its synthesis is again enhanced. The availability of the product in the cells can thereby be maintained at a level appropriate for whatever biochemical function the substance is to serve.

The Nature of the Active Site and the Mechanism of Enzyme Action Within the last 15 years, the three-dimensional molecular structures of more than 100 enzymes and complexes of enzymes with molecules that resemble their usual substrates have been worked out by X-ray diffraction studies of single crystals, and many more such studies are in progress. Because the experimental data are collected over periods of weeks or months, the resulting structural models are of a time-averaged, essentially static, situation. They show the details of the structure of the enzyme and, in some cases, of its complex with a molecule much like its usual substrate but with which the enzyme does not react during the experiment. The mechanism of interaction with the proper substrate must be inferred, but the evidence on which to make this inference, at least for the initial stage of the reaction, is concrete and highly detailed. We are thus in the paradoxical situation of knowing more about the structures of the initial reaction intermediates for some of these hitherto mysterious reactions involving molecules of enormous complexity than about the activated complexes or intermediates for most of the simplest reactions not involving enzymes.

For example, it has been known for some time from studies of chemical reactivity that the closely related digestive enzymes chymotrypsin and trypsin are inactivated if any alteration is made in two particular amino acids that occur in the polypeptide chain of each, a histidine residue that is the 57th amino acid from one end and a serine that is 195th from that end. When the structures of these enzymes were determined in detail, it was found not only that the two molecules are remarkably similar in overall shape and tertiary structure, but that the molecular conformation of each is such that these two critical amino acid residues, although widely separated along the chain, are brought close together by the folding of the chain. A third residue, an aspartic acid at position 102, is also in this same region of space. These three residues are intimately involved in the mechanism by which these enzymes, and some others closely similar to them, hydrolyze proteins. They thus constitute a portion of the active site of each enzyme.

The region around the active site of an enzyme has two functions: recognizing the appropriate substrate or substrates specific for that enzyme and accelerating the appropriate reaction of that substrate. The specificities of trypsin and chymotrypsin are somewhat different, and the difference can be explained in terms of the shape and lining of a cleft surrounding the active site in each molecule. In chymotrypsin, which is specific for splitting peptide bonds adjacent to large nonpolar side chains, this cleft is fairly large and is lined with hydrophobic (nonpolar) amino acids. In trypsin, on the other hand, the cleft has a negatively charged aspartic acid residue in its interior. This is consistent with the fact that trypsin will hydrolyze peptide chains only at positions adjacent to positively charged side groups. Other digestive enzymes closely related to these two are also known, and their specificities can be correlated with their molecular shapes and structural details as well.

While no quantitative explanation has yet been made of the rate enhancement characteristic of an enzyme-catalyzed reaction, qualitative interpretations are now available for a few that have been studied in sufficient detail. They are provided by schemes developed from a careful analysis of the structural details of each molecule and the relative positions of the enzyme and its substrate in the activated complex or intermediate, as well as from other chemical information about the specificity of the enzyme and its inhibitors.

30-4 Bioenergetics and the Role of ATP

In this section we consider some of the principles, compounds, and reactions involved in the flow of energy through organisms. Living cells are open systems. They exchange matter and energy with their environment and are not at equilibrium, although to a good approximation they are in a steady-state condition. A cell is a very efficient engine, operating at nearly constant temperature, pressure, and volume, and is thus unlike most engines contrived by man. Nevertheless, the thermodynamic principles that were developed originally from consideration of heat engines apply to living systems.

Light from the sun is the ultimate energy source upon which all life depends. Photosynthetic plant cells and some bacteria use this energy directly for the synthesis of essential organic molecules by reduction of carbon dioxide, with water or another inorganic compound as the reducing agent. All higher animals, most microorganisms, and even the nonphotosynthetic cells of plants utilize oxidation-reduction reactions to supply the energy they need, chiefly through the oxidation of reduced carbon compounds such as glucose by atmospheric oxygen, producing ultimately CO_2. They are dependent on the photosynthetic cells for the supplies of glucose and other reduced carbon compounds and for returning oxygen to the atmosphere. In turn, the photosynthetic cells are dependent on the others for resupplying carbon dioxide to the atmosphere. This symbiotic (Greek: *symbiosis,* state of living together) interdependence channels solar energy indirectly into the energetic demands of animals—for biosynthesis, for transport of chemical substances throughout the organism, for muscle action, and for nerve-impulse transmission—through the gradual release of the very considerable energy available by combustion of glucose and similar reduced substances.

The fact that this chemical energy is made available gradually is crucial, just as it is for any other useful energy source, such as a battery. The energy from a battery is most useful if it is released under controlled conditions when and where it is needed, rather than by short-circuiting the battery or direct combination of the oxidizing agent and the reducing agent. Similarly, in a normal organism energy is made available in comparatively small amounts at times and places best suited to the needs of the organism. Many of the essential processes in an organism, e.g., the biosyntheses of various molecules, occur with an increase of free energy and thus are not thermodynamically possible unless they are coupled with some process that simultaneously provides sufficient free energy that there is an overall decrease in free energy. Many different molecules, most of them phosphate esters, are used to supply free energy sufficient to drive reactions in living cells essentially to completion. The most widespread of these "high-energy molecules" is ATP, whose formula was given in (30-6), page 683.

Adenosine Triphosphate (ATP)   This molecule appears to be present in every cell of every organism. It is continually being transformed to ADP (adenosine diphosphate) by loss of its terminal phosphate group, either as some inorganic phosphate-ion species (symbolized here as P_i) appropriate to the pH,

$$\text{ATP} + \text{H}_2\text{O} \longrightarrow \text{ADP} + \text{P}_i \tag{30-8}$$

or through transfer of the phosphate to some other molecule.[4] At physiological pH, near 7, both ATP and ADP exist as anions with charges -4 and -3, respectively; the predominant species of each molecule in cells is its $1:1$ complex with Mg^{2+}, an ion present in appreciable concentration in intracellular fluid. It has been estimated that each person synthesizes and breaks down about his or her own body weight of ATP every day; its half-life in typical cells is of the order of 1 s.

Two properties of ATP are critical to its central role in the exchange of energy in organisms. (1) It has a relatively high standard[5] free energy of hydrolysis, $\Delta G^{\ominus\prime}$, about -31 kJ mol^{-1}. (2) Despite its thermodynamic instability, it is kinetically stable—that is, it is not broken down to ADP until an appropriate catalyst is present, so that it can be stored almost indefinitely for use when needed.

Since the breakdown of ATP in (30-8) occurs with a significant decrease in free energy, the reverse process (the synthesis of ATP from ADP and phosphate) requires free energy. Some ATP is made directly during photosynthesis in green leaves, with sunlight supplying the energy; we shall not discuss here the reactions involved. Most of the ATP in both plant and animal cells is made by coupling the reverse of reaction (30-8) with an oxidation reaction that provides sufficient free energy or with the splitting of a phosphate compound that has an even more

---

[4] In (30-8) and subsequent equations, charges on ATP, ADP, and the inorganic phosphate species P_i are omitted.

[5] Standard conditions for hydrolytic reactions of phosphate esters in organisms have been defined as pH $= 7.0$, $t = 37°C$, the presence of an excess of Mg^{2+}, and all reactants and products at $1.0\ M$ concentration. Corresponding values of the standard free-energy change are designated $\Delta G^{\ominus\prime}$. The actual free energy of hydrolysis of ATP under physiological conditions differs from the standard value because of variations in the concentrations of the reactants. It is usually of the order of -50 kJ mol^{-1}.

negative free energy of hydrolysis than does ATP (Sec. 30-5). A typical phosphate compound of this kind, synthesized during the oxidative breakdown of glucose in cells (glycolysis), is 1,3-diphosphoglycerate:

$$\overset{\displaystyle OH}{\underset{\displaystyle}{^{-2}O_3POCH_2\overset{\displaystyle |}{C}HC\underset{\displaystyle \underset{\displaystyle O}{\|}}{O}PO_3^{2-}}} \qquad (30\text{-}9)$$

In the presence of an appropriate enzyme, this anion transfers a phosphate group directly to ADP to form ATP, with a significant decrease in standard free energy. The overall reaction is just the sum of the reverse of (30-8), for which $\Delta G^{\ominus\prime} = +31\text{ kJ mol}^{-1}$, and the hydrolysis reaction for the acyl[6] phosphate group involved, for which $\Delta G^{\ominus\prime} = -49\text{ kJ mol}^{-1}$:

1,3-Diphosphoglycerate + ADP $\longrightarrow$ 3-phosphoglycerate + ATP (30-10)

The corresponding standard free-energy change, $\Delta G_{10}^{\ominus\prime}$, is

$$\Delta G_{10}^{\ominus\prime} = (-49 + 31)\text{ kJ mol}^{-1} = -18\text{ kJ mol}^{-1}$$

and thus ATP can be produced in good yield by this reaction. There are numerous other "high-energy" molecules used for storage of energy in chemical systems. They share with ATP the property of being kinetically stable in the absence of enzymes specific for the reactions in which their energy contribution is needed.

Biological oxidations of carbohydrates, fats, and other reduced molecules that serve as energy sources involve sequences of a dozen or so individual reactions. In each successive oxidizing stage, the oxidizing agent used is one whose oxidation potential is sufficient for the reaction involved but not much greater than needed.

30-5 Metabolic Pathways

The term *metabolism* (Greek: *metabole,* change) refers to the chemical reactions occurring as cells exchange matter and energy with their surroundings. These reactions serve a number of specific purposes, which can be described broadly as (1) the extraction of energy from food or sunlight and its storage in chemical form, as discussed in the previous section; (2) the conversion of nutrients provided to the cell into small molecules or portions of molecules, the precursors for making proteins, lipids, nucleic acids, and other specialized molecules needed by the cell; (3) the synthesis of the macromolecular components of the cell and the other molecules necessary for its proper functioning; (4) the disposal of undesirable materials; and (5) providing for various specialized functions of the cell and the organism.

Metabolic changes occur in a highly regulated manner and are remarkably organized in both space and time. As mentioned in Sec. 30-3, many multienzyme systems exist in cells, assembled in such a way that the product released by one enzyme becomes the substrate for a neighboring enzyme that is specific for the next reaction in a sequence. Many of these reaction sequences or pathways have been studied in great detail and have been found to be closely similar in all forms

---

[6] An acyl group has the general formula $R\overset{\displaystyle}{\underset{\displaystyle \underset{\displaystyle O}{\|}}{C}}\!-$; an example is the acetyl group, for which $R = -CH_3$.

of life. They are often displayed in intricate metabolic flowcharts that show the reactants and products of each change as well as the enzyme specific for it. Our concern is not with the details of any of these pathways but rather with some general comments about them and about metabolic reactions.

Catabolism and Anabolism   Cells can both degrade or break down macromolecules and molecules of intermediate sizes and synthesize molecules from smaller components, as they are needed. These processes are called respectively *catabolism* and *anabolism* (Greek: *kata,* down; *ana,* up). It might seem logical that cells would utilize the same sequences of reactions, traversed in opposite directions, for both the degradation and the synthesis of macromolecules and other necessary cell components. However, there are always some differences in catabolic and anabolic pathways, although some of the individual steps may be catalyzed by the same enzymes and thus involve the identical reaction in opposite directions. The reason for these differences can be understood in terms of the thermodynamics of spontaneous reactions.

Any spontaneous process, whether it be a biosynthesis or a degradation, must occur with a decrease in free energy. Although some of the individual steps in the pathway in either direction may involve such small free-energy changes that they can be effectively reversed by variations in the concentrations of the reactants and products, others involve large decreases in free energy. The products are strongly favored in the latter reactions, which can be reversed only by supplying a large amount of free energy—for example, by coupling with some other reaction or by providing the free energy photochemically. Thus reactions in steps involving large free-energy changes invariably differ in catabolism and anabolism.

Degradative and synthetic pathways differ in other respects as well. They are usually situated in different parts of the cell and their rates are individually dependent on the supplies of various molecules, on the energy available, and on the needs of the cell. Regulatory features seem to operate to prevent oppositely directed anabolic and catabolic sequences from operating in the same tissues simultaneously.

As indicated in the schematic diagram in Fig. 30-9, both catabolism and anabolism can be considered to take place in three general stages. In the first stage of catabolism, stage I in the diagram, the major nutrients of cells are broken down into their major components—proteins into their constituent amino acids, polysaccharides into sugars, and lipids into fatty acids, glycerol, and other molecules. In stage II, the products of stage I are degraded still further; the ultimate product of stage II can be a two-carbon acetyl group, which becomes attached by esterification to an —SH group of a molecule called *coenzyme* A, an adenine nucleotide with additional groups attached. Acetyl coenzyme A acts as an acetyl-transfer agent. It can, for example, convert a four-carbon acid to a six-carbon acid, as indicated in stage III. Stage III is a cyclic stage, in which entering acetyl groups are oxidized to two molecules of CO_2 and simultaneously the nucleotide NAD (Fig. 30-1c) and another similar nucleotide are reduced. A series of coupled oxidations of these nucleotides (not shown in Fig. 30-9) then results in the production of a dozen molecules of ATP for each acetyl group oxidized, so that much of the energy released in the oxidation of the acetyl groups is eventually stored as the energy of ATP molecules.

The citric acid cycle of stage III, sometimes called the Krebs cycle after Hans Krebs, the biochemist who proposed it in 1937, is catalytic with respect to the

Stage I

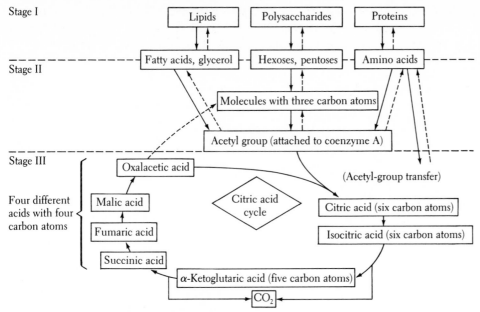

Figure 30-9 The Three Stages of Metabolism
Catabolic pathways are indicated with solid arrows, anabolic pathways with dashed arrows. Stage II and stage III are both catabolic and anabolic. For example, stage III, the citric acid cycle, is the final stage in the breakdown of food to CO_2 and can also furnish small molecules and fragments of molecules for biosynthesis. At physiological pH, citric acid and the other acids of this cycle are present principally in the form of their anions.

degradation of acetyl groups to CO_2 and liberation of hydrogen to NAD. Each acid in the cycle is regenerated in each complete turn of the cycle. These organic acids can play noncatalytic roles as well, serving as the starting materials for the biosyntheses of various essential molecules. For example, as indicated in Fig. 30-9, oxalacetic acid can be converted through three-carbon intermediates into sugars. The five-carbon dicarboxylic acid, α-ketoglutaric acid, is converted by enzymatic reduction in the presence of ammonia into L-glutamic acid (Table 30-3). The latter reaction is an essential one for the biosynthesis of all amino acids in all species, since it is the chief (if not the only) mode of formation of the amino group of an amino acid directly from ammonia. Most other such amino groups are introduced by transfer of the amino group from glutamic acid. This phenomenon of group transfer is common to many biosynthetic reactions. We have already mentioned examples of the transfer of phosphate and acetyl groups and the transfer of hydrogen atoms or electrons in redox reactions. Other commonly transferred groups are methyl, carboxyl, and formyl (—CHO).

The biosynthesis of large molecules involves the addition of small units to molecules that already exist. The units may be, for example, one- or two-carbon groups (such as formyl or acetyl), which are carried by specialized molecules like coenzyme A that act with the enzymes specific for the reaction in question. The added units may also be complete molecules that are, in essence, monomers for the macromolecule—for example, glucose in the buildup of glycogen or amino acids in the synthesis of proteins.

One of the three-carbon intermediates in both the degradation and the synthesis of carbohydrates is 1,3-diphosphoglycerate [Formula (30-9)], mentioned in

$$CH_2COOH$$
$$HOCCOOH$$
$$CH_2COOH$$

Citric acid

the previous section as one of the molecules with enough energy to convert ADP to ATP efficiently [Equation (30-10)]. How is this high-energy diphosphate itself synthesized? By oxidative phosphorylation of glyceraldehyde 3-phosphate,

$$\overset{\displaystyle OH}{\underset{\displaystyle |}{}}$$
$$^{-2}O_3POCH_2CHCHO \qquad (30\text{-}11)$$

that is, by coupling the oxidation of the aldehyde group in (30-11) with a reaction that transfers a phosphate to the resulting carboxylate group. The oxidizing agent is the ubiquitous NAD. The appreciable free energy released in the oxidation is sufficient to provide the energy needed to form the thermodynamically unstable acyl phosphate [Formula (30-9)]:

Glyceraldehyde 3-phosphate + NAD + $H_2O \longrightarrow$
$$3\text{-phosphoglycerate} + \text{NADH} + 2H^+ \qquad (30\text{-}12)$$

with $\Delta G_{12}^{\ominus\prime} = -43\ \text{kJ mol}^{-1}$, and

$$3\text{-Phosphoglycerate} + P_i \longrightarrow 1,3\text{-diphosphoglycerate} \qquad (30\text{-}13)$$

for which $\Delta G_{13}^{\ominus\prime} = 49\ \text{kJ mol}^{-1}$. Thus, the overall standard free-energy change for the oxidative phosphorylation, the sum of (30-12) and (30-13),

Glyceraldehyde 3-phosphate + NAD + P_i + $H_2O \longrightarrow$
$$1,3\text{-diphosphoglycerate} + \text{NADH} + 2H^+ \qquad (30\text{-}14)$$

is $\Delta G_{14}^{\ominus\prime} = +6\ \text{kJ mol}^{-1}$. This positive standard free-energy change is sufficiently small that the corresponding equilibrium constant is not far from unity and the reaction can be made to proceed in either direction with good yield, depending on the concentrations in the cell.

Regulation   The regulation of metabolic pathways, that is, the control of the rates at which the myriad catabolic reactions and biosynthetic reactions in the cell form their products, can occur in several different ways. At the simplest level, the rate of a particular metabolic sequence is a function of the pH and of the concentration within the cells of each enzyme of the pathway, of each substrate of these enzymes, and of all essential coenzymes and metal ions.

Many metabolic pathways, especially those involving biosyntheses, are controlled as well by particular regulatory enzymes, which normally catalyze a reaction at the start of a multienzyme sequence and are inhibited by an end product of the sequence. Thus the cell makes no more of each product than it needs at any time. Some regulatory enzymes are *multivalent,* that is, are stimulated or inhibited by two or more molecules, which might be the products of different pathways. The rate of one sequence can thus be made responsive to the integrated effect of several sequences.

The rates of many catabolic reactions are controlled by the immediate needs of the cell for energy in the form of ATP. The enzymes catalyzing these reactions are inhibited by excess ATP and are stimulated by ADP or AMP.[7] In other words, cells degrade food no more rapidly than necessary to provide the energy they need for their moment-to-moment activities. This is not to imply that an organism always lives "from hand to mouth". Metabolic regulation is such that cells

---

[7] Adenosine monophosphate.

tend to store reduced carbon as glycogen and fats in times when the supply of energy is plentiful and the need for carbon for biosynthesis is minimal, and to call on these stored reserves as conditions change.

Another method of regulation of metabolic pathways is through control of the rate of enzymatic synthesis. Although many enzymes are present in essentially constant amounts in cells, others are synthesized only in response to need. The genes that specify their synthesis are normally repressed or inactive; certain molecules can, however, activate these genes—that is, start the synthesis of the enzymes of some sequence. This mode of control makes it possible for cells to operate economically in expenditure of energy and nutrients, utilizing metabolic pathways only when they are needed. For example, certain bacteria can synthesize all the nitrogenous compounds they need—amino acids, nucleotide bases, and others—from ammonia as the only source of nitrogen if that is all that is provided. However, if these bacteria are fed a mixture of different amino acids and other nitrogenous molecules, the enzyme system that can synthesize these essential compounds directly from ammonia is not needed and the manufacture of the enzymes is suppressed. If the concentrations of the essential compounds in the environment fall below critical levels, the manufacture of the enzymes is resumed.

Finally, in higher animals, hormones secreted internally by various endocrine or ductless glands can either stimulate or inhibit metabolic reactions in other cells. There are many such hormones; they control a great variety of activities, such as growth of bone and muscle, levels of blood sugar, muscle relaxation, secretion of salt and water, and sexual activity. The mechanism of hormone action is still little understood, but hormones clearly operate through regulation of the chemical activities of the cell.

Summary

The chemistry of all living matter shows striking similarities and involves the same amino acids, carbohydrates, nucleic acid fragments, and lipids. Carbohydrates include sugars (polyhydroxyaldehydes and ketones) as well as their oxygen-linked polymers starch, cellulose, and glycogen. Sugars usually exist as heterocyclic rings that include an oxygen atom. Cellulose and starch differ in the way their glucose monomer units are joined. Cellulose is the chief structural element of plants, while starch and glycogen serve as carbohydrate reservoirs for plants and animals, respectively.

The mononucleotides that are components of the nucleic acids DNA and RNA consist of a nitrogen-containing ring compound (a base) bonded to a five-carbon sugar, one OH of which is esterified by phosphoric acid. The five common bases are often specified by the first letters of their names, A, G, C, T, and U. The first four occur in DNA, while in RNA the T is replaced by U. The other primary difference between DNA and RNA is that the five-carbon sugar is deoxyribose in the first and ribose in the second. Most DNA molecules consist of two strands, each of which is a polymer of phosphate-linked mononucleotides with a highly specific sequence of bases that represents a code by which genetic information is stored. The base sequences in the two strands are uniquely related, because the strands are hydrogen-bonded through their bases: A is always bonded to T and G to C. The several forms of RNA are involved in the translation of the genetic code and the subsequent synthesis of proteins.

Lipids are soluble in benzene and ether but not in water. This class of compounds includes fats, which are triesters of fatty acids and glycerol that serve for energy storage in animals and in plant seeds. Phospholipids contain glycerol esterified by phosphoric acid as well as by fatty acids. Together with other lipids and proteins, they are important constituents of membranes inside as well as on the surface of cells. Lipids have a polar end that is hydrophilic and a nonpolar end that is hydrophobic, a property that promotes the arrangement of lipids in membranes and other layers. Membranes are active structures engaged in the highly selective transport of ions and molecules and involved in many biochemical reactions.

Proteins (large polypeptides) consist of at least 50 amino acid residues linked by peptide (amide) bonds. Plants can manufacture all the amino acids they need, but many animals cannot and must obtain some of them through their diets. Proteins serve as structural materials, as enzymes (catalysts for biochemical reactions), as antibodies to counteract foreign materials in organisms, and in other ways. The sequence of amino acids of a protein is called its primary structure. The terms secondary and tertiary structure refer, respectively, to certain hydrogen-bonded arrangements (such as the α helix and several sheetlike structures) and to the detailed arrangement of the remainder of the molecule. Secondary and tertiary structures appear to be determined entirely by the primary structure, and the substitution of one or two amino acids by others may have a drastic effect on the shape of the molecule and may completely change any biological activity associated with the molecule. The amino acid sequence of a protein is determined by the base sequence of the RNA involved in its synthesis (the code for each amino acid being one or several characteristic base triplets) and thus ultimately by the base sequence of the DNA responsible for the RNA in question.

Many enzymes are effective only in the presence of a cofactor. Some cofactors are dinucleotides that are firmly attached to the enzyme. Others are more loosely bound molecules called coenzymes (which often are closely related to vitamins) and ions such as K^+, Ca^{2+}, or Mn^{2+}. Some enzymes are in effect organized groups of subenzymes, capable of catalyzing several reactions in sequence. Enzymes are effective normally only in narrow pH and temperature ranges, outside of which they may become denatured. Their catalytic activity and high specificity are a consequence of the uniquely shaped environments they provide to their substrates. Enzymes can be inhibited by unreactive molecules that compete for the active site, as well as by species that alter the enzyme, often irreversibly. Enzymatic reactions are often regulated by feedback, through competitive inhibition by a reaction product, the synthesis of which is slowed as the product accumulates and accelerated as it is used up.

Photosynthetic plant cells and some bacteria take their energy directly from sunlight. All higher animals, most microorganisms, and the nonphotosynthetic cells of plants utilize oxidation-reduction reactions to supply the energy they need, chiefly through the oxidation of reduced carbon compounds, such as glucose, by atmospheric oxygen. The principles of thermodynamics developed for inanimate systems apply to these processes in living systems as well.

Many of the essential processes in an organism occur with an increase of free energy. This is not possible unless these processes are coupled with others that simultaneously provide sufficient loss of free energy that the sum of the free-energy changes is negative. Adenosine triphosphate (ATP), which appears to be present in the cells of every living organism, provides the free-energy loss that drives many reactions essentially to completion. It is continually being transformed to ADP, adenosine diphosphate, by loss of its terminal phosphate group, a reaction that is accompanied by a large decrease in free energy. Most ATP is made from ADP by coupling the reverse of the hydrolysis with an oxidation reaction that provides sufficient free energy or with the splitting of a phosphate compound that has an even more negative free energy of hydrolysis than does ATP.

The term metabolism refers to the chemical reactions occurring as cells exchange matter and energy with their surroundings. Cells can both degrade or break down macromolecules and molecules of intermediate sizes (catabolism) and synthesize molecules from smaller components, as they are needed (anabolism). Anabolic and catabolic pathways usually take place in different parts of the cell and differ in details because each reaction along either path can proceed only in its spontaneous direction. Both processes can be considered to occur in three general stages, one of which is cyclic.

Metabolic pathways are regulated in a number of ways: (1) by adjustment of pH and of the concentrations of reaction participants; (2) by feedback inhibition through control of enzyme synthesis; (3) in higher animals, by hormonal stimulation and inhibition.

Terms and Concepts

Problems and Questions

30-1 Molecular Weight of Proteins Hemoglobin and myoglobin contain, respectively, 0.344 and 0.330 percent iron. (*a*) Calculate the minimum molecular weight of each protein. (*b*) Each hemoglobin molecule contains four Fe atoms, and each myoglobin molecule contains one Fe atom. Calculate the actual molecular weight of each protein.

30-2 Acid-Base Properties of Amino Acids A representative protein-forming amino acid that has one amino group and one carboxyl group, such as alanine (Table 30-3), is a diprotic acid; that is, it has two acid dissociation constants. The usual form of the amino acid, represented in Table 30-3, picks up a proton at very low pH and loses one at very high pH. The pK's for alanine are about 2.3 and 9.6. Write structural formulas for the low-pH and high-pH forms of this amino acid, and suggest why the values of pK_1 and pK_2 are, respectively, in the ranges 2 to 3 and 9 to 10. (*Hint:* Compare the forms before and after dissociation of each proton with related species, such as acetic acid and ammonia.)

30-3 Important Atoms in Living Systems Table 23-1 lists the 20 most abundant elements on the earth's surface in the order of their abundance. Nine of them are nonmetals, all of which, except silicon, are found in Table 30-1. Show how this striking difference in the behavior of silicon and carbon and the unimportance of Si—Si, Si—C, and Si—H linkages in living systems is consistent with bond energies (Table 17-2). In particular, contrast the energies of C—C and C—H bonds with those of C—O, C—N, C—F, and C—Cl. Then contrast the energies of Si—C and Si—H bonds with those of Si—O, Si—N, Si—F, and Si—Cl. (The bond energy for a Si—N bond is not well established, but it is probably within 10 percent of 440 kJ mol^{-1}.) What do the differences suggest?

30-4 Properties of Water Water is the most important compound in all known organisms, typically constituting 60 to 90 percent of the gross weight of the organism. It is also important in the environment of most organisms. Three physical properties of water, among others, are important for many forms of life: the high heat of vaporization, the high heat capacity, and the fact that water has its maximum density at about 4°C. Discuss reasons for the importance of these properties, and consider the bearing of the hydrogen-bonding tendencies of water on these properties.

30-5 Acidity of Nucleic Acids Nucleic acids are diesters of phosphoric acid. Esters are not usually acidic. Why are nucleic acids acidic, or is this name inappropriate?

30-6 Genetic Code The codons in the genetic code are sequences of three bases in the messenger RNA molecules. Explain why sequences of only two bases could not be used to code for the 20 different amino acids commonly found in proteins.

30-7 DNA for Hemoglobin S What base change must have occurred in the corresponding DNA to produce the messenger RNA described in the text as responsible for the genetic change from hemoglobin A to hemoglobin S (pages 694–695)?

30-8 Anions of Citric Acid Citric acid is a triprotic acid, $C_3H_4(OH)(COOH)_3$, with pK values 3.13, 4.77, and 6.40. Verify the assertion in the legend of Fig. 30-9 that the acids of the citric acid cycle exist principally in the form of their anions by calculating the fraction of citric acid in the form of the -3 ion and the fraction in the form of the -2 ion at pH = 6.9, the pH of a typical liver cell.

30-9 Biochemical Molecules Indicate the important structural features and a principal biological role of the following kinds of molecules: monosaccharides, amino acids, fats, disaccharides, ribonucleotides, enzymes, polysaccharides, fibrous proteins.

30-10 Metal Ions Discuss the biochemical function of various metal ions of both representative metals and transition metals. Indicate why many heavy-metal ions are poisons with cumulative effects (that is, the ions are not effectively eliminated from the system even over comparatively long periods of time).

30-11 Length of a DNA Molecule The average distance between base pairs measured parallel to the axis of a double-helical DNA molecule is 3.4 Å. The average molecular weight of a pair of nucleotides is about 650. What is the approximate length in millimeters of a single DNA molecule of molecular weight 2.8×10^9 (a value typical for the DNA of some bacteria)? About how many base pairs does this DNA contain?

30-12 Hydrogen Bonding Indicate the number and type of hydrogen bonds formed per amino acid unit in the alpha-helical and the sheet-structure portions of proteins. Indicate also the number and type of hydrogen bonds formed between the complementary base pairs in DNA.

Units

"Methanol has a heating power about half that of gasoline. However, it is denser than gasoline, so that 2 gallons of methanol do not require quite as much storage space as 2 gallons of gasoline."
NEWSPAPER ACCOUNT

"As the result of a great and continuing international effort there now exists an internationally agreed language of physical science."
M. L. MCGLASHAN, 1967

The following related topics are discussed in this appendix: SI units, equivalent energies, electrostatic energy and force, units in intermediate steps.

SI Units The SI system, discussed briefly in Chap. 1, Sec. 1-2, is an outgrowth of the mks (meter-kilogram-second) system and is based on seven units considered to be independent. Among them are the meter (m), kilogram (kg), second (s), kelvin (K), mole (mol), and ampere (A).[1] Units related to these by division or multiplication by powers of 10 are denoted by the prefixes in Table A-1.[2] The kilogram is anomalous in that it includes the prefix *kilo-* but has nevertheless been chosen as a basic unit. Prefixes that do not represent powers of 10^3 are to be used sparingly, an example being *centi-*, meaning 10^{-2}: cm = 10^{-2} m. Units may be raised to powers, the exponent applying to the prefix also, so that $km^2 = (1000\,m)^2 = 10^6\,m^2$, and $\mu m^3 = (10^{-6}\,m)^3 = 10^{-18}\,m^3$.

Table A-2 presents various units derived by combining basic units. Several of the derived SI units have special names and symbols. A few of the units used in this text are not SI units. The most common are noted in Table A-3.

Equivalent Energies Although the SI unit of energy is the joule, many other energy units are commonly used in chemistry. Some important interrelations of these units are given in Tables A-4 and A-5; values given on the same line are

---

[1] The remaining unit is the candela (cd), a unit of luminous intensity, which is not used in this book.
[2] Tables A-1, A-4, and A-5 are also given inside the back cover.

Factor	Prefix	Symbol	Factor	Prefix	Symbol
10^{-1}	deci	d	10	deka	da
10^{-2}	centi	c	10^2	hecto	h
10^{-3}	milli	m	10^3	kilo	k
10^{-6}	micro	μ	10^6	mega	M
10^{-9}	nano	n	10^9	giga	G
10^{-12}	pico	p	10^{12}	tera	T
10^{-15}	femto	f			
10^{-18}	atto	a			

Table A-1
Prefixes Denoting Powers of 10

Physical quantity	Unit	Special symbol and name
Speed	$m\,s^{-1}$	
Acceleration	$m\,s^{-2}$	
Force	$kg\,m\,s^{-2} = J\,m^{-1}$	N, newton
Pressure	$kg\,s^{-2}\,m^{-1} = N\,m^{-2}$	Pa, pascal
Energy	$kg\,m^2\,s^{-2} = N\,m$	J, joule
Power	$kg\,m^2\,s^{-3} = J\,s^{-1}$	W, watt
Electric charge	$A\,s$	C, coulomb
Electric potential difference	$kg\,m^2\,A^{-1}\,s^{-3}$ $= J\,A^{-1}\,s^{-1}$ $= W\,A^{-1}$ $= J\,C^{-1}$	V, volt
Electric resistance	$V\,A^{-1}$	Ω, ohm

*For further details on the physical quantities involved, see the Study Guide.

equivalent. For example, by Table A-4, 1 J is equivalent to 0.239 cal, so that multiplication by 0.239 converts an energy in joules to one in calories.

The energy units in Table A-5 require additional comment. The first two columns concern quantities that apply to *one mole* of substance rather than to an isolated atomic or molecular event. On the other hand, energies in electronvolts (eV) usually refer to individual atomic and molecular events, 1 eV being the kinetic energy acquired by one electron accelerated by an electric field of 1 V. Hence the energy unit in this column is *per molecule* not *per mole*. Similarly, the energy unit joule in the table refers to one molecule. The fifth column in Table A-5 shows the temperature T associated with the energy RT per mole (where R is the gas constant, discussed on page 85) or kT per atom or molecule (where k is Boltzmann's constant, $R/N_A = 1.3807 \times 10^{-23}$ J K^{-1}). The significance of these quantities of energy is discussed in Sec. 4-4.

The final column is the wave number equivalent to the energy by the relation $\tilde{\nu} = 1/\lambda = E/hc$, where h is Planck's constant and c is the speed of light.

A table of physical constants is also presented inside the back cover. The values given are based on a report by the U.S. National Bureau of Standards (*Journal of Physical and Chemical Reference Data,* vol. 2, no. 4, p. 741, 1973).

Electrostatic Energy and Force By the laws of electrostatics the potential energy ϵ_p (coulomb energy) associated with two point charges Q_1 and Q_2 at a distance r in a vacuum is

$$\epsilon_p = \frac{Q_1 Q_2}{\alpha r} \tag{A-1}$$

For historical reasons the proportionality constant α is sometimes written as

**Table A-3
Additional Units**

Physical quantity	Unit	Special name
Volume	$liter = 1000\,cm^3 = 10^{-3}\,m^3$ $ml = cm^3 = 10^{-6}\,m^3$	Liter Milliliter
Density	$g\,cm^{-3} = 10^3\,kg\,m^{-3}$	
Molar concentration	$M = mol\,liter^{-1} = 10^{-3}\,mol\,m^{-3}$	Molarity
Length	$Å = 10^{-10}\,m = 0.1\,nm$	Angstrom
Pressure	$atm = 760\,torr = 101.325 \times 10^3\,Pa$ $torr = 1.316 \times 10^{-3}\,atm$ $= 133.3\,Pa$	Atmosphere Torr

Calorie (cal)	Joule (J)	Liter atmosphere (liter atm)
1	4.1840	0.041293
0.23901	1	9.8692×10^{-3}
24.217	101.325	1

$\alpha = 4\pi\epsilon_0$, where ϵ_0 is called the permittivity constant. The force F between the two charges (coulomb force) is given by

$$F = \frac{\mathbf{Q}_1\mathbf{Q}_2}{\alpha r^2} \tag{A-2}$$

When SI units are used ($\mathbf{Q}_1$ and $\mathbf{Q}_2$ in C $=$ s A; r in m; ϵ_p in J $=$ kg m^2 s^{-2}; and F in N $=$ kg m s^{-2}), the proportionality constant α has the value[3]

$$\alpha = 1.11265 \times 10^{-10} \text{ C}^2 \text{ J}^{-1} \text{ m}^{-1} \tag{A-3}$$

The units of α are seen to be those required by (A-1) or (A-2); reduced to basic units, C^2 J^{-1} m^{-1} $=$ kg^{-1} m^{-3} s^4 A^2.

Units in Intermediate Steps   It is essential to formulate a problem in terms of all units involved, but it is often cumbersome to express the units at each of the intermediate steps. Proper units must, however, be affixed to all final results.

A widespread habit of deleting units in intermediate steps is illustrated by the relation

$$x = 0.10 - [\text{H}^+] \tag{A-4}$$

which is more properly written as

$$x = 0.10 - \frac{[\text{H}^+]}{M} \tag{A-5}$$

However, the meaning of (A-4) is unambiguous and division of [H$^+$] by M (meaning mol liter^{-1}) is generally omitted.

A related procedure concerns equilibrium constants, such as

$$K_{\text{N}_2\text{O}_4} = \frac{(P_{\text{NO}_2})^2}{P_{\text{N}_2\text{O}_4}} = 0.141 \text{ atm} \tag{A-6}$$

---

[3] The numerical value of α is related to c, the velocity of light (in a vacuum):

$$\frac{\alpha}{\text{C}^2 \text{ J}^{-1} \text{ m}^{-1}} = 10^7 \left(\frac{c}{\text{m s}^{-1}}\right)^{-2}$$

$E/\text{kJ mol}^{-1}$	$E/\text{kcal mol}^{-1}$	E/eV	E/J	T/K	$\tilde{\nu}/\text{cm}^{-1}$
1	0.23901	0.010364	1.6606×10^{-21}	120.27	83.59
4.1840	1	4.3364×10^{-2}	6.948×10^{-21}	503.2	349.8
96.48	23.060	1	1.6022×10^{-19}	11.605×10^3	8066
6.022×10^{20}	1.4393×10^{20}	6.241×10^{18}	1	7.243×10^{22}	5.034×10^{22}
8.314×10^{-3}	1.9872×10^{-3}	8.617×10^{-5}	1.3807×10^{-23}	1	0.6950
1.196×10^{-2}	2.859×10^{-3}	1.240×10^{-4}	1.986×10^{-23}	1.439	1

[*] Some discretion is required with units that are not strictly energy units. For example, the energy kT characteristic of a temperature of 1 K corresponds also to 8.617×10^{-5} eV, which is sometimes expressed by saying that 1 K "equals" 8.617×10^{-5} eV. Only when both units are interpreted as energy units does 8.617×10^{-5} eV/K represent unity. Note also that the energies in the last four columns might be considered as energies per molecule.

for the reaction

$$N_2O_4(g) \rightleftharpoons 2NO_2(g) \qquad \text{(A-7)}$$

and

$$K_w = [H^+][OH^-] = 1.0 \times 10^{-14} \, M^2 \qquad \text{(A-8)}$$

for the reaction

$$H_2O \rightleftharpoons H^+ + OH^- \qquad \text{(A-9)}$$

(both values at $25°C$). As discussed in Sec. 11-3, equilibrium constants are closely associated with the corresponding balanced chemical equations. Hence, when the chemical equation is explicitly stated or known from the context, the units of equilibrium constants, such as atm and M^2 (in A-6) and (A-8), are often dropped. We usually follow this practice.

Finally, consider logarithmic expressions such as

$$pH = -\log \frac{[H^+]}{M}$$

where $[H^+]$ is divided by M because only logarithms of dimensionless (and nonnegative) quantities have simple definitions. Other examples are $\log (K_w/M^2)$ and $\log (T/K)$. It is often customary to omit the division by the units; that is, one writes $-\log [H^+]$, $\log K_w$, and $\log T$, with the understanding that the units are simply dropped when taking logarithms. Caution must be exercised when there may be uncertainty about the appropriate units, as in $\log P$, which could imply, for example, $\log (P/\text{atm})$ or $\log (P/\text{torr})$. The units *must* be specified in ambiguous cases.

Names and Formulas of Compounds

"The Chemical signs ought to be letters, for the greater facility of writing, and not to disfigure a printed book. . . . I shall take, therefore, for the chemical sign, the initial letter of the Latin name of each elementary substance: but as several have the same initial letter, I shall distinguish them in the following manner: 1. In the class which I call metalloids, I shall employ the initial letter only, even when this letter is common to the metalloid and some metal. 2. In the class of metals, I shall distinguish those that have the same initials with another metal, or metalloid, by writing the first two letters of the word. 3. If the first two letters be common to two metals, I shall, in that case, add to the initial letter the first consonant which they have not in common."

J. J. BERZELIUS, 1814[1]

This appendix has three sections. Sections B-1 and B-2 give the names and formulas of many common inorganic ions and compounds, and discuss the simpler rules of inorganic chemical nomenclature. Students must become familiar with these names, formulas, and rules. Section B-3 contains some of the rules for naming organic compounds.

An important aspect of the systematic naming of compounds is that knowledge of the name must permit the reconstruction of the correct molecular or ionic formula. Names of this kind are called *chemical* or *systematic* names and stand in contrast to trivial names, which are often used for important compounds, such as "aspirin", "vitamin A", "indigo" (a dye), or "muriatic acid" (HCl). Trivial names have little or no relationship to the formula.

The names and rules given here suffice to illustrate the general principles of systematic nomenclature and should be adequate for most of the compounds discussed in this text—particularly in the derivation of a formula from a given name, which is easier than the reverse procedure. An incidental but important point in chemical nomenclature is that a given compound may have a number of acceptable chemical names. All these must, of course, relate to the same formula.

B-1 Simple Inorganic Ions and Compounds

Many simple inorganic compounds are formed by the combination of ions. Their names are based on those of the ions, names that must be memorized, with the corresponding symbols.

An essential part of the formula of an ion is its charge. For example, there is an enormous difference in properties between hydrogen atoms (H) and hydrogen

[1]Quoted by Leicester and Klickstein, "Source Book in Chemistry, 1400–1900", McGraw-Hill Book Co., New York, 1952; *Annals of Philosophy,* vol. 3, p. 362, 1814.

ions (H^+), between sulfur trioxide (SO_3) and sulfite ion (SO_3^{2-}), and between the hydroxyl radical (OH) and the hydroxide ion (OH^-). It is best to study and remember the charge of an ion as part of its formula. Differences such as that between ammonia (NH_3) and ammonium ion (NH_4^+) must also be noted and remembered.

The following subsections also indicate the geometric structure of some ions. These structures need not be memorized; they can be deduced from the electronic structure.

Cations

1. The following metals form only one stable positive ion (cation). The names of these ions are the same as those of the metals, such as potassium ion and magnesium ion.

Alkali metals:	Li^+, Na^+, K^+, Rb^+, Cs^+, Fr^+
Alkaline-earth metals:	Be^{2+}, Mg^{2+}, Ca^{2+}, Sr^{2+}, Ba^{2+}, Ra^{2+}
Others:	Al^{3+}, Zn^{2+}, Cd^{2+}, Ni^{2+}, Pb^{2+}, Ag^+

2. Many metallic elements form two positive ions with different charges. The ion with the smaller charge per atom is often denoted by the ending *-ous*, that with the greater charge by the ending *-ic*. For example,

Cr^{2+}	chromous	Cr^{3+}	chromic
Mn^{2+}	manganous	Mn^{3+}	manganic
Fe^{2+}	ferrous	Fe^{3+}	ferric
Co^{2+}	cobaltous	Co^{3+}	cobaltic
Cu^+	cuprous	Cu^{2+}	cupric
Sn^{2+}	stannous	Sn^{4+}	stannic
Hg_2^{2+}	mercurous[2]	Hg^{2+}	mercuric

    More complicated situations are discussed in Sec. B-2.

3. *Some polyatomic cations*

H_3O^+	hydronium or oxonium (pyramidal)
NH_4^+	ammonium (tetrahedral)
PH_4^+	phosphonium (tetrahedral)
NO_2^+	nitryl (linear)

Anions   Anions containing only one atom (such as F^- and S^{2-}) are given the ending *-ide*. A few polyatomic anions also have the ending *-ide*, as indicated in the lists of names given below. The names of anions containing oxygen with other atoms often end in *-ate*, as in carbonate ion, CO_3^{2-}. However, when two anions of a given element contain different numbers of oxygen atoms, the endings *-ite* and *-ate* are used, as in nitrite (NO_2^-) and nitrate (NO_3^-) ions. The ending *-ite* is used for the ion with fewer oxygen atoms. Sometimes there exist more than two such anions, and prefixes as well as the endings just mentioned are used for naming them. For example, the ions ClO^-, ClO_2^-, ClO_3^-, and ClO_4^- are named, respectively, hypochlorite, chlorite, chlorate, and perchlorate. Further examples can be found in the lists given below. Note that the prefixes and endings do not indicate the actual number of oxygen atoms involved but serve only to order a sequence of related ions.

---

[2] Note the diatomic nature of the Hg_2^{2+} (mercurous) ion.

1. *Halides*

   F^- fluoride Br^- bromide
   Cl^- chloride I^- iodide

2. *Chalcogenides*

   O^{2-} oxide S^{2-} sulfide
   O_2^{2-} peroxide S_2^{2-} disulfide
   OH^- hydroxide HS^- hydrogen sulfide

3. *Planar trigonal oxyanions and related species*

   CO_3^{2-} carbonate NO_3^- nitrate
   HCO_3^- bicarbonate or
            hydrogen carbonate

4. *Tetrahedral oxyanions and related species*

   SiO_4^{4-} silicate HSO_4^- hydrogen sulfate
              (orthosilicate)   or bisulfate
   PO_4^{3-} phosphate $S_2O_3^{2-}$ thiosulfate[3]
              (orthophosphate) ClO_4^- perchlorate
   HPO_4^{2-} hydrogen CrO_4^{2-} chromate
              phosphate $HCrO_4^-$ hydrogen chromate
   $H_2PO_4^-$ dihydrogen MnO_4^- permanganate
              phosphate MnO_4^{2-} manganate
   SO_4^{2-} sulfate

5. *Condensed polyanions linked through oxygen atoms:* Each of the individual
   units linked is tetrahedral (for example, $O_3SOSO_3^{2-}$ for $S_2O_7^{2-}$):

   $S_2O_7^{2-}$ disulfate or pyrosulfate
   $Cr_2O_7^{2-}$ dichromate
   $P_2O_7^{4-}$ diphosphate or pyrophosphate
   $(PO_3)_n^{n-}$ polyphosphate ring, also called metaphosphate
   $Si_2O_7^{6-}$ disilicate
   $(SiO_3)_n^{2n}$ polysilicate ring or (if *n* is very large) chain, also called meta-
                   silicate

6. *Some other important anions*

   H^- hydride OCN^- cyanate (linear)
   CN^- cyanide SCN^- thiocyanate (linear)
   BH_4^- borohydride (tetrahedral) N_3^- azide (linear)
   ClO^- hypochlorite SO_3^{2-} sulfite (pyramidal)
   ClO_2^- chlorite (bent) NO_2^- nitrite (bent)
   ClO_3^- chlorate (pyramidal) NH_2^- amide (bent)

---

[3]The relationship between SO_4^{2-} and $S_2O_3^{2-}$ is that one oxygen atom in sulfate has been replaced by
a sulfur atom in thiosulfate.

Compounds Formed from Ions These are named by suitable combinations of the names of the ions, with the cation(s) named first—e.g., ferric chloride ($FeCl_3$), calcium hydride (CaH_2), sodium hydroxide (NaOH), and lead dichromate ($PbCr_2O_7$).

Compounds of hydrogen with just one other element have names such as hydrogen fluoride (HF) and hydrogen sulfide (H_2S), except that the hydrogen halides (HF, HCl, HBr, HI) are often referred to (especially when in solution) by names such as hydrochloric acid to emphasize their acid character.

The names of oxyacids parallel those of the ions but the endings *-ous* and *-ic* are used rather than *-ite* and *-ate*, as exemplified in the table below.

Name of ionic species	Name of acid species
Carbonate ion	Carbonic acid
Nitrite ion	Nitrous acid
Nitrate ion	Nitric acid
Hypochlorite ion	Hypochlorous acid
Chlorite ion	Chlorous acid
Chlorate ion	Chloric acid
Perchlorate ion	Perchloric acid

Almost all positively charged ions are hydrated in aqueous solution, even though this is not usually indicated in their formulas (see Secs. 7-5 and 9-5). Negatively charged ions (anions) are hydrated also, but the structures are less well defined than those for most cations. Water of crystallization (or hydration) in solids is indicated explicitly in formulas, usually with a dot separating it from the formula for the remainder of the compound, as in $CuSO_4 \cdot 5H_2O$ (Fig. 7-8).

B-2 Oxidation States and Inorganic Nomenclature

An important feature of inorganic nomenclature is the way in which oxidation states (Chap. 10) are specified for each part of the compound, unless there is only one possibility, as with cations of the alkali and alkaline-earth metals.

The system used in the names listed in Sec. B-1 is one of endings and prefixes. In that system the lower and the higher of two oxidation states of a cation are distinguished by the endings *-ous* and *-ic*, respectively, as in the examples under "Cations", item 2, Sec. B-1. The specific oxidation states indicated by these suffixes vary from one element to the next, and thus an alternative and unambiguous means of denoting oxidation states has been developed. They are shown by positive and negative roman numerals enclosed in parentheses after the name of the element or by (0) for oxidation state zero. The system is used most commonly for metallic elements that exhibit several stable oxidation states. Thus $FeCl_2$ is called iron(II) chloride as well as ferrous chloride, and SnS_2 may be referred to as tin(IV) sulfide as well as stannic sulfide.

The endings *-ite* and *-ate* for the anions of oxygen acids (*-ous* and *-ic* for the acids themselves) and the prefixes *hypo-* and *per-* are also related to the oxidation state of the central atom of the species, as illustrated in the following table:

Ionic species	Oxidation state of central atom	Name	Corresponding acid	Name
IO^-	I	Hypoiodite ion	HIO	Hypoiodous acid
IO_2^-	III	Iodite ion	HIO_2	Iodous acid
IO_3^-	V	Iodate ion	HIO_3	Iodic acid
IO_4^-	VII	Periodate ion	HIO_4	Periodic acid
MnO_4^{2-}	VI	Manganate ion	H_2MnO_4	Manganic acid
MnO_4^-	VII	Permanganate ion	$HMnO_4$	Permanganic acid
SO_3^{2-}	IV	Sulfite ion	H_2SO_3	Sulfurous acid
SO_4^{2-}	VI	Sulfate ion	H_2SO_4	Sulfuric acid

Coordination Compounds (compounds containing complex ions or similar neutral species)   The following additional rules apply to coordination compounds:

1. The oxidation state of the central atom is indicated by a roman numeral or zero in parentheses, unless it is unambiguous.
2. Names of complex anions end in *-ate*, while for complex cations and neutral molecules no special ending is given.
3. Neutral ligands are named as the molecule, negative ligands end in *-o* (common endings are *-ido*, *-ito*, and *-ato*), and positive ligands (rare) end in *-ium*. Examples are

$NH_2CH_2CH_2NH_2$	ethylenediamine	ONO^-	nitrito
Cl^-	chloro	NO^-	nitroso
O^{2-}	oxo	CH_3COO^-	acetato
OH^-	hydroxo	SO_4^{2-}	sulfato
CN^-	cyano	NH_2^-	amido
NO_2^-	nitro		

Exceptions to these rules are H_2O (aquo) and NH_3 (ammine; note the double *m*).

The different names given to the NO_2^- group reflect different possibilities in linkage: with the nitro group the nitrogen atom is bonded to the central atom of the coordination compound, whereas for the nitrito group one of the oxygen atoms forms the bond. Such *linkage isomerism* can also occur for other ligands. In the list above, the coordination bond normally involves the atom given first in the formula, except for SO_4^{2-} and acetate, which are written in the conventional way and for each of which an oxygen atom is bonded to the central atom of the coordination compound.

4. The sequence of the ligands in naming a complex is negative, neutral, and positive; no hyphens are used. Within each category the order is that of increasing complexity.
5. The prefixes *di-*, *tri-*, *tetra-*, *penta-*, *hexa-*, . . . are used before such simple names as bromo, nitrito, amido. The prefixes *bis-*, *tris-*, *tetrakis-*, . . . (Greek: *kis*, times) are used in front of more complex names, such as ethylenediamine, particularly if these names contain other prefixes such as *mono-*, *di-*, *tri-*, ... Examples are

$Li[AlH_4]$	lithium tetrahydridoaluminate[4]
$K_2[PtCl_4]$	potassium tetrachloroplatinate(II)
$K_3[Fe(CN)_6]$	potassium hexacyanoferrate(III)
$Na_4[Ni(CN)_4]$	sodium tetracyanonickelate(0)
$CoCl_2(en)_2$	dichlorobis(ethylenediamine)cobalt(II)
$Co(NH_3)_3(NO_2)_3$	trinitrotriammine cobalt(III)
$[Pt(NH_3)_4(ONO)Cl]SO_4$	chloronitritotetrammineplatinum(IV) sulfate

B-3   Modern Organic Nomenclature

The complexity of many organic compounds necessitates an elaborate set of rules so that any compound may be given a name from which the structural formula of that compound can be reconstructed. Furthermore, the rules must, ideally at least, be sufficiently definite that a given compound cannot be named in several different ways, because the names of organic substances are often so complex that the different names would not always be recognized as applying to the same compound. We do not deal with these complexities here, but rather present only a few rules and examples to show how the system works. Examples of the naming of compounds containing the common functional groups are given in Chap. 29.

Alkanes and Alkyl Groups (Secs. 29-1 and 29-2)   Alkanes are hydrocarbons with the general formula C_mH_{2m+2}. They contain no multiple bonds and no rings; every carbon atom forms four bonds and every hydrogen atom one. In molecules of the "straight-chain"[5] or normal alkanes (Table 29-2), no carbon atom has more than two other carbon atoms bonded to it. The simplest alkyl groups, corresponding formally to alkanes from which one hydrogen atom has been removed, are listed in Table 29-3.

Branched Alkanes   Branched alkanes are named as follows: (1) Find and name the longest carbon chain in the molecule; (2) mentally attach the numbers 1, 2, 3, . . . to successive carbon atoms in this chain, the direction being chosen so as to label the side chains by the lowest numbers possible; (3) name the alkyl groups that represent the side chains; (4) put these pieces of information together as shown by the examples below:

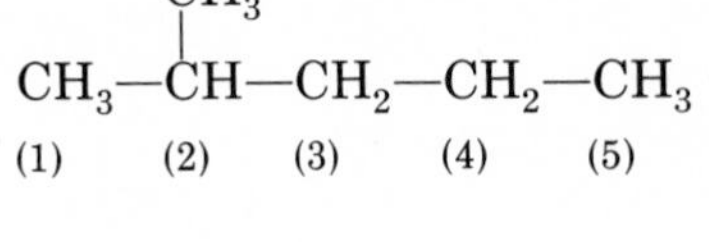

2-Methylpentane

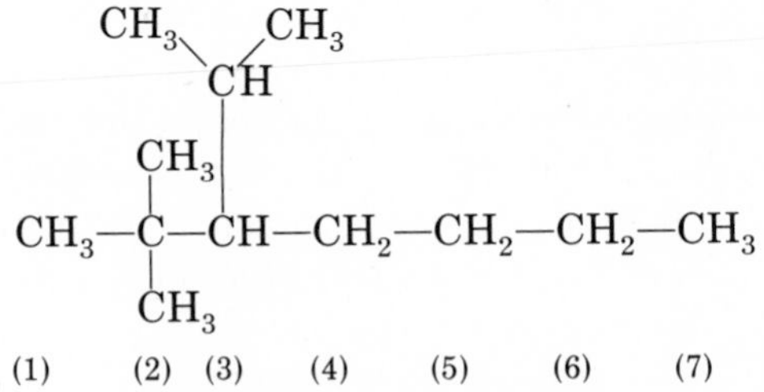

2,2-Dimethyl-3-isopropyl heptane

---

[4] This compound, a powerful and frequently used reagent, is commonly called lithium aluminum hydride.

[5] The C-C-C bond angle in these chains is about 112°, so the chains are far from linear, but topologically they are "straight", which in this context implies "without branches".

Alkenes and Alkynes Hydrocarbons with double bonds are called alkenes or olefins. They are named by the suffix *-ene*. The numbering system introduced earlier is used to show the position of the double bond when this is necessary:

$$CH_2{=}CH_2 \qquad CH_2{=}CHCH_3 \qquad CH_2{=}CHCH_2CH_3$$
$$\phantom{CH_2{=}CHCH_2}{(1)}\quad{(2)}\;{(3)}\;\;{(4)}$$

Ethylene   Propene   1-Butene

$$CH_3CH{=}CHCH_3 \qquad CH_2{=}CH{-}CH{=}CH_2$$
$${(1)}\;\;\;{(2)}\;\;\;\;{(3)}\;{(4)} \qquad\quad {(1)}\;\;\;\;\;{(2)}\;\;\;\;\;{(3)}\;\;\;\;{(4)}$$

2-Butene   1,3-Butadiene

Hydrocarbons with triple bonds are called alkynes. Examples are

$$HC{\equiv}CH \qquad CH_3C{\equiv}CCH_3$$
$$\phantom{HC{\equiv}CH \qquad}{(1)}\;\;\;{(2)}\;\;\;{(3)}{(4)}$$

Ethyne   2-Butyne
(acetylene)

Ring Compounds Cyclic hydrocarbons are denoted systematically by the prefix *cyclo-*:

Cyclopropane   Cyclohexane   Cyclohexene   3-Ethylcyclopentene

Note in the last example that the double bond is assumed to be between C-1 and C-2; the substituent is then given the lowest number consistent with this assumption.

Polyfunctional Compounds Many common polyfunctional compounds have trivial names, but they can be named systematically also. Examples are

1,2-Ethanediol   3-Methyl-2-butene-1-ol   3-Chloronitrobenzene   3-Aminocyclopentanone
(ethylene glycol)

Problems and Questions

B-1 Formulas of Inorganic Compounds and Ions Give the correct formula, including charges if any, for the following compounds and ions: sulfuric acid; sodium sulfate; potassium bromide; ammonium ion; ammonium nitrate; nitric acid; carbonate ion; calcium bicarbonate; calcium phosphate; silver iodide.

B-2 Formulas of Inorganic Compounds and Ions Give the correct formula, including charges if any, for the following: ammonium bicarbonate; barium perchlorate; iron(III) nitrate hexahydrate; calcium fluoride; aluminum sulfate; hydroxide ion; strontium bromide; barium phosphate; sulfite ion; copper(II) sulfate pentahydrate.

B-3   Names of Inorganic Compounds and Ions
Name the following compounds and ions: BaO_2; $CaHPO_4$; $Cr(OH)_3$; HCN; $MgSO_3$; $LiSCN$; S^{2-}; $FeCl_3$; SrI_2; $NiSO_4$; HBr; OCl^-; NH_3.

B-4   Names of Inorganic Compounds and Ions
Name the following compounds and ions: $LiHSO_4$; HS^-; $FePO_4$; H_2CO_3; $BaCrO_4$; $HClO_4$; $Na_2S_2O_3$; $Be(CN)_2$; NH_4ClO_4; Cu_2S; K_2SO_4; HSO_3^-.

B-5   Formulas of Inorganic Complexes   Give chemical formulas for the following compounds and ions: sodium bis(thiosulfato)argentate(I); ammonium tetrathiocyanatodiammine chromate(III); sodium hexanitrocobaltate(III); hydroxopentaaquoaluminum ion; tris(ethylenediamine)cobalt(III) sulfate; dihydroxotetraaquochromium(III) chloride; chlorotriamminecobalt(II) chloride; potassium tetrachloroaurate(III); tetrakis(trichlorophosphine)nickel(0); dichlorotetramminecobalt(III) chloride.

B-6   Formulas of Hydrocarbons   Give formulas for the following hydrocarbons: pentane; 2,3-dimethyl-1-hexene; 3-heptyne; 3-ethyloctane; cyclobutane; 3-methylcyclohexene; 1,3-dimethylbenzene.

Derivation of Some Equations

"Mathematics has a light and wisdom of its own, above any possible application to science, and it will richly reward any intelligent human being to catch a glimpse of what mathematics means to itself."

E. T. BELL[1]

C

In several places in the text, equations that can be derived by elementary calculus procedures have been presented without proof. The derivations are presented here in detail. Equations that appear elsewhere in the book are identified by their original numbers as well as by numbers appropriate to their position in the appendix. Some material on exponents, logarithms, and calculus can be found in the Study Guide.

P,V Work (Sec. 19-2) The work performed on a system when an external pressure P_{surr} causes an infinitesimal change in volume dV is given by

$$Dw = -P_{surr}\, dV \qquad (C\text{-}1)$$

We write Dw rather than dw to indicate that work is not a state function. To calculate the work for a finite change in volume, we must integrate (C-1) along the path by which the change occurs:

$$w = -\int_{V_1}^{V_2} P_{surr}\, dV \qquad (C\text{-}2)$$

Equations have been given in the text for two specific kinds of changes in state.

1. *Constant pressure*:

$$w = -\int_{V_1}^{V_2} P_{surr}\, dV = -P_{surr}\int_{V_1}^{V_2} dV$$

$$= -P_{surr}\, V\Big|_{V_1}^{V_2} = -P_{surr}\,(V_2 - V_1) \qquad (C\text{-}3;\ 19\text{-}2)$$

2. *Isothermal reversible expansion of an ideal gas*: Since the expansion is reversible, $P_{sys} = P_{surr}$. The specification of an ideal gas allows us to make the substitution $P = nRT/V$ while the restriction to an isothermal process means that T is constant and thus can be taken outside the integral:

$$w = -\int_{V_1}^{V_2} P\, dV = -\int_{V_1}^{V_2} \frac{nRT}{V}\, dV = -nRT\int_{V_1}^{V_2} \frac{dV}{V}$$

$$= -nRT\ln V\Big|_{V_1}^{V_2} = -nRT\ln\frac{V_2}{V_1} \qquad (C\text{-}4;\ 19\text{-}3)$$

---

[1]"Men of Mathematics", Simon and Schuster, New York, 1937.

Heat Capacity (Sec. 19-2) A more general definition of heat capacity than (19-4a) and (19-4c) is

$$C = \frac{Dq}{dT} \quad \text{or} \quad Dq = CdT \tag{C-5}$$

where we have omitted the subscript V or P.

Here again the symbol Dq is a warning that heat is not a state function. The heat that must be transferred to the system to produce a change in temperature from T_1 to T_2 can be calculated by integrating (C-5) along a specified path:

$$q = \int Dq = \int_{T_1}^{T_2} C\, dT \tag{C-6a}$$

$$= C \int_{T_1}^{T_2} dT = C\, T \Big|_{T_1}^{T_2} = C(T_2 - T_1) = C\, \Delta T \tag{C-6b; 19-4a; 19-4c}$$

In (C-6b) it has been assumed that C does not depend on T.

Entropy Change on Heating (Sec. 20-2) For any reversible change in state of a system

$$\Delta S = \int_1^2 \frac{Dq_{\text{rev}}}{T}$$

Substituting (C-5) gives

$$\Delta S = \int_{T_1}^{T_2} \frac{C}{T}\, dT \tag{C-7}$$

For a constant heat capacity

$$\Delta S = C \int_{T_1}^{T_2} \frac{dT}{T} = C \ln T \Big|_{T_1}^{T_2} = C \ln \frac{T_2}{T_1} \tag{C-8; 20-8}$$

Rates of Chemical Reactions (Sec. 22-3) For a reaction

$$a\text{A} + b\text{B} \longrightarrow c\text{C} + d\text{D}$$

the instantaneous rate can be defined in terms of any of the components or reactants:

$$\text{Rate} = -\frac{1}{a}\frac{d[\text{A}]}{dt} = -\frac{1}{b}\frac{d[\text{B}]}{dt} = \frac{1}{c}\frac{d[\text{C}]}{dt} = \frac{1}{d}\frac{d[\text{D}]}{dt}$$

which, by the definition of a derivative, is equivalent to (22-6).

1. *First-order reaction*: If the concentration of a substance X is altered by a first-order reaction

$$-\frac{d[\text{X}]}{dt} = k[\text{X}] \tag{C-9}$$

or

$$-\frac{d[\text{X}]}{[\text{X}]} = kdt \tag{C-10}$$

Let the concentration of X at time t_0 be $[\text{X}]_0$ and that at a later time t be $[\text{X}]$. Then to find $[\text{X}]$ we integrate (C-10)

$$-\int_{[X]_0}^{[X]} \frac{d[X]}{[X]} = \int_{t_0}^{t} kdt \tag{C-11}$$

$$-\ln[X] \Big|_{[X]_0}^{[X]} = kt \Big|_{t_0}^{t}$$

$$-(\ln[X] - \ln[X]_0) = k(t - t_0) \tag{C-12}$$

If $t_0 = 0$, we get

$$-\ln \frac{[X]}{[X]_0} = kt \tag{C-13; 22-13}$$

or

$$\frac{[X]}{[X]_0} = e^{-kt}$$

2. *Second-order reaction*: The rate expression for a reaction that is second-order in a single reactant

$$-\frac{d[X]}{dt} = k[X]^2 \tag{C-14}$$

can be rewritten

$$-\frac{d[X]}{[X]^2} = kdt \tag{C-15}$$

and then integrated:

$$-\int_{[X]_0}^{[X]} \frac{d[X]}{[X]^2} = \int_{t_0}^{t} kdt \tag{C-16}$$

$$-\left\{ -\frac{1}{[X]} \Big|_{[X]_0}^{[X]} \right\} = kt \Big|_{t_0}^{t}$$

$$\frac{1}{[X]} - \frac{1}{[X]_0} = k(t - t_0) \tag{C-17}$$

Again, setting $t_0 = 0$ gives

$$\frac{1}{[X]} - \frac{1}{[X]_0} = kt \tag{C-18; 22-17}$$

Tables

"There is measure in all things."
HORACE, 35 B.C.

This appendix contains the following tables:

D-1. Vapor pressure of water at various temperatures
D-2. Solubility-product constants
D-3. Acid constants K_a
D-4. Values of $\widetilde{H}_f^\circ$ and $\widetilde{G}_f^\circ$
D-5. Standard electrode potentials

Table D-1
Vapor Pressure of Water at Various Temperatures

$t/°C$	$P_{vap,H_2O}/torr$	$t/°C$	$P_{vap,H_2O}/torr$
−78 (ice)	0.00056	30	31.8
−20 (ice)	0.78	35	42.2
−10 (ice)	1.95	40	55
0	4.58	50	93
5	6.54	60	149
10	9.2	70	234
15	12.8	80	355
18	15.5	90	526
20	17.5	100	760
22	19.8	110	1075
24	22.4	120	1489
26	25.2	150	3570
28	28.3	200	11,659

$Ag \cdot Ag(CN)_2$*	5×10^{-12}	$Fe(OH)_3$	10^{-36}
$AgBr$	5.2×10^{-13}	FeS	5×10^{-18}
$AgBrO_3$	5.2×10^{-5}	Hg_2Br_2†	1×10^{-21}
$AgCl$	1.8×10^{-10}	Hg_2Cl_2†	1.3×10^{-18}
Ag_2CrO_4	2.0×10^{-12}	Hg_2I_2†	7×10^{-29}
AgI	8.3×10^{-17}	Hg_2SO_4†	6.8×10^{-7}
$AgIO_3$	3.0×10^{-8}	$La(IO_3)_3$	6.2×10^{-12}
$AgOH$	2×10^{-8}	$MgCO_3$	5.6×10^{-6}
$AgSCN$	1.0×10^{-12}	MgF_2	6.6×10^{-9}
Ag_2SO_4	1.6×10^{-5}	$Mg(OH)_2$	1.8×10^{-11}
$Al(OH)_3$	2.0×10^{-33}	$Mn(OH)_2$	1.6×10^{-13}
$BaCO_3$	5.5×10^{-10}	$Ni(OH)_2$	6.3×10^{-18}
$BaCrO_4$	3×10^{-10}	NiS	2.0×10^{-21}
BaF_2	1.7×10^{-6}	$PbCO_3$	3×10^{-14}
$BaSO_4$	1.1×10^{-10}	$PbCl_2$	1.7×10^{-5}
$CaCO_3$	5×10^{-9}	$PbCrO_4$	2×10^{-14}
CaC_2O_4 (oxalate)	2.6×10^{-9}	$Pb(IO_3)_2$	2.6×10^{-13}
CaF_2	3.4×10^{-11}	$PbSO_4$	1.7×10^{-8}
$Ca(IO_3)_2$	3.3×10^{-7}	$SrCO_3$	1.1×10^{-10}
$CaSO_4$	2×10^{-5}	$SrCrO_4$	3×10^{-5}
$Ce(IO_3)_3$	3.2×10^{-10}	SrF_2	2.9×10^{-9}
$Cr(OH)_3$	10^{-30}	$SrSO_4$	2.8×10^{-7}
$CuBr$	4×10^{-8}	$TlCl$	1.9×10^{-4}
$CuCl$	1×10^{-6}	$ZnCO_3$	3×10^{-8}
CuI	4×10^{-12}	$Zn(OH)_2$	7×10^{-18}
$Cu(IO_3)_2$	7.4×10^{-8}	$ZnS(\alpha)$	2.5×10^{-22}
CuS	8×10^{-36}	$ZnS(\beta)$	1.6×10^{-24}

* When $AgCN(s)$ is shaken with water, the preponderant species that go into solution are Ag^+ and $Ag(CN)_2^-$, because of the great stability of the complex $Ag(CN)_2^-$. For this reason $AgCN(s)$ is often formulated as $Ag \cdot Ag(CN)_2(s)$, even though all the silver atoms in the crystal are equivalent. The solubility product of silver cyanide is usually given the form

$$[Ag^+][Ag(CN)_2^-] = K_{sp} = 5 \times 10^{-12}$$

rather than

$$[Ag^+][CN^-] = \text{const} (= 7 \times 10^{-17})$$

† Solutions contain predominantly the species Hg_2^{2+} rather than Hg^+, and the solubility products are formulated as $[Hg_2^{2+}][SO_4^{2-}]$ and $[Hg_2^{2+}][Br^-]^2$.

Table D-3
Acid Constants K_a
(see also Table 12-2)

Name	Formula	K_a
Acetic	HOAc (or CH_3COOH)	1.8×10^{-5}
Ammonium ion	NH_4^+	5.5×10^{-10}
		$(K_b = 1.8 \times 10^{-5})$
Arsenic	H_3AsO_4	$K_1 = 5.6 \times 10^{-3}$
		$K_2 = 1.7 \times 10^{-7}$
		$K_3 = 3 \times 10^{-12}$
Arsenious	H_3AsO_3	$K_1 = 6 \times 10^{-10}$
		$K_2 = 3 \times 10^{-14}$
Benzoic	C_6H_5COOH	6×10^{-5}
Boric	H_3BO_3	$K_1 = 6.4 \times 10^{-10}$
Carbonic	H_2CO_3	$K_1 = 4.4 \times 10^{-7}$
		$K_2 = 4.8 \times 10^{-11}$
Chlorous	$HClO_2$	1.0×10^{-2}
Chromic	H_2CrO_4	$K_1 = 1.2$
		$K_2 = 3.2 \times 10^{-7}$
Dichromic	$H_2Cr_2O_7$	K_1 large
		$K_2 = 8.5 \times 10^{-1}$
	$2HCrO_4^- \rightleftharpoons Cr_2O_7^{2-} + H_2O$	$K = 40$
Isocyanic	HNCO	2.0×10^{-4}
Formic	HCOOH	1.7×10^{-4}
Hydrazinium ion	$^+H_3NNH_2$	1.0×10^{-8}
		$(K_b = 1.0 \times 10^{-6})$
Hydrazoic	HN_3	1.2×10^{-5}
Hydrocyanic	HCN	2×10^{-9}
Hydrofluoric	HF	6.7×10^{-4}
Hydrogen sulfide	H_2S	$K_1 = 9.1 \times 10^{-8}$
		$K_2 = 1.2 \times 10^{-15}$
Hydroxylammonium ion	^+H_3NOH	8.2×10^{-7}
		$(K_b = 1.2 \times 10^{-8})$
Hypobromous	HBrO	2.0×10^{-9}
Hypochlorous	HClO	1.1×10^{-8}
Hypoiodous	HIO	3×10^{-11}
Iodic	HIO_3	2×10^{-1}
Nitrous	HNO_2	4.5×10^{-4}
Oxalic	HOOCCOOH	$K_1 = 5.6 \times 10^{-2}$
		$K_2 = 7.2 \times 10^{-5}$
Phosphoric	H_3PO_4	$K_1 = 7.1 \times 10^{-3}$
		$K_2 = 6.2 \times 10^{-8}$
		$K_3 = 4.4 \times 10^{-13}$
Phosphorous	H_3PO_3	$K_1 = 1.0 \times 10^{-2}$
		$K_2 = 2.6 \times 10^{-7}$
Sulfuric	H_2SO_4	K_1 large
		$K_2 = 1.2 \times 10^{-2}$
Sulfurous	H_2SO_3	$K_1 = 1.0 \times 10^{-2}$
		$K_2 = 5.0 \times 10^{-8}$
Isothiocyanic	HNCS	1.4×10^{-1}

*Some K_b values are given for the corresponding conjugate bases.

	$\widetilde{H}_f^\circ$/kJ mol^{-1}	$\widetilde{G}_f^\circ$/kJ mol^{-1}		$\widetilde{H}_f^\circ$/kJ mol^{-1}	$\widetilde{G}_f^\circ$/kJ mol^{-1}
Ag(s)	0.00	0.00	HI(g)	25.9	1.3
Ag$^+$(aq)	105.9	77.1	H$_2$(g)	0.00	0.00
AgCl(s)	−127.0	−109.7	H$_2$O(g)	−241.8	−228.6
Ba(s)	0.00	0.00	H$_2$O(l)	−285.8	−237.2
Ba^{2+}(aq)	−538.4	−560	H$_2$O$_2$(l)	−187.6	−114.0
Br(g)	111.8	82.4	H$_2$O$_2$(aq)	−191.1	−131.7
Br$^-$(aq)	−120.9	−102.8	H$_2$S(g)	−20.2	−33.0
Br$_2$(g)	30.7	3.1	HS$^-$(aq)	−17.7	12.6
Br$_2$(l)	0.00	0.00	S^{2-}(aq)	41.8	83.7
C(g)	716.7	671.3	Hg(g)	60.8	31.8
C (diamond)	1.9	2.8	Hg(l)	0.00	0.00
C (graphite)	0.00	0.00	Hg$_2$Cl$_2$(s)	−264.9	−210.7
CCl$_4$(g)	−102.9	−60.6	I(g)	106.6	70.2
CCl$_4$(l)	−135.4	−65.3	I$^-$(aq)	−55.9	−51.7
CHCl$_3$(l)	−134.5	−73.7	I$_2$(g)	62.3	19.4
CH$_4$(g)	−74.9	−50.8	I$_2$(s)	0.00	0.00
CO(g)	−110.5	−137.3	K(s)	0.00	0.00
CO$_2$(g)	−393.5	−394.4	K$^+$(aq)	−251.2	−282.3
CO$_3^{2-}$(aq)	−676.3	−528.1	KCl(s)	−435.9	−408.3
H$_2$CO$_3$(aq)	−698.7	−623.4	KNO$_3$(s)	−492.7	−393.1
HCO$_3^-$(aq)	−691.1	−587.1	Mg(s)	0.00	0.00
C$_2$H$_2$(g)	226.7	209.2	Mg^{2+}(aq)	−462.0	−457.9
C$_2$H$_4$(g)	52.3	68.1	MgCl$_2$(s)	−641.8	−592.3
C$_2$H$_6$(g)	−84.7	−32.9	Mn(s)	0.00	0.00
C$_3$H$_8$(g)	−103.8	−23.5	Mn^{2+}(aq)	−223.0	−227.6
CH$_3$OH(l)	−238.6	−166.2	MnO$_2$(s)	−519.7	−464.8
C$_2$H$_5$OH(l)	−277.7	−174.8	MnO$_4^-$(aq)	−542.7	−449.4
CH$_3$COOH(l)	−484.5	−389.9	N(g)	472.7	455.6
CH$_3$COOH(aq)	−485.8	−396.6	NH$_3$(g)	−46.2	−16.7
CH$_3$COO$^-$(aq)	−486.0	−369.4	NH$_4^+$(aq)	−132.8	−79.5
C$_6$H$_6$(l)	49.0	124.5	NO(g)	90.4	86.7
C$_6$H$_6$(g)	82.9	129.7	NO$_2$(g)	33.8	51.8
Ca(s)	0.00	0.00	N$_2$(g)	0.00	0.00
Ca^{2+}(aq)	−543.0	−553.0	N$_2$O(g)	81.5	103.6
CaCO$_3$ (calcite)	−1206.9	−1128.8	N$_2$O$_4$(g)	9.7	98.3
CaCO$_3$ (aragonite)	−1207.0	−1127.7	Na(s)	0.00	0.00
CaO(s)	−635.5	−604.2	Na$^+$(aq)	−239.7	−261.9
Ca(OH)$_2$(s)	−986.6	−896.8	NaCl(s)	−411.0	−384.0
Cl(g)	121.4	105.4	NaHCO$_3$(s)	−947.7	−851.9
Cl$^-$(aq)	−167.4	−131.2	Na$_2$CO$_3$(s)	−1130.9	−1047.7
Cl$_2$(g)	0.00	0.00	ONCl(g)	52.6	66.4
Cr(s)	0.00	0.00	O(g)	247.5	230.1
Cr^{3+}(aq)	−256.1	−215.5	OH$^-$(aq)	−230.0	−157.3
Cr$_2$O$_7^{2-}$(aq)	−1523.0	−1319.6	O$_2$(g)	0.00	0.00
Cu(s)	0.00	0.00	O$_3$(g)	142.3	163.4
Cu^{2+}(aq)	64.4	65.0	Pb(s)	0.00	0.00
CuCl(s)	−134.7	−118.8	PbCl$_2$(s)	−359.2	−314.0
Fe(s)	0.00	0.00	S (s, rhombic)	0.00	0.00
Fe^{2+}(aq)	−87.9	−84.9	S (s, monoclinic)	0.30	0.10
Fe^{3+}(aq)	−47.7	−10.5	SO$_2$(g)	−296.9	−300.4
Fe$_2$O$_3$(s)	−822.2	−741.0	SO$_3$(g)	−395.2	−370.4
Fe$_3$O$_4$(s)	−1117.1	−1014.2	SO$_4^{2-}$(aq)	−907.5	−742.0
H(g)	217.9	203.3	Zn(s)	0.00	0.00
H$^+$(aq)	0.00	0.00	Zn^{2+}(aq)	−152.4	−147.2
HBr(g)	−36.2	−53.2	ZnCl$_2$(s)	−415.9	−369.3
HCl(g)	−92.3	−95.3	ZnO(s)	−348.0	−318.2

*The standard states associated with these values are described on page 408. The values for ions are based on the convention that $\widetilde{H}_f^\circ$ and $\widetilde{G}_f^\circ$ are zero for H$^+$(aq) in its standard state (1 M, 1 atm, and 25°C, unless another temperature is explicitly specified).

Reaction	$E°$
$Ag^+ + e^- \rightleftharpoons Ag(s)$	0.799
$Ag_2O(s) + H_2O + 2e^- \rightleftharpoons 2Ag(s) + 2OH^-$	0.342
$AgCl(s) + e^- \rightleftharpoons Ag(s) + Cl^-$	0.222
$AgBr(s) + e^- \rightleftharpoons Ag(s) + Br^-$	0.071
$AgI(s) + e^- \rightleftharpoons Ag(s) + I^-$	-0.152
$Al^{3+} + 3e^- \rightleftharpoons Al(s)$	-1.66
$Al(OH)_4^- + 3e^- \rightleftharpoons Al(s) + 4OH^-$	-2.35
$H_3AsO_4 + 2H^+ + 2e^- \rightleftharpoons H_3AsO_3 + H_2O$	0.56
$Ba^{2+} + 2e^- \rightleftharpoons Ba(s)$	-2.90
$Be^{2+} + 2e^- \rightleftharpoons Be(s)$	-1.85
$Br_2(l) + 2e^- \rightleftharpoons 2Br^-$	1.065
$Ca^{2+} + 2e^- \rightleftharpoons Ca(s)$	-2.87
$Cd^{2+} + 2e^- \rightleftharpoons Cd(s)$	-0.402
$Cl_2(g) + 2e^- \rightleftharpoons 2Cl^-$	1.359
$2HClO + 2H^+ + 2e^- \rightleftharpoons Cl_2(g) + 2H_2O$	1.63
$HClO_2 + 2H^+ + 2e^- \rightleftharpoons HClO + H_2O$	1.64
$ClO_2(g) + H^+ + e^- \rightleftharpoons HClO_2$	1.27
$ClO_3^- + 2H^+ + e^- \rightleftharpoons ClO_2(g) + H_2O$	1.15
$ClO_4^- + 2H^+ + 2e^- \rightleftharpoons ClO_3^- + H_2O$	1.19
$Co^{2+} + 2e^- \rightleftharpoons Co$	-0.28
$Co^{3+} + e^- \rightleftharpoons Co^{2+}$	1.82
$Cr^{3+} + 3e^- \rightleftharpoons Cr(s)$	-0.74
$Cr_2O_7^{2-} + 14H^+ + 6e^- \rightleftharpoons 2Cr^{3+} + 7H_2O$	1.33
$Cs^+ + e^- \rightleftharpoons Cs(s)$	-2.952
$Cu^+ + e^- \rightleftharpoons Cu(s)$	0.521
$Cu^{2+} + 2e^- \rightleftharpoons Cu(s)$	0.337
$Cu^{2+} + e^- \rightleftharpoons Cu^+$	0.153
$Cu^{2+} + I^- + e^- \rightleftharpoons CuI(s)$	0.85
$F_2(g) + 2e^- \rightleftharpoons 2F^-$	2.87
$Fe^{2+} + 2e^- \rightleftharpoons Fe(s)$	-0.440
$Fe^{3+} + e^- \rightleftharpoons Fe^{2+}$	0.771
$2H^+ + 2e^- \rightleftharpoons H_2(g)$	0.0000
$Hg_2^{2+} + 2e^- \rightleftharpoons 2Hg(l)$	0.792
$Hg_2Cl_2(s) + 2e^- \rightleftharpoons 2Hg(l) + 2Cl^-$	0.268
$2Hg^{2+} + 2e^- \rightleftharpoons Hg_2^{2+}$	0.907
$I_2(s) + 2e^- \rightleftharpoons 2I^-$	0.534
$2IO_3^- + 12H^+ + 10e^- \rightleftharpoons I_2(s) + 6H_2O$	1.19
$K^+ + e^- \rightleftharpoons K(s)$	-2.925
$Li^+ + e^- \rightleftharpoons Li(s)$	-3.03
$Mg^{2+} + 2e^- \rightleftharpoons Mg(s)$	-2.37
$Mn^{2+} + 2e^- \rightleftharpoons Mn(s)$	-1.190
$MnO_2(s) + 4H^+ + 2e^- \rightleftharpoons Mn^{2+} + 2H_2O$	1.23
$MnO_4^- + 8H^+ + 5e^- \rightleftharpoons Mn^{2+} + 4H_2O$	1.51
$HNO_2 + H^+ + e^- \rightleftharpoons NO(g) + H_2O$	0.99
$NO_3^- + 3H^+ + 2e^- \rightleftharpoons HNO_2 + H_2O$	0.94
$Na^+ + e^- \rightleftharpoons Na(s)$	-2.698
$Ni^{2+} + 2e^- \rightleftharpoons Ni(s)$	-0.23
$H_2O_2 + 2H^+ + 2e^- \rightleftharpoons 2H_2O$	1.77
$O_2(g) + 4H^+ + 4e^- \rightleftharpoons 2H_2O$	1.229
$O_2(g) + 2H^+ + 2e^- \rightleftharpoons H_2O_2$	0.69
$H_3PO_4 + 2H^+ + 2e^- \rightleftharpoons H_3PO_3 + H_2O$	-0.276
$Pb^{2+} + 2e^- \rightleftharpoons Pb(s)$	-0.126
$PbSO_4(s) + 2e^- \rightleftharpoons Pb(s) + SO_4^{2-}$	-0.356
$PbO_2(s) + 4H^+ + 2e^- \rightleftharpoons Pb^{2+} + 2H_2O$	1.47
$PbO_2(s) + SO_4^{2-} + 4H^+ + 2e^- \rightleftharpoons PbSO_4(s) + 2H_2O$	1.685
$Rb^+ + e^- \rightleftharpoons Rb(s)$	-2.93
$S(s) + 2H^+ + 2e^- \rightleftharpoons H_2S(g)$	0.141
$HSO_4^- + 9H^+ + 8e^- \rightleftharpoons H_2S(g) + 4H_2O$	0.316
$HSO_4^- + 3H^+ + 2e^- \rightleftharpoons SO_2(g) + 2H_2O$	0.14

$S_4O_6^{2-} + 2e^- \rightleftharpoons 2S_2O_3^{2-}$	0.09
$Sn^{2+} + 2e^- \rightleftharpoons Sn(s)$	-0.140
$Sn(OH)_6^{2-} + 2e^- \rightleftharpoons HSnO_2^- + H_2O + 3OH^-$	-0.90
$Sn(IV) + 2e^- \rightleftharpoons Sn(II)$	0.14 (in 1 M HCl)
$Sr^{2+} + 2e^- \rightleftharpoons Sr(s)$	-2.89
$Tl^+ + e^- \rightleftharpoons Tl(s)$	-0.336
$Tl^{3+} + 2e^- \rightleftharpoons Tl^+$	1.28
$Zn^{2+} + 2e^- \rightleftharpoons Zn(s)$	-0.763

Answers to Exercises and Selected Problems

E

Exercises, Chapter 1

1-1 2.6×10^2 dollar ton^{-1} **1-2** 88 km hour^{-1};
25 m s^{-1} **1-3** (a) pure substance; (b) mixture;
(c), (d) homogeneous; (e) heterogeneous
1-4 (a) 8.40 g cm^{-3}; (b) 1.587 cm^3

Problems, Chapter 1

1-5 2.38 g cm^{-3} **1-7** 58 km hour^{-1}
1-9 9.4 liter (100 km)$^{-1}$

Exercises, Chapter 2

2-1 10.814 **2-2** 17.0 for both **2-3** 310.18
2-4 145.15 **2-5** (a) 32.1, 64.1, 192.4; (b) 16.0 g, 32.1 g,
96.2 g; (c) 144, 288, 864; (d) 1.20×10^{24}, 2.40×10^{24},
7.2×10^{24}; (e) 10.0, 5.0, 1.67 **2-6** (a) 1344 g; (b) 340 ml;
(c) 134 g; (d) 6.70×10^{-21} g; (e) 95.7 g

Problems, Chapter 2

2-5 14.0068 **2-7** 76, 24 **2-11** 10^7 km, or more km of
beach than are on earth; about 250
2-13 2.51×10^{26}
2-15 (a) 147.0; (b) 4.88×10^{-20} g; (c) 551 g; (d) 2.4 ml
2-17 (a) 167.1; (b) 6.04 g; (c) 7.9×10^{-21} g; (d) 0.93 mol
2-19 (a) 0.57 mol; (b) 168 ml; (c) 1.7×10^{24}; (d) 280 g
2-21 (a) 444 g; (b) 296 ml; (c) 37 g; (d) 2.1 g;
(e) 1.23×10^{-21} g

Exercises, Chapter 3

3-1 7.3×10^{-4} **3-2** $4 + 4.6 \times 10^{-10}$ **3-3** $C_6H_{12}O_6$
3-4 $C_6H_5NO_2$ **3-5** (a) The coefficients are 3 and 2 on
the left, and 1 and 6 on the right. (b) The coefficients of O_2,
CO_2, and H_2O are all 6 **3-6** Weight percentages: H, 2.06;
S, 32.69; O, 65.25 **3-7** Two quarters, six dimes, three
nickels, or $Q_2D_6N_3$ **3-8** $CoK_3N_6O_{12}$ **3-9** 81 g H_2O,
3 mol NO, 120 g O_2 **3-10** 48.3 g **3-11** 72 percent
3-12 (a) FeS + 2HCl → H_2S + $FeCl_2$; (b) 0.15 mol FeS,
0.35 mol $FeCl_2$, 0.35 mol H_2S; (c) 19.4 g H_2S, 8.5 g HCl
3-13 (a) 0.025 mol Fe, 2.6 g Cr, 0.075 mol $FeCr_2O_4$;
(b) 0.46 ton Cr **3-14** (a) 0.0345; (b) 0.61, 63 lb Fe_2O_3
3-15 P_4S_7 **3-16** C_3H_8O

Problems, Chapter 3

3-1 (a) 0.86; (b) −0.144; (c) 57; (d) 8.3×10^3; (e) 74.20
3-3 (a) 1.7×10^2; (b) 1428; 142.9×10^2 **3-5** The
coefficients are, in sequence (a) 6,1,7; (b) 1,3,1,2; (c) 1,11,7,8;
(d) 1,1,3,1; (e) 1,4,3,4; (f) 1,2,1,1; (g) 1,4,1,2
3-7 (a) CaH_2; (b) $PbBr_2$ **3-9** $Ag_4V_2O_7$
3-11 C_5H_5N **3-13** (a) C_2H_6N; (b) $C_4H_{12}N_2$
3-15 O/H = 21.3 **3-17** 144 **3-19** C_5H_{12}
3-21 (a) 1.20 g C, 0.10 g H, 0.70 g N; (b) C_2H_2N;
(c) $C_8H_8N_4$ **3-23** (a) 50.1(n/m), where n and m are
positive integers; (b) 75.2 **3-25** 24.31
3-27 (a) 1.4×10^2; (b) M_2O_3 **3-29** 95.6 n, with n an
integer **3-31** 1.50 mol; 2.82 mol **3-33** 1.33 mol
3-35 10.7 mol **3-37** 7.8×10^3 ton **3-39** 0.100 g Ne,
0.299 g MgO, 0.031 g O_2 **3-41** 33.0 lb Pb;
12.3 lb $NaNO_3$ **3-43** 23 percent **3-45** 35.5 ± 0.2
percent **3-47** 17.3 percent

Exercises, Chapter 4

4-1 0.0426 ft^3 (lb in^{-2}) mol^{-1}K^{-1} **4-2** 125 ft^3
4-3 202 ml **4-4** 0.50 mol **4-5** 0.73 g liter^{-1}
4-6 51.7 (sum of atomic weights is 50.5) **4-7** 286, P_4O_{10}
(sum of atomic weights = 284) **4-8** 121 ml
4-9 20.4 liter; 79.5 percent **4-10** 50/3 ml CO and
100/3 ml CH_4 **4-11** 0.542 g liter^{-1} **4-12** In weight
percentage: H_2O, 29; SO_2, 46; CO_2, 25; Av. MW = 35

4-13 (a) $673 \text{ ms}^{-1} = 2.42 \times 10^3 \text{ km hour}^{-1} = 1.51 \times 10^3 \text{ mile}$ hour^{-1} (b) 32.1 K   **4-14**   About 10^{-10} s; 10^{10} s^{-1}
4-15   0.80; 35 mg

Problems, Chapter 4

4-1   (a) 37.0°C; (b) 6×10^3°C; (c) 0.971 atm; (d) 1.6 atm
4-3   (a) 0.080 liter or 80 ml; (b) 1.33×10^{-2} atm; 10.1 torr
4-5   (b) 0.400 g; (c) 0.362 g; (d) 144°C   **4-7**   41 mol;
7.4×10^2 g   **4-9**   36.6 g (which is 1.81 mol)
4-11   586 K; 313°C   **4-13**   0.597 liter
4-15   (a) 0.066 g liter^{-1}; (b) 109 liter; (c) 0.100 atm;
(d) 1.55 mol; (e) 90 liter   **4-17** 9.5×10^2 g
4-19   76.1   **4-21** 9×10^{16} **4-23**   (a) 87; (b) more in
winter by 5 percent   **4-25**   (a) 85; (b) two; 86.4
4-27   (a) 121; (b) 3; (c) $72.0/n$, $n = $ integer
4-29   0.078 g; 1.02 liter   **4-31**   (a) $2n$; (b) $3m$; (c) at least
two   **4-33**   CO, 0.60   **4-35** -29 percent
4-37   (a) 1.73 atm; (b) F_2, 0.89 atm; Ar, 0.84 atm;
(c) 2.54 g liter^{-1}; (d) 0.975   **4-39**   0.989 atm
4-41   (a) 4.9 g liter^{-1}; (b) 12.0 atm; (c) 0.67; (d) 10.0 atm
4-43   (a) 96; (b) 10.6 K

Exercises, Chapter 5

5-1   471 ml   **5-2**   (a) s; g; s,g; l,g (b) g until $\sim$0.32 atm
where g,l; l until 0.6 atm where l,s; s above 0.6 atm

Problems, Chapter 5

5-3 1.15×10^3 atm; 3.3×10^3 atm   **5-5**   (a) liquid;
(b) 5.5°C   **5-7**   The warmer day   **5-9**   128 torr at
25°C   **5-11**   (a) 0.040 mol; (b) 38 torr; (c) 90 percent
5-13   (a) 1.08 mol; 19.5 g; (b) 23 percent; (c) 25 percent

Problems, Chapter 6

6-3   coulomb: 0.50, 0.33, 0.25, 0.20; dispersion: 1.6×10^{-2},
1.4×10^{-3}, 2.4×10^{-4}, 6.4×10^{-5} **6-5**   (a) -1.2×10^{-18} J
(b) -6.9×10^2 kJ mol^{-1} **6-7**   (a) $C_2H_4(OH)_2$, more polar,
forms hydrogen bonds to water molecules; (b) LiF, ionic
compound whereas SiH_4 is nonpolar covalently bonded
molecule; (c) CBr_4, more electrons and higher MW
6-9   (a) Cl_2, Xe, $GeCl_4$. Each has more electrons, higher
MW than the other in its pair and thus is less volatile. All
shown are nonpolar. (b) CH_3Cl, PH_3. Each forms hydrogen
bonds much less strongly than the other member of its pair.

Problems, Chapter 7

7-11   0, 2, 3, and 4   **7-13**   (a) KBr; (b) I_2; (c) ZnS

Exercises, Chapter 8

8-1   31.7 volume percent; 32.5 weight percent
8-2   0.930 M **8-3**   2.69 M **8-4** $[Ca^{2+}] = 0.0220\ M$,
$[Cl^-] = 0.0200\ M$, $[NO_3^-] = 0.0240\ M$
8-5   0.966 mol kg^{-1} **8-6** SO_4^{2-}: 0.011 M, 0.011 mol kg^{-1},
0.00019; HSO_4^-: 0.515 M, 0.525 mol kg^{-1}, 0.00928; H^+:
0.537 M, 0.548 mol kg^{-1}, 0.00968   **8-7**   0.278
8-8   3.81 g   **8-9**   93   **8-10**   1.38 g

Problems, Chapter 8

8-1   0.259   **8-3**   (a) 29 liter; (b) 0.84
8-5   0.91 g ml^{-1} **8-7**   The more stable form in both
cases   **8-9**   3.7 ml, 306 g   **8-11**   (a) 98, (b) 96.1
8-13   MW $= 2.9 \times 10^2$, hence $HgCl_2$ is undissociated
8-15   (a) 86 g liter^{-1}; 1.86 M; (b) 5.5 m; (c) 2.4 m; 3.1 m; 40
8-17   (a) 293 g; (b) 71 g; (c) -8 percent   **8-19**   9.6 mg
8-21   109   **8-23**   0.46 atm   **8-25**   (a) 0.31 torr;
(b) 196.5 torr   **8-27**   (a) heptane, 0.53; (b) $P_H = 53$ torr,
$P_O = 19$ torr; (c) $x_H = 0.74$, $P_H = 74$ torr, $P_O = 10$ torr
8-29   (a) 125 torr; 0.60 (b) 0.43   **8-31**   (a) 1.5 M;
(b) 1.0 M; (c) 4.5 M; (d) 24.4 torr

Exercises, Chapter 9

9-1   (a) 0.07; (b) 3.82; (c) -0.54; (d) 5.26   **9-2**   (a) 2.8 M
(b) $2.1 \times 10^{-5}\ M$ **9-3**   0.110 M **9-4**   97.0 percent
9-5 $[NH_4^+] = 0.045\ M$; $[Al^{3+}] = 0.015\ M$; $[Cl^-] = 0.065\ M$;
$[SO_4^{2-}] = 0.0125\ M$

Problems, Chapter 9

9-1   (a) pOH: 9.3, 5.6, 2.78, 13.48, 14.30; $[H^+]$: 2×10^{-5},
4×10^{-9}, 6.0×10^{-12}, 0.30, 2.0; (b) -0.78, 5.82, 9.15, 1.5,
13.3   **9-3** -0.78, 14.78; 5.82, 8.18; 9.2, 4.8; 1.5, 12.5
9-5   (a) 0.30 mol Mg^{2+}, 0.60 mol NO_3^-; fewer of each;
(b) 19 g   **9-7**   acids: HOAc, HCN, HSO_4^-, $H_2PO_4^-$; NH_3,
OH$^-$; bases: CN$^-$, OAc$^-$; HPO$_4^{2-}$, SO$_4^{2-}$; O^{2-}, NH$_2^-$
9-9   1.026 liter   **9-11**   0.500 M **9-13**   32 g; 16 g
9-15   Two   **9-17**   (a) 333; 333 n (b) $n = 3$
9-19   (a) 0.12 mol Ba^{2+}, 0.24 mol Cl$^-$; (b) 0.20 M Ba^{2+},
0.40 M Cl$^-$, 0.17 M Na$^+$, 0.17 M NO$_3^-$ **9-21**   (a) 0.50 M;
(b) 1.06 g ml^{-1}; (c) 143 g   **9-23**   4.8 percent
9-25   (a) 160; (b) four   **9-27**   11.70   **9-29** Ba^{2+},
0.070 M; Cl$^-$, 0.080 M; NO$_3^-$, 0.060 M; Ag$^+$, 0.000 M
9-31 NH_3, 0.28 M; NH_4^+, 0.20 M; SO_4^{2-}, 0.10 M;
47 g $BaSO_4$ **9-33**   (a) 0.122 mol; (b) 0.090 mol;
(c) $P_{HCl} = 0.74$ atm, $P_{He} = 0.26$ atm   **9-35**   0.709 g

Exercises, Chapter 10

10-1 (a) $+3, -1, +1, 0, +7, +5, +7, +7$; (b) $+4, +6, +3, +2, +3, +8/3$ **10-2** $2I^- + Sn^{4+} \rightarrow I_2 + Sn^{2+}$
10-3 The coefficients of the reactants H^+, O_2, and I^- are, respectively, 4, 1, and 4 **10-4** The coefficients of the reactants H^+, MnO_4^-, and H_2CO are, respectively, 6, 2, and 5 **10-5** The coefficients are, in order, 4, 3, 4, 3, 4, 8, 5
10-6 The coefficients are, in order, 6, 1, 3, 3, 1, 3
10-7 Oxidation: each CH_3OH gives up $2e^-$; reduction: each $Cr_2O_7^{2-}$ reacts with $14H^+$ and $6e^-$; to balance, combine oxidation and reduction half-reactions in the ratio 3:1
10-8 $0.360\ M$

Problems, Chapter 10

10-3 (a) $3, 3, 5, 0, 5, 1, 5, 5, -3$; (c) $2, 3, \frac{8}{3}, 7, 7, 6, 4$
10-5 Reducing agents: (a) N_2H_4, 4; (c) NO_2^-, 2. Oxidizing agents: (b) $HBiO_3$, 2; (d) MnO_4^-, 5
10-7 $0.0300; 0.0150$ **10-9** 45; 135

Exercises, Chapter 11

11-1 $(P_{SO_2})(P_{O_2})^{1/2}/P_{SO_3} = 0.196\ atm^{1/2}$ **11-2** (a) 0.046; 0.0093 M; (b) 3.0×10^{-3} **11-3** 2.63 g liter^{-1} **11-4** 0.54
11-5 1.00 atm **11-6** 0.87 **11-7** $P_{SO_2Cl_2} = 0.022$ atm; $P_{SO_2} = 0.915$ atm; $P_{Cl_2} = 0.063$ atm
11-8 24.5 **11-9** $(K_1/K_2)^{1/2} = 8.38$

Problems, Chapter 11

11-1 (a) no effect; more HBr; (b) less N_2O; less N_2O; (c) less NOCl; more NOCl **11-3** (b) increase; (c) decrease, no effect
11-5 $K = P^2_{NO_2}/(P^2_{NO}P_{O_2}) = 1.0 \times 10^6$ atm^{-1}
11-7 (a) yes, formation; (b) decomposition; (c) no effect
11-9 (a) 10^{-26} atm (b) increase **11-11** 1.81 atm
11-13 0.34 **11-15** $K = 41$ (dimensionless)
11-19 3.1×10^{-5}

Exercises, Chapter 12

12-1 (a) 3 percent (b) 4 percent **12-2** (a) 1.5×10^{-6} (b) 9×10^{-7} **12-3** 2.6×10^{-4} **12-4** 3.17
12-5 4.3 **12-6** 1.23 **12-7** 9.74 **12-8** 1.0×10^{-3}
12-9 (a) 8.8 (8.78) (b) 9.7 (9.73) **12-10** 1(a) pH = 1.6 (decrease of 4.9 units), (b) pH = 12.4 (increase of 5.9 units); 2(a) pH = 6.3, (b) pH = 6.7 (decrease and increase of 0.2 units each) **12-11** $c_a/c_b = 2.8$ for HOAc, 0.29 for HCOOH. For example, 0.28 M acetic acid, 0.10 M sodium acetate, or 0.29 M formic acid (HCOOH), 1.00 M sodium formate **12-12** 3.9, 5.5, 6.5, 7.5, 8.5, 9.6, 10.7, 11.7

12-13 H_3PO_3; H_3PO_3 and $H_2PO_3^-$; $H_2PO_3^-$; HPO_3^{2-}
12-14 (a) 6.9 (b) 7.4 **12-15** 7.9

Problems, Chapter 12

12-1 e, b, d, c, a **12-3** (a) $1.0 \times 10^{-9}\ M$; (b) 1.0×10^{-4}; (c) 0.025 M **12-5** $1.0 \times 10^{-4}\ M$ **12-7** (a) 2.09; (b) 1.30; (c) 12.40; (d) 8.24; (e) 9.54 **12-9** $a - 0.7$ **12-11** OH^-, NH_3, OAc^-, H_2O, Cl^- **12-15** $4.5 \times 10^{-5}\ M$
12-17 About 8.15 **12-21** 0.20 M, 0.20 M, 0.001 M, $1 \times 10^{-11}\ M$, 0.001 M; 0.005 **12-23** 1.4×10^{-11}
12-25 4.82 **12-27** (a) $4 \times 10^{-4}\ M$, 10.6; (b) 5.4
12-29 9.0 **12-31** (a) $5 \times 10^{-5}\ M$; (b) 0.50 M; 0.25 M; (c) 8.7 **12-33** pH changes by -0.04 units, $+0.04$ units **12-35** (a) dissolve 1.0 mol NH_3 and 0.64 mol (monoprotic) strong acid in enough water to make total volume 1 liter; (b) 8.95 **12-37** $\frac{1}{3}$ liter
12-39 (a) 10 ml; (b) 91 ml, 99.9 ml **12-41** (a) 9.3; (b) 3.15

Exercises, Chapter 13

13-1 $[Ca^{2+}][F^-]^2$; $[La^{3+}]^2[CO_3^{2-}]^3$
13-2 $3.4 \times 10^{-11}\ M^3$ **13-3** $1.7 \times 10^{-4}\ M$; 4.9 mg per 100 ml **13-4** $5.3 \times 10^{-4}\ M$; $1.06 \times 10^{-3}\ M$
13-5 $1.5 \times 10^{-7}\ M$ **13-6** $K_{sp} = 1.5 \times 10^{-11}\ M^3$; pH $\geqslant$ 12.3 **13-7** $[Ba^{2+}] = 2 \times 10^{-5}\ M$, $[SO_4^{2-}] = 6 \times 10^{-6}\ M$, $[CrO_4^{2-}] = \frac{3}{2} \times 10^{-5}\ M$
13-8 0.17 M **13-9** 0.19 mol HOAc
13-10 (a) 2.12×10^4 torr^2 (b) $P_{PH_3} = 46.7$ torr, $P_{HBr} = 453$ torr

Problems, Chapter 13

13-1 (a) S^2; (b) $4S^3$; (c) $4S^3$; (d) $27S^4$; (e) S^2
13-3 (a) $3 \times 10^{-9}\ M^3$; (b) $2 \times 10^{-49}\ M^3$; (c) $4 \times 10^{-12}\ M^2$; (d) $1.8 \times 10^{-18}\ M^4$; (e) $2 \times 10^{-14}\ M^2$ **13-5** (a) $9 \times 10^{-6}\ M$ (b) $7 \times 10^{-9}\ M$ **13-7** (a) 5×10^1 mg per 100 ml; (b) 2×10^{-4} mol liter^{-1}; (c) Hint: Note that HF is a weak acid **13-9** K_{sp} for $BaCO_3 = 5 \times 10^{-10}\ M^2$
13-11 Yes **13-13** $3.8 \times 10^{-4}\ M$ **13-15** AgCl precipitates first; a little less than 99 percent
13-17 (a) $P_{NH_3} = Q$; (b) and (c) no effect; (d) P_{NH_3} decreased **13-19** 5.3 percent **13-21** (a) 0.040 torr; (b) exothermic; (c) a little over 8 hour

Exercises, Chapter 14

14-1 396 m s^{-1} **14-2** 9.35 kJ mol^{-1}; 0.0969 eV
14-3 0.35×10^{-20} J
14-4 7.9×10^{-12} m = 7.9×10^{-3} nm
14-5 0.050 Å

Problems, Chapter 14

14-3 (*a*) 10^6 s^{-1} (1 MHz) (radio); 3×10^{10} s^{-1} (microwave); 6×10^{14} s^{-1} (visible); 3×10^{18} s^{-1} (X ray); (*b*) 3×10^{-4} m (0.3 mm) (far infra-red); 3×10^{-7} m (300 nm) (ultraviolet); 3×10^{-11} m (0.03 nm) (X ray) **14-5** 3.00×10^{-19} J **14-7** 6.2×10^3 K **14-9** 4.0 ms^{-1} **14-11** 1.46×10^{-10} m; 2.32×10^{-11} m; 1.28×10^{-11} m **14-15** 6.6×10^{-24} J; 10^{10} s^{-1} **14-17** 5×10^{21} cm^{-2} s^{-1}

Exercises, Chapter 15

15-1 4.085×10^{-19} J; 2.550 eV; 246.0 kJ mol^{-1}
15-2 5.25×10^3 kJ mol^{-1}; 54.4 eV; 8.72×10^{-18} J

Problems, Chapter 15

15-3 54.4 eV, 122.4 eV, 217.6 eV **15-5** $n_1 = 1$, none; $n_1 = 2$, $n_2 = 3$ to $n_2 = 6$, $\lambda = 410$ nm for $n_2 = 6$
15-7 yes; 798 nm **15-9** 315 nm **15-11** 7.83 eV
15-15 $r = 1.01$ Å, 3.76 Å

Exercises, Chapter 16

16-1 Five

Problems, Chapter 16

16-1 Would resemble Pb **16-7** (*a*) Li$^+$, He; NH$_4^+$, CH$_4$, Ne; Ca^{2+}, Cl$^-$; (*b*) Li$^+$, H$^-$; H$_3$O$^+$, NH$_3$, Na$^+$; S^{2-}, K$^+$
16-11 (*b*) Li; (*c*) 50.4 nm **16-13** (*a*) Mg^{2+}, Ca^{2+}, Ar, Br$^-$; (*b*) Na, O, Ne, Na$^+$; (*c*) Al, H, O, F
16-15 (*b*) 5.03×10^{-19} J; (*c*) 1.38×10^{-19} J
16-17 (*a*) Ar; (*b*) Ca; (*c*) Cl; (*d*) K; (*e*) K; (*f*) Cl; (*g*) Si
16-19 (*c*), (*a*), (*b*), (*d*)

Exercises, Chapter 17

17-1 (*a*) 90 kJ; (*b*) -70 kJ **17-2**

17-3

17-4
The carbon-oxygen distances in HCOO$^-$ are equal to each other and intermediate in length, shorter than the C—OH bond and longer than the C—O bond in HCOOH

Problems, Chapter 17

17-1 (*a*) -180 kJ; (*b*) 470 kJ; (*c*) -120 kJ; (*d*) 110 kJ
17-5 (*a*) $d\sqrt{3}/(2\sqrt{2})$ (*c*) $b/a = 0.225$ **17-7** (*a*) 8Cs$^+$, 3.57 Å, cube; (*b*) 6Cl$^-$, 4.12 Å, octahedron; (*c*) 8Cl$^-$, 3.57 Å, cube; 6Cs$^+$, 4.12 Å, octahedron **17-9** (*a*) 8F$^-$, 2.36 Å, cube; (*b*) 4Ca^{2+}, 2.36 Å, tetrahedron; 6F$^-$, 2.73 Å, octahedron

Exercises, Chapter 18

18-1 PF$_3$: 3s, three 3p; sp^3; $<109°$. BF$_3$: 2s, three 2p; sp^2; 120°. GeCl$_4$: 4s, three 4p; sp^3; 109.5°. SF$_2$: 3s, three 3p, five 3d; sp^3; $<109°$ **18-2** Square planar
18-3 $\sigma_s^2\,(\sigma_s^*)^2(\pi_{y,z})^4\sigma_p^2$, 3 bonds; $\sigma_s^2(\sigma_s^*)^2(\pi_{y,z})^4\sigma_p^2\pi_y^*\pi_z^*$, 2 bonds

Problems, Chapter 18

18-5 Three of the molecules are paramagnetic: NO$_2$, S$_2$, and ClO$_2$

Exercises, Chapter 19

19-1 (*a*) $w = -6.1 \times 10^3$ J; (*b*) $\Delta U_{gas} = -6.1 \times 10^3$ J; (*c*) $\Delta U_{surr} = 6.1 \times 10^3$ J **19-2** $w = 205$ J
19-3 (*a*) 17.5 J; (*b*) 16.8 J; (*c*) -96 J; (*d*) -79 J
19-4 -2759.5 kJ **19-5** 926 kJ **19-6** 903 kJ
19-7 -241.8 kJ mol^{-1} **19-8** -904.4 kJ
19-9 -1605.2 kJ **19-10** 28.5 kJ mol^{-1}; 59.8 kJ mol^{-1}
19-11 Final temperature 0.0°C with 31.3 g snow melted

Problems, Chapter 19

19-3 $w = -5.61$ kJ $= -q$; $\Delta U = \Delta H = 0$ **19-5** -191 liter atm; -19.3 kJ **19-7** 21 kJ **19-9** 25.14°C
19-11 94 g **19-13** 55.9 kJ, -3.3 kJ, 55.9 kJ, 52.6 kJ
19-15 37 percent **19-17** $\Delta H = -890.2$ kJ; $\Delta U = -885.2$ kJ **19-19** (*a*) -7844 kJ; (*b*) -7829 kJ
19-21 -1266 kJ; exothermic **19-23** (*a*) -644 kJ mol^{-1}; (*b*) more **19-25** -1307 kJ **19-27** (*a*) -13 kJ mol^{-1}; (*b*) about 1300°C

Exercises, Chapter 20

20-1 $93 \text{ J mol}^{-1} \text{ K}^{-1}$ **20-2** $0.81 \text{ J mol}^{-1} \text{ K}^{-1}$
20-3 $\Delta U = \Delta H = 0, \Delta S = -6.7 \text{ J K}^{-1}$ **20-4** -13 J K^{-1}
20-5 $\Delta G = 28.3 \text{ kJ}$; at the conditions given $I_2(s)$ is the
stable phase   **20-6** -60.2 kJ **20-7** 91.3 kJ
20-8 $\Delta G^{\circ} = -474.4 \text{ kJ}; K = 2 \times 10^{83} \text{ atm}^{-3}$
20-9 $\Delta G^{\circ} \doteq -228.8 \text{ kJ}; K = 1.3 \times 10^{40} \text{ atm } M^{-8}$
20-10 $0.0145 \, M^{-1}$

Problems, Chapter 20

20-1 $104 \text{ J K}^{-1} \text{ mol}^{-1}$ **20-5** -5.76 J K^{-1}
20-7 $28.7 \text{ J K}^{-1} \text{ mol}^{-1}$ **20-9** $\tfrac{1}{8}$; 0.05 J K^{-1} **20-11** B, B
20-13 $232 \text{ kJ}; 195 \text{ kJ}; \text{Fe}_2\text{O}_3$ more stable   **20-15** value
of K: (a) 1.2×10^{40}; (b) 1.2×10^5; (c) 1.8×10^{36};
(d) 6.4×10^{60} **20-17** $\Delta H^{\circ} = -2073.8 \text{ kJ}$;
$\Delta G^{\circ} = -1705.5 \text{ kJ}$ **20-19** (a) P, T constant, only P, V
work; (b) an adiabatic process; (c) T constant; (d) V
constant; (e) $w = 0$, V constant; (f) negative; (g) positive
20-21 (a) I is more stable; B has the higher entropy;
(b) 2.3; 30 percent B   **20-23** $3 \times 10^{-19} \text{ atm}$
20-25 $-61.5 \text{ kJ}; 112.8 \text{ kJ};$ to left   **20-27** $\text{Br}_2, 0.29 \text{ atm},$
$2.2 \times 10^2 \text{ torr}; I_2, 4.0 \times 10^{-4} \text{ atm}, 0.30 \text{ torr}$ **20-29** $55.8 \text{ kJ},$
$2.6 \times 10^{-14}; 1.0 \times 10^{-12}$ **20-31** (a) $\Delta H^{\circ}_{\text{vap}} = 33.9 \text{ kJ},$
$\Delta G^{\circ}_{\text{vap}} = 5.2 \text{ kJ};$ (b) $0.122 \text{ atm};$ (c) $79°\text{C}$
20-33 (a) $q = \Delta H = 51.2 \text{ kJ}, w = -5.0 \text{ kJ}, \Delta U = 46.2 \text{ kJ};$
(b) $97.5 \text{ J K}^{-1} \text{ mol}^{-1}; 4.1 \text{ atm}$

Exercises, Chapter 21

21-1 0.50 A **21-2** $1.055 \text{ g Cu}; 3.58 \text{ g Ag}$ **21-3** 7 kJ
21-4 $E = E^{\circ} - (0.0592/3) \log([\text{Al}^{3+}]/[\text{Ag}^+]^3)$ **21-5** (a) 2
(b) 2 (c) 3 (d) 6   **21-6** -2.00 V **21-7** 0.37 V
21-8 0.34 V **21-9** $\sim 1 \times 10^{54}$ **21-10** $\sim 10^{57}$
21-11 1.7×10^{-8}

Problems, Chapter 21

21-1 (a) 132 C; (b) 1.37×10^{-3}; (c) 8.2×10^{20} **21-3** (a)
1.9×10^{20}; (b) 0.59 g hour^{-1} **21-5** $15.3 \text{ cm}^3; 1.06 \times 10^3 \text{ s}$
21-7 (b) 0.170 A **21-9** 0.203 A **21-11** (a) Al
dissolves at anode, Cu plates out at cathode; (b)
2.00 V; (c) from Al to Cu; (d) 5.4 hour; (e) decreased
21-13 (a) 0.331 V; (b) Zn anode   **21-15** $1.25 \, \mathcal{F}$
21-19 (a) 0.37 V; (b) $5 \times 10^{16} \, M^{-3}$ **21-21** (a) $[\text{Ag}^+] =$
$4 \times 10^{-18} M, [\text{Ni}^{2+}] = 1 M$; (b) -199 kJ per mole
of nickel   **21-23** 9×10^{-17} **21-25** In all three
cases $2\text{Tl} + \text{Tl}^{3+} \to 3\text{Tl}^+, \Delta G^{\circ} = -313 \text{ kJ}$. For (a), (b), (c) in
sequence, $\mathbf{n} = 2, 6, 3$; $E^{\circ} = 1.62 \text{ V}, 0.54 \text{ V}, 1.08 \text{ V}$
21-27 (a) $\text{Mg(OH)}_2(s), -833.8 \text{ kJ mol}^{-1}$; (b) $\text{Mg}^{2+}(aq),$
$-457.9 \text{ kJ mol}^{-1}$; (c) -2.37 V **21-29** $\text{Ag}_2\text{S}(s) \to 2\text{Ag} +$
S^{2-}; $\text{Al} \to \text{Al}^{3+}$

Exercises, Chapter 22

22-1 The rate is (a) halved (b) increased by a factor of 10
(c) decreased by a factor of 9   **22-2** (a) rate =
$k'[\text{H}_2\text{O}_2][\text{I}^-]$; (b) $k' = 7 \times 10^{-2} \, M^{-1} \text{ min}^{-1}$;
(c) $1.8 \times 10^{-5} \text{ mol min}^{-1}$ **22-3** $18.6 \text{ hour}; 43 \text{ hour}$
22-4 102 kJ

Problems, Chapter 22

22-1 (a) second order in NO, first order in Cl_2;
(b) $k = 8.0 \, M^{-2} \text{ s}^{-1}$; (c) about 0.024 mol **22-3** $0.115 \, M$
22-5 1, 2, 1   **22-7** (a) first order; (b) $k = 0.017 \text{ min}^{-1}$
22-11 0.45 **22-13** (a) $0.096 \text{ hour}^{-1}; 0.96 \text{ hour}^{-1}$;
(b) $1.2 \times 10^2 \text{ kJ}$; (c) $0.045 \, M; 0.91 \, M$
22-15 114 kJ mol^{-1} **22-17** (a) $d[\text{D}]/dt =$
$(k_1 k_2 / k_{-1})[\text{A}][\text{B}][\text{C}]; d[\text{E}]/dt = (k_1 k_3 / k_{-1})[\text{A}][\text{B}][\text{C}]$;
(b) k_2 is about $10 k_3$ **22-19** rate constant $=$
$1 \times 10^{10} \, M^{-1} \text{ s}^{-1}$ **22-21** (a) 2; (b) 85 kJ;
(c) 4.1×10^{-5}; 100

Exercises, Chapter 23

23-1 No for both

Problems, Chapter 23

23-1 $2, 3 : 2; 1; 2, 6 : 1$ **23-3** 300 kJ mol^{-1},
142 kJ mol^{-1} **23-5** -24 kJ **23-9** (a) $1.9 \times 10^7 \text{ kg}$;
(b) $9.5 \times 10^6 \text{ kg}$ **23-11** 15; $K_a K_b = 1 \times 10^{-30}$
23-13 (d) $[\text{H}^+]^3[\text{Br}^-]^2[\text{NO}_3^-]/[\text{HNO}_2] = 1.7 \times 10^4 \, M^5$
23-15 $10.5; 0.05 \, M$ **23-17** at around 10 K

Exercises, Chapter 24

24-1 All three elemental species are stable   **24-2** 10^{-2},
which is also the measured value

Problems, Chapter 24

24-1 (b) $7 \times 10^{-11} \text{ atm}$ **24-5** $0.036 \, M$
24-7 $0.36 \text{ g liter}^{-1}, 8 \times 10^9 \text{ kg year}^{-1}$ (for pH = 6,
$1.8 \text{ g liter}^{-1}, 4 \times 10^{10} \text{ kg year}^{-1}$)   **24-13** (a) 0.6, (b) 0.005

Problems, Chapter 25

25-1 In sequence $\Delta \tilde{S}/(\text{J mol}^{-1} \text{ K}^{-1})$ and $T^{\text{equil}}_{\text{estim}}/\text{K}$, (a) CO:
542, 905; CO_2: 277, 837; (b) CO: 705, 958; CO_2: 351, 940;

(c) CO: 364, 820; CO_2: 187, 674   **25-11**   (a) 4.7×10^{-8}; (b) $1.6 \times 10^{-8}\,M$; (c) $1.8 \times 10^{26}\,M$

Problems, Chapter 26

26-1 The overall enthalpy change for Li is 21 kJ less positive than that for Cs   **26-3**   (a) Al_2O_3, MgO, Na_2O; (b) CrO_3, Cr_2O_3, CrO   **26-7**   (b) $K = 2 \times 10^1\,M^3$; $[MnO_4{}^{2-}] = 0.02\,M$ **26-9**   11.0   **26-11** 1.1×10^{-41} **26-15**   (b) $1.7 \times 10^6\,M^{-1}$

Exercises, Chapter 27

27-1 d^{10}, d^3, d^5 **27-2**   (a) Same number of unpaired electrons in weak- or strong-field case ($t_{2g}{}^6 e_g{}^2$); (b) strong-field case (t_{2g})^5; (c) weak-field case ($t_{2g}{}^3 e_g{}^2$); (d) strong-field case (t_{2g})^4

Problems, Chapter 27

27-9   18 for all   **27-13** Mn^{2+}, Sc^{3+}, Cu^+, Hf^{4+} **27-15**   strong field: 2, 2, 2, 0; weak field: 4, 2, 4, 4 **27-17**   The e_g states are split into d_{z^2} and $d_{x^2-y^2}$, of which the second has the higher energy. The t_{2g} states are split into d_{xy} and the doubly degenerate d_{zx} and d_{yz}, where d_{xy} has the higher energy

Exercises, Chapter 28

28-1 $^{233}_{93}Pa$; $^{229}_{90}Th$ **28-2** 4×10^7 year **28-3** 3.633×10^{12} J   **28-4** 3.30×10^3 year

Problems, Chapter 28

28-3 $(\frac{1}{2})^{50} \approx (10^{-3})^5 \approx 10^{-15}$ **28-5** 6.36×10^{-9} g; 1.77×10^{-8} percent   **28-7** 17.3×10^3 year **28-9** 3.48×10^{-12} g   **28-11**   102   **28-13**   (a) $N/N_0 = 1 - 4.2 \times 10^{-11}$ (b) 1.33×10^9 year

Exercises, Chapter 29

29-1   (a) and (c) same molecule; (d) and (e) incorrect **29-2**   heptane; 2-methylhexane; 3-methylhexane; 2,2-

dimethylpentane; 2,3-dimethylpentane; 3,3-dimethylpentane; 2,4-dimethylpentane; 3-ethylpentane; 2,2,3-trimethylbutane

29-3   (a) CH_3CHO; $O{<}{\begin{smallmatrix}CH_2\\CH_2\end{smallmatrix}}$ (b) $O{<}{\begin{smallmatrix}CH_2\\CH_2\end{smallmatrix}}$

29-4   Reduce part of the ethanal with $LiAlH_4$ or $NaBH_4$; then proceed as in Example 29-4

Problems, Chapter 29

29-1 $CH_3CH_2CH_2CH_2CH_2CH_2CH_3$; $H_2NCH_2CH_2OH$;

Cl_2CHCH_2Cl; $CH_3\underset{\underset{OH}{|}}{CH}\underset{\underset{OH}{|}}{CH}CH_2CH_2CH_3$

$CH_3COCH_2\underset{\underset{CH_3}{|}}{\overset{\overset{H_3C}{|}}{CH}}CH_3$; $CH_3\overset{\overset{CH_3}{|}}{CH}\overset{\overset{CH_3}{|}}{C}CH_2\underset{\underset{CH_2CH_3}{|}}{CH}CH_3$; $CH_3\overset{\overset{CH_3}{|}}{CH}CH_2CHO$;

$(CH_3)_2CHOC(CH_3)_3$; $CH_3CH_2OOCCH_2CH_2CH_3$; $CH_3C{\equiv}CCH_2CH$

$BrCH{-}CH_2$
$|\qquad\quad|$
$CH_2{-}CHOH$

29-3   four structures   **29-7**   only one formula is consistent: $(CH_3)_2CHOCH_3$ **29-9**   (a) one; (b) three; (c) three; (d) three; (e) six   **29-11**   (a) two; (b) one possible structure for X is $HOCH_2\underset{\underset{OH}{|}}{CH}CH_2CHO$

29-13   (b), (e), (f)

Problems, Chapter 30

30-1   (a) 1.62×10^4, 1.69×10^4; (b) 6.5×10^4, 1.69×10^4 **30-7**   thymine to adenine   **30-11**   1.5 mm; 4.3 million

Problems, Appendix B

B-1 H_2SO_4; Na_2SO_4; KBr; $NH_4{}^+$; NH_4NO_3; HNO_3; $CO_3{}^{2-}$; $Ca(HCO_3)_2$; $Ca_3(PO_4)_2$; AgI   **B-3**   barium peroxide; calcium hydrogen phosphate; chromic hydroxide or chromium(III) hydroxide; hydrogen cyanide or hydrocyanic acid; magnesium sulfite; lithium thiocyanate; sulfide ion; ferric chloride or iron(III) chloride; strontium iodide; nickel(II) sulfate; hydrogen bromide or hydrobromic acid; hypochlorite ion; ammonia   **B-5** $Na_3[Ag(S_2O_3)_2]$; $NH_4[Cr(NH_3)_2(SCN)_4]$; $Na_3[Co(NO_2)_6]$; $Al(H_2O)_5OH^{2+}$; $[Co(en)_3]_2(SO_4)_3$; $[Cr(H_2O)_4(OH)_2]Cl$; $[Co(NH_3)_3Cl]Cl$; $K[AuCl_4]$; $Ni(PCl_3)_4$; $[Co(NH_3)_4Cl_2]Cl$

Index

Index

Prefixes Denoting Powers of 10

Factor	Prefix	Symbol	Factor	Prefix	Symbol
10^{-1}	deci	d	10	deka	da
10^{-2}	centi	c	10^2	hecto	h
10^{-3}	milli	m	10^3	kilo	k
10^{-6}	micro	μ	10^6	mega	M
10^{-9}	nano	n	10^9	giga	G
10^{-12}	pico	p	10^{12}	tera	T
10^{-15}	femto	f			
10^{-18}	atto	a			

Some Equivalent Energies

Calorie (cal)	Joule (J)	Liter atmosphere (liter atm)
1	4.1840	0.041293
0.23901	1	9.8692×10^{-3}
24.217	101.325	1

Additional Equivalent Energies

$E/\text{kJ mol}^{-1}$	$E/\text{kcal mol}^{-1}$	E/eV	E/J	T/K	$\tilde{\nu}/\text{cm}^{-1}$
1	0.23901	0.010364	1.6606×10^{-21}	120.27	83.59
4.1840	1	4.3364×10^{-2}	6.948×10^{-21}	503.2	349.8
96.48	23.060	1	1.6022×10^{-19}	11.605×10^3	8066
6.022×10^{20}	1.4393×10^{20}	6.241×10^{18}	1	7.243×10^{22}	5.034×10^{22}
8.314×10^{-3}	1.9872×10^{-3}	8.617×10^{-5}	1.3807×10^{-23}	1	0.6950
1.196×10^{-2}	2.859×10^{-3}	1.240×10^{-4}	1.986×10^{-23}	1.439	1